Business Mathematics and Statistics

Business Mathematics and Statistics

B M Aggarwal
MA (Maths.)
Member visiting faculty : *NIRC of ICWA, ICS, ICFAI Centres at Gurgaon and Delhi*
Nodal Centres of AIMA, Guru Nanak Institute of Management

Ane Books Pvt. Ltd.
New Delhi ◆ Chennai

Business Mathematics and Statistics
B M Aggarwal

Published by
Ane Books Pvt. Ltd.
4821, Parwana Bhawan, 1st Floor, 24 Ansari Road,
Darya Ganj, **New Delhi** - 110 002, India
Tel.: +91(011) 23276843-44, Fax: +91(011) 23276863
e-mail: kapoor@anebooks.com, Website: www.anebooks.com

Branch Office :

- Avantika Niwas, 1st Floor, 19 Doraiswamy Road, T. Nagar,
Chennai - 600 017, Tel.: +91(044) 28141554, 28141209
Email : rathinam@anebooks.com

ISBN : 978-81-8052-285-7

Printed at : Sagar Colourscan, New Delhi.

Dedication

With extreme reverence

to

my most loving and affectionate mother

Preface

The book has been written strictly as per the revised syllabus of B.Com II. Delhi University and also caters to the needs of other Universities equally well. No doubt the subject is known for its complicacies but still I have made efforts to deal with the subject conceptually to make it within the grasp of even weak students.

I express my thanks and gratitude to the publishers M/s Ane Books Pvt. Ltd. and their staff for utmost co-operation and patience.

I owe a word of thanks to my family members and my son-in-law Mr. Gaurav Gulati (CA, ICWA, CS) who always inspires me for such ventures.

Any suggestions for improvement of the book will be thankfully acknowledged.

B M AGGARWAL

Contents

UNIT II : BUSINESS STATISTICS

UNIT-I

Business Mathematics

1. Matrices and Determinants
2. Applications of Matrices and Determinants to Business and Economics
3. Functions, Limits and Continuity
4. Differentiation
5. Applications of Differentiation to Business and Economics
6. Integral Calculus
7. Applications of Integration to Business and Economics
8. Basic Mathematics of Finance

1 Matrices and Determinants

1.1 MATRICES

A matrix is a rectangular or a square array of numbers arranged systematically into rows and columns enclosed in a pair of brackets. The array is said to be ordered, meaning that the position of each number is important. Therefore, a matrix is an ordered array of numbers consists of *m* rows and *n* columns that are enclosed in pair of brackets. A matrix is designated by a capital letter. Some authors enclosed the array by Parentheses ( ). But in this book pair of Brackets has been used. The general form of a matrix is

$$A = \begin{pmatrix} a_{11} & a_{12} & a_{13} & \ldots & a_{1n} \\ a_{21} & a_{22} & a_{23} & \ldots & a_{2n} \\ a_{31} & a_{32} & a_{33} & \cdots & a_{3n} \\ \vdots & \vdots & \vdots & \cdots & \vdots \\ a_{m1} & a_{m2} & a_{m3} & \cdots & a_{mn} \end{pmatrix}$$

The matrix A consists of $m \times n$ components or elements a_{ij}. The number in the array are called the elements of the matrix. An element a_{ij} is designated by its position in the matrix, i referring to the row and j to the column. An element a_{24} thus refers to the an element at the intersection of the second row and fourth column.

The matrix A is said to be of order m by n or, alternatively, is said to have dimensions of m by n, if it has m rows and n columns.

1.2 ROW MATRIX

A matrix having single row is called a row matrix or row vector.

1.3 COLUMN MATRIX

A matrix having a single column is called a column matrix or column vector.

Example 1.

$$A = \begin{pmatrix} 3 & 1 \\ 2 & 4 \\ 1 & 6 \end{pmatrix} \text{ is a 3 by 2 matrix}$$

$$B = \begin{pmatrix} 2 & 4 & 1 \\ 3 & 7 & 0 \end{pmatrix} \text{ is a 2 by 3 matrix}$$

$$C = \begin{pmatrix} 3 & 2 & 6 \\ 1 & 5 & -2 \\ -1 & 6 & 1 \end{pmatrix} \text{ is a 3 by 3 square matrix}$$

$$D = (3, 5, 2) \text{ is a 1 by 3 matrix or row matrix}$$

$$E = \begin{pmatrix} 2 \\ 1 \end{pmatrix} \text{ is a 2 by 1 matrix or column matrix}$$

Matrices provide a convenient method of storing large quantities of data. As an example, consider a firm with multiple retail outlets. The inventory of the firm can be represented by the matrix G, where the columns represent the outlets and the rows represent products.

$$\begin{array}{cc} & \text{Outlets} \\ & \begin{array}{ccc} 1 & 2 & 3 \end{array} \;\; \text{Products} \\ G = & \begin{pmatrix} 50 & 60 & 40 \\ 175 & 200 & 125 \\ 40 & 25 & 30 \\ 205 & 235 & 275 \end{pmatrix} \begin{array}{c} 1 \\ 2 \\ 3 \\ 4 \end{array} \end{array}$$

Component a_{21} of matrix G is 175 units of products 2 at outlet 1. Component a_{42} is 235 units of product 4 at outlet 2.

1.4 SUBMATRIX

A matrix obtained by deleting some rows or columns or both of a given matrix is called a **submatrix** of a given matrix, *e.g.*,

$$\text{Matrix B} = \begin{bmatrix} a_{11} & a_{12} & a_{13} \\ a_{21} & a_{22} & a_{23} \end{bmatrix} \text{ is a submatrix of the matrix } A = \begin{bmatrix} a_{11} & a_{12} & a_{13} & a_{14} \\ a_{21} & a_{22} & a_{23} & a_{24} \\ a_{31} & a_{32} & a_{33} & a_{34} \end{bmatrix}$$

[Matrix B is obtained by deleting 4th column and third row of Matrix A]

1.5 EQUAL MATRICES

Two matrices A and B are said to be equal if

(i) The number of rows and columns of A is equal to the number of rows and columns of B and

(ii) The corresponding elements in the two matrices are equal, *e.g.*, the matrices

$$\begin{bmatrix} 2 & 3 & 4 \\ 3 & 6 & 8 \\ 3 & 2 & 1 \end{bmatrix} \text{ and } \begin{bmatrix} 2 & 3 & 4 \\ 3 & 6 & 8 \\ 3 & 2 & 1 \end{bmatrix} \text{ are equal matrices.}$$

Equal matrices may or may not be square matrices. The matrices

$$\begin{bmatrix} 2 & 3 & 1 \\ 1 & 0 & 5 \end{bmatrix} \text{ and } \begin{bmatrix} 2 & 3 & 1 \\ 1 & 0 & 5 \end{bmatrix} \text{ are also equal matrices}$$

1.6 ADDITION AND SUBTRACTION OF MATRICES

Addition or substraction of matrices is defined only if the matrices are of the same order, *i.e.,* if the two matrices have the same number of rows and columns. In the sum/difference matrix, the elements are the sum/difference of the corresponding elements of the two matrices.

If $$A = \begin{bmatrix} a_{11} & a_{12} & a_{13} \\ a_{21} & a_{22} & a_{23} \\ a_{31} & a_{32} & a_{33} \end{bmatrix}, \quad B = \begin{bmatrix} b_{11} & b_{12} & b_{13} \\ b_{21} & b_{22} & b_{23} \\ b_{31} & b_{32} & b_{33} \end{bmatrix}$$

Then $$[A + B] = \begin{bmatrix} a_{11}+b_{11} & a_{12}+b_{12} & a_{13}+b_{13} \\ a_{21}+b_{21} & a_{22}+b_{22} & a_{23}+b_{23} \\ a_{31}+b_{31} & a_{32}+b_{32} & a_{33}+b_{33} \end{bmatrix}$$

$$[A - B] = \begin{bmatrix} a_{11}-b_{11} & a_{12}-b_{12} & a_{13}-b_{13} \\ a_{21}-b_{21} & a_{22}-b_{22} & a_{23}-b_{23} \\ a_{31}-b_{31} & a_{32}-b_{32} & a_{33}-b_{33} \end{bmatrix}$$

If $$A = \begin{bmatrix} 3 & 4 & 5 \\ 6 & 8 & 9 \\ 4 & 2 & 1 \end{bmatrix}, \quad B = \begin{bmatrix} 4 & 8 & 6 \\ 3 & 1 & 5 \\ 4 & 1 & 2 \end{bmatrix}$$

Then $$A + B = \begin{bmatrix} 3+4 & 4+8 & 5+6 \\ 6+3 & 8+1 & 9+5 \\ 4+4 & 2+1 & 1+2 \end{bmatrix} = \begin{bmatrix} 7 & 12 & 11 \\ 9 & 9 & 14 \\ 8 & 3 & 3 \end{bmatrix}$$

$$A - B = \begin{bmatrix} 3-4 & 4-8 & 5-6 \\ 6-3 & 8-1 & 9-5 \\ 4-4 & 2-1 & 1-2 \end{bmatrix} = \begin{bmatrix} -1 & -4 & -1 \\ 3 & 7 & 4 \\ 0 & 1 & -1 \end{bmatrix}$$

Example 2. Let

$$A = \begin{pmatrix} 3 & 4 \\ 6 & 5 \\ 4 & 0 \end{pmatrix} \text{ and } B = \begin{pmatrix} 2 & 5 \\ 3 & 6 \\ 4 & 2 \end{pmatrix}$$

Determine $A + B$ and $A - B$

$$A + B = \begin{pmatrix} 3+2 & 4+5 \\ 6+3 & 5+6 \\ 4+4 & 0+2 \end{pmatrix} = \begin{pmatrix} 5 & 9 \\ 9 & 11 \\ 8 & 2 \end{pmatrix}$$

$$A - B = \begin{pmatrix} 3-2 & 4-5 \\ 6-3 & 5-6 \\ 4-4 & 0-2 \end{pmatrix} = \begin{pmatrix} 1 & -1 \\ 3 & -1 \\ 0 & -2 \end{pmatrix}$$

Example 3. Let

$$C=\begin{pmatrix}1 & 3 & 4\\ 2 & 6 & 5\\ 1 & -1 & 6\end{pmatrix},\ D=\begin{pmatrix}2 & 3 & 5\\ 4 & 1 & 3\\ 5 & 2 & 1\end{pmatrix},\ E=\begin{pmatrix}1 & -2 & 5\\ 3 & -1 & 4\\ 6 & 5 & 3\end{pmatrix}$$

Show that $C+(D+E)=(C+D)+E$

$$C+(D+E)=\begin{pmatrix}1 & 3 & 4\\ 2 & 6 & 5\\ 1 & -1 & 6\end{pmatrix}+\begin{pmatrix}3 & 1 & 10\\ 7 & 0 & 7\\ 11 & 7 & 4\end{pmatrix}=\begin{pmatrix}4 & 4 & 14\\ 9 & 6 & 12\\ 12 & 6 & 10\end{pmatrix}$$

$$(C+D)+E=\begin{pmatrix}3 & 6 & 9\\ 6 & 7 & 8\\ 6 & 1 & 7\end{pmatrix}+\begin{pmatrix}1 & -2 & 5\\ 3 & -1 & 4\\ 6 & 5 & 3\end{pmatrix}=\begin{pmatrix}4 & 4 & 14\\ 9 & 6 & 12\\ 12 & 6 & 10\end{pmatrix}$$

The examples illustrate the commutative and associative laws of addition. The commutative law states that $A+B=B+A$. The associative law states that $C+(D+E)=(C+D)+E$. These examples show that these laws apply to the addition and substraction of matrices.

1.7 MATRIX MULTIPLICATION

Matrix multiplication is divided into two cases.

(1) The multiplication of a matrix by a scalar and

(2) The multiplication of a matrix by a matrix.

The simplest of these cases is the multiplication of a matrix by a scalar.

Multiplication of a Matrices by a Scalar

A matrix may be multiplicated by a scalar number by multiplying each component in the matrix by that scalar. The resulting matrix will be of the same order as the original matrix.

Example 4. Multiply the matrix $Y=\begin{pmatrix}2 & 3 & 5\\ 4 & 6 & -2\end{pmatrix}$ by the scalar $k=3$.

$$kY=3\begin{pmatrix}2 & 3 & 5\\ 4 & 6 & -2\end{pmatrix}=\begin{pmatrix}6 & 9 & 15\\ 12 & 18 & -6\end{pmatrix}$$

Example 5. Multiply the matrix $B=\begin{pmatrix}2 & 1\\ 1 & -3\end{pmatrix}$ by the scalar $k=-2$.

$$kB=-2\begin{pmatrix}2 & 1\\ 1 & -3\end{pmatrix}=\begin{pmatrix}-4 & -2\\ -2 & 6\end{pmatrix}$$

Example 6. Multiply the matrix $A=\begin{pmatrix}1 & 3 & -2\\ 4 & 7 & -1\end{pmatrix}$ by the scalar $k=0.1$.

$$kA=0.1\begin{pmatrix}1 & 3 & -2\\ 4 & 7 & -1\end{pmatrix}=\begin{pmatrix}0.1 & 0.3 & -0.2\\ 0.4 & 0.7 & -0.1\end{pmatrix}$$

Multiplication of a Matrix by a Matrix

Matrix multiplication is somewhat more compliated than the multiplication of a matrix by a scalar. Two matrices may be multiplied together only if the **number of columns in the first matrix is equal as of the number of rows in the second matrix**. If matrix A is of the order a by b and matrix B is of c by d, the matrix product AB is defined only if b is equal to c. Thus, a 2 by 3 matrix can be multiplied by a 3 by 4 matrix but not, for example, by a 2 by 4 matrix. The matrix resulting from the multiplication has dimensions equivalent to the number of rows in the first matrix and the number of column in the second matrix. To summarize, matrix A of order a by b may be multiplied by B of order c by d to form matrix AB of order a by d, provided that b equal c. If the number of columns in the first matrix equals the number of rows in the second matrix, the matrices are said to be **conformable for multiplication**.

If A is a $m \times n$ matrix and B is a $n \times p$ matrix then AB is $m \times p$ matrix.

Example 7. If $A = \begin{vmatrix} 2 & 3 & 4 \\ 5 & 6 & 7 \\ 4 & 7 & 6 \end{vmatrix}, B = \begin{vmatrix} 1 \\ 3 \\ 5 \end{vmatrix}$, find AB.

$$\text{Then AB} = \begin{vmatrix} \text{Sum of the products along thin lines} \\ \text{Sum of the products along medium lines} \\ \text{Sum of the products along thick lines} \end{vmatrix}$$

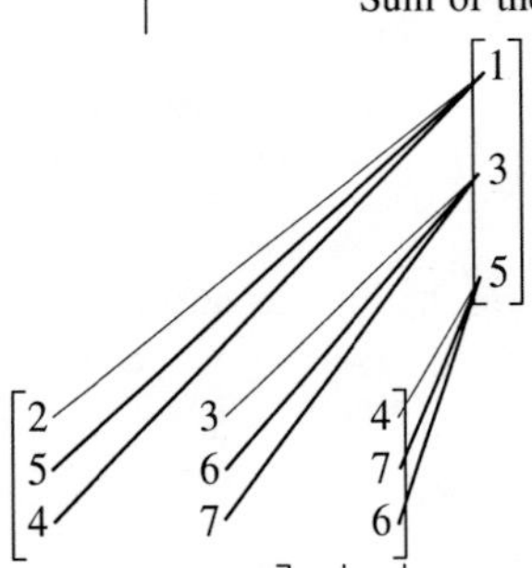

Solution. $AB = \begin{bmatrix} 2 \times 1 + 3 \times 3 + 4 \times 5 \\ 5 \times 1 + 6 \times 3 + 7 \times 5 \\ 4 \times 1 + 7 \times 3 + 6 \times 5 \end{bmatrix} = \begin{vmatrix} 31 \\ 58 \\ 55 \end{vmatrix}$, *i.e.*, AB is a 3×1 matrix.

Example 8. With the same A as in Example 7, find AB if $B = \begin{bmatrix} 1 & 2 \\ 3 & 4 \\ 5 & 6 \end{bmatrix}$

1st column **2nd Column**

Solution. Then $$AB = \begin{vmatrix} \text{1st column} & 2 \times 2 + 3 \times 4 + 4 \times 6 \\ \text{Same as in} & 5 \times 2 + 6 \times 4 + 7 \times 6 \\ \text{Example 7} & 4 \times 2 + 7 \times 4 + 6 \times 6 \end{vmatrix}$$

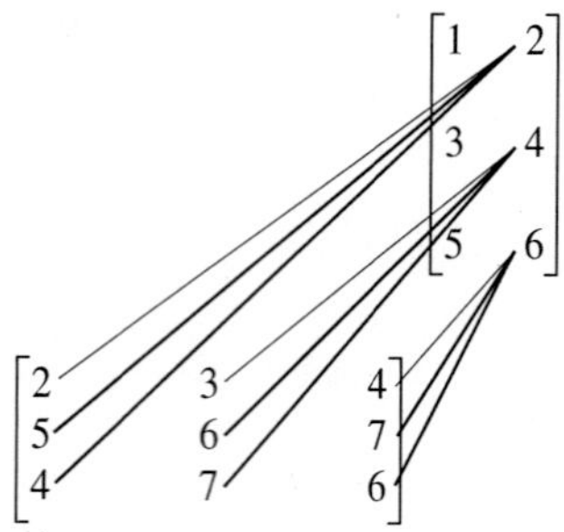

i.e., AB is a 3×2 matrix.

In the above example 8 if B is a matrix of order 3×3, then AB will also be a 3×3 matrix. Process of multiplication is symbolically represented as

$$AB = \begin{bmatrix} 2 & 3 & 4 \\ 5 & 6 & 7 \\ 4 & 7 & 6 \end{bmatrix}_{3 \times 3} \downarrow \begin{bmatrix} 1 \\ 3 \\ 5 \end{bmatrix}_{3 \times 1}$$

Arrows indicate that, for multiplication rows are considered for the first matrix and columns for the 2nd matrix. **(Important)**

$$A = \begin{bmatrix} a_{11} & a_{12} & a_{13} \\ a_{21} & a_{22} & a_{33} \end{bmatrix} \text{ by } B = \begin{bmatrix} b_{11} & b_{12} & b_{13} \\ b_{21} & b_{22} & b_{23} \\ b_{31} & b_{32} & b_{33} \end{bmatrix}$$

Since A is a 2 by 3 matrix and B is 3 by 3 matrix, the matrices are conformable for multiplication. The resulting matrix AB has dimensions 2 by 3. The multiplication is performed as follows :

$$AB = \begin{pmatrix} a_{11} & a_{12} & a_{13} \\ a_{21} & a_{22} & a_{23} \end{pmatrix} \begin{pmatrix} b_{11} & b_{12} & b_{13} \\ b_{21} & b_{22} & b_{23} \\ b_{31} & b_{32} & b_{33} \end{pmatrix}$$

$$AB = \begin{pmatrix} a_{11}b_{11} + a_{12}b_{21} + a_{13}b_{31} & a_{11}b_{12} + a_{12}b_{22} + a_{13}b_{32} & a_{11}b_{13} + a_{12}b_{23} + a_{13}b_{33} \\ a_{21}b_{11} + a_{22}b_{21} + a_{23}b_{31} & a_{21}b_{12} + a_{22}b_{22} + a_{23}b_{32} & a_{21}b_{13} + a_{22}b_{23} + a_{23}b_{33} \end{pmatrix}$$

Matrix AB in the preceding example is a 2 by 3 matrix. The component of the matrix are formed by the sum of the products of the rows of the first matrix and columns of the second matrix. The sum of the products of the components in the first row of A and the first column of B give the first row, first column component of AB. Similarly, the sum of the products of the components in the first row of A and second column of B gives the first row, second column component of AB. This procedure is continued until the matrix AB is formed.

Example 9. Determine the product of $A = \begin{pmatrix} 6 & 2 & 4 \\ 1 & 2 & 2 \end{pmatrix}$ and $B = \begin{pmatrix} 3 & 2 \\ 2 & 4 \\ 4 & 5 \end{pmatrix}$.

$$AB = \begin{pmatrix} 6 & 2 & 4 \\ 1 & 2 & 2 \end{pmatrix} \begin{pmatrix} 3 & 2 \\ 2 & 4 \\ 4 & 5 \end{pmatrix} = \begin{pmatrix} 6 \times 3 + 2 \times 2 + 4 \times 4 & 6 \times 2 + 2 \times 4 + 4 \times 5 \\ 1 \times 3 + 2 \times 2 + 2 \times 4 & 1 \times 2 + 2 \times 4 + 2 \times 5 \end{pmatrix}$$

$$AB = \begin{pmatrix} 18+4+16 & 12+8+20 \\ 3+4+8 & 2+8+10 \end{pmatrix} = \begin{pmatrix} 38 & 40 \\ 15 & 20 \end{pmatrix}$$

Example 10. Determine the product of $A = \begin{pmatrix} 6 & 1 \\ 3 & 7 \end{pmatrix}$ and $B = \begin{pmatrix} 2 & 4 & 7 \\ 3 & 1 & 5 \end{pmatrix}$.

$$AB = \begin{pmatrix} 6 & 1 \\ 3 & 7 \end{pmatrix}\begin{pmatrix} 2 & 4 & 7 \\ 3 & 1 & 5 \end{pmatrix}$$

$$AB = \begin{pmatrix} 6 & 1 \\ 3 & 7 \end{pmatrix}\begin{pmatrix} 2 & 4 & 7 \\ 3 & 1 & 5 \end{pmatrix}$$

$$= \begin{pmatrix} 6\times 2+1\times 3 & 6\times 4+1\times 1 & 6\times 7+1\times 5 \\ 3\times 2+7\times 3 & 3\times 4+7\times 1 & 3\times 7+7\times 5 \end{pmatrix} = \begin{pmatrix} 15 & 25 & 47 \\ 27 & 19 & 56 \end{pmatrix}$$

In the first example, a 2 by 3 matrix is multiplied by a 3 by 2 matrix. The matrices are comfortable for multiplications, since, the number of columns in the first matrix equals the number of rows in the second. The dimensions of the resulting matrix are equal to the number of rows of the first matrix and the number of columns of the second matrix. The second example shows that a 2 by 2 matrix and a 2 by 3 matrix are comfortable for multiplication. The product of the matrices is a 2 by 3 matrix.

Example 11. For $A = (6, 2, 3)$ and $B = \begin{pmatrix} 3 \\ 5 \\ 4 \end{pmatrix}$, find AB and BA.

The row vector A is a 1 by 3 matrix and the column vector B is a 3 by 1 matrix. The inner product AB gives the 1 by 1 matrix of scalar, $AB = 34$.

$$AB = (6,\ 2,\ 3)\begin{pmatrix} 3 \\ 5 \\ 4 \end{pmatrix} = 12 + 10 + 12 = 34$$

Multiplying B by A is equivalent to multiplying a 3 by 1 matrix by a 1 by 3 matrix. Since, the number of columns in the first matrix equals to the number of rows in the second matrix, the product is defined. The product of an n by 1 vector and a 1 by n vector is termed the **outer product** and has dimension n by n. To illustrate,

$$BA = \begin{pmatrix} 3 \\ 5 \\ 4 \end{pmatrix}(6,\ 2,\ 3) = \begin{pmatrix} 12 & 6 & 9 \\ 30 & 10 & 15 \\ 24 & 8 & 12 \end{pmatrix}$$

The outer product of the vectors B and A is the 3 by 3 matrix BA.

SOLVED EXAMPLES

Example 12. If $A = \begin{bmatrix} 2 & 1 \\ 1 & 2 \end{bmatrix}$, $B = \begin{bmatrix} 1 & 0 \\ 0 & 1 \end{bmatrix}$, $C = \begin{bmatrix} 1 & 2 \\ 3 & 4 \\ 1 & 2 \end{bmatrix}$, $D = \begin{bmatrix} 0 & 1 & 0 \\ 1 & 1 & 1 \end{bmatrix}$, Find $A + B - 2DC$.

Solution.

$$A + B - 2DC = \begin{bmatrix} 2 & 1 \\ 1 & 2 \end{bmatrix} + \begin{bmatrix} 1 & 0 \\ 0 & 1 \end{bmatrix} - 2\begin{bmatrix} 0 & 1 & 0 \\ 1 & 1 & 1 \end{bmatrix}\begin{bmatrix} 1 & 2 \\ 3 & 4 \\ 1 & 2 \end{bmatrix}$$

$$= \begin{bmatrix} 2 & 1 \\ 1 & 2 \end{bmatrix} + \begin{bmatrix} 1 & 0 \\ 0 & 1 \end{bmatrix} - 2\begin{bmatrix} 0+3+0 & 0+4+0 \\ 1+3+1 & 2+4+2 \end{bmatrix}$$

$$= \begin{bmatrix} 2 & 1 \\ 1 & 2 \end{bmatrix} + \begin{bmatrix} 1 & 0 \\ 0 & 1 \end{bmatrix} - 2\begin{bmatrix} 3 & 4 \\ 5 & 8 \end{bmatrix}$$

$$= \begin{bmatrix} 2+1-6 & 1+0-8 \\ 1+0-10 & 2+1-16 \end{bmatrix} = \begin{bmatrix} -3 & -7 \\ -9 & -13 \end{bmatrix}$$

Example 13. If $A = \begin{bmatrix} 1 & -1 \\ 0 & 2 \end{bmatrix}$ and $B = \begin{bmatrix} 1 & 0 \\ 1 & 2 \end{bmatrix}$, write the matrices $(A+B)(A-B)$ and $(A^2 - B^2)$, and determine whether they are equal.

Solution.

$$(A+B)(A-B) = \begin{bmatrix} 2 & -1 \\ 1 & 4 \end{bmatrix}\begin{bmatrix} 0 & -1 \\ -1 & 0 \end{bmatrix} = \begin{bmatrix} 1 & -2 \\ -4 & -1 \end{bmatrix}$$

$$A^2 = \begin{bmatrix} 1 & -1 \\ 0 & 2 \end{bmatrix}\begin{bmatrix} 1 & -1 \\ 0 & 2 \end{bmatrix} = \begin{bmatrix} 1 & -3 \\ 0 & 4 \end{bmatrix}$$

$$B^2 = \begin{bmatrix} 1 & 0 \\ 1 & 2 \end{bmatrix}\begin{bmatrix} 1 & 0 \\ 1 & 2 \end{bmatrix} = \begin{bmatrix} 1 & 0 \\ 3 & 4 \end{bmatrix}$$

$$A^2 - B^2 = \begin{bmatrix} 1 & -3 \\ 0 & 4 \end{bmatrix} - \begin{bmatrix} 1 & 0 \\ 3 & 4 \end{bmatrix} = \begin{bmatrix} 0 & -3 \\ -3 & 0 \end{bmatrix} = -3\begin{bmatrix} 0 & 1 \\ 1 & 0 \end{bmatrix}$$

Hence, $(A+B)(A-B) \neq A^2 - B^2$.

Example 14. If $A = \begin{bmatrix} 1 & 0 & 0 \\ 4 & 2 & 3 \\ 2 & 0 & 1 \end{bmatrix}$, find $A^2 - A - I$.

Solution.

$$A^2 = \begin{bmatrix} 1 & 0 & 0 \\ 4 & 2 & 3 \\ 2 & 0 & 1 \end{bmatrix}\begin{bmatrix} 1 & 0 & 0 \\ 4 & 2 & 3 \\ 2 & 0 & 1 \end{bmatrix}$$

$$= \begin{bmatrix} 1\times1+0\times4+0\times2 & 1\times0+0\times2+0\times0 & 1\times0+0\times3+0\times1 \\ 4\times1+2\times4+3\times2 & 4\times0+2\times2+3\times0 & 4\times0+2\times3+3\times1 \\ 2\times1+0\times4+1\times2 & 2\times0+0\times2+1\times0 & 2\times0+0\times3+1\times1 \end{bmatrix}$$

$$= \begin{bmatrix} 1 & 0 & 0 \\ 18 & 4 & 9 \\ 4 & 0 & 1 \end{bmatrix}$$

$$\therefore \quad A^2 - A - I = \begin{bmatrix} 1 & 0 & 0 \\ 18 & 4 & 9 \\ 4 & 0 & 1 \end{bmatrix} - \begin{bmatrix} 1 & 0 & 0 \\ 4 & 2 & 3 \\ 2 & 0 & 1 \end{bmatrix} - \begin{bmatrix} 1 & 0 & 0 \\ 0 & 1 & 0 \\ 0 & 0 & 1 \end{bmatrix}$$

$$= \begin{bmatrix} 1-1-1 & 0-0-0 & 0-0-0 \\ 18-4-0 & 4-2-1 & 9-3-0 \\ 4-2-0 & 0-0-0 & 1-1-1 \end{bmatrix} = \begin{bmatrix} 1 & 0 & 0 \\ 14 & 1 & 6 \\ 2 & 0 & -1 \end{bmatrix}$$

Example 15. If $A = \begin{bmatrix} 4 & 2 \\ 4 & 1 \end{bmatrix}$ and $B = \begin{bmatrix} a & b \\ 3 & 5 \end{bmatrix}$ find a and b such that $AB = BA$. Compute $3A + 5B$.

Solution.

$$A = \begin{bmatrix} 3 & 2 \\ 4 & 1 \end{bmatrix}, B = \begin{bmatrix} a & b \\ 3 & 5 \end{bmatrix}, AB = \begin{bmatrix} 3 & 2 \\ 4 & 1 \end{bmatrix}\begin{bmatrix} a & b \\ 3 & 5 \end{bmatrix}$$

$$AB = \begin{bmatrix} 3a+6 & 3b+10 \\ 4a+3 & 4b+5 \end{bmatrix} \text{ and } BA = \begin{bmatrix} a & b \\ 3 & 5 \end{bmatrix}\begin{bmatrix} 3 & 2 \\ 4 & 1 \end{bmatrix} = \begin{bmatrix} 3a+4b & 2a+b \\ 29 & 11 \end{bmatrix}$$

$$AB = BA \text{ gives } = \begin{bmatrix} 3a+6 & 3b+10 \\ 4a+3 & 4b+5 \end{bmatrix} = \begin{bmatrix} 3a+4b & 2a+b \\ 29 & 11 \end{bmatrix}$$

On equating the respective elements

$$4a + 3 = 29 \Rightarrow a = \frac{26}{4} = 6.5,\ 4b + 5 = 11 \Rightarrow b = \frac{6}{4} = 1.5$$

$$3A + 5B = 3\begin{bmatrix} 3 & 2 \\ 4 & 1 \end{bmatrix} + 5\begin{bmatrix} 6.5 & 1.5 \\ 3 & 5 \end{bmatrix}$$

$$= \begin{bmatrix} 9 & 6 \\ 12 & 3 \end{bmatrix} + \begin{bmatrix} 32.5 & 7.5 \\ 15 & 25 \end{bmatrix} = \begin{bmatrix} 41.5 & 13.5 \\ 27 & 28 \end{bmatrix}$$

Example 16. If $A = \begin{bmatrix} 1 & 2 \\ 3 & 4 \end{bmatrix}$, $B = \begin{bmatrix} 1 & 0 \\ 2 & -3 \end{bmatrix}$ and $C = \begin{bmatrix} 1 & -1 \\ 0 & 1 \end{bmatrix}$, show that $A(B + C) = AB + AC$.

Solution.

$$B + C = \begin{bmatrix} 1 & 0 \\ 2 & -3 \end{bmatrix} + \begin{bmatrix} 1 & -1 \\ 0 & 1 \end{bmatrix} = \begin{bmatrix} 2 & -1 \\ 2 & -2 \end{bmatrix}$$

$$A(B + C) = \begin{bmatrix} 1 & 2 \\ 3 & 4 \end{bmatrix}\begin{bmatrix} 2 & -1 \\ 2 & -2 \end{bmatrix} = \begin{bmatrix} 2+4 & -1-4 \\ 6+8 & -3-8 \end{bmatrix} = \begin{bmatrix} 6 & -5 \\ 14 & -11 \end{bmatrix}$$

$$AB = \begin{bmatrix} 1 & 2 \\ 3 & 4 \end{bmatrix}\begin{bmatrix} 1 & 0 \\ 2 & -3 \end{bmatrix} = \begin{bmatrix} 1+4 & 0-6 \\ 3+8 & 0/12 \end{bmatrix} = \begin{bmatrix} 5 & -6 \\ 11 & -12 \end{bmatrix}$$

$$AC = \begin{bmatrix} 1 & 2 \\ 3 & 4 \end{bmatrix}\begin{bmatrix} 1 & -1 \\ 0 & 1 \end{bmatrix} = \begin{bmatrix} 1+0 & -1+2 \\ 3+0 & -3+4 \end{bmatrix} = \begin{bmatrix} 1 & 1 \\ 3 & 1 \end{bmatrix}$$

$$AB + AC = \begin{bmatrix} 5 & -6 \\ 11 & -12 \end{bmatrix} + \begin{bmatrix} 1 & 1 \\ 3 & 1 \end{bmatrix} = \begin{bmatrix} 6 & -5 \\ 14 & -11 \end{bmatrix}$$

Hence, $A(B + C) = AB + AC$.

Example 17. The inventory of a firm with multiple retail outlets is given by matrix G.

$$\begin{array}{c} \text{Outlets} \\ \begin{array}{cccl} 1 & 2 & 3 & \text{Products} \end{array} \end{array}$$

$$G = \begin{pmatrix} 50 & 60 & 40 \\ 175 & 200 & 125 \\ 40 & 25 & 30 \\ 205 & 235 & 275 \end{pmatrix} \begin{matrix} 1 \\ 2 \\ 3 \\ 4 \end{matrix}$$

If the cost of inventory is given by the vector $C = (20, 10, 30, 40)$, determine the value of inventory at each outlet.

The value of inventory at each outlet is given by the matrix CG. Since, C is a 1 by 4 matrix and G is a 4 by 3 matrix, the matrices are conformable for multiplication. The product is

$$CG = (20,\ 10,\ 30,\ 40)\begin{pmatrix} 50 & 60 & 40 \\ 175 & 200 & 125 \\ 40 & 25 & 30 \\ 205 & 235 & 275 \end{pmatrix}$$

$$CG = (12150,\ 13350,\ 13950)$$

Inventory has a value of Rs. 12,150 at outelet 1, Rs. 13,350 at outlet 2, and Rs. 13,950 at outlet 3.

1.8 COMMUTATIVE LAW, ASSOCIATIVE LAW AND DISTRIBUTIVE LAW

If the matrices A and B are conformable for multiplication. Then,

$$AB \neq BA$$

It may also be noted that if product AB is defined the Product BA may or may not be defined for example is A is a 3×2 matrix and B is 2×1 matrix then Product AB is defined but BA is not defined. If A and B are square matrices of the same order then AB and BA both will be defined but may or may not be equal.

The associative and distributive laws of multiplication do, however, apply to matrix multiplication. Given three matrices A, B and C that are conformable for multiplication, the **associative law states** and $A(BC) = (AB)C$. **The distributive law**, again if conformability is assumed, states that $A(B + C) = AB + AC$. These proeprties are illustrated by the following two examples.

Example 18. For $A = \begin{pmatrix} 3 & -1 & 4 \\ 6 & 8 & 2 \\ 1 & -5 & 4 \end{pmatrix}$ and $B = \begin{pmatrix} 2 & 6 & -1 \\ 3 & -2 & 4 \\ 5 & 3 & -3 \end{pmatrix}$, find AB and BA.

$$AB = \begin{pmatrix} 3 & -1 & 4 \\ 6 & 8 & 2 \\ 1 & -5 & 4 \end{pmatrix}\begin{pmatrix} 2 & 6 & -1 \\ 3 & -2 & 4 \\ 5 & 3 & -3 \end{pmatrix} = \begin{pmatrix} 23 & 32 & -19 \\ 46 & 26 & 20 \\ 7 & 28 & -33 \end{pmatrix}$$

$$BA = \begin{pmatrix} 2 & 6 & -1 \\ 3 & -2 & 4 \\ 5 & 3 & -3 \end{pmatrix}\begin{pmatrix} 3 & -1 & 4 \\ 6 & 8 & 2 \\ 1 & -5 & 4 \end{pmatrix} = \begin{pmatrix} 41 & 51 & 16 \\ 1 & -39 & 24 \\ 30 & 34 & 14 \end{pmatrix}$$

The preceding example illustrates that the cumulative law of multiplication does not apply to the multiplication of matrices. The commutative law of multiplication states that $AB = BA$. Since, AB is not equal to BA, the **commutative law** of multiplication does not apply to matrix multiplication.

Example 19. Show that the associative law of multiplication applies to the matrices

$$A = \begin{pmatrix} 2 & 3 \\ 3 & 5 \end{pmatrix},\ B = \begin{pmatrix} 3 & 4 \\ 5 & 6 \end{pmatrix},\ C = \begin{pmatrix} 4 & 5 \\ 6 & 7 \end{pmatrix}$$

The associative law of multiplication states that $A(BC) = (AB)C$. To verify the law we multiply A by BC.

$$BC = \begin{pmatrix} 3 & 4 \\ 5 & 6 \end{pmatrix}\begin{pmatrix} 4 & 5 \\ 6 & 7 \end{pmatrix} = \begin{pmatrix} 36 & 43 \\ 56 & 67 \end{pmatrix}$$

$$A(BC) = \begin{pmatrix} 2 & 3 \\ 4 & 5 \end{pmatrix}\begin{pmatrix} 36 & 43 \\ 56 & 67 \end{pmatrix} = \begin{pmatrix} 240 & 287 \\ 424 & 507 \end{pmatrix}$$

Next, we multiply AB by the matrix C.

$$AB = \begin{pmatrix} 2 & 3 \\ 4 & 5 \end{pmatrix}\begin{pmatrix} 3 & 4 \\ 5 & 6 \end{pmatrix} = \begin{pmatrix} 21 & 26 \\ 37 & 46 \end{pmatrix}$$

$$(AB)C = \begin{pmatrix} 21 & 26 \\ 37 & 46 \end{pmatrix}\begin{pmatrix} 4 & 5 \\ 6 & 7 \end{pmatrix} = \begin{pmatrix} 240 & 287 \\ 424 & 507 \end{pmatrix}$$

Since, $A(BC)$ equals $(AB)C$, we conclude that the associative law applies in matrix multiplication.

Example 20. Show that the distributive law of algebra applies to the matrices

$$A = \begin{pmatrix} 2 & 3 \\ 4 & 5 \end{pmatrix}, B = \begin{pmatrix} 2 & 3 & 6 \\ 3 & 5 & 1 \end{pmatrix} \text{ and } C = \begin{pmatrix} 1 & 4 & 5 \\ 6 & 3 & 1 \end{pmatrix}$$

The distributive law states that $A(B + C) = AB + AC$, provided that B ad C are comformable for addition and that AB and AC are conformable for multiplication. The matrix $A(B + C)$ is

$$A(B+C) = \begin{pmatrix} 2 & 3 \\ 4 & 5 \end{pmatrix} \left[\begin{pmatrix} 2 & 3 & 6 \\ 3 & 5 & 1 \end{pmatrix} + \begin{pmatrix} 1 & 4 & 5 \\ 6 & 3 & 1 \end{pmatrix} \right]$$

$$A(B+C) = \begin{pmatrix} 2 & 3 \\ 4 & 5 \end{pmatrix} \left[\begin{pmatrix} 3 & 7 & 11 \\ 9 & 8 & 2 \end{pmatrix} = \begin{pmatrix} 33 & 38 & 28 \\ 57 & 68 & 54 \end{pmatrix} \right]$$

and the matrix $AB + AC$ is

$$AB + AC = \begin{pmatrix} 2 & 3 \\ 4 & 5 \end{pmatrix} \begin{pmatrix} 2 & 3 & 6 \\ 3 & 5 & 1 \end{pmatrix} + \begin{pmatrix} 2 & 3 \\ 4 & 5 \end{pmatrix} \begin{pmatrix} 1 & 4 & 5 \\ 6 & 3 & 1 \end{pmatrix}$$

$$AB + AC = \begin{pmatrix} 13 & 21 & 15 \\ 23 & 37 & 29 \end{pmatrix} + \begin{pmatrix} 20 & 17 & 13 \\ 34 & 31 & 25 \end{pmatrix} = \begin{pmatrix} 33 & 38 & 28 \\ 57 & 68 & 54 \end{pmatrix}$$

Since, $A(B + C)$ equals $AB + AC$, this illustrates that the distributive law applies to matrix algebra.

Example 21. If $A = \begin{bmatrix} 1 & 1 \\ a & 2 \end{bmatrix}$, $B = \begin{bmatrix} 1 & 0 \\ 0 & 1 \end{bmatrix}$ and $C = \begin{bmatrix} b & 1 \\ 3 & 3 \end{bmatrix}$, find the values of a and b for which $(A + B) = BC$.

Solution.

$$A + B = \begin{bmatrix} 1 & 1 \\ a & 2 \end{bmatrix} \begin{bmatrix} 1 & 0 \\ 0 & 1 \end{bmatrix} = \begin{bmatrix} 2 & 1 \\ a & 3 \end{bmatrix}$$

$$BC = \begin{bmatrix} 1 & 0 \\ 0 & 1 \end{bmatrix} \begin{bmatrix} b & 1 \\ 3 & 3 \end{bmatrix} = \begin{bmatrix} b & 1 \\ 3 & 3 \end{bmatrix}$$

$$A + B = BC \Rightarrow \begin{bmatrix} 2 & 1 \\ a & 3 \end{bmatrix} = \begin{bmatrix} b & 1 \\ 3 & 3 \end{bmatrix} \Rightarrow b = 2,\ a = 3$$

Example 22. Find matrix A such that

$$A \begin{bmatrix} 3 & 4 \\ 6 & 2 \end{bmatrix} = \begin{bmatrix} 2 & 8 \\ 9 & 4 \end{bmatrix}$$

Solution. Let

$$B = \begin{bmatrix} 3 & 4 \\ 6 & 2 \end{bmatrix},\ C = \begin{bmatrix} 2 & 8 \\ 9 & 4 \end{bmatrix}$$

Since, B is 2×2 matrix and C is also a 2×2 matrix, then A has to be 2×2 matrix.

Let

$$A = \begin{bmatrix} a & b \\ c & d \end{bmatrix}$$

$\therefore$

$$\begin{bmatrix} a & b \\ c & d \end{bmatrix} \begin{bmatrix} 3 & 4 \\ 6 & 2 \end{bmatrix} = \begin{bmatrix} 3a + 6b & 4a + 2b \\ 3c + 6d & 4c + 2d \end{bmatrix} = \begin{bmatrix} 2 & 8 \\ 9 & 4 \end{bmatrix}$$

$\Rightarrow$ $3a + 6b = 2$ ...(I) $\quad$ $4a + 2b = 8$ ...(II)

$3c + 6d = 9$ ...(III) $\quad$ $4c + 2d = 4$ ...(IV)

Solving I and II, we get $a = \frac{22}{9}, b = \frac{-8}{9}$.

Solving III and IV, we get $c = \frac{1}{3}, d = \frac{4}{3}.$

$$\therefore \qquad A = \begin{bmatrix} \frac{22}{9} & \frac{-8}{9} \\ \frac{1}{3} & \frac{4}{3} \end{bmatrix}$$

Second Method

Determinant $\begin{bmatrix} 3 & 4 \\ 6 & 2 \end{bmatrix} = -18 \neq 0.$

$$\text{Adj.} \begin{bmatrix} 3 & 4 \\ 6 & 2 \end{bmatrix} = \begin{bmatrix} 2 & -4 \\ -6 & 3 \end{bmatrix}$$

$$\text{Inverse of} \qquad \begin{bmatrix} 3 & 4 \\ 6 & 2 \end{bmatrix} = \frac{\begin{bmatrix} 2 & -4 \\ -6 & 3 \end{bmatrix}}{-18}$$

Multiplying both sides of the given equation by inverse of $\begin{bmatrix} 3 & 4 \\ 6 & 2 \end{bmatrix}$ we get

$$A = \begin{bmatrix} 3 & 4 \\ 6 & 2 \end{bmatrix} \frac{\begin{bmatrix} 2 & -4 \\ -6 & 3 \end{bmatrix}}{-18} = \frac{\begin{bmatrix} 2 & 8 \\ 9 & 4 \end{bmatrix} \begin{bmatrix} 2 & -4 \\ -6 & 3 \end{bmatrix}}{-18}$$

$$\Rightarrow \qquad AI = -\frac{1}{18} \begin{bmatrix} -44 & 16 \\ -6 & -24 \end{bmatrix} = \begin{bmatrix} \frac{22}{9} & \frac{-8}{9} \\ \frac{1}{3} & \frac{4}{3} \end{bmatrix} \qquad [\because AA^{-1} = I]$$

$$\Rightarrow \qquad A = \begin{bmatrix} \frac{22}{9} & \frac{-8}{9} \\ \frac{1}{3} & \frac{4}{3} \end{bmatrix} \qquad [\because AI = A]$$

Note : Student, can apply second method after learning about the inverse.

Example 23. A radio manufacturing compay produces two types of radio.

Type	Manufacturing Cost
A	Rs. 280
B	Rs. 310

During a week the company produced 128 radios. The total cost of these radios being Rs. 37,280, using matrix algebra find the number of A type and B type radios that have been produced during the week.

Solution. Let the number of type A radios be x, then the number of type B radios $= 128 - x$

$$[x\ (128 - x)] \begin{bmatrix} 280 \\ 310 \end{bmatrix} = [280x + 310\ (128 - x)]$$

$$= [-\ 30x + 310 \times 128] = 37.280$$

$$\Rightarrow \qquad -30x + 310 \times 128 = 37{,}280$$

$$\Rightarrow \qquad 30x = -37{,}280 + 310 \times 128 = 2{,}400$$

$\Rightarrow \qquad x = 80$

$\Rightarrow$ Type A radio = 80,

Type B radio = 128 – 80 = 48

1.9 IDENTITY MATRIX

The identity matrix is a square matrix that has 1's on the upper left to lower right diagonal *i.e.*, principal diagonal and 0's elsewhere. The 2×2 identify matrix, commonly designated by I, is

$$I = \begin{pmatrix} 1 & 0 \\ 0 & 1 \end{pmatrix}$$

Similarly, the 3 by 3 and by 4 by 4 identity matrices, also designated by I, are

$$I = \begin{pmatrix} 1 & 0 & 0 \\ 0 & 1 & 0 \\ 0 & 0 & 1 \end{pmatrix} \text{ and } I = \begin{pmatrix} 1 & 0 & 0 & 0 \\ 0 & 1 & 0 & 0 \\ 0 & 0 & 1 & 0 \\ 0 & 0 & 0 & 1 \end{pmatrix}$$

The identity matrix has properties in matrix algebra similar to those of the number I in ordinary algebra. Specifically, the identity matrix has the property that the product of any matrix A with a comfortable identity matrix I is equal to the matrix A. Symbolically, this means that $AI = A$, or alternatively, $IA = A$. Thus, although we pointed out that the commutative law of multiplication does not apply in matrix multiplication, it does apply in the case of multiplication of a matrix by an identity matrix.

Please note.

(i) Multiplication by a unit matrix leaves the matrix unchanged, *i.e.*, $IA = AI = A$.

(ii) If the multiplicand is not a square matrix then the unit matrix for premultiplication will be different than the unit matrix for postmultiplication, e.g.,

If $A = \begin{bmatrix} 3 & 2 & 1 \\ 4 & 5 & 8 \end{bmatrix}$, then the unit matrix for premultiplication will be $\begin{bmatrix} 1 & 0 \\ 0 & 1 \end{bmatrix}$ and for postmultiplication will be $\begin{bmatrix} 1 & 0 & 0 \\ 0 & 1 & 0 \\ 0 & 0 & 1 \end{bmatrix}$.

(iii) Unit matrix raised by any power is equal to the unit matrix of the same order, *i.e.*,

$$\mathbf{I} = \mathbf{I}^2 = \mathbf{I}^3 = \ldots = \mathbf{I}^n$$

(iv) A unit matrix is always a square matrix.

Example 24. For $A = \begin{pmatrix} 2 & 3 \\ 4 & 5 \end{pmatrix}$, verify that $AI = A - IA$

$$AI = \begin{pmatrix} 2 & 3 \\ 4 & 5 \end{pmatrix} \begin{pmatrix} 1 & 0 \\ 0 & 1 \end{pmatrix} = \begin{pmatrix} 2 & 3 \\ 4 & 5 \end{pmatrix}$$

$$IA = \begin{pmatrix} 1 & 0 \\ 0 & 1 \end{pmatrix} \begin{pmatrix} 2 & 3 \\ 4 & 5 \end{pmatrix} = \begin{pmatrix} 2 & 3 \\ 4 & 5 \end{pmatrix}$$

therefore, $AI = A = IA$.

Example 25. For $A = \begin{pmatrix} 2 & 3 & 4 \\ 5 & 6 & 7 \\ 8 & 9 & 10 \end{pmatrix}$, verify that $AI = A = IA$

$$AI = \begin{pmatrix} 2 & 3 & 4 \\ 5 & 6 & 7 \\ 8 & 9 & 10 \end{pmatrix}\begin{pmatrix} 1 & 0 & 0 \\ 0 & 1 & 0 \\ 0 & 0 & 1 \end{pmatrix} = \begin{pmatrix} 2 & 3 & 4 \\ 5 & 6 & 7 \\ 8 & 9 & 10 \end{pmatrix}$$

$$IA = \begin{pmatrix} 1 & 0 & 0 \\ 0 & 1 & 0 \\ 0 & 0 & 1 \end{pmatrix}\begin{pmatrix} 2 & 3 & 4 \\ 5 & 6 & 7 \\ 8 & 9 & 10 \end{pmatrix} = \begin{pmatrix} 2 & 3 & 4 \\ 5 & 6 & 7 \\ 8 & 9 & 10 \end{pmatrix}$$

1.10 NULL MATRIX

The null matrix is a matrix in which all elements are 0. Unlike the identity matrix, the null matrix need not be a square matrix. A matrix of any size or shape all of whose elements are zero is called a zero matrix or null matrix. The difference between two equal matrices is a zero matrix. The null matrix is commonly designated by the symbol ϕ (phi). A 2 by 3 null matrix is

$$\phi = \begin{pmatrix} 0 & 0 & 0 \\ 0 & 0 & 0 \end{pmatrix}$$

Similarly, a 1 by 3 null matrix or null row vector is

$$\phi = (0, 0, 0)$$

and a 3 by 1 null matrix or null column vector is

$$\phi = \begin{pmatrix} 0 \\ 0 \\ 0 \end{pmatrix}$$

The product of any matrix with a conformable null matrix is a null matrix, *i.e.*,

$A\phi = \phi$. Thus, if $A = \begin{pmatrix} 1 & 2 & 3 \\ 4 & 5 & 6 \end{pmatrix}$ and $\phi = \begin{pmatrix} 0 & 0 \\ 0 & 0 \\ 0 & 0 \end{pmatrix}$

$$A\phi = \begin{pmatrix} 1 & 2 & 3 \\ 4 & 5 & 6 \end{pmatrix}\begin{pmatrix} 0 & 0 \\ 0 & 0 \\ 0 & 0 \end{pmatrix} = \begin{pmatrix} 0 & 0 \\ 0 & 0 \end{pmatrix}$$

Please note that product of two non-zero matrices can be a zero matrix. **(Important)**

For example if $A = \begin{bmatrix} 1 & 1 \\ 0 & 0 \end{bmatrix}$, $B = \begin{bmatrix} 0 & -1 \\ 0 & 1 \end{bmatrix}$, then $AB = \begin{bmatrix} 0 & 0 \\ 0 & 0 \end{bmatrix}$.

1.11 DIAGONAL MATRIX

A square matrix in which all elements except those in the principal (main) diagonal are zero is called a diagonal matrix.

1.12 SCALAR MATRIX

A square matrix in which

(i) All the elements except on the principal diagonal are zero, and

(ii) All the elements on the principal diagonal (or main diagonal) are equal.

Notes :

1. Unit matrices and scalar matrices are special types of diagonal matrices.
2. A unit matrix is a scalar matrix but may not be vice versa.

Example 26.

$$\begin{bmatrix} 4 & 0 & 0 \\ 0 & 3 & 0 \\ 0 & 0 & 1 \end{bmatrix} \qquad \begin{bmatrix} 1 & 0 & 0 \\ 0 & 1 & 0 \\ 0 & 0 & 1 \end{bmatrix} \qquad \begin{bmatrix} a & 0 & 0 \\ 0 & a & 0 \\ 0 & 0 & a \end{bmatrix}$$

Diagonal matrix of order 3 — Unit matrix of order 3 — Scalar matrix of order 3

1.13 TRANSPOSE OF A MATRIX

The matrix obtained by changing the rows into columns and columns into rows in a given matrix A is called the transpose of A. It is denoted by A', *e.g.*,

$$\text{if } A = \begin{bmatrix} 1 & 2 & 3 \\ 4 & 5 & 6 \\ 7 & 8 & 9 \end{bmatrix}, \text{ then } A' = \begin{bmatrix} 1 & 4 & 7 \\ 2 & 5 & 8 \\ 3 & 6 & 9 \end{bmatrix}$$

Please note :

(i) The elements on the principal diagonal in a matrix and its transpose remain the same.

(ii) The transpose of the transpose of a matrix is the matrix itself, *i.e.*, $(A')' = A$

(iii) A^T also denotes transpose of A.

(iv) If A is a $m \times n$ matrix, then A' (or A^T) is a $n \times m$ matrix.

1.14 ORTHOGONAL MATRIX

A square matrix : A is called an orthogonal matrix if the transpose of A is equal to A^{-1}, *i.e.*, if $A' = A^{-1}$ then $AA' = AA^{-1} = I$.

Example 27. Find the values of x, y, z if the matrix

$$A = \begin{bmatrix} 0 & 2y & z \\ x & y & -z \\ x & -y & z \end{bmatrix}$$

obeys the law $A'.A = I$ where A' is the transpose of A and I is the unit matrix of order 3.

Solution. $A = \begin{bmatrix} 0 & 2y & z \\ x & y & -z \\ x & -y & z \end{bmatrix}$, $A' = \begin{bmatrix} 0 & x & x \\ 2y & y & -y \\ z & -z & z \end{bmatrix}$

$$A'.A = \begin{bmatrix} 0 & x & x \\ 2y & y & -y \\ z & -z & z \end{bmatrix} \begin{bmatrix} 0 & 2y & z \\ x & y & -z \\ x & -y & z \end{bmatrix}$$

$$= \begin{bmatrix} 2x^2 & 0 & 0 \\ 0 & 6y^2 & 0 \\ 0 & 0 & 3z^2 \end{bmatrix} = I \qquad \text{(given)}$$

$$\Rightarrow \quad \begin{bmatrix} 2x^2 & 0 & 0 \\ 0 & 6y^2 & 0 \\ 0 & 0 & 3z^2 \end{bmatrix} = \begin{bmatrix} 1 & 0 & 0 \\ 0 & 1 & 0 \\ 0 & 0 & 1 \end{bmatrix}$$

$$\Rightarrow 2x^2 = 1,\ 6y^2 = 1,\ 3z^2 = 1 \Rightarrow x = \pm\frac{1}{\sqrt{2}},\ y = \pm\frac{1}{\sqrt{6}},\ z = \pm\frac{1}{\sqrt{3}}$$

Example 28. If $A = \frac{1}{3}\begin{bmatrix} -1 & 2 & 2 \\ 2 & -1 & 2 \\ 2 & 2 & -1 \end{bmatrix}$, show that A is an orthogonal matrix.

Solution. A matrix A is said to be an orthogonal matrix if $AA' = I$ where A' is the transpose of A.

$$AA' = \frac{1}{3}\begin{bmatrix} -1 & 2 & 2 \\ 2 & -1 & 2 \\ 2 & 2 & -1 \end{bmatrix} \frac{1}{3}\begin{bmatrix} -1 & 2 & 2 \\ 2 & -1 & 2 \\ 2 & 2 & -1 \end{bmatrix}$$

$$= \frac{1}{9}\begin{bmatrix} 9 & 0 & 0 \\ 0 & 9 & 0 \\ 0 & 0 & 9 \end{bmatrix} = \begin{bmatrix} 1 & 0 & 0 \\ 0 & 1 & 0 \\ 0 & 0 & 1 \end{bmatrix} = I.$$

Hence, A is an orthogonal matrix.

Example 29. Find the values of a, b, c if the matrix A is given by

$$A = \frac{1}{3}\begin{bmatrix} a & 2 & 2 \\ 2 & 1 & b \\ 2 & c & 1 \end{bmatrix}$$

obey the law $AA' = I$ where A' is the transpose of A and I is the unit matrix of order 3.

Solution :

$$A' = \frac{1}{3}\begin{bmatrix} a & 2 & 2 \\ 2 & 1 & c \\ 2 & b & 1 \end{bmatrix}$$

$$\therefore \quad AA' = \frac{1}{3}\begin{bmatrix} a & 2 & 2 \\ 2 & 1 & b \\ 2 & c & 1 \end{bmatrix} \times \frac{1}{3}\begin{bmatrix} a & 2 & 2 \\ 2 & 1 & c \\ 2 & b & 1 \end{bmatrix}$$

$$= \frac{1}{9}\begin{bmatrix} a^2+8 & 2a+2+2b & 2a+2c+2 \\ 2a+2+2b & 4+1+b^2 & 4+c+b \\ 2a+2c+2 & 4+c+b & 4+c^2+1 \end{bmatrix}$$

$$= \frac{1}{9}\begin{bmatrix} a^2+8 & 2a+2b+2 & 2a+2c+2 \\ 2a+2b+2 & b^2+5 & b+c+4 \\ 2a+2c+2 & b+c+4 & c^2+5 \end{bmatrix} = \begin{bmatrix} 1 & 0 & 0 \\ 0 & 1 & 0 \\ 0 & 0 & 1 \end{bmatrix} \qquad [\because AA' = 1]$$

Equating corresponding elements on both sides, we get

$$\frac{1}{9}(a^2+8) = 1 \Rightarrow a^2 = 1 \Rightarrow a = \pm 1$$

$$\frac{1}{9}(b^2+5)=1 \Rightarrow b^2=4 \Rightarrow b=\pm 2$$

$$\frac{1}{9}(c^2+5)=1 \Rightarrow c^2=4 \Rightarrow c=\pm 2$$

Substituting these values of a, b, c in other equations, *i.e.*,

$$\frac{1}{9}(2a+2b+2)=0,\ \frac{1}{9}(2a+2c+2)=0,\ \frac{1}{9}(2a+2b+2)=0$$

$$\frac{1}{9}(b+c+4)=0,\ \frac{1}{9}(2a+2c+2)=0,\ \frac{1}{9}(b+c+4)=0$$

We find that all the equations are satisfies with the values

$$a=1, b=-2, c=-2$$

An inverse matrix performs a function in matrix algebra similar to that of the reciprocal in ordinary algebra. The product of a number a and its reciprocal $1/a$ in ordinary algebra is 1. Similarly, in matrix algebra the product of matrix A and its inverse, A^{-1} (read A inverse) is identity matrix I, that is

$$AA^{-1}=\mathrm{I}$$

1.15 INVERSE OF A MATRIX

If A and B are two square matrices of the square order such that $AB = BA = \mathrm{I}$ where I is a unit matrix, of the same order then A is called the inverse of B and B is called the inverse of A. Inverse of a matrix A is written as A^{-1}. A non-square (*i.e.*, rectangular) matrix can not have **an inverse**. From the definition of inverse we conclude that

$$AA^{-1}=A^{-1}A=\mathrm{I} \text{ (unit matrix)}$$

Example 30. Given $A=\begin{bmatrix}3 & 1\\ 0 & 2\end{bmatrix}$, show that

$A^3+A^2-24A+36I=0$ and hence, (not otherwise) find A^{-1}.

Solution. $A^2=\begin{bmatrix}3 & 1\\ 0 & 2\end{bmatrix}\begin{bmatrix}3 & 1\\ 0 & 2\end{bmatrix}=\begin{bmatrix}9 & 5\\ 0 & 4\end{bmatrix}$

$$A^3=A^2A=\begin{bmatrix}9 & 5\\ 0 & 4\end{bmatrix}\begin{bmatrix}3 & 1\\ 0 & 2\end{bmatrix}=\begin{bmatrix}27 & 19\\ 0 & 8\end{bmatrix}$$

$$A^3+A^2-24A+36I=\begin{bmatrix}27 & 19\\ 0 & 8\end{bmatrix}+\begin{bmatrix}9 & 5\\ 0 & 4\end{bmatrix}-\begin{bmatrix}72 & 24\\ 0 & 48\end{bmatrix}+\begin{bmatrix}36 & 0\\ 0 & 36\end{bmatrix}=\begin{bmatrix}0 & 0\\ 0 & 0\end{bmatrix}$$

$$A^3+A^2-24A+36I=0$$

Postmultiplying the equation by A^{-1}, we get

$$A^3A^{-1}+A^2A^{-1}-24AA^{-1}+36A^{-1}=0$$

$$\Rightarrow \quad A^2+A-24I+36A^{-1}=0$$

$$\Rightarrow \qquad A^{-1} = \frac{24I - A - A^2}{36} = \frac{1}{36}\left[\begin{bmatrix} 24 & 0 \\ 0 & 24 \end{bmatrix} - \begin{bmatrix} 3 & 1 \\ 0 & 2 \end{bmatrix} - \begin{bmatrix} 9 & 5 \\ 0 & 4 \end{bmatrix}\right]$$

$$= \frac{1}{36}\begin{bmatrix} 12 & -6 \\ 0 & 18 \end{bmatrix} = \begin{bmatrix} \frac{1}{3} & \frac{-1}{6} \\ 0 & \frac{1}{2} \end{bmatrix}$$

Example 31. If $A = \begin{bmatrix} 1 & a & -2 \\ 2 & 2 & 4 \\ 0 & 0 & 2 \end{bmatrix}$, find a for which $A^2 + 2I_3 = 3A$ where I_3 is the unit matrix of order 3. Hence (and not otherwise) find A^{-1}.

Hint : $A^2 + 2I_3 = \begin{bmatrix} 2a+3 & 3a & 4a-6 \\ 6 & 2a+6 & 12 \\ 0 & 0 & 6 \end{bmatrix}$

$$A^2 + 2I_3 = 3A \Rightarrow \begin{bmatrix} 2a+3 & 3a & 4a-6 \\ 6 & 2a+6 & 12 \\ 0 & 0 & 6 \end{bmatrix} = \begin{bmatrix} 3 & 3a & -6 \\ 6 & 6 & 12 \\ 0 & 0 & 6 \end{bmatrix}$$

$$\Rightarrow \qquad 2a + 3 = 3 \Rightarrow a = 0$$

$$A^2 + 2I_3 = 3A$$

$$\Rightarrow \qquad A^2A^{-1} + 2I_3A^{-1} = 3AA^{-1}$$

$$\Rightarrow \qquad A^{-1} = \frac{1}{2}[3I_3 - A] = \frac{1}{2}\left[3\begin{bmatrix} 1 & 0 & 0 \\ 0 & 1 & 0 \\ 0 & 0 & 1 \end{bmatrix} - \begin{bmatrix} 1 & 0 & -2 \\ 2 & 2 & 4 \\ 0 & 0 & 4 \end{bmatrix}\right] \qquad (\because a = 0)$$

$$\therefore \qquad A^{-1} = \begin{bmatrix} 1 & 0 & 1 \\ -1 & \frac{1}{2} & -2 \\ 0 & 0 & \frac{1}{2} \end{bmatrix}$$

Example 32. Show that $A = \begin{bmatrix} 1 & 1 & 2 \\ 2 & 1 & 2 \\ 2 & 1 & 1 \end{bmatrix}$ satisfies the relation $A^2 - 4A - 5I = Q$ where Q denotes the zero matrix.

Hence, find the inverse of A.

[**Hint :**

$$A^2 = \begin{bmatrix} 1 & 2 & 2 \\ 2 & 1 & 2 \\ 2 & 2 & 1 \end{bmatrix}\begin{bmatrix} 1 & 2 & 2 \\ 2 & 1 & 2 \\ 2 & 2 & 1 \end{bmatrix} = \begin{bmatrix} 9 & 8 & 8 \\ 8 & 9 & 8 \\ 8 & 8 & 9 \end{bmatrix}$$

$$A^2 - 4A - 5I = \begin{bmatrix} 9 & 8 & 8 \\ 8 & 9 & 8 \\ 8 & 8 & 9 \end{bmatrix} - 4\begin{bmatrix} 1 & 2 & 2 \\ 2 & 1 & 2 \\ 2 & 2 & 1 \end{bmatrix} - 5\begin{bmatrix} 1 & 0 & 0 \\ 0 & 1 & 0 \\ 0 & 0 & 1 \end{bmatrix} = 0$$

$$A^2 - 4A - 5I = 0 \Rightarrow A - 4I = 5A^{-1} = 0 \qquad \text{[Post multiplying by } A^{-1}]$$

$$\Rightarrow \qquad 5A^{-1} = A - 4I \Rightarrow A^{-1} = \frac{A-4I}{5} = \frac{1}{5}\left[\begin{bmatrix} 1 & 2 & 2 \\ 2 & 1 & 2 \\ 2 & 2 & 1 \end{bmatrix} - \begin{bmatrix} 4 & 0 & 0 \\ 0 & 4 & 0 \\ 0 & 0 & 4 \end{bmatrix}\right]$$

$$= \frac{1}{5}\begin{bmatrix} -3 & 2 & 2 \\ 2 & -3 & 2 \\ 2 & 2 & -3 \end{bmatrix}$$

1.16 DETERMINANT

One of the important concepts in matrix algebra is that of the determinant of a matrix. The determinant of a matrix is a number that is associated with a square matrix. This number can be positive, negative, or zero. The number is important in that it is used in determining the solutions to systems of simultaneous equations.

The determinant is used in two alternative solution techniques for calculating the solution set to a system of simultaneous equations.

One approach uses the determinant to calculate the inverse of a matrix. The inverse of the matrix is then used to determine the solution to the system of equation.

The second approach used the determination in a solution technique termed as Cramer's rule.

Steps to find the expansion of a 2×2 determinant :

(1) Multiply the elements along the principal diagonal [diagonal starting from the left hand top to right hand bottom].

(2) Multiply the elements along the other diagonal and change the sign.

(3) Add I and II.

1.16.1 Determinants of 2 By 2 and 3 by 3 Matrices

The determinant of the matrix A is commonly represented by one of two symbols, det A of $|A|$. These symbols are used interchangeably.

The determinant of the 2 by 2 matrix A is given by

$$|A| = \begin{vmatrix} a_{11} & a_{12} \\ a_{21} & a_{22} \end{vmatrix} = a_{11} \times a_{22} - a_{21} \times a_{12}$$

To illustrate the formula, assume that matrix $A = \begin{bmatrix} 1 & 3 \\ -2 & 4 \end{bmatrix}$.

$$|A| = \begin{vmatrix} 1 & 3 \\ -2 & 4 \end{vmatrix} = 1(4) - (-2)\,(3) = 10 = \text{Expansion of the determinant } A.$$

Example 33. Find the determinant of matrix $C = \begin{bmatrix} -2 & 3 \\ 6 & 4 \end{bmatrix}$

$$|C| = \begin{vmatrix} -2 & 3 \\ 6 & 4 \end{vmatrix} = -8 - 18 = -26$$

Example 34. Find the determinant of $D = \begin{pmatrix} 1 & 3 \\ 1 & 3 \end{pmatrix}$.

$$|D| = \begin{vmatrix} 1 & 3 \\ 1 & 3 \end{vmatrix} = 3 - 3 = 0$$

The determinant of a 3 by 3 matrix is calculated by a similar formula. The determinant of the 3 by 3 matrix A is given by

$$|A| = \begin{vmatrix} a_{11} & a_{12} & a_{13} \\ a_{21} & a_{22} & a_{23} \\ a_{31} & a_{32} & a_{33} \end{vmatrix}$$

$$= a_{11}a_{22}a_{33} + a_{12}a_{23}a_{31} + a_{13}a_{21}a_{32} - a_{31}a_{22}a_{13} - a_{32}a_{23}a_{11} - a_{33}a_{21}a_{12}$$

The formula can be easily remembered by employing the following scheme. First, repeat the first two columns of the matrix. Next, draw diagonals through the components as shown below. The products of the components joined by

$$\begin{bmatrix} a_{11} & a_{12} & a_{13} \\ a_{21} & a_{22} & a_{23} \\ a_{31} & a_{32} & a_{33} \end{bmatrix} \begin{matrix} a_{11} & a_{12} \\ a_{21} & a_{22} \\ a_{31} & a_{32} \end{matrix}$$

each diagonal are then determined, for example : $a_{11} \times a_{22} \times a_{33}$. The products of the components on the diagonals from upper left to lower right are summed, and the products of the components on the diagonals from lower left to upper right are subtracted from this sum.

$$|A| = a_{11}a_{22}a_{33} + a_{12}a_{23}a_{31} + a_{13}a_{21}a_{32} - a_{31}a_{22}a_{13} - a_{32}a_{23}a_{11} - a_{33}a_{21}a_{12}$$

Example 35. Find the determinant of $A = \begin{pmatrix} 1 & 3 & -4 \\ 2 & -1 & 6 \\ 3 & 0 & -2 \end{pmatrix}$. Use the method of repeating the first two columns.

$$|A| \begin{bmatrix} 1 & 3 & -4 \\ 2 & -1 & 6 \\ 3 & 0 & -2 \end{bmatrix} \begin{matrix} 1 & 3 \\ 2 & -1 \\ 3 & 0 \end{matrix}$$

$$|A| = 1(-1)(-2) + 3.6.3 + (-4).0 - 3(-1)(-4) - (0)6.1 - (-2)2.3$$

$$|A| = 2 + 54 + 0 - 12 - 0 + 12 = 56$$

Example 36. Find the determinant of $A = \begin{pmatrix} 1 & 5 & 4 \\ -3 & 2 & 3 \\ 6 & -1 & 4 \end{pmatrix}$.

$$|A| = 1.2.4 + 5.3.6 + 4(-3)(-1) - 6.2.4 - (-1)3.1 - 4(-3)5$$
$$|A| = 8 + 90 + 12 - 48 + 3 + 60 = 125$$

Example 37. Find the determinant of $A = \begin{pmatrix} 2 & -1 & 3 \\ 3 & -2 & 6 \\ 2 & 1 & 4 \end{pmatrix}$.

$$|A| = 2(-24)\,4 + (-1)\,6.2 + 3.3.1 - 2\,(-2)\,3 - 1.6.2 - 4.3\,(-1)$$
$$|A| = -16 - 12 + 9 + 12 - 12 + 12 = -7$$

1.16.2 Minors, Cofactors and Cofactor Expansion

The above method of calculation of the determinant of a 2×2 matrix and 3×3 matrix does not, apply to matrices of higher dimensions. This section introduces a procedure for calculating a determinant that applies to any square matrix. This procedure is termed the method of **cofactor expansion.**

Before discussing the method of cofactor expansion, we must define the terms : minor and cofactor. Both of these terms are defined with reference to an element a_{ij} of a square matrix A.

The **minor** of the $[a_{ij}]$ element is defined as the determinant of the submatrix formed by deleting the ith row and the jth column of the matrix. The minor of an element is thus a determinant. It is, again, the determinant of the submatrix resulting from deletion of the ith row and jth column. For the 3 by 3 matrix A shown below, the minor of a_{11}.

$$A = \begin{pmatrix} a_{11} & a_{12} & a_{13} \\ a_{21} & a_{22} & a_{23} \\ a_{31} & a_{32} & a_{33} \end{pmatrix}$$

is the determinant of the submatrix formed by deleting the first row and first column of A :

$$\text{minor of } a_{11} = \begin{vmatrix} a_{22} & a_{23} \\ a_{32} & a_{33} \end{vmatrix}$$

Similarly, the minor of the element a_{21} is $\begin{vmatrix} a_{12} & a_{13} \\ a_{32} & a_{33} \end{vmatrix}$ and the minor of a_{22} is $\begin{vmatrix} a_{11} & a_{13} \\ a_{31} & a_{33} \end{vmatrix}$.

A method of determining the elements included in the submatrix is to line out the row and column to which the a_{ij} element belongs. The minor of a_{32} is thus

$$\text{minor of } a_{32} = \begin{vmatrix} a_{11} & a_{13} \\ a_{21} & a_{23} \end{vmatrix}$$

Example 38. Find the minors of elements of the first row and the third column of the matrix $A = \begin{pmatrix} 2 & 3 & 1 \\ -1 & 3 & 2 \\ 4 & 2 & 3 \end{pmatrix}$. Since, the minor of the a_{ij} element is the determinant of the submatrix formed by eliminating the ith row and jth column, the minors of $a_{11}, a_{12}, a_{13}, a_{23}$ and a_{33} are

$$\text{minor of } a_{11} = \begin{vmatrix} 3 & 2 \\ 2 & 3 \end{vmatrix} = 9 - 4 = 5$$

$$\text{minor of } a_{12} = \begin{vmatrix} -1 & 2 \\ 4 & 3 \end{vmatrix} = -3 - 8 = -11$$

$$\text{minor of } a_{13} = \begin{vmatrix} -1 & 3 \\ 4 & 2 \end{vmatrix} = -2 - 12 = -14$$

$$\text{minor of } a_{23} = \begin{vmatrix} 2 & 3 \\ 4 & 2 \end{vmatrix} = 4 - 12 = -8$$

$$\text{minor of } a_{33} = \begin{vmatrix} 2 & 3 \\ -1 & 3 \end{vmatrix} = 6 - (-3) = 9$$

Example 39. Find the minors of the second row of the matrix.

$$A = \begin{pmatrix} 3 & 1 & 6 & 4 \\ 2 & 3 & -1 & 7 \\ 2 & 0 & 3 & 2 \\ 4 & 2 & -3 & 1 \end{pmatrix}$$

$$\text{minor of } a_{21} = \begin{vmatrix} 1 & 6 & 4 \\ 0 & 3 & 2 \\ 2 & -3 & 1 \end{vmatrix}$$

$$\text{minor of } a_{22} = \begin{vmatrix} 3 & 6 & 4 \\ 2 & 3 & 2 \\ 4 & -3 & 1 \end{vmatrix}$$

$$\text{minor of } a_{23} = \begin{vmatrix} 3 & 1 & 4 \\ 2 & 0 & 2 \\ 4 & 2 & 1 \end{vmatrix}$$

$$\text{minor of } a_{24} = \begin{vmatrix} 3 & 1 & 6 \\ 2 & 0 & 3 \\ 4 & 2 & -3 \end{vmatrix}$$

The cofactor of an element a_{ij} of a square matrix is the product of the minor of the a_{ij} element and $(-1)^{i+j}$. The cofactor is thus, simply the minor multiplied by either +1 or –1, the sign depending upon the location of the a_{ij} element, i is the row number and j is the column number of the element. If $i + j$ is even then $(-1)^{i+j}$ is +1 and if $i + j$ is odd then $(-1)^{i+j}$ is –1. For instance, the cofactor of the element a_{12} would be the product of the minor of this element and $(-1)^{1+2}$, or –1. The cofactor of the a_{22} element would be the product of the minor of the a_{22} element and $(-1)^{2+2}$, or +1. **The cofactor of the a_{ij} element is designated by the symbol** A_{ij}. The general formula is

$$A_{ij} = (-1)^{i+j} \text{ minor } a_{ij}$$

Example 40. Find the cofactors of the elements of the first row and of the third column of the matrix $A = \begin{pmatrix} 2 & 3 & 1 \\ -1 & 3 & 2 \\ 4 & 2 & 3 \end{pmatrix}$. The minors of this matrix were calculated previously. Since, the cofactor A_{ij} is the product of $(-1)^{i+j}$ and the minor of a_{ij}, the cofactors are

$$A_{11} = (-1)^{1+1} \times \text{ minor } a_{11} = +1(5) = 5$$
$$A_{12} = (-1)^{1+2} \times \text{ minor } a_{12} = -1\,(-11) = 11$$
$$A_{13} = (-1)^{4} \times \text{ minor } a_{13} = +1\,(-14) = -14$$
$$A_{23} = (-1)^{5} \times \text{ minor } a_{23} = -1\,(-8) = 8$$
$$A_{33} = (-1)^{6} \times \text{ minor } a_{33} = +1\,(9) = 9$$

Example 41. Determine the cofactors of the elements of the second row of the matrix

$$A = \begin{pmatrix} 3 & 1 & 6 & 4 \\ 2 & 3 & -1 & 7 \\ 2 & 0 & 3 & 2 \\ 4 & 2 & -3 & 1 \end{pmatrix}$$

The minor of the elements of the second row were determined previously in Ex. 39. The cofactors of the elements of the second row are

$$A_{21} = (-1)^3 \times \text{ minor of } a_{21} = -1\,(9) = -9$$

$$A_{22} = (-1)^4 \times \text{minor of } a_{22} = +1\ (-9) = -9$$
$$A_{23} = (-1)^5 \times \text{minor of } a_{23} = -1\ (10) = -10$$
$$A_{24} = (-1)^6 \times \text{minor of } a_{24} = +1\ (24) = 24$$

Expansion of the determinant

We can now introduce a method for calculating the expansion of a determinant that applies to any square matrix. To calculate the expansion of a determinant we multiply the elements of any row or column by their respective cofactors. The sum of these products is the expansion of the determinant of the matrix. The expansion of the determinant of the 3 by 3 matrix,

$$A = \begin{pmatrix} a_{11} & a_{12} & a_{13} \\ a_{21} & a_{22} & a_{23} \\ a_{31} & a_{32} & a_{33} \end{pmatrix}$$

is given by multiplying the elements in the first row by the cofactors of the elements of that row,

$$|A| = a_{11}A_{11} + a_{12}A_{12} + a_{13}A_{13}$$

or alternatively, by multiplying the elements in the third column by the cofactors of the elements of that column

$$|A| = a_{13}A_{13} + a_{23}A_{23} + a_{33}A_{33}.$$

In general the expansion of a determinant is defined as the sum of products of the cofactors and the element of any other row or column. This method of calculating determinants is termed **cofactor expansion.**

Step to calculate the expansion of a determinant :

(1) Take the elements of any row or column.

(2) Multiply the elements by their respective cofactors.

(3) Add the products.

Example 42. Calculate the value of the determinant by cofactor expansion of the matrix $A = \begin{pmatrix} 2 & 3 & 1 \\ -1 & 3 & 2 \\ 4 & 1 & 3 \end{pmatrix}$. Verify that the expansion about the first row or the third columns yields the same value of the determinant.

Solution. The cofactors of the first row and of the third column of the matrix were calculated previously in Ex. 38. Cofactor expansion about the first row gives,

$$|A| = 2A_{11} + 3A_{12} + 1A_{13}$$
$$|A| = 2(5) + 3(11) + (-14) = 29$$

Cofactor expansion about the third column gives

$$|A| = 1A_{13} + 2A_{23} + 3A_{33}$$
$$|A| = 2(-14) + 2(8) + 3(9) = 29$$

The reader can verify that the cofactor expansion about any other row or column gives the same value for the determinant.

Properties of Determinants

The arithmetic calculations required to determine the determinant are reduced in proportion to the number of elements a_{ij} that have values of zero.

Certain properties of determinants are useful in obtaining zeros as elements in row or columns. Many of these properties will also prove useful in more advanced studies of matrix applications. These properties of determinants are offered without proof as follows.

1. Interchanging any two rows or any two columns of a matrix changes the sign of the determinant of the matrix but not the absolute value of the determinant. Thus, if

$$A = \begin{pmatrix} 1 & 3 & 2 \\ 3 & 1 & -3 \\ 2 & 0 & 4 \end{pmatrix} \text{ and } B = \begin{pmatrix} 2 & 3 & 1 \\ -3 & 1 & 3 \\ 4 & 0 & 2 \end{pmatrix}, \text{ then } |A| = -|B|$$

B is obtained by interchanging first column and third column.

2. The determinant of a matrix and of its transpose are equal, that is,

$$|A| = |A'|$$

i.e., if in a matrix, the rows are changed into columns and columns into rows, the determinants of the matrix remains unchanged.

3. If all of the elements if any row or column of a matrix are zero, the determinant of the matrix is zero.
4. If the elements in any two rows or two columns are equal or proportional, the determinant of the matrix is zero.
5. If the elements of any row or column are multiplied by a constant k, the determinant of the matrix is multiplied by the constant k.

$$A = \begin{pmatrix} 1 & 3 & 2 \\ 3 & 1 & -3 \\ 2 & 0 & 4 \end{pmatrix} \text{ and } B = \begin{pmatrix} 1k & 3k & 2k \\ 3 & 1 & -3 \\ 2 & 0 & 4 \end{pmatrix}, \text{ the } |B| = k|A|$$

6. Any non-zero multiple of one row may be added to another row, or any non-zero multiple of one column may be added to another column without changing the value of the determinant.
7. The value of the determinant of any identity matrix is 1.
8. If each element of a row or column of a determinant is expressed as the sum of two (or more) terms, the determinant can be expressed as the sum of two (or more) determinants of the same order.

$$A = \begin{vmatrix} \alpha + a_1 & a_2 & a_3 \\ \beta + b_1 & b_2 & b_2 \end{vmatrix} = \begin{vmatrix} \alpha & a_2 & a_3 \\ \beta & b_2 & b_3 \end{vmatrix} + \begin{vmatrix} a_1 & a_2 & a_3 \\ b_1 & b_2 & b_3 \end{vmatrix}$$

9. If the elements of any row (or column) of a determinant are multiplied with the cofactors of the elements of some other row (or colum) and summed up, the result is zero but if the elements of any row (or column) are multiplied with their respective cofactors and summed up, the result is the determinant itself.

$$\text{If } \Delta = \begin{vmatrix} a_1 & b_1 & c_1 \\ a_2 & b_2 & c_2 \\ a_3 & b_3 & c_3 \end{vmatrix} \text{ then } a_1A_1 + b_1B_1 + c_1C_1 = \Delta, \text{ but } a_1A_2 + b_1B_2 + c_1C_2 = 0 \text{ and so on.}$$

1.17 MULTIPLICATION OF DETERMINANTS

The product of two determinants Δ_1 and Δ_2 of the **same order** is a determinant of the **same order** and the element of the product determinant are obtained by the following procedure.

1. The sum of the products of the elements in first row of Δ_1 with the corresponding elements in the first row of Δ_2 constitute the leading elements (*i.e.*, the first element of first row and first column) for the product determinant.

2. The sum of the products of the elements in the first row of Δ_1 with the corresponding elements in the second row of Δ_2 constitute the second element in the first row (or first element in the second column) of the product determinant.
3. The sum of the products of the elements in the first row of Δ_1 with the corresponding elements in the third row of Δ_2 consitute the third element in the first row (or first element in the third column) of the product determinant.
   Likewise, the sum of the product of the elements in the first row of Δ_1 with the corresponding elements in the fourth, fifth, etc., rows of Δ_2 will constitute other elements in the first row of the product determinant.
4. The sum of the products of the elements in the 2nd row of Δ_1 with the corresponding elements in the 1st, 2nd, 3rd, 4th, etc., rows of Δ_2 will constitute the element in the second row of the product determinant.

Similarly, by taking the elements in the 3rd, 4th, 5th, ..., etc., rows of Δ_1 and forming the sum of the products with the corresponding element in the different rows of Δ_2 we can determine the elements in the 3rd, 4th, 5th, ..., etc. rows of the product determinant, *e.g.*,

if
$$\Delta_1 = \begin{vmatrix} a_1 & b_1 \\ a_2 & b_2 \end{vmatrix}, \Delta_2 = \begin{vmatrix} \alpha_1 & \beta_1 \\ \alpha_2 & \beta_2 \end{vmatrix}$$

Then
$$\Delta_1 \Delta_2 = \begin{vmatrix} a_1\alpha_1 + b_1\beta_1 & a_1\alpha_2 + b_1\beta_2 \\ a_2\alpha_1 + b_2\beta_1 & a_2\alpha_2 + b_2\beta_2 \end{vmatrix}$$

If
$$\Delta_1 = \begin{vmatrix} a_1 & b_1 & c_1 \\ a_2 & b_2 & c_2 \\ a_3 & b_3 & c_3 \end{vmatrix}, \Delta_2 = \begin{vmatrix} \alpha_1 & \beta_1 & \gamma_1 \\ \alpha_2 & \beta_2 & \gamma_2 \\ \alpha_3 & \beta_3 & \gamma_3 \end{vmatrix}$$

$$\Delta_1 \Delta_2 = \begin{vmatrix} a_1\alpha_1 + b_1\beta_1 + c_1\gamma_1 & a_1\alpha_2 + b_1\beta_2 + c_1\gamma_2 & a_1\alpha_3 + b_1\beta_3 + c_1\gamma_3 \\ a_2\alpha_1 + b_2\beta_1 + c_2\gamma_1 & a_2\alpha_2 + b_2\beta_2 + c_2\gamma_2 & a_1\alpha_3 + b_2\beta_3 + c_2\gamma_3 \\ a_3\alpha_1 + b_3\beta_1 + c_3\gamma_1 & a_3\alpha_2 + b_3\beta_2 + c_3\gamma_2 & a_3\alpha_3 + b_3\beta_3 + c_3\gamma_3 \end{vmatrix}$$

1.18 CONJUGATE ELEMENTS

Two elements of a determinant are said to be conjugate if one occupies with reference to the leading element the same position in the rows as the other does in the columns, *i.e.*, the element in the first row and second column and the element in the second row and first column are conjugates.

1.19 SYMMETRIC DETERMINANT

A determinant in which every two conjugate elements are equal is called a symmetric determinant, *e.g.*,

$$\begin{vmatrix} a & h & g \\ h & b & f \\ g & f & c \end{vmatrix} \text{ is a symmetric determinant.}$$

1.20 ORTHO SYMMETRIC DETERMINANT

A symmetric determinant in which all the elements along the principal diagonal are equal is called ortho symmetric determinant, *e.g.*,

$$\begin{vmatrix} a & h & g \\ h & a & f \\ g & f & a \end{vmatrix} \text{ is an ortho symmetric determinant.}$$

1.21 SKEW SYMMETRIC DETERMINANT

A skew symmetric determinant is one in which every element is equal to its conjugate with sign changed. Since, diagonal elements are their own conjugates it follows that in such a determinant all the diagonal elements are zero.

1.22 SKEW DETERMINANT

A determinant in which all except the elements along the principal diagonal, are equal to their conjugates with sign changed is called a skew determinant.

Example 43.

$$\begin{vmatrix} o & h & g \\ -h & o & f \\ -g & -f & o \end{vmatrix} \qquad \begin{vmatrix} a & h & g \\ -h & b & f \\ -g & -f & c \end{vmatrix}$$

(Skew symmetric determinant) (Skew determinant)

1.23 ADJUGATE (OR ADJOINT) OF A DETERMINANT

A determinant Δ' formed by the cofactors of the corresponding elements of a determinant Δ is called the adjugate (or adjoint) of Δ, *i.e.*, if

$$\Delta = \begin{vmatrix} a_1 & b_1 & c_1 \\ a_2 & b_2 & c_2 \\ a_3 & b_3 & c_3 \end{vmatrix}, \text{ then } \Delta' = \begin{vmatrix} A_1 & B_1 & C_1 \\ A_2 & B_2 & C_2 \\ A_3 & B_3 & C_3 \end{vmatrix}$$

$$\Delta' = \Delta^2.$$

1.24 RECIPROCAL DETERMINANT

$$\text{If } \Delta = \begin{vmatrix} a_1 & b_1 & c_1 \\ a_2 & b_2 & c_2 \\ a_3 & b_3 & c_3 \end{vmatrix} \text{ then } \overline{\Delta} = \begin{vmatrix} \frac{A_1}{\Delta} & \frac{B_1}{\Delta} & \frac{C_1}{\Delta} \\ \frac{A_2}{\Delta} & \frac{B_2}{\Delta} & \frac{C_2}{\Delta} \\ \frac{A_3}{\Delta} & \frac{B_3}{\Delta} & \frac{C_3}{\Delta} \end{vmatrix} \text{ is called the reciprocal determinant of } \Delta.$$

$$\overline{\Delta} = \frac{1}{\Delta^3} \times \Delta' = \frac{1}{\Delta^3} \times \Delta^2 = \frac{1}{\Delta} \Rightarrow \overline{\Delta}\Delta = 1$$

Example 44. Prove that

$$\begin{vmatrix} \alpha & \beta & \gamma \\ \alpha^2 & \beta^2 & \gamma^2 \\ \beta+\gamma & \gamma+\alpha & \alpha+\beta \end{vmatrix} = (\alpha+\beta+\gamma)(\beta-\gamma)(\gamma-\alpha)(\alpha-\beta)$$

Solution. Let $\Delta = \begin{vmatrix} \alpha & \beta & \gamma \\ \alpha^2 & \beta^2 & \gamma^2 \\ \beta+\gamma & \gamma+\alpha & \alpha+\beta \end{vmatrix} = \begin{vmatrix} \alpha+\beta+\gamma & \alpha+\beta+\gamma & \alpha+\beta+\gamma \\ \alpha^2 & \beta^2 & \gamma^2 \\ \beta+\gamma & \gamma+\alpha & \alpha+\beta \end{vmatrix}$ $(R_1 \to R_1 + R_3)$

$$= (\alpha+\beta+\gamma) \begin{vmatrix} 1 & 1 & 1 \\ \alpha^2 & \beta^2 & \gamma^2 \\ \beta+\gamma & \gamma+\alpha & \alpha+\beta \end{vmatrix}$$

$$= (\alpha+\beta+\gamma) \begin{vmatrix} 1 & 0 & 0 \\ \alpha^2 & \beta^2-\alpha^2 & \gamma^2-\alpha^2 \\ \beta+\gamma & \alpha-\beta & \alpha-\gamma \end{vmatrix} \qquad \begin{matrix} C_2 \to C_2 - C_1 \\ C_3 \to C_3 - C_1 \end{matrix}$$

$$= (\alpha+\beta+\gamma)(\beta-\alpha)(\gamma-\alpha) \begin{vmatrix} 1 & 0 & 0 \\ \alpha^2 & \beta+\alpha & \gamma+\alpha \\ \beta+\gamma & -1 & -1 \end{vmatrix}$$

$$= (\alpha+\beta+\gamma)(\beta-\alpha)(\gamma-\alpha)[-\beta-\alpha+\gamma+\alpha]$$

$$= (\alpha+\beta+\gamma)(\beta-\gamma)(\gamma-\alpha)(\alpha-\beta).$$

Example 45. Show that $\begin{vmatrix} -x^2 & xy & xz \\ xy & -y^2 & yz \\ xz & yz & -z^2 \end{vmatrix} = (2xyz)^2.$

Solution. Let $\Delta = \begin{vmatrix} -x^2 & xy & xz \\ xy & -y^2 & yz \\ xz & yz & -z^2 \end{vmatrix} = xyz \begin{vmatrix} -x & x & x \\ y & -y & y \\ z & z & -z \end{vmatrix}$

$$= (xyz)(xzy) \begin{vmatrix} -1 & 1 & 1 \\ 1 & -1 & 1 \\ 1 & 1 & -1 \end{vmatrix}$$

$$= (xyz)^2 \begin{vmatrix} -1 & 1 & 1 \\ 0 & 0 & 2 \\ 0 & 2 & 0 \end{vmatrix} \qquad \begin{matrix} R_2 \to R_2 - R_1 \\ R_3 \to R_3 - R_1 \end{matrix}$$

$$= 4(xyz)^2 = (2xyz)^2.$$

Example 46. Evaluate $\begin{vmatrix} 1 & x & x^2-yz \\ 1 & y & y^2-zx \\ 1 & z & z^2-xy \end{vmatrix}$ without direct expansion.

Solution. Let

$$\Delta = \begin{vmatrix} 1 & x & x^2-yz \\ 1 & y & y^2-zx \\ 1 & z & z^2-xy \end{vmatrix}$$

$$= \begin{vmatrix} 1 & x-y-x & x^2-yz \\ 0 & z-x & y^2-zx-x^2+zx \\ 0 & & z^2-xy-x^2+yz \end{vmatrix} \quad \begin{matrix} R_2 \to R_2 - R_1 \\ R_3 \to R_3 - R_1 \end{matrix}$$

$$= \begin{vmatrix} 1 & x & x^2-yz \\ 0 & (y-x) & (y-x)(x+y+z) \\ 0 & (z-x) & (z-x)(x+y+z) \end{vmatrix}$$

$$= (y-x)(z-x) \begin{vmatrix} 1 & x & x^2-yz \\ 0 & 1 & x+y+z \\ 0 & 1 & x+y+z \end{vmatrix}$$

$$= 0 \qquad [\because R_2 \text{ and } R_3 \text{ are identical}]$$

Example 47. Show that $\begin{vmatrix} 1+x & 1 & 1 \\ 1 & 1+y & 1 \\ 1 & 1 & 1+z \end{vmatrix} = xyz\left(1+\frac{1}{x}+\frac{1}{y}+\frac{1}{z}\right)$

Solution. L.H.S. $\begin{vmatrix} 1+x & 1 & 1 \\ 1 & 1+y & 1 \\ 1 & 1 & 1+z \end{vmatrix} = \begin{vmatrix} x+1 & 1 & 1 \\ 1 & y+1 & 1 \\ 1 & 1 & z+1 \end{vmatrix}$

$$= xyz \begin{vmatrix} 1+\frac{1}{x} & \frac{1}{x} & \frac{1}{x} \\ \frac{1}{y} & 1+\frac{1}{y} & \frac{1}{y} \\ \frac{1}{z} & \frac{1}{z} & 1+\frac{1}{z} \end{vmatrix}$$

$$= xyz \begin{vmatrix} 1+\frac{1}{x}+\frac{1}{y}+\frac{1}{z} & 1+\frac{1}{x}+\frac{1}{y}+\frac{1}{z} & 1+\frac{1}{x}+\frac{\frac{1}{yh1}}{z} \\ \frac{1}{y} & 1+\frac{1}{y} & \frac{1}{y} \\ \frac{1}{z} & \frac{1}{z} & 1+\frac{1}{z} \end{vmatrix}$$

$$R_1 \to R_1 + R_2 + R_3$$

$$= xyz\left(1+\frac{1}{x}+\frac{1}{y}+\frac{1}{z}\right) \begin{vmatrix} 1 & 1 & 1 \\ \frac{1}{y} & 1+\frac{1}{y} & \frac{1}{y} \\ \frac{1}{z} & \frac{1}{z} & 1+\frac{1}{z} \end{vmatrix}$$

$$= xyz\left(1+\frac{1}{x}+\frac{1}{y}+\frac{1}{z}\right)\begin{vmatrix} 1 & 0 & 0 \\ \frac{1}{y} & 1 & 0 \\ \frac{1}{z} & 0 & 1 \end{vmatrix} \qquad \text{(Operating)} \begin{matrix} C_2 \to C_2 - C_1 \\ C_3 \to C_3 - C_1 \end{matrix}$$

$$= xyz\left(1+\frac{1}{x}+\frac{1}{y}+\frac{1}{z}\right) = \text{R.H.S.}$$

Example 48. Express the square of

$\begin{vmatrix} x_1 & y_1 & 0 \\ x_2 & y_2 & 0 \\ x_3 & y_3 & 0 \end{vmatrix}$ as a third order determinant without any zero element, what is its value.

Solution. $\begin{vmatrix} x_1 & y_1 & 0 \\ x_2 & y_2 & 0 \\ x_3 & y_3 & 0 \end{vmatrix}^2 = \begin{vmatrix} x_1 & y_1 & 0 \\ x_2 & y_2 & 0 \\ x_3 & y_3 & 0 \end{vmatrix} \times \begin{vmatrix} x_1 & y_1 & 0 \\ x_2 & y_2 & 0 \\ x_3 & y_3 & 0 \end{vmatrix} = \begin{vmatrix} x_1^2+y_1^2 & x_1x_2+y_1y_2 & x_1x_3+y_1y_3 \\ x_2x_1+y_2y_1 & x_2^2+y_2^2 & x_2x_3+y_2y_3 \\ x_3x_1+y_3y_1 & x_3x_2+y_3y_2 & x_3^2+y_3^2 \end{vmatrix}$

Which is a third order determinant without any zero element. Since, one column in the given determinant is zero, the value of determinant is zero, hence, the value of its square is also zero.

Example 49. Show that $\begin{vmatrix} 0 & r & q \\ r & 0 & p \\ q & p & 0 \end{vmatrix}^2 = \begin{vmatrix} -p^2 & pq & pr \\ pq & -q^2 & qr \\ pr & qr & -r^2 \end{vmatrix} = (2pqr)^2$

Solution. Let $\Delta = \begin{vmatrix} 0 & r & q \\ r & 0 & p \\ q & p & 0 \end{vmatrix}$, Then Δ' the determinant formed by the cofactors of the corresponding element of Δ and $\Delta' = \Delta^2$.

$$\Delta' = \begin{vmatrix} -p^2 & pq & pr \\ pq & -q^2 & qr \\ pr & qr & -r^2 \end{vmatrix}$$

$$\therefore \qquad \Delta^2 = \begin{vmatrix} 0 & r & q \\ r & 0 & p \\ q & p & 0 \end{vmatrix}^2 = \begin{vmatrix} -p^2 & pq & pr \\ pq & -q^2 & qr \\ pr & qr & -r^2 \end{vmatrix} \qquad (\because \Delta^2 = \Delta')$$

Also $\begin{vmatrix} 0 & r & q \\ r & 0 & p \\ q & p & 0 \end{vmatrix} = r(pq) \times q(rp) = 2pqr$

Hence, $$\begin{vmatrix} 0 & r & q \\ r & 0 & p \\ q & p & 0 \end{vmatrix}^2 = \begin{vmatrix} -p^2 & pq & pr \\ pq & -q^2 & qr \\ pr & qr & -r^2 \end{vmatrix} = (2pqr)^2$$

Example 50. Without expanding the determinant, prove that

$$\begin{vmatrix} 1 & a & bc \\ 1 & b & ca \\ 1 & c & ab \end{vmatrix} = \begin{vmatrix} 1 & a & a^2 \\ 1 & b & b^2 \\ 1 & c & c^2 \end{vmatrix}.$$

Solution. L.H.S. $\begin{vmatrix} 1 & a & bc \\ 1 & b & ca \\ 1 & c & ab \end{vmatrix} = \dfrac{1}{abc}\begin{vmatrix} a & a^2 & abc \\ b & b^2 & abc \\ c & c^2 & abc \end{vmatrix}$

$$= \frac{abc}{abc}\begin{vmatrix} a & a^2 & 1 \\ b & b^2 & 1 \\ c & c^2 & 1 \end{vmatrix} = -\begin{vmatrix} 1 & a^2 & a \\ 1 & b^2 & b \\ 1 & c^2 & c \end{vmatrix}$$

[Interchanging Ist and third column. This interchange will change the sign of the determinant]

$$= \begin{vmatrix} 1 & a & a^2 \\ 1 & b & b^2 \\ 1 & c & c^2 \end{vmatrix} == \text{R.H.S.}$$

Example 51. Show that $\begin{vmatrix} a-b-c & 2a & 2a \\ 2b & b-c-a & 2b \\ 2c & 2c & c-a-b \end{vmatrix} = (a+b+c)^3.$

Hence, find the values of x from

$$\begin{vmatrix} 3-x & 2x & 2x \\ 2x-2 & 1-x & 2x-2 \\ 2x-4 & 2x-4 & -1-x \end{vmatrix} = 0$$

Solution. $\begin{vmatrix} a-b-c & 2a & 2a \\ 2b & b-c-a & 2b \\ 2c & 2c & c-a-b \end{vmatrix} = \begin{vmatrix} a+b+c & a+b+c & a+b+c \\ 2b & b-c-a & 2b \\ 2c & 2c & c-a-b \end{vmatrix}$ $(R_1 \to R_1 + R_2 + R_3)$

$$= (a+b+c)\begin{vmatrix} 1 & 1 & 1 \\ 2b & b-c-a & 2b \\ 2c & 2c & c-a-b \end{vmatrix}$$

$$= (a+b+c)\begin{vmatrix} 0 & 0 & 1 \\ 0 & -(a+b+c) & 2b \\ a+b+c & a+b+c & c-a-b \end{vmatrix}$$

Operating $C_1 \to C_1 - C_3$, $C_2 \to C_2 - C_3$

$$= (a+b+c)^3\begin{vmatrix} 0 & 0 & 1 \\ 0 & -1 & 2b \\ 1 & 1 & c-a-b \end{vmatrix} = (a+b+c)^3.$$

Putting $a = x, b = x-1, c = x-2$

We have $a-b-c = 3-x$, $2a = 2x$ and so on.

$$\therefore \quad \begin{vmatrix} 3-x & 2x & 2x \\ 2x-2 & 1-x & 2x-2 \\ 2x-4 & 2x-4 & -1-x \end{vmatrix} = 0 \Rightarrow (3x-3)^3 = 0 \Rightarrow x = 1 \quad [\because a+b+c = 3x-3]$$

Example 52. If $\begin{vmatrix} p & b & c \\ a & q & c \\ a & b & r \end{vmatrix} = 0$ then evaluate

$\frac{p}{p-a} + \frac{q}{q-b} + \frac{r}{r-c}$ where $p - a \neq 0, q - b \neq 0, r - c \neq 0$.

Solution. $\begin{vmatrix} p & b & c \\ a & q & c \\ a & b & r \end{vmatrix} = 0$

$$\Rightarrow \quad \begin{vmatrix} p-a & 0 & c-r \\ 0 & q-b & c-r \\ a & b & r \end{vmatrix} = 0 \qquad \text{Operating } \begin{matrix} R_1 \to R_1 - R_3 \\ R_2 \to R_2 \to R_3 \end{matrix}$$

$$\Rightarrow \quad (p-a)\,[r(q-b) - b(c-r)] - (c-r)\,[a(q-b)] = 0$$

$$\Rightarrow \quad r(p-a)\,(q-b) - b(p-a)\,(c-r) - a(c-r)\,(q-b) = 0$$

$$\Rightarrow \quad \frac{r}{c-r} - \frac{b}{q-b} + \frac{a}{p-a} = 0 \quad [\text{Dividing by } [(q-b)\,(c-r)\,(p-a)]]$$

$$\Rightarrow \quad \frac{r}{r-c} + \frac{b}{q-b} + \frac{a}{p-a} = 0$$

$$\Rightarrow \quad \frac{r}{r-c} + \frac{b}{q-b} + 1 + \frac{a}{p-a} + 1 = 2$$

$$\Rightarrow \quad \frac{r}{r-c} + \frac{q}{q-b} + \frac{p}{p-a} = 2.$$

Example 53. If a, b, c, are all positive and the pth, qth, rth terms of a G.P., then prove that

$$\begin{vmatrix} \log a & p & 1 \\ \log b & q & 1 \\ \log c & r & 1 \end{vmatrix} = 0$$

Solution. Let first term $= A$ and common ratio $= R$. Then,

$$a = AR^{(p-1)},\ b = AR^{(q-1)} \text{ and } c = AR^{(r-1)}$$

$\therefore$ $\log a = \log A + (p-1)\log R$; $\log b = \log A + (q-1)\log R$ and $\log c = \log A + (r-1)\log R$

$$\therefore \text{ Given determinant} = \begin{vmatrix} \log A + (p-1)\log R & q & 1 \\ \log A + (q-1)\log R & q & 1 \\ \log A + (r-1)\log R & r & 1 \end{vmatrix}$$

$$= \begin{vmatrix} \log A & p & 1 \\ \log A & q & 1 \\ \log A & r & 1 \end{vmatrix} + \begin{vmatrix} (p-1)\log R & p & 1 \\ (q-1)\log R & q & 1 \\ (r-1)\log R & r & 1 \end{vmatrix}$$

$$= \log A \begin{vmatrix} 1 & p & 1 \\ 1 & q & 1 \\ 1 & r & 1 \end{vmatrix} + \log R \begin{vmatrix} p-1 & p & 1 \\ q-1 & q & 1 \\ r-1 & r & 1 \end{vmatrix}$$

$$= 0 + \log R \begin{vmatrix} 0 & p & 1 \\ 0 & q & 1 \\ 0 & r & 1 \end{vmatrix} = 0 \quad [\because \text{(i) Two columns are identical (ii) operate } R_1 \to R_1 + (R_3 - R_2)]$$

Example 54. Find the value of K in order that the points (5, 5), (K, 1) and (10, 7) lie on a straight line.

Solution. The points (x_1, y_1), (x_2, y_2) and (x_3, y_3) lie on a straight line or are collinear if

$$\begin{vmatrix} x_1 & y_1 & 1 \\ x_2 & y_2 & 1 \\ x_3 & y_3 & 1 \end{vmatrix} = 0$$

The given points will be collinear if

$$\begin{vmatrix} 5 & 5 & 1 \\ K & 1 & 1 \\ 10 & 7 & 1 \end{vmatrix} \Rightarrow \begin{vmatrix} 5-10 & 5-7 & 0 \\ K-10 & 1-7 & 0 \\ 10 & 7 & 1 \end{vmatrix} = 0 \quad [R_1 \to R_1 - R_3, R_2 \to R_2 - R_3]$$

$$= \begin{vmatrix} -5 & -2 & 0 \\ K-10 & -6 & 0 \\ 10 & 7 & 1 \end{vmatrix} = 0$$

$$\Rightarrow \qquad 30 + 2(K - 10) = 0 \Rightarrow K = -5$$

Example 55. Show that

$$\begin{vmatrix} y+z & x & y \\ z+x & z & x \\ x+y & y & z \end{vmatrix} = (x+y+z)\,(x-z)^2.$$

Solution. Let

$$\Delta = \begin{vmatrix} y+z & x & y \\ z+x & z & x \\ x+y & y & z \end{vmatrix}$$

Performing $R_1 \to R_2 + R_3$, we get

$$\Delta = \begin{vmatrix} 2(x+y+z) & x+y+z & x+y+z \\ z+x & z & x \\ x+y & y & z \end{vmatrix}$$

$$= (x+y+z) \begin{vmatrix} 2 & 1 & 1 \\ z+x & z & x \\ x+y & y & z \end{vmatrix}$$

Performing $C_1 \to C_1 - 2C_2$ and $C_2 \to C_2 - C_3$, we get

$$\Delta = (x+y+z) \begin{vmatrix} 0 & 0 & 1 \\ x-z & z-x & x \\ x-y & y-z & z \end{vmatrix}$$

$$= (x+y+z)\,[(x-z)\,(y-z) - (x-y)\,(z-x)] \qquad \text{(Expanding along first row)}$$

$$= (x+y+z)(x-z)[(y-z)+(x-y)]$$
$$= (x+y+z)(x-z)^2$$

Example 56. Show that

$$\begin{vmatrix} a+b+c & -c & -b \\ -c & a+b+c & -a \\ -b & -a & a+b+c \end{vmatrix} = 2(b+c)(c+a)(a+b).$$

Solution. Operating $R_1 \to R_1 + R_2$, we get

$$\begin{vmatrix} a+b & a+b & -(a+b) \\ -c & a+b+c & -a \\ -b & -a & a+b+c \end{vmatrix} = (a+b)\begin{vmatrix} 1 & 1 & -1 \\ -c & a+b+c & -a \\ -b & -a & a+b+c \end{vmatrix}$$

$$= (a+b)\begin{vmatrix} 1 & 1 & -1 \\ -(b+c) & b+c & b+c \\ -b & -a & a+b+c \end{vmatrix} \qquad \text{Operating } R_2 \to R_2 + R_3$$

$$= (a+b)(b+c)\begin{vmatrix} 1 & 1 & -1 \\ -1 & 1 & 1 \\ -b & -a & a+b+c \end{vmatrix}$$

$$= (a+b)(b+c)\begin{vmatrix} 0 & 2 & 0 \\ -1 & 1 & 1 \\ -b & -a & a+b+c \end{vmatrix} \qquad \text{Operating } R_1 \to R_1 + R_2$$

$$= (a+b)(b+c)(-2)[-(a+b+c)+b] = 2(b+c)(c+a)(a+b)$$

1.25 SINGULAR AND NONSINGULAR MATRICES

The terms singular and nonsingular are used with reference to the value of the determinant of a square matrix. A square matrix A is said to be singular if the determinant of A is zero. Conversely, matrices whose determinants are nonzero are termed nonsingular.

Example 57. Determine if the matrix A is singular or nonsingular

$$A = \begin{pmatrix} 2 & 3 \\ 4 & 3 \end{pmatrix}$$

Since $|A| = -6$ is nonzero, the matrix in nonsingular.

Example 58. Determine if the matrix A is singular or nonsingular.

$$A = \begin{pmatrix} 4 & 6 \\ 2 & 3 \end{pmatrix}$$

Since $|A| = 0$ is zero, the matrix is singular.

1.26 THE INVERSE : ADJOINT METHOD

The inverse of the matrix A was defined earlier as the matrix A^{-1} such that

$$AA^{-1} = I$$

where A and A^{-1} are both square matrices and I is the identity matrix. Now we will calculate the inverse by adjoint method.

1.27 COFACTOR MATRIX

The cofactor of the eleemnt a_{ij} of matrix A is designated by the symbol A_{ij}, where i refers to the row location (number) and j the column location (number). The cofactor matrix, of A designated cof. A, is the matrix of cofactors. That is,

$$\text{cof. A} = \begin{pmatrix} A_{11} & A_{12} & \dots & A_{1n} \\ A_{21} & A_{22} & \dots & A_{2n} \\ \vdots & \vdots & \dots & \vdots \\ Am_1 & A_{m2} & \cdots & A_m \end{pmatrix} \qquad \text{...(1.15)}$$

where the A_{ij} in row i and column j is the cofactor of the element a_{ij} of the square matrix A.

Example 59. Determine the cofactor matrix of the matrix

$$A = \begin{pmatrix} 3 & -2 & 4 \\ 6 & 4 & 5 \\ 3 & -1 & 6 \end{pmatrix} = \begin{pmatrix} a_{11} & a_{12} & a_{13} \\ a_{21} & a_{22} & a_{23} \\ a_{31} & a_{32} & a_{33} \end{pmatrix}$$

The minors of the element a_{ij} are

minor $a_{11} = 29$ minor $a_{12} = 21$ minor $a_{13} = -18$

minor $a_{21} = -8$ minor $a_{22} = 6$ minor $a_{23} = 3$

minor $a_{31} = -26$ minor $a_{32} = -9$ minor $a_{33} = 24$

The cofactor of each element is given by the product of the minor and $(-1)^{i+j}$. The cofactor matrix is

$$\text{cof}\,A = \begin{pmatrix} 29 & -21 & -18 \\ 8 & 6 & -3 \\ -26 & 9 & 24 \end{pmatrix}$$

Example 60. Determie the cofactor matrix for the matrix

$$A = \begin{pmatrix} 6 & -2 & -3 \\ -1 & 8 & -7 \\ 4 & -3 & 6 \end{pmatrix}$$

The minors of the element a_{ij} are

minor $a_{11} = 27$ minor $a_{12} = 22$ minor $a_{13} = -29$

minor $a_{21} = -21$ minor $a_{22} = 48$ minor $a_{23} = -10$

minor $a_{31} = 38$ minor $a_{32} = -45$ minor $a_{33} = 46$

The cofactor matrix is

$$\text{cof}\,A = \begin{pmatrix} 27 & -22 & -29 \\ 21 & 48 & 10 \\ 38 & 45 & 46 \end{pmatrix}$$

1.28 ADJOINT MATRIX

The adjoint matrix is the transpose of the cofactor matrix. That is,

$$\text{adj}\,A = (\text{cof.}\,A)'$$

where adj A designates the adjoint matrix of A. This definition is illustrated by the following two examples.

Example 61. Determine the adjoint matrix for

$$A = \begin{pmatrix} 3 & -2 & 4 \\ 6 & 4 & 5 \\ 3 & -1 & 6 \end{pmatrix}$$

The cofactor matrix was calculated above and is

$$\text{cof}\, A = \begin{pmatrix} 29 & -21 & -18 \\ 8 & 6 & -3 \\ -26 & 9 & 24 \end{pmatrix}$$

the adjoint matrix is the transpose of the cofactor matrix :

$$\text{adj}\, A = \begin{pmatrix} 29 & 8 & -26 \\ -21 & 6 & 9 \\ -18 & -3 & 24 \end{pmatrix}$$

Example 62. Determine the adjoint matrix for

$$A = \begin{pmatrix} 6 & -2 & -3 \\ -1 & 8 & -7 \\ 4 & -3 & 6 \end{pmatrix}$$

The cofactor matrix was calculated previously. The adjoint matrix is the transpose of the cofactor matrix.

$$\text{adj}\, A = \begin{pmatrix} 27 & 21 & 38 \\ -22 & 48 & 45 \\ -29 & 10 & 46 \end{pmatrix}$$

1.29 INVERSE OF A MATRIX

The inverse matrix, A^{-1}, is given by

$$A^{-1} = \left(\frac{1}{|A|}\right) \text{adj}\, A$$

provided A is nonsingular. This definition states that the inverse of A exists provided that the determinant of A is nonzero. The inverse is given by the product of the scalar $1/|A|$ and the adjoint matrix. This method of calculating an inverse matrix is illustrated by the following examples.

Example 63. Calculate the inverse of

$$A = \begin{pmatrix} 3 & -2 & 4 \\ 6 & 4 & 5 \\ 3 & -1 & 6 \end{pmatrix}$$

Applying the properties of determinants, we obtain

$$|A| = 3 \begin{vmatrix} 1 & -2 & 4 \\ 2 & 4 & 5 \\ 1 & -1 & 6 \end{vmatrix} = 3 \times 2 \begin{vmatrix} 1 & -2 & 4 \\ 1 & 2 & 2.5 \\ 1 & -1 & 6 \end{vmatrix}$$

$$|A| = 3 \times 2 \begin{vmatrix} 1 & -2 & 4 \\ 2 & 4 & -1.5 \\ 0 & 1 & 2 \end{vmatrix} = (3 \times 2\, [(1)\,(8 + 1.5)]$$

$$|A| = 57$$

$$A^{-1} = \frac{1}{|A|} \text{adj } A$$

$$= \frac{1}{57}\begin{bmatrix} 29 & 8 & -26 \\ -21 & 6 & 9 \\ -18 & -3 & 24 \end{bmatrix}$$

Example 64. Given $A^{-1} = \begin{bmatrix} \frac{5}{7} & \frac{1}{7} \\ \frac{3}{7} & \frac{2}{7} \end{bmatrix}$ evaluate $A^2 + 2A$.

Solution. Note. Please read the question carefully. Here A^{-1} is given and we have to calculate $A^2 + 2A$ so we will first find A

$$A = [A^{-1}]^{-1} = \frac{\text{Adj } A^{-1}}{|A^{-1}|}$$

$$|A^{-1}| = \frac{10}{49} - \frac{3}{39} = \frac{1}{7}$$

$$\text{Adj } A^{-1} = \begin{bmatrix} \frac{2}{7} & \frac{-1}{7} \\ \frac{-3}{7} & \frac{5}{7} \end{bmatrix}$$

$$\therefore \quad A = 7\begin{bmatrix} \frac{2}{7} & \frac{-1}{7} \\ \frac{-3}{7} & \frac{5}{7} \end{bmatrix} = \begin{bmatrix} 2 & -1 \\ -3 & 5 \end{bmatrix}$$

$$A^2 = \begin{bmatrix} 2 & -1 \\ -3 & 5 \end{bmatrix}\begin{bmatrix} 2 & -1 \\ -3 & 5 \end{bmatrix} = \begin{bmatrix} 7 & -7 \\ -21 & 28 \end{bmatrix}$$

$$A^2 + 2A = \begin{bmatrix} 7 & -7 \\ -21 & 28 \end{bmatrix} + 2\begin{bmatrix} 2 & -1 \\ -3 & 5 \end{bmatrix}$$

$$= \begin{bmatrix} 7 & -7 \\ -21 & 28 \end{bmatrix} + \begin{bmatrix} 4 & -2 \\ -6 & 10 \end{bmatrix} = \begin{bmatrix} 11 & -9 \\ -27 & 38 \end{bmatrix}$$

1.30 CALCULATION OF INVERSE BY ROW TRANSFORMATION

The Gaussian elimination method can be used to calculate an inverse of a matrix. The procedures involves augmenting the square matrices A with the identity matrix I :

$$A \mid \text{I}$$

Row operations are then employed to obtain an identity matrix on the left side of the vertical line. Concurrent with the identity matrix on the left of the vertical line, the inverse matrix is obtained on the right of the vertical line. The Gaussian elimination method involves transforming the augmented matrix $A|\text{I}$ to the inverse matrix $\text{I} \mid A^{-1}$ through row operations.

The justification for this procedure can be demonstrated with matrix algebra. First, write the matrix A augmented with I.

$$A \mid \text{I}$$

If the inverse matrix A^{-1} were known, we could multiply the matrices on both sides of the vertical line by A^{-1}, that is,

$$AA^{-1} \mid IA^{-1}$$

This product would give

$$I \mid A^{-1} \tag{1.1}$$

Instead of using the inverse to obtain the identity matrix on the left side of the vertical line, we use row operations. The row operations applied to obtain the identity matrix concurrently yields the inverse matrix on the right side of the vertical line.

The Gaussian elimination method for obtaining the inverse matrix is illustrated by the following examples.

Example 65. Determine the inverse of $A = \begin{pmatrix} 5 & 3 \\ 6 & -2 \end{pmatrix}$

To determine the inverse, we augment the matrix A with the identity matrix I and apply row operations to obtain an identity matrix on the left side of the vertical line. The augmented matrix is

$$A|I = \left(\begin{array}{cc|cc} 5 & 3 & 1 & 0 \\ 6 & -2 & 0 & 1 \end{array}\right)$$

Multiply the first row by $\frac{1}{5}$

$$\left(\begin{array}{cc|cc} 1 & 3/5 & 1/5 & 0 \\ 6 & -2 & 0 & 1 \end{array}\right)$$

Add –6 times the first row to the second row.

$$\left(\begin{array}{cc|cc} 1 & 3/5 & 1/5 & 0 \\ 0 & -28/5 & -6/5 & 1 \end{array}\right)$$

Multiply the second row by $-\frac{5}{28}$

$$\left(\begin{array}{cc|cc} 1 & 3/5 & 1/5 & 0 \\ 0 & 1 & 6/28 & -5/28 \end{array}\right)$$

Add $-\frac{3}{5}$ times the second row to the first row.

$$\left(\begin{array}{cc|cc} 1 & 0 & 2/28 & 3/28 \\ 0 & 1 & 6/28 & -5/28 \end{array}\right)$$

This results is an identity matrix on the left side of the vertical line. From formula (1.1) we know that the inverse of A is given on the right side of the vertical line. The inverse is

$$A^{-1} = \begin{pmatrix} \frac{2}{28} & \frac{3}{28} \\ \frac{6}{28} & -\frac{5}{28} \end{pmatrix}$$

Example 66. Determine the inverse of $A = \begin{pmatrix} 3 & 2 \\ 6 & -2 \end{pmatrix}$

Augment the matrix A with the identity matrix I.

$$\left(\begin{array}{cc|cc} 3 & 2 & 1 & 0 \\ 6 & -2 & 0 & 1 \end{array}\right)$$

Add 1 times the second row to the first row.

$$\left(\begin{array}{cc|cc} 9 & 0 & 1 & 1 \\ 6 & -2 & 0 & 1 \end{array}\right)$$

Multiply the first row by $\frac{1}{9}$.

$$\left(\begin{array}{cc|cc} 1 & 0 & 1/9 & 1/9 \\ 6 & -2 & 0 & 1 \end{array}\right)$$

Add –6 times the first row to the second row.

$$\left(\begin{array}{cc|cc} 1 & 0 & 1/9 & 1/9 \\ 0 & -2 & -2/3 & 1/3 \end{array}\right)$$

Multiply the second row by $-\frac{1}{2}$.

$$\left(\begin{array}{cc|cc} 1 & 0 & 1/9 & 1/9 \\ 0 & 1 & 1/3 & -1/6 \end{array}\right)$$

the inverse of A is given on the right side of the vertical line and is

$$A^{-1} = \begin{pmatrix} \frac{1}{9} & \frac{1}{9} \\ \frac{1}{3} & -\frac{1}{6} \end{pmatrix}$$

Example 67. Determine the inverse of $A = \begin{pmatrix} 2 & 3 \\ -4 & 6 \end{pmatrix}$.

Augment the matrix A with the identity matrix I.

$$\left(\begin{array}{cc|cc} 2 & 3 & 1 & 0 \\ -4 & 6 & 0 & 1 \end{array}\right)$$

Add 2 times the first row to the second row.

$$\left(\begin{array}{cc|cc} 2 & 3 & 1 & 0 \\ 0 & 12 & 2 & 1 \end{array}\right)$$

Multiply the first row by $\frac{1}{2}$ and the second row by $\frac{1}{12}$.

$$\left(\begin{array}{cc|cc} 1 & 3/2 & 1/2 & 0 \\ 0 & 1 & 1/6 & 1/12 \end{array}\right)$$

Add $-\frac{3}{2}$ times the second row to the first row.

$$\left(\begin{array}{cc|cc} 1 & 0 & 1/4 & -3/24 \\ 0 & 1 & 1/6 & 1/12 \end{array}\right)$$

The inverse of A is

$$A^{-1} = \begin{pmatrix} \frac{1}{4} & -\frac{3}{24} \\ \frac{1}{6} & \frac{1}{12} \end{pmatrix}$$

Example 68. Determine the inverse of $A = \begin{pmatrix} 2 & 4 & -6 \\ 4 & 2 & 2 \\ 3 & -3 & 1 \end{pmatrix}$.

Augment the matrix A with I.

$$A|I = \left(\begin{array}{ccc|ccc} 2 & 4 & -6 & 1 & 0 & 0 \\ 4 & 2 & 2 & 0 & 1 & 0 \\ 3 & -3 & 1 & 0 & 0 & 1 \end{array}\right)$$

Multiply the first row by $\frac{1}{2}$.

$$\left(\begin{array}{ccc|ccc} 1 & 2 & -3 & 1/2 & 0 & 0 \\ 4 & 2 & 2 & 0 & 1 & 0 \\ 3 & -3 & 1 & 0 & 0 & 1 \end{array}\right)$$

Add -4 times the first row to the second row and add -3 times the first row to the third row.

$$\left(\begin{array}{ccc|ccc} 1 & 2 & -3 & 1/2 & 0 & 0 \\ 0 & -6 & 14 & -2 & 1 & 0 \\ 0 & -9 & 10 & -3/2 & 0 & 1 \end{array}\right)$$

Multiply the second row by $-\frac{1}{6}$.

$$\left(\begin{array}{ccc|ccc} 1 & 2 & -3 & 1/2 & 0 & 0 \\ 0 & 1 & -7/3 & 1/3 & -1/6 & 0 \\ 0 & -9 & 10 & -3/2 & 0 & -1 \end{array}\right)$$

Add -2 times the second row to the first row and add 9 times the second row to the third row. Multiply the new third row by $-\frac{1}{11}$.

$$\left(\begin{array}{ccc|ccc} 1 & 0 & 5/3 & -1/6 & 1/3 & 0 \\ 0 & -6 & -7/3 & 1/3 & -1/6 & 0 \\ 0 & -9 & 1 & -3/22 & 3/22 & -1/11 \end{array}\right)$$

Add $\frac{7}{3}$ times the third row to the second row and add $-\frac{5}{3}$ times the third row to the first row.

$$\left(\begin{array}{ccc|ccc} 1 & 0 & 0 & 4/66 & 7/66 & 10/66 \\ 0 & 1 & 0 & 1/66 & 10/66 & -14/66 \\ 0 & 0 & 1 & -9/66 & 9/66 & -6/66 \end{array}\right)$$

The inverse of A is

$$A^{-1} = \begin{pmatrix} \frac{4}{66} & \frac{7}{66} & \frac{10}{66} \\ \frac{1}{66} & \frac{10}{66} & -\frac{14}{66} \\ -\frac{9}{66} & \frac{9}{66} & -\frac{6}{66} \end{pmatrix}$$

1.31 TRIANGULAR MATRICES

A square matrix is said to be a triangular matrix if every element above or below the principal diagonal is zero. If every element above the principal diagonal is zero, the matrix is called **lower triangular.** If every element below the principal diagonal is zero the matrix is called **upper triangular.**

$$\begin{bmatrix} 3 & 0 & 0 \\ 4 & 5 & 0 \\ 3 & 2 & 1 \end{bmatrix} \qquad \begin{bmatrix} 3 & 2 & 1 \\ 0 & 5 & 4 \\ 0 & 0 & 3 \end{bmatrix}$$

Lower triangular Upper triangular

1.32 CONJUGATE OF MATRIX

The matrix obtained from a given matrix 'A' by replacing each of its element by their corresponding complex conjugates is called the conjugate of A and is denoted by $\overline{A}$, *e.g.*,

$$\text{If } A = \begin{bmatrix} 1+i & 3 & 2-i \\ 3+4i & 1-2i & 5 \\ 6i & 8 & 3 \end{bmatrix}, \text{ then } \overline{A} = \begin{bmatrix} 1-i & 3 & 2+i \\ 3-4i & 1+2i & 5 \\ -6i & 8 & 3 \end{bmatrix}$$

Please note :

(i) The order of the conjugate of the matrix is the same as that of the matrix.

(ii) If all the element of A are real, then $A = \overline{A}$.

1.33 TRANSPOSED CONJUGATE OF A MATRIX

Transposed conjugate of a matrix A is denoted by A^{θ}.

$$\text{If } A = \begin{bmatrix} 3+i & 3+2i & 4 \\ 2+2i & 5-3i & 7-i \\ 4 & 6 & 8-i \end{bmatrix}, \text{ then } A^{\theta} = \begin{bmatrix} 3-i & 2-2i & 4 \\ 3-2i & 5+3i & 6 \\ 4 & 7+i & 8+i \end{bmatrix}$$

To obtain the transposed conjugate of a matrix A we replace the elements in A' by their complex conjugates.

Please note :

(i) If A is a $m \times n$ matrix then A^{θ} is a $n \times m$ matrix.

(ii) The transpose of the conjugate coincides with the conjugate of transpose, *i.e.*, $(\overline{A})' = \overline{(A)'} = A^{\theta}$. Also $[A^{\theta}]^{\theta} = A$.

1.34 SYMMETRIC MATRIX

A square matrix A is said to the a symmetric matrix if the transpose of A is equal to A, *i.e.,* if $A = A'$, *e.g.*,

$$\begin{bmatrix} a & h & g \\ h & b & f \\ g & f & c \end{bmatrix} \text{ is a symmetric matrix.}$$

Please note :

(i) In a symmetric matrix any element $a_{ij} = a_{ji}$ for all values of i and j.

(ii) A rectangular matrix cannot be a symmetric matrix.

1.35 SKEW SYMMETRIC

A square matrix A is said to be skew symmetric if the transpose of A is equal to $-A$, *i.e.*, if $A' = -A$, *e.g.*, the matrix $\begin{bmatrix} 0 & c & d \\ -c & 0 & f \\ -d & -f & 0 \end{bmatrix}$ is a skew symmetric or a square matrix A is called skew symmetric, if $a_{ij} = -a_{ji}$ for all i and j.

Please note. In a skew symmetric matrix all the diagonal elements are zero.

1.36 HERMITIAN MATRIX

A square matrix 'A' is said to be a Hermitian matrix if $A^{\theta} = \mathrm{A}$, *e.g.*, the matrix $\begin{bmatrix} 1 & 2+i \\ 2-i & 4 \end{bmatrix}$ is a Hermitian matrix.

Please note. All the diagonal elements of a Hermitian matrix are real.

1.37 SKEW HERMITIAN MATRIX

A square matrix A is said to be skew Hermitian if $A^{\theta} = -A$, *e.g.*, the matrix $\begin{bmatrix} 2i & 1+i & 3-2i \\ -1+i & 4i & 4+i \\ -3-2i & -4+i & 0 \end{bmatrix}$ is a skew Hermitian matrix.

1.38 SINGULAR AND NON-SINGULAR MATRICES

A square matrix A is said to be non-singular if determinant of A, *i.e.*, $|A|$ is not equal to zero. If $|A| = 0$, the matrix is said to be singular.

The necessary and sufficient condition for a square matrix A to possess an inverse is that it must be non-singular.

Important note. To find the inverse of matrix, first step should be to prove that the matrix is non singular.

1.39 SELF-RECIPROCAL MATRIX

A matrix A is called self-reciprocal if $A^{-1} = A$, *e.g.*, the matrix

$$\begin{bmatrix} 0 & 0 & 1 \\ 0 & 1 & 0 \\ 1 & 0 & 0 \end{bmatrix} \text{ is a self-reciprocal matrix.}$$

1.40 INVOLUNTARY MATRIX

A matrix whose square equals identity matrix is called an involuntary matrix

For example, the matrix $\begin{bmatrix} 0 & 1 & -1 \\ 4 & -3 & 0 \\ 3 & -3 & 0 \end{bmatrix}$ is an involuntary matrix.

1.41 NILPOTENT MATRIX

A square matrix A is said to be nilpotent if $A^p = 0$ for some positive integer p. For example,

$$A = \begin{bmatrix} 0 & 0 \\ 2 & 0 \end{bmatrix} \text{ is nilpotent as } A^2 = \begin{bmatrix} 0 & 0 \\ 0 & 0 \end{bmatrix}.$$

1.42 IDEMPOTENT MATRIX

A square matrix A si said to be idempotent if $A^2 = A$. For example,

$$\begin{bmatrix} -1 & 3 & 5 \\ 1 & -3 & -5 \\ -1 & 3 & 5 \end{bmatrix}$$

is an idempotent matrix as in this case $A^2 = A$.

Method for finding the Solution of Linear Equations

The following three methods for finding the solution of linear equations will be discussed here

1. Using inverse of a matrix.
2. Cramer's Rule.
3. Gauss elimination method is method of row operations or row transformations.

1. Matrix Inversion Method : Consider the system of equations

$$a_{11}z + a_{12}y = c_1$$

$$a_{21}x + a_{22}y = c_2$$

This system of equations can be written in the form

$$AX = B$$

where $A = \begin{bmatrix} a_{11} & a_{12} \\ a_{21} & a_2 \end{bmatrix}, X = \begin{bmatrix} x \\ y \end{bmatrix}, B = \begin{bmatrix} c_1 \\ c_2 \end{bmatrix}$

Similarly the system of equations

$$a_{11}x + a_{12}y + a_{13}z = d_1$$

$$a_{21}x + a_{22}y + a_{23} = d_2$$

$$a_{31}x + a_{32}y + a_{33} = d_3$$

Can be written in the form

$$AX = B$$

where $$A = \begin{bmatrix} a_{11} & a_{12} & a_{13} \\ a_{21} & a_{22} & a_{23} \\ a_{31} & a_{32} & a_{33} \end{bmatrix}, X = \begin{bmatrix} x \\ y \\ z \end{bmatrix}, B = \begin{bmatrix} c_1 \\ c_2 \\ c_3 \end{bmatrix}$$

$$AX = B \Rightarrow A^{-1}AX = A^{-1}B$$

$$\Rightarrow \qquad IX = A^{-1}B \Rightarrow X = A^{-1}B$$

The right hand matrix and left hand matrix are matrices of the same order, therefore by equation the two matrices we can find the value of x, y, z.

Example 69. Solve the following system of linear euations by determining the inverse of a matrix

$$3x_1 + 3x_2 + 4x_3 = 20$$

$$-2x_1 + 4x_2 - 2x_3 = -6$$

$$4x_1 - 2x_2 + 3x_3 = 16$$

In matrix form, the system of equations $AX = B$.

$$\begin{pmatrix} 3 & 3 & 4 \\ -2 & 4 & -2 \\ 4 & -2 & 3 \end{pmatrix}\begin{pmatrix} x_1 \\ x_2 \\ x_3 \end{pmatrix} = \begin{pmatrix} 20 \\ -6 \\ 16 \end{pmatrix}$$

The inverse of A is calculated by formula (??),

$$A^{-1} = \frac{\text{adj}\, A}{|A|}$$

where $A|A| = \begin{vmatrix} 3 & 3 & 4 \\ -2 & 4 & -2 \\ 4 & -2 & 3 \end{vmatrix}$

$|A| = 3\,(4)\,3 + 3(-2)\,4 + 4(-2)\,(-2) - 4\,(4)\,4 - (-2)\,(-2)\,3 - 3(-2)\,(3)$

$|A| = -30$

The cofactors of A are

$$A_{11} = (-1)^2 \begin{vmatrix} 4 & -2 \\ -2 & 3 \end{vmatrix} = +1\,(12 - 4) = 8$$

$$A_{12} = (-1)^3 \begin{vmatrix} -2 & -2 \\ 4 & 3 \end{vmatrix} =`-1\,(-6 + 8) = -2$$

$$A_{13} = (-1)^4 \begin{vmatrix} -2 & 4 \\ 4 & -2 \end{vmatrix} = +1\,(4 - 16) = -12$$

$$A_{21} = (-1)^3 \begin{vmatrix} 3 & 4 \\ -2 & 3 \end{vmatrix} = -1\,(9 + 8) = -17$$

$$A_{22} = (-1)^4 \begin{vmatrix} 3 & 4 \\ 4 & 3 \end{vmatrix} = +1\,(9 - 16) = -7$$

$$A_{23} = (-1)^5 \begin{vmatrix} 3 & 3 \\ 4 & -2 \end{vmatrix} = +1\,(-6 - 12) = 18$$

$$A_{31} = (-1)^4 \begin{vmatrix} 3 & 4 \\ 4 & -2 \end{vmatrix} = +1\,(-6-16) = -22$$

$$A_{32} = (-1)^5 \begin{vmatrix} 3 & 4 \\ -2 & -2 \end{vmatrix} = -1\,(-6+8) = -2$$

$$A_{33} = (-1)^6 \begin{vmatrix} 3 & 3 \\ -2 & 4 \end{vmatrix} = +1\,(12+6) = 18$$

The cofactor matrix is

$$\text{cof}\, A = \begin{pmatrix} 8 & -2 & -12 \\ -17 & -7 & 18 \\ -22 & -2 & 18 \end{pmatrix}$$

The adjoint of a matrix is given by the transpose of cofactor A,

$$\text{adj}\, A = \begin{pmatrix} 8 & -17 & -22 \\ -2 & -7 & -2 \\ -12 & 18 & 18 \end{pmatrix}$$

The inverse matrix is

$$A^{-1} = -\frac{1}{30}\begin{pmatrix} 8 & -17 & -22 \\ -2 & -7 & -2 \\ -12 & 18 & 18 \end{pmatrix}$$

The solution vector for the three simultaneous equations is

$$X = A^{-1}B = -\frac{1}{30}\begin{pmatrix} 8 & -17 & -22 \\ -2 & -7 & -2 \\ -12 & 18 & 18 \end{pmatrix}\begin{pmatrix} 20 \\ -6 \\ 16 \end{pmatrix}$$

$$X = -\frac{1}{30}\begin{pmatrix} -90 \\ -30 \\ -60 \end{pmatrix} = \begin{pmatrix} 3 \\ 2 \\ 1 \end{pmatrix}$$

Example 70. Examine whether the following system of linear homogeneous equations has a non-trivial solution :

$$x - y + z = 0$$
$$x + 2y - z = 0$$
$$2x + y + 3z = 0$$

Solution. The determinant of the coefficient matrix is :

$$|A| = \begin{vmatrix} 1 & -1 & 1 \\ 1 & 2 & -1 \\ 2 & 1 & 3 \end{vmatrix} = 1(6+1) + 1(3+2) + 1(1-4) = 9 \neq 0$$

Since $|A| \neq 0$, the system possess only the trivial solution : $x = 0$, $y = 0$, $z = 0$.

Example 71. Solve the following system of homogeneous equations using matrices :

$$x + y - z = 0$$
$$x - 2y + z = 0$$
$$3x + 6y - 5z = 0$$

Solution. The given system can be expressed in the matrix form as

$$\underset{\mathbf{A}}{\begin{bmatrix} 1 & 1 & -1 \\ 1 & -2 & 1 \\ 3 & 6 & -5 \end{bmatrix}} \underset{\mathbf{X}}{\begin{bmatrix} x \\ y \\ z \end{bmatrix}} = \underset{\mathbf{B}}{\begin{bmatrix} 0 \\ 0 \\ 0 \end{bmatrix}}$$

Now, $$|A| = \begin{vmatrix} 1 & 1 & -1 \\ 1 & -2 & 1 \\ 3 & 6 & -6 \end{vmatrix} = 1(10-6) - 1(-5-3) - 1(6+6) = 0$$

Since $|\mathbf{A}| = 0$, and $B = 0$ the given system has infinitely many non-trivial solutions.

To find these solutions, we let $z = k$, and express the first two equations as follows :

$$x + y - k = 0 \Rightarrow x + y = k$$

$$x - 2y + k = 0 \Rightarrow x - 2y = -k$$

This system can be written as :

$$\begin{bmatrix} 1 & 1 \\ 1 & -2 \end{bmatrix} \begin{bmatrix} x \\ y \end{bmatrix} = \begin{bmatrix} k \\ -k \end{bmatrix}$$

$$\Rightarrow \quad \begin{bmatrix} x \\ y \end{bmatrix} = \begin{bmatrix} 1 & 1 \\ 1 & -2 \end{bmatrix}^{-1} = \begin{bmatrix} k \\ -k \end{bmatrix} = \frac{1}{3}\begin{bmatrix} -2 & -1 \\ -1 & 1 \end{bmatrix}\begin{bmatrix} k \\ -k \end{bmatrix} = -\frac{1}{3}\begin{bmatrix} -2k+k \\ -k-k \end{bmatrix} = \begin{bmatrix} k/3 \\ 2k/3 \end{bmatrix}$$

$$\Rightarrow \quad x = \frac{k}{3}, \; y = \frac{2k}{3}$$

Since, k is an arbitrarily chosen constant, we can replace k by $3k$ to get $x = k$, $y = 2k$ and $z = 3k$ also satisfy the third equation. Thus, the general solution of the given system is :

$$x = k, \; y = 2k \text{ and } z = 3k$$

where k is an arbitrarily chosen constant.

Example 72. If $A = \begin{bmatrix} 1 & -1 & 0 \\ 2 & 3 & 4 \\ 0 & 1 & 2 \end{bmatrix}$ and $B = \begin{bmatrix} 2 & 2 & -4 \\ -4 & 2 & -4 \\ 2 & -1 & 5 \end{bmatrix}$ be two square matrices of order three, verify that $AB = BA = 6\mathrm{I}$.

Hence, solve the set of linear equations.

$$x - y = 3, \; 2x + 3y + 4z = 17, \; y + 2z = 7$$

[**Hint :** Prove that $AB = BA = 6\mathrm{I}$

$$\therefore \quad BA = 6\mathrm{I} = B = 6\mathrm{I}A^{-1} = 6A^{-1} = A^{-1} = \frac{B}{6}$$

In the given system of equations.

$$\begin{bmatrix} x \\ y \\ z \end{bmatrix} = A^{-1}B_1 \text{ where } A^{-1} = \frac{B}{6}, \; B_1 = \begin{bmatrix} 3 \\ 17 \\ 7 \end{bmatrix}$$

i.e., $$\begin{bmatrix} x \\ y \\ z \end{bmatrix} = \frac{1}{6}\begin{bmatrix} 2 & 2 & -4 \\ -4 & 2 & -4 \\ 2 & -1 & 5 \end{bmatrix}\begin{bmatrix} 3 \\ 17 \\ 7 \end{bmatrix} = \frac{1}{6}\begin{bmatrix} 12 \\ -6 \\ 24 \end{bmatrix} = \begin{bmatrix} 2 \\ -1 \\ 4 \end{bmatrix}$$

Note. This question can also be stated as find the product of the matrices A ad B as defined above and use this product to solve the system of linear Equalise

$$x - y = 3,\ 2x + 3y = 4z = 17$$
$$y = 2z = 7$$

Gauss Jorden elimination method of solving linear equations : This method makes use of elementary row operations. The following row operations are defined

1. Any two row of a matrix may be interchanged.
2. A row may be multiplied by a nonzero constant.
3. A multiple of one row may be added to another row.

Before proceding further let us define an augmented matrix

Let the system of equation be

$$a_{11}x + a_{12}y + a_{13}z = b_1$$
$$a_{21}x + a_{22}y + a_{23}z = b_2$$
$$a_{31}x + a_{32}y + a_{33}z = b_3$$

Let $A = \begin{bmatrix} a_{11} & a_{12} & a_{13} \\ a_{21} & a_{22} & a_{23} \\ a_{31} & a_{32} & a_{33} \end{bmatrix}$ to the coefficient matrix

$B = \begin{bmatrix} b_1 \\ b_2 \\ b_3 \end{bmatrix}$ to the right hand column matrix, *i.e.,* matrix of constant terms. Then

A/B written in the matrix form as

$$\left[\begin{array}{ccc|c} a_{11} & a_{12} & a_{13} & b_1 \\ a_{21} & a_{22} & a_{23} & b_2 \\ a_{31} & a_{32} & a_{33} & b_3 \end{array}\right] \text{ is defined as an augmented matrix}$$

The solution to the system of equations is determined by applying row operations to the rows of the augmented matrix. The row operations are applied with the objetive of obtaining an identity matrix in the position originally occupied by the coefficient matrix A. Once, the identity matrix is obtained, the solution to the system of equations appears in the position originally occupied by the right-hand side vector.

To demonstrate the solution technique, row operations are performed on both the augmented matrix and the system of two equations in two equations in two unknowns. Our purpose is to show that applying row operations to augmented matrices is analogous to the algebraic operations used in the solution of systems of equatiosn and can, therefore, be used to determine the solution for systems of equations. The following steps are involved to solve the system of equations

(1) Write the augmented matrix A/B

(2) Apply the row operations to transform the coefficient matrix A to any identity matrix I

(3) The solution to the system of equations is then read from the right-hand side of the vertical line.

Illustration. Solve the following system of equations

$$3x_1 + 2x_2 = 14$$
$$6x_1 - 2x_2 = 4$$

Solution. The augmented matrix is

$$\left(\begin{array}{cc|c} 3 & 2 & 14 \\ 6 & -2 & 4 \end{array}\right) \qquad \begin{array}{l} 3x_1 + 2x_2 = 14 \\ 6x_1 - 2x_2 = 4 \end{array}$$

The first row operation is to add one time and second row or second equation to the first row or equation. This gives

$$\left(\begin{array}{cc|c} 9 & 0 & 18 \\ 6 & -2 & 4 \end{array}\right) \qquad \begin{array}{l} 9x_1 + 0x_2 = 18 \\ 6x_1 - 2x_2 = 4 \end{array}$$

Next, multiply the first row or equation by $\frac{1}{9}$.

$$\left(\begin{array}{cc|c} 1 & 0 & 2 \\ 6 & -2 & 4 \end{array}\right) \qquad \begin{array}{l} 1x_1 + 0x_2 = 2 \\ 6x_1 - 2x_2 = 4 \end{array}$$

Add –6 times the first row or equation to the second row or equation.

$$\left(\begin{array}{cc|c} 1 & - & 2 \\ 0 & -2 & -8 \end{array}\right) \qquad \begin{array}{l} 1x_1 + 0x_2 = 2 \\ 0x_1 - 2x_2 = -8 \end{array}$$

the last operation is to multiply the second row or equation by $-\frac{1}{2}$.

$$\left(\begin{array}{cc|c} 1 & 0 & 2 \\ 0 & 1 & 4 \end{array}\right) \qquad \begin{array}{l} 1x_1 + 0x_2 = 2 \\ 0x_1 + 1x_2 = 4 \end{array}$$

The solution to the system of equation is read directly as $x_1 = 2$ and $x_2 = 4$. The top element to the right of the vertical line is the solution value of x_1 and the bottom element is the solution value of x_2. This can be demonstrated by writing the augmented matrix as the matrix equation.

$$\begin{pmatrix} 1 & 0 \\ 0 & 1 \end{pmatrix}\begin{pmatrix} x_1 \\ x_2 \end{pmatrix} = \begin{pmatrix} 2 \\ 4 \end{pmatrix}$$

and converting the matrix equation to two equations in the two unknowns.

$$1x_1 + 0x_2 = 2$$

$$0x_1 + 1x_2 = 4$$

Example 73. As a second example of solving matrix equations using row operations, consider the following three equations

$$2x + 1y - 2z = -1$$

$$4x - 2y + 3z = 14$$

$$1x - 1y + 2z = 7$$

Solution. Row operations are performed on the augmented matrix and the system of equations. This will again demonstrate the parallel between the algebraic solution and the matrix solution of systems of equations. The augmented matrix and the system of equations are

$$\left(\begin{array}{ccc|c} 2 & 1 & -2 & -1 \\ 4 & -2 & 3 & 14 \\ 1 & -1 & 2 & 7 \end{array}\right) \qquad \begin{array}{r} 2x + 1y - 2z = -1 \\ 4x - 2y + 3z = 14 \\ 1x - 1y + 2z = 7 \end{array}$$

Add 1 time the third row to the first row.

$$\left(\begin{array}{ccc|c} 3 & 0 & 0 & 6 \\ 4 & -2 & 3 & 14 \\ 1 & -1 & 2 & 7 \end{array}\right) \qquad \begin{aligned} 3x + 0y + 0z &= 6 \\ 4x - 2y + 3z &= 14 \\ 1x - 1y + 2z &= 7 \end{aligned}$$

Multiply the first row by $\frac{1}{3}$.

$$\left(\begin{array}{ccc|c} 1 & 0 & 0 & 2 \\ 4 & -2 & 3 & 14 \\ 1 & -1 & 2 & 7 \end{array}\right) \qquad \begin{aligned} 1x + 0y + 0z &= 2 \\ 4x - 2y + 3z &= 14 \\ 1x - 1y + 2z &= 7 \end{aligned}$$

Add –4 times the first row to the second row and –1 times the first row to the third row.

$$\left(\begin{array}{ccc|c} 1 & 0 & 0 & 2 \\ 0 & -2 & 3 & 6 \\ 0 & -1 & 2 & 5 \end{array}\right) \qquad \begin{aligned} 1x + 0y + 0z &= 2 \\ 0x - 2y + 3z &= 6 \\ 0x - 1y + 2z &= 5 \end{aligned}$$

Multiply the second row by $-\frac{1}{2}$.

$$\left(\begin{array}{ccc|c} 1 & 0 & 0 & 2 \\ 0 & 1 & -3/2 & -3 \\ 0 & -1 & 2 & 5 \end{array}\right) \qquad \begin{aligned} 1x + 0y + 0z &= 2 \\ 0x + 1y - \frac{3}{2}z &= -3 \\ 0x - 1y + 2z &= 5 \end{aligned}$$

Add 1 times the second row to the third row.

$$\left(\begin{array}{ccc|c} 1 & 0 & 0 & 2 \\ 0 & 1 & -3/2 & -3 \\ 0 & 0 & 1/2 & 2 \end{array}\right) \qquad \begin{aligned} 1x + 0y + 0z &= 2 \\ 0x + 1y - \frac{3}{2}z &= -3 \\ 0x + 0y + \frac{1}{2}z &= 2 \end{aligned}$$

Multiply the third row by 2.

$$\left(\begin{array}{ccc|c} 1 & 0 & 0 & 2 \\ 0 & 1 & -3/2 & -3 \\ 0 & 0 & 1 & 4 \end{array}\right) \qquad \begin{aligned} 1x + 0y + 0z &= 2 \\ 0x + 1y - \frac{3}{2}z &= -3 \\ 0x + 0y + 1z &= 4 \end{aligned}$$

Add $\frac{3}{2}$ times the third row to the second row.

$$\left(\begin{array}{ccc|c} 1 & 0 & 0 & 2 \\ 0 & 1 & 0 & 3 \\ 0 & 0 & 1 & 4 \end{array}\right) \qquad \begin{aligned} 1x + 0y + 0z &= 2 \\ 0x + 1y + 0z &= 3 \\ 0x + 0y + 1z &= 4 \end{aligned}$$

The solution to the three equations is $x = 2$, $y = 3$ and $z = 4$.

It has been demonstrated in the preceding two examples that the row operations are applied to the augmented matrix with the objective of obtaining an identity matrix on the left side of the vertical line. Once, the identity matrix is obtained, the solution vector can be read from the right side of the vertical line. This is possible since, the row operations on the augmented matrix correspond to algebraic operations on the system of equations.

Any combination of row operations is acceptable in obtaining the identity matrix to the left of the vertical line. Many persons find it useful to first obtain 1 in the a_{11} position and to then use multiples of the first row to obtain zeros elsewhere in the first column. The student next obtains 1 in the a_{22} position and uses multiples of the second row to obtain zeros elsewhere in the second column. This procedure is followed until the identity matrix is obtained. This method is used in the following examples.

Example 74. Determine the solution vector for the following two equations, using the Gaussian elimination method.

$$-2x_1 + 4x_2 = 4$$
$$3x_1 - 2x_2 = 10$$

Solution. Write the equations in matrix form as $A \mid B$

$$\left(\begin{array}{cc|c} -2 & 4 & 4 \\ 3 & -2 & 10 \end{array}\right)$$

Multiply the first row by $-\frac{1}{2}$.

$$\left(\begin{array}{cc|c} 1 & -2 & -2 \\ 3 & 2 & 10 \end{array}\right)$$

Add −3 times the first row to the second row.

$$\left(\begin{array}{cc|c} 1 & -2 & -2 \\ 0 & 4 & 16 \end{array}\right)$$

Multiply the second row by $\frac{1}{4}$.

$$\left(\begin{array}{cc|c} 1 & -2 & -2 \\ 0 & 1 & 4 \end{array}\right)$$

Add 2 times the second row to the first row.

$$\left(\begin{array}{cc|c} 1 & 0 & 6 \\ 0 & 1 & 4 \end{array}\right)$$

The solution is $x_1 = 6$ and $x_2 = 4$.

Example 75. Determine the solution vector to the following system of equations, using the Gaussian elimination method.

$$2x + 3y - 4z = 9$$
$$3x - 2y + 3z = -15$$
$$1x + 4y - 2x = 12$$

Solution. Write the equations in matrix form as $A|B$.

$$\left(\begin{array}{ccc|c} 2 & 3 & -4 & 9 \\ 3 & -2 & 3 & -15 \\ 1 & 4 & -2 & 12 \end{array}\right)$$

Multiply the first row by $\frac{1}{2}$.

$$\left(\begin{array}{ccc|c} 1 & 3/2 & -2 & 9/2 \\ 3 & -2 & 3 & -15 \\ 1 & 4 & -2 & 12 \end{array}\right)$$

Add –3 times the first row to the second row and add –1 times the first row to the third row.

$$\left(\begin{array}{ccc|c} 1 & 3/2 & -2 & 9/2 \\ 0 & -13/2 & 9 & -57/2 \\ 0 & 5/2 & 0 & 15/2 \end{array}\right)$$

Interchange the second and third row.

$$\left(\begin{array}{ccc|c} 1 & 3/2 & -2 & 9/2 \\ 0 & 5/2 & 0 & 15/2 \\ 0 & -13/2 & 9 & -57/2 \end{array}\right)$$

Multiply the second row by $\frac{2}{5}$.

$$\left(\begin{array}{ccc|c} 1 & 3/2 & -2 & 9/2 \\ 0 & 1 & 0 & 3 \\ 0 & -13/2 & 9 & -57/2 \end{array}\right)$$

Add $-\frac{3}{2}$ times the second row to the first row, and add $\frac{13}{2}$ time the second row to the third row.

$$\left(\begin{array}{ccc|c} 1 & 0 & -2 & 0 \\ 0 & 1 & 0 & 3 \\ 0 & 0 & 9 & -9 \end{array}\right)$$

Adding $\frac{2}{9}$ times the third to the first row we get,

$$\left(\begin{array}{ccc|c} 1 & 0 & 0 & -2 \\ 0 & 1 & 0 & 3 \\ 0 & 0 & 9 & -9 \end{array}\right)$$

Multiple the third row by $\frac{1}{9}$ we get.

$$\left(\begin{array}{ccc|c} 1 & 0 & 0 & -2 \\ 0 & 1 & 0 & 3 \\ 0 & 0 & 1 & -1 \end{array}\right)$$

$\therefore\ x = -2,\ y = 3,\ z = -1$

3rd Procedure of Gauss Jordan Elimination Method

In this procedure coefficient matrix is reduced to a triangular matrix by performing row operation on the coefficient matrix and the matrix of constant terms on the right hand side. The following steps are taken

Step 1. Express the given system of equations in the matrix form as $AX = B$.

Step 2. Transform the coefficient matrix A to the triangular matrix by appropriate row operations on both sides of $AX = B$.

Step 3. Let the transformed form be $CX = D$, where C is the triangular matrix. This matrix form is then converted back to the system of linear equations by multiplying the matrices C and X and equating it to the matrix D.

Step 4. The equations thus obtained are solved to obtain the required solution. The following example will make the method clear. We illustrate the method by suitable example.

Example 76. Solve the following system of linear equatios using Gauss Jordan Elimination method :

$$x + 2y + 3z = 1$$
$$x + 3y + 5z = 2$$
$$2x + 5y + 9z = 3$$

Solution. The given system of equations in matrix form is

or,
$$\begin{bmatrix} 1 & 2 & 3 \\ 1 & 3 & 5 \\ 2 & 5 & 9 \end{bmatrix} \mathbf{X} = \begin{bmatrix} 1 \\ 2 \\ 3 \end{bmatrix}$$

Now we shall transform the coefficient matrix A to the triangular matrix by applying suitable row operations on A and B (Remember any such row operation on the product of two matrices is equivalent to applying the same on the prefactor only) Applying the operations $R_2 \to R_2 - R_1$ and $R_3 \to R_3 - 2R_1$, the given system is equivalent to

$$\begin{bmatrix} 1 & 2 & 3 \\ 0 & 1 & 2 \\ 0 & 1 & 3 \end{bmatrix} \mathbf{X} = \begin{bmatrix} 1 \\ 1 \\ 1 \end{bmatrix}$$

Now applying $R_3 \to R_3 - R_2$, the above system is equivalent to

$$\begin{bmatrix} 1 & 2 & 3 \\ 0 & 1 & 2 \\ 0 & 0 & 1 \end{bmatrix} \mathbf{X} = \begin{bmatrix} 1 \\ 1 \\ 0 \end{bmatrix}$$

The coefficient matrix of the given system has now been reduced to the triangular matrix. Rewriting the above matrix form in terms of equations, we get

$$x + 2y + 3z = 1$$
$$y + 2z = 1$$
$$z = 0$$

Substituting $z = 0$ form (iii) in (ii), we get $y = 1$. Substituting these values of y and z in (i), we get $x = -1$.

Hence, the solution is : $x = -1, y = 1, z = 0$

Example 77. Solve the following system of linear equations using elementary operation :

$$2x + 3y - z = 9$$
$$x + y + z = 9$$
$$3x - y - z = -1$$

Solution. The given system of equations in matrix form is :

$$\mathbf{AX} = \mathbf{B}$$

or,
$$\begin{bmatrix} 2 & 3 & -1 \\ 1 & 1 & 1 \\ 3 & -1 & -1 \end{bmatrix} \mathbf{X} = \begin{pmatrix} 9 \\ 9 \\ -1 \end{pmatrix}$$

Interchanging R_1 and R_2 we get

$$\begin{bmatrix} 1 & 1 & 1 \\ 2 & 3 & -1 \\ 3 & -1 & -1 \end{bmatrix} \mathbf{X} = \begin{bmatrix} 9 \\ 9 \\ -1 \end{bmatrix}$$

Applying the operations $R_2 \to R_2 - 2R_1$ and $R_3 \to R_3 - 3R_1$, then system is equivalent to

$$\begin{bmatrix} 1 & 1 & 1 \\ 0 & 1 & -3 \\ 0 & -4 & -4 \end{bmatrix} \mathbf{X} = \begin{bmatrix} 9 \\ -9 \\ -28 \end{bmatrix}$$

Applying the operations $R_3 \to R_3 + 4R_2$, the system is equivalent to

$$\begin{bmatrix} 1 & 1 & 1 \\ 0 & 1 & -3 \\ 0 & 0 & -16 \end{bmatrix} \mathbf{X} = \begin{bmatrix} 9 \\ -9 \\ -64 \end{bmatrix}$$

Rewriting the above system in terms of equations, we have

$$x + y + z = 9$$

$$y - 3z = 9$$

$$-16z = -64$$

The last equation gives $z = 4$. Substituting $z = 4$ in (ii), we get $y = 3$. Substituting these values of y and z in (i), we get $x = 2$.

Hence, the solution is : $x = 2, y = 3, z = 4$.

Example 78. Using elementary row operations, solve the following system of equations :

$$2x + 4y - z = 9;\ 3x - y + 5z = 5;\ 8x + 2y + 9z = 19$$

Solution. The given system of equations in matrix form is :

$$\mathbf{AX} = \mathbf{B}$$

or,
$$\begin{bmatrix} 2 & 4 & -1 \\ 3 & -1 & 5 \\ 8 & 1 & 9 \end{bmatrix} \mathbf{X} = \begin{bmatrix} 9 \\ 5 \\ 19 \end{bmatrix}$$

Applying the operation $R_1 \to \frac{1}{2} R_1$, the system is equivalent to

$$\begin{bmatrix} 1 & 2 & -1/2 \\ 3 & -1 & 5 \\ 8 & 2 & 9 \end{bmatrix} \mathbf{X} = \begin{bmatrix} 9/2 \\ 5 \\ 19 \end{bmatrix}$$

Applying the operations $R_2 \to R_2 - 3R_1$ and $R_3 \to R_3 - 8R_1$, the system is equivalent to

$$\begin{bmatrix} 1 & 2 & -1/2 \\ 0 & -7 & 13/2 \\ 0 & -14 & 13 \end{bmatrix} \mathbf{X} = \begin{bmatrix} 9/2 \\ -17/2 \\ -17 \end{bmatrix}$$

Applying the operations $R_3 \to R_3 - 2R_2$, the system is equivalent to

$$\begin{bmatrix} 1 & 2 & -1/2 \\ 0 & -7 & 13/2 \\ 0 & 0 & 0 \end{bmatrix} \mathbf{X} = \begin{bmatrix} 9/2 \\ -17/2 \\ 0 \end{bmatrix}$$

Rewriting the above system in terms of equations, we have

$$x+2y-\frac{1}{2}z=\frac{9}{2}$$
$$-7y+\frac{13}{2}z=-\frac{17}{2}$$
$$0=0$$

The last equation, though true, is reductant and the system is equivalent to

$$x+2y-\frac{1}{2}z=\frac{9}{2}$$
$$-7y+\frac{13}{2}z=-\frac{17}{2}$$

Solving these for x and y in terms of z, we get $y=\frac{1}{14}(13z+17)$ and $x=\frac{1}{14}(29-19z)$. By assigning an arbitrary value k to z, the general solution to the given system is given by

$$x=\frac{1}{14}(29-19k),\ y=\frac{1}{14}(13k+17) \text{ and } z=k$$

Since, k is arbitrary, the system has infinite number of solutions.

Example 79. Solve, if consistent, the following system of equations.

$$x-3y+4z=3$$
$$2-5y+7z=6$$
$$3x-8y+11z=11$$

Solution. The given system of equations in matrix form is

$$\mathbf{AX}=\mathbf{B}$$

$$\begin{bmatrix}1 & -3 & 4\\ 2 & -5 & 7\\ 3 & -8 & 11\end{bmatrix}\mathbf{X}=\begin{bmatrix}3\\ 6\\ 11\end{bmatrix}$$

Applying the operation $R_2 \to R_2-2R_1$ and $R_3 \to R_3-3R_1$, the system is equivalent to

$$\begin{bmatrix}1 & -3 & 4\\ 0 & 1 & -1\\ 0 & 1 & -1\end{bmatrix}\mathbf{X}=\begin{bmatrix}3\\ 0\\ 2\end{bmatrix}$$

Applying $R_2 \to R_2-R_1$, we get

$$\begin{bmatrix}1 & -3 & 4\\ 0 & 1 & -3\\ 0 & 0 & 0\end{bmatrix}\mathbf{X}=\begin{bmatrix}3\\ 0\\ 2\end{bmatrix}$$

$\therefore$

$$x-3y+4z=3$$
$$y-3z=0$$

1.43 CRAMER'S RULE FOR SOLVING SIMULTANEOUS EQUATIONS

Cramer's rule can be applied for solving a system of n equation with n variables. This technique bypassses the calculation of the inverse of the coefficient matrix or the reduction of the augmented coefficient matrix through row operation. Instead, Cramer's requires the calculation of determinants for $n+1$ matrices.

Consider the following system of equations

$$\left.\begin{aligned} a_1x+b_1y+c_1z \\ a_2x+b_2y+c_2z \\ a_3x+b_3y+c_3z \end{aligned}\right\} \begin{aligned} &= d_1 \\ &= d_2 \ldots \text{'}A\text{'} \\ &= d_3 \end{aligned}$$

By Cramer's rule

$$x=\frac{\Delta_x}{\Delta},\ y=\frac{\Delta_y}{\Delta},\ z=\frac{\Delta_z}{\Delta}$$

where $\Delta = \begin{vmatrix} a_1 & b_1 & c_1 \\ a_2 & b_2 & c_2 \\ a_3 & b_3 & c_3 \end{vmatrix}$ *i.e.*, Δ is the determinant formed by the coefficients of x, y, z.

Δ_x is the determinant obtained by replacing the column of coefficient of x in Δ by the column of constant terms in the given system of equations.

i.e.,

$$\Delta_x = \begin{vmatrix} d_1 & b_1 & c_1 \\ d_2 & b_2 & c_2 \\ d_3 & b_3 & c_3 \end{vmatrix}$$

i.e., Δ_x is the determinant obtained by replacing the column of coefficients of x in Δ by the column of constant terms in the given system of equations.

$$\Delta_y = \begin{vmatrix} a_1 & d_1 & c_1 \\ a_2 & d_2 & c_2 \\ a_3 & d_3 & c_3 \end{vmatrix}$$

i.e., Δ_y is the determinant obtained by replacing the column of coefficients of y in Δ by the column of constant terms in the given system of equations.

$$\Delta_z = \begin{vmatrix} a_1 & b_1 & d_1 \\ a_2 & b_2 & d_2 \\ a_3 & b_3 & d_3 \end{vmatrix}$$

i.e., Δ_z is the determinant obtained by replacing the column of coefficients of z in Δ by the column of constant terms in the given system of equations.

Caution : Before applying Cramer's rule, students should reduce the equations to the standard form as in '*A*' above. If the system of equations is written as

$$a_1x+b_1y+c_1z+d_1=0$$
$$a_2x+b_2y+c_2z+d_2=0$$
$$a_3x+b_3y+c_3z+d_3=0$$

then

$$x=-\frac{\Delta_x}{\Delta},\ y=-\frac{\Delta_y}{\Delta},\ z=-\frac{\Delta_z}{\Delta}$$ **(Note carefully)**

Example 80. Solve the 3 by 3 system of equations, using Cramer's rule.

$$3x_1-4x_2+8x_3=26$$
$$6x_1-3x_2-5x_3=1$$
$$-x_1+1x_2+3x_3=11$$

$$x_1 = \frac{\begin{vmatrix} 26 & -4 & 8 \\ 1 & 3 & -5 \\ 11 & 1 & 3 \end{vmatrix}}{\begin{vmatrix} 3 & -4 & 8 \\ 6 & 3 & -5 \\ -2 & 1 & 3 \end{vmatrix}} = \frac{340}{170} = 2$$

$$x_2 = \frac{\begin{vmatrix} 3 & 26 & 8 \\ 6 & 1 & -5 \\ -2 & 11 & 3 \end{vmatrix}}{\begin{vmatrix} 3 & -4 & 8 \\ 6 & 3 & -5 \\ -2 & 1 & 3 \end{vmatrix}} = \frac{510}{170} = 3$$

$$x_3 = \frac{\begin{vmatrix} 3 & -4 & 26 \\ 6 & 3 & 1 \\ -2 & 1 & 11 \end{vmatrix}}{\begin{vmatrix} 3 & -4 & 8 \\ 6 & 3 & -5 \\ -2 & 1 & 3 \end{vmatrix}} = \frac{680}{170} = 4$$

The same rule can be applied for a 2×2 system of equations.

Questions

1. If $A = \begin{bmatrix} 3 & 5 \\ 2 & a \end{bmatrix}, B = \begin{bmatrix} 4 & b \\ 2 & 9 \end{bmatrix}, C = \begin{bmatrix} 26 & a \\ 14 & 45 \end{bmatrix}$, find a and b when $2A + 5B = C$.] **(ICWA June, 1996)**

[**Ans.** $a = 0, b = -2$]

2. Show that

$$A = \begin{bmatrix} 1 & 2 & 2 \\ 2 & 1 & 2 \\ 2 & 2 & 1 \end{bmatrix}$$

satisfies the relation.

$A^2 - 4A - 5I = 0$, where I is the identity matrix and 0 is the null matrix. Hence find A^{-1}. **(ICWA Dec., 1996)**

[**Ans.** $A^{-1} = \frac{1}{5}\begin{bmatrix} 3 & 2 & 2 \\ 2 & -3 & 2 \\ 2 & 2 & -3 \end{bmatrix}$]

3. If $A = \begin{bmatrix} 1 & -1 \\ 2 & -1 \end{bmatrix}, B = \begin{bmatrix} x & 1 \\ y & -1 \end{bmatrix}$,

and $(A + B)^2 = A^2 + B^2$, find x and y. **(ICWA June, 1997)**

[**Ans.** $x = 1, y = 4$].

4. If $A = \begin{bmatrix} 2 & 0 & -1 \\ 2 & 4 & -1 \\ 1 & -8 & -3 \end{bmatrix}$ show that

$A\,(\text{Adj.}\,A) = |A|\, I_3$. **(ICWA June, 1998)**

5. If $A = \begin{bmatrix} 0 & 1 \\ -1 & 0 \end{bmatrix}$, find α and β all reals such that $(\alpha I + \beta A)^2 = A$.

6. Which of the following is an orthogonal matrix?

(i) $\begin{pmatrix} 1 & 0 \\ -1 & 0 \end{pmatrix}$ (ii) $\begin{pmatrix} 1 & 0 \\ 1 & -1 \end{pmatrix}$ (iii) $\begin{pmatrix} 1 & -1 \\ -1 & 1 \end{pmatrix}$ **(CA (Foundation) May 2000)**

7. In a certain city there are 5 Colleges and 20 Schools. Each school has three Peons, 1 Clerk and 1 Head Clerk. Whereas a college has 5 Peons, 3 Clerks and I Head Clerk and an additional Staff as a Caretaker. The monthly salary of each of them is as follows :

Peon	Rs. 1,100	Head Clerk	Rs. 3,000
Clerk	Rs. 1,700	Caretaker	Rs. 2,500

Using matrix algebra, find the total monthly salary bill of each school and college.

(CA (Foundation) May 1996)

[**Ans.** Rs. 16,100; Rs. 8,000]

8. Obtain the inverse of the matrix $\begin{bmatrix} 2 & 4 & -1 \\ 3 & 1 & 2 \\ 1 & 3 & -3 \end{bmatrix}$ hence solve the following system of equations

$2x + 4y - z = 9, 3x + y + 2z = 7, x + 3y - 3z = 4$ **(CA (Foundation) May 1995)**

[**Ans.** $x = 1, y = 2, z = 1$]

9. One unit of commodity A, is produced by combining 1 unit of land, 2 units of labour and 5 units of capital. One unit of B is produced by combining 2 units of land, 3 units of labour and 1 units of capital. One unit of C is produced by 3 units of land, 1 unit of labour and 2 units of capital. If the price of A, B and C are Rs. 27, Rs. 16 and Rs. 19 respectively, find the rent R, wage W and rate of interest I.

(D.U., B.Com. (Hons.), 1984 (External))

[**Ans.** $R = 3, W = 2, I = 4$]

[**Hint :** If R denotes the rent, W denotes the wage and I denotes interest Then,

$$\begin{matrix} R + 2W + 5I = 27 \\ 2R + 3W + 1I = 15 \\ 3R + W + 2I = 19 \end{matrix} \Rightarrow \begin{bmatrix} 1 & 2 & 5 \\ 2 & 3 & 1 \\ 3 & 1 & 2 \end{bmatrix} \begin{bmatrix} R \\ W \\ I \end{bmatrix} = \begin{bmatrix} 27 \\ 16 \\ 19 \end{bmatrix}].$$

Solve by matrix method.

Rent 8 is for level, wage for labour and interest on capital.

10. A, B and C have Rs. 1,250 1,700 and 2,100 respectively. They utilized the amount to purchase three types of shares of prices x, y and z respectively. A Purchases 20 shares of price, x, 50 shares of price y and 30 shares of price z. B purchases 44 of price x, 30 of price y and 60 of price x, C purchases 12 of price x, 40 of price y and 100 of price z. Find x, y and z by matrix method. **(D.U., B.Com. (Hons.) 1982 (External))**

[**Ans.** $x = 10, y = 12, z = 15$]

[**Hint :** $\left.\begin{matrix} 20x + 50y + 30z = 1{,}250 \\ 44x + 30y + 60z = 1{,}700 \\ 12x + 40y + 100z = 2{,}100 \end{matrix}\right\}$].

Solve by matrix inversion method.

11. A salesman has the following record of sales during three weeks for the three items, X, Y and Z which have different rates of commission :

Weeks	Unit sold			Total Commission required (in Rs.)
	X	Y	Z	
I	40	30	20	270
II	50	50	40	450
III	60	30	10	260

Find out, using matrix method, rates of commission on items, X, Y and Z.

(D.U., B.Com. (Hons.) 1988 (External))

[**Ans.** 2, 5, 5]

[**Hint :** Let Rs. x_1, x_2, x_3 be the rates of commission per unit for the items X, Y and Z respectively. Then

$$\begin{bmatrix} 40 & 30 & 20 \\ 50 & 50 & 40 \\ 60 & 30 & 10 \end{bmatrix}\begin{bmatrix} x_1 \\ x_2 \\ x_3 \end{bmatrix}=\begin{bmatrix} 270 \\ 450 \\ 260 \end{bmatrix}].$$

12. Explain the concept of adjoint of square matrix. Show that $\frac{\text{Adj. } A}{|A|}$ is the inverse of matrix A where symbols have their usual meaning. **(D.U., B.Com. (Hons.) 1988)**

[**Hint :** Let $A\begin{bmatrix} a_{11} & a_{12} & a_{13} \\ a_{21} & a_{22} & a_{23} \\ a_{31} & a_{32} & a_{33} \end{bmatrix}$. Then Adj. $A=\begin{bmatrix} A_{11} & A_{21} & A_{31} \\ A_{12} & A_{22} & A_{32} \\ A_{13} & A_{22} & A_{33} \end{bmatrix}$].

13. The prices in Rs. of the three commodities X, Y, Z are x, y and z per unit respectively. A purchases 4 units of z and sells 3 units of x and 5 units of y. B Purchases 3 units of Y and sells 2 units of X and 1 unit of Z. C purchases I unit of X and sells 4 units of Y and 6 units of Z. In the process A, B, C Earn Rs. 6,000, 5,000 and 13,000 respectively. Using matrices, find prices per unit of the three commodities [note that selling the unit is positive earning and buying the unit is negative earning]. [**Ans.** $x =$ Rs. 3,000, $y =$ Rs. 1,000 and $z =$ Rs. 2,000]

14. Food I has 3 units of Vitamin A, 9 units of Vitamin B and 12 units of Vitamin C. Food II has 6, 9 and 15 units respectively and Food III has 9, 0, 9 units respectively. 33 units of Vitamin A, 27 of B and 60 of C are required. Find the amount of three foods that will provide exactly these amounts, Use matrix method.

[**Hint :** If x_1 is the amount of Food I, x_2 the amount of Food II, x_3 the amount of Food III necessary to meet the requirement, then

$$\begin{array}{cc} & \text{Food} \\ & \begin{array}{ccc} \text{I} & \text{II} & \text{III} \end{array} \\ \text{Vitamin} \begin{array}{c} \text{A} \\ \text{B} \\ \text{C} \end{array} & \begin{bmatrix} 3 & 6 & 9 \\ 9 & 9 & 0 \\ 12 & 15 & 9 \end{bmatrix} \end{array}\begin{bmatrix} x_1 \\ x_2 \\ x_3 \end{bmatrix}=\begin{bmatrix} 33 \\ 27 \\ 60 \end{bmatrix}$$

Here $\begin{bmatrix} 3 & 6 & 9 \\ 9 & 9 & 0 \\ 12 & 15 & 9 \end{bmatrix}=0$ Hence there is no unique solution]

15. If $A=\begin{bmatrix} 1 & -1 & 0 \\ 2 & 3 & 4 \\ 0 & 1 & 2 \end{bmatrix}$ and $B=\begin{bmatrix} 2 & 2 & -4 \\ -4 & 2 & -4 \\ 2 & -1 & 5 \end{bmatrix}$ be two square matrices of order three, verify that $AB = BA = 6I$.

Hence solve the set of linear equations.

$$x - y = 3,\ 2x + 3y + 4z = 17,\ y + 2z = 7$$

[**Hint :** Prove that $AB = BA = 6I$

$$\therefore \qquad BA = 6I = B = 6IA^{-1} = 6A^{-1} = A^{-1} = \frac{B}{6}$$

In the given system of equation.

$$\begin{bmatrix} x \\ y \\ z \end{bmatrix} = A^{-1}B_1 \text{ where } A^{-1} = \frac{B}{6},\ B_1 = \begin{bmatrix} 3 \\ 17 \\ 7 \end{bmatrix}$$

$$i.e., \begin{bmatrix} x \\ y \\ z \end{bmatrix} = \frac{1}{6}\begin{bmatrix} 2 & 2 & -4 \\ -4 & 2 & -4 \\ 2 & -1 & 5 \end{bmatrix}\begin{bmatrix} 3 \\ 17 \\ 7 \end{bmatrix} = \frac{1}{6}\begin{bmatrix} 12 \\ -6 \\ 24 \end{bmatrix} = \begin{bmatrix} 2 \\ -1 \\ 4 \end{bmatrix} \Rightarrow x = 2, y = -1, z = 4].$$

Note : This question can also be stated as find the product of the matrices A and B as defined above and use this product to solve the system of linear equations

$$x - y = 3,\ 2x + 3y = 4z = 17$$
$$y + 2z = 7$$

16. (i) In a certain city there are 50 colleges and 400 schools. Each school and college has 18 peons, 5 clerks and one cashier. Each college in addition has one section officer and one librarian. The monthly salary of each of them is as follows :

Peon—Rs. 300, clerk—Rs. 500, cashier—Rs. 600, section officer—Rs. 700 and librarian—Rs. 900. Using matrix notation find : (i) Total number of posts of each kind in schools and colleges taken together.

(ii) The total monthly salary bill of the schools and college taken together.

[**Ans.** (i) Peon—8,100, clerk—2,250, cashier—450, section officer—50, librarian—50
(ii) Total salary—Rs. 39,05,000]

17. An amount of Rs. 5,000 is put into three different investments at the rates of interest of 6%, 7% and 8% per annum respectively. The total annual income is Rs. 358. If the combined income from the first two investments is Rs. 70 more than the income from the third, find the amount of each investment by using matrix Algebra.

(D.U., B.Com. (Hosn.) 1989)

[**Ans.** 1,000; 2,000; 1,800]

18. An amount of Rs. 7,000 is placed in three diffrent saving accounts with annual interest rates of 5%, 8% and 8.5%. The total annual yield from these accounts is Rs. 550. Equal amount has been placed in the 5% and 8% saving accounts. You are required to obtain amount placed in each account with the help of matrix Algebra.

(Punjab M.A. Eco. 1990)

[**Ans.** Rs. 1,125, Rs. 1,125, Rs. 4,750]

19. Three products *X*, *Y* and *Z* are produced after being processed through three departments D_1, D_2 and D_3. The following data are available.

Products	Hours required for a unit produced		
	In D_1	In D_2	In D_3
X	2	5	1
Y	1	2	3
Z	2	2	3
Max. time available in hours	1,100	1,800	1,400

Find by matrix the number of units produced for each product to have full utilization of capacity.

[**Ans.** Units produced for product $X = 20$, Units produced for product $Y = 100$,
Units produced for product $Z = 300$]

[**Hint :** Let x_1, x_2 and x_3 denote the number of units of products *X*, *Y* and *Z* respectively to have full utilization of capacity. Then

$$2x_1 + x_2 + 2x_3 = 1{,}100$$

$$5x_1 + 2x_3 + 2x_3 = 1{,}800$$

and $$x_1 + 3x_2 + 3x_3 = 1{,}400$$

The above system of equations can be written in the matrix form as

$$\underset{A}{\begin{bmatrix} 2 & 1 & 2 \\ 5 & 2 & 2 \\ 1 & 3 & 3 \end{bmatrix}} \underset{X}{\begin{bmatrix} x_1 \\ x_2 \\ x_3 \end{bmatrix}} = \underset{B}{\begin{bmatrix} 1{,}100 \\ 1{,}800 \\ 1{,}400 \end{bmatrix}}].$$

20. A truck company owns three types of trucks *X*, *Y*, *Z* which are equipped to carry three different types of machines per load as shown below :

	Trucks		
	Type *X*	Type *Y*	Type *Z*
Machine I	2	3	4
Machine II	1	1	2
Machine III	3	2	1

How many trucks of each type should be used to carry exactly 29 of Type I machines, 13 of Type II machines and 16 of Type III machines? Assume that each truck is fully loaded.

[**Ans.** Type $X = 2$, Type $Y = 3$, Type $Z = 4$]

[**Hint :** Let the trucks required of Type $X = x$, Type $Y = y$, Type $Z = z$.

Then
$$2x + 3y + 4x = 29$$
$$x + y + 2z = 13$$
$$3x + 2y + z = 16].$$

21. Obtain a square matrix X of order 2, which satisfies the matrix equation $CX - AB = 0$. Given that

$$A = \begin{bmatrix} 2 & 1 \\ 0 & 3 \end{bmatrix}, B = \begin{bmatrix} -1 & 1 \\ 2 & 1 \end{bmatrix}, C = \begin{bmatrix} 0 & 1 \\ -1 & 1 \end{bmatrix}$$

and 0 is the null matrix $\begin{bmatrix} 0 & 0 \\ 0 & 0 \end{bmatrix}$.

22. If $A = \begin{bmatrix} i & 0 \\ 0 & -i \end{bmatrix}, B = \begin{bmatrix} 0 & -1 \\ 1 & 0 \end{bmatrix}$, and $C = \begin{bmatrix} 0 & i \\ i & 0 \end{bmatrix}$

Prove that $A^2 = B^2 = C^2 = -I_2$ and $AB = -C = -BA$.

23. Find the product of the two matrices A and B where

$$A = \begin{bmatrix} -5 & 1 & 3 \\ 7 & 1 & -5 \\ 1 & -1 & 1 \end{bmatrix}, B = \begin{bmatrix} 1 & 1 & 2 \\ 3 & 2 & 1 \\ 2 & 1 & 3 \end{bmatrix}$$

and use it for solving the Equations.

$$x + y + 2z = 1, 3x + 2y + z = 7, 2x + y + 3z = 2$$ [**Ans.** $x = 2, y = 1, z = -1$]

[**Hint :** $AB = 4I \Rightarrow \frac{A}{4}B = I \Rightarrow B^{-1} = \frac{A}{4}$

The given equations can be written as

$$\begin{bmatrix} 1 & 1 & 2 \\ 3 & 2 & 1 \\ 2 & 1 & 3 \end{bmatrix}\begin{bmatrix} x \\ y \\ z \end{bmatrix} = \begin{bmatrix} 1 \\ 7 \\ 2 \end{bmatrix}].$$

❑❑❑

2

Applications of Matrices and Determinants to Business and Economics

2.1 INTRODUCTION

The concept of matrices and determinants is applied in Business and Economics in various situations. For Example

(1) Solution of Equations [Cramer's rule and matrix inverse method].

(2) Input-output analysis where we deal with the inter-relationship between different sectors of an economy. (Not discussed in this chapter).

(3) We use matrix addition and multiplication for different types of calculations (Explained in solved examples).

Example 1. The daily cost "C" of operating a hospital is a linear function of the number of in-patients and out-Patients P plus *a* fixed cost '*a*', *i.e.,* $C = a + bP + dI$ (a, b and d are constants). Given the following data for three days, find the values of a, b and d by using matrix algebra.

Day	*Cost (Rs.)*	*Number of in-patients*	*Number of out-patients*
1	6,950	40	10
2	6,725	35	9
3	7,100	40	12

Solution. The three Equations as per the data are

$$a + 10b + 40d = 6{,}950 \text{ or } a + 40d + 10b = 6{,}950$$

$$a + 9b + 35d = 6{,}725 \text{ or } a + 35d + 9b = 6{,}725$$

$$a + 12b + 40d = 7{,}100 \text{ or } a + 40d + 12b = 7{,}100$$

In the matrix form, these Equations can be written as

$$\begin{bmatrix} 1 & 40 & 10 \\ 1 & 35 & 9 \\ 1 & 40 & 12 \end{bmatrix} \begin{bmatrix} a \\ d \\ b \end{bmatrix} = \begin{bmatrix} 6{,}960 \\ 6{,}725 \\ 7{,}100 \end{bmatrix}$$

where $$A=\begin{bmatrix}1 & 40 & 10\\ 1 & 35 & 9\\ 1 & 40 & 12\end{bmatrix};\ X=\begin{bmatrix}a\\ d\\ b\end{bmatrix};\ C=\begin{bmatrix}6{,}950\\ 6{,}725\\ 7{,}100\end{bmatrix}$$

$$A^{-1}=-\frac{1}{10}\begin{bmatrix}60 & -80 & 10\\ -3 & 2 & 1\\ 5 & 0 & -5\end{bmatrix}$$

$$\therefore \quad \begin{bmatrix}a\\ d\\ b\end{bmatrix}=-\frac{1}{10}\begin{bmatrix}60 & -80 & 10\\ -3 & 2 & 1\\ 5 & 0 & -5\end{bmatrix}\begin{bmatrix}6{,}950\\ 6{,}725\\ 7{,}100\end{bmatrix}=\begin{bmatrix}5{,}000\\ 30\\ 75\end{bmatrix}$$

$$\therefore \quad a=5{,}000,\ d=30,\ b=75$$

Example 2. Matrix S and C gives the daily sales and sales price of hot drinks for a shopkeeper.

$$\begin{matrix}\text{Mon.}\\ \text{Tue.}\\ \text{Wed.}\end{matrix}=\begin{bmatrix}\text{Tea} & \text{Coffee} & \text{Chocolate}\\ 20 & 20 & 10\\ 25 & 25 & 12\\ 35 & 15 & 15\end{bmatrix};\quad C=\begin{matrix}\text{Tea}\\ \text{Coffee}\\ \text{Chocolate}\end{matrix}\begin{bmatrix}40\,P\\ 70\,P\\ 80\,P\end{bmatrix}$$

Find by matrix method the total revenue for the three days.

Solution. Total revenue = Sales × Sales price. In this case total revenue is equal to the product SC

$$=\begin{bmatrix}20 & 20 & 10\\ 25 & 25 & 12\\ 35 & 15 & 15\end{bmatrix}\times\begin{bmatrix}40P\\ 70P\\ 80P\end{bmatrix}=\begin{bmatrix}3{,}000P\\ 3{,}170P\\ 3{,}650P\end{bmatrix}\begin{matrix}\text{Monday}\\ \text{Tueesday}\\ \text{Wednesday}\end{matrix}$$

∴ Total Revenue for 3 days [3,000 + 3,710 + 3,650] = 10,360 P = Rs 103.60.

Example 3. An automobile company uses three types of steel S_1, S_2 and S_3 for producing different types of cars C_1, C_2 and C_3. Steel requirements in tons for each type of car and total available steel of all three types is summarised in the following table.

Types of Steel	*Types of Car*			*Total Steel available*
	C_1	C_2	C_3	
S_1	2	3	4	29
S_2	1	1	2	13
S_3	3	2	1	16

Using matrix algebra determine the number of cars of each type which can be produced.

Solution. Let x be the number of cars of type C_1, y the number of cars of type C_2, z the number of cars of type C_3

$$\begin{bmatrix}2 & 3 & 4\\ 1 & 1 & 2\\ 3 & 2 & 1\end{bmatrix}\begin{bmatrix}x\\ y\\ z\end{bmatrix}=\begin{bmatrix}29\\ 13\\ 16\end{bmatrix}$$

$$\begin{bmatrix}x\\ y\\ z\end{bmatrix}=\frac{1}{5}\begin{bmatrix}-3 & 5 & 2\\ 5 & -10 & 0\\ -1 & 5 & -1\end{bmatrix}\begin{bmatrix}29\\ 13\\ 16\end{bmatrix}$$

where $\begin{bmatrix} -3 & 5 & 2 \\ 5 & -10 & 0 \\ -1 & 5 & -1 \end{bmatrix}$ is the inverse of $\begin{bmatrix} 2 & 3 & 4 \\ 1 & 1 & 2 \\ 3 & 2 & 1 \end{bmatrix}$

$$= \frac{1}{5}\begin{bmatrix} -87+65+32 \\ 145-130+0 \\ -29+65-16 \end{bmatrix} = \frac{1}{5}\begin{bmatrix} 10 \\ 15 \\ 20 \end{bmatrix} = \begin{bmatrix} 2 \\ 3 \\ 4 \end{bmatrix}$$

$\Rightarrow$ $$x = 2, y = 3, z = 4.$$

Example 4. A manufacturer produces two types of products X and Y. Each product is first processed in a machine M_1 and then sent to another machine M_2 for finishing. Each unit of X requires 20 minutes time in M_1 and 10 minutes time in M_2 whereas each unit of Y requires 10 minutes time in M_1 and 20 minutes time in M_2. The total time available on each machine is 600 minutes. Calculate the number of units of two types of products produced by constructing a matrix equation of the form AX = B and then solving it by matrix inversion method.

Solution. The given information in tabular form can be written as under :

Machine	M_1 (Time in minutes)	M_2 (Time in minutes)
Product X	20	10
Product Y	10	20
Total time available	600	600

Let x and y denote the number of units manufactured of the type X and Y respectively.

Then $$\begin{bmatrix} 20 & 10 \\ 10 & 20 \end{bmatrix}\begin{bmatrix} x \\ y \end{bmatrix} = \begin{bmatrix} 600 \\ 600 \end{bmatrix} \quad \text{...(I)}$$

If $$A = \begin{bmatrix} 20 & 10 \\ 10 & 20 \end{bmatrix}, X = \begin{bmatrix} x \\ y \end{bmatrix}, B = \begin{bmatrix} 600 \\ 600 \end{bmatrix}$$

Then $A^{-1} = \frac{1}{300}\begin{bmatrix} 20 & -10 \\ -10 & 20 \end{bmatrix}$ [Pl. complete the calculations]

From (I),

$$\begin{bmatrix} x \\ y \end{bmatrix} = A^{-1}B = \frac{1}{300}\begin{bmatrix} 20 & -10 \\ -10 & 20 \end{bmatrix}\begin{bmatrix} 600 \\ 600 \end{bmatrix} = \begin{bmatrix} 20 \\ 20 \end{bmatrix}$$

$\Rightarrow$ $$x = 20, \ y = 20$$

i.e., The number of units produced are 20 each of type X and Y.

Example 5. A man buys 8 dozens of mangoes, 10 dozens of apples and 4 dozens of bananas. Mangoes cost Rs. 18 per dozen, apples Rs. 9 per dozen and bananas Rs. 6 per dozen. Represent the quantities bought by a row matrix and the prices by a column matrix and hence, obtain the total cost.

Solution. Let A the row matrix representing the quantities bought, then A = [8 10 4] is a 1×3 matrix.

Let B be the column matrix representing the cost of mangoes, applies and bananas per dozen, then

$$B = \begin{bmatrix} 18 \\ 9 \\ 6 \end{bmatrix}$$

The total cost is given by the product AB

and
$$AB = [8 \;\; 10 \;\; 4]\begin{bmatrix}18\\9\\6\end{bmatrix} = [144 + 90 + 24] = [258]$$

Hence, the total cost in Rs. 258.

Example 6. There are two families A and B. There are 2 men; 3 women and 1 child in Family A and 1 man, 1 woman and 2 children in Family B. The recommended daily allowance for calories is—Men : 2,400, Women : 1,900, Children : 1,800 and for proteins is—Men : 55 g, Women : 45 g, Children : 33 g.

Solution. Represent the above information by matrices. Using matrix multiplication calculate the total requirements of calories and proteins for each of the two families.

$$\begin{matrix}\text{Families}\\ \begin{bmatrix}A\\B\end{bmatrix}\end{matrix} = \begin{matrix}\text{men} & \text{women} & \text{children}\\ \begin{bmatrix}2 & 3 & 1\\1 & 1 & 2\end{bmatrix}\end{matrix} \begin{bmatrix}\text{Cal.} & \text{Prot.}\\ 2{,}400 & 55\\ 1{,}900 & 45\\ 1{,}800 & 33\end{bmatrix}\begin{matrix}\\ \text{Men}\\ \text{Women}\\ \text{Children}\end{matrix}$$

$$\begin{matrix} & \text{Calories} & \text{Proteins}\end{matrix}$$

$$\begin{bmatrix}A\\B\end{bmatrix} = \begin{bmatrix}4{,}800 + 5{,}700 + 1{,}800 & 110 + 135 + 33\\ 2{,}400 + 1{,}900 + 3{,}600 & 55 + 45 + 66\end{bmatrix}$$

$$= \begin{bmatrix}12{,}300 & 278\\ 7{,}900 & 166\end{bmatrix}$$

Requirements of

$\therefore$ Family A $\rightarrow$ Cal. 12,300 units, Proteins 278 g.

Family B $\rightarrow$ Cal. 7,900 units, Proteins 166.

Example 7. A firm produces three products P_1, P_2 and P_3 requiring the mixup of three materials M_1, M_2 and M_3. The per unit requirement of each product for each material (in units) is as follows :

$$A = \begin{matrix} & M_1 & M_2 & M_3 \\ P_1 & 2 & 3 & 1\\ P_2 & 4 & 2 & 5\\ P_3 & 2 & 4 & 2\end{matrix}$$

Using matrix notations, find :

(i) the total requirements of each material if the firm produces 100 units of each product.

(ii) the per unit cost of production of each product if the per unit costs of materials M_1, M_2 and M_3 are Rs. 5, Rs. 10 and Rs. 5 respectively, and

(ii) the total cost production if the firm produces 200 units of each product.

Solution. Let
$$\begin{matrix} M_1 \; M_2 \; M_3 \\ A = \begin{bmatrix}2 & 3 & 1\\4 & 2 & 5\\2 & 4 & 2\end{bmatrix}\begin{matrix}P_1\\P_2\\P_3\end{matrix}\end{matrix}$$

$$B = \begin{bmatrix} 100 \\ 100 \\ 100 \end{bmatrix} \begin{matrix} P_1 \\ P_2 \\ P_3 \end{matrix} \text{ and } C = \begin{bmatrix} 5 \\ 10 \\ 5 \end{bmatrix} \begin{matrix} M_1 \\ M_2 \\ M_3 \end{matrix}$$

(i) Total requirement of each material = $A^T.B$

$$\begin{matrix} & \begin{matrix} P_1 & P_2 & P_3 \end{matrix} \\ = \begin{matrix} M_1 \\ M_2 \\ M_3 \end{matrix} & \begin{bmatrix} 2 & 4 & 2 \\ 3 & 2 & 4 \\ 1 & 5 & 2 \end{bmatrix} \end{matrix} \begin{bmatrix} 100 \\ 100 \\ 100 \end{bmatrix} \begin{matrix} P_1 \\ P_2 \\ P_3 \end{matrix}$$

$$= \begin{bmatrix} 800 \\ 900 \\ 800 \end{bmatrix} \begin{matrix} M_1 \\ M_2 \\ M_3 \end{matrix} \Rightarrow \begin{matrix} M_1 = 800 \text{ units} \\ M_2 = 900 \text{ units} \\ M_3 = 800 \text{ units} \end{matrix}$$

(ii) Cost of production of each product = A.C

$$= \begin{bmatrix} 2 & 3 & 1 \\ 4 & 2 & 5 \\ 2 & 4 & 2 \end{bmatrix} . \begin{bmatrix} 5 \\ 10 \\ 5 \end{bmatrix} = \begin{bmatrix} 45 \\ 65 \\ 60 \end{bmatrix}$$

unit cost of production, *i.e.,* of P_1 = Rs. 45, P_2 = Rs. 65 and P_3 = Rs. 60

(iii) Total cost of production of 200 units of each product

$$= [200 \ \ 200 \ \ 200] \begin{bmatrix} 45 \\ 65 \\ 60 \end{bmatrix}$$

i.e., total cost of production of 200 units of each product = Rs. 34,00.

Example 8. Mr. X has invested a part of this investment in 10% bond A and a part in 15% bond B. His interest income during risk year is Rs. 4,000. If he invests 20% more in 10% bond A and 10% more in 15% bond B, his income during second year increases by Rs. 500. Find his initial investment and the new investment in bonds A and B, using matrix method.

Solution. Let, the amount invested in 10% bond be Rs. x and in 15% bond be Rs. y

Then $$\frac{10}{100}x + \frac{15}{100}y = 4{,}000 \qquad \text{...(I)}$$

and $$\left(x + \frac{x \times 20}{100}\right)\frac{10}{100} + \left(y + \frac{10y}{100}\right)\frac{15}{100} = 4{,}500$$

$$\Rightarrow \quad 0.12x + 0.165y = 4{,}500$$

$$\Rightarrow \quad 0.24x + 0.330y = 9000 \qquad \text{...(II)}$$

Equations I and II after simplification becomes

$$2x + 3y = 80{,}000 \qquad \text{...(III)}$$

$$8x + 11y = 3{,}00{,}000 \qquad \text{...(IV)}$$

Equations III and IV in matrix form can be written as

$$\begin{bmatrix} 2 & 3 \\ 8 & 11 \end{bmatrix}\begin{bmatrix} x \\ y \end{bmatrix} = \begin{bmatrix} 80{,}000 \\ 3{,}00{,}000 \end{bmatrix}$$

Determinant $\begin{vmatrix} 2 & 3 \\ 8 & 11 \end{vmatrix} = 22 - 24 = -2$

Inverse of $\begin{bmatrix} 2 & 3 \\ 8 & 11 \end{bmatrix} = -\frac{1}{2}\begin{bmatrix} 11 & -3 \\ -8 & 2 \end{bmatrix}$

$\therefore$

$$\begin{bmatrix} x \\ y \end{bmatrix} = -\frac{1}{2}\begin{bmatrix} 11 & -3 \\ -8 & 2 \end{bmatrix}\begin{bmatrix} 80{,}000 \\ 3{,}00{,}000 \end{bmatrix}$$

$$= -\frac{1}{2}\begin{bmatrix} 8{,}80{,}000 - 9{,}00{,}000 \\ -6{,}40{,}000 + 6{,}00{,}000 \end{bmatrix}$$

$$= \begin{bmatrix} 10{,}000 \\ 20{,}000 \end{bmatrix}$$

$\Rightarrow$ $x = \text{Rs. } 10{,}000$ and $y = \text{Rs. } 20{,}000$

$\therefore$ Initial investment in bond A = Rs. 10,000

Initial investment in bond B = Rs. 20,00

New Investment in bond A $= \left(x + \frac{x \times 20}{100}\right)$

$$= 10{,}000 + \frac{10{,}000 \times 20}{100} = \text{Rs. } 12{,}000$$

New investment in bond B $= \text{Rs. } 20{,}000 + \frac{20{,}000 \times 10}{100}$

$= \text{Rs. } 22{,}000$

Example 9. A transport company uses 3 types of trucks T_1, T_2 and T_3, to transport 3 types of vehicles V_1, V_2 and V_3. The capacity of each truck in terms of 3 types of vehicles is given below :

	V_1	V_2	V_3
T_1	1	3	2
T_2	2	2	3
T_3	3	2	2

Using matrix method find :

(i) The number of trucks of each type required to transport 85, 105 and 110 vehicles of V_1, V_2 and V_3 types respectively.

(ii) Find the number of vehicles of each type which can be transported if company has 10, 20 and 30 trucks of each type respectively.

Solution. (i) Let x, y, z be the respective number of trucks of Type T_1, T_2 and T_3

Then

$$x + 2y + 3z = 85$$
$$3x + 2y + 2z = 105$$
$$2x + 3y + 2z = 110$$

In the matrix form

$$\begin{bmatrix} 1 & 2 & 3 \\ 3 & 2 & 2 \\ 2 & 3 & 2 \end{bmatrix}\begin{bmatrix} x \\ y \\ z \end{bmatrix} = \begin{bmatrix} 85 \\ 105 \\ 110 \end{bmatrix}$$

Let
$$A = \begin{bmatrix} 1 & 2 & 3 \\ 3 & 2 & 2 \\ 2 & 3 & 2 \end{bmatrix} = \begin{bmatrix} a_{11} & a_{12} & a_{13} \\ a_{21} & a_{22} & a_{23} \\ a_{31} & a_{32} & a_{33} \end{bmatrix}$$

Cofactor of $a_{11} = -2$ Cofactor of $a_{12} = -2$
Cofactor of $a_{13} = 5$ Cofactor of $a_{21} = 5$
Cofactor of $a_{22} = -4$ Cofactor of $a_{23} = 1$
Cofactor of $a_{31} = -2$ Cofactor of $a_{32} = 7$
Cofactor of $a_{33} = -4$

Determinant
$$A = -2 - 2 \times 2 + 3 \times 5$$
$$= -2 \times 4 + 15 = 9$$

$$A^{-1} = \frac{1}{9}\begin{bmatrix} -2 & 5 & -2 \\ -2 & -4 & 7 \\ 5 & 1 & -4 \end{bmatrix}$$

$$\begin{bmatrix} x \\ y \\ z \end{bmatrix} = \frac{1}{9}\begin{bmatrix} -2 & 5 & -2 \\ -2 & -4 & 7 \\ 5 & 1 & -4 \end{bmatrix}\begin{bmatrix} 85 \\ 105 \\ 110 \end{bmatrix} = \frac{1}{9}\begin{bmatrix} 135 \\ 180 \\ 90 \end{bmatrix}$$

$$= \begin{bmatrix} 156 \\ 20 \\ 10 \end{bmatrix}$$

$\therefore\ x = 5, y = 20, z = 10$

(ii) Number of vehicles of each type are given by

$$\begin{bmatrix} 1 & 2 & 3 \\ 3 & 2 & 2 \\ 2 & 3 & 2 \end{bmatrix}\begin{bmatrix} 10 \\ 20 \\ 30 \end{bmatrix} = \begin{bmatrix} 10+40+90 \\ 30+50+60 \\ 20+60+60 \end{bmatrix}$$

$$= \begin{bmatrix} 140 \\ 130 \\ 130 \end{bmatrix} \begin{matrix} \to V_1 \\ \to V_2 \\ \to V_3 \end{matrix}$$

Example 10. The manager of a car rental company is planning its maintenance programme for the next year and wants to determine the needs for certain repair parts and the expected costs. The company rents out three types of cars : large, medium and small. The matrix N gives the number of each size of cars available for renting in the four regions.

Region

$$\begin{matrix} & \text{I} & \text{II} & \text{III} & \text{IV} & \\ N = & \left(\begin{matrix} 160 \\ 400 \\ 500 \end{matrix} \right. & \begin{matrix} 150 \\ 300 \\ 200 \end{matrix} & \begin{matrix} 100 \\ 10 \\ 150 \end{matrix} & \left. \begin{matrix} 120 \\ 400 \\ 300 \end{matrix} \right) & \begin{matrix} \text{Large} \\ \text{Medium} \\ \text{Small} \end{matrix} \end{matrix}$$

Because of cost and frequency of replacement, the manager is concentrating on four repair parts in particular : fan belts, spark plugs, batteries and tyres. On the basis of past records, the average number of repair parts needed per car during a year are given in matrix R. Further, the matrix C gives the cost per unit for the four parts.

$$R = \begin{pmatrix} & \text{Large} & \text{Medium} & \text{Small} & \\ & 1.7 & 1.6 & 1.5 & \text{Fan belts} \\ & 12.0 & 8.0 & 5.0 & \text{Spark Plugs} \\ & 0.9 & 0.75 & 0.5 & \text{Batteries} \\ & 4.0 & 6.5 & 6.0 & \text{Tyres} \end{pmatrix}$$

$$C = (125 \;\; 80 \;\; 300 \;\; 350)$$

Through matrix operations, compute :

(i) the total demand for each type of car,

(ii) total demand for each repair part needed for the fleet, and

(iii) total combined cost for all repairs parts.

Solution. Matrix N can be broken into four matrices as

Region I

$$N_1 = \begin{bmatrix} 160 \\ 400 \\ 500 \end{bmatrix} \begin{matrix} \text{Large cars} \\ \text{Medium cars} \\ \text{Small cars} \end{matrix}$$

Region II

$$N_2 = \begin{bmatrix} 150 \\ 300 \\ 20 \end{bmatrix} \begin{matrix} \text{Large cars} \\ \text{Medium cars} \\ \text{Small cars} \end{matrix}$$

Region III

$$N_3 = \begin{bmatrix} 100 \\ 100 \\ 150 \end{bmatrix} \begin{matrix} \text{Large cars} \\ \text{Medium cars} \\ \text{Smallcars} \end{matrix}$$

Region IV

$$N_4 = \begin{bmatrix} 120 \\ 400 \\ 300 \end{bmatrix} \begin{matrix} \text{Large cars} \\ \text{Medium cars} \\ \text{Small cars} \end{matrix}$$

$$N_1 + N_2 + N_3 + N_4 = \begin{bmatrix} 160 + 150 + 100 + 120 \\ 400 + 300 + 100 + 400 \\ 500 + 200 + 150 - + 300 \end{bmatrix} \begin{matrix} \text{Large cars} \\ \text{Medium cars} \\ \text{Smallcars} \end{matrix}$$

$$= \begin{bmatrix} 530 \\ 1{,}200 \\ 1{,}150 \end{bmatrix} \begin{matrix} \text{Large cars} \\ \text{Medium cars} = S \\ \text{Small cars} \end{matrix}$$

∴ The above matrix gives the total demand

Changing rows into columns and columns in rows in S we get

$$S^T = [530 \;\; 1{,}200 \;\; 1{,}150]$$

$R^T =$	Fan belts	Spark plug	Batteries	Tyres
Large	1.7	12	0.9	4
Medium	1.6	8	0.75	6.5
Small	1.5	5	0.5	6

$$S^T \times R^T = \begin{matrix} \text{Large} & \text{Medium} & \text{Small} \\ [530 & 1{,}200 & 1{,}150] \end{matrix} \begin{bmatrix} 1.7 & 12 & 0.9 & 4 \\ 1.6 & 8 & 0.75 & 6.5 \\ 1.5 & 5 & 0.5 & 6 \end{bmatrix}$$

$$\begin{matrix} & \text{Fanbelt} & \text{Spark plugs} & \text{Batteries} & \text{Tyres} \\ = [& 4{,}546 & 21{,}710 & 1{,}952 & 16{,}820] \end{matrix}$$

$$= [4{,}546 \;\; 21{,}710 \;\; 1{,}952 \;\; 16{,}820] \begin{bmatrix} 125 \\ 80 \\ 300 \\ 350 \end{bmatrix}$$

$$= 5{,}68{,}250 + 17{,}36{,}800 + 5{,}85{,}600 + 58{,}87{,}000$$

$$= \text{Rs. } 87{,}77{,}650 = \text{Total Cost.}$$

Example 11. A firm produces two products P_1 and P_2, passing through two machines, M_1 and M_2 before completion. M_1 can produce either 10 units of P_1 or 15 units of P_2 per hour. M_2 can produce 15 units of either product per hour. Find daily production of P_1 and P_2 if time available is 12 hours of machine M_1 and 10 hours of M_2 per day using matrix inversion.

Solution. Time taken to produce one unit of P_1 in machine $M_1 = \frac{10}{10}$ hr.

Time taken to produce one unit of P_2 in machine $M_1 = \frac{1}{15}$ hr.

Time taken to produce one unit of P_1 in machine $M_2 = \frac{1}{15}$ hr.

Time taken to produce one unit of P_2 in machine $M_2 = \frac{1}{15}$

If x unit of P_1 and y units of P_2 are produced

Then $$\frac{x}{10} + \frac{y}{15} = 12 \Rightarrow 3x + 2y = 360 \qquad \text{...(I)}$$

and $$\frac{x}{15} + \frac{y}{15} = 10 \Rightarrow x + y = 150 \qquad \text{...(II)}$$

Equations (I) and (II) can be represented in the matrix from as

$$\begin{bmatrix} 3 & 2 \\ 1 & 1 \end{bmatrix} \begin{bmatrix} x \\ y \end{bmatrix} = \begin{bmatrix} 360 \\ 150 \end{bmatrix}$$

$$\begin{bmatrix} x \\ y \end{bmatrix} = \text{Inverse of} \begin{bmatrix} 3 & 2 \\ 1 & 1 \end{bmatrix} \times \begin{bmatrix} 360 \\ 150 \end{bmatrix}$$

$$= \begin{bmatrix} 1 & -2 \\ -1 & 3 \end{bmatrix} \begin{bmatrix} 360 \\ 150 \end{bmatrix} = \begin{bmatrix} 60 \\ 90 \end{bmatrix}$$

$\Rightarrow$ $$x = 60, \; y = 90$$

Example 12. A firm has two machines M_1 and M_2 costing Rs. 45,000 and Rs. 30,000. Each has 5 years life with scrap value nil. Find depreciation of each machine for each year using matrix notations if (i) both are depreciated by sum of the years digit method, (ii) first is depreciated by the sum of the year's digit method and second by straight line method.

Solution. (i) According to the sum of the years digit method, the depreciation rate for five years would be in ratio 5 : 4 : 3 : 2 : 1 or 5/15, 4/15, 3/15, 2/15 and 1/5 respectively. Now depreciation (D) = cost (C) × depreciation rate (R). Thus if

$$C = \begin{bmatrix} 45{,}000 \\ 30{,}000 \end{bmatrix} \text{ and } R = \begin{matrix} \text{1st} & \text{2nd} & \text{3rd} & \text{4th} & \text{5th} \\ [5/15 & 4/14 & 3/15 & 2/15 & 1/5] \end{matrix}$$

then

$$D = CR = \begin{bmatrix} 45{,}000 \\ 30{,}000 \end{bmatrix} [5/15,\ 4/15,\ 3/15,\ 2/15,\ 1/15]$$

$$\begin{array}{c} \\ = \end{array} \begin{array}{ccccc} \text{1st} & \text{2nd} & \text{3rd} & \text{4th} & \text{5th} \\ \end{array}$$

$$= \begin{bmatrix} 15{,}000 & 12{,}000 & 9{,}000 & 6{,}000 & 3{,}000 \\ 10{,}000 & 8{,}000 & 6{,}000 & 4{,}000 & 2{,}000 \end{bmatrix} \begin{matrix} M_1 \\ M_2 \end{matrix} \begin{matrix} \text{Depreciation of machine I for different years} \\ \text{Depreciation of machine II for different years} \end{matrix}$$

(ii) Here

$$C = \begin{bmatrix} 45{,}000 & 0 \\ 0 & 30{,}000 \end{bmatrix}$$

$$R = \begin{bmatrix} 5/15 & 4/15 & 3/15 & 2/15 & 1/15 \\ 1/5 & 1/5 & 1/5 & 1/5 & 1/5 \end{bmatrix}$$

$$D = CR = \begin{bmatrix} 45{,}000 & 0 \\ 0 & 30{,}00 \end{bmatrix} \begin{bmatrix} 5/15 & 4/15 & 3/15 & 2/15 & 1/15 \\ 1/5 & 1/5 & 1/5 & 1/5 & 1/5 \end{bmatrix}$$

$$= \begin{bmatrix} 15{,}000 & 12{,}000 & 9{,}000 & 6{,}000 & 3{,}00 \\ 6{,}000 & 6{,}000 & 6{,}000 & 6{,}000 & 6{,}000 \end{bmatrix} \begin{matrix} M_1 \\ M_2 \end{matrix} \begin{matrix} \text{Depreciation of machine I for different years} \\ \text{Depreciation of machine II for different years} \end{matrix}$$

Example 13. A finance company has offices loated in every division, every district and every taluka in a certain state in India. Assume that there are five divisions, 30 districts and 200 talukas in the state. Each office has 1 head clerk, 1 cashier, 1 clerk and 1 peon. A division office has, in addition, an office superintendent, 2 clerks, 1 typist and 1 peon. A district office has, in addition, 1 clerk and 1 peon. The basic monthly salaries are as follows : Office Superintendent Rs 500, Head Clerk Rs. 200, Cashier Rs. 175, Clerks and Typist Rs. 150 and Peons Rs. 100. Using matrix notation find :

(a) the total number of posts of each kind in all the offices taken together,

(b) the total basic monthly salary bill of each kind of office, and

(c) the total basic monthly salary bill of all the offices, taken together.

Solution. The number of offices can be arranged as elements of a row matrix.

$$A = \begin{bmatrix} \text{Division} & \text{District} & \text{Taluka} \\ 5 & 30 & 200 \end{bmatrix}$$

Staff composition can be arranged in 3×6 matrix

$$\begin{matrix} & \text{O} & \text{H} & \text{C} & \text{T} & \text{Cl} & \text{P} \end{matrix}$$

$$\text{B} = \begin{bmatrix} 1 & 1 & 1 & 1 & 3 & 2 \\ 0 & 1 & 1 & 0 & 2 & 2 \\ 0 & 1 & 1 & 0 & 1 & 1 \end{bmatrix} \begin{matrix} \text{Division} \\ \text{District} \\ \text{Taluka} \end{matrix}$$

where O = Office Superintendent.

H = Head Clerk, C = Cashier,

T = Typist, Cl = Clerk, P = Peon.

Column matrix D will have the elements that correspond to basic monthly salaries is :

$$D = \begin{bmatrix} 500 \\ 200 \\ 175 \\ 150 \\ 150 \\ 100 \end{bmatrix} \begin{matrix} \text{O} \\ \text{H} \\ \text{C} \\ \text{T} \\ \text{Cl} \\ \text{P} \end{matrix}$$

(a) The total number of posts of each kind in all the offices are the elements of the matrix

$$AB = [5 \quad 30 \quad 200] \begin{bmatrix} 1 & 1 & 1 & 1 & 3 & 2 \\ 0 & 1 & 1 & 0 & 2 & 2 \\ 0 & 1 & 1 & 0 & 1 & 1 \end{bmatrix} = \begin{bmatrix} \text{O} & \text{H} & \text{C} & \text{T} & \text{Cl} & \text{P} \\ 5 & 235 & 235 & 5 & 275 & 270 \end{bmatrix}$$

(b) The total basic monthly salary bill of each kind of office are the elements of the matrix

$$BD = \begin{bmatrix} 1 & 1 & 1 & 1 & 3 & 2 \\ 0 & 1 & 1 & 0 & 2 & 2 \\ 0 & 1 & 1 & 0 & 1 & 1 \end{bmatrix} \begin{bmatrix} 500 \\ 200 \\ 175 \\ 150 \\ 150 \\ 100 \end{bmatrix} = \begin{bmatrix} 1{,}675 \\ 875 \\ 625 \end{bmatrix}$$

(c) The total basic monthly salary bill of all the offices taken together are the elements of the matrix

$$ABD = [5 \ 30 \ 200] \begin{bmatrix} 1{,}675 \\ 875 \\ 625 \end{bmatrix} = [1{,}59{,}625]$$

Example 14. A company is considering which of the three methods of production it should use in producing three goods A, B and C. The amount of each good produced by each method is shown in the matrix :

$$\begin{matrix} & \begin{matrix} A & B & C \end{matrix} \\ \begin{matrix} \text{Method I} \\ \text{Method II} \\ \text{Method III} \end{matrix} & \begin{bmatrix} 4 & 8 & 2 \\ 5 & 7 & 1 \\ 5 & 3 & 9 \end{bmatrix} \end{matrix}$$

The vector [10 4 6] represents the profit per unit for the goods A, B and C in that order. Using matrix multiplication, find which method maximizes total profit.

Solution. The matrix giving the amount of each good produced by each method is

$$\begin{matrix} & \begin{matrix} A & B & C \end{matrix} \\ \begin{matrix} \text{Method I} \\ \text{Method II} \\ \text{Method III} \end{matrix} & \begin{bmatrix} 4 & 8 & 2 \\ 5 & 7 & 1 \\ 5 & 3 & 9 \end{bmatrix} \end{matrix}$$

and that giving the profit per unit for the goods A, B and C is

$$\begin{matrix} & \text{Profit} \\ \begin{matrix} A \\ B \\ C \end{matrix} & \begin{bmatrix} 10 \\ 4 \\ 6 \end{bmatrix} \end{matrix}$$

Then, the total profit from each method is given by the matrix product

$$\begin{bmatrix} 4 & 8 & 2 \\ 5 & 7 & 1 \\ 5 & 3 & 9 \end{bmatrix}\begin{bmatrix} 10 \\ 4 \\ 6 \end{bmatrix} = \begin{bmatrix} 84 \\ 84 \\ 116 \end{bmatrix}$$ **(Pl. note)**

116 is the largest element in the column (profit) matrix, and it appears in the third row that corresponds to Method III. Hence, the Method III maximizes the total profit.

Example 15. A company in the wholesale trade, sells sportwear and stocks in two brands, A and B, of football kit, each consisting of a shirt, a pair of shorts and a pair of socks. The costs for brand A are £5.75 for a shirt, £3.99 for a pair of shorts and £1.85 for a pair of socks and those for brand B are £6.25 for a shirt, £4.48 for a pair of shorts and £1.97 for a pair of socks. Three customers X, Y and Z demand the following combinations of Brands : X, 36 kits of brand A and 48 kits of brand B; Y, 24 kits of brand A and 72 kits of brand B; Z, 60 kits of brand A.

(i) Express the costs of brands A and B in matrix form, then the demands of the customers X, Y and Z in matrix form.

(ii) By forming the product of the two matrices that you obtain in the previous part, deduce the detailed costs to each of the customers.

(iii) As a result of recent price increases, Z has ceased to be the same. The company no longer stocks socks of either brand.

Re-express the demands of X and Y in matrix form and find its inverse matrix.

Solution. (i) If C is the cost matrix and D is the demand matrix, then :

$$C = \begin{array}{r} \\ \text{shirts} \\ \text{shorts} \\ \text{socks} \end{array}\begin{array}{c} \begin{array}{cc} A & B \end{array} \\ \begin{bmatrix} 5.75 & 6.25 \\ 3.99 & 4.48 \\ 1.85 & 1.97 \end{bmatrix} \end{array}$$

and

$$D = \begin{array}{r} \\ A \\ B \end{array}\begin{array}{c} \begin{array}{ccc} X & Y & Z \end{array} \\ \begin{bmatrix} 36 & 24 & 60 \\ 48 & 72 & 0 \end{bmatrix} \end{array}$$

(ii) $C \times D = \begin{bmatrix} 5.75 & 6.25 \\ 3.99 & 4.48 \\ 1.85 & 1.97 \end{bmatrix} \times \begin{bmatrix} 36 & 24 & 60 \\ 48 & 72 & 0 \end{bmatrix}$

$$= \begin{array}{r} \\ \text{shorts} \\ \text{shorts} \\ \text{socks} \end{array}\begin{array}{c} \begin{array}{ccc} X & Y & Z \end{array} \\ \begin{bmatrix} 507 & 588 & 345 \\ 358.68 & 418.32 & 239.40 \\ 161.16 & 186.24 & 111 \end{bmatrix} \end{array}$$

(iii) The new demand matrix is :

$$D = \begin{array}{r} \\ A \\ B \end{array}\begin{array}{c} \begin{array}{cc} X & Y \end{array} \\ \begin{bmatrix} 36 & 24 \\ 48 & 72 \end{bmatrix} \end{array}$$ and its inverse can be calculated as follows

$$D^{-1} = \frac{1}{(36)(72)-(48)(24)}\begin{bmatrix} 72 & -24 \\ -48 & 36 \end{bmatrix} = \begin{bmatrix} \frac{72}{1,440} & \frac{-24}{1,440} \\ \frac{-48}{1,440} & \frac{36}{1,440} \end{bmatrix}$$

$$= \begin{bmatrix} 0.05 & -0.017 \\ -0.033 & 0.025 \end{bmatrix}$$

Example 16. A shop sells 20 table fans, 30 ceiling fans and 10 pedestal fans in a month. The price of a table fan is Rs. 300, that of a ceiling fan is Rs. 400 and of pedestal fan is Rs. 500. The cost to the shop is Rs. 220 for one table fan, Rs, 325 for a ceiling fan and Rs. 400 for a pedestal fan. Find the monthly profits of the shop.

Solution. Consider the three matrices A, B and C

$$A = [20 \;\; 30 \;\; 10],\; B = \begin{bmatrix} 300 \\ 400 \\ 500 \end{bmatrix} \text{ and } C = \begin{bmatrix} 220 \\ 325 \\ 400 \end{bmatrix} \text{ respectively,}$$

where A represents the quantity of goods sold, B represents the selling price of three types of fans and C represents the cost price of three types of fans.

$$\text{Then, total revenue } R = AB = [20 \;\; 30 \;\; 10]\begin{bmatrix} 300 \\ 400 \\ 500 \end{bmatrix} = [23,000]$$

and,
$$\text{total cost } TC = AC = [20 \;\; 30 \;\; 10]\begin{bmatrix} 220 \\ 325 \\ 400 \end{bmatrix} = [18,150]$$

Thus, the profit $= R - TC = 23,000 - 18,150 = 4,850$.

Hence, the monthly profit of the shop is Rs. 4,850.

Example 17. In a certain city there are 50 colleges and 400 schools. Each school and college has 18 peons, 5 clerks and 1 cashier. Each college in addition has 1 section officer and one librarian. The monthly salary of each of them is as follows :

Peon—Rs. 3000, Clerk—Rs. 5000, Cashier—Rs. 6000, Section Officer—Rs. 7000 and Librarian—Rs. 9000.

Using matrix notation find (a) total number of posts of each kind in schools and colleges taken together, (b) the total monthly salary bill of all the schools and colleges taken together.

Solution. Consider the three matrices A, B and C as

$$A = \begin{matrix} \text{College} & \text{School} \\ [50 & 400] \end{matrix},\; B = \begin{bmatrix} P & Cl & C & SO & L \\ 18 & 5 & 1 & 1 & 1 \\ 18 & 5 & 1 & 0 & 0 \end{bmatrix} \begin{matrix} \\ \text{College,} \\ \text{School} \end{matrix}\; C = \begin{bmatrix} 3000 \\ 5000 \\ 6000 \\ 7000 \\ 9000 \end{bmatrix} \begin{matrix} P \\ Cl \\ C \\ SO \\ L \end{matrix}$$

where the row matrix A represents the number of colleges and schools in that order, the columns of matrix B represents the number of peons, clerks, cashier, section officer, librarian while the rows of B represent college and schools in that order and the column matrix C represents the monthly salary of peon, clerk, cashier, section officer and librarian in that order.

(a) Now,
$$AB = [50 \ \ 400]\begin{bmatrix} 18 & 5 & 1 & 1 & 1 \\ 18 & 5 & 1 & 0 & 0 \end{bmatrix}$$
$$= [8,100 \ \ 2,250 \ \ 450 \ \ 50 \ \ 50]$$

where first element represents total number of peons, second element represents total number of clerks, third elements represents number of cashiers, fourth element represents total number of section officers and fifth element represents total number of librarians.

(b) The total monthly salary bill of each college and school is given by the matrix product

$$BC = \begin{bmatrix} 18 & 5 & 1 & 1 & 1 \\ 18 & 5 & 1 & 0 & 0 \end{bmatrix}\begin{bmatrix} 3000 \\ 5000 \\ 6000 \\ 7000 \\ 9000 \end{bmatrix} = \begin{bmatrix} 1,01,000 \\ 85,000 \end{bmatrix}$$

Now, the total monthly salary bill of all colleges and schools taken together is given by the matrix product

$$A\,(BC) = [50 \ \ 400]\begin{bmatrix} 1,01,000 \\ 85,000 \end{bmatrix} = [3,90,50,000]$$

Example 18. The following matrix gives the proportionate mix of constituents used for three fertilisers :

		Constituents			
		A	B	C	D
	1	0.5	0	0.5	0
Fertiliser	2	0.2	0.3	0	0.5
	3	0.2	0.2	0.1	0.5

(a) If sales are 1,000 tins (of one kilogram) per weeek, 20% being fertiliser 1, 30% being fertiliser 2, and 50% being fertiliser 3, how much of each constituent is used?

(b) If the cost of each constituent is 50 paise, 60 paise, 75 paise and 100 paise per 100 grams, respectively, how much does a one kilogram tin of each fertiliser cost?

(c) What is the total cost per week?

Express the calculations and answers in matrix form.

Solution. (a) The sales of fertiliser per week can be expressed as the following matrix :

$$1,000\,[0.2 \ \ 0.3 \ \ 0.5] = [200 \ \ 300 \ \ 500]$$

The requirement of each constituent is given by the matrix product

$$[200 \ \ 300 \ \ 500]\begin{bmatrix} 0.5 & 0 & 0.5 & 0 \\ 0.2 & 0.3 & 0 & 0.5 \\ 0.2 & 0.2 & 0.1 & 0.5 \end{bmatrix} = [260 \ \ 190 \ \ 150 \ \ 400]$$

i.e., the requirements of constituents are : A—260, B—190, C—150, D—400.

(b) Cost of each constituent per 100 gm are 50 paise, 60 paise, 75 paise and 100 paise, *i.e.*, 500 paise, 600 paise, 750 paise and 1,000 paise per 1,000 grams (one kilogram) of each constituent, respectively.

The cost of one kilogram tin of each fertiliser in paise is given by the matrix product

$$\begin{bmatrix} 0.5 & 0 & 0.5 & 0 \\ 0.2 & 03 & 0 & 0.5 \\ 0.2 & 0.2 & 0.1 & 0.5 \end{bmatrix} \begin{bmatrix} 500 \\ 600 \\ 750 \\ 1{,}000 \end{bmatrix} = \begin{bmatrix} 625 \\ 780 \\ 795 \end{bmatrix}$$

i.e., costs per 1 kilogram tin of fertiliser are :

1—Rs. 6.25, 2—Rs. 7.80, 3—Rs. 7.95.

(c) The total cost in paise of fertiliser if 1,000 tins (of one kilogram) are needed per week may be calculated by

either : $$[200 \quad 300 \quad 500] \begin{bmatrix} 625 \\ 780 \\ 795 \end{bmatrix} = [7{,}56{,}500]$$

or : $$[260 \quad 190 \quad 150 \quad 400] \begin{bmatrix} 500 \\ 600 \\ 750 \\ 1{,}000 \end{bmatrix} = [7{,}56{,}500]$$

Hence, the total cost per week is Rs. 7,565.

Example 19. A holding company H has two subsidiaries S_1 and S_2. The subsidiaries also own a portion of the share capital of H. The percentage ownership of the group is given below :

Owners of Shares	*Percentage of Share*	*Capital held in Company*	
	H	S_1	S_2
H	0	60	80
S_1	3	0	10
S_2	6	10	0
Outside Shareholders	91	30	10
Total	100	100	100

If the separately earned net profits of H, S_1 and S_2 are Rs. 30,000, Rs. 17,500 and Rs. 5,000 respectively, find the total net profit of each company (separately earned profit plus share in the profits of the other companies) by using matrix algebra. Also, verify that the sum of the profits allocated to the outside shareholders equals the sum of the separately earned profits of all the companies.

(D.U., B.Com. (Hons.) 1990)

Solution. Suppose that the total net profit of H, S_1 and S_2 is x, y and z respectively. Then we have the following system of linear equations :

$$x = 30{,}000 + \frac{60}{100}y + \frac{80}{100}z \Rightarrow 5x - 3y - 4z = 1{,}50{,}000$$

$$y = 17{,}500 + \frac{3}{100}x + \frac{10}{100}z \Rightarrow -3x + 100y - 10z = 17{,}50{,}000$$

$$z = 5{,}000 + \frac{6}{100}x + \frac{10}{100}y \Rightarrow -3x - 5y + 50z = 2{,}50{,}000$$

On simplification, the above system of equation can be written in the following matrix form

$$\begin{bmatrix} 5 & -3 & -4 \\ -3 & 100 & -10 \\ -3 & -5 & 50 \end{bmatrix} \begin{bmatrix} x \\ y \\ z \end{bmatrix} = \begin{bmatrix} 1{,}50{,}000 \\ 17{,}50{,}000 \\ 2{,}50{,}000 \end{bmatrix}$$

Performing the operations $R_2 \to 3R_1 + 5R_2$ and $R_3 \to 3R_1 + 5R_3$, we obtain

$$\begin{bmatrix} 5 & -3 & -4 \\ 0 & 491 & -62 \\ 0 & -34 & 238 \end{bmatrix} \begin{bmatrix} x \\ y \\ z \end{bmatrix} = \begin{bmatrix} 1{,}50{,}000 \\ 92{,}00{,}000 \\ 17{,}00{,}000 \end{bmatrix}$$

Performing the operations $R_3 \to (1/34)\, R_3$, we obtain

$$\begin{bmatrix} 5 & -3 & -4 \\ 0 & 491 & -62 \\ 0 & -1 & 7 \end{bmatrix} \begin{bmatrix} x \\ y \\ z \end{bmatrix} = \begin{bmatrix} 1{,}50{,}000 \\ 92{,}00{,}000 \\ 50{,}000 \end{bmatrix}$$

Performing the operation $R_3 \to 491R_3$, we obtain

$$\begin{bmatrix} 5 & -3 & -4 \\ 0 & 491 & -62 \\ 0 & -491 & 3{,}437 \end{bmatrix} \begin{bmatrix} x \\ y \\ z \end{bmatrix} = \begin{bmatrix} 1{,}50{,}000 \\ 92{,}00{,}000 \\ 2{,}45{,}50{,}000 \end{bmatrix}$$

Performing the operation $R_3 \to R_3 + R_2$, we get

$$\begin{bmatrix} 5 & -3 & -4 \\ 0 & 491 & -62 \\ 0 & 0 & 3{,}375 \end{bmatrix} \begin{bmatrix} x \\ y \\ z \end{bmatrix} = \begin{bmatrix} 1{,}50{,}000 \\ 92{,}00{,}000 \\ 3{,}37{,}50{,}000 \end{bmatrix} \Rightarrow \begin{aligned} 5x - 3y - 4z &= 1{,}50{,}000 \\ 0x + 491y - 62z &= 92{,}00{,}000 \\ 0x + 0y + 3375z &= 3{,}37{,}50{,}000 \end{aligned}$$

These equation gives

$$z = 10{,}000,\ y = 20{,}000 \text{ and } x = 50{,}000$$

Verification : The sum of the profits of the outside shareholders

$$= \frac{91}{100}x + \frac{30}{100}y + \frac{10}{100}z$$

$$= 45{,}500 + 6{,}000 + 1{,}000 = 52{,}500$$

Also the sum of the separately earned profits of all the companies

$$= 30{,}000 + 17{,}500 + 5{,}000 = 52{,}500$$

Example 20. A firm produces three products A, B and C which it sells in two markets. Annual sales in units are given below :

Market	*Units Sold*		
	A	B	C
I	8,000	4,000	16,000
II	7,000	18,000	19,000

If the prices per unit of A, B and C are Rs. 2.50, Rs. 1.25 and Rs. 1.50 respectively and the costs per unit are Rs. 1.70, Rs. 1.20 and Re. 0.80 respectively, find the total profit in each market by using matrix algebra.

Solution. Let P denote the matrix of the profit of the firm from the products A, B and C. The

$$P = \begin{bmatrix} 2.50 - 1.70 \\ 1.25 - 1.20 \\ 1.50 - 0.80 \end{bmatrix} = \begin{bmatrix} 0.80 \\ 0.05 \\ 0.70 \end{bmatrix}$$

Now the matrix of the profit of the firm in Market I $= P_I$

$$= [8{,}000 \quad 4{,}000 \quad 16{,}000] \begin{bmatrix} 0.80 \\ 0.05 \\ 0.70 \end{bmatrix}$$

$$= [6{,}400 + 200 + 11{,}200] = [17{,}800]$$

And the matrix of the profit of the firm in Market II $= P_{II}$

$$= [7{,}000 \quad 18{,}000 \quad 19{,}000] \begin{bmatrix} 0.80 \\ 0.05 \\ 0.70 \end{bmatrix}$$

$$= [5{,}600 + 900 + 6{,}300] = [12{,}800]$$

Therefore, total profit in Market I = Rs. 17,800 and total profit in Market II = Rs. 12,800.

Example 21. In an engineering workshop there are 10 machines for drilling, 8 machines for turning and 7 machines for grinding. Three types of brackets are made. Type I brackets require 0 minutes for drilling, 5 minutes for turning and 4 minutes for grinding. The corresponding times for type II and II brackets are 3, 3, 2 and 3, 2, 2 minutes respectively. How many brackets of each type should be produced per hour so that all the machines remain fully occupied during an hour? Solve by matrix algebra. **(D.U., B.Com. (Hons.) 2008)**

Solution. Let the number of Type I brakcets by x, Type II brackets by y and Type III brackets be z.

	Type I	*Type II*	*Type III*
Drilling	0	3	3
Turning	5	3	2
Grinding	4	2	2

The following matrix equation will represent the problem

$$\begin{bmatrix} 0 & 3 & 3 \\ 5 & 3 & 2 \\ 4 & 2 & 2 \end{bmatrix} \begin{bmatrix} x \\ y \\ z \end{bmatrix} = \begin{bmatrix} 10 \times 60 = 600 \\ 8 \times 60 = 480 \\ 7 \times 60 = 420 \end{bmatrix}$$

$$\begin{bmatrix} x \\ y \\ z \end{bmatrix} = \begin{bmatrix} 0 & 3 & 3 \\ 5 & 3 & 2 \\ 4 & 2 & 2 \end{bmatrix}^{-1} \begin{bmatrix} 600 \\ 480 \\ 420 \end{bmatrix}$$

$$= -\frac{1}{12} \begin{bmatrix} 2 & 0 & -3 \\ -2 & -12 & 15 \\ -2 & 12 & -15 \end{bmatrix} \begin{bmatrix} 600 \\ 480 \\ 420 \end{bmatrix}$$

$$= -\frac{1}{12} \begin{bmatrix} 1200 + 0 + 1260 \\ -1200 + 5760 + 6300 \\ -1200 + 5760 - 6300 \end{bmatrix}$$

$$= -\frac{1}{12}\begin{bmatrix} -60 \\ -660 \\ -1740 \end{bmatrix} = \begin{bmatrix} 5 \\ 55 \\ 145 \end{bmatrix}$$

$\Rightarrow x = 5, y = 55, z = 145.$

Example 22. A company is manufacturing two types of autocycles for gents and ladies separately, which are assembled and finished in two workshops W_1 and W_2. Each type takes 15 hours and 10 hours for assembly and 5 hours and 2 hours for finishing in the respective shops. If total number of hours available are 400 and 120 in workshops W_1 and W_2 respectively, calculate the number of units of autocycles produced using matrix method. **(CA (Foundation) May, 1992)**

Solution. Let x be the number of autocycles for gents and y the number of autocycle for ladies, then

$$15x + 10y = 400$$
$$5x + 2y = 120$$

The equations in the matrix form can be written as

$$\begin{bmatrix} 15 & 10 \\ 5 & 2 \end{bmatrix}\begin{bmatrix} x \\ y \end{bmatrix} = \begin{bmatrix} 400 \\ 120 \end{bmatrix}$$

Let $A = \begin{bmatrix} 15 & 10 \\ 5 & 2 \end{bmatrix}$, $|A| = 30 - 50 = -20$

Adj. $A = \begin{bmatrix} 2 & -10 \\ -5 & 15 \end{bmatrix}$, $A^{-1} = \frac{-1}{20}\begin{bmatrix} 2 & -10 \\ -5 & 15 \end{bmatrix}$

$$\therefore \quad \begin{bmatrix} x \\ y \end{bmatrix} = \frac{-1}{20}\begin{bmatrix} 2 & -10 \\ -5 & 15 \end{bmatrix}\begin{bmatrix} 400 \\ 120 \end{bmatrix}$$

$$= \frac{-1}{20}\begin{bmatrix} 2 \times 400 - 10 \times 120 \\ -5 \times 400 + 15 \times 120 \end{bmatrix}$$

$$= \frac{-1}{20}\begin{bmatrix} -400 \\ -200 \end{bmatrix} = \begin{bmatrix} 20 \\ 10 \end{bmatrix} \Rightarrow x = 20, y = 10$$

Example 23. There are two families X and Y. Family X has 2 men, 3 women and one child, while family Y has one men, one women and two children. Their individual daily requirements are as follows :

Man : 2,400 calories and 55 gms protein, Women : 1,900 calories and 45 gms protein, Child : 1,800 calories and 33 gms protein.

Present the above information in the form of matrices. Using matrix multiplication, calculate the total daily requirements of calories and protein for each of the two families.

(CA (Foundation) Nov., 2000)

Solution. Let two families X and Y of men, women and children be represented by 2×3 matrix A and their individual daily requirements in calories and gms by 3×2 matrix B.

$$A = \begin{matrix} \\ X \\ Y \end{matrix}\begin{bmatrix} \text{men} & \text{women} & \text{children} \\ 2 & 3 & 1 \\ 1 & 1 & 2 \end{bmatrix} 4$$

$$B = \begin{matrix} \\ M \\ W \\ Ch \end{matrix}\begin{bmatrix} \text{Cal} & \text{gm} \\ 2{,}400 & 55 \\ 1{,}900 & 45 \\ 1{,}800 & 33 \end{bmatrix} 4$$

$$AB = \begin{matrix} X \\ Y \end{matrix}\begin{bmatrix} 2\times 2400+3\times 1900+1\times 1800 & 12\times 55+3\times 45+1\times 33 \\ 1\times 2400+1\times 1900+2\times 1800 & 1\times 55+1\times 45+2\times 33 \end{bmatrix}_{2\times 2}$$

$$B = \begin{matrix} \\ X \\ Y \end{matrix}\begin{matrix} \text{Cal} & \text{gm} \\ \begin{bmatrix} 12{,}300 & 278 \\ 7{,}900 & 166 \end{bmatrix} \end{matrix}$$

Family X requires daily 12,300 calories and 278 gms of protein

Family Y requires daily 7,900 calories and 166 gms of protein.

Example 24. A student has 4 places where he can have his lunch. The College Canteen charges Rs. 9 for a cold drink, Rs. 6 for a cutlet and Rs. 5 for a sandwich. The Coffee House charges Rs. 10, Rs. 8 and Rs. 9 for the same items, while fast food joint charges Rs. 12, Rs. 15, Rs. 15 and the Restaurant charges Rs. 15, Rs. 25 and Rs. 20 for the above items respectively. The student wants to have one cold drink, two cutlets and one sandwich. Where should he have his lunch so that the lunch costs him the least? (Use of Matrix-Algebra is required). **(CA (Foundation) May, 2002)**

Solution. The description in the problem may be put in matrix form as follows :

$$\begin{matrix} \\ \text{Canteen} \\ \text{Cofee House} \\ \text{Fast food joint} \\ \text{Restaurant} \end{matrix}\begin{pmatrix} \text{Cold drink} & \text{Cutlets} & \text{Sandwiches} \\ 9 & 6 & 5 \\ 10 & 8 & 9 \\ 12 & 15 & 15 \\ 15 & 25 & 20 \end{pmatrix}_{4\times 3}$$

The order is given by the following 3×1 matrix :

$$\begin{matrix} \text{Cold drink} \\ \text{Cutlet} \\ \text{Sandwich} \end{matrix}\begin{pmatrix} 1 \\ 2 \\ 1 \end{pmatrix}_{3\times 1}$$

Using matrix-multiplication, the cost of lunch at four places are given by :

$$\begin{matrix} \text{Canteen} \\ \text{Cofee House} \\ \text{Fast food joint} \\ \text{Restaurant} \end{matrix}\begin{pmatrix} 9 & 6 & 5 \\ 10 & 8 & 9 \\ 12 & 15 & 15 \\ 15 & 25 & 20 \end{pmatrix}\begin{pmatrix} 1 \\ 2 \\ 1 \end{pmatrix} = \begin{bmatrix} 9\times 1+\ 6\times 2+\ 5\times 1 \\ 10\times 1+\ 8\times 2+\ 9\times 1 \\ 12\times 1+15\times 2+15\times 1 \\ 15\times 1+25\times 2+20\times 1 \end{bmatrix} = \begin{pmatrix} 26 \\ 35 \\ 57 \\ 85 \end{pmatrix}$$

$\therefore$ The cost will be least if be takes his lunch in the canteen.

Example 25. A company is to employ 60 labourers from either of the party A and B, comprising persons in different age groups as under :

Category	*I (20-25 years)*	*II (26-30 years)*	*III (31-40 years)*
Party *A*	25	20	15
Party *B*	20	30	10

Rate of labour applicable to categories I, II and III are Rs. 1,200, Rs. 1,000 and Rs. 600 respectively. Using Matrices, find which party is economically preferable over the other.

(CA (Foundation) November, 1997)

Solution.

$$\begin{matrix} \text{Party } A \\ \text{Party } B \end{matrix}\overset{\begin{matrix}\text{I} & \text{II} & \text{III}\end{matrix}}{\begin{bmatrix} 25 & 20 & 15 \\ 20 & 30 & 10 \end{bmatrix}}\begin{bmatrix} 1200 \\ 1000 \\ 600 \end{bmatrix} = \begin{bmatrix} 30000 + 20000 + 9000 \\ 24000 + 30000 + 6000 \end{bmatrix}$$

$$= \begin{bmatrix} 59000 \\ 60000 \end{bmatrix} \begin{matrix} \rightarrow \text{Total wages for party } A \\ \rightarrow \text{Total wages for party } B \end{matrix}$$

Since total wages for party A are less, party A should be preferred.

Example 26. Consider an-investor who wants to invest Rs. 3000 in bank deposits, equity Shares and Unit Trust. Inview of the risk involved in buying equity shares he wants to invest in equity shares equal to 20% of his total investments in bank deposits and unit trust. Because of certain tax exemptions available to him he would also like to maintain a 3 : 2 ratio between investments in Bank deposits and unit trust. Determine the amount he would invest in each of the three forms of investments. Use matrix Algebra method to solve Problem. **(AIMA May, 1992)**

Solution. Let Rs. x be invested in equity shares, Rs. y in Bank deposits and Rs. z in unit trust, then

$$x + y + z = 3000 \qquad \ldots\text{(i)}$$

$$x = \frac{y+z}{5} \Rightarrow 5x - y - z = 0 \qquad \ldots\text{(ii)}$$

$$y = \frac{3}{2}z \Rightarrow 0x + 2y - 3z = 0 \qquad \ldots\text{(iii)}$$

Equations (i), (ii), (iii) in the matrix form can be written as

$$\begin{bmatrix} 1 & 1 & 1 \\ 5 & -1 & -1 \\ 0 & 2 & -3 \end{bmatrix}\begin{bmatrix} x \\ y \\ z \end{bmatrix} = \begin{bmatrix} 3000 \\ 0 \\ 0 \end{bmatrix}$$

Now we will find the inverse of

$$\begin{bmatrix} 1 & 1 & 1 \\ 5 & -1 & -1 \\ 0 & 2 & -3 \end{bmatrix} = A$$

let

$$\begin{bmatrix} 1 & 1 & 1 \\ 5 & -1 & -1 \\ 0 & 2 & -3 \end{bmatrix} = \begin{bmatrix} a_{11} & a_{12} & a_{13} \\ a_{21} & a_{22} & a_{23} \\ a_{31} & a_{32} & a_{33} \end{bmatrix}$$

Cofactor of $a_{11} = 5$, Cofactor of $a_{12} = 15$

Cofactor of $a_{13} = 10$, Cofactor of $a_{21} = 5$

Cofactor of $a_{22} = -3$, Cofactor of $a_{23} = -2$

Cofactor of $a_{31} = 0$, Cofactor of $a_{32} = 6$

Cofactor of $a_{33} = -6$

$$\text{Adj. } A = \begin{bmatrix} 5 & 5 & 0 \\ 15 & -3 & 6 \\ 10 & -2 & -6 \end{bmatrix}, \ \text{Det. } A = 30$$

Inverse of $$A = \frac{1}{30}\begin{bmatrix} 5 & 5 & 0 \\ 15 & -3 & 6 \\ 10 & -2 & -6 \end{bmatrix} = \begin{bmatrix} 1/6 & 1/6 & 0 \\ 1/2 & -1/10 & 1/5 \\ 1/3 & -1/15 & -1/5 \end{bmatrix}$$

$$\therefore \quad \begin{bmatrix} x \\ y \\ z \end{bmatrix} = \begin{bmatrix} 1/6 & 1/6 & 0 \\ 1/2 & -1/10 & 1/5 \\ 1/3 & -1/15 & -1/5 \end{bmatrix}\begin{bmatrix} 3000 \\ 0 \\ 0 \end{bmatrix} = \begin{bmatrix} 500 \\ 1500 \\ 1000 \end{bmatrix}$$

$\Rightarrow x = 500, y = 15000, z = 1000.$

Example 27. An amount of Rs. 5000 is put into three different investment at the rates of interest of 6%, 7% and 8% per annum respectively. The total annual income is Rs. 358. If the combined income from first two investments is Rs. 70 more than income from third. Find the amount of each income by using matrix algebra. **(AIMA, Nov. 1992)**

Solution. Let the income from 6%, 7% and 8% Interest be Rs. x, y and z respectively.*

Then $$x + y + z = 358 \qquad \ldots(i)$$

$$\frac{100}{6}x + \frac{100}{7}y + \frac{100}{8}z = 5000 \Rightarrow \frac{x}{6} + \frac{y}{7} + \frac{z}{8} = 50$$

$$\Rightarrow \quad 28x + 24y + 21z = 8400 \qquad \ldots(ii)$$

$$x + y - 70 = z \Rightarrow x + y - z = 70 \qquad \ldots(iii)$$

Equations (i), (ii), (iii) in matrix form can be written as

$$\begin{bmatrix} 1 & 1 & 1 \\ 28 & 24 & 21 \\ 1 & 1 & -1 \end{bmatrix}\begin{bmatrix} x \\ y \\ z \end{bmatrix} = \begin{bmatrix} 358 \\ 8400 \\ 70 \end{bmatrix}$$

Let $$A = \begin{bmatrix} 1 & 1 & 1 \\ 28 & 24 & 21 \\ 1 & 1 & -1 \end{bmatrix} = \begin{bmatrix} a_{11} & a_{12} & a_{13} \\ a_{21} & a_{22} & a_{23} \\ a_{31} & a_{32} & a_{33} \end{bmatrix}$$

$A_{11} = -45, A_{12} = 49, A_{13} = 4, A_{21} = 2, A_{22} = -2, A_{23} = 0, A_{31} = -3, A_{32} = 7, A_{33} = -4$

$$\text{Adj. } A = \begin{bmatrix} -45 & 2 & -3 \\ 49 & -2 & 7 \\ 4 & 0 & -4 \end{bmatrix}, |A| = 8$$

$$A^{-1} = \frac{1}{8}\begin{bmatrix} -45 & 2 & -3 \\ 49 & -2 & 7 \\ 4 & 0 & -4 \end{bmatrix}$$

$$\therefore \quad x = A^{-1}B = \frac{1}{8}\begin{bmatrix} -45 & 2 & -3 \\ 49 & -2 & 7 \\ 4 & 0 & -4 \end{bmatrix}\begin{bmatrix} 358 \\ 8400 \\ 70 \end{bmatrix}$$

$$= \frac{1}{8}\begin{bmatrix} 480 \\ 1232 \\ 1152 \end{bmatrix} = \begin{bmatrix} 60 \\ 154 \\ 144 \end{bmatrix} \Rightarrow x = 60, y = 154, z = 144$$

$\therefore$ The incomes are Rs. 60, Rs. 154 and Rs. 144 respectively.

* Interest $= \frac{Prn}{100} \Rightarrow x = \frac{P \times 6}{100} \Rightarrow P = \frac{100}{6}x$, where P is the amount borrowed at 6% interest.

❑❑❑

3 Functions, Limits and Continuity

3.1 CONSTANTS AND VARIABLES

In mathematics, different operations like addition, multiplication, division, etc. are performed on the things know as quantities. These quantities are capable of measurement and division into parts. The classification of quantities is as follows :

- Quantities
  - Constants
    - Absolute constants
    - Arbitrary constants
  - Variables
    - Independent variables
    - Dependent variables

In general, a constant is a quantity that does not vary, *i.e.*, it retains the same value all through the investigation.

Absolute constants are those quantities which retain the same value (*i.e.*, which are unchangeable) in every set of mathematical operations, .*e.g*, π, e, real numbers, etc.

Arbitrary constants are those quantities which retain the same value only in one set of mathematical operations. Arbitrary constants may have different values in different sets of mathematical operations.

For example, in the equation of the straight line $y = mx + c$, m and c are constants but have different values for different straight lines. Constants are normally denoted by the letters $a, b, c, d, \ldots$, etc.

Variable quantities are those quantities which take up different values during any set of operations. Variables are normally denoted by u, v, x, y, z, etc.

In the equation of the straight line $y = mx + c, x$ and y are variables.

Independent variables are those which have any arbitrarily assigned value and the **dependent variables** are those whose value depends on the value assumed by the independent variable. For example,

1. The circumference of a circle $= 2\pi r$. If we give different values to r the circumference will have different values, *i.e.*, we can say that radius is an independent variable and circumference is a dependent variable.

2. The volume of a sphere $= \frac{4}{3}\pi r^3$. Here, again r is the independent variable and volume is the dependent variable.
3. The distance travelled by a falling body in any time t [starting from rest] is given by $S = \frac{1}{2}gt^2$. Here, t is the independent variable and distance (S) is dependent variable.
4. Let the total cost function be given by

$$c = ax^2 + bx + k$$

where x is quantity of output. Here, x is independent variable and the total cost (c) is dependent variable.
5. Let the demand function be $p = 12 - 4x$, where p is the price and x the quantity demanded, then x is independent variable and p is dependent variable.

In all the above examples, we have seen that the dependent and independent variables are related by constants. In other words, constants are used to relate dependent and independent variables.

3.2 FUNCTION OR MAPPING

Let A and B be two non-empty sets. If to each element $x \in A$ there corresponds a unique element $y \in B$, by means of a rule denoted by f then we say a function or mapping f is defined from the set A into the set B, *i.e.*, a mapping from the set A to the set B is a rule that assigns to each element x of A an element y of B.

Symbolically it is written as

$$y = f(x) \qquad \text{...(I)}$$

'y is called the function of x'.

Suppose 'the cube of' is a mapping from the set A = {1, 2, 3, 4} to set B, then the element of set B will be {1, 8, 27, 64}.

In (I), x is called the independent variable and y is called the dependent variable.

y is also called the image of x under f and x is called the pre-image of y under f.

The set of values which an independent variable can assume is called the **domain** of the function and the set of corresponding values of y, *i.e.*, dependent variable is called the **range** of the function. In other words, the set of all the images of the independent variable under a mapping is called the **range** of the function.

The **mapping** or function f from the set A to the set B is called '**onto**' if B has no element other than the images of the elements of A.

The mapping or function f from the set A to the set B is called '**into**' if B, besides having all the images of the elements of A, has at least one element which is not an image of any element in A.

In general, if a set B contains all the images of the elements of the set A under the mapping (or function) f, then the set B is called the **co-domain** for the function f. If the mapping is 'onto', then

Co-domain = Range.

If the mapping is 'into', then set of the elements in the range is a proper subset of the elements in the co-domain. For example.

If A = {1, 2, 3, 4} and B = {1, 8, 27, 64, 70} and if the mapping f is 'the cube of' then B is the co-domain of f and the range of f = {1, 8, 27, 64}.

3.3 TYPES OF FUNCTIONS

(1) Linear Functions : A function of the form $y = f(x)$, *e.g.*, $y = mx + c$; m and c being real constants (real numbers) is known as a linear function. The graph of linear function is always a straight line. The graph of a linear function $y = mx + c$ is a straight line with **slope** m and **intercept** c on y-axis.

If $m = 0, y = c$, *i.e.*, for all values of x, the value of the function remains the same, *i.e.*, the function $y = mx + c$ becomes a constant function when $m = 0$. The function $y = f(x)$ is said to be a **constant function** if for every value of x, the value of the function $f(x)$ remains the same constant *e.g.*, in the function $y = mx + c$, if $m = 0$ then $y = c$ for all real values of x].

When $m = 1$ and $c = 0$, $y = f(x) = x$, *i.e.*, the function $y = f(x) = mx + c$ become an identity function when $m = 1$ and $c = 0$. [The function $y = f(x) = x$ for all real values of x is called the identity function defined on the set of real numbers].

Note : The domain and range of linear function is the set of real numbers.

(2) Quadratic Functions : A function of the form $y = f(x) = ax^2 + bx + c$ where a, b and c are real constants [real numbers] is called a **quadratic function**. The graph of quadratic function is always a **parabola.**

(3) Cubic Functions : A function of the form $y = f(x) = ax^3 + bx^2 + cx + d$ where a, b, c, d are real constants is called a cubic function.

(4) Exponential Functions : In general a function whose independent variable appears as exponent (or Power) is called an exponential function. $y = a^x$ is an exponential function with base a.

The function defined by $y = e^x$ for all real values of x is called an exponential function with **base** e. The domain of exponential function is the set of real numbers.

(5) Logarithmic Functions : A function defined by $y = \log x$ for $x > 0$ is called a natural logarithmic function or simply logarithmic function. The domain of logarithmic function is the set of positive real numbers.

Note : $y = \log x \Rightarrow y = \log_e x$ because in calculus we deal only with base e.

(6) Even and Odd Functions : A function $y = f(x)$ is said to be an even function of x if $f(-x) = f(x)$, *e.g.*, $f(x) = x^4 + x^2$ is an even function of x because $f(-x) = (-x)^4 + (-x)^2 = x^4 + x^2 = f(x)$.

A function $y = f(x)$ is said to be an odd function of x if $f(-x) = -f(x)$, *e.g.*, $f(x) = x^3 - x$ is an odd function of x.

(7) Greatest Integer Functions : If x is a real number then $[x]$ denotes the largest integer less than or equal to x, *e.g.*, if $x = 2.999 \ldots$ then $[x] = 2$ and if $x = 3.000 \ldots. 1$ then $[x] = 3$.

If $x = -\frac{5}{2}$, then $[x] = -3$, In general greatest integer function is denoted as $y = [x]$ = largest integer $\leq x$.

Note : A function $y = f(x)$ is said to be defined at $x = a$ if it has finite value at $x = a$, *i.e.*, if $f(a)$ is finite. If the function $f(x)$ takes one of the meaningless forms $\frac{0}{0}, \frac{\infty}{\infty}$, $0 \times \infty$, $\infty - \infty$, etc. for a particular value of x, the function is said to be **'not defined'** for that value of x.

(8) Algebraic (Or Polynomial) Functions : Any expression of the variable x containing different powers of x is known as an algebraic function or Polynomial function, *e.g.*,

$$f(x) = x^2 - 2x - 2 \text{ is an algebraic function of degree 2.}$$

In general a function of the form $f(x) = a_0x^n + a_1x^{n-1} + \ldots + a_n$

Here n is a non-negative integer and $a_0, a_1, \ldots, a_n$ are constants with $a \neq 0$ is called a polynomial function in x of degree n. a_0 is called the leading coefficient.

A linear function is a polynomial fucntion of degree 1 and a quadratic function is a polynomial function of degree 2. The domain of any polynomial function is the set of all real numbers.

(9) Rational functions : A function defined by the quotient of two polynomial $\left[i.e., \frac{f(x)}{g(x)}\right]$ in x is called a rational function, *e.g.*, the function $f(x) = \frac{x^2 - 3x + 1}{x - 2}$ is a rational function provided $x \neq 2$.
[It should be noted that the ratio of two polynomials defines a rational function only for those values of x, which do not make denominator zero.]

(10) Irrational Functions : An algebraic function which is not rational is known is irrational, *e.g.*, $f(x) = \frac{8x + 5}{\sqrt{3 - 5x - 4x^2}}$ is a irrational function.

(11) Reciprocal or identity function : An algebraic function of the form $y = f(x) = \frac{1}{x}$ is called the reciprocal of identity function. The domain is the set of real numbers except zero and range is the set of real numbers.

(12) Absolute value fuction : The function $f(x) = |x|$ is called the absolute value function. The domain of the absolute value function is the set of real numbers while the range is the set of non-negative real numbers. If x is possitive $f(x) = x$ and if x is negative $f(x) = -x$ so that $f(x)$ is always positive. For graph refer Example 1.

(13) Explicit Function : y is said to be an explicit function of x if y can be directly expressed in terms of x, *e.g.*, $y = \log x + e^x$ is an explicit function of x.

(14) Implicit Function : y is said to be an implicit function of x when x and y both occur together in an equation and y is not capable of being directly expressed in terms of x, *e.g.*, $x^2 + 2y + xy^2 + x^2y + y^2 = 9$ (or some constant) is an implicit function of x.

(15) Parametric Functions : In parametric functions, both the variables, *i.e.*, dependent and independent, are expressed as functions of a third variable called **Parameter,** *e.g.*, $x = f_1(t)$, $y = f_2(t)$ defines a function in terms of parameter t.

(16) Increasing and Decreasing Functions : A function $y = f(x)$ defined in the closed interval $[a, b]$ is said to be increasing if for any x_1, x_2 in the interval, $x_1 < x_2 \Rightarrow f(x_1) \leq f(x_2)$. The function is said to be decreasing if $x_1 < x_2 \Rightarrow f(x_1) \geq f(x_2)$.

If $x_1 < x_2 \Rightarrow f(x_1) < f(x_2)$ the function is said to be strictly increasing. If $x_1 < x_2 \Rightarrow f(x_1) > f(x_2)$ the function is said to be strictly decreasing.

Functions relating to Business like Demand function, Supply function, Revenue function, Profit function, Consumption function, etc., have been discussed while discussing maxima-minima and demand.

3.4 GRAPHS

The graph of a function $y = f(x)$ is the pictorial representation of the function and is obtained by locating on a rectangular co-ordinate system the points defined by the number pairs (x, y) or $[x, f(x)]$.

By giving different values of the independent variable, corresponding value of the dependent variable are obtained and pairs of corresponding values of two variables formed are plotted on a graph paper. The points so plotted are joined approximately by a smooth curve (curve may be a straight line).

Example 1. Draw the graphs represented by the following functions :

(i) $y = |x|$ (ii) $y^2 = x$ (iii) $y = x^2$ (iv) $y = \frac{x^2}{x}$ (v) $y = \frac{1}{x}$

Solution : (i) $y = |x| \Rightarrow y = x$ when $x > 0 = -x$ when $x < 0$, *i.e.*, y is always positive

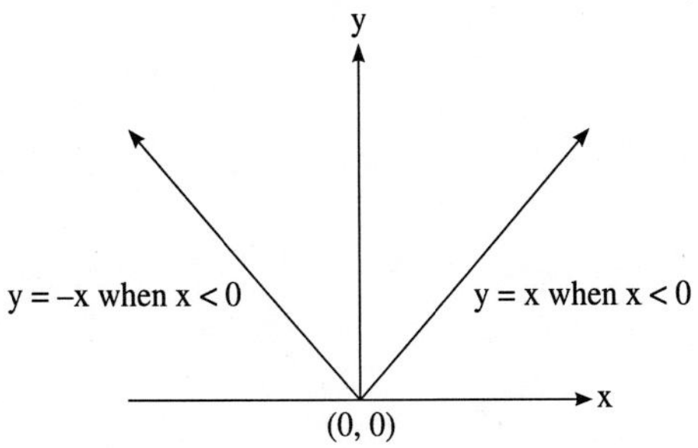

(ii) The given functions is $y^2 = x$

$x = 0, \quad y = 0$
$x = 1, \quad y = \pm 1$
$x = 2, \quad y = \pm 1.4$
$x = 3, \quad y = \pm 1.73$
$x = 4, \quad y = \pm 2$
$x = 5, \quad y = \pm 2.2$
$x = 6, \quad y = \pm 2.45$

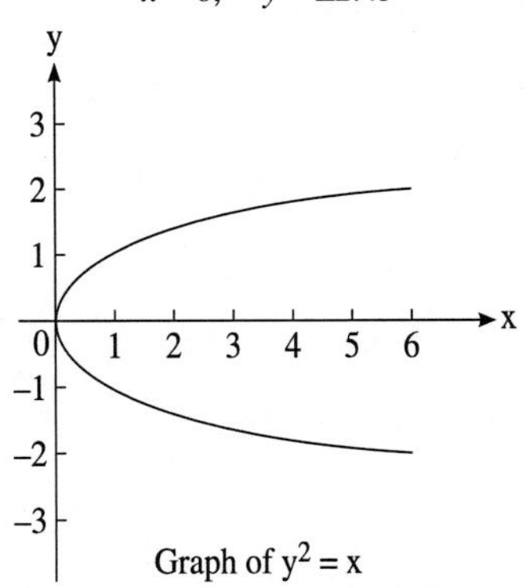

Graph of $y^2 = x$

On the positive side of x-axis the graph is symmetrical about x-axis. For negative values of x no real values of y exist and hence, no graph exists on the negative side of x-axis.

Note : Sometimes distinction is made between a **function** and a **relation.** By function we generally mean single valued function, *i.e.,* for each value of the independent variable there is only one value of the dependent variable. A relation is a multi-value function, *i.e.,* for each value of the independent variable then can be more then one values of the dependent variable. Hence, $y = \sqrt{x}$ is a relation.

(iii) The given function is $y = x^2$

$x = 0, \quad y = 0$
$x = 1, \quad y = 1$

$x = 1.5,\quad y = 2.25$
$x = 2,\quad y = 4$
$x = 2.5,\quad y = 6.25$
$x = -1,\quad y = 1$
$x = -1.5,\quad y = 2.25$
$x = -2,\quad y = 4$
$x = -2.5,\quad y = 6.25$

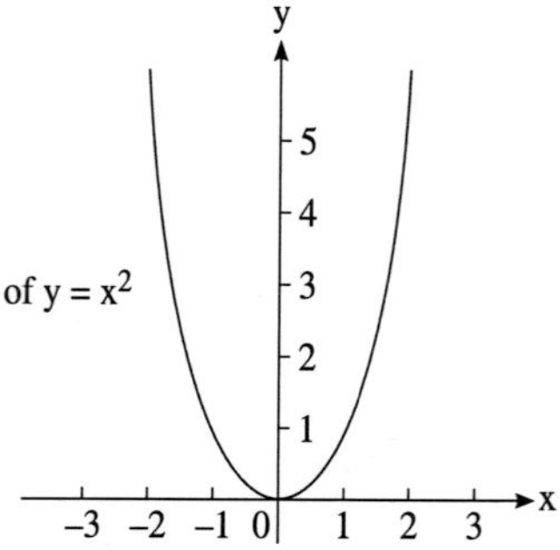

Graph of $y = x^2$

In the above example, on the positive side of y-axis, the graph is symmetrical about y-axis. Since, for no value of x (positive or negative) negative values of y exist, the graph of the function $y = x^2$ does not extend on the negative side of y-axis.

Note : In both the above examples we have seen that for every finite value of x there is a finite value of y and the graph plotted by set of points (x, y) is continuous. Sometimes for a certain value of x, y may not be defined, *i.e.,* the value of y may not exist, *e.g.,* in the function $y = \frac{x^2}{x}$, $y = \frac{0}{0}$ when $x = 0$. Such a point cannot be represented on the graph and is called a **hole** or a point of discontinuity. When $x \neq 0$ the function becomes $y = x$ which is a straight line. Hence, a function $y = \frac{x^2}{x}$ represents a straight line $y = x$ with a hole (*i.e.,* discontinuity) at $x = 0$. It may be noted that when $x = 0$, $y = \frac{x^2}{x}$ and when $x \neq 0$, $y = x$.

$$y = \frac{0}{0} \quad \text{(indeterminate) when } x = 0$$

(iv) The given function is $y = \frac{x^2}{x}$

$x = 0,\quad y = \frac{0}{0}$
$x = 1,\quad y = 1$
$x = -1,\quad y = -1$
$x = 1.5,\quad y = 1.5$
$x = -1.5,\quad y = -1.5$
$x = 2,\quad y = 2$
$x = 2.5,\quad y = 2.5$
$x = -2,\quad y = -2$
$x = -2.5,\quad y = -2.5$

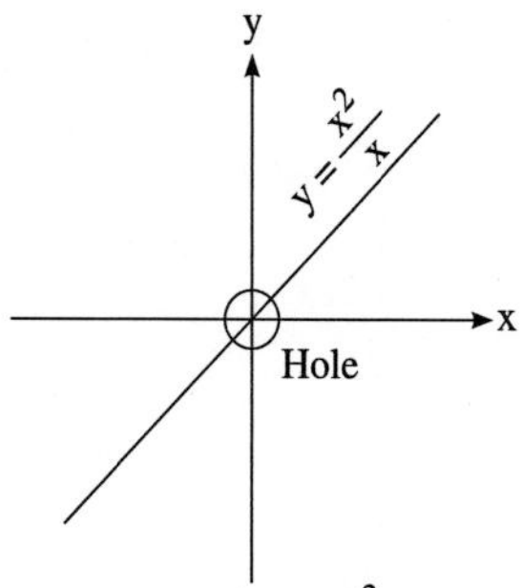

Graph of $y = \frac{x^2}{x}$

Since, the given function is not defined at $x = 0$, therefore, $x = 0$ is a point of discontinuity, *i.e.,* hole.

(v) The given function is $y = \frac{1}{x}$

When $x = 0, y = \frac{1}{0}$

i.e., the function is not defined at $x = 0$. Therefore $x = 0$ is a point of discontinuity **(hole)**.

For positive values of x except zero, y decreases as x increases and y increases as x decreases, *i.e.,* $x \to \infty, y \to 0$.

For negative values of x except zero, y increases as x decreases, *i.e.,* as $x \to -\infty, y = 0$.

Here the curve is of the form shown in the figure.

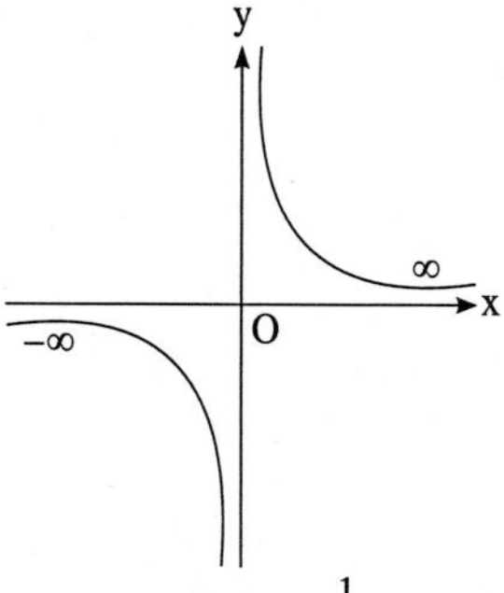

Graph of $y = \frac{1}{x}$

Example 2. Plot the graph of : $y = x + |x|$

Solution : $y = x + |x|$

(i) When $x > 0$ $\quad y = x + x = 2x \quad [\because$ When $x > 0, |x| = x]$

and the graph will be a straight line as follows :

$x = 0, \quad y = 0$
$x = 0.5, \quad y = 1$
$x = 1, \quad y = 2$
$x = 1.5, \quad y = 3$
$x = 2, \quad y = 4$

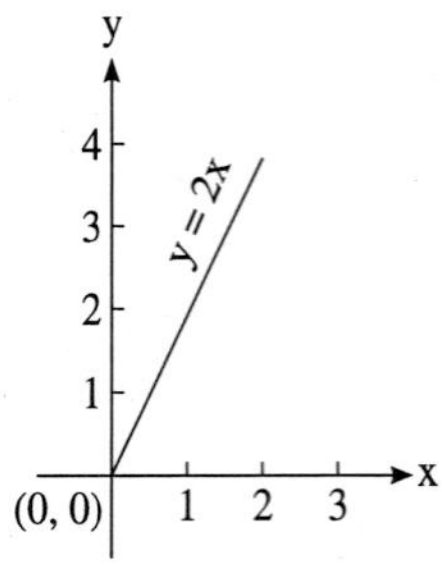

Graph of y = 2x

(ii) When $x < 0$

$$f(x) = 0, \qquad [\because \text{ when } x \text{ is negative } |x| = -x]$$

Hence graph will simply be the portion of x-axis for which x is negative. [$\because$ $y = 0$ for all $x < 0$]

$$x = 0, y = 0 \text{ is origin.}$$

Graph the function

$$f(x) = \begin{cases} x & 0 \le x \le 1 \\ 2 - x & 1 < x \le 2 \end{cases}$$

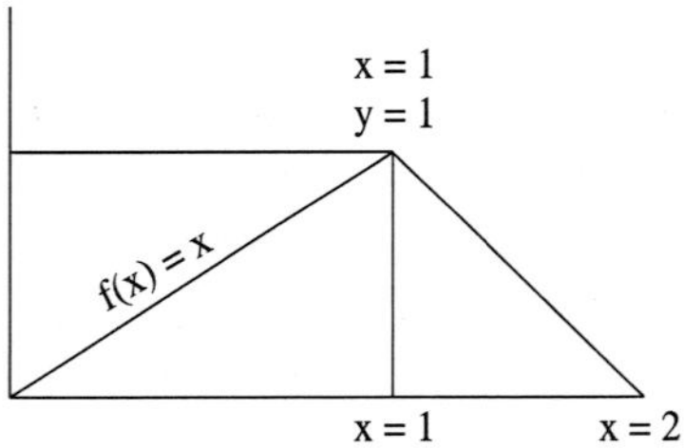

3.5 LINE OF SYMMETRY

If in a figure, when folded along the x-axis the part of the curve below the x-axis coincides exactly with the part of the curve above the x-axis, then x-axis is called the **line of symmetry**. In general, if a curve is folded along any line and the part of the curve below that line coincides exactly with the part of the curve above that line, then that line is called the **line of symmetry.**

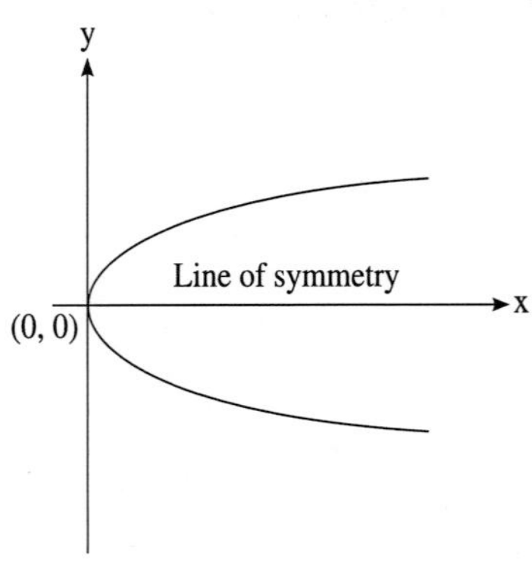

Example 3. Draw the graph of the following function

$$y = x\,|x| : -\infty \leq x \leq \infty$$

Solution : When $x > 0$, *i.e.,* when x is positive

$$y = x^2$$

when $x < 0$, *i.e.,* when x is negative

$$y = -x^2$$

Table for $y = x^2$

x	1	2	3
y	1	4	9

Table for $y = -x^2$

x	–1	–2	–3
y	–1	–4	–9

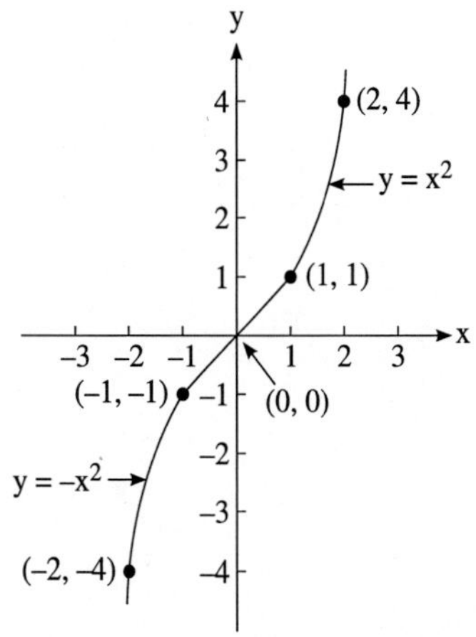

Example 4. Draw the graph of $f(x)$ as defiend below and examine from it whether $f(x)$ is continuous at $x = 0$

$$fx) = -x \text{ for } x \leq 0, f(x) = x \text{ for } 0 < x < 1$$
$$f(x) = 2 - x \text{ for } x \geq 1$$

Solution : Table for $f(x) = 2 - x = y$

x	1	2	3	4
$f(x)$	1	0	–1	–2

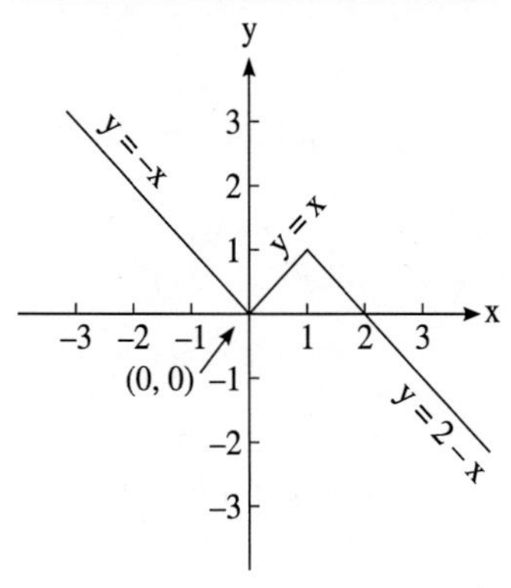

The given function is continuous at $x = 0$

Note : In this example please note that when $x = 1, f(x) = 2 - 1$ and when $x = 0, f(x) = 0$. So, we can also define $f(x) = x$ for $0 \le x \le 1$.

Example 5. Draw the graph of the function $y = \frac{|x|}{x}$.

Solution : For all positive values of x

$$y = \frac{|x|}{x} = \frac{x}{x} = 1$$

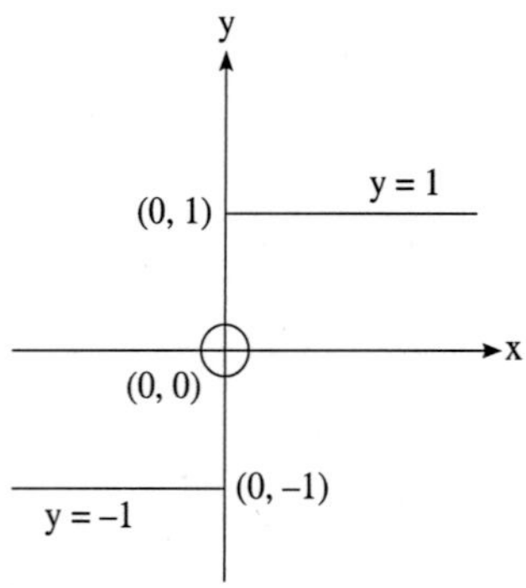

When $x = 0$, y is not defined.

For all negative values of x

$$y = \frac{|x|}{x} = \frac{-x}{x} = -1$$

$\therefore$ The graph of $y = \frac{|x|}{x}$ is as shown in the above Fig.

3.6 LIMIT OF A FUNCTION

(i) Meaning of x ends to a : When the difference $|x - a|$, (*i.e.*, numerical difference between x and a) can be made less than any positive quantity, however small, we say x tends to a and is written as $x \to a$.

(ii) Meaning of x tends to zero : When x varies in such a way that its numerical value becomes and remains less than any positive number that we may choose, no matter how small, we say x tends to zero and is written as $x \to 0$.

(iii) Meaning of x tends to infinity : When x becomes and remains greater than any positive number that we may choose, however great that number may be, we say x tends to infinity and is written as $x \to \infty$.

3.7 NEIGHBOURHOOD OR PROXIMITY OF A POINT

Let c be any real number, then any open interval around c is called the neighbourhood of c, *e.g.*,

$]c - \varepsilon, c + \varepsilon[, \varepsilon > 0$ is the neighbourhood of c.

Any neighbourhood from which the point c is excluded is called **deleted neighbourhood of** c. Geometrically it means set of all those points which are within an ε distance from c on either side except for the point c.

3.8 LIMIT OF A FUNCTION f(x)

A number l is said to be the limit of $f(x)$ as $x \to a$ written as

$$\lim_{x \to a} f(x) = l \text{ if}$$

(i) The function is defined and single valued in the **deleted neighbourhood** of a.

(ii) For every positive number ε, however small, there exists a positive number δ (usually dependent on ε), such that

$$|f(x) - l| < \varepsilon$$

whenever $$0 < |x - a| < \delta$$

From the above definition it is interesting to note that $\lim_{x \to a} f(x)$ may exist, even if the function $f(x)$ is not defined at $x = a$. Sometimes both the things may happen, *i.e.*,

(i) The function is defined at $x = a$, and (ii) $\lim_{x \to a} f(x)$ also exist.

3.9 RIGHT HAND AND LEFT HAND LIMITS

$$\text{LEFT } [x < a] \quad | \quad \text{RIGHT } [x > a]$$

$$\downarrow$$

$$x = a$$

The variable point x can approach a either from the left of a or from the right of a. The respective approaches are indicated by writing

$$x \to a-$$

and $$x \to a+$$

If $$\lim_{x \to a-} f(x) = l_1 \text{ (left hand limit)}$$

and $$\lim_{x \to a+} f(x) = l_2 \text{ (right hand limit)},$$

we say $$\lim_{x \to a} f(x) = l \text{ if and only if}$$

$$l_1 = l_2 = l$$

It will be proved later that the limit of a function, if exists, is unique.

Illustration : (i) Consider a function

$$f(x) = \frac{x^2 - 9}{x - 3}$$

Clearly, the function is not defined at $x = 3$ but is defined in the deleted neighbourhood of 3.

If $x \neq 3$, then $\dfrac{x^2 - 9}{x - 3} = \dfrac{(x-3)(x+3)}{x - 3} = x + 3$

Let us give different values to x in the deleted neighbourhood of 3. Say $x = 3.1, 3.01, 3.001, 3.0001$, etc., from right side and $x = 2.9, 2.99, 2.999, 2.9999$, etc. from left side. Then,

Left hand side		*Right hand side*	
$x = 2.9$	$f(x) = 5.9$	$x = 3.1$	$f(x) = 6.1$
$x = 2.99$	$f(x) = 5.99$	$x = 3.01$	$f(x) = 6.01$
$x = 2.999$	$f(x) = 5.999$	$x = 3.001$	$f(x) = 6.001$
$x = 2.9999$	$f(x) = 5.9999$	$x = 3.0001$	$f(x) = 6.0001$

i.e., as x gets nearer and nearer to 3 from either side, $f(x)$ gets close to 6, *i.e.,* $|f(x) - 6|$ can be made as small as we wish (less than any Pre-assigned positive quantity however small) by taking x sufficiently close to 3 (but not equal to 3). In this case $f(x)$ has a limiting value 6 as $x \to 3$. Symbolically written as

$$\lim_{x \to 3} f(x) = \lim_{x \to 3} \frac{x^2 - 9}{x - 3} = 6$$

Note that even if a function is not defined for some value x, the limit of the function as x approaches that value may exist.

Now consider the function

$$f(x) = \frac{|x - 3|}{x - 3}, \; x \neq 3$$
$$= 0, \qquad x = 3$$

When $x > 3, x - 3$ is > 0 and $|x - 3|$ is also > 0.

When $x < 3, x - 3$ is < 0 (Negative) and $|x - 3|$ is positive *i.e.,* > 0 .

$\therefore$ When $x > 3$, $|x - 3| = (x - 3)$ and when $x < 3$, $|x - 3| = -(x - 3)$

$$\lim_{x \to 3^+} f(x) = \lim_{x \to 3^+} \frac{|x - 3|}{x - 3} = 1 \qquad \text{(Right hand limit)}$$

and when $x < 3$

$$\lim_{x \to 3^-} f(x) = \lim_{x \to 3^-} \frac{|x - 3|}{x - 3} = -1 \qquad \text{(Left hand limit)}$$

Since, Left hand limit $\neq$ Right hand limit,

$$\lim_{x \to 3} \frac{|x - 3|}{x - 3} \text{ do not exist.}$$

3.10 METHODS OF FINDING THE LIMIT OF A FUNCTION $f(x)$ AS x TENDS TO A FINITE QUANTITY SAY 'a'

There are three methods for finding limit of a function $f(x)$ as x tends to finite quantity, *i.e.,* 'a' :

(i) Method of factors

(ii) Method of substitution

(iii) Method of rationalisation

(iv) L-Hospital's Rule.

(i) Methods of factors

In method 1, if $f(x)$ is of the form $\frac{g(x)}{h(x)}$, factorise $g(x)$ and $h(x)$, cancel the common factors and then put the value of x.

(ii) Find the value of $\lim_{x \to 1} \frac{x^3 - 1}{x^2 - 1}$.

Solution : Now,
$$\frac{x^3 - 1}{x^2 - 1} = \frac{(x - 1)(x^2 + x + 1)}{(x - 1)(x + 1)} = \frac{x^2 + x + 1}{x + 1} \qquad \text{when } x \neq 1$$

$$\therefore \quad \lim_{x \to 1} \frac{x^3 - 1}{x^2 - 1} = \lim_{x \to 1} \frac{x^2 + x + 1}{x + 1} = \frac{1 + 1 + 1}{1 + 1} = \frac{3}{2}$$

(ii) Method of Substitution

The following steps are involved :

(i) Put $x = a + h$ where h is very small but $\neq 0$, *i.e.*, $x \to a, h \to 0$

(ii) Simplify numerator and denominator and cancel common powers of h.

(iii) Put $h = 0$.

The result is the required limit.

Example 6. Evaluate $\lim\limits_{x \to a} \dfrac{x^n - a^n}{x - a}$.

Solution : Put $x = a + h$ where h is very small, then

$$\frac{x^n - a^n}{x - a} = \frac{(a+h)^n - a^n}{a + h - a} = \frac{(a+h)^n - a^n}{h}$$

$$= \frac{a^n\left[\left(1 + \frac{h}{a}\right)^n - 1\right]}{h}$$

$$= \frac{a^n\left[n\frac{h}{a} + \frac{n(n-1)}{2}\frac{h^2}{a^2} + \ldots\right]}{h}$$

$$= a^n\left[\frac{n}{a} + \frac{n(n-1)}{2}\frac{h}{a^2} + \ldots\right]$$

$$\therefore \quad \lim_{x \to a} \frac{x^n - a^n}{x - a} = \lim_{x \to 0} a^n\left[\frac{n}{a} + \frac{n(n-1)}{2}\frac{h}{a^2} + \ldots\right]$$

$$= \boldsymbol{a^n . \frac{n}{a} = na^{n-1}}$$

(iii) Method of Rationalisation

This method is useful where radical signs are involved either in the numerator or in the denominator. The numerator or denominator (as required) is rationalised and limit taken. The following example will make the method clear.

Example 7. Evaluate $\lim\limits_{x \to 0} \dfrac{\sqrt{1+x} - \sqrt{1-x}}{x}$

Solution : Rationalising the numerator, we get

$$\frac{\sqrt{1+x} - \sqrt{1-x}}{x} = \frac{\sqrt{1+x} - \sqrt{1-x}}{x} \times \frac{\sqrt{1+x} + \sqrt{1-x}}{\sqrt{1+x} + \sqrt{1-x}}$$

$$= \frac{2x}{x[\sqrt{1+x} + \sqrt{1-x}]} = \frac{2}{\sqrt{1+x} + \sqrt{1-x}}$$

$$\lim_{x \to 0} \frac{\sqrt{1+x} - \sqrt{1-x}}{x} = \lim_{x \to 0} \frac{2}{\sqrt{1+x} + \sqrt{1-x}} = 1$$

L-Hospital's Rule

If $f(x)=\dfrac{g(x)}{h(x)}$ takes the form $\dfrac{0}{0}$ when $x \to a$, differentiate the numerator and denominator separately with respect to x and then take the limit of $\dfrac{g'(x)}{h'(x)}$ as $x \to a$. If $\dfrac{g'(x)}{h'(x)}$ also takes the form $\dfrac{0}{0}$ when $x \to a$, again differentiate the numerator and denominator w.r.t. x and take the limit of $\dfrac{g''(x)}{h''(x)}$ as $x \to a$ and so on. Sometimes several repetitions of the rule are necessary to evaluate the limit. This rule is called **L-Hospital's Rule**. If it is possible to factorise the numerator and denominator, the method of factors can also be used where $\dfrac{g(x)}{h(x)}$ take the form $\dfrac{0}{0}$ when $x \to a$.

3.11 INFINITE LIMITS (IMPORTANT)

For finding the limit of $f(x)=\dfrac{g(x)}{h(x)}$ as $x \to \infty$, we divide the numerator and denominator by highest power of x occurring in $f(x)$ (numerator or denominator whichever power is higher) and then use $\dfrac{1}{x}, \dfrac{1}{x^2}$, etc. $\to 0$ as $x \to \infty$.

Example 8. Evaluate $\lim\limits_{x \to \infty} \dfrac{(x+1)(2x+3)}{(x+2)(3x+4)}$

Solution : Now, $\dfrac{(x+1)(2x+3)}{(x+2)(3x+4)} = \dfrac{2x^2+5x+3}{3x^2+10x+8}$

$$= \frac{2+\frac{5}{x}+\frac{3}{x^2}}{3+\frac{10}{x}+\frac{8}{x^2}} \quad \text{[Dividing the Numerator and Denominator by } x^2]$$

$$\therefore \quad \lim_{x \to \infty} \frac{(x+1)(2x+3)}{(x+2)(3x+4)} = \lim_{x \to \infty} \frac{2+\frac{5}{x}+\frac{3}{x^2}}{3+\frac{10}{x}+\frac{8}{x^2}} = \frac{2}{3} \quad \left[\because \begin{array}{r}\text{when } x \to \infty \\ 1/x \to 0\end{array}\right]$$

Otherwise

$$\lim_{x \to \infty} \frac{(x+1)(2x+3)}{(x+2)(3x+4)} = \lim_{x \to \infty} \frac{\left(1+\frac{1}{x}\right)\left(2+\frac{3}{x}\right)}{\left(1+\frac{2}{x}\right)\left(3+\frac{4}{x}\right)} = \frac{2}{3}$$

Note : It may be noted that $\lim\limits_{x \to \alpha} f(x)$ and $f(a)$ are not same. $\lim\limits_{x \to \alpha} f(x)$ stands for the limiting value of $f(x)$ when x has any arbitrarily chosen value in the deleted neighbourhood of a excluding a. But $f(a)$ stands for the value of $f(x)$ when x is exactly equal to a [*i.e.*, the value of $f(x)$ obtained by substituting $x = a$ in $f(x)$].

3.12 THE FOLLOWING CASES MAY ARISE IN LIMITS

1. $\lim_{x \to a} f(x)$ exists but $f(a)$ does not exist. For example $\lim_{x \to 1} \frac{x^2 - 1}{x - 1} = 2$ but $f(1) = \frac{0}{0}$, *i.e.*, $f(1)$ does not exist.

2. $\lim_{x \to a} f(x)$ does not exist but $f(a)$ exists, *e.g.*, let $f(x)$ be defined as

$$f(x) = \begin{cases} 1 & \text{when } x > 0 \\ 0 & \text{when } x = 0 \\ -1 & \text{when } x < 0 \end{cases}$$

   Here left hand limit ≠ Right hand limit hence, $\lim_{x \to 0} f(x)$ does not exist but $f(0)$ exists.

3. $f(a)$ and $\lim_{x \to a} f(x)$ both may exist and may be equal *i.e.*,

$$\lim_{x \to 2} (x + 2) = 4 \text{ and } f(2) = 4.$$

4. $f(a)$ and $\lim_{x \to a} f(x)$, both may exist and may be unequal, *e.g.*, let a function $f(x)$ be defined as

$$f(x) = \begin{cases} 0 & \text{when } x \neq 0 \\ 1 & \text{when } x = 0 \end{cases}$$

   Here, $\lim_{x \to 0} f(x) = 0$ and $f(0) = 1$

5. $f(a)$ and $\lim_{x \to a} f(x)$ both may not exist

   e.g., if $f(x) = \frac{1}{x}$, then $f(0) = \infty$

   Right hand limit $= +\infty$.

   Left hand limit $= -\infty$.

   i.e., neither $f(0)$ exists nor $\lim_{x \to 0} f(x)$ exists.

3.13 SOME RESULTS ON LIMITS

(i) If $f(x) = k$, a constant, then $\lim_{x \to a} f(x) = k$

(ii) If $\lim_{x \to a} f(x) = A$ and $\lim_{x \to a} g(x) = B$.

then $\lim_{x \to a} kf(x) = k \lim_{x \to a} f(x) = kA$ and $\lim_{x \to a} k\,[g(x)] = kB$

(iii) $\lim_{x \to a} [f(x)\, g(x)] = \lim_{x \to a} f(x) \times \lim_{x \to a} g(x) = A \times B$

(iv) $\lim_{x \to a} \frac{f(x)}{g(x)} = \frac{\lim_{x \to a} f(x)}{\lim_{x \to a} g(x)} = \frac{A}{B}$ provided $B \neq 0$.

i.e., the limit of the quotient of the functions is equal to the quotient of the limits of the functions provided that the limit of the divisor is not equal to zero.

(v) $\lim\limits_{x \to a} [f(x)]^{1/n} = [A]^{1/n}$, provided $[A]^{1/n}$ is a real number.

(vi) $\lim\limits_{x \to a} [f(x) \pm g(x)] = \lim\limits_{x \to a} f(x) \pm \lim\limits_{x \to a} g(x) = A \pm B$.

In general, the limit of any number of functions connected by plus or minus sign is equal to the limits of the separate functions, connected by the same sign. **(Important)**

3.14 SOME IMPORTANT LIMITS (CTM)

1. $\lim\limits_{x \to \infty} \left(1 + \frac{1}{x}\right)^x = e$
2. $\lim\limits_{x \to 0} (1 + x)^{1/x} = e$
3. $\lim\limits_{x \to 0} \frac{a^x - 1}{x} = \log_e a$
4. $\lim\limits_{x \to 0} \frac{\log(1 + x)}{x} = 1$
5. $\lim\limits_{x \to 0} \left(\frac{e^x - 1}{x}\right) = 1$ [$\because \log_e e = 1$]
6. $\lim\limits_{x \to 0} \frac{(1 + x)^n - 1}{x} = n$
7. $\lim\limits_{x \to a} \frac{x^n - a^n}{x - a} = xa^{n-1}$

Example 9. Prove that $\lim\limits_{x \to 0} \frac{a^x - 1}{x} = \log_e a.$

Solution : Put $a^x - 1 = y$

i.e., $a^x = 1 + y \Rightarrow x = \log_a (1 + y)$ $[\because \log_a a = 1]$

$\because$ $\frac{a^x - 1}{x} = \frac{y}{\log_a (1 + y)}$

When $x \to 0, y \to a^0 - 1$, *i.e.,* $y \to 1 - 1$, *i.e.,* $y \to 0$

Now,
$$\lim_{x \to 0} \frac{a^x - 1}{x} = \lim_{x \to 0} \frac{y}{\log_a (1 + y)}$$
$$= \lim_{y \to 0} \frac{1}{\frac{1}{y} \log_a (1 + y)} = \lim_{y \to 0} \frac{1}{\log_a (1 + y)^{1/y}}$$
$$= \frac{1}{\log_e a} \qquad \left[\because \lim_{y \to 0} (1 + y)^{1/y} = e\right]$$
$$= \log_e a \qquad \left[\because \log_a b = \frac{1}{\log_b a}\right]$$

Example 10. Give one example in each of the following cases :

(i) $f(a)$ does not exist but $\lim\limits_{x \to a} f(x)$ exists.

(ii) $f(a)$ exists but $\lim\limits_{x \to a} f(x)$ does not exist.

(iii) $f(a)$ and $\lim\limits_{x \to a} f(x)$ both exist but they are unequal.

(iv) $f(a)$ and $\lim_{x \to a} f(x)$ both exist and are equal.

(v) Neither $f(a)$ nor $\lim_{x \to a} f(x)$ exists.

Solution : (i) If $f(x) = \dfrac{x^2 - 9}{x - 3}$ then $f(3)$ does not exist but $\lim_{x \to 3} f(x)$ exists.

(ii) If $f(x) = 1$ when $x > 0, f(x) = 0$ when $x = 0$ and $f(x) = -1$ when $x < 0$. Here, $f(0)$ exists but $\lim_{x \to 0} f(x)$ does not exist.

(iii) $f(x) = x$ when $x \neq 0, f(x) = 1$ when $x = 0$. Here, $\lim_{x \to 0} f(x) = 0$, but $f(0) = 1$, *i.e.*, $f(0)$ and $\lim_{x \to 0} f(x)$ both exist but are not equal.

(iv) If $f(x) = |x|$, then $f(0) = 0$ and $\lim_{x \to 0} f(x) = 0$

(v) If $f(x) = \dfrac{1}{x}$ then neither $f(0)$ exist nor $\lim_{x \to 0} f(x)$ exists.

Example 11. Evaluate
$$\lim_{x \to \frac{1}{2}} \frac{2x^2 + 3x - 2}{2x^2 + 5x - 3}$$

Solution : Let
$$y = \frac{2x^2 + 3x - 2}{2x^2 + 5x - 3}$$

when $x = \dfrac{1}{2}, y = \dfrac{0}{0}$, *i.e.*, indeterminate form, *i.e.*, the function is not defined at $x = \dfrac{1}{2}$.

$$2x^2 + 3x - 2 = 2x^2 + 4x - x - 2 = 2x(x + 2) - 1(x + 2)$$
$$= (2x - 1)(x + 2)$$
$$2x^2 + 5x - 3 = 2x^2 + 6x - x - 3 = 2x(x + 3) - 1(x + 3)$$
$$= (2x - 1)(x + 3)$$

$$\therefore \quad \frac{2x^2 + 3x - 2}{2x^2 + 5x - 3} = \frac{(2x - 1)(x + 2)}{(2x - 1)(x + 3)} = \frac{x + 2}{x + 3} \qquad \text{when } x \neq \frac{1}{2}$$

$$\therefore \quad \lim_{x \to \frac{1}{2}} \frac{2x^2 + 3x - 2}{2x^2 + 5x - 3} = \lim_{x \to \frac{1}{2}} \frac{x + 2}{x + 3} = \frac{5}{7}$$

Example 12. Evaluate the limit
$$\lim_{x \to \frac{1}{3}} \frac{9x^2 - 1}{3x - 1}$$

Solution : The function is not defined at $x = \dfrac{1}{3}$

$$f(x) = \frac{9x^2 - 1}{3x - 1} = \frac{(3x-1)(3x+1)}{(3x-1)} = 3x + 1 \qquad \text{when } x \neq \frac{1}{3},$$

$$\lim_{x \to \frac{1}{3}} \frac{9x^2 - 1}{3x - 1} = \lim_{x \to \frac{1}{3}} (3x + 1) = 2$$

Example 13. Evaluate

$$\lim_{x \to 0} \frac{4x + |x|}{3x - 2|x|}$$

Solution : Let

$$y = \frac{4x + |x|}{3x - 2|x|}$$

when x is positive $|x| = x$

$$\therefore \qquad y = \frac{4x + x}{3x - 2x} = \frac{5x}{x} = 5$$

when x is negative, $|x| = -x$

$$\therefore \qquad y = \frac{4x + |x|}{3x - 2|x|} = \frac{4x - x}{3x + 2x} = \frac{3}{5}$$

$$\text{R.H.L.} = \lim_{x \to 0^+} \frac{4x + |x|}{3x - 2|x|} = \frac{5x}{x} = 5$$

$$\text{L.H.L.} = \lim_{x \to 0^-} \frac{4x + |x|}{3x - 2|x|} = \frac{3x}{5x} = \frac{3}{5}$$

Since, R.H.L. ≠ L.H.L., the given limit does not exist.

Example 14. Evaluate

$$\lim_{x \to 0} \frac{a^x - b^x}{x}.$$

Solution :

$$\lim_{x \to 0} \frac{a^x - b^x}{x} = \lim_{x \to 0} \frac{a^x - b^x + 1 - 1}{x}$$

$$= \lim_{x \to 0} \left[\left(\frac{a^x - 1}{x}\right) - \left(\frac{b^x - 1}{x}\right)\right]$$

$$= \log a - \log b = \log \frac{a}{b}$$

Example 15. Evaluate

$$\lim_{m \to \infty} P\left(1 + \frac{r}{m}\right)^{-mt}$$

Solution :

$$\lim_{m \to \infty} P\left(1 + \frac{r}{m}\right)^{-mt} = \lim_{m \to \infty} P\left[\left[1 + \frac{r}{m}\right]^{\frac{m}{r}}\right]^{-rt}$$

when $m \to \infty$, $\frac{r}{m} \to 0$, Put $\frac{r}{m} = x$

$$\therefore \qquad \lim_{m \to \infty} P\left[\left(1+\frac{r}{m}\right)^{\frac{m}{r}}\right]^{-rt} = P \lim_{x \to 0} [(1+x)^{1/x}]^{-rt} = Pe^{-rt}$$

$$\left[\because \lim_{x \to 0} (1+x)^{1/x} = e \text{ and } \lim_{x \to a} [f(x)]^{1/x} = [\lim_{x \to a} f(x)]^{1/x}\right]$$

Example 16. For the demand function $p = \frac{a}{x+b}$, (where p is price, x is quatity demanded and a and b are positive constants), show by using the concept of limit that the demand increases to infinitely large amount as the price falls. Also, show that total revenue reaches a limiting value as the quantity demaded increases.

Solution :

$$p = \frac{a}{x+b} \Rightarrow p(x+b) = a$$

$$\Rightarrow \qquad px = a - bp \Rightarrow x = \frac{a}{p} - b$$

As the price falls, the quantity $\frac{a}{p}$ increases, therefore, x increases. [$\because$ b is constant]

$$\lim_{p \to 0} \frac{a}{p} - b = \infty, \text{ i.e., } x \text{ increases to infinity as } p \to 0$$

$$p = \frac{a}{x+b}$$

$$\text{Revenue} = px = \frac{ax}{x+b} = \frac{a}{1+\frac{b}{x}}$$

when $x \to \infty$, $\frac{b}{x} \to 0$ and Revenue $\to a$

Note : The above question can also be stated as

For the demand function $x = \frac{a}{p} - b$ (where p is a price, x is quantity demanded and a, b are positive constants). Show, by using the concept of limit, that the demand increases to infinitely large amount as the price falls. Also show that the total revenue reaches a limiting value as the quantity demanded increases.

Example 17. Find $\lim \frac{3x+1}{4x-3}$ as $x \to \infty$. What percentage error is involved by taking $\frac{3}{4}$ as the value of this expression when $x = 10$? How large must x be for the expression to differ from $\frac{3}{4}$ by less than 0.05?

Solution :

$$\lim_{x \to \infty} \frac{3x+1}{4x-3} = \lim_{x \to \infty} \frac{3+\frac{1}{x}}{4-\frac{3}{x}} = \frac{3}{4} = 0.750 \qquad \text{[Dividing the sum and den. by } x\text{]}$$

when $x = 10, \dfrac{3x+1}{4x-3} = \dfrac{31}{37} = 0.838$ (approx.)

$$\text{Error} = \text{Actual value} - \text{Limiting value}$$
$$= 0.838 - 0.750 = 0.888$$
$$\% \text{ error} = \frac{\text{error}}{\text{actual value}} \times 100$$
$$= \frac{0.088}{0.838} \times 100 = \frac{8800}{8}38 = 10.50\% \text{ of actual value.}$$

For the expression $\dfrac{3x+1}{4x-3}$ to differ from 0.750 by less than 0.005, the value of x must be such that

$$\left(\frac{3x+1}{4x-3} - \frac{3}{4}\right) < 0.005$$

$\Rightarrow$ $$\frac{3x+1}{4x-3} < \frac{3}{4} + 0.005 = 0.755$$

$\Rightarrow$ $$3x + 1 < 0.755\,(4x - 3)$$

$\Rightarrow$ $$(3x + 1) < 3.020\,x - 2.265$$

$\Rightarrow$ $$1 + 2.265 < 3.020x - 3x \Rightarrow 0.02x > 3.265 \textbf{ (Pl. note)}$$

$\therefore$ $$x > \frac{3.265}{0.020} = 163.25$$

Thus, x must be greater than 163.25 for the given expression to differ from $\dfrac{3}{4}$ by less than 0.005.

Example 18. Give the function : $y = \dfrac{x^2 + x - 56}{x - 7}$, $(x \neq 7)$, find the left-hand side limit and the right side limit of y as x approaches 7. Can we conclude from these answers that y has a limit as x approaches 7?

Solution : $$y = \frac{x^2 + x - 56}{x - 7} = \frac{(x-7)(x+8)}{x-7}$$
$$= x + 8 \text{ when } x \neq 7$$
$$\text{L.H.L.} = \lim_{x \to 7^-} \left[\frac{x^2 + x - 56}{x - 7}\right] = \lim_{x \to 7^-} (x + 8) = 15$$
$$\text{R.H.L.} = \lim_{x \to 7^+} \left(\frac{x^2 + x - 56}{x - 7}\right) = \lim_{x \to 7^+} (x + 8) = 15$$

Since, R.H.L. = L.H.L., the limit as $x \to 7$ exists. [The limit of a function as x approaches a certain value may exist even if the function is not defined at that point.]

Example 19. A lead shot is allowed to fall in vacuum. It falls $16x^2$ feet in x seconds. Find an expression for the average number of feet fallen per second between the ath and $(a + x)$th second. Find the limit of the expression as $x \to 0$ and give a meaning to this limit.

Solution : In x seconds lead shot falls $16x^2$ feet.

In x seconds lead falls $16x^2$, in a seconds it will fall $16a^2$ feet and in $(a+x)$ seconds lead shot falls $16(a+x)^2$ feet.

Distance travelled between ath and $(a+x)$th seconds

$$= 16(a+x)^2 - 16a^2 = 32ax + 16x^2$$

Average number of feet fallen per second $= \dfrac{\text{Distance travelled}}{\text{Total time taken}}$

$$\frac{32ax - 16x^2}{a+x-a} = \frac{32ax + 16x^2}{x} = 16(x+2a) \text{ ft.}$$

$$\lim_{x\to 0} 16(x+2a) = 16 \times 2a = 32a$$

The limit of the expression indicates the speed of lead shot in seconds and is equal to the gravitational pull.

Example 20. A firm has purchased a machinery for Rs. c being depreciated a r per cent per annum on diminishing balance method. Using limit show that sum of depreciation of all the years is equal to initial cost. (r may be used as in relation interest).

Solution : Depreciation in the first year $= \dfrac{cr}{100}$

Value of the machinery in the beginning of second year

$$= c - \frac{cr}{100}$$

$$= c\left(1 - \frac{r}{100}\right)$$

Depreciation in the second year $= c\left(1 - \dfrac{r}{100}\right)\dfrac{r}{100}$

Value of the machinery in the beginning of third year

$$= c\left(1 - \frac{r}{100}\right) - c\left(1 - \frac{r}{100}\right)\frac{r}{100}$$

$$= c\left(1 - \frac{r}{100}\right)^2$$

Total depreciation during the first n years

$$= \frac{cr}{100} + \left(1 - \frac{r}{100}\right)\frac{cr}{100} + \left(1 - \frac{r}{100}\right)^2 \frac{cr}{100} + \ldots + \left(1 - \frac{r}{100}\right)^{n-1} \frac{cr}{100}$$

$$= \frac{cr}{100}\left[1 + \left(1 - \frac{r}{100}\right) + \left(1 - \frac{r}{100}\right)^2 + \ldots + \left(1 - \frac{r}{100}\right)^{n-1}\right]$$

The series in the bracket is a G.P. with 1 as first term and $\left(1 - \dfrac{r}{100}\right)$ as common ratio.

∴ Total depreciation during first n years

$$= \frac{cr}{100}\left[\frac{1-\left(1-\frac{r}{100}\right)^n}{1-\left(1-\frac{r}{100}\right)}\right]$$

$$= c\left[1-\left(1-\frac{r}{100}\right)^n\right]$$

$$\lim_{n\to\infty}\left[1-\left(1-\frac{r}{100}\right)^n\right] = c \text{ because } \left(1-\frac{r}{100}\right) < 1 \text{ and therefore } \lim_{x\to\infty}\left(1-\frac{r}{100}\right)^n = 0$$

Example 21. It is known that income Y_t is growing with time t and is given by $Y_t = a + bt + ct^2$. The population N_t is known to be increasing with t and is given by $N_t = p + qt + rt^2$. Find the limiting value of per capita income as t tends to infinity.

Solution : Let I_t = per capita income at time t.

Then $$I_t = \frac{\text{Income}}{\text{Population}} = Y_t/N_t$$

$$= \frac{a+bt+ct^2}{p+qt+rt^2},$$ we have to find $\lim_{x\to\infty} I_t$.

Dividing both the numerator and the denominator by the highest power of t in the fraction, *i.e.*, t^2, we get

$$\lim_{t\to\infty} I_t = \lim_{t\to\infty}\frac{Y_t}{N_t} = \lim_{t\to\infty}\left[\frac{a+bt+ct^2}{p+qt+rt^2}\right]$$

$$= \lim_{t\to\infty}\left[\frac{(a/t^2)+(b/t)+c}{(p/t^2)+(q/t)+r}\right] = \frac{c}{r}$$

Example 22. Evaluate $\lim_{x\to 0}\frac{e^{1/x}-1}{e^{1/x}+1}$

Solution :

$$\text{R.H.L.} = \lim_{x\to 0+}\frac{e^{1/x}-1}{e^{1/x}+1} = \lim_{x\to 0+}\frac{1-\frac{1}{e^{1/x}}}{1+\frac{1}{e^{1/x}}} = 1$$

$$\text{L.H.L.} = \lim_{x\to 0-}\frac{e^{1/x}-1}{e^{1/x}+1} = \lim_{h\to 0}\frac{e^{1/0-h}-1}{e^{1/0-h}+1}$$

$$= \lim_{h\to 0}\frac{e^{-1/h}-1}{e^{-1/h}+1}$$

$$= \lim_{h \to 0} \frac{\frac{1}{e^{1/h}} - 1}{\frac{1}{e^{1/h}} + 1} = -1$$

R.H.L. ≠ L.H.L. ⇒ the given limit does not exist.

Example 23. Evaluate

$$\lim_{x \to 1} \frac{x^2 - 1}{\sqrt{3x+1} - \sqrt{5x-1}}$$

Solution : Let $f(x) = \frac{x^2 - 1}{\sqrt{3x+1} - \sqrt{5x-1}}$, the function is not defined at $x = 1$

$$f(x) = \frac{x^2 - 1}{\sqrt{3x+1} - \sqrt{5x-1}} \left[\frac{\sqrt{3x+1} + \sqrt{5x-1}}{\sqrt{3x+1} + \sqrt{5x-1}}\right]$$

$$= \frac{(x^2 - 1)[\sqrt{3x+1} + \sqrt{5x-1}]}{3x + 1 - 5x + 1}$$

$$= \frac{(x-1)(x+1)[\sqrt{3x+1} + \sqrt{5x-1}]}{-2(x-1)}$$

$$= \frac{(x+1)}{-2}[\sqrt{3x+1} + \sqrt{5x-1}] \qquad \text{when } x \neq 1$$

$$\lim_{x \to 1} f(x) = \lim_{x \to 1} \left(\frac{x+1}{-2}\right)(\sqrt{3x+1} + \sqrt{5x-1}) = -4$$

Example 24. Evaluate

$$\lim_{x \to 0} \frac{1}{x}[(1+x)^3 - 1],$$

Generalise your result by finding

$$\lim_{x \to 0} \frac{1}{x}[(a+x)^3 - a^3]$$

Solution :
$$\frac{1}{x}[(1+x)^3 - 1] = \frac{1}{x}[1 + x^3 + 3x + 3x^2 - 1] = (x^2 + 3 + 3x)$$

$$\therefore \quad \lim_{x \to 0} \frac{1}{x}[(1+x)^3 - 1] = \lim_{x \to 0} (x^2 + 3x + 3) = 3 \qquad \text{...(I)}$$

Now
$$\lim_{x \to 0} \frac{1}{x}[(a+x)^3 - a^3] = \lim_{x \to 0} \frac{1}{x}[a^3 + x^3 + 3ax^2 + 3a^2x - a^3]$$

$$= \lim_{x \to 0} [x^2 + 3ax + 3a^2] = 3a^2 \qquad \text{...(II)}$$

If we put $a = 1$ in (II) we get (I)

Example 25. Show that $\lim_{x \to 1} \frac{x^3 - 1}{x - 1} = 3.$

Solution : $\dfrac{x^3-1}{x-1} = \dfrac{(x-1)(x^2+x+1)}{x-1} = x^2+x+1$ when $x \neq 1$

$$\lim_{x \to 1} \frac{x^3-1}{x-1} = \lim_{x \to 1} (x^2+x+1) = 3$$

Example 26. Use the formula

$$\lim_{x \to 0} \frac{a^x-1}{x} = \log_e a$$

to show that $\displaystyle\lim_{x \to 0} \left(\frac{2^x-1}{(1+x)^{1/2}-1} \right) = 2 \log 2$

Solution : $\dfrac{2^x-1}{\sqrt{1+x}-1} = \dfrac{(2^x-1)[\sqrt{1+x}+1]}{[\sqrt{(1+x)}-1][\sqrt{1+x}+1]}$

$$= \frac{2^x-1}{x} [\sqrt{1+x}+1]$$

$$\lim_{x \to 0} \frac{(2^x-1)}{\sqrt{1+x}-1} = \lim_{x \to 0} \frac{2^x-1}{x} \lim_{x \to 0} (\sqrt{1+x}+1)$$

$$= \lim_{x \to 0} [\sqrt{1+x}+1].\lim_{x \to 0} \frac{2^x-1}{x} = 2 \log_e 2$$

Example 27. Does the limit

$$\lim_{x \to 0} \frac{|x|}{x} \text{ exist? Explain.}$$

Solution : The function $f(x) = \dfrac{|x|}{x}$ is not defined at $x = 0$. In the deleted neighbourhood of $x = 0$, when $x > 0$, $\dfrac{|x|}{x}$ is equal to $\dfrac{x}{x} = 1$ and when $x < 0$, $\dfrac{|x|}{x} = \dfrac{-x}{x} = -1$

$\therefore$ Right hand limit $= \displaystyle\lim_{x \to 0^+} f(x) = \lim_{x \to 0} \frac{|x|}{x} = 1$

and left hand limit $= \displaystyle\lim_{x \to 0^-} f(x) = \lim_{x \to 0^-} \frac{|x|}{x} = -1$

Since, R.H.L. $\neq$ L.H.L. $\displaystyle\lim_{x \to 0} \frac{|x|}{x}$ does not exist.

Example 28. If $[x]$ = largest integer $\leq x$,

Evaluate ; (i) $\displaystyle\lim_{x \to 2^+} \{x-[x]\}$, (ii) $\displaystyle\lim_{x \to 2^-} \{x-[x]\}$ and give the conclusion.

Or

Does the $\displaystyle\lim_{x \to 2} \{x-[x]\}$ exist? Explain your answer.

Solution : (Note the solution carefully)

Now $[x]$ is given to be the largest integer less than or equal to x, therefore, when x approaches 2 from the right hand side $[x]$ will be equal to 2 and when x approaches 2 from the left hand side $[x]$, it will be equal to 1 ($\because$ when x approaches 2 from the left hand side x always remains less than 2 and $[x] = 1$), when $x \to 2^+$ (*i.e.,* when x approaches 2 from the right hand side) the difference $\{x - [x]\}$ can be made as near to zero as we like and we can say

$$\lim_{x \to 2^+} \{x - [x]\} = 0$$

When x approaches 2 from left hand side, the difference $\{x - [x]\}$ can be made as near to 1 as we like and we can say

$$\lim_{x \to 2^-} \{x - [x]\} = 1$$

Since, L.H.L. $\neq$ R.H.L. $\lim_{x \to 2} \{x - [x]\}$ does not exist.

Example 29. Find $\lim_{n \to \infty} \left[\frac{1}{n^3}(1^2 + 2^2 + 3^2 + \ldots + n^2)\right]$

Solution :
$$\lim_{n \to \infty} \left[\frac{1}{n^3}(1^2 + 2^2 + 3^2 + \ldots + n^2)\right] = \lim_{x \to \infty} \frac{n(n+1)(2n+1)}{6n^3}$$

$$= \lim_{n \to \infty} \frac{n(n+1)(2n+1)}{6n^2} = \lim_{n \to \infty} \left[\frac{2n^2}{6n^2} + \frac{3n}{6n^2} + \frac{1}{6n^2}\right] = \frac{1}{3}$$

$$\left[\because \text{ when } n \to \infty, \frac{1}{n} \to 0\right]$$

Example 30. Evaluate $\lim_{h \to 0} \frac{\sqrt[3]{8+h} - 2}{h}$

Solution : Put $8 + h = x^3 \Rightarrow h = x^3 - 8$

$$\sqrt[3]{8+h} = x, \text{ when } h \to 0, x \to 2$$

$$\therefore \quad \lim_{h \to 0} \frac{\sqrt[3]{8+h} - 2}{h} = \lim_{x \to 0} \frac{x-2}{x^3 - 8} = \lim_{x \to 2} \frac{1}{x^2 + 2x + 4} = \frac{1}{12} \quad \text{(Pl. note)}$$

$$[\because x^3 - 8 = (x-2)(x^2 + 2x + 4)]$$

Second Method

$$\sqrt[3]{8+h} = (8+h)^{1/3} = 2\left(1 + \frac{h}{8}\right)^{1/3}$$

Apply Binomial theorem.

Example 31. Evaluate

$$\frac{1 - \sqrt{1 - x^2}}{x^2}$$

Solution : Let $f(x) = \dfrac{1-\sqrt{1-x^2}}{x^2}$ which is not defined at $x = 0$

$$f(x) = \frac{1-\sqrt{1-x^2}}{x^2}\left[\frac{1+\sqrt{1-x^2}}{1+\sqrt{1-x^2}}\right]$$

$$= \frac{1-(1-x^2)}{x^2[1+\sqrt{1-x^2}]} = \frac{1}{1+\sqrt{1-x^2}} \qquad \text{when } x \neq 0$$

$$\therefore \qquad \lim_{x \to 0} f(x) = \frac{1}{2}$$

Example 32. Obtain $\lim\limits_{x \to 0} \dfrac{10^x - 2^x - 5^x + 1}{x^2}$

Solution :

$$\lim_{x \to 0} \frac{10^x - 2^x - 5^x + 1}{x^2} = \lim_{x \to 0} \frac{5^x \times 2^x - 2^x - 5^x + 1}{x^2}$$

$$= \lim_{x \to 0} \frac{2^x[5^x - 1] - 1[5^x - 1]}{x^2}$$

$$= \lim_{x \to 0}\left[\left(\frac{2^x - 1}{x}\right) \times \left(\frac{5^x - 1}{x}\right)\right]$$

$$= \lim_{x \to 0} \frac{2^x - 1}{x} \times \lim_{x \to 0} \frac{5^x - 1}{x} = \log_e 2 \times \log_e 5$$

Example 33. Evaluate $\lim\limits_{x \to \infty} \sqrt{x}\,(\sqrt{x+4} - \sqrt{x})$

Solution :

$$\sqrt{x+4} - \sqrt{x} = \frac{(\sqrt{x+4} - \sqrt{x})(\sqrt{x+4} + \sqrt{x})}{\sqrt{x+4} + \sqrt{x}}$$

$$= \frac{4}{\sqrt{x+4} + \sqrt{x}}$$

$$\lim_{x \to \infty} \sqrt{x}\,(\sqrt{x+4} - \sqrt{x}) = \lim_{x \to \infty} \frac{4\sqrt{x}}{\sqrt{x+4} + \sqrt{x}}$$

$$= \lim_{x \to \infty} \frac{4}{\sqrt{1 + \frac{4}{x}} + 1} = \frac{4}{2} = 2 \qquad \left[\because \text{ when } x \to \infty, \frac{4}{x} \to 0\right]$$

Example 34. Evaluate $\lim\limits_{x \to -2} \dfrac{x^3 + 8}{|x| - 2}$

Solution : Whether x approaches -2 from right or from left $|x| = -x$

$$\therefore \qquad \lim_{x \to -2} \frac{x^3 + 8}{|x| - 2} = \lim_{x \to -2} \frac{x^3 + 8}{-x - 2} \qquad \textbf{(Pl. note)}$$

$$= \lim_{x \to -2} \frac{(x+2)(x^2 - 2x + 4)}{-(x+2)}$$

$$= \lim_{x \to -2} -(x^2 - 2x + 4) = -12$$

Example 35. Find $\lim_{x \to 0} e^x$

Solution :

$$e^x = \left(\frac{e^x - 1 + 1}{x}\right)x \quad \textbf{(Pl. note)}$$

$$= \left(\frac{e^x - 1}{x}\right)x + 1$$

$$\lim_{x \to 0} e^x = \lim_{x \to 0} \left(\frac{e^x - 1}{x}\right)x + 1 = 1 \qquad \left[\because \lim_{x \to 0} \frac{e^x - 1}{x} = 1\right]$$

Example 36. If $\lim_{t \to 2} \frac{mt^2 - n}{t - 2} = 4$, find m and n.

Solution :

$$\frac{mt^2 - n}{t - 2} = (mt + 2m) + \frac{4m - n}{t - 2} \qquad \text{[By actual Division]}$$

$$\lim_{t \to 2} \left(\frac{mt^2 - n}{t - 2}\right) = \lim_{t \to 2} (mt + 2m) + \lim_{t \to 2} \left(\frac{4m - n}{t - 2}\right) \qquad ...(1)$$

When $t \to 2$, $\frac{4m - n}{t - 2} \to \infty$. But the limit on the left hand side of (I) is a finite limit, therefore $\frac{4m - n}{t - 2}$ can be taken to be equal to zero which $\Rightarrow n = 4m$.

When $n = 4m$, $\lim_{t \to 2} \left(\frac{mt^2 - n}{t - 2}\right) = \lim_{t \to 2} (mt + 2m) = 4m = 4 \Rightarrow m = 1$ (given)

$$\therefore \quad m = 1, n = 4 \qquad [\because n = 4m]$$

Example 37. Evaluate $\lim_{h \to 0} \frac{\sqrt{x+h} - \sqrt{x}}{h}$

Solution :

$$\frac{\sqrt{x+h} - \sqrt{x}}{h} = \frac{\sqrt{x+h} - \sqrt{x}}{h} \times \left[\frac{\sqrt{x+h} + \sqrt{x}}{\sqrt{x+h} + \sqrt{x}}\right]$$

$$= \frac{x + h - x}{h(\sqrt{x+h} + \sqrt{x})} = \frac{1}{\sqrt{x+h} + \sqrt{x}}$$

$$\therefore \quad \lim_{h \to 0} \frac{\sqrt{x+h} - \sqrt{x}}{h} = \lim_{h \to 0} \frac{1}{\sqrt{x+h} + \sqrt{x}} = \frac{1}{2\sqrt{x}}$$

2nd Method

By Binomial theorem

$$\sqrt{x+h} = \sqrt{x}\left(1 + \frac{h}{x}\right)^{1/2} = \sqrt{x}\left(1 + \frac{h}{2x}\right)$$

[neglecting higher powers of h]

$$\lim_{h \to 0} \frac{\sqrt{x+h} - \sqrt{x}}{h} = \lim_{h \to 0} \frac{\sqrt{x}\left[1 + \dfrac{h}{2x}\right] - \sqrt{x}}{h} = \frac{1}{2\sqrt{x}}$$

Example 38. Find the $\lim_{x \to 1} \frac{x-1}{\sqrt{x-1} + \sqrt{x+1}}$

$$\text{R.H.L.} = \lim_{x \to 1^+} \frac{x-1}{\sqrt{x-1} + \sqrt{x+1}} = 0$$

$$\text{R.H.L.} = \lim_{x \to 1^-} \frac{x-1}{\sqrt{x-1} + \sqrt{x+1}}$$

$$= \frac{\text{A negative quantity}}{\text{an imaginary quantity + a positive quantity}}$$

∴ L.H.L. does not exist.

i.e., R.H.L. ≠ L.H.L. Hence the limit does not exist.

Example 39. Show that

$$\lim_{x \to 4}\left[\frac{1}{\log(x-3)} - \frac{1}{x-4}\right] = \frac{1}{2}$$

Solution :

$$\lim_{x \to 4}\left[\frac{1}{\log(x-3)} - \frac{1}{x-4}\right]$$

$$= \lim_{x \to 4} \frac{(x-4) - \log(x-3)}{[\log(x-3)](x-4)} = \lim_{x \to 4} \frac{f(x)}{g(x)}$$

By L-Hospital's Rule

$$\lim_{x \to 4} \frac{f(x)}{g(x)} = \lim_{x \to 4} \frac{f'(x)}{g'(x)} = \lim_{x \to 4} \frac{f''(x)}{g''(x)} = \ldots$$

$$\frac{f'(x)}{g'(x)} = \frac{1 - \dfrac{1}{(x-3)}}{\log(x-3) + \dfrac{x-4}{x-3}} = \frac{x-4}{(x-3)\log(x-3) + x - 4}$$

when $x \to 4$, $\frac{f'(x)}{g'(x)} \to \frac{0}{0}$

$$\frac{f''(x)}{g''(x)} = \frac{1}{\dfrac{(x-3)}{x-3} + \log(x-3) + 1}$$

when $x \to 4$, $\frac{f''(x)}{g''(x)} \to \frac{1}{2}$ [∵ $\log 1 = 0$]

∴ The required limit $= \frac{1}{2}$

Example 40. Find $\lim_{x \to 0} \frac{\sqrt{(2+x^2)} - \sqrt{(2-x^2)}}{x^2}$.

Solution : $\lim_{x \to 0} \dfrac{\sqrt{(2+x^2)} - \sqrt{(2-x^2)}}{x^2}$

$$= \lim_{x \to 0} \frac{\sqrt{(2+x^2)} - \sqrt{(2-x^2)}}{x^2} \times \left(\frac{\sqrt{2+x^2} + \sqrt{2-x^2}}{\sqrt{2+x^2} + \sqrt{2-x^2}} \right)$$

$$= \lim_{x \to 0} \frac{2x^2}{x^2 [\sqrt{2+x^2} + \sqrt{2-x^2}]}$$

$$= \lim_{x \to 0} \frac{2}{\sqrt{2+x^2} + \sqrt{2-x^2}} = \frac{2}{2\sqrt{2}} = \frac{1}{\sqrt{2}}$$

$\therefore$ The required limit $= \dfrac{1}{\sqrt{2}}$.

Example 41. Find the limit as $x \to 1$ of

$$f(x) = \frac{x^3 - 3x + 2}{x^3 + x^2 - 5x + 3}$$

Solution :

$$x^3 - 3x + 2 = x^3 - 4x + x + 2$$
$$= x(x^2 - 4) + 1(x + 2)$$
$$= [x(x-2)(x+2) + 1(x+2)]$$
$$= (x+2)[x(x-2) + 1]$$
$$= (x+2)(x^2 - 2x + 1) = (x+2)(x-1)^2$$

$$x^3 + x^2 - 5x + 3 = (x^3 - x) + (x^2 - x) - 3x + 3$$
$$= x(x^2 - 1) + x(x-1) - 3(x-1)$$
$$= x(x-1)(x+1) + x(x-1) - 3(x-1)$$
$$= (x-1)[x(x+1) + x - 3]$$
$$= (x-1)[x^2 + x + x - 3]$$
$$= (x-1)(x^2 + 2x - 3)$$
$$= (x-1)(x^2 + 3x - x - 3) = (x-1)[x(x+3) - 1(x+3)]$$
$$= (x-1)(x-1)(x+3)$$
$$= (x-1)^2(x+3)$$

$$\therefore \quad \lim_{x \to 1} \frac{x^3 - 3x + 2}{x^3 + x^2 - 5x + 3} = \lim_{x \to 1} \frac{(x-1)^2(x+2)}{(x-1)^2(x+3)} = \lim_{x \to 1} \frac{x+2}{x+3} = \frac{3}{4}$$

Second Method [By L-Hospital's rule]

$$\lim_{x \to 1} \frac{x^3 - 3x + 2}{x^3 + x^2 - 5x + 3} = \lim_{x \to 1} \frac{\frac{d}{x}(x^3 - 3x + 2)}{\frac{d}{dx}(x^3 + x^2 - 5x + 3)}$$

$$= \lim_{x \to 1} \left[\frac{3x^2 - 3}{3x^2 + 2x - 5} \right]$$

$$= \lim_{x \to 1} \left[\frac{\frac{d}{dx}(3x^2 - 3)}{\frac{d}{dx}(3x^2 + 2x - 5} \right]$$

$$= \lim_{x \to 1} \frac{6x}{6x + 2} = \frac{3}{4}$$

3.15 CONTINUITY

A function $f(x)$ is said to be continuous at $x = a$, if

(i) $f(x)$ is defined as $x = a$, *i.e.*, $f(a)$ exists, and

(ii) $\lim_{x \to a} f(x)$ exists, and

(iii) $\lim_{x \to a} f(x) = f(a)$, *i.e.*, the value of the function at $x = a$ coincides with the limit of the function as $x \to a$.

For a function $f(x)$ to be continuous at $x = a$, the above three conditions should be simultaneously satisfied.

Note : $f(x)$ is defined at $x = a$ means the function $f(x)$ has a finite value at $x = a$. The function $f(x)$ ceases to be continuous for those values of x which make $f(x)$ infinite.

Points where $f(x)$ fails to be continuous are called discontinuities of $f(x)$ and $f(x)$ is said to be discontinuous at these points.

A function $f(x)$ is said to be **continuous** in an open interval $]a, b[$ if it is continuous at every point in the interval but if the domain of definition of the function is a closed interval condition (ii) fails at end points ($\because$ the function will not be defined at any point on the left of a and right of b) and the function is said to be continuous in the closed interval $[a, b]$ if

(i) It is continuous in the open interval $a < x < b$

(ii) $\lim_{x \to a^+} f(x) = f(a)$, and (ii) $\lim_{x \to b^-} f(x) = f(b)$.

Properties of Continuous Functions

1. A polynomial function is continuous everywhere.
2. The sum, difference and product of any finite number of continuous functions are continuous.
3. The quotient of two functions which are continuous at a point is continuous at that point, provided the denominator is not zero at that point.
4. The *n*th root of a polynomial function is continuous whenever it is defined.
5. The exponential and the logarithmic functions are continuous in their respective domains.

Example 42. Examine the continuity of the function

$$f(x) = |x| + |x - 1| + |x - 2| \text{ at } x = 2.$$

Solution : When $x = 2$

$$f(2) = 2 + 1 + 0 = 3 \qquad \text{(By actual calculations)}$$

i.e., the function is defined at $x = 2$.

On the right hand side, *i.e.,* when $x > 2$

$$f(x) = x + x - 1 + x - 2 = 3x - 3$$

and on the left hand side, *i.e.,* when $x < 2$,

$$|x| = x, |x - 1| = x - 1, \text{ and } |x - 2| = -(x - 2).$$

$\therefore$ when $x < 2$

$$f(x) = x + x - 1 - x + 2 = x + 1$$

Now $$\lim_{x \to 2^+} f(x) = \lim_{x \to 2^+} (3x - 3) = 3$$

and $$\lim_{x \to 2^-} f(x) = \lim_{x \to 2^-} (x + 1) = 3$$

Right hand limit = Left hand limit = value of the function at $x = 2$.

$\therefore$ The function $f(x) = |x| + |x - 1| + |x - 2|$ is continuous at $x = 2$.

Example 43. A function is defined as below

$$\begin{aligned} f(x) &= x && \text{when } 0 < x < 1 \\ &= 2 - x && \text{when } x \geq 1. \end{aligned}$$

Show that it is continuous at $x = 1$.

Solution : Step I. The function is defined at $x = 1$ and its value $= 2 - 1 = 1$, *i.e.,* $f(1) = 1$.

Step II. On the right side, *i.e.,* when $x > 1$

$$f(x) = 2 - x$$

$$\lim_{x \to 1^+} f(x) = \lim_{x \to 1^+} (2 - x) = 1$$

Step III. On the left side, *i.e.,* when $x < 1$

$$f(x) = x$$

$$\lim_{x \to 1^-} f(x) = \lim_{x \to 1^-} x = 1$$

i.e., Right hand limit = Left hand limit = value of the function at $x = 1$

Therefore the function is continuous at $x = 1$.

Example 44. If $[x]$ = greatest integer $\leq x$,, investigate the continuity of $f(x) = x - [x]$ in the interval :

(a) $1 < x < 2$, (b) $1 \leq x \leq 2$.

Solution : (a) For every x in the interval $1 < x < 2$

$$[x] = 1, \text{ and } x - [x] = x - 1$$

If c be any point in the given interval, then

$$f(c) = \lim_{x \to c+} f(x) = c - 1$$

i.e., $f(x)$ is continuous in the interval $1 < x < 2$.

(b) In the interval $1 \leq x \leq 2$, the given function will be continuous if

(i) $f(x)$ is continuous in $1 < x < 2$

(ii) $\lim_{x \to 1^+} f(x) = f(1)$

(iii) $\lim_{x \to 2^-} f(x) = f(2)$

It has already been proved that $f(x)$ is continuous in the interval

$$1 < x < 2$$

Now
$$\lim_{x \to 1^+} f(x) = 0 = f(1)$$

and
$$\lim_{x \to 2^-} f(x) = 2 - 1 = 1 \neq f(2)$$

[Because at $x = 2$, $[x] = 2$ and $f(x) = 2 - 2 = 0$ and when $x \to 2$ from left hand side $[x]$ will remain equal to 1]

Hence, the function is not continuous in the closed interval [1, 2].

Example 45. Discuss the continuity of

$$f(x) = \frac{|x|}{x} \text{ at } x = 0.$$

Solution :
$$f(x) = \frac{|x|}{x}$$

The function is not defined at $x = 0$ and when x approaches zero from right hand side. Then,

$$\frac{|x|}{x} = 1 \qquad \therefore \text{ R.H.L.} = \lim_{x \to 0^+} \frac{|x|}{x} = 1$$

and when x approaches zero from left hand side. Then

$$\frac{|x|}{x} = -1, \text{ i.e., L.H.L.} = \lim_{x \to 0^-} \frac{|x|}{x} = -1$$

Hence, the function is discontinuous at $x = 0$.

Example 46. Discuss the continuity of the function

$$f(x) = \begin{bmatrix} 5x - 4 & \text{when } 0 < x \le 1 \\ 4x^2 - 3x & \text{when } 1 < x < 2 \\ 3x - 4 & \text{when } x \ge 2 \end{bmatrix}$$

at points $x = 1$ and 2.

Solution : Consider the point $x = 1$

$$f(1) = 5 - 4 = 1$$

$$\text{RHL} = \lim_{x \to 1^+} f(x) = \lim_{x \to 1^+} (4x^2 - 3x) = 1$$

$$\text{LHL} = \lim_{x \to 1^-} f(x) = \lim_{x \to 1^-} (5x - 4) = 1$$

$\therefore f(1) = \lim_{x \to 1} f(x)$. Hence, the function is continuous at $x = 1$.

Consider the point $x = 2$

$$f(x) = 3x - 4$$

$$f(2) = 3 \times 2 - 4 = 2$$

$$\text{RHL} = \lim_{x \to 2^+} f(x) = \lim_{x \to 2^+} (3x - 4) = 2$$

$$\text{LHL} = \lim_{x \to 2^-} f(x) = \lim_{x \to 2^-} (4x^2 - 3x) = 10$$

LHL $\neq$ RHL. Hence, the function is not continuous at $x = 2$.

Example 47. Discuss the continuity at $x = 2$ of the function

$$\begin{aligned} f(x) &= 5x - 4 && : \quad 0 < 1 \leq 1 \\ &= 4x^2 - 3x && : \quad 1 < x < 2 \\ &= 3x + 4 && : \quad x \geq 2 \end{aligned}$$

Solution : $f(2)$ = value of the function at $x = 2$

$$= 3 \times 2 + 4 = 10$$

$$\text{L.H.L.} = \lim_{x \to 2^-} f(x) = \lim_{x \to 2^-} 4x^2 - 3x = 10$$

$$\text{R.H.L.} = \lim_{x \to 2^+} f(x) = \lim_{x \to 2^+} (3x + 4) = 10$$

$\Rightarrow$ L.H.S. = R.H.L. = Value of the function at $x = 2$

$\Rightarrow$ The given function is continuous at $x = 2$.

Example 48. Show that the function $f(x)$ as defined below is discontinuous at $x = \frac{1}{2}$.

$$\begin{aligned} f(x) &= x && \text{when } 0 \leq x < \frac{1}{2} \\ &= 1 && \text{when } x = \frac{1}{2} \\ &= 1 - x && \text{when } \frac{1}{2} < x < 1. \end{aligned}$$

Solution : Please note that for taking right hand limit we have to take $f(x) = 1 - x$ and for taking left hand limit we have to take $f(x) = x$.

Now $$f\left(\frac{1}{2}\right) = 1$$ [Given]

Right hand limit

$$\lim_{x \to \frac{1}{2}+} f(x) - \lim_{h \to 0} \left[1 - \left(\frac{1}{2} + h\right)\right] = \lim_{h \to 0} \left(\frac{1}{2} - h\right) = \frac{1}{2}$$

Left hand limit

$$\lim_{x \to \frac{1}{2}-} f(x) - \lim_{x \to \frac{1}{2}-} x - \lim_{h \to 0} \left(\frac{1}{2} + h\right) - \frac{1}{2}$$

Though Left hand limit = Right hand limit but each is not equal to the value of the function at $x - \frac{1}{2}$. Hence, the given function is not continuous at $x - \frac{1}{2}$. **(Please note)**

Example 49. Suppose $f(x) = \begin{cases} 2, & x \le 0 \\ 3x+2, & 0 < x \le 1 \\ \dfrac{x}{x-1}, & 1 < x \end{cases}$

Is $f(x)$ continuous at $x = 1$ Give reasons.

Solution :

$$\text{LHL} = \lim_{x \to 1^-} f(x) = \lim_{x \to 1^-} (3x+2) = 5$$

$$\text{RHL} = \lim_{x \to 1^+} f(x) = \lim_{x \to 1^+} \frac{x}{x-1} = \infty$$

RHL $\ne$ LHL $\Rightarrow \lim\limits_{x \to 1} f(x)$ does not exist. Hence, the given function is not continuous at $x - 1$.

Example 50. A function $f(x)$ is defined as follows :

$$\begin{aligned} f(x) &= \frac{9x}{x+2} && \text{for } x < 1 \\ &= 3 && \text{for } x = 1, \\ &= \frac{x+3}{x} && \text{for } x > 1 \end{aligned}$$

Examine the continuity of $f(x)$ in the interval $(-3, 3)$.

Solution : [Pl. note the solution carefully]

When $x = -2, x + 2$ becomes equal to zero and the value of the function $f(x) = \dfrac{9x}{x+2}$ goes to infinity. Hence, the given function is discontinuous at $x = -2$.

At $x = 1, f(x) = 3$

Left hand limit

$$\lim_{x \to 1^-} f(x) = \lim_{h \to 0} \frac{9(1-h)}{(1-h)+2} = 3$$

Right hand limit

$$\lim_{x \to 1^+} f(x) = \lim_{h \to 0} \frac{(1+h)+3}{1+h} = 4$$

i.e., L.H.L. $\ne$ R.H.L.

Hence, the given function is not continuous at $x = 1$.

Since, the values -2, 1 belong to the given interval, we can say that the given function is not continuous in the interval $(-3, 3)$.

Example 51. Show that $f(x) = 3x^2 + 2x - 1$ is continuous at $x = 2$. Hence, prove that $f(x)$ is continuous for all values of x.

Solution : $f(2) = 3 \times 2^2 + 2 \times 2 - 1 = 15$

Right hand limit

$$\lim_{x \to 2^+} (3x^2 + 2x - 1) = \lim_{h \to 0} [3(2 + h)^2 + 2(2 + h) - 1]$$

$$= \lim_{h \to 0} (3h^2 + 14h + 15) = 15 \qquad ...(II)$$

Left hand limit

$$\lim_{x \to 2^-} (3x^2 + 2x - 1) = \lim_{h \to 0} [3(2 - h)^2 + 2(2 - h) - 1]$$

$$= \lim_{h \to 0} (3h^2 - 14h + 15) = 15 \qquad ...(III)$$

Since, $f(2)$ = Right hand limit = Left hand limit, the given function is continuous at $x = 2$.

To prove that the function $f(x) = 3x^2 + 2x - 1$ is continuous for all values of x, we will examine the behaviour of the function at any value $x = k$ arbitrarily chosen.

Now, $$f(k) = 3k^2 + 2k - 1 \qquad ...(IV)$$

$$\lim_{x \to k^+} f(x) = \lim_{h \to 0} [3(k + h)^2 + 2(k + h) - 1]$$

$$= \lim_{h \to 0} (3k^2 + 3h^2 + 6kh + 2k + 2h - 1) \qquad ...(V)$$

$$= 3k^2 + 2k - 1$$

$$\lim_{x \to k^-} f(x) = \lim_{h \to 0} [3(k - h)^2 + 2(k - h) - 1] = 3k^2 + 2k - 1 \qquad ...(VI)$$

From (IV), (V) and (VI) we find that the given function is continuous as $x = k$, but k is an arbitrarily chosen value of x. Hence, the function $f(x) = 3x^2 + 2x - 1$ is continuous for all values of x.

[Please note the second part of solution carefully]

Example 52. Show that $f(x) = \frac{e^{-1/x}}{1 + e^{1/x}}$, when $x \neq 0$

$f(x) = 0$, when $x = 0$,

is discontinuous at $x = 0$

Solution : [Please note the solution carefully]

$$f(0) = 0$$

Right hand limit

$$\lim_{x \to 0^+} f(x) = \lim_{h \to 0} \frac{e^{-1/(0 + h)}}{1 + e^{1/(0 + h)}} = \lim_{h \to 0} \frac{e^{-1/h}}{1 + e^{1/h}}$$

When $h \to 0$, $\frac{1}{h} \to \infty$ and $e^{1/h} \to \infty$ and $e^{-1/h} = e^{1/h} \to 0$

$$\therefore \qquad \lim_{x \to 0^+} f(x) = 0$$

Left hand limit

$$\lim_{x \to 0^-} f(x) = \lim_{h \to 0} \frac{e^{-1/0-h}}{1+e^{1/0-h}} = \lim_{h \to 0} \frac{e^{1/h}}{1+e^{-1/h}} = \infty$$

i.e., Right hand limit ≠ Left hand limit.

Hence, the given function is discontinuous at $x = 0$.

Example 53. Examine the continuity at $x = 0$ of the function $f(x)$ defined as under :

$$f(x) = \frac{x}{1+e^{1/x}} \; x \neq 0 = 0 \; x - 0$$

Solution : R.H.L. $\lim_{x \to 0^+} f(x) = \lim_{h \to 0} \frac{0+h}{1+e^{1/h}} = \lim_{h \to 0} \frac{h}{1+e^{1/h}} = 0$

L.H.L. $\lim_{h \to 0^-} f(x) = \lim_{h \to 0} \frac{0-h}{1+e^{-1/h}} = \lim_{h \to 0} \frac{-h}{1+\frac{1}{e^{1/h}}} = 0$

R.H.L. = L.H.L. = 0

Hence, the given function is continuous at $x = 0$.

Example 54. A function is defined as under

$$f(x) = \frac{e^{1/x^2}}{e^{1/x^2} - 1} \quad \text{when } x \neq 0$$

$$= 1 \quad \text{when } x = 0$$

Examine the continuity at $x = 0$

Solution : R.H.L. $= \lim_{x \to 0^+} \frac{e^{1/x^2}}{e^{1/x^2} - 1} = \lim_{h \to 0^+} \frac{e^{1/h^2}}{e^{1/h^2} - 1} = \lim_{h \to 0^+} \frac{1}{1 - \frac{1}{e^{1/h^2}}} = 1$

L.H.L. $= \lim_{x \to 0^-} f(x) = \lim_{h \to 0^-} \frac{e^{1/h^2}}{e^{1/h^2} - 1} = \lim_{x \to 0^-} \frac{1}{1 - \frac{1}{e^{1/h^2}}} = 1$

R.H.L. = L.H.L. = value of the function at $x = 0$

∴ The function is continuous at $x = 0$

Example 55. Given $y = f(x) = \frac{x^2 - x - 12}{x - 4}$

(i) Evaluate the limit of y as $x \to 4$.

(ii) Is this function continuous at $x = 4$? Why ?

Solution : $x^2 - x - 12 = (x - 4)(x + 3)$

$$y = \frac{(x-4)(x+3)}{x-4} = x + 3 \qquad \text{when } x \neq 4$$

$$\therefore \quad \lim_{x \to 4} \frac{x^2 - x - 12}{x - 4} = \lim_{x \to 4} (x + 3) = 7$$

The function is not continuous at $x = 4$ because the function is not defined at $x = 4$. However, if the function is defined as under, then it will become continuous at $x = 4$.

$$y = \frac{x^2 - x - 12}{x - 4} \qquad \text{when } x \neq 4$$
$$= 7 \qquad \text{when } x = 4.$$

Example 56. Find the points of discontinuity of the function

$$\frac{2x^2 + 6x - 5}{12x^2 + x - 20}$$

Solution : The given function will be discontinuous at the points where the denominator is equal to zero

$$\begin{aligned} 12x^2 + x - 20 = 0 &\Rightarrow 12x^2 + 16x - 15x - 20 = 0 \\ &\Rightarrow 4x(3x + 4) - 5(3x + 4) = 0 \\ &\Rightarrow (4x - 5)\,(3x + 4) = 0 \\ &\Rightarrow x = \frac{5}{4} \text{ or } \frac{-4}{3} \end{aligned}$$

which are the points of discontinuity.

Example 57. Find the limit of

$$\lim_{x \to 0} f(x) = \frac{e^{1/x} - e^{-1/x}}{e^{1/x} + e^{-1/x}},$$

Solution :

$$\lim_{x \to 0} \frac{e^{1/x} - e^{-1/x}}{e^{1/x} + e^{-1/x}} = \lim_{x \to 0} \frac{1 - \dfrac{e^{-1/x}}{e^{1/x}}}{1 + \dfrac{e^{-1/x}}{e^{1/x}}} = \lim_{x \to 0} \frac{1 - \dfrac{1}{e^{2/x}}}{1 + \dfrac{1}{e^{2/x}}} = 1$$

$$\left[\because \text{ when } x \to 0, \frac{2}{x} \to \infty,\ e^{2/x} \to \infty \text{ and } \frac{1}{e^{2/x}} \to 0\right]$$

Questions

1. Define limit and continuity of a function at a given point. Find the limits

   (i) $\lim_{x \to -1} \left(\dfrac{x^2 + 1}{x^2 + x + 1} \right)$ (ii) $\lim_{x \to \infty} \left(\dfrac{2}{x^2 + 1} \right)$ **[Ans.** (i) 2, (ii) 0]

2. Find the limit as $n \to \infty$ of $F(n) - \dfrac{3n^4 - 2n^2 - 5}{2n^4 - 5n + 7}$ **[Ans.** 3/2]

3. Evaluate $\lim_{x \to 0} \left(\dfrac{\sqrt{1 + x} - \sqrt{1 - x}}{x} \right)$ **[Ans.** 1]

4. Find the limit $x \to \infty$ $\dfrac{3x^2 - 4x + 1}{7x^2 + 5}$ **[Ans.** 3.7}

5. If $A(x) = \dfrac{e^x - e^{-x}}{2}$. Find the limit of $A(x)_{x \to \infty}$. **[Ans.** ∞]

6. Is it true to state that if the limit of the denominator function is not equal to zero, the limit of the ratio of two functions equals the ratio of their limits? Correct the statement if you think that it is not true? [**Ans.** True]

7. If $MR = AR\left[1-\frac{1}{\eta}\right]$, where η is the price electricity of demand. Find the limit of the MR when the elasticity tends to infinity and interpret phenomenon. (When $n \to \infty$ then AR and MR are equal).

8. Verify that $\frac{1}{x}(1-\sqrt{1-x})$ is equal to the reciprocal of $(1+\sqrt{1-x})$ as long as $x \neq 0$. Deduce that as $x \to 0$, $\frac{1}{x}(1-\sqrt{1-x}) \to \frac{1}{2}$. By a similar method, show that $\frac{1}{x}(\sqrt{1+x}-\sqrt{1-x}) \to 1$ as $x \to 0$

9. If a is a fixed positive number and if $n \to \infty$ through integral values, show that $a^n \to 0$ if $a < 1$ and $a^n \to \infty$ if $a > 1$. What can be said of the case if $a = 1$? Deduce the limit of $\frac{1}{1+a^n}$ as $n \to \infty$ for values of $a \geq 1$.

[**Ans.** 0, 1/2]

10. Examine the continuity at $x = 1$ and $x = 2$ of the function

$$F(x) = \begin{cases} 2x & \text{if } 0 \leq x \leq 1 \\ 2 - x & \text{if } 1 \leq x \leq 2 \\ x^2 - 2x & \text{if } x > 2 \end{cases}$$

[**Ans.** (i) The function is not continuous at $x = 1$,
(ii) The function is continuous at $x = 2$]

11. Is the function $F(x)$ continuous at $x = 1$, given

$$\begin{aligned} F(x) &= 3x - 2 \text{ when } 1/2 \leq x \leq 1 \\ &= 1 \text{ at } x - 1 \\ &= x + 2 \text{ when } 1 < x \leq 2? \end{aligned}$$

[**Ans.** The give function is not continuous at $x = 1$]

12. Examine for continuity at $x = 2$ for

$$\begin{aligned} F(x) &= 3x + 5 \text{ when } 1 \leq x < 2 \\ &= 8 \text{ at } x = 2 \\ &= 4x - 1 \text{ when } 2 < x \leq 3 \end{aligned}$$

[**Ans.** $F(x)$ is not continuous at $x = 2$]

13. What type of demand curve corresponds to the demand law

$$P = \frac{a}{x} - c$$

where a and c are positive constants? Show that demand is positive no matter how large the price. Is there any limit to the extent of the demand as price tends to zero? Show that total revenue falls steadily as output rises.

14. Evaluate $\lim_{x \to a} \frac{x^3 - a^3}{x - a}$

[**Ans.** $3a^2$]

❑❑❑

4 Differentiation

4.1 DERIVATIVES

Let $y = f(x)$ be a function of x. Let there be a small change Δy in the value of y corresponding to a small change Δx in the value of x, then

$$y + \Delta y = f(x + \Delta x)$$

$\Delta y =$ change in the value of $y -= f(x + \Delta x) - f(x)$

$$\therefore \qquad \frac{\Delta y}{\Delta x} = \frac{f(x + \Delta x) - f(x)}{\Delta x}$$

= Average rate of change of y in the interval between x and $x + \Delta x$.

The $\lim\limits_{\Delta x \to 0} \frac{\Delta y}{\Delta x}$ if it exists (*i.e.*, if this limit is finite) is called the **derivative** of y, w.r.t. x and is written as

$$\frac{dy}{dx} = \lim_{\Delta x \to 0} \frac{f(x + \Delta x) - f(x)}{\Delta x}$$

= Instantaneous rate of change of y, w.r.t. x.

i.e., the instantaneous rate of change of the function with respect to its independent variable is called the **derivative of the function** and the process of finding the derivative is called **differentiation.**

Notes : (i) Δx (or Δy or both) may be positive or negative. It may happen that Δx is positive and Δy is negative or vice versa. But for the derivation of all standard results Δx, Δy are presumed to be positive, *i.e.*, incremments.

(ii) $\frac{dy}{dx}$ does not mean $dy \div dx$. It stands for $\frac{d}{dx}(y)$ and the notation $\frac{d}{dx}$ **stands** as a symbol to denote the **operation of differentiation**.

(iii) Derivative of a function is also known as **differential coefficient** or **first derivative.** If $y = f(x)$, $\frac{dy}{dx}$ is also written as $f'(x)$.

(iv) The degree of the derivative of the polynomial function is one less than the degree of function **(Please note).**

(v) A function that possesses a derivative at a point is said to be differentiable at that point. A function is said to be differentiable **in an interval** if it is differentiable at every point of the interval.

4.2 THEOREM

If the derivative of a function $f(x)$ exists at any point say $x = x_0$, then the function must be continuous at that point.

Proof : Since, the derivative of function $f(x)$ exists at $x = x_0$, the limit

$$= \lim_{\Delta x_0 \to 0} \frac{f(x_0 + \Delta x_0) - f(x_0)}{\Delta x_0} \text{ exists.}$$

Now $$f(x_0 + \Delta x_0) - f(x_0) = \left[\frac{f(x_0 + \Delta x_0) - f(x_0)}{\Delta x_0}\right] \times \Delta x_0$$

$$\therefore \lim_{\Delta x_0 \to 0} [f(x_0 + \Delta x_0) - f(x_0)] = \lim_{\Delta x_0 \to 0} \frac{f(x_0 + \Delta x_0) - f(x_0)}{\Delta x_0} \times \lim_{\Delta x_0 \to 0} \Delta x_0$$

$$= f'(x_0) \times 0 = 0$$

i.e., $$\lim_{\Delta x_0 \to 0} f(x_0 + \Delta x_0) = \lim_{\Delta x_0 \to 0} f(x_0)$$

or $$\lim_{\Delta x_0 \to 0} f(x) = f(x_0) \text{ i.e., } f(x) \text{ is continuous at } x = x_0.$$

But the converse of this theorem is not true, *i.e,* a function $f(x)$ may be continuous at a point without being differentiable at that point.

For example, consider the following functions :

(i) $f(x) = |x|$, (ii) $f(x) = x^{1/3}$

(i) Now, $f(0) = |0| = 0$

$$\lim_{x \to 0^+} f(x) = \lim_{x \to x^-} f(x) = 0 = f(0)$$

i.e., The function $f(x) = |x|$ is continuous at $x = 0$. But

$$\lim_{\Delta x \to 0} \frac{f(0 + \Delta x) - f(0)}{\Delta x} = \lim_{\Delta x \to 0} \frac{|\Delta x|}{\Delta x}$$

Now $$\lim_{\Delta x \to 0^+} \frac{|\Delta x|}{\Delta x} = 1 \qquad \left[\because \text{ when } \Delta x > 0, \frac{|\Delta x|}{\Delta x} = 1\right]$$

and $$\lim_{\Delta x \to 0^-} \frac{|\Delta x|}{\Delta x} = -1 \qquad \left[\because \text{ when } \Delta x < 0, \frac{|\Delta x|}{\Delta x} = -1\right]$$

$\therefore$ The $\lim_{\Delta x \to 0} \frac{f(0 + \Delta x) - f(0)}{\Delta x} = \lim_{\Delta x \to 0} \frac{|\Delta x|}{\Delta x}$ does not exist hence the function $f(x) = |x|$ is not derivable at $x = 0$.

(ii) $$f(0) = (0)^{1/3} = 0$$

$$\lim_{x \to 0^+} f(x) = \lim_{x \to 0^+} x^{1/3} = 0 = \lim_{x \to 0^-} x^{1/3}$$

i.e., the function $f(x) = x^{1/3}$ is continuous at $x = 0$.

Now

$$\lim_{\Delta x \to 0} \frac{f(0 + \Delta x)^{1/3} - f(0)}{\Delta x} = \lim_{\Delta x \to 0} \frac{(\Delta x)^{1/3}}{\Delta x} = \lim_{\Delta x \to 0} \frac{1}{(\Delta x)^{2/3}} = \infty$$

i.e., the $\lim\limits_{\Delta x \to 0} \dfrac{f(0+\Delta x) - f(0)}{\Delta x}$ does not exist. Hence, the function is not differentiable at $x = 0$.

4.3 RULES OF DIFFERENTIATION

1. The derivative of a constant is zero : Since, derivative (differential coefficient) measures the instantaneous rate of the change of a variable, and a constant has no change, the differential coefficient of a constant must be zero. Mathematically,

let $y = f(x)$ is a constant function.

let $f(x) = c$ (a constant)

Then $f(x + \Delta x)$ is also equal to c*

$\therefore$ $$f(x+\Delta x) - f(x) = 0$$

Hence, $$\lim_{\Delta x \to 0} \frac{f(x+\Delta x) - f(x)}{\Delta x} = 0$$

i.e, if $$f(x) = c$$

Then $$\frac{d}{dx}[f(x)] = 0$$

2. The derivative of a variable with respect to itself is unity

Let $$y = f(x) = x$$

Thus, $$f(x+\Delta x) = x + \Delta x$$

and $$\frac{f(x+\Delta x) - f(x)}{\Delta x} = \frac{(x+\Delta x) - x}{\Delta x} = 1$$

$\therefore$ $$\lim_{\Delta x \to 0} \frac{f(x+\Delta x) - f(x)}{\Delta x} = 1$$

i.e., if $y = f(x) = x$ then $\dfrac{dy}{dx} = 1$

But if $y = u$ then

$$\frac{dy}{dx} = \frac{dy}{du}\cdot\frac{du}{dx} = 1 \times \frac{du}{dx} = \frac{du}{dx}$$ **(Pl. note)**

3. The derivative of x^n with respect to $x = n\,x^{n-1}$

Let $$y = x^n = f(x)$$

$$f(x+\Delta x) = (x+\Delta x)^n$$ **(Pl. note)**

$$\lim_{\Delta x \to 0} \frac{f(x+\Delta x) - f(x)}{\Delta x} = \lim_{\Delta x \to 0} \frac{(x+\Delta x)^n - x^n}{\Delta x} = \lim_{\Delta x \to 0} \frac{x^n\left(1 + \dfrac{\Delta x}{x}\right)^n - x^n}{\Delta x}$$

$$= \lim_{\Delta x \to 0} x^n \left[\frac{1 + x\left(\dfrac{\Delta x}{x}\right)}{\Delta x}\right] - x^n$$

* If $y = f(x) = a$ constant then for any change in the value of x, y remains constant.

$$= \lim_{\Delta x \to 0} \frac{x^n + nx^{n-1}\Delta x - x^n}{\Delta x} = nx^{n-1}$$

(By Binomial theorem rejecting higher powers of Δx)

4. Derivative of u^n with respect to

let $$y = u^n = f(u)$$

$$\lim_{\Delta x \to 0} \frac{f(u+\Delta u) - f(u)}{\Delta x} = \lim_{\Delta x \to 0} \frac{(u+\Delta u)^n - u^n}{\Delta u} \cdot \frac{\Delta u}{\Delta x}$$

$$= \lim_{\Delta x \to 0} nu^{n-1} \frac{\Delta u}{\Delta x} = nu^{n-1} \frac{du}{dx}$$

5. The derivative of a constant times a differentiable function is equal to constant times the derivative of the function

i.e., $y = cf(x)$, then

$$\frac{dy}{dx} = c\frac{d}{dx}[f(x)] = cf'(x) \qquad \left[f'(x) = \frac{d}{dx}f(x)\right]$$

6. The additive constant disappears on differentiation, i.e.,

$$\frac{d}{dx}[f(x) + K] = \frac{d}{dx}f(x) = f'(x).$$

[$\because$ If $\phi(x) = f(x) + k$, then $\phi(x+\Delta x) = f(x+\Delta x) + k$ and $\phi(x+\Delta x) - \phi(x) = f(x+\Delta x) - f(x)$]

7. Prove that the differential coefficient of the algebraic sum of any number of functions is the algebraic sum of their differential coefficients.

Proof : Let $$y = u + v + w \ldots$$

where $u, v, w \ldots$ are functions of x.

Let x change to $x + \Delta x$ and correspondingly y change to $y + \Delta y$, and $u, v, w \ldots$ change to $u + \Delta u$, $v + \Delta u$, $w + \Delta w \ldots$ respectively

Then $$y + \Delta y = (u + \Delta u) + (v + \Delta v) + (w + \Delta w) + \ldots$$

$$\Delta y = \Delta u + \Delta v + \Delta w + \ldots$$

and $$\frac{\Delta y}{\Delta x} = \frac{\Delta u}{\Delta x} + \frac{\Delta v}{\Delta x} + \Delta\frac{w}{\Delta x} + \ldots$$

Taking limits when $\Delta x \to 0$, we get*

$$\frac{dy}{dx} = \lim_{\Delta x \to 0} \frac{\Delta u}{\Delta x} + \lim_{\Delta x \to 0} \frac{\Delta v}{\Delta x} + \lim_{\Delta x \to 0} \frac{\Delta w}{\Delta x} + \ldots$$

$$= \frac{du}{dx} + \frac{dv}{dx} + \frac{dw}{dx} + \ldots$$

If $y = u - v$, then

$$\frac{dy}{dx} = \frac{du}{dx} - \frac{dv}{dx}$$

and if $y = u + v - w + z$, then

$$\frac{dy}{dx} = \frac{du}{dx} + \frac{dv}{dx} - \frac{dw}{dx} + \frac{dz}{dx}$$

8. Differential coefficient of product of two functions. [Important]

Prove that

$$\frac{d}{dx}(uv) = u\frac{dv}{dx} + v\frac{du}{dx}$$

Let $y = uv$ where u and v are functions of x.

If x changes to $x + \Delta x$ and correspondingly y changes to $y + \Delta y$, and u and v change to $u + \Delta u$ and $v + \Delta v$ respectively, then

$$y + \Delta y = (u + \Delta u)(v + \Delta v)$$

$$= uv + u\,\Delta v + v\,\Delta u + \Delta u.\Delta v$$

$$\Delta y = u.\Delta v + v.\Delta u + \Delta u.\Delta v$$

$$\frac{\Delta y}{\Delta x} = u\frac{\Delta v}{\Delta x} + v\frac{\Delta u}{\Delta x} + \Delta u\frac{\Delta v}{\Delta x}$$

Taking limits when $\Delta x \to 0$ and consequently $\Delta y \to 0$, $\Delta u \to 0$, $\Delta v \to 0$ we get

$$\frac{dy}{dx} = u\frac{dv}{dx} + v\frac{du}{dx}$$

9 (a). Rule to find the differential coefficient of product of two functions, *i.e.*,

$\frac{d}{dx}$ [Product of two functions] = first function × differential coefficient of second function + second function × differential coefficient of first function.

$$\frac{d}{dx}(uvw) = vw\frac{du}{dx} + wu\frac{dv}{dx} + uv\frac{dw}{dx}$$

i.e., $\frac{d}{dx}$ (Product of three functions) = Product of second and third function $\times \frac{d}{dx}$ (first function) + product of third and first functions $\times \frac{d}{dx}$ second function + Product of first and second functions $\times \frac{d}{dx}$ (third function).

9 (b). Important Rule

There is an other form of the Product Rule which is easy to remember

If $y = uv$; Then

$$\frac{\frac{d}{dx}(uv)}{uv} = \frac{\frac{du}{dx}}{u} + \frac{\frac{dv}{dx}}{v} \Rightarrow \frac{d}{dx}(uv) = v\frac{du}{dx} + u\frac{dv}{dx}$$ **[Pl. note]**

this Rule can be extended to the product of any finite number of functions.

$$\frac{\frac{d}{dx}(uvw....)}{(uvw...)} = \frac{\frac{du}{dx}}{u} + \frac{\frac{dv}{dx}}{v} + \frac{\frac{dw}{dx}}{w} + ...$$ **[Pl. note]**

10. Differential coefficient of quotient of two functions.

Prove that $\frac{d}{dx}\left(\frac{u}{v}\right) = \frac{v\frac{du}{dx} - u\frac{dv}{dx}}{v^2}$ where u and v are functions of x.

Proof : Let $y = \frac{u}{v}$ where u and v are functions of x.

Let x be changed to $x + \Delta x$ and corresponding to this change let u, v change to $u + \Delta u$, and $v + \Delta v$ respectively and y to $y + \Delta y$.

$\therefore$ $$y + \Delta y = \frac{u + \Delta u}{v + \Delta v}$$

i.e., $$\Delta y = \frac{u + \Delta u}{v + \Delta v} - \frac{u}{v} = \frac{v\,\Delta u - u\,\Delta v}{(v + \Delta v)\,v}$$

$$\frac{\Delta y}{\Delta x} = \frac{v\frac{\Delta u}{\Delta x} - u\frac{\Delta v}{\Delta x}}{(v + \Delta v)\,v}$$

Taking limits when $\Delta x \to 0$ [when $\Delta x \to 0$, Δy, Δu, Δv also tend to zero], we get

$$\frac{dy}{dx} = \frac{v\frac{du}{dx} - u\frac{dv}{dx}}{v^2}$$

In other words, we can express it as

Diff. coeff. of the quotient of two functions

$$\frac{\text{Denominator [Diff. coeff. of the numerator]} - \text{Numerator [Diff. coeff. of the denominator]}}{[\text{Denominator}]^2}$$

1. Chain Rule (Derivative of the function of function) :

(1) If $y = f(u)$ and $u = g(x)$ [*i.e.,* $y = f\{g(x)\}$, *i.e.,* y is a function of a function of x]

Then, $$\frac{dy}{dx} = \frac{dy}{du} \times \frac{du}{dx}$$

i.e., Differential coefficient (Derivative) of y w.r.t. $x =$ (Differential coefficient of y with respect to u) $\times$ (Differential coefficient of u, with respect to x)

(2) If $y = f(u)$, $u = g(v)$ and $v = h(x)$, then

$$\frac{dy}{dx} = \frac{dy}{du}.\frac{du}{dv}.\frac{dv}{dx}$$

(3) $\frac{dy}{dx}.\frac{dx}{dy} = 1$

4.4 PARAMETRIC FUNCTIONS

If $x = f(t)$, $y = g(t)$, *i.e.,* both the variables x and y and y are functions of a third variable called the parameter, then

$$\frac{dy}{dx} = \frac{dy}{dt}\Big/\frac{dx}{dt} = \frac{g'(t)}{f'(t)}$$

4.5 DIFFERENTIATION OF IMPLICIT FUNCTIONS

A relation connecting x and y in such a way that none of them can be expressed explicitly (directly) in terms of the other is called an **implicit function** of x and y, *e.g.,* $ax^2 + 2hxy + by^2 = 0$,

$x^2 + y^2 + 2xy = 15$ are implicit functions of x and y. $\frac{dy}{dx}$ in such a function can be obtained by differentiating each term w.r.t. x, considering y as a function of x and then solving the resulting equation to $\frac{dy}{dx}$. For diferentiating different terms, general rule of differentiation, *i.e.,* sum of product rule, chain rule, etc., are used.

While dealing with implicit function following formula will be frequently useed :

If $y = u^n$, then $$\frac{dy}{dx} = nu^{n-1}\frac{du}{dx}$$

For example, $$\frac{d}{dx}(y^2) = 2y\frac{dy}{dx} \text{ and } \frac{d}{dx}(u^3) = 3u^2\frac{du}{dx}$$

$$\frac{d}{dx}(xy^2) = 2xy\frac{dy}{dx} + y^2, \frac{d}{dx}(x^3) = 3x^2 \text{ etc.}$$

4.6 GEOMETRICAL MEANING OF THE DERIVATIVE

Let $y = f(x)$ be the equation of the curve APQ referred to OX, OY as the co-ordinate axes.

Let $P(x, y)$ and $Q(x + \Delta x, y + \Delta y)$ be two neighbouring points on the curve. Draw PM, QN perpendicular on the x-axis. Draw $Pl \perp QN$, join QP and produce it to meet the x-axis at R.

Let $\angle XRP = \theta$. $\angle LPQ$ is also equal to θ.

$$\therefore \quad \tan\theta = \frac{\Delta y}{\Delta x}$$

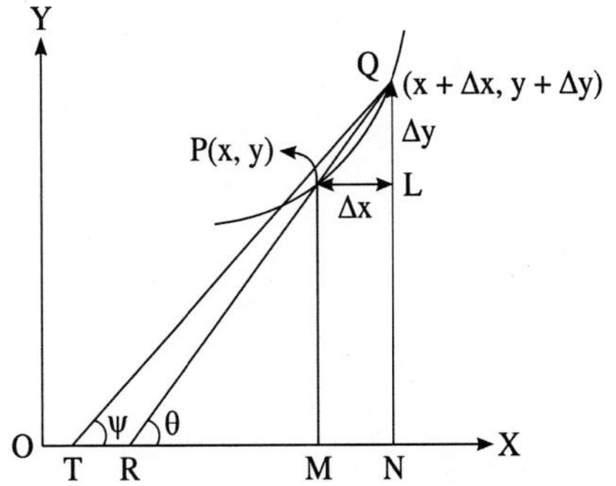

Keeping P fixed, when $Q \to P$, the line $PQ \to PT$, the tangent at the point P and the angle $\theta \to \psi$ (ψ is the angle which the tangent at P makes with the positive direction of x-axis). Also $\Delta x \to 0$, $\Delta y \to 0$.

In the limiting position when $\Delta y \to 0$, $\Delta x \to 0$ and $\theta \to \psi$, we get

$$\lim_{\Delta x \to 0} \frac{\Delta y}{\Delta x} = \lim_{\theta \to \psi} \tan\theta$$

i.e., $\frac{dy}{dx} = \tan\psi =$ slope of tangent at P.

If the tangent at point P is parallel to x-axis (Horizontal tangent) then $\psi = 0$ and $\frac{dy}{dx} = 0$. If the tangent at the point P is parallel to y-axis, *i.e.*, perpendicular to x-axis (vertical tangent), then $\psi = 90^\circ$ and $\tan \psi = \infty$. **(Important)**

STANDARD RESULTS

(i) $\frac{d}{dx}(x^n) = nx^{n-1}$ (ii) $\frac{d}{dx}(\text{constant}) = 0$

(iii) $\frac{d}{dx}(u^n) = nu^{n-1}\frac{du}{dx}$ (iv) $\frac{d}{dx}(\log_a)x = \frac{\log_a e}{x}$

(v) $\frac{d}{dx} = (\log_e x) = \frac{1}{x}$ (vi) $\frac{d}{dx}(\log_e u) = \frac{\log_e e}{x}\frac{du}{dx} = \frac{1}{u}\frac{du}{dx}$

(vii) $\frac{d}{dx} = (\log_a u) = \left(\frac{\log_a e}{u}\right)\frac{du}{dx}$ (viii) $\frac{d}{dx}(a^x) = a^x \log_e a$ **[Important]**

(ix) $\frac{d}{dx}(a^u) = (a^u \log_e a)\frac{du}{dx}$ (x) $\frac{d}{dx}(e^x) = e^x$

(xi) $\frac{d}{dx}(e^u) = \frac{du}{dx}$ (xii) $\frac{d}{dx}(e^{ax}) = ae^{ax}$

Rule to find the different coefficient of x^n, * $n \in R$: Reduce the power of x by unity and multiply the function so obtained by the original power of x. For example,

$$\frac{d}{dx}(x^2) = 2x,\ \frac{d}{dx}(x^3) = 3x^2,\ \frac{d}{dx}(x^4) = 4x^3$$

Rule to find the differential coefficient of $(ax + b)^n$ with respect to x where n is positive or negative integer or a fraction :

$$\frac{d}{dx}(ax + b)^n = n(ax + b)^{n-1} \times a$$

For finding the differential coefficient of $(ax + b)^n$, reduce the power by 1, multiply the function so obtained by the original power and also by the coefficient of x. For example,

$$\frac{d}{dx}(2x + 3)^2 = 2(2x + 3) \times 2$$

$$\frac{d}{dx}(ax + 4)^6 = 6(ax + 4)^5 \times a = 6a(ax + 4)^5$$

Note. $\frac{d}{dx}\left(\frac{dy}{dx}\right) = \frac{d^2y}{dx^2}$ = second derivative of y w.r.t. x. **[Pl. note]**

and $\frac{d}{dx}\left(\frac{d^2y}{dx^2}\right) = \frac{d^3y}{dx^3}$ = Third derivative of y w.r.t. x.

SOLVED EXAMPLES

Example 1. Find the differential coefficient of $\log_a x$ where a and x are both positive.

Solution : Let

$$y = \log_a x$$

$$y + \Delta y = \log_a (x + \Delta x)$$

$$\therefore \quad \Delta y = \log_a (x + \Delta x) - \log_a x$$

$$= \log_a \frac{x + \Delta x}{x} \qquad [\because \log_a m - \log_a n = \log_a \frac{m}{n}]$$

$$= \log_a \left[1 + \frac{\Delta x}{x}\right]$$

Dividing both sides by Δx, we get

$$\frac{\Delta y}{\Delta x} = \frac{1}{\Delta x} \log_a \left(1 + \frac{\Delta x}{x}\right) = \frac{1}{x} \frac{x}{\Delta x} \log_a \left(1 + \frac{\Delta x}{x}\right)$$

$$= \frac{1}{x} \log_a \left(1 + \frac{\Delta x}{x}\right)^{\frac{x}{\Delta x}}$$

$$= \frac{1}{x} \log_a (1 + p)^{1/p} \qquad \text{where } p = \frac{\Delta x}{x}.$$

Now, when $\Delta x \to 0, p$ also $\to 0$

$$\therefore \quad \lim_{\Delta x \to 0} \frac{\Delta y}{\Delta x} = \frac{dy}{dx} = \frac{1}{x} \lim_{\Delta x \to 0} \log_a \left(1 + \frac{\Delta x}{x}\right)^{\frac{x}{\Delta x}}$$

$$= \frac{1}{x} \lim_{p \to 0} \log_a (1 + p)^{1/p} = \frac{1}{x} \log_a e \qquad [\because \lim_{x \to 0} (1 + x)^{1/x} = e]$$

(when $a = e, \frac{dy}{dx} = \frac{1}{x} \log_e e = \frac{1}{x}$), *i.e., if* $y = \log_e x$ *then* $\frac{dy}{dx} = \frac{1}{x}$

Example 2. If $y = \log_a u$ find $\frac{dy}{dx}$.

Solution : Suppose when x changes from x to $x + \Delta x$, u changes to $u + \Delta u$ and y to $y + \Delta y$. Then,

$$\frac{\Delta y}{\Delta x} = \left[\log_a \left(1 + \frac{\Delta u}{u}\right)\right] \frac{1}{\Delta x} \qquad \left[\because \Delta y = \log_a (u + \Delta u) - \log_a u = \log_a \left(\frac{u + \Delta u}{u}\right)\right]$$

$$= \frac{1}{u} \left[\frac{u}{\Delta u} \log_a \left(1 + \frac{\Delta u}{u}\right)\right] \frac{\Delta u}{\Delta x}$$

$$= \frac{1}{u} \left[\log_a \left(1 + \frac{\Delta u}{u}\right)^{\frac{u}{\Delta u}}\right] \frac{\Delta u}{\Delta x}$$

$$\lim_{\Delta x \to 0} \frac{\Delta y}{\Delta x} = \frac{dy}{dx}$$

$$\therefore \quad \frac{dy}{dx} = \frac{1}{u} \lim_{\Delta u \to 0} \log_a \left[1 + \frac{\Delta u}{u}\right]^{\frac{u}{\Delta u}} \lim_{\Delta u \to 0} \frac{\Delta u}{\Delta x}$$

$$= \frac{1}{u} [\log_a e] \frac{du}{dx}$$

$$= \frac{1}{u} \frac{du}{dx} \qquad \text{when } a = e. \text{ [Please note } \log_e e = 1]$$

Example 3. If $y = \log 5x$, then

$$\frac{dy}{dx} = \frac{1}{5x}.5 = \frac{1}{x}.$$

Example 4. Differentiate $\sqrt{x}$ from first principles.

Solution : Let $y = \sqrt{x}$

$$y + \Delta y = \sqrt{x + \Delta x} = (x + \Delta x)^{1/2}$$

$$= \sqrt{x}\left(1 + \frac{\Delta x}{x}\right)^{1/2}$$

$$= \sqrt{x}\left[1 + \frac{1}{2}\frac{\Delta x}{x}\right]$$

expanding by Binomial theorem and neglecting higher power of Δx

$$\Delta y = y + \Delta y - y = (x + \Delta x)^{1/2} - \sqrt{x}$$

$$= \sqrt{x}\left[1 + \frac{1}{2}\frac{\Delta x}{x}\right] - \sqrt{x} = \frac{1}{2}\frac{\Delta x}{\sqrt{x}}$$

$$\Rightarrow \quad \frac{\Delta y}{\Delta x} = \frac{1}{2\sqrt{x}}$$

$$\lim_{\Delta x \to 0} \frac{\Delta y}{\Delta x} = \frac{dy}{dx} = \frac{1}{2\sqrt{x}}$$

Example 5. Differentiate x^2 from first principles.

Solution : Let $y = x^2$

$$y + \Delta y = (x + \Delta x)^2$$

$$\Delta y = (x + \Delta x)^2 - x^2 = (\Delta x)^2 + 2x\,\Delta x$$

$$\frac{\Delta y}{\Delta x} = 2x + \Delta x$$

$$\lim_{\Delta x \to 0} \frac{\Delta y}{\Delta x} = \frac{dy}{dx} = 2x$$

Example 6. Differentiate $\frac{1}{x}$ w.r.t. x from first principles.

Solution : Let $y = \frac{1}{x}$, $y + \Delta y = \frac{1}{x + \Delta x}$

$$\Delta y = \frac{1}{x + \Delta x} - \frac{1}{x} = \frac{x - x - \Delta x}{x(x - \Delta x)} = -\frac{\Delta x}{(x^2 - x\,\Delta x)}$$

$$\frac{\Delta y}{\Delta x} = -\frac{1}{(x^2 - x\,\Delta x)} = \lim_{\Delta x \to 0} \frac{\Delta y}{\Delta x} = \frac{-1}{x^2}$$

Example 7. Differentiate $\frac{1}{\sqrt{x}}$ w.r.t. x from first principles.

Solution : Let $y = \frac{1}{\sqrt{x}}$, $y + \Delta y = \frac{1}{\sqrt{x + \Delta x}}$

$$\Delta y = \frac{1}{\sqrt{x+\Delta x}} - \frac{1}{\sqrt{x}} = \frac{\sqrt{x} - \sqrt{x+\Delta x}}{\sqrt{x}\,\sqrt{x+\Delta x}}$$

$$= \frac{\sqrt{x} - (x+\Delta x)^{1/2}}{\sqrt{x}\,\sqrt{x+\Delta x}} = \frac{\sqrt{x} - \sqrt{x}\left(1+\frac{\Delta x}{x}\right)^{1/2}}{\sqrt{x}\,\sqrt{x+\Delta x}}$$

$$= \frac{\sqrt{x} - \sqrt{x}\left(1+\frac{1}{2}\frac{\Delta x}{x}\right)}{\sqrt{x}\,\sqrt{x+\Delta x}} \quad \text{neglecting higher powers of } x$$

$$= \frac{-1}{2x\sqrt{x+\Delta x}}\Delta x$$

$$\frac{\Delta y}{\Delta x} = -\frac{1}{2x\sqrt{x+\Delta x}}$$

$$\lim_{\Delta x \to 0} \frac{\Delta y}{\Delta x} = \frac{dy}{dx} = -\frac{1}{2x^{3/2}}$$

Example 8. Using first principle and not otherwise, find the derivative of the function

$$f(x) = x^2 + \frac{1}{x} + 7$$

Solution :

$$f(x+\Delta x) = (x+\Delta x)^2 + \frac{1}{x+\Delta x} + 7$$

$$\Delta y = f(x+\Delta x) - f(x) = (x+\Delta x)^2 - x^2 + \frac{1}{x+\Delta x} - \frac{1}{x}$$

$$= x^2 + \Delta x^2 + 2x\,\Delta x - x^2 + \frac{x - x - \Delta x}{x(x+\Delta x)}$$

$$= \Delta x^2 + 2x\,\Delta x - \frac{\Delta x}{x(x+\Delta x)}$$

$$\frac{\Delta y}{\Delta x} = \Delta x + 2x - \frac{1}{x(x+\Delta x)}$$

$$\lim_{\Delta x \to 0} = \frac{\Delta y}{\Delta x} = \frac{dy}{dx} = 2x - \frac{1}{x^2}$$

Example 9. Find the differential coefficient of

$$7x^5 + 9x^4 - 17x^3 + 11x^2 - \frac{5}{x^5} + \frac{3}{x^3} + 100$$

Solution :

$$y = 7x^5 + 9x^4 - 17x^3 + 11x^2 - \frac{5}{x^5} + \frac{3}{x^3} + 100$$

$$= 7x^5 + 9x^4 - 17x^3 + 11x^2 - 5x^{-5} + 3x^{-3} + 100$$

$$\frac{dy}{dx} = 7 \times 5x^{5-1} + 9 \times 4x^{4-1} - 17 \times 3x^{3-1} + 11 \times 2x^{2-1}$$

$$- 5 \times (-5)\,x^{-5-1} + 3 \times (-3)\,x^{-3-1}$$

$$= 35x^4 + 36x^3 - 51x^2 + 22x + 25x^{-6} - 9x^{-4}$$

$$= 35x^4 + 36x^3 - 51x^2 + 22x + \frac{25}{x^6} - \frac{9}{x^4}$$

Example 10. (i) Find the derivative of $y = 2x^3 - 3\log x + 6e^{2x} + \frac{35}{x}$.

(ii) Find the first derivative of $y = x^2 e^x$.

(iii) Find the derivative of $(1 - x - x^2)^{-3}$.

(iv) Find $\frac{dy}{dx}$ if $x^2 - 2xy + y^2 = 8$.

Solution : (i) $\frac{dy}{dx} = 6x^2 - \frac{3}{x} + 6e^{2x} \times 2 - \frac{35}{x^2} = 6x^2 - \frac{3}{x} + 12e^{2x} - \frac{35}{x^2}$.

(ii) $\frac{dy}{dx} = x^2 e^x + e^x \times 2x = xe^x (x + 2)$ [By Product Rule taking x^2 as first function]

(iii) Let $y = (1 - x - x^2)^{-3}$

$$\frac{dy}{dx} = -3(1 - x - x^2)^{(-3-1)} \frac{d}{dx}(1 - x - x^2) \qquad \left[\because \frac{d}{dx}(u^n) = nu^{n-1}\frac{du}{dx}\right]$$

$$= -3(1 - x - x^2)^{-4}(-1 - 2x) = \frac{3(1 + 2x)}{(1 - x - x^2)^4}$$

(iv) Differentiating both sides of the given function, we get

$$2x - 2x\frac{dy}{dx} - 2y + 2y\frac{dy}{dx} = 0$$

$$\therefore \qquad \frac{dy}{dx} = \frac{2(x - y)}{2(x - y)} = 1$$

Example 11. Differentiate w.r.t. x

$$4x^2 - 7x + 8 - \frac{4}{\sqrt{x}} + \frac{6}{\sqrt{6}}$$

Solution : Let

$$y = 4x^2 - 7x + 8 - \frac{4}{\sqrt{x}} + \frac{6}{\sqrt{6}}$$

$$= 4x^2 - 7x - 4x^{-1/2} + 8 + \frac{6}{\sqrt{6}}$$

$$\frac{dy}{dx} = 4 \times 2x - 7 - 4\left(-\frac{1}{2}\right)x^{-\frac{1}{2} - 1} = 8x - 7 + 2x^{-3/2} = 8x + \frac{2}{x^{3/2}} - 7$$

Example 12. Differentiate w.r.t. x

$$\frac{x^3 + 3x^2 - 4}{x}$$

Solution : Let

$$y = \frac{x^3 + 3x^2 - 4}{x} = x^2 + 3x - \frac{4}{x} = x^2 + 3x - 4x^{-1}$$

$$\frac{dy}{dx} = 2x + 3 + 4x^{-2} = 2x + 3 + \frac{4}{x^2} = \frac{2x^3 + 3x^2 + 4}{x^2}$$

Second Method [By Quotient Rule]

$$\frac{dy}{dx} = \frac{x\frac{d}{dx}(x^3 + 3x^2 - 4) - (x^3 + 3x^2 - 4)\frac{d}{dx}(x)}{x^2}$$

$$= \frac{x(3x^2 + 6x) - (x^3 + 3x^2 - 4)}{x^2} = \frac{2x^3 + 3x^2 + 4}{x^2}$$

Example 13. Differentiate $2x^4 + 3x^3 - 6x^{2/3} + \frac{1}{\sqrt{x}}$.

Solution : Let

$$y = 2x^4 + 3x^3 - 6x^{2/3} + \frac{1}{\sqrt{x}}$$

$$= 2x^4 + 3x^3 - 6x^{2/3} + x^{-1/2}$$

$$\frac{dy}{dx} = 8x^3 + 9x^2 - 6 \times \frac{2}{3} x^{\frac{2}{3} - 1} - \frac{1}{2} x^{-\frac{1}{2} - 1}$$

$$= 8x^3 + 9x^2 - 4x^{-1/3} - \frac{1}{2} x^{-3/2} = 8x^3 + 9x^2 - \frac{4}{x^{1/3}} - \frac{1}{2x^{3/2}}.$$

Example 14. Differentiate w.r.t. x

(i) $5x^4 + 3x^3 - 7x^2 + 9x - 100$ (ii) $\frac{e^x + 1}{e^x - 1}$.

Solution : (i) Let

$$y = 5x^4 + 3x^3 - 7x^2 + 9x - 100$$

$$\frac{dy}{dx} = 5 \times 4x^3 + 3 \times 3x^2 - 7 \times 2x + 9$$

$$= 20x^3 + 9x^2 - 14x + 9$$

(ii) Let $y = \frac{e^x + 1}{e^x - 1}$

$$\frac{dy}{dx} = \frac{(e^x - 1)\, e^x - (e^x + 1)\, e^x}{(e^x - 1)^2}$$

$$= \frac{e^{2x} - e^x - e^{2x} - e^x}{(e^x - 1)^2} = -\frac{2e^x}{(e^x - 1)^2}$$

Example 15. Find $\frac{dy}{dx}$ if $y = x^n e^{ax}$.

Solution :

$$\frac{dy}{dx} = x^n \frac{d}{dx}(e^{ax}) + e^{ax} \frac{d}{dx}(x^n)$$

$$= x^n.a.e^{ax} + e^{ax}.n.x^{n-1}$$

$$= (ax^n + nx^{n-1})\, e^{ax} = e^{ax}\, x^{n-1}\,(ax + n)$$

Example 16. Find $\frac{dy}{dx}$ of (i) $y = \frac{\sqrt{x} - 1}{\sqrt{x} + 1}$ (ii) $y = x^3 (\log x)^2$ (iii) $y = (5 - 2x)(2x^3 + 3)$.

Solution : (i) $\frac{dy}{dx}$ (By Quotient Rule)

$$= \frac{(\sqrt{x}+1)\frac{d}{dx}(\sqrt{x}-1) - (\sqrt{x}-1)\frac{d}{dx}(\sqrt{x}+1)}{(\sqrt{x}+1)^2}$$

$$= \frac{(\sqrt{x}+1)\frac{1}{2}x^{\frac{1}{2}-1} - (\sqrt{x}-1)\frac{1}{2}x^{\frac{1}{2}-1}}{(\sqrt{x}+1)^2}$$

$$= \frac{\frac{1}{2}+\frac{1}{2\sqrt{x}}-\frac{1}{2}+\frac{1}{2\sqrt{x}}}{(\sqrt{x}+1)^2} = \frac{1}{\sqrt{x}(\sqrt{x}+1)^2}$$

(ii) By Product Rule

$$\frac{d}{dx}[x^3(\log x)^2] = x^3 \times \frac{2\log x}{x} + (\log x)^2 \times 3x^2 \qquad \text{(Refer Rule 9(b) Page 2.5)}$$

$$= 2x^2 \log x + 3x^2(\log x)^2$$

Second Method (Product Rule)

$$\frac{\frac{d}{dx}[x^3(\log x)^2]}{x^3(\log x)^2} = \frac{\frac{d}{dx}(x^3)}{x^3} + \frac{\frac{d}{dx}(\log x)^2}{(\log x)^2} \qquad \text{(Refer Rule 9(b) Page 2.5)}$$

(iii) By Product Rule

$$\frac{\frac{d}{dx}[(5-2x)(2x^3+3)]}{(5-2x)(2x^3+3)} = \frac{\frac{d}{dx}(5-2x)}{(5-2x)} + \frac{\frac{d}{dx}(2x^3+3)}{2x^3+3}$$

$$= \frac{2}{5-2x} + \frac{6x^2}{2x^3+3}$$

$\therefore$
$$\frac{d}{dx}[(5-2x)(2x^3+3)] = -2(2x^3+3) + 6x^2(5-2x)$$

[Pl. try by the method $\frac{d}{dx}(uv) = u\frac{dv}{dx} + v\frac{du}{dx}$]

Example 17. If $y = x^{e^{-x^2}}$, find $\frac{dy}{dx}$.

Solution : $\log y = e^{-x^2} \log x$

$$\frac{1}{y}\frac{dy}{dx} = \frac{e^{-x^2}}{x} + \log x\,(e^{-x^2})(-2x)$$

$$= e^{-x^2}\left[\frac{1}{x} - 2x\log x\right]$$

$\therefore$
$$\frac{dy}{dx} = x^{e^{-x^2}} e^{-x^2}\left[\frac{1}{x} - 2x\log x\right]$$

Example 18. Differentiate the following functions w.r.t. x :

(i) $\frac{ax^2 - bx + c}{\sqrt{x}}$ (ii) $10^x + 10e^x$

(iii) $x^e + e^x - ex$ (iv) $\log(\sqrt{x}\, e^x)$

(v) $\log(x^2 e^{mx})$ (vi) $e^{3 \log x}$

Solution : (i) Let $y = \frac{ax^2 + bx + c}{\sqrt{x}} = ax^{3/2} - bx^{1/2 + cx^{-1/2}}$

$$\frac{dy}{dx} = \frac{3}{2} ax^{1/2} - \frac{1}{2} bx^{-1/2} - \frac{1}{2} cx^{-3/2}$$

$$= \frac{3}{2} a\sqrt{x} - \frac{b}{2\sqrt{x}} - \frac{c}{2x^{3/2}}$$

(ii) Let $y = 10^x + 10e^x$

$$\frac{dy}{dx} = \frac{d}{dx}(10^x) + \frac{d}{dx} 10e^x = \frac{d}{dx}(10^x) + 10e^x$$

To differentiate 10^x, let $y_1 = 10^x$

Taking log of both sides, we get, $\log y_1 = x \log 10$

Differentiating w.r.t. x, we get

$$\frac{1}{y_1}\frac{dy_1}{dx} = \log 10 \text{ or } \frac{dy_1}{dx} = y_1 \log 10 = 10^x \log_e 10$$

$\therefore$
$$\frac{dy}{dx} = 10^x \log_e 10 + 10e^x$$

(iii) Let
$$y = x^e + e^x - ex$$

$$\frac{dy}{dx} = ex^{e-1} + e^x - e$$

(iv) Let
$$y = \log(\sqrt{x}\, e^x)$$

$$\frac{dy}{dx} = \frac{1}{\sqrt{x}\, e^x} . \frac{d}{dx}(\sqrt{x}\, e^x)$$

$$= \frac{1}{\sqrt{x}\, e^x}\left[\sqrt{x}\, e^x + \frac{1}{2}\frac{e^x}{\sqrt{x}}\right]$$

$$= \frac{1}{\sqrt{x}}\left[\sqrt{x} + \frac{1}{2\sqrt{x}}\right] = \frac{2x+1}{2x}$$

Second Method

$$\log \sqrt{x}\, e^x = \frac{1}{2}\log x + x \log e = \frac{1}{2}\log x + x,$$

$$\frac{d}{dx}\log[\sqrt{x}e^x] = \frac{1}{2x} + 1 = \frac{2x+1}{2x}$$

(v) Please try. [**Ans.** $\frac{2}{x} + m$]

(vi) Let
$$y = e^{3\log x} = e^{\log x^3}$$
$$\frac{dy}{dx} = e^{\log x^3} \cdot \frac{d}{dx} \log (x^3)$$
$$= e^{\log x^3} \left[\frac{1}{x^3} \cdot \frac{d}{dx} (x)^3 \right]$$
$$= e^{\log x^3} \left[\frac{1}{x^3} .3x^2 \right] = e^{\log x^3} \left[\frac{3}{x} \right]$$
$$= x^3 \times \frac{3}{x} = 3x^2 \qquad [\because e^{\log x} = x \text{ and } e^{\log x^3} = x^3]$$

Second Method

$$y = e^{\log x^3}$$
$$\log y = \log x^3 \log e = \log x^3$$
$$y = x^3, \frac{dy}{dx} = 3x^2$$

Example 19. If $f(x + y) = f(x) f(y)$ for all x and y and if $f(5) = 2$ and $f'(0) = 3$ find $f'(5)$.

Solution :
$$f(x + y) = f(x) f(y)$$
$\therefore$
$$f(x + a) = f(x) f(a)$$
and
$$f'(x + a) = f'(x) f(a) \qquad \text{[Differentiating w.r.t. } x]$$

Now let $x = 0$ and $a = 5$, then
$$f'(5) = f'(0) f(5)$$
$$= 3 \times 2 = 6 \qquad [\because f'(0) = 3 \text{ and } f(5) = 2]$$

Example 20. Differentiate
$$(a\sqrt{x} + b)(b\sqrt{x} + a)$$

Solution : Let
$$y = (a\sqrt{x} + b)(b\sqrt{x} + a)$$
$$\frac{dy}{dx} = (a\sqrt{x} + b) \frac{1}{2} \frac{b}{\sqrt{x}} + (b\sqrt{x} + a) \frac{1}{2} \frac{a}{\sqrt{x}}$$
$$= ab + \frac{1}{2\sqrt{x}} (a^2 + b^2)$$

4.7 LOGARITHMIC DIFFERENTIATION

When a function of x is raised to a power which is also a function of x or when the given function consists of the product of a number of functions, we first take logarithm of both sides and then differentiate. We use following properties of logarithms :

(i) $\log_a mn = \log_a m + \log_a n$ (ii) $\log_a (m/n) = \log_a m - \log_a n$

(iii) $\log_a m^n = n \log_a m$ (iv) $\log_e e = 1$.

(v) $\log_b a = \dfrac{1}{\log_a b}$

4.8 SUCCESSIVE DIFFERENTIATION

If $y = f(x)$ is a differentiable function of x, then its derivative $\frac{dy}{dx}$ is called the **first derivative** of y or $f(x)$ and is denoted by y' or $f'(x)$ or y_1. If the first derivative is differentiable, its derivative is called **second derivative** of the original function and is denoted by $\frac{d^2y}{dx^2}$ or y'' or $f''(x)$ or y_2. If second derivative is differentiable, its derivative is called the **third derivative** of the original function and is denoted by $\frac{d^3y}{dx^3}$ or y''' or $f'''(x)$ or y_3 and so on ... The nth derivative is denoted by $\frac{d^ny}{dx^n}$ or $f^n(x)$.

It has already been pointed out that the degree of the derivative of the function is exactly one less than the degree of the function, *i.e.*, if $y = f(x)$ is a polynomial function of degree n, then $\frac{dy}{dx}$ will be a polynomial function of degree $n - 1$, $\frac{d^2y}{dx^2}$ of degree $n - 2$ and so on. $\frac{d^ny}{dx^n}$, *i.e.*, nth derivative will be a constant function and $(n + 1)$th derivative will be zero. **(Pl. note)**

Example 21. If $y = x^x$, find $\frac{dy}{dx}$.

Solution : $$y = x^x \Rightarrow \log y = x \log x$$

Differentiating both sides w.r.t. x

$$\frac{1}{y}\frac{dy}{dx} = x \times \frac{1}{x} + (\log x) \times 1 = 1 + \log x \qquad \left[\because \frac{d}{dx}(\log x) = \frac{1}{x}\right]$$

$$\therefore \qquad \frac{dy}{dx} = y(1 + \log x) = x^x(1 + \log x)$$

Example 22. If $y = x^{\log x}$, find $\frac{dy}{dx}$.

Solution : $\log y = (\log x) \log x = (\log x)^2$

Differentiating both sides w.r.t. x

$$\frac{1}{y}\frac{dy}{dx} = 2(\log x).\frac{d}{dx}(\log x) = \frac{2 \log x}{x}$$

$$\therefore \qquad \frac{dy}{dx} = y\left(\frac{2 \log x}{x}\right) = \frac{x^{\log x}\, 2 \log x}{x} = 2x^{(\log x - 1)} \log x$$

Example 23. Differentiate :

(i) e^{e^x} (ii) $(\sqrt{x})^{\sqrt{x}}$ (iii) $[x^x]^x$ (iv) x^{x^x}

(v) If $x = y^{y^{y^{\ldots\alpha}}}$, find $\frac{dy}{dx}$ (vi) If $y = \log_y x$, find $\frac{dy}{dx}$

Solution : (i) let $y = e^{e^x}$

Taking log of both sides

$$\log y = e^x \log e = e^x \qquad [\because \log_e e = 1]$$

Differentiating both sides w.r.t. x we get

$$\frac{1}{y}\frac{dy}{dx} = e^x \Rightarrow \frac{dy}{dx} = ye^x = e^{e^x}.e^x$$

(ii) Let $y = [\sqrt{x}]^{\sqrt{x}}$

Taking log of both sides, we get $\log y = \sqrt{x} \log \sqrt{x}$

Differentiating both sides w.r.t. x

$$\frac{1}{y}\frac{dy}{dx} = \sqrt{x}\,\frac{1}{\sqrt{x}}.\frac{1}{2\sqrt{x}} + (\log \sqrt{x})\,\frac{1}{2\sqrt{x}}$$

$$= \frac{1}{2\sqrt{x}}[1 + \log \sqrt{x}]$$

i.e.,

$$\frac{dy}{dx} = y\,\frac{1}{2\sqrt{x}}[1 + \log \sqrt{x}]$$

$$= (\sqrt{x})^{\sqrt{x}}\left[\frac{1}{2\sqrt{x}} + \frac{1}{2\sqrt{x}}\log x^{1/2}\right]$$

$$\frac{dy}{dx} = (\sqrt{x})^{\sqrt{x}}\left[\frac{1}{2\sqrt{x}} + \frac{1}{4\sqrt{x}}\log x\right]$$

Second Method

$$\log y = \sqrt{x}\log\sqrt{x} = \frac{1}{2}x^{1/2}\log x$$

$$\frac{1}{y}\frac{dy}{dx} = \frac{1}{2}\left(\frac{1}{2}\right)x^{-1/2}\log x + \frac{1}{2}\frac{x^{1/2}}{x}$$

$$= \frac{\log x}{4\sqrt{x}} + \frac{1}{2\sqrt{x}}$$

(iii) Let $y = [x^x]^x$

Taking log of both sides, we get

$$\log y = x \log x^x = x^2 \log x$$

Differentiating both sides w.r.t. x we get

$$\frac{1}{y}\frac{dy}{dx} = x^2\frac{1}{x} + 2x\log x = x(1 + 2\log x)$$

$$\frac{dy}{dx} = [x^x]^x\, x(1 + 2\log x) = x[x^x]^x\,[1 + 2\log x]$$

(iv) Let $y = x^{x^x}$

Taking log of both sides, we get $\log y = x^x \log x$

Differentiating both sides w.r.t. x

$$\frac{1}{y}\frac{dy}{dx} = x^x\frac{1}{x} + \log x\,\frac{d}{dx}(x^x) \qquad ...(I)$$

For finding $\frac{d}{dx}(x^x)$, let $u = x^x \Rightarrow \log u = x \log x$

$$\frac{1}{u}\frac{du}{dx} = x\frac{1}{x} + \log x + 1 + \log x$$

i.e., $$\frac{du}{dx} = x^x(1 + \log x)$$

Substituting in (i), we get

$$\frac{1}{y}\frac{dy}{dx} = x^x\frac{1}{x} + \log x\ \{x^x(1 + \log x)\}$$

$$\frac{dy}{dx} = x^{x^x}.x^x\left[\frac{1}{x} + (\log x)\,(1 + \log x)\right]$$

$$= x^{x^x}.x^x \log x\left[1 + \log x + \frac{1}{x\log x}\right]^*$$

(v) $x = y^{y^{y\ldots\infty}} \Rightarrow x = y^x$

Taking log of both sides

$$\log x = x\log y$$

Differentiating $$\frac{1}{x} = x.\frac{1}{y}\frac{dy}{dx} + \log y$$

i.e., $$\frac{1}{x} - \log y = \frac{x}{y}.\frac{dy}{dx}$$

i.e., $$\frac{dy}{dx} = (1 - x\log y)\,\frac{y}{x^2}$$

(vi) $y = \log_y x \Rightarrow y^y = x$

Taking log of both sides

$$y\log y = \log x$$ [Also $\log_y x = \dfrac{\log_e x}{\log_e y}$]

Differentiating both sides w.r.t. x

$$y.\frac{1}{y}\frac{dy}{dx} + (\log y)\,\frac{dy}{dx} = \frac{1}{x}$$

$$\frac{dy}{dx}(1 + \log y) = \frac{1}{x} \Rightarrow \frac{dy}{dx} = \frac{1}{x(1 + \log y)}$$

Example 24. If $y = 1 + x + \dfrac{x^2}{2!} + \dfrac{x^3}{3!} + \ldots\infty$ show that $\dfrac{dy}{dx} = y.$

Solution : $$\frac{dy}{dx} = \frac{d}{dx}(1) + \frac{d}{dx}(x) + \frac{d}{dx}\left(\frac{x^2}{2!}\right) + \frac{d}{dx}\left(\frac{x^3}{3!}\right) + \ldots\infty$$

$$= 0 + 1 + \frac{2x}{2} + \frac{3x^2}{3!} + \ldots\infty$$

$$= 1 + x + \frac{x^2}{2!} + \frac{x^3}{3!} + \ldots\infty = y$$ which proves the result.

Example 25. If $y = x^{1/2} + x^{-1/2}$ prove that $2x\frac{dy}{dx} + y = 2\sqrt{x}$.

Solution :
$$\frac{dy}{dx} = \frac{d}{dx}(x^{1/2}) + \frac{d}{dx}(x^{-1/2})$$
$$= \frac{1}{2}x^{\frac{1}{2}-1} + \left(\frac{-1}{2}\right)x^{\left(-\frac{1}{2}-1\right)} = \frac{1}{2\sqrt{x}} - \frac{1}{2(x)^{3/2}}$$
$$= \frac{\sqrt{x}}{2x} - \frac{1}{2x.\sqrt{x}} = \frac{1}{2x}(x^{1/2} - x^{-1/2})$$

$\Rightarrow$
$$2x\frac{dy}{dx} = x^{1/2} - x^{-1/2} + x^{1/2} - x^{1/2} = 2x^{1/2} - (x^{1/2} + x^{-1/2})$$
$$= 2\sqrt{x} - y$$

$\Rightarrow$
$$2x\frac{dy}{dx} + y = 2\sqrt{x}$$
[Pt. note]

Example 26. If $2x = y^{\frac{1}{m}} + y^{-\frac{1}{m}}$ prove that $(x^2 - 1)\left(\frac{dy}{dx}\right)^2 = m^2y^2$.

Solution : Differentiating
$$2x = y^{\frac{1}{m}} + y^{-\frac{1}{m}}$$
w.r.t. x, we get
$$2 = \frac{1}{m}[y^{\frac{1}{m}-1} - y^{-\frac{1}{m}-1}]\frac{dy}{dx}$$

$\Rightarrow$
$$2my = [y^{\frac{1}{m}} - y^{-\frac{1}{m}}]\frac{dy}{dx}$$

$\Rightarrow$
$$4m^2y^2 = \left[\frac{dy}{dx}\right]^2 [y^{\frac{1}{m}} - y^{-\frac{1}{m}}]^2$$
$$= \left[\frac{dy}{dx}\right]^2 [[y^{\frac{1}{m}} + y^{-\frac{1}{m}}]^2 - 4] = \left[\frac{dy}{dx}\right]^2 [4x^2 - 4]$$

$\Rightarrow$
$$m^2y^2 = (x^2 - 1)\left[\frac{dy}{dx}\right]^2$$

Example 27. If $x^y = e^{x-y}$ prove that $\frac{dy}{dx} = \frac{\log x}{(1 + \log x)^2}$.

Solution : $x^y = e^{x-y} \Rightarrow y \log x = (x - y) \log e \Rightarrow y \log x = x - y$

$\Rightarrow$
$$y = \frac{x}{1 + \log x}$$
$$\frac{dy}{dx} = \frac{(1 + \log x) - \frac{x}{x}}{(1 + \log x)^2} = \frac{\log x}{(1 + \log x)^2}$$

Example 28. Differentiate $\log[\sqrt{x-a}+\sqrt{x-b}]$ w.r.t. x.

Solution : Let $$y=\log[\sqrt{x-a}+\sqrt{x-b}]$$

$$\frac{dy}{dx}=\frac{\left[\frac{1}{2}(x-a)^{-1/2}+\frac{1}{2}(x-b)^{-1/2}\right]}{\sqrt{x-a}+\sqrt{x-b}}$$

$$=\frac{\frac{1}{2}\left[\frac{1}{\sqrt{x-a}}+\frac{1}{\sqrt{x-b}}\right]}{\sqrt{x-a}+\sqrt{x-b}}=\frac{1}{2\sqrt{(x-a)(x-b)}}$$

Example 29. If $y=ae^{mx}+be^{-mx}$, prove that $\frac{d^2y}{dx^2}-m^2y=0$.

Solution : Differentiating both sides of $y=ae^{mx}+be^{-mx}$ w.r.t. x, we get

$$\frac{dy}{dx}=mae^{mx}-mbe^{-mx} \qquad ...(I)$$

Differentiating (I) again w.r.t. x, we get

$$\frac{d^2y}{dx^2}=m^2ae^{mx}+m^2be^{-mx}=m^2(ae^{mx}+be^{-mx})=m^2y$$

$\Rightarrow \frac{d^2y}{dx^2}-m^2y=0$ which proves the result.

Example 30. Find $\frac{dy}{dx}$ if $y=\log e^x\left(\frac{x-2}{x+2}\right)^{3/4}$.

Solution :
$$y=\log e^x+\frac{3}{4}\log(x-2)-\frac{3}{4}\log(x+2)$$

$$\frac{dy}{dx}=\frac{1}{e^x}.e^x+\frac{3}{4(x-2)}-\frac{3}{4(x+2)}$$

$$=1+\frac{\frac{3}{4}(x+2-x+2)}{x^2-4}=1+\frac{3}{x^2-4}.$$

Example 31. If $y=[x+\sqrt{1+x^2}]^m$ show that $(1+x^2)\frac{d^2y}{dx^2}+x\frac{dy}{dx}-m^2y=0$.

Solution : $y=[x+\sqrt{1+x^2}]^m \Rightarrow \log y=m\log[x+\sqrt{1+x^2}]$

Differentiating both sides w.r.t. x, we get

$$\frac{1}{y}\frac{dy}{dx}=\frac{m\left[1+\frac{x}{\sqrt{1+x^2}}\right]}{x+\sqrt{1+x^2}}=\frac{m}{\sqrt{1+x^2}}\Rightarrow\frac{dy}{dx}=\frac{my}{\sqrt{1+x^2}}$$

Differentiating again w.r.t. x we get

$$\frac{d^2y}{dx^2} = m\frac{dy}{dx} \times \frac{1}{\sqrt{1+x^2}} = my\left[-\frac{1}{2}(1+x^2)^{-3/2}\right]2x$$

$$= m\frac{dy}{dx} \times \frac{1}{\sqrt{1+x^2}} - \frac{mxy}{(1+x^2)^{3/2}}$$

$$\Rightarrow \quad \frac{d^2y}{dx^2}(1+x^2) = m\sqrt{1+x^2}\frac{dy}{dx} - x\frac{my}{\sqrt{1+x^2}}$$

$$= m^2y - x\frac{dy}{dx} \qquad \text{[By (I)]}$$

$$\Rightarrow \quad (1+x^2)\frac{d^2y}{dx^2} + x\frac{dy}{dx} - m^2y = 0$$

Example 32. Find $\frac{dy}{dx}$ when $y = \log\sqrt{2x+\sqrt{4x^2+a^2}}$.

Solution :

$$y = \frac{1}{2}\log(2x+\sqrt{4x^2+a^2})$$

$$\frac{dy}{dx} = \frac{1}{2}\left[\frac{2+\frac{1}{2}(4x^2+a^2)^{-1/2}(8x)}{2x+\sqrt{4x^2+a^2}}\right]$$

$$= \frac{2x+\sqrt{4x^2+a^2}}{\sqrt{4x^2+a^2}\,\{2x+\sqrt{4x^2+a^2}\}} = \frac{1}{\sqrt{4x^2+a^2}}.$$

Example 33. Differentiate

$$\frac{1}{\sqrt{x^2+a^2}+\sqrt{x^2+b^2}} \text{ w.r.t. } x.$$

Solution :

$$y = \frac{1}{\sqrt{x^2+a^2}+\sqrt{x^2+b^2}}$$

$$= \frac{\sqrt{x^2+a^2}-\sqrt{x^2+b^2}}{[\sqrt{x^2+a^2}+\sqrt{x^2+b^2}]\,[\sqrt{x^2+a^2}-\sqrt{x^2+b^2}]}$$

$$= \frac{\sqrt{x^2+a^2}-\sqrt{x^2+b^2}}{a^2-b^2}$$

$$\frac{dy}{dx} = \frac{d}{dx}\left[\frac{\sqrt{x^2+a^2}}{a^2-b^2}\right] - \frac{d}{dx}\left[\frac{\sqrt{x^2+b^2}}{a^2-b^2}\right]$$

$$= \frac{\frac{1}{2}(x^2+a^2)^{-1/2}\,2x}{a^2-b^2} - \frac{\frac{1}{2}(x^2+b^2)^{-1/2}\,2x}{a^2-b^2}$$

$$= \frac{x}{(a^2-b^2)}\left[\frac{1}{\sqrt{x^2+a^2}} - \frac{1}{\sqrt{x^2+b^2}}\right]$$

Example 34. Obtain the derivative of $x^2 \log_3 x$ with respect to x.

Solution : $$y = x^2 \log_3 x = x^2 \frac{\log_e x}{\log_e 3} = x^2 (\log_e x) \log_3 e \qquad \left[\because \log_b a = \frac{1}{\log_a b}\right]$$

$$\frac{dy}{dx} = \log_3 e \left[\frac{x^2}{x} + 2x \log_e x\right] = \log_3 e\, [1 + 2 \log_e x]\, x$$

Example 35. If $y = \dfrac{x}{\sqrt{1+x^2}}$, prove that $x^3 \dfrac{dy}{dx} = y^3$.

Solution : $$\frac{dy}{dx} = \frac{\sqrt{1+x^2 - x \,.\, \frac{1}{2}(1+x^2)^{-1/2}\, 2x}}{(1+x^2)} \qquad \text{[Quotient Rule]}$$

$$= \frac{\sqrt{1+x^2} - \dfrac{x^2}{\sqrt{1+x^2}}}{1+x^2} = \frac{1}{(1+x^2)^{3/2}}$$

$$= \frac{x^3}{x^3 [\sqrt{1+x^2}]^3} = \frac{1}{x^3}\left[\frac{x}{\sqrt{1+x^2}}\right]^3 = \frac{y^3}{x^3} \Rightarrow x^3 \frac{dy}{dx} = y^3 \qquad \left[\because y = \frac{x}{\sqrt{1+x^2}}\right]$$

Example 36. If $f(x) = \left(\dfrac{c+x}{d+x}\right)^{c+d+2x}$, show that

$$f'(0) = \left(\frac{c}{d}\right)^{c+d}\left[2 \log \frac{c}{d} + \frac{d^2 - c^2}{cd}\right]$$

Solution : $$f(x) = \left(\frac{c+x}{d+x}\right)^{c+d+2x}$$

$\Rightarrow$ $$\log [f(x)] = (c+d+2x)\, [\log (c+x) - \log (d+x)]$$

Differentiating both sides w.r.t. x, we get

$$\frac{1}{f(x)} f'(x) = 2\, [\log (c+x) - \log (d+x)] + (c+d+2x)\left[\frac{1}{c+x} - \frac{1}{d+x}\right]$$

$\Rightarrow$ $$f'(0) = f(0)\left[2\, (\log c - \log d) + (c+d)\left(\frac{1}{c} - \frac{1}{d}\right)\right]$$

$$= \left(\frac{c}{d}\right)^{c+d}\left[2 \log \frac{c}{d} + \frac{d^2 - c^2}{cd}\right] \qquad \left[\because f(0) = \left[\frac{c}{d}\right]^{c+d}\right]$$

Example 37. If $y = x^3 \log \left(\dfrac{1}{x}\right)$, prove that $x \dfrac{d^2y}{dx^2} - 2\dfrac{dy}{dx} + 3x^2 = 0$.

Solution : $y = x^3 \log \left(\dfrac{1}{x}\right) = x^3 (\log 1 - \log x) = -x^3 \log x$

Differentiating both sides w.r.t. x, we get

$$\frac{dy}{dx} = -x^3 \frac{d}{dx}(\log x) - \log x \frac{d}{dx}(x^3)$$

$$= -\frac{x^3}{x} - (\log x)\, 3x^2 = -x^2 - 3x^2 \log x \quad \text{...(I)}$$

$$x\frac{dy}{dx} = -3x^3 \log x - x^3 = 3y - x^3 \qquad \left[\because\ y = x^3 \log \frac{1}{x} = -x^3 \log (x)\right] \text{...(II)}$$

Differentiating (II), w.r.t. x, we get

$$x\frac{d^2y}{dx^2} + \frac{dy}{dx} = 3\frac{dy}{dx} - 3x^2 \Rightarrow x\frac{d^2y}{dx^2} - 2\frac{dy}{dx} + 3x^2 = 0$$

Example 38. If $y = x^2 + \frac{1}{x^2}$, show that $x^2\frac{d^2y}{dx^2} + x\frac{dy}{dx} = 4y$.

Solution : Differentiating $y = x^2 + \frac{1}{x^2}$ w.r.t. x we get

$$\frac{dy}{dx} = 2x + \left[\frac{-2}{x^3}\right] = 2\left[x - \frac{1}{x^3}\right] = \frac{2}{x}\left[x^2 - \frac{1}{x^2}\right]$$

$$\Rightarrow \qquad x^2\left(\frac{dy}{dx}\right)^2 = 4\left[x^2 - \frac{1}{x^2}\right]^2 = 4\left[\left(x^2 + \frac{1}{x^2}\right)^2 - 4\right]$$

$$= 4\,[y^2 - 4] = 4y^2 - 16 \quad \text{...(II)}$$

Differentiating (II), w.r.t. x we get

$$(x^2)\, 2\left(\frac{dy}{dx}\right)\frac{d^2y}{dx^2} + 2x\left(\frac{dy}{dx}\right)^2 = 8y\frac{dy}{dx}$$

$$\Rightarrow \qquad x^2\frac{d^2y}{dx^2} + x\frac{dy}{dx} = 4y$$

Example 39. If $x^y + y^x = a^b$, find $\frac{dy}{dx}$.

Solution : Let $u = x^y$, then $\log u = y \log x$.

$$\frac{1}{u}\frac{du}{dx} = \frac{y}{x} + (\log x)\frac{dy}{dx} \Rightarrow \frac{du}{dx} = u\left[\frac{y}{x} + (\log x)\frac{dy}{dx}\right]$$

$$\Rightarrow \qquad \frac{d}{dx}(x^y) = x^y\left[\frac{y}{x} + \frac{dy}{dx}(\log x)\right] \quad \text{...(I)}$$

Let $v = (y^x)$, then $\log v = x \log y$

Differentiating w.r.t. x, we get $\frac{1}{v} \times \frac{dv}{dx} = \frac{x}{y}\frac{dy}{dx} + \log y$

$$\Rightarrow \qquad \frac{d}{dx}(y^x) = y^x\left[\frac{x}{y}\frac{dy}{dx} + \log y\right]$$

Differentiating both sides of $x^y + y^x = a^b$, w.r.t. x, we get

$$\frac{d}{dx}(x^y) + \frac{d}{dx}(y^x) = 0$$

$$\Rightarrow \qquad x^y\left(\frac{y}{x}+\frac{dy}{dx}\log x\right)+y^x\left(\frac{x}{y}\frac{dy}{dx}+\log y\right)=0$$

$$\Rightarrow \qquad \frac{dy}{dx}\left[x^y\log x+\frac{x}{y}y^x\right]=-\left[y^x\log y+\frac{y}{x}x^y\right]$$

$$\Rightarrow \qquad \frac{dy}{dx}=-\frac{\left[y^x\log y+\frac{y}{x}x^y\right]}{x^y\log x+\frac{x}{y}y^x}$$

Example 40. If $x^y y^x = c$, find $\frac{dy}{dx}$ at the point $(x = e, y = e)$, c being constant.

Solution : $x^y y^x = c \Rightarrow y\log x + x\log y = \log c$ ...(I)

(Taking log of both sides)

Differentiating (I), w.r.t. x, we get

$$\frac{y}{x}+\log x\frac{dy}{dx}+\frac{x}{y}\frac{dy}{dx}+\log y=0$$

$$\Rightarrow \qquad \frac{dy}{dx}\left(\log x+\frac{x}{y}\right)=-\left(\log y+\frac{y}{x}\right)\Rightarrow\frac{dy}{dx}=-\left[\frac{\log y+\frac{y}{x}}{\log x+\frac{x}{y}}\right]$$

$$\Rightarrow \qquad \left[\frac{dy}{dx}\right]_{y=e}=-\left[\frac{\log e+1}{\log e+1}\right]=-1 \qquad [\because \log_e e=1]$$

Example 41. If $x^y = y^x$, show that $\frac{dy}{dx}=\frac{y(x\log y-y)}{x(y\log x-x)}$.

Solution : Refer Ex. 39, Page 24.

Example 42. Differentiate $x^x + x^{1/x}$ w.r.t. x.

Solution : Let $u = x^x$, $\log u = x\log x$

$$\frac{1}{u}\frac{du}{dx}=\frac{x}{x}+\log x=1+\log x$$

$$\Rightarrow \qquad \frac{du}{dx}=u(1+\log x)\Rightarrow\frac{du}{dx}=x^x(1+\log x)=\frac{d}{dx}(x^x)$$

$$\Rightarrow \qquad \frac{du}{dx}=4(1+\log x)$$

Let $v = x^{1/x}$, then $\log v=\frac{1}{x}\log x$

Differentiating w.r.t. x, we get

$$\frac{1}{v}\frac{dv}{dx}=\frac{1}{x}.\frac{1}{x}+\left(-\frac{1}{x^2}\right)\log x=\frac{1}{x^2}(1-\log x)$$

$$\Rightarrow \qquad \frac{dv}{dx}=v\left[\frac{1-\log x}{x^2}\right]\Rightarrow\frac{dv}{dx}=x^{1/x}\left[\frac{1-\log x}{x^2}\right]=\frac{d}{dx}[x^{1/x}]$$

$$\therefore \qquad \frac{d}{dx}[x^x + x^{1/x}] = \frac{d}{dx}(x^x) + \frac{d}{dx}[x^{1/x}]$$

$$= x^x(1 + \log x) + x^{1/x}\left[\frac{1 - \log x}{x^2}\right]$$

Example 43. Differentiate w.r.t. x (i) $(1 + x)^{1+x}$ (ii) $(\log x)^{\log x}$.

Solution : (i) Let $y = (1 + x)^{1+x} \Rightarrow \log y = (1 + x)\log(1 + x)$

Differentiating w.r.t. x, we get

$$\frac{1}{y}\frac{dy}{dx} = (1 + x)\frac{1}{1 + x} + \log(1 + x) = 1 + \log(1 + x)$$

$$\Rightarrow \qquad \frac{dy}{dx} = (1 + x)^{1+x}[1 + \log(1 + x)]$$

Note : Since, $\log_e e = 1$, the above result can be written as

$$\frac{dy}{dx} = (1 + x)^{1+x}[\log_e e + \log_e(1 + x)] = (1 + x)^{1+x}[\log e(1 + x)]$$

(ii) Let $y = (\log x)^{\log x} \Rightarrow \log y = \log x \log[\log x]$

Differentiating w.r.t. x, we get

$$\frac{1}{y}\frac{dy}{dx} = (\log x)\frac{1}{\log x}.\left[\frac{1}{x}\right] + \log[\log x]\frac{1}{x}$$

$$= \frac{1}{x}[1 + \log(\log x)]$$

$$\Rightarrow \qquad \frac{dy}{dx} = (\log x)^{\log x}\left[\frac{1 + \log(\log x)}{x}\right]$$

Note : $\frac{d}{dx}[\log\{\log(x)\}] = \left[\frac{1}{\log x}\right]\frac{d}{dx}(\log x) = \frac{1}{\log x}\left(\frac{1}{x}\right) = \frac{1}{x}\left[\frac{1}{\log x}\right]$

Example 44. If $y = \sqrt{x + y}$ find $\frac{dy}{dx}$.

Solution : $y = \sqrt{x + y} \Rightarrow y^2 = x + y$

Differentiating both sides w.r.t. x, we get

$$2y\frac{dy}{dx} = 1 + \frac{dy}{dx} \Rightarrow \frac{dy}{dx}(2y - 1) = 1 \Rightarrow \frac{dy}{dx} = \frac{1}{2y - 1}$$

Example 45. If $y = x^{x^{x\ldots\infty}}$ prove that $x\frac{dy}{dx} = \frac{y^2}{1 - y\log x}$.

Solution : $y = x^{x^{x\ldots\infty}} \Rightarrow y = x^y$

Taking log of both sides, we get, $\log y = y\log x$

Differentiating both sides w.r.t. x, we get

$$\frac{1}{y}\frac{dy}{dx} = \frac{y}{x} + [\log x]\frac{dy}{dx}$$

$$\Rightarrow \qquad \left(\frac{1}{y} - \log x\right)\frac{dy}{dx} = \frac{y}{x} \Rightarrow x\frac{dy}{dx} = \frac{y^2}{1 - y\log x}$$

Example 46. Find the slope of the curve $x^2 + xy + 2y^2 = 28$ at $x = 2, y = 3$.

Solution : $x^2 + xy + 2y^2 = 28$

Differentiating both sides w.r.t. x, we get

$$2x + x\frac{dy}{dx} + y + 4y\frac{dy}{dx} = 0 \Rightarrow \frac{dy}{dx}(x + 4y) = -(y + 2x)$$

$$\Rightarrow \qquad \frac{dy}{dx} = -\frac{(y + 2x)}{x + 4y}$$

When $x = 2, y = 3$ we get $\frac{dy}{dx} = -\frac{7}{14} = -\frac{1}{2}$

Example 47. If $y = \sqrt{25 - 3x}$ find $\frac{dy}{dx}$ and $\frac{d^2y}{dx^2}$ at $x = 3$.

Solution :

$$y = (25 - 3x)^{1/2}$$

$$\frac{dy}{dx} = \frac{1}{2}(25 - 3x)^{-1/2}(-3)$$

$$= \frac{-3/2}{\sqrt{25 - 3x}} = \frac{-3}{8} \text{ at } x = 3$$

$$\frac{d^2y}{dx^2} = -\frac{3}{2} \times \frac{-1}{2} \times (-3) \times \frac{1}{(25 - 3x)^{3/2}}$$

$$= \frac{-9}{4} \times \frac{1}{(25 - 3x)^{3/2}} = \frac{-9}{256} \text{ at } x = 3$$

Example 48. If $y = \cfrac{\sqrt{x}}{1 + \cfrac{\sqrt{x}}{1 + \cfrac{\sqrt{x}}{1 + \ldots \infty}}}$ show that $\frac{dy}{dx} = \frac{1}{2\sqrt{x}\,(1 + 2y)}$.

Solution : $y = \cfrac{\sqrt{x}}{1 + \cfrac{\sqrt{x}}{1 + \cfrac{\sqrt{x}}{1 + \ldots \infty}}} \Rightarrow y = \frac{\sqrt{x}}{1 + y}$

i.e., $y + y^2 = \sqrt{x}$

Differentiating both sides w.r.t. x, we get

$$\frac{dy}{dx} + 2y\frac{dy}{dx} = \frac{1}{2\sqrt{x}}$$

$$\frac{dy}{dx}(1 + 2y) = \frac{1}{2\sqrt{x}} \Rightarrow \frac{dy}{dx} = \frac{1}{2\sqrt{x}}(1 + 2y)^{-1}$$

Example 49. If $x^p y^q = (x + y)^{p+q}$, show that $\frac{dy}{dx} = \frac{y}{x}$. What is the value of $\frac{dy}{dx}$ when $\sqrt{x}\,\sqrt[3]{y} = (x + y)^{5/6}$?

Solution : Take log of both sides,

$$p \log x + q \log y = (p+q) \log (x+y)$$

Differentiating both sides w.r.t. x, we get

$$\frac{p}{x} + \frac{q}{y}\frac{dy}{dx} = \frac{p+q}{x+y}\left(1 + \frac{dy}{dx}\right) \Rightarrow \frac{dy}{dx}\left[\frac{q}{y} - \frac{p+q}{x+y}\right] = \frac{p+q}{x+y} - \frac{p}{x}$$

$$\Rightarrow \quad \frac{dy}{dx}\left[\frac{qx+qy-py-qy}{y(x+y)}\right] = \frac{px+qx-px-py}{x(x+y)}$$

$$\Rightarrow \quad \frac{dy}{dx}\frac{(qx-py)}{y} = \frac{qx-py}{x} \Rightarrow \frac{dy}{dx} = \frac{y}{x}$$

Second Part

$$\sqrt{x}\,\sqrt[3]{y} = (x+y)^{5/6}$$

$$\Rightarrow \quad x^{1/2}\,y^{1/3} = (x+y)^{(1/2+1/3)}$$

This is of the form $\quad x^p.y^q = (x+y)^{p+q}$

$$\therefore \quad \frac{dy}{dx} = \frac{y}{x}$$

Example 50. Find $\frac{dy}{dx}$ when $x^y = e^{x+y}$.

Solution : Refer Ex. 27, Page 20.

Example 51. Differentiate w.r.t. $x\sqrt{x^2+1} + \log[\log(x+\sqrt{x^2+1})]$

Solution : Let $\quad y = x\sqrt{x^2+1} + \log[\log(x+\sqrt{x+1})]$

$$\frac{dy}{dx} = \frac{d}{dx}[x\sqrt{x^2+1}] + \frac{d}{dx}\log[\log(x+\sqrt{x^2+1})]$$

$$= \sqrt{x^2+1} + x.\frac{1}{2}(x^2+1)^{-1/2}.2x + \frac{1}{\log(x+\sqrt{x^2+1})}.\frac{\left[1+\frac{1}{2}(x^2+1)^{-1/2}.2x\right]}{x+\sqrt{x^2+1}}$$

$$= \frac{x^2+1+x^2}{\sqrt{x^2+1}} + \frac{x+\sqrt{x^2+1}}{(x+\sqrt{x^2+1})(\sqrt{x^2+1})}.\frac{1}{\log(x+\sqrt{x^2+1})}$$

$$= \frac{2x^2+1}{\sqrt{x^2+1}} + \frac{1}{(\sqrt{x^2+1})\log(x+\sqrt{x^2+1})}$$

Example 52. Examine $y = \sqrt{|x|}$ to find the values of x for which the graph has a vertical tangent.

Solution : $y = \sqrt{|x|}$

$y = \sqrt{x}$ when $x \geq 0$

$= \sqrt{-x}$ when $x < 0$

$\frac{dy}{dx} = \frac{1}{2\sqrt{x}}$ when $x \geq 0$, [-ve value of x is ignored]

[If the tangent is vertical at $x = a$, then $\frac{dy}{dx} \Rightarrow \infty$ as $x \to a$]

$\therefore$ The tangent will be vertical when $x = 0$. When $x = 0$, y is also equal to zero therefore, the tangent is vertical at (0, 0).

Example 53. If $\sqrt{y+x}+\sqrt{y-x}=c$, show that

$$\frac{dy}{dx}=\frac{y}{x}-\sqrt{\frac{y^2}{x^2}-1}.$$

Solution :

$$\sqrt{y+x}+\sqrt{y-x}=c$$

Differentiating both sides, w.r.t. x, we get

$$\frac{1}{2}(y+x)^{-1/2}\left(\frac{dy}{dx}+1\right)+\frac{1}{2}(y-x)^{-1/2}\left(\frac{dy}{dx}-1\right)=0$$

i.e.,
$$\frac{1}{2}\frac{dy}{dx}\left[\frac{1}{(y+x)^{1/2}}+\frac{1}{(y+x)^{1/2}}\right]+\frac{1}{2}\left[\frac{1}{(y+x)^{1/2}}-\frac{1}{(y-x)^{1/2}}\right]=0$$

i.e.,
$$\frac{dy}{dx}\frac{(y-x)^{1/2}+(y+x)^{1/2}}{(y^2-x^2)^{1/2}}=\frac{(y+x)^{1/2}-(y-x)^{1/2}}{(y^2-x^2)^{1/2}}$$

i.e.,
$$\frac{dy}{dx}=\frac{[(y+x)^{1/2}-(y-x)^{1/2}]}{[(y+x)^{1/2}+(y-x)^{1/2}]}\times\frac{[(y+x)^{1/2}-(y-x)^{1/2}]}{[(y+x)^{1/2}-(y-x)^{1/2}]}$$

$$=\frac{y}{x}-\frac{2(y^2-x^2)^{1/2}}{2x}=\frac{y}{x}-\sqrt{\frac{y^2}{x^2}-1}$$

Second Method : Squaring both sides of the equation $\sqrt{y+x}+\sqrt{y-x}=c$, we get

$$(y+x)+(y-x)+2\sqrt{y^2-x^2}=c^2$$

$\Rightarrow$
$$2\sqrt{y^2-x^2}=c^2-2y \qquad \text{...(I)}$$

Squaring again

$$4(y^2-x^2)=c^4+4y^2-4c^2y$$

$\therefore$
$$4x^2-4c^2y+c^4=0$$

Differentiating w.r.t. x

$$8x-4c^2\frac{dy}{dx}=0 \Rightarrow \frac{dy}{dx}=\frac{2x}{c^2} \qquad \text{...(II)}$$

$$=\frac{x}{\sqrt{y^2-x^2}+y}=\frac{x}{y+\sqrt{y^2-x^2}} \qquad \text{(By I)}$$

$$=\frac{x\,[y-\sqrt{y^2-x^2}]}{x^2}=\frac{[y-\sqrt{y^2-x^2}]}{x}$$

[Multiplying the numerator and denominator by $(y-\sqrt{y^2-x^2})$]

$$=\frac{y}{x}-\sqrt{\frac{y^2}{x^2}-1}$$

Example 54. If $x^3y^4=(x+y)^7$ show that $\frac{dx}{dy}=\frac{x}{y}$.

Solution : $x^3y^4=(x+y)^7$

Differentiating both sides w.r.t. y, we get

$$3x^2y^4\frac{dx}{dy} + 4x^3y^3 = 7(x+y)^6\left(1+\frac{dx}{dy}\right)$$

$$\Rightarrow \quad \frac{dx}{dy}[3x^2y^4 - 7(x+y)^6] = 7(x+y)^6 - 3x^3y^3$$

$$\Rightarrow \quad \frac{dx}{dy}\left[3x^2y^4 - \frac{7(x+y)^7}{x+y}\right] = \frac{7(x+y)^7}{x+y} - 3x^3y^3$$

$$\Rightarrow \quad \frac{dx}{dy}\left[x^2y^4 - \frac{7x^3y^4}{x+y}\right] = \frac{7x^3y^4}{x+y} - 4x^3y^3 \qquad [\because (x+y)^7 = x^3y^4]$$

$$\Rightarrow \quad \frac{dx}{dy}\left[3y^2 - \frac{7xy^2}{x+y}\right] = \left[\frac{7xy^2}{x+y} - 4xy\right]$$

$$\Rightarrow \quad \frac{dx}{dy}[3xy^2 + 3y^3 - 7xy^2] = [7xy^2 - 4x^2y - 4xy^2]$$

$$\Rightarrow \quad \frac{dx}{dy}(3y^3 - 4xy^2) = (3xy^2 - 4x^2y)$$

$$\Rightarrow \quad \frac{dx}{dy}y(4y^2 - 4xy) = x(3y^2 - 4xy) \Rightarrow \frac{dx}{dy} = \frac{x}{y}$$

Example 55. If $x^5y^3 = (x+y)^8$ show that $y = x\dfrac{dy}{dx}$.

Solution : $x^5y^3 = (x+y)^8$

Differentiating both sides w.r.t. x

$$5x^4y^3 + 3x^5y^2\frac{dy}{dx} = 8(x+y)^7\left(1+\frac{dy}{dx}\right)$$

$$\Rightarrow \quad \frac{dy}{dx}[3x^5y^2 - 8(x+y)^7] = 8(x+y)^7 - 5x^4y^3$$

$$\Rightarrow \quad \frac{dy}{dx}\left[3x^5y^2 - \frac{8(x+y)^8}{x+y}\right] = \frac{8(x+y)^8}{x+y} - 5x^4y^3$$

$$\Rightarrow \quad \frac{dy}{dx}\left[3x^5y^2 - \frac{8x^5y^3}{x+y}\right] = \frac{8x^5y^3}{x+y} - 5x^4y^3 \qquad [\because (x+y)^8 = x^5y^3]$$

$$\Rightarrow \quad \frac{dy}{dx}\left[3x - \frac{8xy}{x+y}\right] = \left[\frac{8xy}{x+y} - 5y\right]$$

$$\Rightarrow \quad \frac{dy}{dx}[3x^2 + 3xy - 8xy] = 8xy - 5xy - 5y^2$$

$$x\frac{dy}{dx}[3x - 5y] = y[3x - 5y] \Rightarrow x\frac{dy}{dx} = y$$

Example 56. Evaluate $\sqrt[3]{25}$ approximately by use of differentials.

Solution : If $y = f(x)$ is a function of x and if Δx is a small increment in the value of x then Δy the corresponding increment in the value of y is approximately given by

$$\Delta y = f'(x)\,\Delta x$$

Now if $f(x) = \sqrt[3]{x}$

then $$\Delta y = \sqrt[3]{(x+\Delta x)} - \sqrt[3]{x} = \frac{d}{dx}\sqrt[3]{x}\,\Delta x = \frac{1}{3}x^{-2/3}\,\Delta x \qquad ...(I)$$

If we put $x = 27, \Delta x = -2$, then from (I), we get

$$\sqrt[3]{25} - \sqrt[3]{27} \approx \frac{1}{3}(27)^{-2/3}(-2)$$

i.e.,

$$\sqrt[3]{25} \approx 3 - \frac{2}{27}$$

$$= \frac{79}{27} = 2.926$$

i.e., $\sqrt[3]{25} \approx 2.926$ [≈ indicates approximately equal to]

Example 57. If $y = \sqrt{x+1} - \sqrt{x-1}$ prove that

$$(x^2-1)\frac{d^2y}{dx^2} + x\frac{dy}{dx} - \frac{1}{4}y = 0$$

Solution :

$$\frac{dy}{dx} = \frac{1}{2}(x+1)^{-\frac{1}{2}} - \frac{1}{2}(x-1)^{-\frac{1}{2}} = \frac{1}{2}\left[\frac{1}{\sqrt{x+1}} - \frac{1}{\sqrt{x-1}}\right]$$

$$= \frac{1}{2}\frac{(x-1)^{1/2} - (x+1)^{1/2}}{\sqrt{x^2-1}} = -\frac{y}{2\sqrt{x^2-1}}$$

$$\Rightarrow \quad \frac{dy}{dx}\sqrt{x^2-1} = -\frac{y}{2} \qquad \text{...(I)}$$

Differentiating (I), we get

$$\frac{d^2y}{dx^2}\sqrt{x^2-1} + \frac{1}{2}(x^2-1)^{-1/2}(2x)\frac{dy}{dx} = -\frac{1}{2}\left(\frac{dy}{dx}\right) = \frac{y}{4\sqrt{x^2-1}}$$

$$\Rightarrow \quad \frac{d^2y}{dx^2}(x^2-1) + x\frac{dy}{dx} - \frac{1}{4}y = 0$$

Example 58. If $xy = ax^2 + \frac{b}{x}$ show that $x^2y_2 + 2(xy_1 - y) = 0$.

Solution : $xy = ax^2 + \frac{b}{x}$

Differentiating w.r.t. x we get

$$x\frac{dy}{dx} + y = 2ax - \frac{b}{x^2} \qquad \text{...(I)}$$

Differentiating (I) again w.r.t. x, we get

$$x\frac{d^2y}{dx^2} + \frac{dy}{dx} + \frac{dy}{dx} = 2a + 2\frac{b}{x^3}$$

$$\Rightarrow \quad x^2\frac{d^2y}{dx^2} + 2x\frac{dy}{dx} = 2\left(ax + \frac{b}{x^2}\right) = 2y \quad \left[\because xy = ax^2 + \frac{b}{x} \Rightarrow y = ax + \frac{b}{x^2}\right]$$

$$\Rightarrow \quad x^2y_2 + 2(xy_1 - y) = 0 \quad \left[\because \frac{d^2y}{dx^2} = y_2, \frac{dy}{dx} = y_1\right]$$

Example 59. If $y = 2x + \frac{4}{x}$, prove that

$$x^2 \frac{d^2y}{dx^2} + x\frac{dy}{dx} - y = 0$$

Solution :

$$y = 2x + \frac{4}{x}$$

$$\frac{dy}{dx} = 2 - \frac{4}{x^2}; \frac{d^2y}{dx^2} = \frac{8}{x^3}$$

$$x^2 \frac{d^2y}{dx^2} + x\frac{dy}{dx} - y = x^2 \times \frac{8}{x^3} + 2x - \frac{4}{x} - 2x - \frac{4}{x}$$

$$= \frac{8}{x} + 2x - \frac{4}{x} - 2x - \frac{4}{x} = 0$$

Example 60. If $y = \log [x + \sqrt{x^2 + a^2}]$, prove that

$$(a^2 + x^2)\frac{d^2y}{dx^2} + x^2\frac{dy}{dx} = 0$$

Solution :

$$y = \log (x + \sqrt{a^2 + x^2})$$

$$\frac{dy}{dx} = \frac{1}{x + \sqrt{a^2 + x^2}}\left[1 + \frac{x}{\sqrt{a^2 + x^2}}\right]$$

$$= \frac{x + \sqrt{a^2 + x^2}}{x + \sqrt{a^2 + x^2}} \times \frac{1}{\sqrt{a^2 + x^2}} = \frac{1}{\sqrt{a^2 + x^2}}$$

$$\Rightarrow \qquad (\sqrt{x^2 + a^2})\frac{dy}{dx} = 1$$

Differentiating again, we get

$$\sqrt{x^2 + a^2}\frac{d^2y}{dx^2} + \frac{dy}{dx} \times \frac{1}{2}(x^2 + a^2)^{-1/2}(2x) = 0$$

$$\Rightarrow \qquad (x^2 + a^2)\frac{d^2y}{dx^2} + x\frac{dy}{dx} = 0$$

Example 61. If $y = \log\left[\frac{\sqrt{x^2 + a^2} + x}{\sqrt{x^2 + a^2} - x}\right]$ find $\frac{dy}{dx}$.

Solution : Now

$$y = \log_e \frac{\sqrt{x^2 + a^2} + x}{\sqrt{x^2 + a^2} - x}$$

$$= \log_e [\sqrt{x^2 + a^2} + x] - \log_e [\sqrt{x^2 + a^2} - x]$$

Further Refer Ex. 61.

Example 62. Differentiate :

(i) $\log(xe^x)$ w.r.t. $x \log x$ (ii) $\dfrac{ax+b}{cx+d}$ w.r.t. $\dfrac{a'x+b'}{c'x+d'}$ (iii) $x^{1/3}$ w.r.t. $x^{2/5}$

Solution : (i) Let $y = \log(xe^x)$ and $u = x \log x$

Then we are to find $\dfrac{dy}{du}$

Now $$\frac{dy}{dx} = \frac{1}{xe^x}(xe^x + e^x) = \frac{1+x}{x}$$

and $$\frac{du}{dx} = x.\frac{1}{x} + \log x = 1 + \log x$$

$$\frac{dy}{du} = \frac{dy}{dx}.\frac{dx}{du} = \frac{dy}{dx}\Big/\frac{du}{dx}$$

$$= \frac{1+x}{x}.\frac{1}{1+\log x} = \frac{(1+x)}{x(1+\log x)}$$

(ii) Let $y = \dfrac{ax+b}{cx+d}$ and $u = \dfrac{a'x+b'}{c'x+d'}$

The we are to find $\dfrac{dy}{du}$

$$y = \frac{ax+b}{cx+d}$$

$$\frac{dy}{dx} = \frac{(cx+d)\,a - (ax+b)\,c}{(cx+d)^2} = \frac{ad-bc}{(cx+d)^2}$$

Similarly $$\frac{du}{dx} = \frac{a'd'-b'c'}{(c'x+d')^2}$$

$\therefore$ $$\frac{dy}{du} = \frac{dy}{dx} \div \frac{du}{dx}$$

$$= \frac{(ad-bc)\,(c'x+d')^2}{(a'd'-b'c')\,(cx+d)^2}$$

(iii) Please try. **Ans.** $\dfrac{5}{6x^{1/15}}$

Example 63. If $x = \dfrac{1-t^2}{1+t^2}$, $y = \dfrac{2t}{1+t^2}$ prove that $\dfrac{dy}{dx} + \dfrac{x}{y} = 0.$

Solution : $$\frac{dx}{dt} = \frac{(1+t^2)(-2t) - (1-t^2)\,2t}{(1+t^2)^2}$$

$$= \frac{-2t\,[1+t^2+1-t^2]}{(1+t^2)^2} = \frac{-4t}{(1+t^2)^2}$$

$$\frac{dy}{dt} = \frac{(1+t^2)\times 2 - 2t\times 2t}{(1+t^2)^2} = \frac{2(1-t^2)}{(1+t^2)^2}$$

$$\therefore \qquad \frac{dy}{dx} = \frac{dy}{dt} \div \frac{dx}{dt} = \frac{2(1-t^2)}{(1+t^2)^2} \div \frac{-4t}{(1+t^2)^2}$$

$$= \frac{2(1-t^2)}{-4t} = -\frac{(1-t^2)}{2t} = -\frac{x}{y}$$

i.e.,

$$\frac{dy}{dx} + \frac{x}{y} = 0$$

Example 64. If $x^2 + y^2 = t - \frac{1}{t}$, $x^4 + y^4 = t^2 + \frac{1}{t^2}$, prove that $\frac{dy}{dx} = xy^3$.

Solution : $$(x^2 + y^2)^2 = \left(t - \frac{1}{t}\right)^2 \Rightarrow x^4 + y^4 + 2x^2y^2 = t^2 + \frac{1}{t^2} - 2$$

$$\Rightarrow \qquad 2x^2y^2 = -2 \qquad \left[\because x^4 + y^4 = t^2 + \frac{1}{t^2}\right]$$

$$\Rightarrow \qquad x^2y^2 = -1 \qquad ...(I)$$

Differentiating both sides w.r.t. x we get

$$(x^2)\, 2y \frac{dy}{dx} + y^2(2x) = 0 \Rightarrow \left(\frac{dy}{dx}\right)\frac{x^2y}{y^2x} = -1$$

$$\Rightarrow \qquad \frac{dy}{dx}\frac{x^2y^2}{xy^3} = -1 \Rightarrow \frac{dy}{dx} = xy^3 \qquad [\because x^2y^2 = -1 \text{ by (I)}]$$

Questions

1. Differentiate from first principles :

(i) x^2 (ii) x^n (iii) $x^3 + x$ (iv) $\sqrt{\frac{x+1}{x+2}}$

(v) $\frac{1}{\sqrt{ax+b}}$ (vi) $\sqrt{x^2+1}$

[**Ans.** (i) $2x$ (ii) nx^{n-1} (iii) $3x^2 + 1$] (iv) $\frac{1}{2(x+1)^{1/2}(x+2)^{3/2}}$ (v) $\frac{-a}{2(ax+b)^{3/2}}$ (vi) $\frac{x}{\sqrt{x^2+1}}$]

2. Differentiate the following w.r.t. x :

(i) $5x^4 + x^3 + 3x^2 + 9$ (ii) $\frac{x^4 + 3x^2}{\sqrt{x}}$ (iii) $x^5 + x^4 + \log x + \frac{1}{x} + \frac{1}{x^2}$

(iv) $(2x+3)(x+1)$ (v) $\frac{5 \log x}{x}$ (vi) $\frac{\log x}{x^2}$

(vii) $x^2 \log x$ (viii) $(\log x)^2$ (ix) $x(\log x)^3$

(x) $\frac{2+x^2}{3+x^3}$ (xi) $(\log x)\, x^5$ (xii) $(x+1)(x+2)\log x$

(xiii) $(2x^2+1)^{7/2}$ (xiv) $[5x^2+3x^3+4]^6$ (xv) $(\log x)\dfrac{1}{x}$

[**Ans.** (i) $20x^3+3x^2+6x$ (ii) $\dfrac{7}{2}x^{5/2}+\dfrac{9}{2}x^{1/2}$ (iii) $5x^4+4x^3+\dfrac{1}{x}-\dfrac{1}{x^2}-\dfrac{2}{x^3}$ (iv) $4x+5$ (v) $5\left[\dfrac{1-\log x}{x^2}\right]$

(vi) $\dfrac{1-2\log x}{x^3}$ (vii) $x(1+2\log x)$ (viii) $\dfrac{2\log x}{x}$ (ix) $(\log x)^2(3+\log x)$ (x) $\dfrac{8x^4+6x^2+6x}{(3+x^3)^2}$

(xi) $x^4+5x^4\log x$ (xii) $(x+2)\log x+(x+1)\left[\dfrac{x+2}{x}\right]+\log x$ (xiii) $14(2x^2+1)^{5/2}$

(xiv) $6(5x^2+3x^3+4)(10x+9x^2)$ (xv) $\dfrac{1}{x^2}[1-\log x]$]

3. Differentiate w.r.t. x :

(i) $x^5+5^x+\log_5(x)$ (ii) x^{a+b} (iii) $\dfrac{1}{\sqrt{x}}+3^x$ (iv) $3^x\,2^{-2x}$

[**Ans.** (i) $5x^4+5^x\log 5+\dfrac{1}{x}\log_5 e$ (ii) $(a+b)\,x^{a+b-1}$ (iii) $-\dfrac{1}{2}x^{-3/2}+3^x\log 3$ (iv) $\left(\dfrac{3}{4}\right)^x\log\left(\dfrac{3}{4}\right)$]

4. Differentiate w.r.t. x :

(i) $\dfrac{x+e^x}{1+\log x}$ (ii) $\dfrac{e^x-e^{-x}}{e^x+e^{-x}}$ (iii) $\dfrac{\sqrt{x}}{a+bx}$

(iv) $(3x-1)^4(x+2)^5(2x-3)^3$ (v) $x^3e^{(1-x)}$ (vi) $(x+e^x)\,e^{-x}+\log x$

[**Ans.** (i) $\dfrac{x(e^x+\log x+e^x\log x)-e^x}{x(1+\log x)^2}$ (ii) $\left(\dfrac{2}{e^x+e^{-x}}\right)^2$ (iii) $\dfrac{a-bx}{2\sqrt{x}\,(a+bx)^2}$

(iv) $(3x-1)^4(x+2)^5(2x-3)^3\left(\dfrac{12}{3x-1}+\dfrac{5}{x+2}+\dfrac{6}{2x-3}\right)$

(v) $x^2e^{1-x}(3-x)$ (vi) $\left(e^{-x}(1-x)+\dfrac{1}{x}\right)$]

5. Find $\dfrac{dy}{dx}$ for the function $x^yy^x=k^2$ where k is a constant.

[**Ans.** $-y(x\log y+y)/x(y\log x+x)$]

6. Find $\dfrac{dy}{dx}$ if (i) $x^y+y^x=1$ (ii) $x^y+xy=8$

[**Ans.** (i) $\dfrac{yx^{y-1}+y^x\log y}{x^y\log x+xy^{x-1}}$ (ii) $\dfrac{-y(x^{y-1}+1)}{x(x^{y-1}\log x+1)}$]

7. If $y=\sqrt{\log x+\sqrt{\log x+\sqrt{\log x+\ldots}}}$ prove that

$$x\frac{dy}{dx}=\frac{1}{2y-1}.$$

8. If $x^{1/2}y^{3/2}=(x+y)^2$, then prove that $\dfrac{dy}{dx}=\dfrac{y}{x}$.

9. If $x^3+y^3+3axy=0$, then prove that $\dfrac{dy}{dx}=-\left[\dfrac{x^2+ay}{y^2+ax}\right]$.

10. If $y=\dfrac{x^2}{e^x}$, find $\dfrac{dy}{dx}$.

[**Ans.** $\dfrac{dy}{dx}=\dfrac{x}{e^x}(2-x)$]

11. If $y=(5x+1)(x^2-4)$, find $\dfrac{dy}{dx}$.

[**Ans.** $15x^2+2x-20$]

12. If $y = 2e^{a^{2x}}$ find $\frac{dy}{dx}$.

[**Ans.** $2e^{a^{2x}} e^{2x}$]

13. Find $\frac{dy}{dx}$ if $\log\left(\frac{x}{y}\right) = x + y$.

[**Ans.** $\frac{y(1-x)}{x(1+y)}$]

14. Find $\frac{dy}{dx}$ if

(i) $y = (ax + b)\, e^{-x}$ (ii) $y = \frac{1+e^x}{1-e^x}$ (iii) $y = \log(ax^2 + bx + c)$

(iv) $y = (x^2 + 1)\log(x^2 + 1)$ (v) $y = x^{-2} e^{-x^2}$ (vi) $y = e^{x^2 - 3x + 7}$

(vii) $x^{2/3} + y^{2/3} = a^{2/3}$ (viii) $y = [3x^3 + 2x^2 - 6x^{-4}]^{-1}$ (ix) $y = \frac{1+\sqrt{x}}{1-\sqrt{x}}$

(x) $y = 7x^{1/3} + \frac{3}{x} - \frac{5}{5\sqrt{x}}$ (xi) $y = a + bx + cx^2 + e^{3x} + e^{-x^2}$

(xii) $y = e^{5x^2 + 2x + 2}$ (xiii) $y = e^x(6x^3 + 5x + 2)$ (xiv) $y = e^{\sqrt{x}}$

(xv) $y = \sqrt{x} + 3\sqrt{x + \frac{2}{x}}$ (xvi) $y = e^{3x} + e^{x^2 + 1}$ (xvii) $y = a^{x^2} + 2^x$

(xviii) $y = xe^{-x} + (x + 4)\, e^x$ (xix) $y = e^{3x} + \sqrt{x^2 - 4 + 5}$ (xx) $y = \frac{3x + \log x}{3x - 4}$

(xxi) $y = \frac{e^x + x^2}{x - \log x}$

[**Ans.** (i) $e^{-x}[a(1-x) + b]$ (ii) $\frac{2e^x}{(1-e^x)^2}$ (iii) $\frac{2ax + b}{ax^2 + bx + c}$ (iv) $2x[1 + \log(x^2 + 1)]$

(v) $-2x^2 e^{-x^2}\left[\frac{x^2+1}{x}\right]$ (vi) $(2x - 3)\, e^{x^2 - 3x + 7}$ (vii) $(y/x)^{1/3}$ (viii) $[9x^2 + 4x + 24x^{-5}]\,[3x^3 + 2x^2 - 6x^{-4}]^{-}2$

(ix) $\frac{x^{-1/2}}{(1-\sqrt{x})^2}$ (x) $\frac{7}{3}x^{-2/3} - 3x^{-2} + x^{-6/5}$ (xi) $b + cx + 3e^{3x} - 2xe^{-x^2}$ (xii) $2e^{5x^2 + 2x + 2}(5x + 1)$

(xiii) $(6x^3 + 18x^2 + 5x + 7)\, e^x$ (xiv) $\frac{1}{2\sqrt{x}} e^{\sqrt{x}}$ (xv) $\frac{1}{2}x^{-1/2} + \frac{1}{3}x^{-2/3} - \frac{1}{x^2}$ (xvi) $3e^{3x} + 2xe^{x^2+1}$

(xvii) $2xa^{x^2}\log_e a + 2^x \log_e 2$ (xviii) $(1 - x)\, e^{-x} + (x + 5)\, e^x$ (xix) $3e^{3x} + \frac{x}{\sqrt{x^2 - 4}}$

(xx) $\frac{(3x-4)(3x+1) - 3x(3x + \log x)}{(3x-4)^2}$ (xxi) $\frac{(x - \log x)(e^x + 2x)\,x - (e^x + x^2)(x - 1)}{x(x - \log x)^2}$]

15. If $x\sqrt{1+y} + y\sqrt{1+x} = 0$ prove that $\frac{dy}{dx} = -(1+x)^{-2}$.

16. If $y = \left(\frac{1+x}{1-x}\right)^r$ prove that $(1 - x^2)\frac{dy}{dx} = 2ry$
where r is a given positive integer.

17. If $y = \sqrt{x} - \frac{1}{\sqrt{x}} + \frac{x+2}{2-x}$ and $\frac{dy}{dx}$ at $x = 4$.

[**Ans.** 21/16]

18. Find $\frac{dy}{dx}$ if $y = x\log x \log\log x$.

[**Ans.** $1 + (1 + \log x)\log\log x$]

19. If $y = a^{\log(\log x)} + b^{x \log x}$ find $\frac{dy}{dx}$.

[**Ans.** $\frac{a^{\log \log x}(\log a)}{x \log x} b^{x \log x} (\log b)(1 + \log x)$]

20. If $y = \log[\sqrt{x-a} + \sqrt{x-b}]$ find $\frac{dy}{dx}\left[\frac{1}{2\sqrt{(x-a)(x-b)}}\right]$.

21. If $y = \log_a[(x-1)(x^2+1)^{-1/4}]$ find $\frac{dy}{dx}$.

[**Ans.** $\log_a e \frac{1}{x-1} - \frac{x}{2(x+1)}$]

22. If $y = e^{\sqrt{x}+1} - e^{\sqrt{x+2}}$ find $\frac{dy}{dx}$.

[**Ans.** $\frac{e^{\sqrt{x}+2}}{2\sqrt{x}} - \frac{e^{\sqrt{x+2}}}{2\sqrt{x+2}}$]

23. Find $\frac{dy}{dx}$ when $x^2 + y^2 = \log(x^2 - y^2)$.

[**Ans.** $\frac{x}{y}\left[\frac{1-x^2+y^2}{x^2-y^2+1}\right]$]

24. If $x = y(1 + \log x)$ show that $\frac{dy}{dx} = \frac{\log x}{(1+\log x)^2}$.

25. If $y = \frac{x}{\log x}$, find $\frac{dy}{dx}$.

[**Ans.** $\frac{\log x - 1}{(\log x)^2}$]

26. Differentiate $\frac{(x-1)^{4/3}(3x+7)^{2/5}}{\sqrt{x-2}\,(2x+1)^3}$

[**Ans.** $\frac{(x-1)^{4/3}(3x+7)^{2/5}}{\sqrt{x-2}\,(2x+1)^3}\left(\frac{4}{3(x-1)} + \frac{6}{5(3x+7)} - \frac{1}{2(x-2)} - \frac{6}{2x+1}\right)$]

27. If $y\sqrt{x^2+1} = \log[x + \sqrt{x^2+1}]$

show that $(x^2+1)\frac{dy}{dx} + xy - 1 = 0$.

28. If $y = (\log x)^2$, prove that

$$x^2 \frac{d^2y}{dx^2} + x\frac{dy}{dx} = 2.$$

29. (i) If $y = \frac{\sqrt{x^2+1} + \sqrt{x^2-1}}{\sqrt{x^2+1} - \sqrt{x^2-1}}$

prove that $\frac{dy}{dx} = 2x + \frac{2x^3}{\sqrt{x^4-1}}$.

(ii) Differentiate w.r.t. x

$$\log e^x \left(\frac{x-2}{x+2}\right)^{3/4}$$

[**Ans.** $\frac{x^2-1}{x^2-4}$]

[**Hint :** $y = \log e^x \left(\frac{x-2}{x+2}\right)^{3/4} = x \log_e e + \frac{3}{4}[\log(x-2) - \log(x+2)]$

$= x + \frac{3}{4}[\log(x-2) - \log(x+2)]$

(iii) If $\frac{x^2}{a^2} + \frac{y^2}{b^2} = 1$, find $\frac{d^2y}{dx^2}$ and show that $a^2y^3 \frac{d^2y}{dx^2} + b^4 = 0$.

[**Ans.** $\frac{-b^4}{a^2y^3}$]

30. Find $\frac{dy}{dx}$ when

(i) $\log xy = x^2 + y^2$ [**Ans.** $\frac{y(2x^2-1)}{x(2y^2-1)}$]

(ii) $x = y \log (xy)$ [**Ans.** $\frac{x-y}{x(1+\log xy)}$]

(iii) $\sqrt{\frac{y}{x}} + \sqrt{\frac{x}{y}} = 6$ [**Ans.** $\frac{\sqrt{y}\,(\sqrt{x} - 3\sqrt{y})}{\sqrt{x}\,(3\sqrt{x} - \sqrt{y})}$]

(iv) $ax^2 + 2hxy + by^2 = 1$ [**Ans.** $\left(\frac{ax+hy}{hx+by}\right)$]

31. If $y = x^{y^x}$ show that $\frac{dy}{dx} = \frac{y \log y\,(x \log x \log y + 1)}{(1 - x \log y)\, x \log x}$.

[**Hint :** $y = x^{y^x} \Rightarrow \log y = y^x \log x \Rightarrow \log(\log y) = x \log y + \log(\log x)$
Differentiating both sides

$$\frac{1}{\log y} \cdot \frac{1}{y}\frac{dy}{dx} = \frac{x}{y}\frac{dy}{dx} + \log y + \frac{1}{x \log x}$$

32. If $y^3 - 3ax^2 + x^3 = 0$, show that

$$\frac{d^2y}{dx^2} = -\frac{2a^2x^2}{y^5}.$$

33. If $x^2 + 2xy + 3y^2 = 1$, show that

$$(x+3y)^3 \frac{d^2y}{dx^2} + 2 = 0.$$

34. If $y = [x + \sqrt{x^2 - 1}\,]^m$ show that

$$(x^2 - 1)\frac{d^2y}{dx^2} + x\frac{dy}{dx} - m^2y = 0$$

35. If $x = at^2$, $y = 2at$, find $\frac{d^2y}{dx^2}$.

[**Ans.** $\frac{-1}{2at^3}$]

36. Differentiate $\frac{\log x}{e^x}$.

[**Ans.** $\frac{1 - x \log x}{xe^x}$]

37. Differentiate w.r.t. x

(i) $\frac{\sqrt{x+1} + \sqrt{x-1}}{\sqrt{x+1} - \sqrt{x-1}}$

[**Ans.** $1 + \frac{x}{\sqrt{x^2-1}}$]

(ii) $\sqrt{\dfrac{x^2-2ax}{a^2-2ab}}$

[**Ans.** $\dfrac{x-a}{(\sqrt{a^2-2ab})\,(\sqrt{x^2-2ax})}$]

(iii) $e^{\log[x+\sqrt{x^2+a^2}]}$

[**Ans.** $e^{\log(x+\sqrt{x^2+a^2})}\dfrac{1}{\sqrt{x^2+a^2}}$]

38. If $\sqrt{1-x^2}+\sqrt{1-y^2}=a(x-y)$

prove that $\dfrac{dy}{dx}=\dfrac{\sqrt{1-y^2}}{\sqrt{1-x^2}}$.

39. If $x\sqrt{1+y}+y\sqrt{1+x}=0$,

show that $\dfrac{dy}{dx}=\dfrac{-1}{(1+x)^4}$.

40. Differentiate $(\log x)^{\log x}$

[**Ans.** $\dfrac{(\log x)^{\log x}}{x}\log(\log x+1)$]

41. Prove that the function $y=\log\dfrac{1}{1+x}$ satisfies the relationship $x\dfrac{dy}{dx}+1=e^y$.

[**Hint :** $y=\log_e\dfrac{1}{1+x}\Rightarrow e^y=\dfrac{1}{1+x}$]

42. If $\dfrac{x}{x-y}=\log\dfrac{a}{x-y}$, prove that $\dfrac{dy}{dx}=2-\dfrac{x}{y}$.

43. If $y=\dfrac{x}{\sqrt{1+x^2}}$ prove that $x^3\dfrac{dy}{dx}=y^2$.

44. Differentiate $x^{e^x}+e^{x^x}$.

[**Ans.** $\dfrac{x^{e^x}e^x}{x}(1+x\log x)+e^{x^x}.x^x(1+\log x)$]

45. If $ye^y=x$, prove that $\dfrac{dy}{dx}=\dfrac{y}{x(1+y)}$.

46. If $y=\log[x+\sqrt{1+x^2}]$ show that

$$(1+x^2)\,y_2-xy_1=0$$

47. If $y=\sqrt{x+1}+\sqrt{x-1}$ show that

$$(x^2-1)\,y_2+xy_1=\frac{1}{4}y.$$

48. If $y=A\,[x+\sqrt{x^2-1}]^n$ show that

$$(x^2-1)\,y_2+xy_1-n^2y=0$$

49. If $f(x)=e^{x^2}-2$ show that $f'(0)=0$.

50. If $y=\sqrt{1-x^2}$ show that

$$(1-x^2)\,y_2-xy=\frac{x}{y_1}$$

51. If $y = x \log \dfrac{x}{a+x}$ show that

$$x^3 \frac{d^2y}{dx^2} = \left[x\frac{dy}{dx} - y\right]^2$$

[**Hint :** $y = x \log \dfrac{x}{a+x} = x \log x - x \log (a+x)$

$$\frac{dy}{dx} = 1 + \log x - \frac{x}{a+x} - \log (a+x) = 1 + \log \frac{x}{a+x} - \frac{x}{a+x}$$

$$= \log \frac{x}{a+x} + \frac{a}{a+x} = \frac{y}{x} + \frac{a}{a+x}$$

$$\left[x\frac{dy}{dx} - y\right] = \frac{ax}{a+x} \Rightarrow \left[x\frac{dy}{dx} - y\right]^2 = \left[\frac{ax}{a+x}\right]^2$$

$$\frac{d^2y}{dx^2} = \frac{d}{dx}\left[1 + \log \frac{x}{a+x} - \frac{x}{a+x}\right] = \frac{a+x}{x}\frac{d}{dx}\left[\frac{x}{a+x}\right] - \frac{d}{dx}\left[\frac{x}{a+x}\right]$$

$$= \left[\frac{a+x}{x} - 1\right]\frac{d}{dx}\left[\frac{x}{a+x}\right]$$

$$= \left[\frac{a}{x}\right]\frac{a}{(a+x)^2} = \frac{a^2}{x(a+x)^2}$$

$$x^3\left[\frac{d^2y}{dx^2}\right] = \frac{a^2x^2}{(a+x)^2} = \left[\frac{ax}{a+x}\right]^2].$$

❑❑❑

5

Applications of Differentiation to Business and Economics

5.1 INTRODUCTION

Maxima and Minima

Let the figure below represent the graph of the function $y = f(x)$, continuous in the closed interval (a, b).

From the graph it is clear that the given function has maximum value at B (*i.e.*, at $x = b$) and minimum value at A (*i.e.*, at $x = a$). *These are the maximum and minimum values in the domain of definition of the given function.*

Now consider the point P on the graph corresponding to $x = p$. If we take a small neighbourhood of P, say $p - \delta, p + \delta$, then for every x, such that $p - \delta < x < p + \delta$, $f(x) \leq f(p)$, *i.e.*, $f(p)$ is the maximum value in the small neighbourhood of the point P. Here the student must distinguish between $f(b)$ and $f(p)$, $f(b)$ is the maximum value of the function in the domain of definition whereas $f(p)$ is the maximum value of the function in a small interval in the neighbourhood of P.

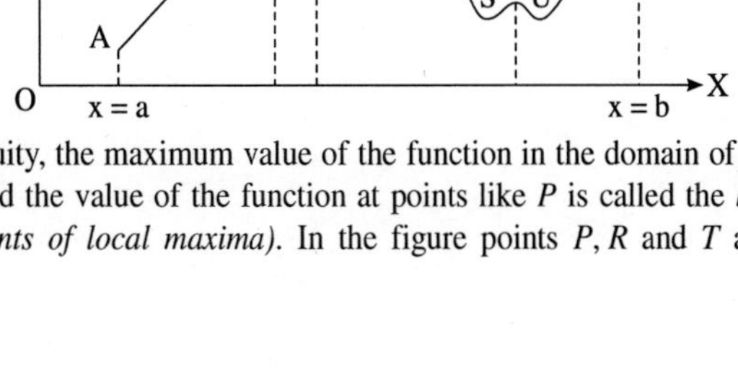

To avoid ambiguity, the maximum value of the function in the domain of definition is called the *absolute maximum* and the value of the function at points like P is called the *local maximum (points like P are called points of local maxima).* In the figure points P, R and T are the points of local maxima.

It may be noted that in a given interval there is only one point representing the absolute maximum value of the function but there can be more than one points representing the local maxima.

Now consider the point Q corresponding to $x = q$. If we take small neighbourhood of Q say $q - \delta, q + \delta$, then for every x such that

$$q - \delta < x < q + \delta, f(x) \geq f(q)$$

i.e., $f(q)$ is the minimum value of the function in the small neighbourhood of Q. The points like Q are called points of *local minima*. The minimum value of the function in the domain of definition is called **absolute minimum**. In a given interval there is only one point representing the absolute minimum value of the function but there can be more than one points representing the local minimum (Q, S, U, in the figure represent local minima).

To summarise, we may say that in a given interval, there can be only one point representing absolute maximum and only one point representing absolute minimum but there can be many points representing local maxima and local minima.

It may further be noted that :

1. The value of the function at a point of local minima may sometimes be more than the value of the function at a point of local maxima (points Q and T in the figure).
2. The points of local maxima and local minima occur alternatively.
3. The absolute maximum and absolute minimum values of a function may occur either at end points or at a point of local maxima and local minima respectively. See figure below.

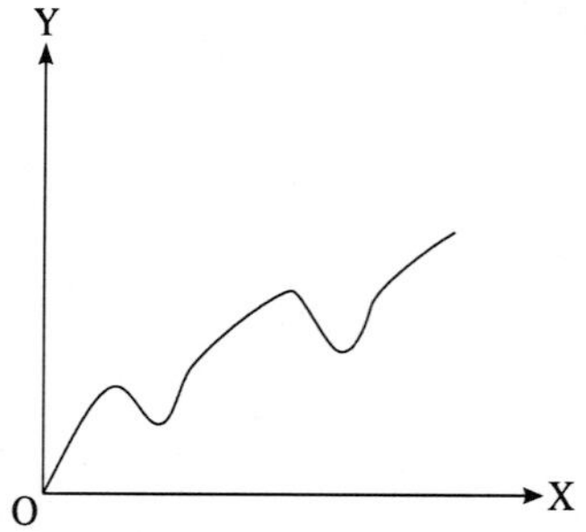

Absolute maximum and absolute minimum at end points.

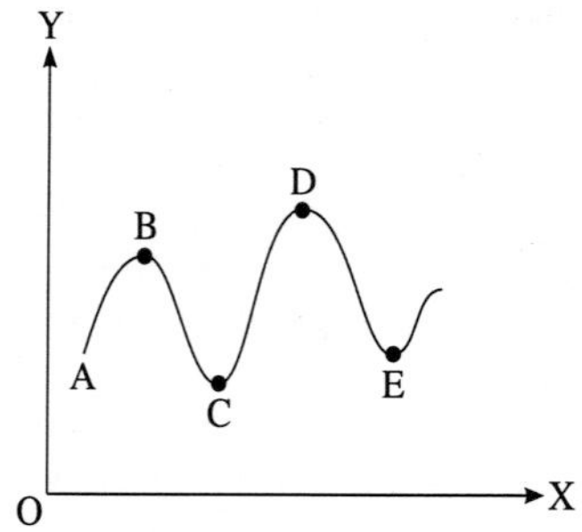

Absolute maximum and absolute minimum coinciding with local maximum and local minimum D and C respectively.

4. At least one local maxima or local minima must lie between two equal values of the given function unless it is a constant function.

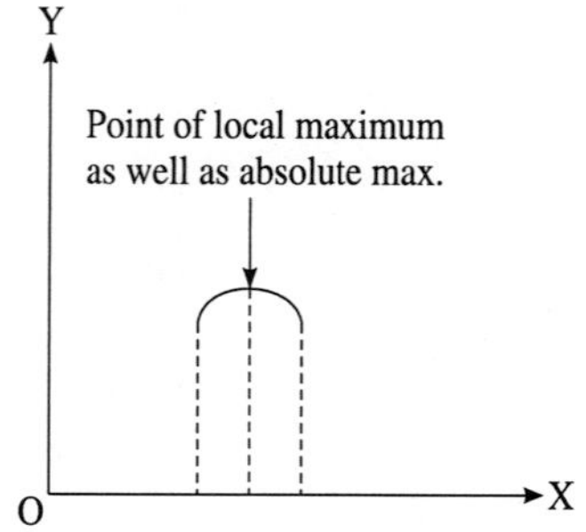

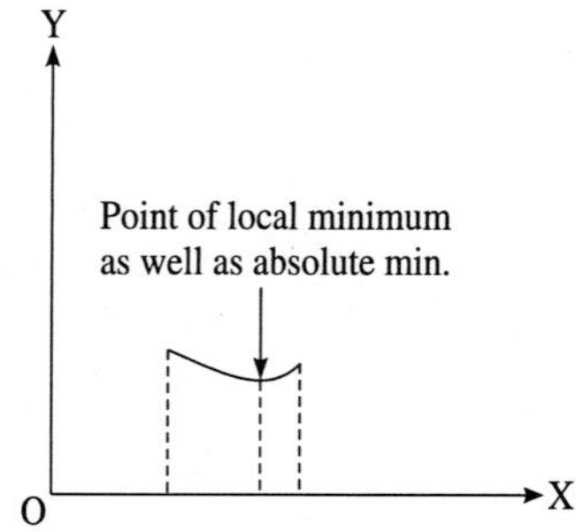

5. The local maximum and local minimum values taken together are called **extreme values** and the points at which these occurs are called **turning points.**
6. If a function has a single local maximum or local minimum in an interval then that is also the absolute maximum or the absolute minimum value respectively of the function in that interval.

5.2 DETERMINATION OF LOCAL MAXIMA AND LOCAL MINIMA

When $f(x)$ increases with x to a certain value and then diminishes, the function attains a local maximum at that turning point.

Similarly when $f(x)$ dimnishes with x increasing to a certain value and then increases, the function attains a local minima at that turning point. When $f(x)$ increases the curve is said to be **rising** and when $f(x)$ diminishes the curve is said to be **falling.**

We know that the slope of the curve at any point is represented by $\frac{dy}{dx}$ and is equal to the tangent of the angle made by the tangent at that point with the positive direction of x-axis. The tangent at any point on the rising curve makes an acute angle with the positive direction of x-axis, *i.e.*, $\frac{dy}{dx}$ is positive

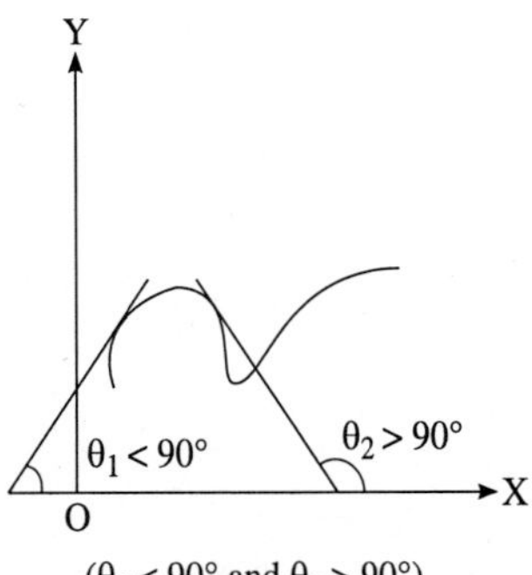

$(\theta_1 < 90°$ and $\theta_2 > 90°)$

at such points, and the tangent at any point on the falling curve makes an obtuse angle with the positive direction of x-axis. Hence, $\frac{dy}{dx}$ is negative at such points. But at the points of local maxima (or local minima) the tangent is parallel to x-axis, *i.e.*, $\frac{dy}{dx} = 0$. Hence we conclude :

$\frac{dy}{dx}$ changes from positive to negative in passing through a point of local maxima and from negative to positive in passing through a point of local minima. At the points of local maxima or local minima $\frac{dy}{dx} = 0$ [if $\frac{dy}{dx}$ exists].

But if, at any point, $f'(x)\left[i.e., \frac{dy}{dx}\right]$ is equal to zero, the point may or may not represent local maxima or local minima, *i.e., all the roots of the equation $f'(x) = 0$ do not correspond to local maxima or local minima.* **(Important)**

In order to determine the roots of $f'(x) = 0$ correspond to local maxima or local minima, we have to take the help of second derivative and the rule is :

Those roots of the equation $f'(x) = 0$ which make $f''(x)$, i.e., $\frac{d^2y}{dx^2}$ negative represent the points of local maxima and the roots which make $f''(x)$ i.e., $\frac{d^2y}{dx^2}$ positive represent points of local minima.

If any root of the equation $f'(0) = 0$ makes $\frac{d^2y}{dx^2} = 0$, we have to make further investigations which will be discussed later.

A very interesting case arises when the given function is continuous at a point but the derivative of the function at that point is infinitely large [we normally say that the derivative does not exist at that point], the function may have local maxima or minima at such points. These points are obtained by solving $\frac{1}{f'(x)} = 0$. In such cases we note the behaviour of $f'(x)$ in the deleted neighbourhood of the point, if $f'(x)$ changes from positive to negative while passing over the point, it represents ***local maxima****. If $f'(x)$ changes negative to positive while passing over the point it represents* ***local minima.*** *Second derivative will not be used in such cases.*

Let
$$y = f(x) = a - b(x-c)^{2/3}$$
$$\frac{dy}{dx} = -\frac{2b}{3}(x-c)^{-1/3} = \frac{2b}{3(x-c)^{1/3}}$$
at $x = c$,
$$\frac{dy}{dx} = \infty$$
$$\frac{1}{f'(x)} = \frac{3(x-c)^{1/3}}{2b}$$
and
$$\frac{1}{f'(x)} = 0, \text{ give } x = c$$

When x is slightly less than $c, f'(x)$ is positive and when x is slightly more than c, $f'(x)$ is negative. Hence, the function $f(x) = a - b\,(x-c)^{2/3}$ has local maxima at $x = c$.

5.3 CONCAVITY, CONVEXITY AND POINTS OF INFLEXION

Let $y = f(x)$ be a function of x, then the graph of $f(x)$ is said to be **concave upwards (or downwards)** at a point P if in the neighbourhood of P, the curve lies above (or below) the tangent at P on both

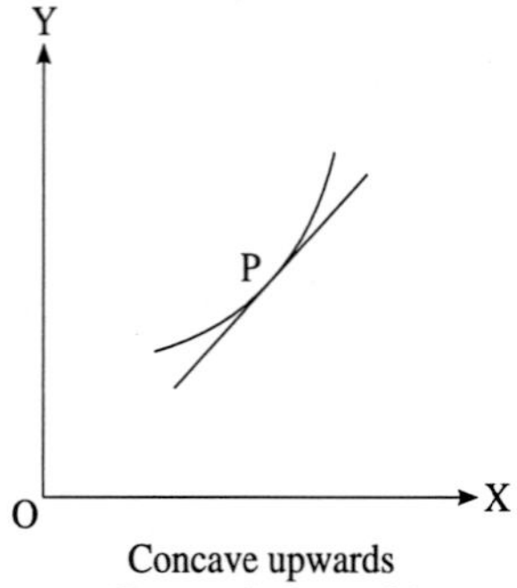

Concave upwards
(Convex downwards)

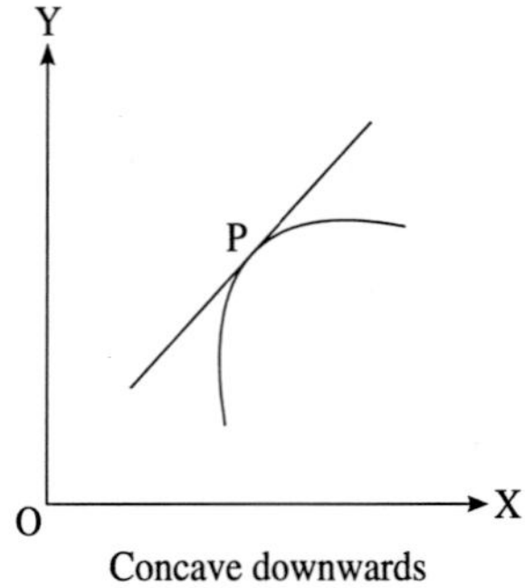

Concave downwards
(Convex upwards)

the sides. The graph which is concave upwards is called **convex downwards** [$f''(x)$ is Positive] and the graph which is concave downwards is called **convex upwards** [$f''(x)$ is negative].

A point Q on the graph of $f(x)$ is said to be a point of inflexion if the graph changes the sense of concavity at this point.

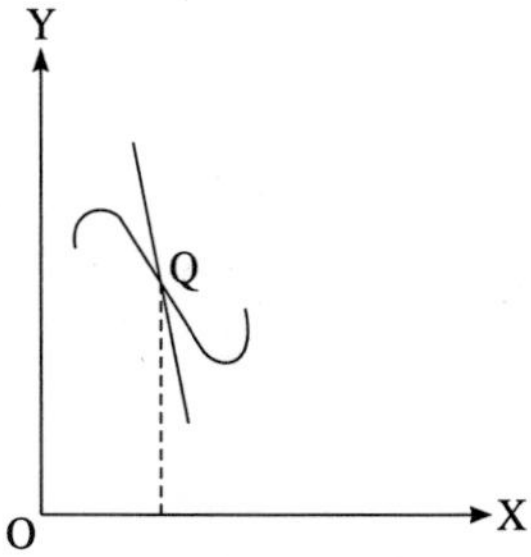

At Q, graph changes from concave downwards to concave upwards.

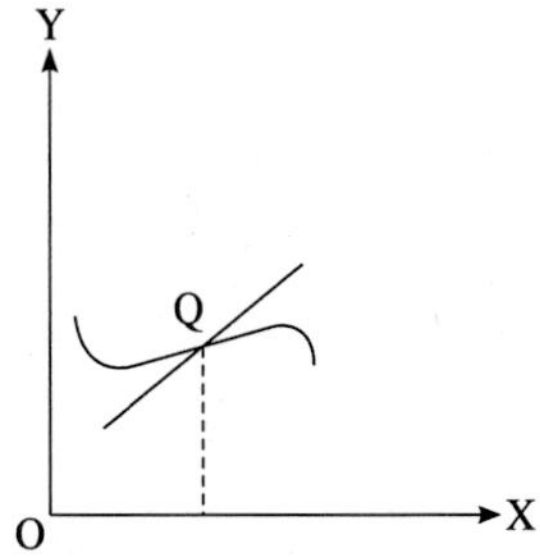

At Q, graph changes from concave upwards to concave downwards.

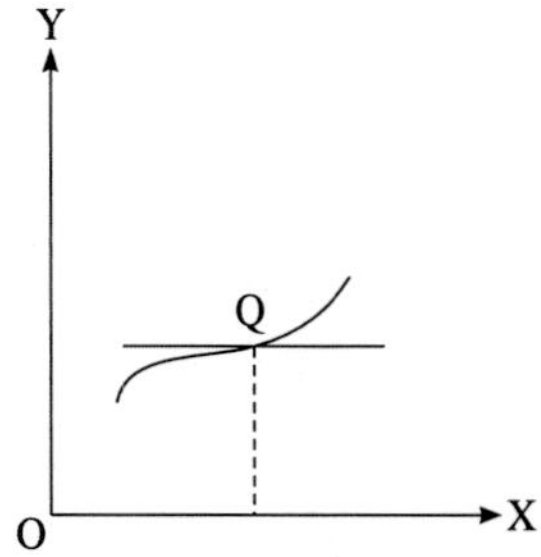

At Q, graph changes from concave downwards to concave upwards.

At the point of inflexion if $f''(x)$ exists, it is equal to zero. If $f''(x)$ does not exist, the it must change sign, *i.e.,* from positive to negative or *vice versa.*

The students should note a characteristic difference between the turning point and a point of inflexion that at the turning point, the function is increasing on the left and decreasing on the right (or *vice versa*) in the close neighbourhood of the point. But in the close neighbourhood of the point of inflexion, the function is either increasing or decreasing on both sides. **(Please note)**

Notes :

1. The only points where $f(x)$ can have maximum or minimum values are those at which the derivative is zero or does not exist. When we say derivative does not exist, we refer to those points at which the derivative is infinitely large but the function is defined and continuous at these points. All points where $f(x)$ can have maximum or minimum values are also called **critical points.**
2. If $f'(x)$ does not exist at $x = c$, then $f(x)$ will have maximum or minimum value at $x = c$, if and only if $f'(x)$ changes sign as x passes through the value c. The value of $f(c)$ being a maximum or minimum at $x = c$, according as the change of sign in $f'(c)$ is from plus to minus, *i.e.,* +ve to −ve or from minus to plus, *i.e.,* −ve to +ve.
3. If $f'(x) = 0$ at $x = c$ (say) then second derivative test (already explained) will determine whether the point at $x = c$ is a point of maxima or minima.
   It may again be noted that whether $f'(x)$ exists or not at a certain point say $x = c$ but $f'(x)$ will change sign while passing over the point of local maxima and local minima.
4. The curve $y = f(x)$ has a point of inflexion at $x = c$ if (1) $\frac{d^2y}{dx^2} = 0$ when $x = c$, and (2) $\frac{d^2y}{dx^2}$ changes sign as passes through the value $x = c$. If $\frac{d^2y}{dx^2}$ does not exist, *i.e.,* if $\frac{d^2y}{dx^2}$ is infinitely large at $x = c$ (say) but changes sign while passing through the point $x = c$, then also the function is said to have a point of inflexion at $x = c$. **(Important)**

For example, examine the curve $(x-2)^{1/3}$ for inflexion points.

Let $$y = (x-2)^{1/3}$$

$$\frac{dy}{dx} = \frac{1}{3}(x-2)^{-2/3}$$

$$\frac{d^2y}{dx^2} = -\frac{1}{3}\cdot\frac{2}{3}(x-2)^{-5/3} = -\frac{2}{9}(x-2)^{-5/3}$$

Now $$\frac{d^2y}{dx^2} = \infty \text{ at } x = 2.$$

But when x is slightly less than 2, $\frac{d^2y}{dx^2} > 0$, and when x is slightly greater than 2, $\frac{d^2y}{dx^2} < 0$, *i.e.*, $\frac{d^2y}{dx^2}$ changes from positive to negative as x passes as through 2, *i.e.*, $x = 2$ represents a point of inflexion at $x = 2$, $y = 0$.

$\therefore$ (2, 0) is the point of inflexion.

5. If $f'(c) = 0$ (*i.e.*, if $f'(x) = 0$ at $x = c$ and $f'''(c) \neq 0, f(c)$ is neither maximum nor minimum and then c is said to be a point of inflexion. If $f'''(c)$ is also zero, $f^{iv}(c) \neq 0$, $f(c)$ is maximum or minimum according as $f^{iv}(c)$ is negative or positive. Generally, if $f'(c) = f'''(c) = f^{iv}(c) = \ldots = f^{n-1}(c) = 0$ and $f^n(c) \neq 0$, c is a point of inflexion if n is odd and $f(c)$ is maximum or minimum if n is even according as $f^n(c)$ is negative or positive.

Rule : For points of local maxima $\frac{dy}{dx} = 0, \frac{d^2y}{dx^2} < 0$

For points of local minima $\frac{dy}{dx} = 0, \frac{d^2y}{dx^2} > 0.$

SOLVED EXAMPLES

Example 1. Show that the height of a closed cylinder of given surface and maximum volume, is equal to the diameter of its base.

Solution : Let S be the fixed surface area of a closed cylinder of radius r and height h. Let V be its volume.

Then, $$S = (2\pi r^2 + 2\pi rh) \Rightarrow h = \left(\frac{S - 2\pi r^2}{2\pi r}\right)^* \quad \text{...(i)}$$

And, $$V = \pi r^2 h = \pi r^2\left(\frac{S - 2\pi r^2}{2\pi r}\right) = \left(\frac{rS - 2\pi r^3}{2}\right) \quad \text{[using (i)]}$$

* Total surface area of a cylinder = curved surface area (= $2\pi rh$) + Area of the top (= πr^2) + Area of the base (= πr^2).

$$\therefore \qquad \frac{dV}{dr} = \left(\frac{S - 6\pi r^2}{2}\right) \text{ and } \frac{d^2V}{dr^2} = -6\pi r$$

Now, $$\frac{dV}{dr} = 0 \Rightarrow (S - 6\pi r^2) = 0$$

$$\Rightarrow \qquad S = 6\pi r^2 \Rightarrow (2\pi r^2 + 2\pi rh) = 6\pi r^2 \Rightarrow h = 2r, \ i.e., \ r = \frac{1}{2}h.$$

And $$\left[\frac{d^2V}{dr^2}\right]_{r=(h/2)} = -6\pi \times \frac{1}{2}h = -3\pi h < 0$$

So, V is maximum, when $h = 2r$.

Example 2. A closed right circular cylinder has volume 2156 cubic units. What should be the radius of its base, so that its total surface area may be maximum ?

Solution : Let r be the radius and h the height of the cylinder.

Then, $$\pi r^2 h = 2156 \Rightarrow h = \left(\frac{2156}{\pi r^2}\right)$$

Let S be its total surface area.

Then, $$S = (2\pi r^2 + 2\pi rh) = \left(2\pi r^2 + 2\pi r \times \frac{2156}{\pi r^2}\right) = \left(2\pi r^2 + \frac{4312}{r}\right)$$

$$\therefore \qquad \frac{dS}{dr} = \left(4\pi r - \frac{4312}{r^2}\right) \text{ and } \frac{d^2S}{dr^2} = \left(4\pi + \frac{8624}{r^3}\right)$$

Now, $$\frac{dS}{dr} = 0 \Rightarrow \left(4\pi r - \frac{4312}{r^2}\right) = 0$$

$$\Rightarrow \qquad 4\pi r^3 = 3212 \Rightarrow r = \left(\frac{4312}{4\pi}\right)^{1/3} = \left(\frac{4312 \times 7}{5 \times 22}\right)^{1/3} = 7$$

$\frac{d^2S}{dr^2}$ is positive for all positive values for r.

$\therefore$ S is minimum, when $r = 7$.

Example 3. A firm has revenue function given by $R = 8D$ (where R is gross revenue and D is quantity sold and production cost function given by

$$C = 1{,}50{,}000 + 60\left(\frac{D}{900}\right)^2$$

Find the total profit function and the number of units to be sold to get the maximum profit.

(ICWA Dec. 1997)

Solution : Profit = Revenue – Cost **(Please note)**

i.e., Profit $$(P) = 8D - 1{,}50{,}000 - 60\left(\frac{D}{900}\right)^2$$

$$\frac{dP}{dD} = 8 - 120\left[\frac{D}{900}\right] \times \frac{1}{900}$$

Profit will be maximum or minimum when $\frac{dP}{dD} = 0$

$$\frac{dP}{dD} = 0 \Rightarrow 8 - \frac{120}{900 \times 900} \times D = 0 \Rightarrow D = 54{,}000 \text{ units.}$$

$$\frac{d^2P}{dD^2} = -\frac{1}{900 \times 900} \times 120$$

which is negative. Hence, the profit will be maximum when the number of units sold = 54,000.

Example 4. The total cost function (in Rupees) of x units of a product is $c(x) = x^2 + 78x + 2{,}500$ and the demand function is $p = 600 - 8x$, when the price is Rs. p per unit. Show that the maximum net revenue (*i.e.*, profit) is obtained when 29 units are produced. Also find the price at which profit is maximum. **(ICWA Dec. 1994)**

Solution : Total Revenue $= px$

$$\begin{aligned} \text{Net Revenue} = \text{Profit} &= px - c(x) \\ &= x(600 - 8x) - (x^2 + 78x + 2{,}500) \\ &= 600x - 8x^2 - x^2 - 78x - 2{,}500 \\ &= -9x^2 + 522x - 2{,}500 \end{aligned}$$

$$\frac{d\,(\text{Profit})}{dx} = -18x + 522$$

$$\frac{d^2}{dx^2}(\text{Profit}) = -18$$

For Profit to be maximum or minimum $\frac{d\,(\text{Profit})}{dx} = 0$

$\Rightarrow$ $$18x = 522 \Rightarrow x = 29$$

Since $\frac{d^2\,(\text{Profit})}{dx^2}$ is negative for all values of x, therefore Profit is maximum when 29 units are produced. When $x = 29$, $p = 600 - 8 \times 29 = 600 - 232 = \text{Rs. } 368$.

Example 5. For a certain establishment the total cost function C and the total revenue function R are given by $C = x^3 - 12x^2 + 48x + 11$ and $R = 83x - 4x^2 - 21$ where $x =$ output. Obtain the output for which profit is maximum and the maximum profit. **(ICWA June 1986)**

Solution :

$$\begin{aligned} \text{Profit } (P) &= \text{Revnue} - \text{Cost} \\ &= 83x - 4x^2 - 21 - x^3 + 12x^2 - 48x - 11 \\ &= -x^3 + 8x^2 + 35x - 32 \end{aligned}$$

$$\frac{dP}{dx} = -3x^2 + 16x + 35$$

$$\frac{dP}{dx} = 0 \Rightarrow -3x^2 + 16x + 35 = 0 \Rightarrow 3x^2 - 16x - 35 = 0$$

$$x = \frac{16 \pm \sqrt{256 + 420}}{6} = \frac{16 \pm \sqrt{676}}{6}$$

i.e., $$x = \frac{16 \pm 26}{6} = 7, \text{ or } -5/3$$

when $x = 7$, Profit $= -49 \times 7 + 8 \times 49 + 35 \times 7 - 32$

$$= -343 + 392 + 245 - 32 = 262$$

$$\frac{d^2P}{dx^2} = -6x + 16 = -26 \text{ when } x = 7$$

$$= 26 \text{ when } x = -5/3$$

Hence, the profit will be maximum at $x = 7$ and the maximum profit = Rs. 262.

Example 6. A company charges Rs. 550 for a transistor set on orders of 50 or less sets. The charge is reduced by Rs. 5 per set for each set ordered in excess of 50. Determine the largest size of the order the company should allow so as to receive a maximum revenue.

Solution : Let the order be for x sets, then when $x > 50$ the revenue from the sale of each set $= 550 - (x - 50)\,5$

Total Revenue received from the sale of x sets

$$= R = x\,[550 - (x - 50)\,5]$$

$$\frac{dR}{dx} = 550 - (x - 50)\,5 + x\,[-5]$$

$$\frac{d^2R}{dx^2} = -5 - 5 = -10$$

For maxima or minima $$\frac{dR}{dx} = 0$$

$$\Rightarrow \quad 550 - 5x + 250 - 5x = 0$$

$$\Rightarrow \quad 10x = 800 \Rightarrow x = 80$$

Since $\frac{d^2R}{dx^2}$ is negative for all values of x, the revenue is maximum when the size of the order is 80 sets.

Example 7. A company has a demand curve given by the function $2Q + 3P = 160$. The average cost curve of the firm is given by the relation

$$AC = 3Q^2 - 18Q + 63 + \frac{5}{Q}.$$ Find the level of output which.

(i) maximise total revnue,

(ii) minimises marginal cost. **(ICWA QM June 2000)**

Solution : $2Q + 3P = 160 \Rightarrow P = \frac{160 - 2Q}{3}$

Total Revenue $(R) = PQ = \left[\frac{160 - 2Q}{3}\right] Q$

$$\frac{dR}{dQ} = \frac{160 - 2Q}{3} - \frac{2}{3}Q$$

$$= \frac{160}{3} - \frac{4Q}{3}$$

$$\frac{d^2R}{dQ^2} = \frac{-4}{3}$$

For total revenue to be maximum $\frac{dR}{dQ} = 0$

$$\frac{dR}{dQ} = 0 \Rightarrow \frac{160}{3} - \frac{4Q}{3} = 0$$

$$\Rightarrow Q = 40$$

$\frac{d^2R}{dQ^2}$ is negative for all values of Q therefore the total Revenue is maximum when $Q = 40$

Total cost $(C) = Q\,(AC)$

$$= 3Q^2 - 18Q^2 + 63Q + 5$$

Marginal cost $= \frac{dC}{dQ} = 9Q^2 - 36Q + 63 = MC$

$$\frac{d}{dQ}(MC) = 18Q - 36, \quad \frac{d^2}{dQ^2}(MC) = 18$$

For marginal cost to be minimum, $\frac{d}{dQ}(MC) = 0$

$$\frac{d\,(MC)}{dQ} = 0 \Rightarrow 18Q^2 - 36 = 0$$

$$\Rightarrow Q = \sqrt{2}$$

when $Q = \sqrt{2}$, $\frac{d^2(MC)}{dQ^2}$ is +ve, Hence marginal cost is minimum when $Q = \sqrt{2}$.

Example 8. If n be the number of workers employed, the average cost of production is given by

$$C = 24n + \frac{3}{2(n-4)}$$

Show that $n = 4\frac{1}{4}$ will make C minimum. Will you then advise to employ 4 or 5 workers?

(ICWA June 1990)

Solution :

$$\frac{dC}{dn} = 24 - \frac{3}{2}(n-4)^{-2}$$

$$\frac{d^2C}{dn^2} = 3(n-4)^{-3}$$

For C to be maximum or minimum

$$\frac{dC}{dn} = 0 \Rightarrow 24 - \frac{3}{2(n-4)^2} = 0$$

$$\Rightarrow \qquad 24 = \frac{3}{2(n-4)^2} \Rightarrow (n-4)^2 = \frac{1}{16}$$

$$\Rightarrow \qquad n - 4 = \frac{1}{4} \Rightarrow n = 4 + \frac{1}{4} = 4\frac{1}{4}$$

$\frac{d^2C}{dn^2}$ is positive when $n = 4\frac{1}{4}$

$\therefore$ The value of C is minimum at $n = 4\frac{1}{4}$

when $n = 4,\ C = 24 \times 4 + \frac{3}{2(4-4)} = 96 + \infty = \infty$

when $n = 5,\ C = 24 \times 5 + \frac{2}{2(5-4)} = 121\frac{1}{2} = \text{Rs. } 121.5$

$\therefore$ 5 workers should be employed because cost canot be infinite.

Example 9. A firm has the following functions :

$$p = 100 - 0.01x$$

$$TC = 50x + 30{,}000 \text{ where } P = \text{Price},\ x = \text{No. of units}$$

How many units are to be manufactured to maximise the profit ? If a tax of Rs. 10 per unit is levied, what should be the number of units to be manufactured to maximise the profit ?

Solution : Number of units to be manufactured before tax :

Given cost function $TC = 50x + 30{,}000$

$$\text{Revenue} = \text{Demand} \times \text{Price} = x(100 - 0.01x) = 100x - 0.01x^2$$

$$\text{Profit} = \text{Total Revenue} - \text{Total Cost} = p \text{ (say)}$$

$$= 100x - 0.01x^2 - 50x - 30{,}000 = 50x - 0.01x^2 - 30{,}000$$

$$\frac{dp}{dx} = 50 - 0.02x$$

$$\frac{dp}{dx} = 0 \Rightarrow 50 - 0.02x = 0 \Rightarrow x = \frac{50}{0.02} = 2{,}500 \text{ units,}$$

$$\frac{d^2p}{dx^2} = -0.02 \text{ which is } -\text{ve for all } x.$$

The firm can maximise its profit by manufacturing 2,500 units

$$\text{Price per unit} = 100 - 0.01x = 100 - 0.01\ (2{,}500) = \text{Rs. } 75$$

$\therefore$ The firm can price the product at Rs. 75 per unit.

Units to be manufactured after tax :

Cost function (after tax) $= 50x + 30{,}000 + 10x = 60x + 30{,}00$

$$\text{Price} = 100 - 0.01x$$

$$\text{Revenue} = \text{Quantity} \times \text{Price} = x(100 - 0.01x) = 100x - 0.01x^2$$

$$\text{Profit} = \text{Revenue} - \text{Cost}$$

$$= [100x - 0.01x^2] - [60x + 30{,}000]$$

$$= 100x - 0.01x^2 - 60x - 30{,}000$$

$$= 40x - 0.01x^2 - 30{,}000$$

$$\frac{dp}{dx} = \frac{d}{dx}(40x - 0.01x^2 - 30{,}000) = 40 - 0.02x$$

$$\therefore \quad \frac{dp}{dx} = 0 \Rightarrow x = \frac{40}{0.02} = 2{,}000 \text{ units}$$

$$\frac{d^2p}{dx^2} = -0.02 \text{ which is negative for all } x$$

$\therefore$ The firm has to manufacture 2,000 units to maximise its profit.

$$\text{Price per unit} = 100 - 0.01x = 100 - 0.01\ (2{,}000) = 100 - 20 = \text{Rs. } 80.$$

Profit maximising price is Rs. 80 per unit.

Example 10. A sitar manufacturer can sell x sitars per week at Rs. P each where $5x = 375 - 3P$. The cost of production is $500 + 13x + \frac{1}{5}x^2$ rupees. Find how many sitars he should manufacture for maximum profit and what is that profit ?

Solution : Profit is the difference between total revenue and total cost.

Now $\quad 5x = 375 - 3P \Rightarrow 3P = 375 - 5x \Rightarrow P = \dfrac{375 - 5x}{3}$

$$\text{Total Revenue} = \frac{x(375 - 5x)}{3} \qquad [\text{Revenue} = px]$$

$$\text{Total cost} = 500 + 13x + \frac{1}{5}x^2$$

$$\text{Profit } (F) = \text{Total Revenue} - \text{Total Cost}$$

$$= \frac{x(375 - 5x)}{3} - \left(500 + 13x + \frac{x^2}{5}\right)$$

$$= 125x - \frac{5x^2}{3} - 500 - 13x - \frac{x^2}{5}$$

$$= 112x - \frac{28x^2}{15} - 500$$

$$\frac{dF}{dx} = 112 - \frac{56x}{15}$$

Now $$\frac{dF}{dx} = 0 \Rightarrow 112 = \frac{56x}{15} \Rightarrow x = 30$$

$$\frac{d^2F}{dx^2} = \frac{-56}{15}$$

which is negative for all values of x. Hence, profit will be maximum when $x = 30$, *i.e.,* when 30 sitars are produced per week.

Maximum profit $= 112 \times 30 - \frac{28 \times 900}{15} - 500 =$ Rs. 1,180.

Example 11. A firm produces x units of output per week at a total cost of Rs. $\left(\frac{1}{3}x^3 - x^2 + 5x\right)$, find the level at which the marginal cost and the average variable cost attain their minimum.

Solution : Marginal cost $= \frac{d}{dx}$ (Total cost) $= \frac{d}{dx}\left(\frac{1}{3}x^3 - x^2 + 5x\right) = x^2 - 2x + 5$

$$\frac{d}{dx}\text{(Marginal cost)} = \frac{d}{dx}\text{(M.C.)} = 2x - 2$$

$$\frac{d}{dx}\text{(M.C.)} = 0 \Rightarrow 2x - 2 = 0 \Rightarrow x = 1$$

$\frac{d^2\,\text{(M.C.)}}{dx^2} = 2$ which is positive for all values of x.

$\therefore$ Marginal cost is minimum when output is 1 unit per week and the minimum marginal cost $1 - 2 + 5 =$ Rs. 4. **(Please note)**

Total variable cost = Rs. $\left[\frac{1}{3}x^3 - x^2 + 5x\right]$

Average variable cost (A.C.) $= \frac{\text{Total variable cost}}{\text{output}} = \frac{1}{3}x^2 - x + 5$

$$\frac{d}{dx}\text{(A.C.)} = \frac{2}{3}x - 1$$

$$\frac{d\,\text{(A.C.)}}{dx} = 0 \Rightarrow x = \frac{3}{2}$$

$$\frac{d^2}{dx^2}[\text{(A.C.)}] = \frac{2}{3}$$

which is positive for all values of x. Hence, average variable cost is minimum at $x = \frac{3}{2}$.

Example 12. A monopolist has a demand curve

$$x = 106 - 2p$$

and average cost curve A.C. $= 5 + \frac{x}{50}$ where, p is the price per uint output and x is the number of units of output.

If the total revenue is $R = xp$, determine the cost profitable output and the maximum profit.

(ICWA Q.M. June 1995)

Solution :
$$\text{A.C.} = 5 + \frac{x}{50}$$

$$\text{Total Cost} = 5x + \frac{x^2}{50}$$

$$\text{Profit} = \frac{x(106-x)}{2} - \left[5x + \frac{x^2}{50}\right] \quad \left[\because x = 106 - 2p \Rightarrow p = \frac{106-x}{2}\right]$$

$$= \frac{1}{2}\left[106x - 10x - x^2 - \frac{2x^2}{50}\right]$$

$$= \frac{1}{2}\left[96x - \frac{52x^2}{50}\right]$$

$$\frac{d\,(\text{Profit})}{dx} = \frac{1}{2}\left[96 - \frac{104x}{50}\right], \frac{d^2\,(\text{Profit})}{dx^2} = -\frac{52}{50}$$

$$\frac{d(\text{Profit})}{dx} = 0 \Rightarrow 96 = \frac{104x}{50} \Rightarrow x = \frac{96 \times 50}{104} = 46.15 = 46$$

$$\frac{d^2\,(\text{Profit})}{dx^2} < 0 \Rightarrow \text{Profit is maximum at } x = 46 \text{ units.}$$

Most Profitable output = 46 units

$$\text{Profit} = \frac{1}{2}\left[96 \times 46 - \frac{52}{50} \times 46 \times 46\right] = \text{Rs. } 1{,}107.68.$$

Example 13. A monopolist has the following demand and cost functions :

$$P = 1{,}620 - 3x$$

$$TC = \frac{1}{3}x^3 - 3x^2 + 15x + 500$$

The government levies tax at the rate of Rs. 5 per unit sold. Find the profit maximising price and quantity after levy of tax.

Solution : Given demand function is $P = 1{,}620 - 3x$

$$\text{Total revenue} = \text{Quantity demanded} \times \text{Price} = x(1{,}620 - 3x) = 1{,}620x - 3x^2$$

$$\text{Given Cost Function} = \text{TC} = \frac{1}{3}x^3 - 3x^2 + 15x + 500$$

$$\text{Cost function after levy of tax} = \frac{1}{3}x^3 - 3x^2 + 15x + 500 + 5x$$

$$= \frac{1}{3}x^3 - 3x^2 + 20x + 500$$

[∵ Tax is considered as the cost and added to the cost]

Profit = Total revenue − Total cost

$$= [1{,}620x - 3x^2] - \left[\frac{1}{3}x^3 - 3x^2 + 20x + 500\right]$$

$$= 1{,}620x - 3x^2 - \frac{1}{3}x^3 + 3x^2 - 20x - 500 = 1{,}600x - \frac{1}{3}x^3 - 500$$

$$\frac{dP}{dx} = 1{,}600 - x^2,\ \frac{dP}{dx} = 0 \Rightarrow 1{,}600 - x^2 = 0 \Rightarrow x = \pm 40$$

$(d^2P/dx^2) = -2x$ which is −ve for $x > 0$

Since the number of units maufactured cannot be −ve we reject the negative value of x.

∴ The firm earns maximum profit when it produces 40 units.

Price per unit at the production level of 40 units.

$$P = 1{,}620 - 3x = 1{,}620 - 3(40) = \text{Rs. } 1{,}500.$$

Exmaple 14. A piece of machinery costs Rs. 12,000. The total cost of operation from the time of purchase upto a time t is given by the function $20t^2 + 15t$. If the machine is sold as scrap after t years, the resale value is given by the function $6{,}880 - 60t^2$. Find the optimal time for replacement of the machine. **(D.U. B. Com. (Hons.) 2000)**

Solution : Let the machine be replaced after t years

$$\text{Net cost} = 12{,}000 + 20t^2 + 15t - 6{,}880 + 60t^2 \text{ *}$$

$$\text{Average Net Cost} = \frac{12{,}000}{t} + \frac{20t^2}{t} + \frac{15t}{t} - \frac{6{,}800}{t} + \frac{60t^2}{t}$$

$$= \frac{5{,}120}{t} + 80t + 15$$

Here, we have to find the value of t, so that Average Net Cost [or Average Cost] is minimum

$$\frac{d}{dt}(\text{AC}) = -\frac{5{,}120}{t^2} + 80$$

$\dfrac{d^2}{dt^2}(\text{AC}) = \dfrac{10{,}240}{t^3}$ which is positive for all positive values of t

For maxima or minima $\quad \dfrac{d}{dt}(\text{AC}) = 0 \Rightarrow 5{,}120 = 80t^2$

$\Rightarrow \quad t^2 = 64 \Rightarrow t = 8$ years

Since, 2nd derivative, *i.e.*, $\dfrac{d^2}{dt^2}(\text{AC})$ is positive for all positive values of t, the average cost is minimum if the machine is replaced after 8 years.

* Total cost = Cost of machine + Cost of operation

Net cost = Total cost − Scrap

= Cost of machine + Cost of operation − Scrap.

Example 15. A company makes a profit of Rs. 5 on each unit of its product it sells. If it spends an amount of Rs. A per week on advertising, then the number of items per week it sells is given by $x = 2{,}000\,(1 - e^{-KA})$, where $k = 0.001$.

Find the value of A that will maximize the net profit. **(D.U., B. Com. (Hons.) 1997)**

Solution : Net Profit

$$(P) = 5 \times 2{,}000\,(1 - e^{-KA}) - A$$

$$= 10{,}000\,[1 - e^{-KA}] - A$$

$$\frac{dP}{dA} = K\,[10{,}000\, e^{-KA}] - 1$$

$$= 10e^{00.001A - A} - 1$$

$$\frac{d^2P}{dA^2} = -[10\, e^{-0.001A}]\, 0.001$$

which is negative for all Positive values of A

For maxima or minima $\frac{dP}{dA} = 0$

$$\Rightarrow \quad 10e^{-0.001A} = 1 \Rightarrow e^{-0.001A} = \frac{1}{10}$$

$$\Rightarrow \quad 0.001A = \log_e 10$$

$$\Rightarrow \quad A = 1000 \log_e 10$$

$\Rightarrow$ The profit will be maximum when $A = 1{,}000 \log_e 10$.

Example 16. A shoe manufacturer produces both the high grade and the low grade shoes in the units of X hundreds of pairs and Y hundreds of pairs respectively. The plant is capable of producing the shoes according to the relationship

$$Y = \frac{42 - 5X}{10 - X}, X < 10$$

If high grade shoes are sold at a price twice the low grade shoes, determine the number of pairs of both which should be produced to maximize total revenue. **(D.U., B. Com. (Hons.) 1998)**

Solution : Let Rs. p per hundred be the selling price of low grade shoes and Rs. $2p$ per hundred the selling price of high grade shoes

Total Revenue

$$(TR) = Yp + 2Xp$$

$$= p\left[\frac{42 - 5X}{10 - X}\right] + 2Xp$$

$$\frac{d\,(TR)}{dX} = p\left[\frac{(10 - X)\,(-5) + (42 - 5X)}{(10 - X)^2}\right] + 2p$$

$$= p\,\frac{(5X - 50 + 42 - 5X)}{(10 - X)^2} + 2p$$

$$= \frac{-8p}{(10 - X)^2} + 2p$$

$$\frac{d^2(TR)}{dX^2} = -\frac{16p}{(10-X)^3} \text{ which is negative if } X < 10 \text{ and positive if } X > 10$$

For maxima or minima $\frac{d(TR)}{dX} = 0$

$$\Rightarrow \qquad \frac{8p}{(10-X)^2} = 2p \Rightarrow (10-X)^2 = 4$$

$$\Rightarrow \qquad 10 - X = \pm 2$$

$$\Rightarrow \qquad X = 8 \text{ or } 12$$

when $X = 8$, $\frac{d^2(TR)}{dX^2}$ is negative

$\therefore$ TR will be maximum when $X = 8$

when $X = 8$, $Y = \frac{42-40}{10-8} = 1$

$\therefore$ The manufacturer should produce 800 pairs of high grade shoes and 100 pairs of low grade shoes to maximize his total revenue. When $X = 10$, Y becomes infinite, hence, we reject this value.

Example 17. The cost of producing x units of TV sets by a monopolist is given by $TC = \frac{x^2}{25} + 3x + 100$ and the demand function is given by $x = 75 - 3p$ (where p is price). If a tax of k per set is imposed by the government, determine the monopolist's output and price under the new situation. Find the value of k that maximizes tax revenue to the government.

(D.U., B. Com. (Hons.) 1996)

Solution : Total tax = kx

After the imposition of tax, the new cost function, say

$$TC_n = \frac{1}{25}x^2 + 3x + 100 + kx$$

Profit, say $\pi = TR - TC_n$

$$= px - \frac{1}{25}x^2 - 3x - 100 - kx$$

$$= 25x - \frac{x^2}{3} - \frac{1}{25}x^2 - 3x - 100 - kx \left[\because x = 75 - 3p \Rightarrow p = \frac{75-x}{3}, \ \therefore \ px = 25x - \frac{x^2}{3}\right]$$

$$= -\frac{28}{75}x^2 + x(22-k) - 100,$$

which is to be maximized.

$$\frac{d\pi}{dx} = -\frac{56}{75}x + (22-k)$$

$$\frac{d^2\pi}{dx^2} = -\frac{56}{75} \text{ which is negative}$$

For maxima or minima

$$\frac{d\pi}{dx}=0 \Rightarrow -\frac{56}{75}x+22-k=0$$

$$\Rightarrow x=\frac{75}{56}(22-k)$$

Since, $\frac{d^2\pi}{dx^2}$ is negative for all values of x, the profit will be maximum when $x=\frac{75}{56}(22-k)$ and price where profit is maximum is given by

$$p=25-\frac{25}{56}(22-k)=\frac{425}{28}-\frac{25}{56}k$$

Total tax $T=kx=k\frac{75}{56}(22-k)$

$$=\frac{75}{56}(22k-k^2)$$

$$\frac{dT}{dk}=\frac{75}{56}(22-2k)$$

$$\frac{d^2T}{dk^2}=-\frac{75}{28}$$

For maximum or minimum value of k, $\frac{dT}{dk}=0$

$\Rightarrow$ $22-2k=0 \Rightarrow k=11$

Since, $\frac{d^2T}{dk^2}$ is –ve for all k, the tax will be maximum when $k=11$.

Example 18. A manufacturing company has a contract to supply 4,000 units of an item per year at a uniform rate. The storage cost per unit per year amounts to Rs. 50 and the set-up cost per production run is Rs. 16. If the production run can be started instantaneously and shortages are not permitted, determine the number of units which should be produced per run to minimize the total cost. Also calculate the cost. **(D.U., B. Com. (Hons.) 1995)**

Solution : Set-up cost per production run can be considered as ordering cost.

$\therefore$ Total cost of storage and set-up $=\frac{x}{2}\times 50+\frac{4000}{x}\times 160$*

where x is the number of units manufactured per production run.

$$\frac{d}{dx}(\text{Total cost})=25-\frac{4{,}000\times 160}{x^2}$$

$\frac{d^2}{dx^2}(\text{Total cost})=\frac{2\times 4{,}000\times 160}{x^3}$ which is positive for all positive values of x.

* While calculating cost, material cost has been ignored because the cost per unit is not given otherwise. It is to be noted that total material cost does not have any effect on the order size/production run to minimize the total cost.

For maxima or minima, $\frac{d}{dx}$ (Total cost) $= 0$

$$\Rightarrow \qquad 25 - \frac{4{,}000 \times 160}{x^2} = 0 \Rightarrow 25x^2 = 4{,}000 \times 160$$

$$\Rightarrow x = 160$$

Since, $\frac{d^2}{dx^2}$ (Total cost) is +ve, the cost will be minimum when 160 units are produced in each productio run.

Minimum cost of storage and set-up $= \left(\frac{160}{2}\right) 50 + \frac{4{,}000}{160} \times 160 = \text{Rs. } 8{,}000.$

Example 19. A firm's annual sales are S units of a product which the firm buys from a supplier. If the replenishment cost is Rs. r per order and the holding cost Rs. h per unit per year, find the economic order quantity. **(D.U., B. Com. (Hons.) 1991)**

Solution : Total variable cost $(TVC) = \left(\frac{S}{x}\right) r + \frac{x}{2} h$

where x is the number of units ordered in each order.

$$\frac{d}{dx}(\text{TVC}) = -\frac{Sr}{x^2} + \frac{h}{2}$$

$\frac{d^2}{dx^2}(\text{TVC}) = 2\frac{Sr}{x^3}$ which is positive for all positive values of x

For maxima and minima

$$\frac{d}{dx}(\text{TVC}) = 0$$

$$\Rightarrow \qquad -\frac{Sr}{x^2} + \frac{h}{2} = 0 \Rightarrow \frac{Sr}{x^2} = \frac{h}{2}$$

$$\Rightarrow \qquad x^2 = \frac{2Sr}{h} \Rightarrow x = \sqrt{\frac{2Sr}{h}}.$$

Example 20. A chemical company acquires its requirement in 20 orders of the size 250 drums each. Each drum costs Rs. 80. The replenishment cost is Rs. 40 irrespective of the order size, and the annual carrying cost is 2% of the average rupee inventory. Find how much can company save if it buys according to economic order quantity, derive using calculus. **(D.U., B. Com. (Hons.) 1999)**

Solution : Total annual requirement $= 250 \times 20 = 5{,}000$ drums

Let x units be the size of each order. We will find the value of x so that the total cost is minimum. This value of x will be economic order quantity.

Total Cost (TC) = Material Cost + Ordering Cost + Holding Cost

$$= 5{,}000 \times 80 + \frac{5{,}000}{x} \times 40 + \frac{x}{2}\left(\frac{2}{100} \times 80\right)^*$$

* Carrying cost is always calculated on the average inventory.

$$\frac{d}{dx}(\text{TC}) = \frac{-5{,}00}{x^2} \times 40 + 0.8$$

$$\frac{d^2(\text{TC})}{dx^2} = \frac{5{,}00 \times 80}{x^3} \text{ which is +ve for all positive values of } x$$

$$\frac{d}{dx}(\text{TC}) = 0 \Rightarrow 0.8 = \frac{5{,}000 \times 40}{x^2}$$

$$\Rightarrow \quad 0.8x^2 = 5{,}000 \times 40$$

$$\Rightarrow \quad x^2 = 50 \times 5{,}000 \Rightarrow x = 500 \text{ units}$$

Since $\frac{d^2(\text{TC})}{dx^2}$ is positive, cost will be minimum, when $x = 5{,}000$

$$\text{TC} = 5{,}000 \times 80 + \frac{5{,}000}{500} \times 40 + \frac{500}{2} \times \left(\frac{2}{100} \times 80\right)$$

$$= 4{,}00{,}000 + 400 + 400$$

$$= \text{Rs. } 4{,}00{,}800$$

Total cost when order is placed for 250 drums each

$$= 5{,}000 \times 80 + \frac{5{,}00}{250} \times 40 + \frac{250}{2} \times \frac{2}{100} \times 80$$

$$= 4{,}00{,}000 + 800 + 200 = \text{Rs. } 4{,}01{,}000$$

$$\text{Net saving} = 4{,}01{,}000 - 4{,}00{,}800$$

$$= \text{Rs. } 200.$$

Example 21. A grain trader foresees an increase in demand for a particular variety of rice over the next 3 months and wishes to enter the market at such a time that his profit from trading is maximized. Until the time the enters the market he shall only be procuring the rice from the producers. According to his assessment he can enter the market now with a supply of 10,000 kgs of rice and sell the entire lot at a profit of Rs. 2.60 per kg. He further feels that if, instead of entering the market now, he waits for the market to pickup then he can increase his stock by 200 kgs of rice per day; however his profit will be reduced by 2 paise per kg per day.

You are required to answer the following questions :

(a) Using differential calculus find out the number of days the trader should wait before entering the market in order to maximize his profit.

(b) Calculate the maximum profit.

Solution : Stock at x days from now = $(10{,}000 + 200x)$ kgs

Profit per kg at x days from now = Rs. $(2.60 - 0.02x)$

Total Profit

$$P = (10{,}000 + 200x)(2.60 - 0.02x)$$

$$= 26{,}000 - 200x + 520x - 4x^2$$

$$= 26{,}000 + 320x - 4x^2$$

$$\frac{dP}{dx} = 320 - 8x, \frac{d^2P}{dx^2} = -16$$

$$\frac{dP}{dx} = 0 \Rightarrow 320 - 8x = 0 \Rightarrow x = 40$$

Since $\frac{d^2P}{dx^2}$ is always negative, we may say profit will be maximum, when $x = 40$ days, *i.e.*, trader should wait for 40 days before entering the market in order to earn maximum profit.

$$\begin{aligned}\text{Maximum Profit} &= 26{,}000 + 320 \times 40 - 4 \times (40)^2 \\ &= 26{,}000 + 12{,}800 - 6{,}400 \\ &= 6{,}000 + 6{,}400 \\ &= \text{Rs. } 32{,}400.\end{aligned}$$

Example 22. A firm requires 10,000 units of a material per annum. The cost of purchasing is Re. 1 per unit, the cost of replenishment of stock of material is Rs. 25 and the cost of storing material is 12.5% per annum of the average rupee inventory. Find the optimum order size and the corresponding total cost using calculus. **(D.U., B. Com. (Hons.) 1998)**

Solution : Let x units be ordered in one order

Then

$$\text{Total Cost (TC)} = 10{,}000 + \frac{10{,}000}{x} \times 25 + \frac{x}{2} \times \left(\frac{12.5}{100}\right)$$

$$\frac{d}{dx}(\text{TC}) = -\frac{10{,}000}{x^2} \times 25 + \frac{12.5}{200}$$

$$\frac{d^2(\text{TC})}{dx^2} = \frac{20{,}000 \times 25}{x^3}$$

which is +ve for all positive values of x

For maxima and minima

$$\frac{d}{dx}(\text{TC}) = 0 \Rightarrow \frac{10{,}000 \times 25}{x^2} = \frac{12.5}{200}$$

$$\Rightarrow 12.5x^2 = 200 \times 10{,}000 \times 25$$

$$\Rightarrow x = \sqrt{\frac{200 \times 10{,}000 \times 25}{12.5}}$$

$$= 2{,}000 \text{ units.}$$

Since $\frac{d^2}{dx^2}$(TC) is positive for all positive values of x, the total cost will be minimum when $x = 2{,}000$.

When $x = 2{,}000$

$$\text{TC} = 10{,}000 + \frac{10{,}000}{2{,}000} \times 25 + \frac{2{,}000}{2} \times \left[\frac{12.5}{100}\right]$$

$$= \text{Rs. } 10{,}250.$$

Example 23. A departmental store sells 2,500 top-of-the-line refrigerators per year, with sales occurring at a relatively constant rate. The annual holding cost per refrigerator is Rs. 10. To reorder there is a basic Rs. 20 service fee per reorder plus a Rs. 9 insurance charge for each refrigerator ordered. How many times per year and in what lot size should the store reorder to minimize the total annual inventory cost?

Solution : Total Cost (TC) $= 2{,}500y + \frac{2{,}500}{x} \times 20 + \frac{x}{2} \times 10 + 2{,}500 \times 9*$ where Rs. y is the cost per refrigerator and x is the number of units ordered in each order

$$\frac{d}{dx}(\text{TC}) = -\frac{2{,}500}{x^2} \times 20 + \frac{1}{2} \times 10$$

$$\frac{d^2}{dx^2}(\text{TC}) = \frac{5{,}000}{x^3} \times 20 \text{ which is positive for all positive values of } x.$$

For maxima or minima

$$\frac{d}{dx}(\text{TC}) = 0 \Rightarrow \frac{2{,}500}{x^2} \times 20 = 5$$

$$\Rightarrow \qquad x^2 = 10{,}000 \Rightarrow x = 100{**}$$

Since, 2nd derivative, *i.e.*, $\frac{d^2}{dx^2}$(TC) is positive for all positive values of x, the total cost will be minimum when $x = 100$.

Number of times the store should reorder

$$= \frac{2{,}500}{100} = 25$$

∴ The store should reorder 25 times in a year and each order should be placed for 100 units.

Example 24. A famous Hamburger Palace requires 45,000 hamburger patties every quarter (three months) based on estimated daily sales of 475 hamburgers plus an additional 25 that are either burned or eaten by employees each day. These hamburgers are packed in boxes of 10 patties each. The manager of the restaurant, for each order he places, incurs order charges of 45 paise due to handling and delivery. He is also aware that, if he places a large order, he must be prepared to store the patties in a freezer. He estimates the cost of storing each box of patties to be 2 paise per box. How many boxes of patties should be ordered and how many orders should be placed to minimized the cost of storage and of ordering?

Solution : Let Rs. y be the cost of each box and x the number of boxes ordered in each order. Then

$$\text{Total Quarterly Cost (TC)} = \left[\frac{45{,}000}{10}\right]y + \frac{45{,}000}{10} \times \frac{1}{x} \times 0.45 + \frac{x}{2} \times \frac{2}{100}$$

$$\frac{d}{dx}(\text{TC}) = -\frac{450 \times 45}{10x^2} + \frac{1}{100}$$

$$\frac{d^2}{dx^2}(\text{TC}) = \frac{2 \times 450 \times 45}{10x^3}$$

which is positive for all positive values of x.

* Total Insurance Cost = Number of Refrigerators ordered per year × Insurance Cost per refrigerator = 2,500 × 9.

** Since quantity ordered cannot be negative, negative value is rejected.

For maxima or minima $\frac{d}{dx}(\text{TC}) = 0 \Rightarrow \frac{450 \times 45}{10x^2} = \frac{1}{100} \Rightarrow x^2 = 4500 \times 45 \Rightarrow x = 450$

$\therefore$ The cost will be minimum when 450 units are manufactured.

5.4 ELASTICITY OF DEMAND

Elasticity of demand

- Price elasticity of demand
- Income elasticity of demand
- Cross elasticity of demand
- Substitution dleasticity of demand

In general elasticity of demand or demand elasticity is a measure of relative responsiveness of quantity demanded to change in any one of the determinants of demand, assuming other determinants to be constant (*i.e.*, unchanged).

Mathematically

$$E_d \text{ (Elasticity of demand)} = \frac{\text{Percentage change in quantity demanded of a good, say } X}{\text{Percentage change in the determinant under consideration, say } Z}$$

where Z can be any of the determinants of demand, *viz.*, price, income, etc.

5.5 PRICE ELASTICITY OF DEMAND DENOTED BY e_p or η or E_d

Price elasticity of demand is the degree of responsiveness of the demand for a commodity to a change in its price.

According to Marshall, "*Elasticity of demand may be defined as the percentage change in the quantity demanded divided by percentage change in the price*".

According to Anatol Murrad, "*Elasticity of demand is the ratio of relative change in quantity to relative change in price.*"

Price Elasticity of Demand

$$= \frac{\text{Percentage change in quantity demanded}}{\text{Percentage change in price}}$$

$$= \frac{\text{Proportionate change in quantity demanded}}{\text{Proportionate change in price}}$$

$$= \frac{\dfrac{\text{Change in quantity demanded}}{\text{Quantity demanded at original price}}}{\dfrac{\text{Change in price}}{\text{Original price}}}$$

where 'x' is the quantity demanded at original price and p is the original price per unit and the demand function is given by $x = f(p)$.

It may further be noted that if the price increases, quantity demanded will decrease, *i.e.*, corresponding to any change in price, quantity demanded changes in the opposite direction, *i.e.*, $\frac{\Delta x}{\Delta p}$ is **always negative**. But some times we take only numerical value and hence ignore the sign.

Price elasticity of demand is denoted by η or e_d or e_p etc.

$$\eta = -\frac{p}{x}\left(\frac{\Delta x}{\Delta p}\right) = \frac{p}{x}\left(\frac{\Delta x}{\Delta p}\right) \text{ (numerically)}$$

In the words of Alfred Marshal, "*The elasticity (or responsiveness) of demand in a market is great or small according as the amount demanded increases much or little for a given fall in price and diminishes much or little for a given rise in price*".

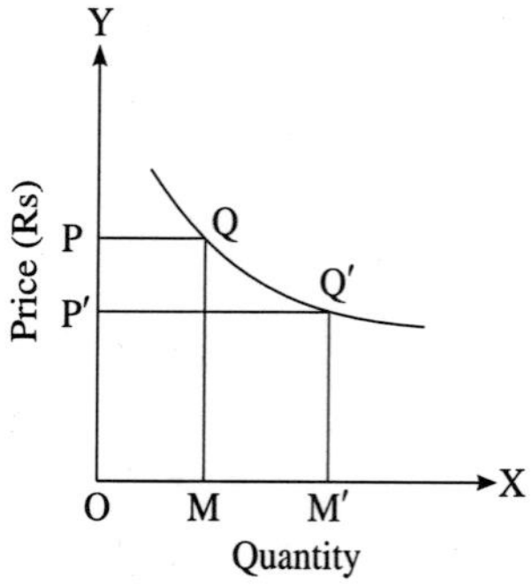

Demand Curve for good A.

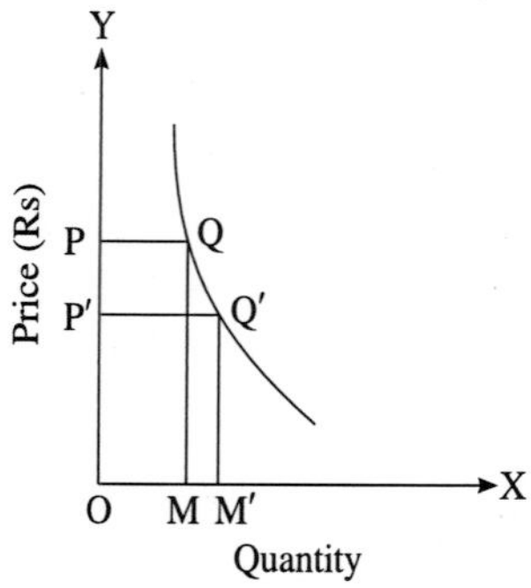

Demand Curve for good B.

Demand for a good is said to be *elastic* if the elasticity of demand for it is greater than 1. The demand for a good is said to be *inelastic* if the elasticity of demand for it is less than 1. The elasticity of demand equal to one, *i.e.*, unit elasticity of demand represents the dividing line between elastic and inelastic demands an in this case the demand is said to be **moderately elastic**.

If there are two goods say *A* and *B*, then the demand for good *A* is said to be more elastic than the demand for good *B* if corresponding to a fixed change in price, the change in quantity demanded of *A* is more than that of *B*.

As *PP′* in curves I and II is equal and *MM′* in curve I is greater than *MM′* in curve II, the demand for good A is more elastic than for good *B*.

5.6 TYPES OF PRICE ELASTICITY OF DEMAND

(i) Perfectly elastic demand,
(ii) Perfectly inelastic demand,
(iii) Demand with unit elasticity,
(iv) Relatively elastic demand and
(v) Relatively inelastic demand

(i) Perfectly elastic demand : The demand for a good is said to be perfectly elastic if for a very small change in price, there is a very large (approximately infinite) change in the quantity demanded. It is purely an imaginary concept and the value of the elasticity of demand is very large and may go up to infinity. The graph is a horizontal line parallel to *x*-axis. This type of demand curve is found under Perfect Competition.

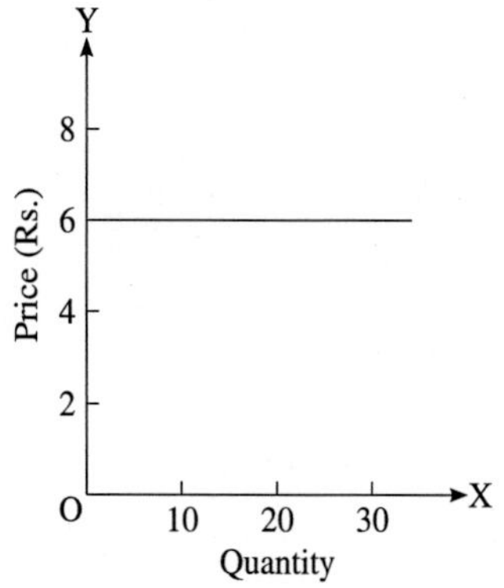

(ii) Perfectly inelastic demand : Demand for a commodity or good is said to be a perfectly inelastic if the quantity demanded does not change at all in response to any change in price. In case of inelastic demand, even substantial changes in price leaves the demand unaffected. For example, salt is a necessity which fulfils a basic human need in a way in which no other good can do. The consumption of salt remains almost unchanged for any change in price.

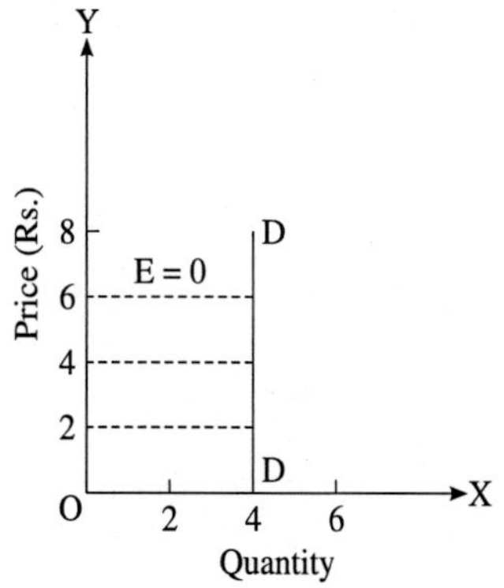

(iii) Demand with unit elasticity :

$$h = \frac{\Delta x/x}{\Delta p/p} = 1 \Rightarrow \frac{\Delta x}{x} = \frac{\Delta p}{p}$$

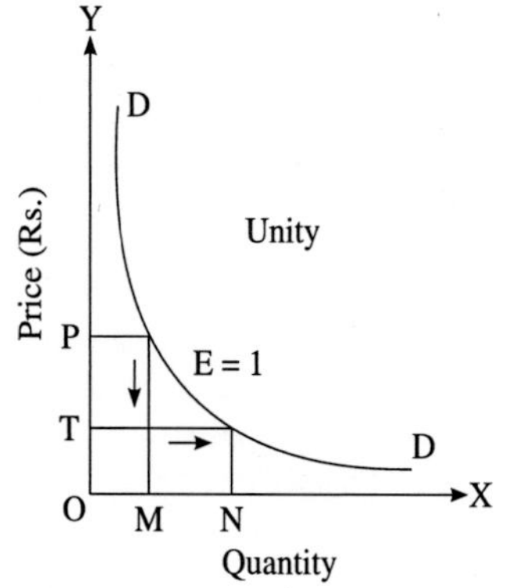

⇒ when the elasticity of demand is unity, any proportionate change in the price causes an equal proportionate change in the quantity demanded. Here, the shape of the demand curve is a rectangular hyperbola.

(iv) Relatively elastic demand or elastic demand : Relatively elastic demand ⇒ the proportionate change in quantity demanded is more than the proportionate change in price. Here, the shape of the demand curve is almost flat.

(v) Relatively inelastic demand or inelastic demand : Relatively inelastic demand ⇒ the proportionate change in demand is less than the proportionate chage in price. Here, the shape of the demand curve is steep.

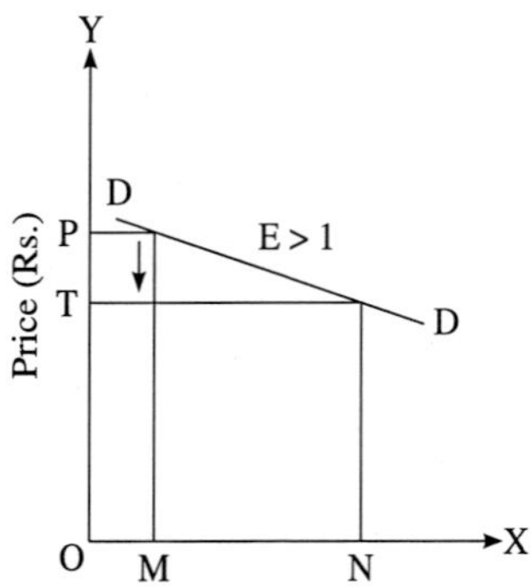

Relatively elastic demand.

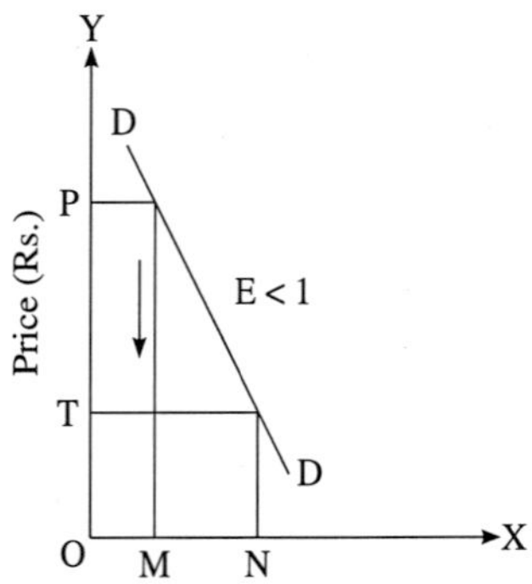

Relatively inelastic demand.

5.7 SUMMARY OF THE CHARACTERISTICS OF VARIOUS TYPES OF PRICE ELASTICITY OF DEMAND

Type	*Numerical Expression*	*Description*	*Shape of Curve*
1. Perfectly Elastic	∞	Infinite	Horizontal
2. Perfectly Inelastic	0	Zero	Vertical
3. Unit Elasticity	1	One	Rectangular Hyperbola
4. Relatively Elastic	> 1	More than one	Flat
5. Relatively Inelastic	< 1	Less than one	Steep

Important Notes :

(1) The value of elasticity of demand (or elasticity coefficient) varies between zero and infinity, *i.e.,* the range of the values of elasticity coefficient is

$$0 \le E_d \le \infty$$

When the elasticity of demand is zero, the demand curve is a vertical straight line and when the elasticity of demand is infinite, the demand curve is a horizontal straight line.

(2) The main reason for difference in responsiveness of various goods to change in their price is the availability and non-availability of competing substitutes.

(3) A rectangular hyperbola is a curve such that the rectangles formed by the two axes and the abscissa (x coordinate) and the ordinate (y coordinate) of points on the curve have constant area.

(4) It may be pointed out here that the price elasticity of demand for a certain commodity and price elasticity of demand for a certain brand of that commodity may be radically different, *e.g.*, The price elasticity of demand for a toothpaste and price elasticity of demand for a particular brand of toothpaste may be different.

5.8 CROSS ELASTICITY OF DEMAND (e_C)

The effect of a change in the prices of related goods upon the demand for a particular commodity may be determined by measuring the 'cross elasticity of demand'. Cross elasticity of demand may be defined as "*the proportionate (or percentage) change in the quantity demanded of a particular commodity in response to a change in the price of another related commodity.*" Thus the price of one commodity, say Y is the independent variable whereas the quantity of another commodity, say X, is the dependent variable. The cross elasticity of demand can be measured by the following formula, *viz.*,

$$e_c = \frac{\text{Proportionate change in the quantity demanded or purchased of good } X}{\text{Proportionate change in the price charged for a commodity (or good) } Y}$$

$$= \frac{\dfrac{\text{Change in Quantity demanded of } X}{\text{original quantity demanded of } X} \times 100}{\dfrac{\text{Change in Price of } Y}{\text{Original Price of } Y} \times 100}$$

$$= \frac{\dfrac{\Delta Q_x}{Q_x}}{\dfrac{\Delta P_Y}{P_Y}} = \frac{\Delta Q_X}{Q_X} \times \frac{P_Y}{\Delta P_Y} \quad \Rightarrow \quad e_c = \frac{P_Y}{Q_x} \times \frac{\Delta Q_X}{\Delta P_Y}$$

$\dfrac{\Delta Q_X}{\Delta P_Y}$ is determined with the help of partial derivatives as $\dfrac{\Delta Q_X}{\Delta P_Y}$ = Partial derivative of Quantity demanded of good X w.r.t. Price of Good Y.

Example 25. The demand curve for a commodity is given by $p = 20 - \sqrt{x}$ where p and x are the price and amount of the commodity respectively. Find the marginal quantity demand, average quantity demand. Also find the elasticity of demand at a price $p = 2$ to state whether the demand is elastic or inelastic. **(ICWA Dec. 1997)**

Solution : Marginal demand $= \dfrac{dx}{dp} = 2p - 40*$

Average demand $= \dfrac{x}{p} = \dfrac{400}{p} - 40 + p**$

---

* $p = 20 - \sqrt{2} \Rightarrow x = (20 - p)^2, \dfrac{dx}{dp} = 2\,(20 - p)\,(-1).$

** $x = (20 - p)^2 = 400 + p^2 - 40p \Rightarrow \dfrac{x}{p} = \dfrac{400}{p} - 40 + p.$

$$\text{Elasticity of demand} = -\frac{p}{x}\frac{dx}{dp}$$

$$= -\frac{2}{9} \text{ when } p = 2$$

Since $|\eta| < 1$ the demand is enelastic.

Example 26. If demand function $P = 10e^{-\frac{x}{4}}$, find elasticity of demand. Also find out the demand at which elasticity is unity. **(Rajasthan Univ. M.A. Eco. 1997)**

Solution :

$$P = 10e^{-\frac{x}{4}}$$

$$\frac{dp}{dx} = -\frac{1}{4}10e^{-\frac{x}{4}}$$

$$e_d = -\frac{P}{x}\frac{dx}{dp} = -\frac{10e^{-\frac{x}{4}}}{x} \times \frac{1}{\left(-\frac{1}{4}\times 10\right)e^{-\frac{x}{4}}} = \frac{4}{x}$$

If $e_d = 1$, then $\frac{4}{x} = 1 \Rightarrow x = 4$.

Example 27. Show that in general slope of average cost curve [*i.e.,* AC curve]

$= \frac{1}{x}(\text{MC} - \text{AC})$ where x is the output. Verify this result when

$$C = 100 + x + 2x^2$$

Solution : $\text{Average cost} = \frac{\text{Total cost}}{\text{No. of units manufactured}} = \frac{C}{x}$

If Y = Average cost $= \frac{C}{x}$ then $\frac{dY}{dx}$ is the slope of the average cost curve.

$$\frac{dY}{dx} = \frac{x\frac{dC}{dx} - C}{x^2} = \frac{1}{x}\left[\frac{dC}{dx} - \frac{C}{x}\right] = \frac{1}{x}[\text{MC} - \text{AC}]$$

$$Y = \frac{C}{x} = \frac{100}{x} + 1 + 2x = \text{Average Cost}, \quad \frac{dY}{dx} = -\frac{100}{x^2} + 2$$

$$\text{Marginal cost} = \frac{dC}{dx} = 1 + 4x$$

$$\frac{1}{x}(\text{MC} - \text{AC}) = \frac{1}{x}\left[(1 + 4x) - \left(\frac{100}{x} + 1 + 2x\right)\right]$$

$$= \frac{1}{x} + 4 - \frac{100}{x^2} - \frac{1}{x} - 2 = \frac{-100}{x^2} + 2 = \frac{dY}{dx}.$$

Example 28. The total revenue from the sale of a good X is given by the equation $R = 60Q - Q^2$ for $0 \le Q \le 60$ where R is the total revenue and Q is the quantity bought at a price P. Calculate the value of MR when the point price elasticity of demand is 2. **(D.U. B.A. Eco. (Hons.) 1988)**

Solution : $R = 60Q - Q^2$, $AR = \dfrac{R}{Q} = 60 - Q$

$$\frac{dR}{dQ} = MR = 60 - 2Q$$

$$MR = AR\left(1 - \frac{1}{e}\right) = (60 - Q)\left(1 - \frac{1}{2}\right)$$

$$\Rightarrow \quad 60 - 2Q = \frac{1}{2}(60 - Q) \Rightarrow 60 = 3Q \Rightarrow Q = 20$$

$$\therefore \quad MR = (60 - 2 \times 20) = 20.$$

Example 29. Suppose that a producer sells the same product of two groups that have constant elasticity demand curves with elasticities, e_1 and e_2. The marginal cost of production is constant at level C. What price is charged to each group. Which of the two prices will be higher ?

(D.U. B.A. Eco. (Hons.) 1991)

Solution : Let Rs. p_1 be the price per unit charged from one group and Rs. p_2 the price per unit charged from the other group.

For maximum profit

$$MC = MR$$

$$\therefore \quad MC = MR_1 = MR_2 \qquad \text{...(I)}$$

where R_1 is the revenue from the first group. R_2 the revenue from the second group

$$\Rightarrow \quad p_1\left(1 - \frac{1}{e_1}\right) = p_2\left(1 - \frac{1}{e_2}\right) \qquad \text{...(II)} \left[\because \; MR = p\left(1 - \frac{1}{e}\right)\right]$$

where e_1 is the price elasticity for the first group and e_2 the price elasticity for the second group.

If $MC = C$ then from (II)

$$\Rightarrow \quad C = p_1\left(1 - \frac{1}{e_1}\right) = p_2\left(1 - \frac{1}{e_2}\right)$$

$$p_1 = \frac{C}{1 - \frac{1}{e_1}}, \; p_2 = \frac{C}{1 - \frac{1}{e_2}},$$

$\Rightarrow$ If $e_1 > e_2$ then $p_1 < p_2$, *i.e.,* the group for which price elasticity of demand is less will be charged more.

Example 30. The demand curve for a commodity is given by

$$x = 20 - 2p - p^2$$

where P and x are the price and the quantity in demand respectively.

Find the elasticity of demand for $p = 2.5$. For which demand the elasticity will be unity?

(ICWA QM Dec., 1994)

Solution : Elasticity of demand $= -\dfrac{dx}{dp}\dfrac{p}{x}, \dfrac{dx}{dp} = -2 - 2p$

$$\therefore \qquad e_d = \frac{(2+2p)}{20-2p-p^2} \times p \qquad \text{...(I)}$$

$$= \frac{2(1+2.5)\,(2.5)}{20-2\times 2.5-(2.5)^2} \text{ when } p = 2.5$$

$$= \frac{17.5}{8.75} = 2$$

If $e_d = 1$, then from (I)

$$\frac{(2+2p)p}{20-2p-p^2} = 1 \Rightarrow (2+2p)\,p = 20-2p-p^2$$

$$\Rightarrow \qquad 2p+2p^2 = 20-2p-p^2$$

$$\Rightarrow \qquad 3p^2+4p-20 = 0$$

$$\Rightarrow \qquad 3p^2+10p-6p-20 = 0$$

$$\Rightarrow \qquad p(3p+10)-2(3p+10) = 0$$

$$\Rightarrow \qquad (p-2)\,(3p+10) = 0 \Rightarrow p = 2 \text{ or } p = -10/3$$

Since, p cannot be –ve, we reject the negative value. Hence, $p = 2$ when $e_d = 1$.

when $p = 2$ $\qquad x = 20 - 4 - 4 = 12$

Example 31. Show that the elasticity of demand with respect to price, for the demand function $x = 3p^{-2}$ is constant, p and x being the price and quantity in demand respectively.

(ICWA QM June, 1995)

Solution : The given demand function is

$$x = 3p^{-2}$$

$$\frac{dx}{dp} = -6p^{-3}$$

Elasicity of demand $= \dfrac{-p}{3p^{-2}}(-6p^{-3}) = 2$

which is independent of p and hence is constant.

Example 32. If a demand law is given by $p = 10e^{-x/2}$, find elasticity of demand. Also show that if the demand law is in the form $p = ae^{-kx}$, then $\eta_d = 1/kx$ (p being price and x being demand).

Solution : For the demand law $p = 10e^{-x/2}$

$$1 = 10e^{-x/2}\left(-\frac{1}{2}\right)\frac{dx}{dp} \Rightarrow \frac{dx}{dp} = \frac{-1}{5e^{-x/2}}$$

$$\therefore \qquad \eta_d = -\frac{p}{x}\frac{dx}{dp} = -\frac{10e^{-x/2}}{x}\left[\frac{-1}{5e^{-x/2}}\right] = \frac{2}{x}$$

Again, for the demand law $p = ae^{-kx}$,

$$1 = ae^{-kx}(-k)\frac{dx}{dp} \text{ or } \frac{dx}{dp} = \frac{-1}{kae^{-kx}}$$

$$\therefore \qquad n_d = -\frac{p}{x}\frac{dx}{dp} = -\frac{ae^{-kx}}{x}\left(\frac{-1}{ka^{-kx}}\right) = \frac{1}{kx}.$$

Example 33. The demand curve is $p = 200 - 5\log_e x$. Prove that MR $= p\left[1 - \frac{1}{\eta_d}\right]$ where η_d is the price elasticity of demand.

Solution : For the demand curve, $p = 200 - 5\log_e x$,

$$1 = \frac{d}{dp}(200 - 5\log_e x) = -\frac{5}{x}\frac{dx}{dp} \Rightarrow \frac{dx}{dp} = -\frac{x}{5}$$

$$\therefore \qquad \eta_d = -\frac{p}{x}\frac{dx}{dp} = -\left(\frac{200 - 5\log_e x}{x}\right)\left(-\frac{x}{5}\right)$$

$$= \frac{200 - 5\log_e x}{5}$$

$$R = px = (200 - 5\log_e x)\,x = 200x - 5x\log_e x$$

$$MR = 200 - 5\left(\log_e x + x\cdot\frac{1}{x}\right) = 195 - 5\log_e x \qquad \text{...(I)}$$

Now,

$$p\left(1 - \frac{1}{\eta_d}\right) = (200 - 5\log_e x)\left[1 - \frac{5}{(200 - 5\log_e x)}\right]$$

$$= 195 - 5\log_e x \qquad \text{...(II)}$$

From (I) and (II), we have

$$MR = p\left(1 - \frac{1}{\eta_d}\right).$$

Example 34. If η_1 and η_2 be the price elasticity of demand for the demand laws $p = e^x$ and $p = e^x/x$, show that $\eta_1\eta_2 = \eta_1 - \eta_2$.

Solution : Differentiating the demand function $p = e^x$ w.r. to p, we have

$$1 = e^x\frac{dx}{dp} \Rightarrow \frac{dx}{dp} = \frac{1}{e^x}$$

$$\therefore \qquad \eta_1 = -\frac{p}{x}\cdot\frac{dx}{dp} = -\frac{e^x}{x}\left(\frac{1}{e^x}\right) = -\frac{1}{x} \qquad \text{...(I)}$$

Again, differentiating the demand function $p = e^x/x$ w.r. to p, we have

$$1 = \frac{x.e^x - 1.e^x}{x^2}\cdot\frac{dx}{dp} \Rightarrow \frac{dx}{dp} = \frac{x^2}{(x-1)\,e^x}$$

$$\therefore \qquad \eta_2 = -\frac{p}{x}\cdot\frac{dx}{dp} = \frac{e^x/x}{x}\left\{\frac{x^2}{(x-1)\,e^x}\right\} = \frac{(-1)}{x-1}$$

$$= \left(-\frac{1}{x}\right) \cdot \frac{1}{\left\{1 + \left(-\frac{1}{x}\right)\right\}} = \eta_1 \cdot \frac{1}{1 + \eta_1} \qquad \text{[on using (I)]}$$

$\Rightarrow \eta_2 + \eta_1 \eta_2 = \eta_1$. Hence, $\eta_1 \eta_2 = \eta_1 - \eta_2$.

Questions

1. A watermelon grower wishes to send his produce to the city market as early as possible in the season to catch the prices. He can send now 6 tons at a price of Rs. 60 per ton. By waiting he estimates that he can add 3 tons per week to what he can send now but then the price would be reduced by Rs. 5 per ton per week. How long should he wait for the maximum revenue. Also determine the price at that level that maximises the total revenue.
[**Hint :** Exactly similar to example 1.] [**Ans.** (i) 5 units (ii) Rs. 735 per unit]

2. The total cost function for producing a commodity in x quantity is $C = 60 - 12x + 2x^2$. Find the average cost function and the level of output at which this function is minimum. Also verify that $MC = AC$ at the minimum point of AC curve.

3. A cheap gramophone sells for Rs. p and it is known that the demand, x hundred machines per year is given by $x = [90/(p+5)] - 6$. Determine at what price and output the total revenue is greatest.
[**Ans.** Price Rs $5(\sqrt{3} - 1)$ hundred and output $6(\sqrt{3} - 1)$ hundred]

4. A company notices that higher sale of a particular item which it produces are achieved by lowering the price charged. As a result the total revenue from the sales at first rises as the number of units sold increases, reaches a maximum and then falls off. This pattern of the total revenue is described by the relation :

$$y = 40{,}00{,}000 - (x - 2{,}000)^2,$$

where y is the total revenue and x the number of units sold :
(i) Find what number of units sold maximizes total revenue.
(ii) What is the amount of maximum revenue ?
(iii) What would be the total revenue if 2,500 units are sold ? [**Ans.** (i) 2,000, (ii) 40,00,000, (iii) 37,50,000]

5. The demand function for a commodity X is given by

$$x = 300 - 0.5p_X^2 + 0.02p_0 + 0.05y$$

where x is the quantity demanded of X, p_X the price of X, p_x the price of a related commodity and y is the constants income. Compute :
(i) the price elasticity of demand for X,
(ii) the income elasticity of demand for X, and
(iii) cross elasticity of demand for x w.r.t. p_0,
when $p_x = 12$, $p_0 = 10$ and $y = 200$. [**Ans.** (i) 0.6, (ii) 0.04, (iii) 0.0008]

6. The utility function of the consumer is given by $U = x_1 x_2 - 10x_1$, where x_1 and x_2 are the quantities of two commodities consumed. Find the optimal value of x_1 and x_2 if his income is 116, and product prices are 2 and 8 respectively. [**Ans.** $x_1 = 18, x_2 = 10$]

7. A tour operator charges Rs. 136 per passenger for 100 passengers up to a discount of Rs. 4 for each 10 passengers in excess of 100. Determine the number of passengers that will maximize the amount of money the tour operator receives. [**Ans.** 220]

❑❑❑

6 Integral Calculus

INDEFINITE INTEGRATION

6.1 INTRODUCTION

In differential calculus we have studied the method of finding out the derivative (differential coefficient) of a given function with respect to its variable. Now, we shall discuss the 'reverse of differentiation' (generally called **integration** or more specifically **indefinite integration**) which deals with the situation where the derivative is given and we have to find the primitive function.

If $f(x)$ represents the differential coefficient of $g(x)$, *i.e.*, $\frac{d}{dx}[g(x)] = f(x)$, then the problem of indefinite integration is to find $g(x)$ when $f(x)$ is given, $g(x)$ is called the **indefinite integral** of $f(x)$ with respect to x.

Since, the derivative of constant is zero, all functions differing by constants will have the same derivative.

i.e., $$\frac{d}{dx}[g(x)] = \frac{d}{dx}[g(x) + c]$$

where c is an arbitrarily chosen constant. Therefore, indefinite integral of a function will have infinite values differing by constants. For example, if $f(x)$ is the integral of $g(x)$, then $f(x) + c$ is also the integral of $g(x)$, *i.e.,* the indefinite integral of a function is not unique. *While dealing with indefinite integrals, the word indefinite is generally omitted.*

The function to be integrated called **integrand** and is inserted between the symbols $\int$ and ***dx*** **where** $\int$ is the symbol of integration and dx indicates the variable x with respect to which the integration is to be performed. The symbol $\int$ which is the integral sign is elongated S, the first letter of the word 'Sum'. $\int$ is read as integral of $f(x)$, w.r.t. x [w.r.t. is the standard symbol for 'with respect to'].

Example :

(i) $\int x^3\, dx = \frac{x^4}{4}$ $\qquad \left(\frac{d}{dx}\left(\frac{x^4}{4}\right) = x^3\right)$

(ii) If $\frac{d}{dx}[g(x)] = f(x)$, then $\int f(x)\, dx = g(x) + c$, where c is an arbitrary constant called *constant of integration* and is determined from the conditions given in the question.

Since, by definition, integration and differentiation are inverse operations, they neutralise the effect of each other, *i.e.,*

$$\frac{d}{dx}\left[\int f(x)\,dx\right] = f(x).$$

But it should be noted that $\int\left[\frac{d}{dx}f(x)\right]dx = f(x) + c \qquad \left[\because \frac{d}{dx}[f(x)] = \frac{d}{dx}[f(x)+c]\right]$

So far, we have discussed only one aspect of integration, *i.e.,* indefinite integration, but there is another aspect of integration, which views an integral as the limiting value of an algebraic summation. This aspect is called *definite integration* and the integral obtained is called *definite integral.* Definite integral has a *unique value.* This approach will be discussed later.

6.2 GENERAL RULE FOR INTEGRATION

In the integrand add one to the power of the variable and divide by the increased power and then add a constant (general rule for differentiation is to multiply the function with the power of the variable and decrease the power of variable by one), i.e.,

$$\int x^n\,dx = \frac{x^{n+1}}{n+1} + c, \quad \frac{d}{dx}\left[\frac{x^{n+1}}{n+1}\right] = \frac{(n+1)\,x^n}{n+1} = x^n$$

Examples :

(i) $\int 2x\,dx = 2.\frac{x^{1+1}}{2} + c = x^2 + c$

(ii) $\int x^5\,dx = \frac{x^{5+1}}{6} + c = \frac{x^6}{6} + c$

6.3 EXCEPTION TO THE GENERAL RULE OF INTEGRATION

The above-mentioned result holds for all values of n except for $= -1$. When $n = -1$, we have another formula

$$\int \frac{1}{x}\,dx = \int x^{-1}\,dx = \log_e x + c$$

because $\qquad \frac{d}{dx}(\log_e x) = \frac{1}{x}$

This result holds good if $x > 0$, if $x < 0$, *i.e.,* if x is negative, $\log x$ is not defined.

Now when $\qquad x < 0$

Put $\qquad x = -y \; (y > 0)$

$$dx = -dy$$

$$\frac{d}{dx}\{\log_e(-x)\} = -\frac{d}{dy}(\log_e y) = -\frac{1}{y} = \frac{1}{x}$$

$\therefore$ $$\int \frac{1}{x}\,dx = \log_e(-x) + c \qquad [\log_e(-x) \text{ is defined when } x < 0]$$

Hence, whether $x > 0$ or $x < 0$

$$\int \frac{1}{x}\,dx = \log_e |x| + c \qquad \textbf{(Please note)}$$

In the fundamental formula (general rule) which we have discussed, there can be a particular case

$$\int c\,ds = \int cx^0\,dx = \frac{cx^{0+1}}{0+1} = cx \qquad \textbf{(Important)}$$

which gives another result.

The integral of a constant with respect to x is equal to constant multiplied by x, e.g.,

$$\int 2dx = 2x.$$

Two important results

(i) $\int \frac{\text{Diff. coeff. of any function}}{\text{The function}}\,dx = \log_e(\text{the function}) + c$

For example,

(i) $\int \frac{a}{ax}\,dx = \log ax + c$ (ii) $\int \frac{b}{a+bx}\,dx = \log(a+bx) + c$

(iii) $\int \frac{\text{Diff. coeff. of any function}}{\sqrt{\text{The function}}}\,dx = 2\sqrt{\text{The function}} + c$

Also refer result (xxii) given below.

Important Note : Most of the results of integration are obtained from the corresponding results of differentiation. Therefore, in integration the result can always be checked. If the function obtained by integration be differentiated, we should get the original function.

6.4 SOME STANDARD RESULTS

(i) $\int x^n\,dx = \frac{x^{n+1}}{n+1} + c.$ [Except when $n = -1$]

(ii) $\int \frac{1}{x}\,dx = \log |x| + c$

(iii) $\int 1\,dx = x + c$

(iv) $\int 0\,dx = c$

(v) $\int a^x \, dx = \frac{a^x}{\log_e a} + c$*

(vi) $\int a^{mx} \, dx = \frac{a^{mx}}{m \log_e a} + c$

(vii) $\int a^{mx+b} \, dx = \frac{a^{mx+b}}{m \log_e a} + c$

(viii) $\int e^x \, dx = e^x + c$

(ix) $\int e^{mx} \, dx = \frac{e^{mx}}{m} + c$

(x) $\int (ax+b)^n \, dx = \frac{(ax+b)^{n+1}}{a(n+1)} + c$

(xi) $\int \frac{1}{ax+b} \, dx = \frac{1}{a} \log (ax+b) + c$

(xii) $\int a dx = ax + c$

(xiii) $\int kf(x) \, dx = k \int f(x) \, dx + c$

(xiv) $\int \frac{dx}{x^2 - a^2} = \frac{1}{2a} \log_e \frac{x-a}{x+a} + c, \ (x > a)$

(xv) $\int \frac{1}{a^2 - x^2} \, dx = \frac{1}{2a} \log_e \frac{a+x}{a-x} + c, \ (x < a)$

(xvi) $\frac{1}{\sqrt{x^2 + a^2}} \, dx = \log_e [x + \sqrt{x^2 + a^2}] + c$

(xvii) $\int \frac{1}{\sqrt{(x^2 - a^2)}} \, dx = \log_e [x + \sqrt{x^2 - a^2}] + c$

(xviii) $\int \sqrt{x^2 \pm a^2} \, dx = \frac{1}{2} x \sqrt{x^2 \pm a^2} \pm \frac{a^2}{2} \log_e [x + \sqrt{x^2 \pm a^2}]$

(xix) $\int e^x [f(x) + f'(x)] \, dx = e^x f(x)$

(xx) $\int [f(x) \pm g(x)] \, dx = \int f(x) \, dx \pm \int g(x) \, dx$

(xxi) $\int \frac{f'(x)}{f(x)} \, dx = \log_e f(x)$

(xxii) $\int [f(x)]^n f'(x) \, dx = \frac{[f(x)]^{n+1}}{n+1}$

---

* If we put $a = e$ in result (v) we get result (viii) similarly if we put $a = e$ in result (vi) we get result (ix).

Illustration : Find the curve whose slope is

$$\frac{dy}{dx} = \frac{2y}{x}, x > 0, y > 0$$

and which passes through the point (1, 1). $\left(\frac{dy}{dx} = \text{slope}\right)$

Solution : $\frac{dy}{dx} = \frac{2y}{x} \Rightarrow \frac{dy}{y} = \frac{2}{x}dx$

Integrating both sides, we get $\int \frac{dy}{y} = 2\int \frac{dx}{x}$

$$\log y = 2 \log x + \log c$$

[Constant of integration has been taken as log c for convenience.]

$$= \log x^2 c$$

i.e., $y = x^2 c$

Since, it passes through (1, 1) we have $1 = c$

∴ The equation of the required curve is $y = x^2$.

6.5 PLEASE REMEMBER THE FOLLOWING RULES OF INTEGRATION ALSO

(i) The integral of constant is equal to the constant multiplied by the independent variable, *i.e.,*

$$\int c\,dx = cx$$

(ii) The integral of the product of a constant and a function is equal to the product of the constant and the integral of the function, *i.e.,*

$$\int c\,g(x)\,dx = c\int g(x)\,dx$$

i.e., a constant factor before integration remain so after integration.

(iii) The integral of the sum (or difference) of two or more functions is equal to the sum (or difference) of the integrals of the functions taken separately, *e.g.,*

i.e., $\int [f_1(x) \pm f_2(x) \pm \ldots \pm f_n(x)]\,dx = \int f_1(x)\,dx \pm \int f_2(x)\,dx \pm \ldots \pm \int f_n(x)\,dx$

(iv) If in a fraction the numerator (after suitable adjustment of constant, if necessary) is the differential coefficient of the denominator, then the integral of the function is equal to the logarithm of the modulus of the denominator, 7.4.

Illustrations. (i) $\int \frac{dx}{ax} = \frac{1}{a}\int \frac{dx}{x} = \frac{1}{a}\log|x| + c$ or $\int \frac{dx}{ax} = \frac{1}{a}\int \frac{a}{ax}dx = \frac{1}{a}\log|ax| + c$

$$= \frac{1}{a}\log|x| + \frac{1}{a}\log a + c = \frac{1}{a}\log|x| + c' \qquad \left[\frac{1}{a}\log a + c = c'\right]$$

(ii) $\int \frac{dx}{ax+b} = \frac{1}{a}\int \frac{a\,dx}{(ax+b)} = \frac{1}{a}\log|(ax+b)| + c$

(iii) $\int \frac{x\,dx}{2x^2} + 3 = \frac{1}{4}\int \frac{4x\,dx}{2x^2+3} = \frac{1}{4}\log|2x^2+3| + c$

$$= \frac{1}{4}\log(2x^2+3) + c$$

(iv) $\int 0\,dx = c$ where c is a constant.

Note. Integration of $f(x)$ w.r.t. x is also called **anti-derivative** of $f(x)$ w.r.t. x.

SOLVED EXAMPLES

Example 1. Integrate the following w.r.t. x :

(a) 7	(b) x^3	(c) $\sqrt{x}$	(d) $x^{3/2}$	(e) $\frac{3}{\sqrt{x}}$
(f) $\frac{1}{225}$	(g) $\frac{1}{2x}$	(h) $\frac{1}{x^n}$	(i) $\sqrt{3}$	(j) $\frac{4}{7}$
(k) 0	(l) $x+2$	(m) $(3x^2-4)^3\,(6x)$		(n) $\sqrt{3x^2-4}\,(6x)$

Solution : (a) $\int 7\,dx = \int 7x^0\,dx = \frac{7x^{0+1}}{0+1} + c = 7x + c$

(b) $\int x^3\,dx = \frac{x^4}{4} + c$

(c) $\int \sqrt{x}\,dx = \int x^{1/2}\,dx = \frac{x^{\frac{1}{2}+1}}{\frac{1}{2}+1} + c = \frac{2}{3}x^{3/2} + c$

(d) $\int x^{3/2}\,dx = \frac{x^{\frac{3}{2}+1}}{\frac{3}{2}+1} + c = \frac{2}{5}x^{5/2} + c$

(e) $\int \frac{3}{\sqrt{x}}\,dx = \int 3x^{-1/2}\,dx = 3 \times 2x^{1/2} + c = 6\sqrt{x} + c$

(f) $\int \frac{1}{225}\,dx = \frac{x}{225} + c$

(g) $\int \frac{1}{2x}\,dx = \frac{1}{2}\int \frac{1}{x}\,dx = \frac{1}{2}\log x + c$

(h) $\int \frac{1}{x^n} dx = \int x^{-n} dx = \frac{x^{-n+1}}{-n+1} + c$

(i) $\int \sqrt{3}\, dx = \sqrt{3}x + c,$

(j) $\int \frac{4}{7} dx = \frac{4}{7}x + c$

(k) $\int 0\, dx = c$

(l) $\int (x+2)\, dx = \frac{x^2}{2} + 2x + c$

(m) $\int (3x^2 - 4)^3 (6x)\, dx = \frac{(3x^2-4)^4}{4} + c$ $\left[\because \frac{d}{dx}(3x^2 - 4) = 6x\right]$

(n) $\int (\sqrt{3x^2 - 4})\, 6x\, dx = \frac{(3x^2-4)^{1/2+1}}{3/2} = \frac{2}{3}(3x^2 - 4)^{3/2}$

Example 2. Integrate [Or Find the anti-derivative of]

(i) $\int \frac{1+x}{(2+x)^2} e^x dx$ (ii) $\int \frac{xe^x}{(x+1)^2} dx$

(iii) $\int \frac{e^x(1 + x\log x)}{x} dx$ (iv) $\int \frac{(1-x)\, e^x}{x^2} dx$

Solution : (i) $\int \frac{1+x}{(2+x)^2} e^x dx = \int \frac{(2+x-1)}{(2+x)^2} e^x dx$

$$= \int \left(\frac{e^x}{2+x} - \frac{e^x}{(2+x)^2}\right) dx$$

$$= \int \left[\frac{1}{2+x} - \frac{1}{(2+x)^2}\right] e^x dx$$

$$= \frac{e^x}{2+x} + c$$

[Refer Page 6.4 (xix)]

$\left[\because \text{Let } f(x) = \frac{1}{2+x}, \text{ then } f'(x) = -\frac{1}{(2+x)^2}\right]$

$\int \left[\frac{1}{2+x} - \frac{1}{(2+x)^2}\right] e^x dx$ is of the form $\int [f(x) + f'(x)]\, e^x dx]$

(ii) [**Hint :** $\frac{xe^x}{(x+1)^2} = \left[\frac{x+1}{(x+1)^2} - \frac{1}{(x+1)^2}\right] e^x = \left[\frac{1}{x+1} - \frac{1}{(x+1)^2}\right] e^x$.] **Ans.** $\frac{e^x}{x+1} + c$

(iii) $e^x\left(\frac{1+x\log x}{x}\right)=e^x\left(\frac{x\log x+1}{x}\right)=e^x\left(\log x+\frac{1}{x}\right)$ **Ans.** $e^x \log x + c$

(iv) [**Hint :** $\frac{1-x}{x^2}e^x=\left[\frac{-1}{x}+\frac{1}{x^2}\right]e^x$] **Ans.** $-\frac{1}{x}e^x+c$

6.6 METHOD OF SUBSTITUTION

This is the most useful method. By suitable substitution, we simplify the integrand so that standard formulae may be applied and the integration becomes easier. In this method the independent variable is changed in the following manner.

If the variable x in $\int f(x)\,dx$ be changed to t by means of the relation $x=\phi(t)$ where $\phi(t)$ possesses a continuous derivative $\phi'(t)$ then

$$\int f(x)\,dx=\int f[\phi(t)]\,\phi'(t)\,dt \qquad [\because \text{ if } x=\phi(t),\ dx=\phi'(t)\,dt]$$

i.e., to carry out the substitution $x=\phi(t)$ in $\int f(x)\,dx$, we replace x by $\phi(t)$ and dx by $\phi'(t)\,dt$. The resulting integral in t is then evaluated. After integration, t is replaced by its value in terms of x obtained from the relation $x=\phi(t)$. This gives the value of

$\int f(x)\,dx$ in terms of x. Solved examples will make the rule clear.

6.7 PARTIAL FRACTIONS

If a fraction* is expressed in terms of simpler fractions connected by plus or minus sign then the simpler fractions are called partial fractions. For example,

$$\frac{1}{(x+1)(x+2)}=\frac{1}{x+1}-\frac{1}{x+2}$$

Then $\frac{1}{x+1}$ and $\frac{1}{x+2}$ are called the partial fractions corresponding to $\frac{1}{(x+1)(x+2)}$.

Please note that the degree of denominator of each partial fraction is lower than the degree of the denominator of the given fraction.

---

* By fraction we mean an algebraic proper fraction [An expression of the form P/Q where P and Q are polynomials in x is called a rational fraction. If P is of lower degree than Q, P/Q is called a proper fraction. If P is of higher degree than Q, P/Q is called an improper fraction. An improper fraction can always be expressed as the sum of an integral function and a proper function. For example,

$\frac{x^2+1}{x+1}$ is an improper fraction because the degree of the numerator is 2 and that of the denominator is 1 and $2>1$.

$$\frac{x^2+1}{x+1}=(x-1)+\frac{2}{x+1}$$

$$= \text{Integral function} + \text{a proper fraction}$$

6.8 RULES TO RESOLVE A GIVEN PROPER FRACTION INTO PARTIAL FRACTIONS

1. Examine the factors in the denominator of the given proper fraction.
2. To each non-repeated factor of the type $x - a$ in the denominator, there corresponds a fraction of the type $\frac{A}{x-a}$ where A is a constant. For example, examine the proper fraction

$$\frac{x^2+1}{(x+1)(x+2)(x+3)}$$

If we are to break this fraction into partial fractions we can say, let

$$\frac{x^2+1}{(x+1)(x+2)(x+3)} = \frac{A}{x+1} + \frac{B}{x+2} + \frac{C}{x+3}$$

where A, B, C are constants.

3. To a factor of the type $(x-b)^n$ in the denominator, there corresponds a group of fractions of the form

$$\frac{A_1}{x-b} + \frac{A_2}{(x-b)^2} + \frac{A_3}{(x-b)^3} + \ldots + \frac{A_n}{(x-b)^n}$$

i.e., to any linear factor $x - b$ occurring twice, *i.e.*, $(x-b)^2$ in the denominator there corresponds two partial fractions

$$\frac{A_1}{x-b} + \frac{A_2}{(x-b)^2}$$

and if $x - b$ occurs three times, *i.e.*, if we have $(x-b)^3$ in the denominator, there corresponds three partial fractions of the type

$$\frac{A_1}{x-b} + \frac{A_2}{(x-b)^2} + \frac{A_3}{(x-b)^3}$$

If $(x - b)$ occurs n *times*, *i.e.*, if we have $(x-b)^n$ in the denominator then there corresponds n partial fractions of the form given above.

4. To a non-repeated quadratic factor of the type $(ax^2 + bx + c)$ in the denominator, there corresponds a fraction of the form

$$\frac{Bx+C}{ax^2+bx+c}$$

For example, if the given fraction is

$$\frac{1}{(x+1)(x^2+2x+3)}$$

then there correspond two partial fractions of the form

$$\frac{A}{x+1} + \frac{Bx+C}{x^2+2x+3}$$

where A, B, C are constants.

A repeated quadratic factor of the form $(ax^2+bx+c)^n$ in the denominator corresponds to n partial fractions of the form

$$\frac{B_1x+C_1}{ax^2+bx+c}+\frac{B_2x+C_2}{(ax^2+bx+c)^2}+\frac{B_3x+C_3}{(ax^2+bx+c)^3}+\ldots+\frac{B_nx+C_n}{(ax^2+bx+c)^n}.$$

6.9 SOME STANDARD SUBSTITUTIONS

1. In integrals of the form

$$\int (ax+b)\,dx \text{ and } \int \frac{dx}{ax+b}, \text{ put } ax+b=t$$

2. In integrals of the form

$$\sqrt{ax+b}\,dx \text{ and } \int \frac{dx}{\sqrt{ax+b}}, \text{ put } ax+b=t^2$$

3. In expressions containing fractional powers of x, put $x=t^r$ where r is the L.C.M. of the denominators of the fractional indices.
4. In integrals of the type

$$\int \text{(function of } x^2)\, x\,dx, \text{ put } x^2=t$$

5. In integrals of the type

$$\int \frac{1}{x(px^m+q)}\,dx, \text{ put } x^m=\frac{1}{z}$$

6. Sometimes the substitution is made by inspection.

Example 3. Integrate $\int \frac{1-x}{\sqrt{1+x}}\,dx$

Solution : $1-x=2-(1+x)$

$$\int \frac{1-x}{\sqrt{1+x}}\,dx=\int\left[\frac{2}{\sqrt{1+x}}-\frac{\sqrt{1+x}}{1}\right]dx$$

$$=\frac{2(1+x)^{-\frac{1}{2}+1}}{\frac{-1}{2}+1}-\frac{[1+x]^{\frac{1}{2}+1}}{\frac{1}{2}+1}$$

$$=4\sqrt{1+x}-\frac{2}{3}(1+x)^{3/2}$$

Example 4. $\int \frac{dx}{lx+m}$.

Solution : Put $lx+m=t$

i.e.,
$$ldx=dt \Rightarrow dx=\frac{1}{l}\,dt$$

$$\int \frac{dx}{lx+m}=\frac{1}{l}\int \frac{1}{t}\,dt=\frac{1}{l}\log t+c=\frac{1}{l}\log(lx+m)+c$$

Example 5. Find $\int (4x+5)^6\, dx$.

Solution : Put $4x+5 = t$

i.e., $$4dx = dt \Rightarrow dx = \frac{1}{4}\, dt$$

$\therefore$ $$\int (4x+5)^6\, dx = \frac{1}{4}\int t^6\, dt = \frac{1}{4}\left(\frac{t^7}{7}\right) + c$$

$$= \frac{1}{28}(4x+5)^7 + c$$

Example 6. Integrate $\int \frac{x+a}{x-a}\, dx$.

Solution : $$\frac{x+a}{x-a} = \frac{x-a+2a}{x-a} = 1 + \frac{2a}{x-a}$$

$\therefore$ $$\int \frac{x+a}{x-a}\, dx = \int 1\, dx + \int \frac{2a}{x-a}\, dx = x + 2a \log(x-a) + c$$

Example 7. $\int \frac{x^2+ax+b}{ax+b}\, dx$.

Solution : $$\int \frac{x^2+ax+b}{ax+b}\, dx = \int \frac{x^2}{ax+b}\, dx + \int 1\, dx$$

$$= \int \left[\frac{x}{a} - \frac{b}{a^2} + \frac{b^2}{a^2(ax+b)}\right] dx + \int 1\, dx\ ^*$$

$$= \frac{x^2}{2a} - \frac{b}{a^2}x + \frac{b^2}{a^2}\log(ax+b) + x + c$$

Example 8. Assuming $\int e^x\, dx = e^x$, show that

$$\int x^m\, dx = \frac{x^{m+1}}{m+1} \text{ where } m \neq -1.$$

Solution : Changing x to ax (dx to $a\,dx$) in the given relation, we get

$$\int e^{ax}(a\, dx) = e^{ax} \text{ or } \int e^{ax}\, dx = \frac{1}{a}e^{ax} \qquad \text{...(I)}$$

Put $x = e^z$, then $dx = e^z dz$

---

* $ax+b)\ x^2\ \left(\frac{x}{a} - \frac{b}{a^2}\right.$

$x^2 + \frac{bx}{a}$

$-\quad -$

$-\frac{bx}{a}$

$-\frac{bx}{a} - \frac{b^2}{a^2}$

$+\quad +$

$\frac{b^2}{a^2}$

Now $$\int x^m\,dx = \int e^{mz}.e^z dz = \int e^{(m+1)z}\,dz$$

$$= \frac{e^{(m+1)z}}{m+1} \qquad \text{[From (I)]}$$

$$= \frac{(e^z)^{m+1}}{m+1} = \frac{x^{m+1}}{m+1}$$

Example 9. Find $\int \left[\frac{2a}{\sqrt{x}} - \frac{b}{x^2} + 3c\,\sqrt[3]{x^2}\right] dx.$

Solution : Now $\int \left[\frac{2a}{\sqrt{x}} - \frac{b}{x^2} + 3c\sqrt[3]{x^2}\right] dx = \int \frac{2a}{\sqrt{x}}\,dx - \int \frac{b}{x^2}\,dx + 3c\int x^{2/3}\,dx$

$$= 2a\frac{x^{1/2}}{\frac{1}{2}} - b\frac{x^{-1}}{-1} + 3c\frac{x^{5/3}}{\frac{5}{3}} + k$$

$$= 4ax^{1/2} + \frac{b}{x} + \frac{9c}{5}x^{5/3} + k$$

where k is the constant of integration.

Example 10. Find the value of $\int \frac{lx+m}{nx+p}\,dx.$

Solution : Now $$\frac{lx+m}{nx+p} = \frac{l}{n}\left[\frac{x+\frac{m}{l}}{x+\frac{p}{n}}\right] = \frac{l}{n}\left[\frac{x+\frac{m}{l}+\frac{p}{n}-\frac{p}{n}}{x+\frac{p}{n}}\right]$$

$$= \frac{l}{n}\left[1+\frac{\frac{m}{l}-\frac{p}{n}}{x+\frac{p}{n}}\right]$$

$$= \frac{l}{n}\left[1+\frac{mn-lp}{ln}\left(\frac{n}{nx+p}\right)\right]$$

$\therefore$ $$\int \frac{lx+m}{x+p}\,dx = \int \frac{l}{n}\,dx + \frac{mn-lp}{n^2}\int \frac{n}{nx+p}\,dx$$

$$= \frac{l}{n}x + \frac{mn-lp}{n^2}\log(nx+p) + c$$

Example 11. Evaluate $\int \frac{x^4 + 3x^2 + 1}{x^3}\, dx.$

Solution : $\int \frac{x^4 + 3x^2 + 1}{x^3}\, dx = \int x\, dx + 3\int \frac{1}{x}\, dx + \int \frac{1}{x^3}\, dx$

$$= \frac{x^2}{2} + 3 \log x - \frac{1}{2x^2} + c.$$

Example 12. $\int (x^2 + 1)(2x^3 - 3)\, dx$

(Mysore Univ. B.Com. 1995)

Solution : $\int (x^2 + 1)(2x^3 - 3)\, dx = \int (2x^5 - 3x^2 + 2x^3 - 3)\, dx$

$$= 2\frac{x^6}{6} - \frac{3x^3}{3} + \frac{2x^4}{4} - 3x + c$$

$$= \frac{x^6}{3} - x^3 + \frac{x^4}{2} - 3x + c$$

$$= \frac{x^6}{3} + \frac{x^4}{2} - x^3 - 3x + c$$

Example 13. Evaluate $\int \sqrt{(ax + b)}\, dx.$

Solution :

$$\int \sqrt{(ax + b)}\, dx = \frac{2}{3}\frac{(ax + b)^{3/2}}{a} + c = \frac{2}{3a}(ax + b)^{3/2} + c.$$

Example 14. Evaluate $\int (ax + b)\sqrt{cx + d}\, dx.$

Solution : Put

$$cx + d = t^2 \Rightarrow x = \frac{t^2 - d}{c}$$

$$c\, dx = 2t\, dt \Rightarrow dx = \frac{2}{c} t\, dt$$

$$\therefore \quad \int (ax + b)\sqrt{cx + d}\, dx = \int \left[a\frac{t^2 - d}{c} + b\right] t.\frac{2}{c} t\, dt$$

$$= \frac{2a}{c^2}\int (t^4 - t^2 d)\, dt + \frac{2b}{c}\int t^2\, dt$$

$$= \frac{2a}{c^2}\left[\frac{t^5}{5} - \frac{t^3}{3} d\right] + \frac{2b}{c}\frac{t^3}{3} + k$$

$$= \frac{2a}{c^2}\left[\frac{(cx + d)^{5/2}}{5} - \frac{(cx + d)^{3/2}}{3} d\right] + \left(\frac{2b}{c}\right)\frac{(cx + d)^{3/2}}{3} + k$$

Student may simplify further.

Example 15. $\int \frac{dx}{\sqrt{1-e^{2x}}}$.

Solution : Put $1 - e^{2x} = t^2 \Rightarrow e^{2x} = 1 - t^2$

i.e.,

$$2e^{2x}\,dx = -2t\,dt \Rightarrow e^{2x}\,dx = -t\,dt \Rightarrow dx = \frac{-t}{e^{2x}}\,dt = \frac{-t}{1-t^2}\,dt$$

$$\Rightarrow \qquad dx = \frac{t\,dt}{t^2-1}$$

$$\therefore \qquad \frac{1}{\sqrt{1-e^{2x}}}\,dx = \int \frac{1}{t}\left[\frac{t}{t^2-1}\right] dt$$

$$= \int \frac{dt}{t^2-1} = \int \frac{dt}{(t+1)\,(t-1)}$$

$$= \frac{1}{2}\int \left[\frac{1}{t-1} - \frac{1}{t+1}\right] dt$$

$$= \frac{1}{2}\,[\log\,(t-1) - \log\,(t+1)] + \log c$$

$$= \frac{1}{2}\log c\left(\frac{t-1}{t+1}\right) \qquad \text{where } t = \sqrt{1-e^{2x}}.$$

Example 16. $\int \frac{dx}{\sqrt{x}+1}$.

Solution : Put $x = t^2$. Then $dx = 2t\,dt$

$$\therefore \qquad \int \frac{dx}{\sqrt{x}+2} = \int \frac{2t\,dt}{t+2} = 2\int \frac{(t+2-2)}{t+2}\,dt$$

$$= 2\int\left(1 - \frac{2}{t+2}\right)dt = 2t - 4\log\,(t+2) + c$$

$$= 2\sqrt{x} - 4\log\,(\sqrt{x}+2) + c$$

Example 17. Evaluate $\int [e^{a \log x} + e^{x \log a}]\,dx$.

Solution : $\int [e^{a\log x} + e^{x\log a}]\,dx = \int e^{a\log x}\,dx + \int e^{x\log a}\,dx$

$$= \int e^{\log x^a}\,dx + \int e^{\log a^x}\,dx \qquad \text{...(I)}$$

Now let $\qquad y = e^{\log x^a}$

$$\log y = \log e^{\log x^a} = \log x^a \Rightarrow y = x^a$$

Similarly $e^{\log a^x} = a^x$

$$\therefore \qquad \int (e^{a \log x} + e^{x \log a})\, dx = \int x^a\, dx + \int a^x\, dx$$

$$= \frac{x^{a+1}}{a+1} + c_1 + \frac{a^x}{\log a} + c_2$$

$$= \frac{x^{a+1}}{a+1} + \frac{a^x}{\log a} + c \qquad \text{[Taking } c = c_1 + c_2]$$

Example 18. Evaluate $\int \frac{dx}{x(x^4+1)}$.

Solution : Let $x^4 = \frac{1}{z}$, $\therefore\ 4x^3\, dx = \frac{-1}{z^2}\, dz$

$$\Rightarrow \qquad \frac{4x^4}{x}\, dx = \frac{-1}{z^2}\, dz \Rightarrow \frac{4}{zx}\, dx = \frac{-1}{z^2}\, dz$$

$$\Rightarrow \qquad \frac{4}{x}\, dx = \frac{-1}{z}\, dz \Rightarrow \frac{dx}{x} = \frac{-1}{4z}\, dz$$

$$\therefore \qquad \int \frac{dx}{x(x^4+1)} = -\frac{1}{4} \int \frac{1}{z\left(\frac{1}{z}+1\right)}\, dz = -\frac{1}{4} \int \frac{dz}{z+1}$$

$$= -\frac{1}{4} \log (z+1) + c$$

$$= -\frac{1}{4} \log \left(\frac{1}{x^4} + 1\right) + c = -\frac{1}{4} \log \frac{x^4+1}{x^4} + c$$

If we take the arbitrary constant c as $-\frac{1}{4} \log c$, the result can be further simplified as

$$\int \frac{dx}{x(x^4-1)} = -\frac{1}{4} \log \frac{x^4+1}{x^4} - \frac{1}{4} \log c$$

$$= -\frac{1}{4} \log \frac{x^4+1}{x^4}$$

Example 19. Find the integral of $\frac{1}{x(x^n-1)}$.

(Punjab, B.Com. 1989)

Hint : Put $x^n = \frac{1}{z}$ **Ans.** $\left[-\frac{1}{n} \log \left(\frac{x^n+1}{x^n}\right) + c\right]$

Example 20. Find $\int \frac{\sqrt{1+\log x}}{x}\,dx.$

Solution : Put $1 + \log x = t \Rightarrow \frac{1}{x}\,dx = dt$

$$\int \frac{\sqrt{1+\log x}}{x}\,dx = \int \sqrt{t}\,dt = \frac{2}{3}t^{3/2} + c = \frac{2}{3}(1+\log x)^{3/2} + c$$

Example 21. Evaluate $\int \frac{\sqrt{1-x^2}}{x^4}\,dx.$ **(Note the solution)**

Solution : Put $x = \frac{1}{t}$, $\therefore\ dx = -\frac{1}{t^2}\,dt$

i.e., $$\int \frac{\sqrt{1-x^2}}{x^4}\,dx = -\int t\sqrt{t^2-1}\,dt. \text{ Again put } t^2 - 1 = z^2$$

$2t\,dt = 2z\,dz \Rightarrow t\,dt = z\,dz$

$\therefore$ $$-\int t\sqrt{t^2-1}\,dt = -\int z^2\,dz = -\frac{z^3}{3} + c$$

$$= \frac{(t^2-1)^{3/2}}{3} + c = -\frac{(1-x^2)^{3/2}}{3x^3} + c$$

6.10 INTEGRATION BY PARTS

If u and v be two functions of x, then

$$\frac{d}{dx}(uv) = u\frac{dv}{dx} + v\frac{du}{dx}$$

Integrating both sides with respect to x, we get

$$uv = \int u\frac{dv}{dx}\,dx + \int v\frac{du}{dx}\,dx$$

i.e., $$\int u\frac{dv}{dx}\,dx = uv - \int v\frac{du}{dx}\,dx \qquad \text{...(I)}$$

If $u = f(x)$, $\frac{dv}{dx} = g'(x)$

Then $$\int \frac{dv}{dx}\,dx = \int g'(x)\,dx, \text{ i.e., } v = \int g'(x)\,dx$$

$\therefore$ (I) can be written as

$$\int f(x)\,g'(x)\,dx = f(x)\int g'(x)\,dx - \int\left[\int (g'(x)\,dx)\,\frac{d}{dx}f(x)\right]dx$$ **(Importat)**

In words, we can put the **rule** as under :

Integral of the product of two functions is equal to :

[First function × Integral of the second function –

Integral of (integral of the second function × Differential coefficient of the first function)]

While solving problems of integration by parts, following rules may be followed :

(i) Choose the first and second functions in such a way that the differential coefficient of the first function and the integral of the second function can be easily found.

(ii) Take logarithmic functions and the powers of x, viz., x, x^2, x^3 etc. as the first function.

(iii) In case of integrals of the form $\int (\log x)^n \, dx$ take 1 as the second function and $(\log x)^n$ as the first function.

(iv) Rule of integration by parts can be used repeatedly if required.

Example 22. Evaluate $\int \log x \, dx$.

Solution : Taking unity as the second function, and $\log x$ as the first function and integrating by parts, we get

$$\int \log x \, dx = x \log x - \int x.\frac{1}{x} \, dx + c$$

$$= x \log x - x + c = x(\log x - 1) + c$$

Example 23. Evaluate : $\int t^2 e^t dt$.

Solution : Integrating by parts taking t^2 as first function and e^t as second function

$$\int t^2 e^t dt = [t^2 e^t] - \int 2te^t dt = t^2 e^t - 2te^t + 2\int e^t \, dt$$

$$= t^2 e^t - 2te^t + 2e^t + c$$

Example 24. $\int (2x + 3) \log (x^2 + 3x + 4) \, dx$.

Solution : Put $x^2 + 3x + 4 = z$, $\quad \therefore (2x + 3) \, dx = dz$

$$\therefore \quad \int (2x + 3) \log (x^2 + 3x + 4) \, dx = \int \log z \, dz$$

$$= (\log z) \, z - \int z \times \frac{1}{z} \, dz$$

$$= z \log z - z = z(\log z - 1)$$

$$= (x^2 + 3x + 4) [\log (x^2 + 3x + 4) - 1].$$

Example 25. Evaluate $\int (\log t)^2 \, dt$.

Solution : Integrating by parts taking $(\log t)^2$ as the first function and 1 as second function, we get

$$\int (\log t)^2 \, dt = (\log t)^2 \, t - \int \frac{t \times 2 \log t}{t} \, dt = t(\log t)^2 - 2\int \log t \, dt$$

Integrating $\int \log t \, dt$ by taking $\log t$ as first function and 1 as second function, we get

$$\int (\log t) \, dt = (\log t) \, t - \int \frac{1}{t} \times t \, dt = t \log t - t$$

$\therefore$
$$\int (\log t)^2 \, dt = t(\log t)^2 - 2t \log t + 2t$$

$$= t[(\log t)^2 - 2 \log t + 2] + c.$$

Example 26. Evaluate $\int x \, e^{ax} \, dx$.

Solution : Integrating by parts by taking x as the first function and e^{ax} as the second function, we get

$$\int x \, e^{ax} \, dx = x \frac{e^{ax}}{a} - \int \frac{e^{ax}}{a} \, dx = \frac{xe^{ax}}{a} - \frac{e^{ax}}{a^2} + c$$

$$= \frac{1}{a} e^{ax} \left[x - \frac{1}{a} \right] + c.$$

Example 27. $\int \frac{\log x}{x^2} \, dx$.

Solution : Integrating by parts taking $\log x$ as first function and $\frac{1}{x^2}$ as second function, we get

$$\int \frac{\log x}{x^2} \, dx = \log x \left[-\frac{1}{x} \right] + \int \frac{1}{x} \cdot \frac{1}{x} \, dx$$

$$= -\frac{\log x}{x} + \int \frac{1}{x^2} \, dx = -\frac{\log x}{x} - \frac{1}{x} + c = -\left[\frac{\log x}{x} + \frac{1}{x} \right] + c$$

$$= -\frac{1}{x} [1 + \log x] + c.$$

Example 28. $\int x^3 \, e^{x^2} \, dx$.

Solution : Put $x^2 = z$, $2x \, dx = dz$

$$\int x^3 \, e^{x^2} \, dx = \int \frac{x^2}{2} 2xe^{x^2} \, dx = \int \frac{x^2}{2} e^{x^2} (2x) \, dx = \int \frac{z}{2} e^z \, dz = \frac{1}{2} \int ze^z \, dz$$

Integrating by parts taking e^z as second function and z as first function we get

$$\frac{1}{2}\int ze^z\,dz = \frac{1}{2}[ze^z - e^z] = \frac{1}{2}e^z(z-1)$$

$$\therefore \qquad \int x^3 e^{x^2}\,dx = \frac{1}{2}e^{x^2}(x^2-1).$$

Example 29. Evaluate $\int \frac{\log x}{(x+1)^2}\,dx.$

Solution : Integrating by parts, taking $\log x$ as first function and $(x+1)^{-2}$ as second function, we get

$$\int \frac{\log x}{(x+1)^2}\,dx = \log x\,\frac{(x+1)^{-1}}{-1} + \int \frac{1}{x(x+1)}\,dx$$

Now let $\qquad \frac{1}{x(x+1)} = \frac{A}{x} + \frac{B}{x+1} \qquad$ [Method of Partial Fractions]

$$1 = A(x+1) + Bx$$

Put $x = 0$ then $A = 1$

Again putting $x = 1$, we get $\qquad 1 + 2A + B \Rightarrow B = -1 \qquad [\because A = 1]$

[Else for getting the value of B, put $x = -1$]

Note : We could have obtained the value of A and B by equating coefficients of equal powers of x on both sides also.

$$\therefore \qquad \int \frac{1}{x(x+1)}\,dx = \int \frac{1}{x}\,dx - \int \frac{1}{x+1}\,dx$$

$$= \log x - \log(x+1) + c = \log\frac{x}{x+1} + c$$

Thus

$$\int \frac{\log x}{(x+1)^2}\,dx = -\frac{\log x}{x+1} + \log\frac{x}{x+1} + c = \log\frac{x}{x+1} - \frac{\log x}{x+1} + c.$$

Example 30. Evaluate $\int \frac{(x^2+1)\,e^x}{(x+1)^2}\,dx.$

Solution : Now $\qquad \frac{(x^2+1)\,e^x}{(x+1)^2} = \frac{(x^2+1+2x-2x)\,e^x}{(x+1)^2}$

$$= \frac{(x+1)^2 e^x}{(x+1)^2} - \frac{2xe^x}{(x+1)^2} = e^x - \frac{2xe^x}{(x+1)^2}$$

Now we will evaluate $\int \frac{2xe^x}{(x+1)^2}\,dx$

$$\int \frac{2xe^x}{(x+1)^2} = \int \frac{2e^x}{(x+1)}\,dx - \int \frac{2e^x}{(x+1)^2}\,dx \qquad [\because\ 2x = 2x+2-2 = 2(x+1)-2]$$

$$= \frac{2}{x+1}\,e^x + \int \frac{2e^x}{(x+1)^2}\,dx - \int \frac{2e^x}{(x+1)^2}\,dx + c_1$$

$$= \frac{2}{x+1}\,e^x + c_1$$

$$\therefore \quad \int\left[e^x - \frac{2xe^x}{(x+1)^2}\right]dx = e^x - \frac{2e}{(x+1)} + c = e^x\left(\frac{x-1}{x+1}\right) + c.$$

Example 31. Evaluate $\int e^x(1+x)\log(xe^x)\,dx$.

Solution : Put $xe^x = t$

Differentiating both sides, we get $(xe^x + e^x)\,dx = dt$

$$\Rightarrow \qquad e^x(1+x)\,dx = dt$$

$$\therefore \qquad \int e^x(1+x)\log(xe^x)\,dx = \int \log(xe^x)\,e^x(1+x)\,dx$$

$$= \int \log t\,dt = t(\log t - 1) + c = xe^x(\log xe^x - 1) + c$$

Example 32. Evaluate $\int \frac{xe^x}{(1+x)^2}\,dx$.

Solution : Integrating by parts taking xe^x as the first function and $\frac{1}{(1+x)^2}$ as the second function, we get

$$\int \frac{xe^x}{(1+x)^2}\,dx = -\frac{xe^x}{1+x} + \int \frac{1}{1+x}\frac{d}{dx}(xe^x)\,dx$$

$$= -\frac{xe^x}{1+x} + \int \frac{1}{1+x}[xe^x + e^x]\,dx$$

$$= -\frac{xe^x}{1+x} + \int e^x \frac{1+x}{1+x}\,dx = -\frac{xe^x}{1+x} + \int e^x\,dx = -\frac{xe^x}{1+x} + e^x + c$$

$$= \frac{e^x}{1+x} + c \text{ where } c \text{ is the arbitrary constant.}$$

Example 33. Evaluate $\int x^n \log x\,dx$.

Solution : Integrating by parts taking $\log x$ as first function and x^n as second function, we get

$$\int (\log x)\,x^n\,dx = \log x\,\frac{x^{n+1}}{n+1} - \int \frac{x^{n+1}}{n+1}\cdot\frac{1}{x}\,dx$$

$$= (\log x)\frac{x^{n+1}}{n+1} - \int \frac{x^n}{n+1}\,dx$$

$$= (\log x)\frac{x^{n+1}}{n+1} - \frac{x^{n+1}}{(n+1)^2} + c$$

Example 34. Integrate $\dfrac{dx}{x[6(\log x)^2 + 7\log x + 2]}$

Solution : Put $\log x = z,\ \dfrac{1}{x}dx = dz$

$$\therefore \quad \int \frac{dx}{x[6(\log x)^2 + 7\log x + 2]} = \int \frac{1}{6z^2 + 7z + 2}\,dz$$

$$= \frac{1}{6}\int \frac{1}{z^2 + \frac{7}{6}z + \frac{1}{3}}\,dz$$

$$= \frac{1}{6}\int \frac{1}{\left(z + \frac{7}{12}\right)^2 - \left(\frac{1}{12}\right)^2}$$

$$= \frac{1}{6} \times 6\left[\log \frac{z + \frac{7}{12} - \frac{1}{12}}{z + \frac{7}{12} + \frac{1}{12}}\right] + c$$

$$= \log \frac{\left(z + \frac{1}{2}\right)}{\left(z + \frac{2}{3}\right)} + c \quad \text{where } z = \log x.$$

Example 35. If $f'(x) = 3x^2 + 2$ and $f(0) = 0$, find $f(2)$.

$$f'(x) = 3x^2 + 2$$

Solution : Integrating both sides w.r.t. x, we get

$$f(x) = \frac{3x^3}{3} + 2x + c$$

when $x = 0, f(x) = 0$ $\quad \therefore c = 0$

Hence, $f(x) = x^3 + 2x$

$\therefore$ $f(2) = (2)^3 \times 2 \times 2 = 12.$

Example 36. Integrate $\int \frac{dx}{(e^x - 1)^2}$.

Solution : Put $e^x - 1 = t$, *i.e.,* $e^x dx = dt$

$$\therefore \quad \int \frac{dx}{(e^x-1)^2} = \int \frac{dt}{(t+1)\, t^2}$$

$$\left[\because\ e^x - 1 = t \Rightarrow e^x = 1 + t \text{ and } e^x\, dx = dt,\ dx = \frac{dt}{e^x} = \frac{dt}{1+t}\right]$$

Let
$$\frac{1}{(t+1)\, t^2} = \frac{A}{t+1} + \frac{B}{t} + \frac{C}{t^2}$$

i.e.,
$$1 = A(t^2) + B(t)\,(t+1) + C(t+1)$$

Putting $t = -1$ we get $A = 1$

Comparing coefficients of equal powers of t on both sides we get

$$1 = C \Rightarrow C = 1$$

and
$$A + B = 0 \Rightarrow B = -1 \qquad [\because\ A = 1]$$

$$\frac{1}{(t+1)t^2} = \frac{1}{t+1} - \frac{1}{t} + \frac{1}{t^2}$$

i.e.,
$$\int \frac{dx}{(e^x-1)^2} = \int \frac{dt}{(t+1)\, t^2} = \int \frac{1}{t+1}\, dt - \int \frac{1}{t}\, dt + \int \frac{1}{t^2}\, dt$$

$$= \log\,(t+1) - \log t - \frac{1}{t} = \log\left(\frac{t+1}{t}\right) - \frac{1}{t}$$

$$= \log\left(\frac{e^x}{e^x - 1}\right) - \frac{1}{e^x - 1} + c.$$

Example 37. Integrate $\int \frac{x^2 + 5x + 2}{x+2}\, dx$.

Solution :
$$\frac{x^2 + 5x + 2 + 4 - 4}{x+2} = \frac{x^2 + 5x + 6}{x+2} - \frac{4}{x+2}$$

$$= (x+3) - \frac{4}{x+2} \qquad [\because\ x^2 + 5x + 6 = (x+2)\,(x+3)]$$

$$\therefore \quad \int \frac{x^2 + 5x + 2}{x+2}\, dx = \int (x+3)\, dx - 4 \int \frac{1}{x+2}\, dx$$

$$= \frac{x^2}{2} + 3x - 4 \log\,(x+2) + c.$$

Example 38. Integrate $\int \frac{x^3}{(x^2+1)^3}\,dx.$

Solution : Put $x^2+1=t \Rightarrow x^2=t-1$ and $2x\,dx=dt$

$$\therefore \quad \int \frac{x^3}{(x^2+1)^3}\,dx = \int \frac{x^2 \times x\,dx}{(x^2+1)^3}$$

$$= \frac{1}{2}\int \frac{(t-1)\,dt}{t^3} = \frac{1}{2}\int \left(\frac{1}{t^2}-\frac{1}{t^3}\right)dt$$

$$= \frac{1}{2}\left[-\frac{1}{t}+\frac{1}{2t^2}\right] = \frac{1}{2}\left(\frac{(1-2t)}{2t^2}\right)+c$$

$$= \frac{1}{2}\frac{[1-2(x^2+1)]}{2(x^2+1)^2} = -\frac{1}{4}\frac{(2x^2+1)}{(x^2+1)^2}+c.$$

Example 39. $\int \frac{1+x\log x}{x}e^x\,dx.$

Solution : $\int \left(\frac{1+x\log x}{x}\right)e^x\,dx = \int \frac{1}{x}e^x\,dx + \int (\log x)\,e^x\,dx$

$$= \int e^x.\frac{1}{x}\,dx + \int (\log x)\,e^x\,dx$$

$$= e^x \log x - \int (\log x)\,e^x\,dx + \int (\log x)\,e^x\,dx + c = e^x \log x + c.$$

Second Method

$$\int \frac{1+x\log x}{x}e^x\,dx = \int e^x\left(\log x + \frac{1}{x}\right)dx = e^x \log x + c$$

$$\left[\because \int e^x[f(x)+f'(x)]\,dx = e^x f(x)\right]$$

Example 40. Determine $\int \frac{6-x}{(x-3)\,(2x+5)}\,dx$ by the method of partial fractions.

Solution : Let $\quad \frac{6-x}{(x-3)\,(2x+5)} = \frac{A}{x-3}+\frac{B}{2x+5}$

$$\Rightarrow \quad 6-x = A(2x+5)+B(x-3)$$

Comparing coefficients of equal powers of x on both sides, we get

$$2A+B=-1$$

$$5A-3B=6$$

i.e., $\quad A=\frac{3}{11}, B=-\frac{17}{11}$

$$\therefore \quad \int \frac{(6-x)\,dx}{(x-3)\,(2x+5)} = \int \frac{3}{11(x-3)}\,dx - \int \frac{17}{11(2x+5)}\,dx$$

$$= \frac{3}{11}\log(x-3) - \frac{17}{22}\log(2x+5)+c.$$

Example 41. Evaluate $\int \frac{dx}{(1+e^x)(1-e^{-x})}$.

Solution : $\int \frac{dx}{(1+e^x)(1-e^{-x})} = \int \frac{dx}{e^x - e^{-x}} = \int \frac{e^x\,dx}{e^{2x}-1}$

Put $e^x = t$, *i.e.*, $e^x\,dx = dt$

$$\therefore \qquad \int \frac{e^x\,dx}{e^{2x}-1} = \int \frac{dt}{t^2-1}$$

$$= \frac{1}{2}\log\left(\frac{t-1}{t+1}\right) = \frac{1}{2}\log\left(\frac{e^x-1}{e^x+1}\right) + c.$$

Example 42. Evaluate $\int \frac{e^x + e^{-x}}{e^x - e^{-x}}\,dx$.

(P.U., B.Com. Dec. 1984, G.N.D.U. B.A. 1999)

Solution : $\int \frac{e^x + e^{-x}}{e^x - e^{-x}}\,dx = \int \frac{e^{2x}+1}{e^{2x}-1}\,dx$

Put $e^{2x} - 1 = t$, *i.e.*, $e^{2x}\,dx = dt$, *i.e.*, $dx = \frac{dt}{2e^{2x}} = \frac{dt}{2(t+1)}$

$$\therefore \qquad \int \frac{e^{2x}+1}{e^{2x}-1}\,dx = \frac{1}{2}\int \frac{t+2}{t(t+1)}\,dt = \frac{1}{2}\int \frac{1}{t+1}\,dt + \int \frac{1}{t(t+1)}\,dt$$

$$= \frac{1}{2}\log(t+1) + \int \frac{1}{t(t+1)}\,dt$$

$$= \frac{1}{2}\log(t+1) + \int \frac{1}{t}\,dt - \int \frac{1}{t+1}\,dt$$

$$= \frac{1}{2}\log(t+1) + \log t - \log(t+1) + c$$

where $t = e^{2x} - 1$.

Questions

1. Integrate

(i) $\int \frac{2x^2 + 3x^3 + 4}{x}\,dx$ (ii) $\int \frac{(x+2)^2}{\sqrt{x}}\,dx$ (iii) $\int e^{4x}\,dx$

(iv) $\int \frac{e^{5x} + e^{3x}}{e^{4x}}\,dx$ (v) $\int x\sqrt{x+1}\,dx$

[**Ans.** (i) $x^2 + x^3 + 4\log x + c$, (ii) $\frac{2}{5}x^{5/2} + \frac{8}{3}x^{3/2} + 2x^{1/2} + c$, (iii) $\frac{e^{4x}}{4} + c$, (iv) $e^x - e^{-x} + c$,

(v) $\frac{2}{3}x(x+1)^{3/2} - \frac{4}{15}(x+1)^{5/2} + c$]

2. Integrate

(i) $\int x\sqrt{x}\,dx$ (ii) $\int \frac{x^2-1}{x-1}\,dx$ (iii) $\int \frac{x^2-1}{x+1}\,dx$

(iv) $\int \frac{x^2+3x+2}{x+1}\,dx$ (v) $\int \frac{x^2-x-6}{x+2}\,dx$ (vi) $\int \sqrt{x}\,e^x\,dx$

[**Ans.** (i) $\frac{2}{5}x^{5/2}+c$, (ii) $\frac{x^2}{2}+x+c$, (iii) $\frac{x^2}{2}-x+c$ (iv) x^2+2x+c, (v) $(x^2-3x)+c$, (vi) $e^x\left(\sqrt{x}-\frac{1}{2\sqrt{x}}\right)+c$]

3. $\int e^x\left(\frac{1+x\log x}{x}\right)dx$

[**Ans.** $e^x \log x + c$]

4. $\int \frac{dx}{x(\log x)^2}$

[**Ans.** $\frac{-1}{\log x}+c$]

[**Hint :** Put $\log x = z$]

5. $\int e^x\left(\frac{1}{x}-\frac{1}{x^2}\right)dx$

[**Ans.** $\frac{e^x}{x}+c$]

6. $\int \frac{\log\sqrt{x}}{3x}\,dx$

[**Ans.** $\frac{1}{3}(\log\sqrt{x})^2+c$]

[**Hint :** $\log\sqrt{x}=\frac{1}{2}\log x$]

7. $\int \frac{(x+1)\,(x+2\log x)^2}{x}\,dx.$

[**Ans.** $\frac{1}{3}(x+\log x)^3+c$]

[**Hint :** Put $x+\log x = z$]

8. $\int (2x+5)\sqrt{x^2+5x}\,dx.$

[**Ans.** $\frac{2}{3}(x^2+5x)^{3/2}+c$]

[**Hint :** Apply $\int [f(x)]^n f'(x)\,dx = \frac{[f(x)]^{n+1}}{n+1}$]

9. $\int \frac{dx}{x+\sqrt{x}}.$

[**Ans.** $2\log(1+\sqrt{x})+c$]

[**Hint :** Put $1+\sqrt{x}=z$]

10. $\int \frac{dx}{\sqrt{x}+\sqrt{1+x}}.$

[**Ans.** $\frac{2}{3}[(1+x)^{3/2}-x^{3/2}]+c$]

[**Hint :** Rationalise the denominator by multiplying numerator and denominator by $\sqrt{x}-\sqrt{1+x}$.]

11. (i) $\frac{dx}{\sqrt{x+1}-\sqrt{x}}$ (ii) $\frac{dx}{\sqrt{x+2}-\sqrt{x+3}}.$

[**Ans.** (i) $\frac{2}{3}[(x+1)^{3/2}+x^{3/2}]+c$ (ii) $\frac{-2}{3}[(x+2)^{3/2}+(x+3)^{3/2}]+c$]

12. $\int \frac{(x+1)(x-4)}{\sqrt{x}}\, dx.$

[**Ans.** $\frac{2}{5}x^{5/2} - 2x^{3/2} - 8x^{1/2} + c$]

13. $\int \frac{x\, dx}{\sqrt{x}+1}.$

[**Ans.** $\frac{2}{3}x^{3/2} - x + 2\sqrt{x} - 2\log|\sqrt{x}+1| + c$]

14. Evaluate $\int x\, e^x\, dx.$

[**Ans.** $e^x(x-1)+c$]

15. Evaluate $\int \log x^2\, dx.$

[**Ans.** $x\log x^2 - 2x + c$]

16. Evaluate $\int x(\log x)^2\, dx.$

[**Ans.** $\frac{x^2}{2}(\log x)^2 - \frac{x^2}{2}\log x - \frac{x^2}{4}$]

17. $\int \frac{\log x}{x^2}\, dx.$

[**Ans.** $\frac{-\log x}{x} - \frac{1}{x}$]

18. $\int \frac{1}{x}\log(\log x)\, dx.$

[**Ans.** $\log x\,[\log(\log x) - 1]$]

19. $\int \left(\frac{1}{\log x} - \frac{1}{(\log x)^2}\right) dx.$

[**Ans.** $\frac{x}{\log x}$]

20. Evaluate $\int xe^{ax}\, dx.$

[**Ans.** $\frac{e^{ax}}{a^2}(ax-1)$]

DEFINITE INTEGRATION

6.11 INTRODUCTION

After dealing with indefinite (general) integrals, we now come to another type of integrals which have unique value and are called **definite integrals.** Strictly speaking, the definite integral $\int_a^b f(x)\, dx$ is defined as the limiting sum of an infinite series. By fundamental theorem of Integral Calculus the sum is proved to be equal to $F(b) - F(a)$

where $$\int f(x)\, dx = F(x) + c.$$

The definite integral of a function $f(x)$ in the closed interval (a, b) [denoted by $\int_a^b f(x)\, dx$] is equal to the difference in the value of the integral of $f(x)\, dx$ for two assigned values say a and b of the independent variable x and is written as $\int_a^b f(x)\, dx = [F(x)+c]_a^b = F(b) - F(a)$ where

$F(x) + c = \int f(x)\, dx$, a is called the lower limit and b is called the upper limit of integration $(a < b)$ although the word limit does not carry the same meaning as in differential calculus. The interval (a, b) is called the **range of integration.**

To evaluate a definite integral such as $\int_a^b f(x)\, dx$ process is as under :

(i) Find the indefinite integral $\int f(x)\, dx = F(x) + c$ (say).

(ii) Put $x = b$ (upper limit), and get $F(b) + c$.

(iii) Put $x = a$ (lower limit), and get $F(a) + c$.

(iv) Subtract (iii) from (ii), and get $F(b) - F(a)$.

Thus $\int_a^b f(x)\, dx$ means the value of the integral of $f(x)$ with respect to x when $x = b$ minus the value of this integral when $x = a$.

Notes :

(i) $f(x)$ is assumed to be continuous in the closed interval (a, b).

(ii) In definite integrals, the variable is assumed to be increasing from the lower limit to the upper limit.

(iii) Be careful while dealing with negative limits, *e.g.*, if the limits are -2 and 0. The integral will be represented as

$$\int_{-2}^{0} f(x)\, dx \qquad [\because \text{ Zero is greater than any negative number.}]$$

Here, the terms upper and lower limits simply mean the values of the variable x at the end of the range of values a to b (closed interval (a, b)) over which we are finding the values of the definite integral.

6.12 PROPERTIES OF DEFINITE INTEGRALS

(i) the interchange of the limits of integration changes only the sign of the definite integral, *i.e.*,

$$\int_a^b f(x)\, dx = -\int_b^a f(x)\, dx$$

$$\int_a^b f(x)\, dx = F(b) - F(a) \qquad \text{...(I)}$$

and

$$\int_b^a f(x)\, dx = F(a) - F(b)$$

$$= -[F(b) - F(a)] \qquad \text{...(II)}$$

From (I) and (II) result follows.

(ii) Change of variable does not change the value of the integral, *i.e.,*

$$\int_a^b f(x)\,dx = \int_a^b f(x)\,dz.$$

When the variable is changed, new limits of integration have to be determined, otherwise the final integral has to be changed in terms of the original variable before applying the limits.

(iii) When the lower limit is zero, the integral $f(x)$ can be changed to $f(a-x)$ where a is the upper limit, *i.e.,*

$$\int_0^a f(x)\,dx = \int_0^a f(a-x)\,dx.$$

Proof : Let $$I = \int_0^a f(x)\,dx$$

Put $$a - x = t \Rightarrow x = a - t$$

$$dx = -dt$$

when $x = 0, t = a; x = a, t = 0$

$$\therefore \quad \int_0^a f(x)\,dx = -\int_a^0 f(a-t)\,dt$$

$$= -\int_a^0 f(a-x)\,dx \quad \text{[By property (ii)]}$$

$$= \int_0^a f(a-x)\,dx \quad \text{[By property (i)]}$$

(iv) Any definite integral can be expressed as the sum of any number of definite integrals by splitting the range of integration.

Proof : Suppose in $\int_a^b f(x)\,dx$, the range (a to b) is split as

$$(c_1 - a) + (c_2 - c_1) + (c_3 - c_2) + \ldots + (b - c_n) = b - a$$

Then $$\int_a^b f(x)\,dx = \int_a^{c_1} f(x)\,dx + \int_{c_1}^{c_2} f(x)\,dx + \int_{c_2}^{c_3} f(x)\,dx + \ldots + \int_{c_n}^{b} f(x)\,dx = F(b) - F(a)$$

(v) When $f(x)$ is periodic function such that $f(x) = f(a + x)$

then $$\int_0^{na} f(x)\,dx = n\int_0^a f(x)\,dx$$

Proof : By splitting the range of integration, we get

$$\int_0^{na} f(x)\,dx = \int_0^a f(x)\,dx + \int_a^{2a} f(x)\,dx + \ldots + \int_{(n-1)a}^{na} f(x)\,dx$$

Let us consider

$$\int_a^{2a} f(x)\,dx$$

Put $x = a + t \Rightarrow dx = dt$

when $x = a,\ t = 0$ and when $x = 2a,\ t = a$

$$\therefore \quad \int_a^{2a} f(x)\,dx = \int_0^a f(a+t)\,dt + \int_0^a f(a+x)\,dx = \int_0^a f(x)\,dx \qquad \text{(Given)}$$

Similarly, all other integrals can each be shown as equal to $\int_0^a f(x)\,dx$.

$$\therefore \quad \int_0^{na} f(x)\,dx = n\int_0^a f(x)\,dx$$

(vi) For any function $f(x)$

$$\int_0^a f(x)\,dx = \int_0^{a/2} f(x)\,dx + \int_0^{a/2} f(a-x)\,dx.$$

Proof : By property (iv)

$$\int_0^a f(x)\,dx = \int_0^{a/2} f(x)\,dx + \int_{a/2}^a f(x)\,dx \qquad \text{...(I)}$$

In $$\int_{a/2}^a f(x)\,dx$$

Put $a - x = y$, *i.e.*, $-dx = dy$, $x = a - y$

when $x = \frac{a}{2}$, $y = \frac{a}{2}$ when $x = a$, $y = 0$

$$\text{Then} \quad \int_{a/2}^a f(x)\,dx = -\int_{a/2}^0 f(a-y)\,dy = \int_0^{a/2} f(a-y)\,dy \qquad \text{[Property (i)]}$$

$$= \int_0^{a/2} f(a-x)\,dx \qquad \text{[Property (ii)]}$$

$$\therefore \quad \int_0^a f(x)\,dx = \int_0^{a/2} f(x)\,dx + \int_a^{a/2} f(a-x)\,dx.$$

(vii) $\int_a^b f(x)\,dx = \int_a^b f(a+b-x)\,dx.$

Proof : Put $t = a + b - x$, $dt = -dx$

Also when $x = a$, $t = b$ and when $x = b$, $t = a$

$$\therefore \quad \int_a^b f(x)\,dx = -\int_b^a f(a+b-t)\,dt$$

$$= \int_a^b f(a+b-t)\,dt \qquad \text{[Property (i)]}$$

$$= \int_a^b f(a+b-x)\,dx$$

(viii) $$\int_0^{2a} f(x)\,dx = 2\int_0^a f(x)\,dx \text{ if } f(2a-x) = f(x)$$

and $$\int_0^{2a} f(x)\,dx = 0 \text{ if } f(2a-x) = -f(x)$$

Now $$\int_0^{2a} f(x)\,dx = \int_0^a f(x)\,dx + \int_a^{2a} f(x)\,dx$$

In $\int_a^{2a} f(x)\,dx$, put $2a - x = t \Rightarrow -dx = dt,\ 2a - t = x.$

When $x = a, t = a$ and when $x = 2a, t = 0$

$$\therefore \quad \int_a^{2a} f(x)\,dx = -\int_a^0 f(2a-t)\,dt = \int_0^a f(2a-t)\,dt$$

$$= \int_0^a f(2a-x)\,dx$$

$$\therefore \quad \int_a^{2a} f(x)\,dx = \int_0^a f(x)\,dx + \int_0^a f(2a-x)\,dx$$

$$= 2\int_0^a f(x)\,dx \text{ if } f(2a-x)\,dx = f(x)$$

$$= 0 \text{ if } f(2a-x) = -f(x)$$

(ix) $\int_{-a}^a f(x)\,dx = 2\int_0^a f(x)\,dx$ if $f(x)$ is an even function of x

$= 0$ if $f(x)$ is an odd function of x.

$$\int_{-a}^0 f(x)\,dx = \int_{-a}^0 f(x)\,dx = \int_0^a f(x)\,dx$$ [Property (iv)]

In the first integral on the right hand side put $x = -t$, then

$$\int_{-a}^0 f(x)\,dx = -\int_a^0 f(-t)\,dt = \int_0^a f(-t)\,dt = \int_0^a f(-x)\,dx$$

Hence, $$\int_{-a}^a f(x)\,dx = \int_0^a f(-x)\,dx + \int_0^a f(x)\,dx$$

$$= 2\int_0^a f(x)\,dx \text{ if } f(-x) = f(x), \textit{ i.e.,} \text{ if } f(x) \text{ is an even function of } x$$

and $$\int_{-a}^a f(x)\,dx = 0 \text{ if } f(-x) = -f(x) \textit{ i.e.,} \text{ if } f(x) \text{ is an odd function of } x.$$

Example 43. Evaluate $\int_0^1 \frac{1-t}{1+t}\,dt.$

Solution : $$\int_0^1 \frac{1-t}{1+t}\,dt = \int_0^1 \frac{1-t+1-1}{1+t}\,dt = \int_0^1 \frac{2-(1+t)}{1+t}\,dt$$

$$= \int_0^1 \frac{2}{1+t}\,dt - \int_0^1 dt = [2\log(1+t) - t]_0^1$$

$$= 2\log 2 - 1$$

Example 44. Prove that

$$\int_a^b \frac{\log x}{x}\,dx = \frac{1}{2}\log\left(\frac{b}{a}\right)\log ab.$$

Solution : Put $\log x = t$ then $\frac{1}{x}\,dx = dt$;

when $x = a$, $t = \log a$ and when $x = b$, $t = \log b$

$$\therefore \quad \int_a^b \frac{\log x}{x}\,dx = \int_{\log a}^{\log b} t\,dt = \left[\frac{t^2}{2}\right]_{\log a}^{\log b}$$

$$= \frac{1}{2}[(\log b)^2 - (\log a)^2]$$

$$= \frac{1}{2}[(\log b - \log a)(\log b + \log a)]$$

$$= \frac{1}{2}\log\left(\frac{b}{a}\right)\log ab.$$

Method II

Integrating by parts, taking $\log x$ as the first function and $\frac{1}{x}$ as the second function, we get

$$\int (\log x)\frac{1}{x}\,dx = (\log x)\log(x) - \int(\log x)\frac{1}{x}\,dx$$

i.e.,

$$2\int (\log x)\frac{1}{x}\,dx = (\log x)^2 \Rightarrow \int(\log x)\frac{1}{x}\,dx = \frac{1}{2}(\log x)^2$$

$$\int_a^b (\log x)\frac{1}{x}\,dx = \frac{1}{2}|(\log x)^2|_a^b = \frac{1}{2}[(\log b)^2 - (\log a)^2]$$

$$= \frac{1}{2}(\log b - \log a)(\log b + \log a) = \frac{1}{2}\log\left(\frac{b}{a}\right)\log ab$$

Method III

$$\int_a^b (\log x)\frac{1}{x}\,dx = \left[\frac{(\log x)^2}{2}\right]_a^b \qquad \left[\because \int [f(x)]^n f'(x)\,dx = \frac{[f(x)]^{n+1}}{n+1}\right]$$

$$= \frac{(\log b)^2 - (\log a)^2}{2} = \frac{1}{2}(\log b - \log a)(\log b + \log a)$$

$$= \frac{1}{2}\log\left(\frac{b}{a}\right)\log ab.$$

Example 45. Show that $\int_0^{1/3} \sqrt{1-3x}\, dx = \frac{2}{9}$.

Solution : Put $$1 - 3x = z, -3\, dx = dz \Rightarrow dx = -\frac{dz}{3}$$

when $x = 0, z = 1$, when $x = 1/3, z = 0$

$$\therefore \quad \int_0^{1/3} \sqrt{1-3x}\, dx = -\frac{1}{3}\int_1^0 \sqrt{z}\, dz = \frac{1}{3}\int_0^1 \sqrt{z}\, dz = \frac{1}{3}\left[\frac{z^{\frac{1}{2}+1}}{\frac{1}{2}+1}\right]_0^1$$

$$= \frac{1}{3} \times \frac{2}{3} [z^{3/2}]_0^1 = \frac{2}{9}.$$

Example 46. Evaluate $\int_4^8 \frac{dx}{(x-3)(x+1)^2}$.

Solution : Let $$\frac{1}{(x-3)(x+1)^2} = \frac{A}{x-3} + \frac{B}{x+1} + \frac{C}{(x+1)^2}$$

Then $$1 = A(x+1)^2 + B(x-3)(x+1) + C(x-3) \quad \text{...(I)}$$

Putting $x = -1$, in (I) we get $-4C = 1$, *i.e.,* $C = -1/4$

Putting $x = 3$, in (I) we get $I = 16A$, *i.e.,* $A = \frac{1}{16}$

Equating constant terms on both sides

$$1 = A - 3B - 3C$$

$$= \frac{1}{16} - 3B + \frac{3}{4} = \frac{13}{16} - 3B \Rightarrow B = -\frac{1}{16}$$

$$\therefore \quad \frac{1}{(x-3)(x+1)^2} = \frac{1}{16(x-3)} - \frac{1}{16(x+1)} - \frac{1}{4(x+1)^2}$$

$$\therefore \quad \int_4^8 \frac{1}{(x-3)(x+1)^2} dx = \frac{1}{16}\int_4^8 \frac{1}{x-3} dx - \frac{1}{16}\int_4^8 \frac{1}{x+1} dx - \frac{1}{4}\int_4^8 \frac{1}{(x+1)^2} dx$$

$$= \frac{1}{16} [\log (x-3)]_4^8 - \frac{1}{16} [\log (x+1)]_4^8 + \frac{1}{4}\left[\frac{1}{(x+1)}\right]_4^8$$

$$= \frac{1}{16}\left(\log \frac{25}{9}\right) - \frac{1}{45} = \frac{1}{8} \log \frac{5}{3} - \frac{1}{45}$$

Example 47. Evaluate $\int_e^{e^2} \frac{1}{x \log_e x}$.

Solution : Put $\log_e x = t$; when $[x = e, t = \log_e e = 1$ and when $x = e^2, t = \log_e e^2 = 2 \log_e e = 2]$

$$\log_e x = t \Rightarrow \frac{1}{x} dx = dt$$

$$\therefore \qquad \int_e^{e^2} \frac{1}{x \log x}\, dx = \int_1^2 \frac{1}{t}\, dt = [\log t]_1^2$$

$$= \log 2 - \log 1 = \log 2.$$

Example 48. Evaluate $\int_1^2 \frac{dx}{\sqrt{(2x-1)}}$.

Solution : Put $2x - 1 = t^2 \Rightarrow 2dx = 2t\, dt \Rightarrow dx = t\, dt.$

When $x = 1$,'$t = 1$ and when $x = 5, t = 3$

$$\therefore \qquad \int_1^5 \frac{dx}{\sqrt{(2x-1)}} = \int_1^3 dt = [t]_1^3 = 2.$$

Example 49. Evaluate $\int_1^{e^2} \frac{dx}{x(1 + \log_e x)}$.

Solution : Put $\log_e x = t \Rightarrow \frac{1}{x} dx = dt$

when $x = 1, t = \log_e 1 = 0$

$x = e^2, t = \log_e (e^2) = 2$

$$\therefore \qquad \int_1^{e^2} \frac{dx}{x(1 + \log_e x)} = \int_0^2 \frac{dt}{1+t} = [\log (1+t)]_0^2$$

$$= \log 3 - \log 1 = \log 3 \qquad [\because \log 1 = 0]$$

Example 50. Evaluate $\int_1^e x \log_e x\, dx$.

Solution : Now integrating by parts by taking $\log x$ as the first function and x as second function.

$$\int (\log_e x)\, dx = (\log_e x) \frac{x^2}{2} - \int \frac{x^2}{2} . \frac{1}{x}\, dx = \frac{x^2}{2} \log_e x - \frac{x^2}{4}$$

$$\therefore \qquad \int_1^e x \log_e x\, dx = \left(\frac{x^2}{2} \log_e x - \frac{x^2}{4} \right)_1^e$$

$$= \frac{e^2}{2} \log_e e - \frac{e^2}{4} - \frac{1}{2} \log_e 1 + \frac{1}{4}$$

$$= \frac{e^2}{4} + \frac{1}{4} = \frac{e^2 + 1}{4} \qquad [\because \log_e e = 1]$$

Example 51. Evaluate $\int_1^2 x(x^2 + 1)^3\, dx$.

Solution : Here we put $\qquad x^2 + 1 = t \qquad$ ...(I)

When $\qquad x = 1,\ t = 2$

$x = 2,\ t = 5$

Differentiating (I), we get $2x\,dx = dt \Rightarrow x\,dx = \dfrac{dt}{2}$ ...(II)

$$\therefore \quad \int_1^2 x(x^2+1)^3\,dx = \frac{1}{2}\int_2^5 t^3\,dt = \frac{1}{8}[t^4]_2^5$$

$$= \frac{1}{8}(625-16) = \frac{609}{8}$$

Example 52. Evaluate $\displaystyle\int_0^2 \frac{\sqrt{x}}{\sqrt{2-x}+\sqrt{x}}\,dx.$

Solution : Let $\displaystyle I = \int_0^2 \frac{\sqrt{x}}{\sqrt{2-x}+\sqrt{x}}\,dx$

$$= \int_0^2 \frac{\sqrt{2-x}}{\sqrt{2-2+x}+\sqrt{2-x}}\,dx \quad \left[\text{Property (iii), } i.e., \int_0^a f(x)\,dx = \int_0^a f(a-x)\,dx\right]$$

$$= \int_0^2 \frac{\sqrt{2-x}}{\sqrt{x}+\sqrt{2-x}}\,dx$$

$$\therefore \quad 2I = \int_0^2 \frac{\sqrt{x}}{\sqrt{2-x}+\sqrt{x}}\,dx + \int_0^2 \frac{\sqrt{2-x}}{\sqrt{x}+\sqrt{2-x}}\,dx$$

$$= \int_0^2 \frac{\sqrt{x}+\sqrt{2-x}}{\sqrt{2-x}+\sqrt{x}}\,dx = \int_0^2 1\,dx = [x]_0^2 = 2$$

$$\therefore \quad I = 1.$$

Example 53. Prove that $\displaystyle\int_2^e \left\{\frac{1}{\log x} - \frac{1}{(\log x)^2}\right\} dx = e - \frac{2}{\log 2}.$

Solution : $\displaystyle\int \frac{1}{\log x}\,dx = \frac{x}{\log x} + \int \frac{x}{(\log x)^2} \times \frac{1}{x}\,dx = \frac{x}{\log x} + \int \frac{1}{(\log x)^2}\,dx$

[Integrating by parts taking $\dfrac{1}{\log x}$ as first function and 1 as second function.]

$$\therefore \quad \int_2^e \left[\frac{1}{\log x} - \frac{1}{(\log x)^2}\right] dx = \left[\frac{x}{\log x}\right]_2^e + \int_2^e \frac{dx}{(\log x)^2} - \int_2^e \frac{dx}{(\log x)^2}$$

$$= \left[\frac{x}{\log x}\right]_2^e = \frac{e}{\log_e e} - \frac{2}{\log_e 2} = e - \frac{2}{\log_e 2}.$$

Questions

1. Evaluate $\int_0^1 \frac{1-x}{1+x}\,dx.$ [**Ans.** $2 \log 2 - 1$]

[**Hint :** $1 - x = -(x - 1 + 1 - 1) = -(x + 1) + 2$]

2. Evaluate (i) $\int_6^7 \frac{x}{x-4}.$ [**Ans.** (i) $\log \frac{3}{2}$]

3. Evaluate $\int_0^9 \left[\sqrt{x} - \frac{1}{\sqrt{x}}\right] dx.$ [**Ans.** 12]

4. Evaluate $\int_{-1}^1 (x+1)^2\,dx.$ [**Ans.** $\frac{8}{3}$]

5. Evaluate $\int_1^2 x^2 e^x\,dx.$ [**Ans.** $2e^2 - e$]

6. Evaluate $\int_1^2 \frac{x^2-1}{x^2} e^{x+\frac{1}{x}}\,dx.$ [**Ans.** $e^{\frac{5}{2}} - e^2$]

[**Hint :** Put $x + \frac{1}{x} = z$]

7. Evaluate $\int_0^2 \frac{6x+5}{3x^2+5x+1}\,dx.$ [**Ans.** $\log 23$]

8. Prove that $\int_0^2 x(x-1)(x-2)\,dx = 0.$

❑❑❑

7

Applications of Integration to Business and Economics

7.1 INTRODUCTION

The process of integration has very useful applications in the following fields of Business and Economics :

(1) **Marginal cost** $= \frac{d}{dx}$ (Total cost)

$\therefore \int$ (Marginal cost) dx = total variable cost + Fixed cost = Total cost.

In the integration of marginal cost the constant of integration will represent **Fixed cost.**

(2) **Marginal revenue** $= \frac{d}{dx}$ (Total Revenue)

Total revenue $= \int$ (Marginal revenue) dx

When production $= 0$, Revenue $= 0$

(3) **To find the total cost when marginal cost is given**

Total cost for producing x units $= C(x) = \int MC(x)\, dx + K$ where K is the constant of integration and is determined from the initial conditions given in the question. $C(x)$ is also called **Total Cost Function** or **Cost Function**.

(4) **To find the total Revenue for selling x units when Marginal revenue is given**

Total Revenue $= \int$ (Marginal Revenue) $dx + K$

$TR(x)$ [or simply $R(x)$] $= \int MR(x)\, dx + K$

where K is the constant of integration. $TR(x)$ is also called **total revenue function**.

(5) **To find the maximum profit when Marginal Revenue and Marginal Cost Functions are given,**

$$\text{Profit} = \text{Revenue} - \text{Cost}$$

$$\frac{d}{dx}(\text{Profit}) = \frac{d}{dx}(\text{Revenue}) - \frac{d}{dx}(\text{Cost})$$

$\Rightarrow$ Marginal Profit = Marginal Revenue – Marginal Cost

$\therefore$ If Marginal Revenue and Marginal Cost Functions are given, then

$$\text{Total profit} = \int [(\text{Marginal Revenue}) - (\text{Marginal Cost})]\, dx + K$$

(6) **To find demand function when the elasticity of demand is given.**

$$\text{Price elasticity of demand} = e_p = -\frac{p}{x}\frac{dx}{dp}$$

$$\Rightarrow \quad e_p \frac{dp}{p} = -\frac{dx}{x}$$

$$\Rightarrow \quad e_p \int \frac{dp}{p} = -\int \frac{dx}{x} + \log K$$

$$\Rightarrow \quad e_p \log p = -\log x + \log K \ *$$

$$\Rightarrow \quad \log p^{(e_p)} + \log x = \log K$$

$$\Rightarrow \quad \log xp^{e_p} = \log K$$

$$\Rightarrow \quad xp^{(e_p)} = K$$

(7) To find the Consumption Function if marginal propensity to consume *i.e.*, $\frac{dC}{dI}$ is given.

(8) To find Producer's Surplus and Consumer's Surplus.

(9) In compound interest calculations when compounding is done continuously.

Example 1. Obtain the demand function for a commodity whose elasticity of demand is given as

$$e_{xp} = a - bp$$

where a, b are constants and p denotes the price per unit of the commodity.

(D.U., B.Com. (Hons.) 1985)

Solution : Elasticity of demand $= -\frac{p}{x}\frac{dx}{dp} = a - bp$

$$\Rightarrow \quad -\frac{dx}{x} = \left[\frac{a - bp}{p}\right] dp$$

$$= \left[\frac{a}{p} - b\right] dp$$

Integrating both sides we get

$$-\log x = a \log p - bp + c$$

$$\Rightarrow \quad \log x = -\log p^a + \log_e e^{-bp+e} \qquad [\because \log_e e^x = x \log_e e = x]$$

* log K is a constant, depending on the situation. Constant of integration can be selected in any convenient way.

$$= \log p^{-\alpha} e^{-bp+e}$$

$\Rightarrow$ $x = p^{-a} e^{-bp+e}$ where c is constant of integration.

Example 2. Marginal revenue function of a firm is $\dfrac{ab}{(x-b)^2} - c$. Prove that the demand law is

$$p = \frac{a}{b-x} - c.$$

(D.U., B.Com. (Hons.) 1986)

Solution : MR $= \dfrac{dR}{dx} = \dfrac{ab}{(x-b)^2} - c \Rightarrow dR = \dfrac{ab}{(x-b)^2} - c$

Integrating both sides with respect to x, we get

$$R = \int \frac{ab}{(x-b)^2}\, dx - c \int dx + A$$

where A is constant of integration

$$R = -\frac{ab}{(x-b)} - cx + A$$

When $x = 0, R = 0$

$\therefore$ $$0 = -\frac{ab}{-b} + A \Rightarrow A = -a$$

$\therefore$ Revenue function $$R = \frac{-ab}{x-b} - cx - a$$

$\therefore$ The demand law is given by the relation

$$R = px = \frac{-ab}{x-b} - cx - a$$

$\therefore$ $$p = \frac{-ab}{x(x-b)} - \frac{cx}{x} - \frac{a}{x} \Rightarrow p = \frac{-ab}{x(x-b)} - c - \frac{a}{x}$$

$$= \frac{-ab - cx(x-b) - a(x-b)}{x(x-b)}$$

$$= \frac{-ab - cx(x-b) - ax + ab}{x(x-b)} = \frac{[-c(x-b) - a]\, x}{x(x-b)}$$

$$= -c - \frac{a}{x-b} = \frac{a}{b-x} - c$$

Hence the demand law is given by the relation

$$p = \frac{a}{b-x} - c.$$

Example 3. The marginal cost $C'(x)$ and the marginal revenue $R'(x)$ are given by

$$C'(x) = 20 + \frac{x}{20} \text{ and } R'(x) = 30$$

The fixed cost is Rs. 200. Determine the maximum profit and the number of items produced for this profit. **(D.U., B.Com. (Hons.) 1987)**

Solution : $$C'(x) = 20 + \frac{x}{20} \text{ and } R'(x) = 30$$

Fixed cost = 200

i.e., $C(0)$ = fixed cost = 200 (Cost of the producing zero units).

$C(x) = \int C'(x)\,dx + K$ where K is a constant of integration

$$= 20x + \frac{x^2}{40} + K$$

when $x = 0$, $C(0) = K = 200$ = Fixed cost

Hence cost function is $C = 20x + \frac{x^2}{40} + 200$

$R(x) = \int R'(x)\,dx + K'$ ... where K' is a constant of integration

$$= 30x + K'$$

when $x = 0, R = 0$, therefore $K' = 0$.

Hence Revenue function $R = 30x$

$$\therefore \text{ Profit} = \pi = R - C = 30x - \left(20x + \frac{x^2}{40} + 200\right)$$

$$= 10x - \frac{x^2}{40} - 200$$

$$\frac{d\pi}{dx} = 10x - \frac{2x}{40} = 10 - \frac{x}{20}$$

On putting $\frac{d\pi}{dx} = 0$, we get $x = 200$

$\frac{d^2\pi}{dx^2} = -\frac{1}{20}$ which is negative

$\therefore$ $x = 200$ gives maximum profit

And the maximum profit is given by

$$\pi = 10 \times 200 - \frac{(200)^2}{40} - 200 = 800$$

$\therefore$ x = No. of items produced = 200 and Max. Profit π = Rs. 800.

Example 4. A company suffers a loss of Rs. 1,000 if its product does not sell at all. Marginal revenue and marginal cost functions for the product are given by :

$$\text{MR} = 50 - 4x, \text{MC} = -10 + x$$

Determine the total profit function, break-even points and the profit maximising level of output.

(D.U., B.Com. (Hons.) 1989)

Solution : Profit (P) = Revenue (R) – Cost (C)

$$\frac{dP}{dx} = \frac{dR}{dx} - \frac{dC}{dx}$$

$$= 50 - 4x + 10 - x = 60 - 5x$$

Integrating both sides with respect to x, we get

$$P = 60x - \frac{5x^2}{2} + K \qquad ...(I)$$

where K is the constant of integration

If $x = 0$, then loss = Rs. 1,000

i.e., If $x = 0$, then $P = -1{,}000$

Putting these values in (I), we get $K = -1000$

$\therefore$ Profit function is given by

$$P = 60x - \frac{5x^2}{2} - 1{,}000 \qquad ...(II)$$

$$\frac{dP}{dx} = 60 - 5x$$

Also $\frac{d^2P}{dx^2} = -5 < 0$ for all values of x

$$\frac{dP}{dx} = 0 \Rightarrow 60 - 5x = 0 \Rightarrow x = 12$$

$\therefore$ $x = 12$ gives maximum profit

Break even points* : Put $P = 0$ in (II)

$$\therefore \qquad 60x - \frac{5x^2}{2} - 1{,}000 = 0$$

$$\Rightarrow \qquad \frac{5x^2}{2} - 60x + 1{,}000 = 0$$

$\Rightarrow x^2 - 24x + 400 = 0 \Rightarrow x = \frac{24 \pm \sqrt{576 - 1{,}600}}{2}$ which gives imaginary values of x.

$\therefore$ Break even points cannot be obtained.

Example 5. The marginal cost function of a firm is given by MC = 3,000 $e^{0.3x}$ + 50, when x is quantity produced. If fixed cost is Rs. 80,000, find the total cost function of the firm.

(D.U. B.Com. (Hons.) 1990)

Solution : It is given that

$$\text{MC} = 3{,}000\, e^{0.3x} + 50 \text{ and } C(0) = 80{,}000$$

Integrating the marginal cost function (MC), we obtain

$$C(x) = \int (3{,}000\, e^{0.3x} + 50)\, dx + k$$

where k is constant of integration

$$= 10{,}000\, e^{0.3x} + 50x + k$$

$$\Rightarrow \qquad C(0) = 10{,}000 + k = 80{,}000 \Rightarrow k = 70{,}000 \qquad \textbf{(Pl. note)}$$

* At break even point there is no profit, *i.e.*, Revenue = Cost.

Thus, the total cost function of the firm is give by

$$C(x) = 10{,}000\, e^{0.3x} + 50x + 70{,}000$$

Example 6. A firm's marginal revenue function is

$$\text{MR}(x) = 20\, e^{-x/10}\left(1 - \frac{x}{10}\right)$$

Find the corresponding demand function. **(D.U. B.Com. (Hons.) 1998)**

Solution :

$$\text{MR} = 20\, e^{-x/10}\left(1 - \frac{x}{10}\right)$$

$$= 20\left(1 - \frac{x}{10}\right)e^{-x/10}$$

$$\text{Revenue} = \int 20\left(1 - \frac{x}{10}\right)e^{-x/10}\, dx$$

$$= 20\left(1 - \frac{x}{10}\right)e^{-x/10}(-10) + 10\int 20\, e^{-x/10}\, \frac{(-1)}{10}\, dx \;*$$

$$= -200\left(1 - \frac{x}{10}\right)e^{-x/10} + 200\, e^{-x/10} + c$$

$$= 20x\, e^{-x/10} + c$$

where c is constant of integration

when $\qquad x = 0,\ R = 0$

$\therefore \qquad 0 = 20 \times 0 \times e^{-0} + c \Rightarrow c = 0$

$\therefore \qquad R = 20x\, e^{-x/10}$

$$px = 20x\, e^{-x/10} \qquad [\because\ R = px]$$

$\Rightarrow p = 20\, e^{-x/10}$. Which is the required demand function.

Example 7. The elasticity of demand of a commodity with respect to price is calculated to be $\dfrac{5p}{(p+3)(p-2)}$ where p is price. Find the demand function if it is known that the quantity demanded is 5 units as $p = 3$. **(D.U. B.Com. (Hons.) 1996)**

Solution : Let e_d denote the elasticity demand. Then

$$e_d = -\frac{p}{x}\frac{dx}{dp}$$

---

* (1) Take $e^{-x/10}$ as second function and $\left(1 - \dfrac{x}{10}\right)$ as first function.

(2) The integral can be done as under also

$$\text{Revenue} = \int 20e^{-x/10}\left(1 - \frac{x}{10}\right)dx = \int (20 - 2x)\, e^{-x/10}\, dx$$

$$= 20\int e^{-x/10}\, dx - 2\int x\, e^{-x/10}\, dx$$

$$\Rightarrow \qquad \frac{5p}{(p+3)(p-2)} = -\frac{p}{x}\frac{dx}{dp}$$

$$\Rightarrow \qquad \int \frac{5}{(p+3)(p-2)}\,dp = -\int \frac{1}{x}\,dx$$

$$\Rightarrow \qquad \int \frac{-1}{p+3}\,dp + \int \frac{1}{p-2}\,dp = -\int \frac{1}{x}\,dx^{*}$$

$$\Rightarrow \qquad -\log(p+3) + \log(p-2) = -\log x + \log k,$$

where k is an arbitrary constants. Thus

$$\log \frac{p-2}{p+3} = \log \frac{k}{x}$$

$$\Rightarrow \qquad \frac{p-2}{p+3} = \frac{k}{x} \Rightarrow x = \frac{k(p+3)}{p-2}$$

It is given that $x = 5$ at $p = 3$. Hence

$$5 = \frac{k(3+3)}{3-2} \Rightarrow k = \frac{5}{6}$$

Thus, the required demand function is given by

$$x = \frac{5(p+3)}{6(p-2)}.$$

Example 8. If the marginal revenue of a firm is given by MR $= 9 - 4x^2$, find the total revenue R in terms of x. What can be maximum revenue possible ?

Solution : MR $= (9 - 4x^2)$ where x is the number of units produced

$$\int (\text{MR})\,dx = \text{Total revenue}$$

$$\therefore \qquad \int (9 - 4x^2)\,dx = 9x - \frac{4x^3}{3} + c = \text{R}$$

When $x = 0$, R $= 0$, $\therefore\ c = 0$

$\therefore$ Total Revenue $\qquad (\text{R}) = 9x - \frac{4}{3}x^3$

Revenue is maximum when $\frac{dR}{dx} = 0$ and $\frac{d^2R}{dx^2} < 0$, *i.e.*, $\frac{d^2R}{dx^2}$ is negative

---

* Let $\qquad \frac{5}{(p+3)(p-2)} = \frac{A}{p+3} + \frac{B}{p-2}$

$\Rightarrow \qquad 5 = A(p-2) + B(p+3)$

Putting $p = 2$ we get $\qquad 5 = 5p \Rightarrow p = 1$

Putting $p = -3$, we get $\qquad 5 = -5A \Rightarrow A = -1$

$\therefore \qquad \frac{5}{(p+3)(p-2)} = \frac{-1}{p+3} + \frac{1}{p-2}$

$$\frac{dR}{dx}=0 \Rightarrow 9-4x^2=0 \Rightarrow x=\pm\frac{3}{2} \qquad [\because \text{ Marginal Revenue} = \frac{dR}{dx}]$$

Since, the production cannot be negative, we reject the negative sign. Therefore, $x=\frac{3}{2}$

$\frac{d^2R}{dx^2}=-8x$ which is negative for all positive values of x. Hence, Revenue is maximum when $x=\frac{3}{2}$.

When $x=\frac{3}{2}$, maximum revenue $=9\times\frac{3}{2}-\frac{4}{3}\left(\frac{3}{2}\right)^3=\frac{27}{2}-\frac{9}{2}=9.$

Example 9. Marginal cost of production (in Rs.) is $3+\frac{x}{3,00}+3^{-0.03x}$ where x denotes the number of units. Find the cost of producing 100 units. It is given that $e^{-3}=0.05$ (approx.)

Solution : $\text{MC}=3+\frac{x}{3,000}+e^{-0.03x}$

Total cost of producing 100 units.

$$\int_0^{100}(\text{MC})\,dx=\int_0^{100}\left(3+\frac{x}{3,000}+e^{-0.03}x\right)dx$$

$$=\left[3x+\frac{x^2}{6,000}-\frac{e^{-0.03x}}{0.03}\right]_0^{100}$$

$$=300+\frac{10,000}{6,000}-\frac{e^{-3}}{0.03}-\left(-\frac{e^0}{0.03}\right)$$

$$=300+\frac{5}{3}-\frac{5}{3}+\frac{1}{0.03}=300+\frac{10}{3}=\text{Rs. } 333.33 \qquad [\because e^{-3}=0.05]$$

Example 10. The cost of a truck engine overhaul is Rs. A and the operating costs are at rate of Rs. $a+bx$ per km, where the engine has run x kms since the last overhaul. If the engine is overhauled at M km, find the total cost function. **(D.U., B.Com. (Hons.) 2000)**

Solution : Let K = Total Cost

= Cost of truck engine overhaul + Operating cost

Let C = Operating cost

$\therefore$ $$\frac{dC}{dx}=a+bx$$

$$dC=(a+bx)\,dx$$

Total operating cost in running the engine for M km is given by

$$\int dC=C=\int_0^M (a+bx)\,dx$$

$$= \left[ax + \frac{bx^2}{2} \right]_0^M = aM + \frac{bM^2}{2} - 0$$

$$= aM + \frac{bM^2}{2}$$

Since, the engine is overhauled at M kms, the total cost function is given by

Total cost $= A + aM + \frac{1}{2} bM^2$.

Example 11. If the marginal revenue is given by

$$\text{MR} = 27 - 12x + x^2$$

Find the total revenue function.

Solution : Given, $\text{MR} = 27 - 12x + x^2$

Integrating both sides w.r.t. x, we have

$$R = \int (\text{MR})\, dx = \int (27 - 12x + x^2)\, dx$$

$$= 27x - 6x^2 + (x^3/3) + k \qquad \text{where } k \text{ is constant}$$

when $x = 0, R = 0$, then $k = 0$. Hence,

$$R = 27x - 6x^2 + (x^3/3)$$

which gives total revenue function.

Example 12. Find the total revenue between 0 to 10 units of output (x) from the marginal revenue given by

$$\text{MR} = 3(x^2/20) - 10x + 100$$

Solution : Given, $\text{MR} = 3(x^2/20) - 10x + 100$, then the total revenue between output 0 to 10 units is

$$R = \int_0^{10} (\text{MR})\, dx = \int_0^{10} \left(\frac{3}{20} x^2 - 10x + 100 \right) dx$$

$$= \left| \frac{1}{20} x^3 - 5x^2 + 100x \right|_0^{10}$$

$$= \frac{1}{20} (10)^3 - 5(10)^2 + 100(10) = 50 - 500 + 1{,}000 = \text{Rs. } 550.$$

Example 13. The elasticity of demand w.r.t. price of commodity is $\frac{3-x}{x}$, p being the price when the demand is x. Find the demand function when the price is 2, the demand is 1.

Solution : Given, $-\frac{p}{x}\frac{dx}{dp} = e_d = \frac{3-x}{x}$, then $\frac{-1}{3-x}\, dx = \frac{dp}{p}$

$$\Rightarrow \qquad \int \frac{-dx}{3-x} = \int \frac{dp}{p} \Rightarrow \log(3-x) = \log p + \log k,$$

where $\log k$ is constant,

$$\Rightarrow \qquad \log(3-x) = \log pk \Rightarrow 3 - x = pk.$$

Now $x = 1$ when $p = 2$, then $3 - 1 = 2k \Rightarrow k = 1$

Hence, the demand function is $p = 3 - x$.

Example 14. Determine the equation of a demand curve which has price elasticity of demand constant 'c' throughout.

Solution : Given, $-\frac{p}{x}\frac{dx}{dp} = e_d = c$, then $\frac{dx}{x} + c\frac{dp}{p} = 0$

Integrating both sides, we have

$$\log x + c \log p = \log k, \text{ where } \log k \text{ is constant of integration}$$

$$\Rightarrow \qquad \log x + \log p^c = \log k \Rightarrow \log(xp^c) = \log k \Rightarrow xp^c = k,$$

which is the required demand curve.

The above question can also be stated as 'Derive the demand function for which the elasticity of demand is a constant.' **(D.U. B.Com. (Hons.) 1994)**

Example 15. The average cost of manufacturing 5,000 slippers is Rs. 70. The marginal cost function of the same is given by MC = 20 – 0.04x. Find the total cost function and average cost function.

Solution : Given $\qquad \text{MC} = 20 - 0.04x$

It is known that

$$\text{TC} = \int \text{MC}\, dx = \int [20 - 0.04x]\, dx = 20x - \frac{0.04x^2}{2} + c$$

$$\text{TC} = 20x - 0.02x^2 + c$$

$$\text{Average cost} = \frac{20x}{x} - \frac{0.02x^2}{x} + \frac{c}{x} = 20 - 0.02x + \frac{c}{x}$$

It is given in the problem that when output is 5,000 units the average cost is Rs. 70. **(Pl. note)**

i.e.,
$$70 = 20 - 0.02(5{,}000) + \frac{c}{5{,}000}$$

$$70 = 20 - 100 + \frac{c}{5{,}000} \Rightarrow 70 = -80 + \frac{c}{5{,}000}$$

$$\Rightarrow \qquad \frac{c}{5{,}000} = 70 + 80 = 150, \quad \therefore c = 7{,}50{,}000$$

$$\text{TC} = 20x - 0.02x^2 + 7{,}50{,}000$$

$$\text{AC} = 20 - 0.02x + \frac{7{,}50{,}000}{x}.$$

Example 16. The marginal reveue function of a product is give by MR = 500 – 0.01x and the marginal cost function is given by MC = 100 + 0.006x. The fixed cost is Rs. 1,50,000. Find the profit function.

Solution : Given, $\quad MR = 500 - 0.01x$

Total reveue $= \int MR\,dx = \int (500 - 0.01x)\,dx = 500x - \dfrac{0.01x^2}{2} + c$

When $x = 0$, TR = 0, $\therefore\ c = 0$

Hence, $TR = 500x - \dfrac{0.01x^2}{2} = 500x - 0.005x^2$

Given, $\quad MC = 100 + 0.06x$

Total Cost $= \int MC\,dx = \int (100 + 0.006x)\,dx$

$$= 100x + 0.006\frac{x^2}{2} + c$$

$$= 100x + 0.003x^2 + c$$

The fixed cost is Rs. 1,50,000, *i.e.,* when $x = 0$, TC = Rs. 1,50,000.

Therefore, $\quad c = 1{,}50{,}000$

$\therefore$ Total cost $= 100x + 0.003x^2 + 1{,}50{,}000$

Profit = Total revenue – Total cost

$$= 500x - 0.005x^2 - [100x + 0.003x^2 + 1{,}50{,}000]$$

$$= 500x - 0.005x^2 - 100x - 0.003x^2 - 1{,}50{,}000$$

$$= 400x - 0.008x^2 - 1{,}50{,}000$$

$\therefore$ Profit function $= 400x - 0.008x^2 - 1{,}50{,}000$.

Example 17. The marginal cost function of a product is given as $MC = 6 + 10x - 6x^2$. The total cost of a manufacturing one unit is Rs. 54. Find the total cost function and average cost function.

Solution : $\quad MC = 6 + 10x - 6x^2$

Total $= TC = \int (MC)\,dx = \int (6 + 10x - 6x^2)\,dx$

$$= 6x + \frac{10x^2}{2} - \frac{6x^3}{3} + c$$

when $x = 1$, TC = 54

i.e., $54 = 6 + 5 - 2 + c \Rightarrow c = 45$

Total cost function $= 6x + 5x^2 - 2x^3 + 45$

Average cost function $= \dfrac{TC}{x} = 6 + 5x - 2x^2 + \dfrac{45}{x}$.

Example 18. The marginal cost of producing a product is given by MC = $10 + 0.02x$. The marginal revenue function, MR = $20 - 0.002x$. The fixed cost is Rs. 150. Find (a) the total cost function and (b) revenue function.

Solution : $\quad \text{MC} = 10 + 0.02x$

Total cost $\quad = \int (10 + 0.02x)\, dx = 10x + \dfrac{0.02x^2}{2} + c$

When $x = 0$, Total cost = 150

$\therefore \quad 150 = c$

Total cost function = $10x + 0.01x^2 + 150$

$$\text{MR} = 20 - 0.002x$$

Total revenue = $\int (20 - 0.002x)\, dx = 20x - \dfrac{0.002x^2}{2} + c$

When $x = 0$, Total revenue = 0, Hence, $c = 0$

$\therefore$ Total Reveue function = $20x - 0.001x^2$.

Example 19. If the elasticity of demand

$$e_d = \frac{-(5P + 2P^2)}{Q}$$

And $Q = 500$ when $P = 10$. Find the demand function. **(Rajasthan Univ. M.A. (Eco.) 1998)**

Solution :
$$\frac{-P}{Q} \cdot \frac{dQ}{dP} = -\left(\frac{5P + 2P^2}{Q}\right)$$

$\Rightarrow \quad PdQ = P(5 + 2P)\, dp \Rightarrow dQ = (5 + 2P)\, dP$

Integrating both sides we get

$Q = 5P + P^2 + C$ where C is the constant of integration

when $P = 10$, $Q = 500$

$\therefore \quad 500 = 50 + 100 + C$

$\Rightarrow \quad C = 350$

$\therefore \quad Q = 5P + P^2 = 350$ is the demand function.

Example 20. The price elasticity of demand of a commodity is $E_p = \dfrac{3p}{(p-1)(p+2)}$. Find the corresponding demand function if quantity demanded is 8 units when the price is Rs. 2. **(D.U., B.Com. (Hons.) 1997)**

Solution :
$$-\frac{p}{x}\frac{dx}{dp} = \frac{3p}{(p-1)(p+2)}$$

$\Rightarrow \quad \dfrac{dx}{x} = \dfrac{3}{(p-1)(p+2)}\, dp \qquad \text{...(I)}$

Let $$\frac{3}{(p-1)(p+2)} = \frac{A}{(p-1)} + \frac{B}{p+2}$$

$$\Rightarrow \qquad 3 = A(p+2) + B(p-1)$$

Putting $p = 1$ we get $3 = 3A \Rightarrow A = 1$

Putting $p = -2$ we get $3 = -3B \Rightarrow B = -1$

(1) becomes $-\frac{dx}{x} = \left(\frac{1}{p-1} - \frac{1}{p+2}\right) dp$

Integrating both sides, we get

$$-\log x = \log (p-1) - \log (p+2) - \log c$$

$$\Rightarrow \qquad x^{-1} = \frac{(p-1)}{c(p+2)} \Rightarrow x = \frac{c(p+2)}{p-1}$$

Example 21. The marginal cost function when output is x units is given by

$$\text{MC} = x^2 - 2x + 5$$

Find the total cost function C and average cost function AC if the fixed cost is Rs. 30. Also show that the slope of the average cost curve is $\frac{\text{MC} - \text{AC}}{x}$.

(ICWA QM June 1997)

Solution : $$\frac{dC}{dx} = x^2 - 2x + 5$$

$$\int dC = \int x^2 - 2x + 5\, dx$$

$$\Rightarrow \qquad C = \frac{x^3}{3} - x^2 + 5x + K = \text{Total cost function}$$

when $x = 0$, $C = K = 30$ (given)

$$\therefore \qquad c = \frac{x^3}{3} - x^2 + 5x + 30$$

$$\text{AC} = \frac{x^2}{3} - x + 5 = \frac{30}{x}$$

Slope of $$\text{AC} = \frac{d}{dx}(\text{AC}) = \frac{2x}{3} - 1 - \frac{30}{x^2} \qquad \text{...(I)}$$

$$\frac{\text{MC} - \text{AC}}{x} = \frac{x^2 - 2x + 5 - \frac{x^2}{3} + x - 5 - \frac{30}{x}}{x} = \frac{\frac{2x^2}{3} - x - \frac{30}{x}}{x}$$

$$= \frac{2x}{3} - 1 - \frac{30}{x^2} \qquad \text{...(II)}$$

(I) = (II) Proves the result.

7.2 CONSUMERS' SURPLUS

In the words of Dr. Marshall "The excess of price which a person would be willing to pay rather than go without the thing over that which he actually does pay is the economic measure of this surplus of satisfaction. It may be called **consumer's surplus".** It is estimated by the area between the demand curve and the price which is actually paid.

Let $p = f(x)$ be a demand function where y is the price per unit and x, the number of units. Let a buyer purchase x_0 units at a price p_0, then the total amount paid by the consumer $= x_0 p_0$. But the total satisfaction derived by the consumer = Area under the demand curve between $x = 0$ and $x = x_0$.

In the figure an area *OABC* represents the area under the demand curve between $x = 0$ and $x = x_0$ and gives the total satisfaction derived by the consumer from the purchase of x_0 units.

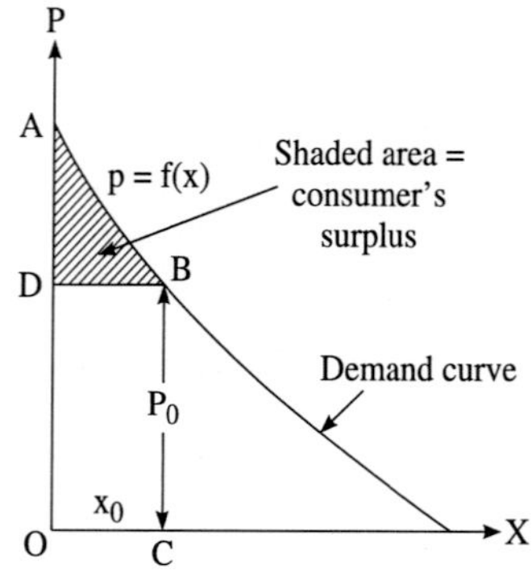

The area under the demand curve from $x = 0$ to $x = x_0 = \int_0^{x_0} f(x)\, dx$

$\therefore$ Consumers' surplus $= \int_0^{x_0} f(x)\, dx - x_0 p_0 =$ **what a consumer is prepared to pay – what he actually pays**.

There is an another method to find the consumers' surplus by integratig the demand function (as a function of price) with respect to p :

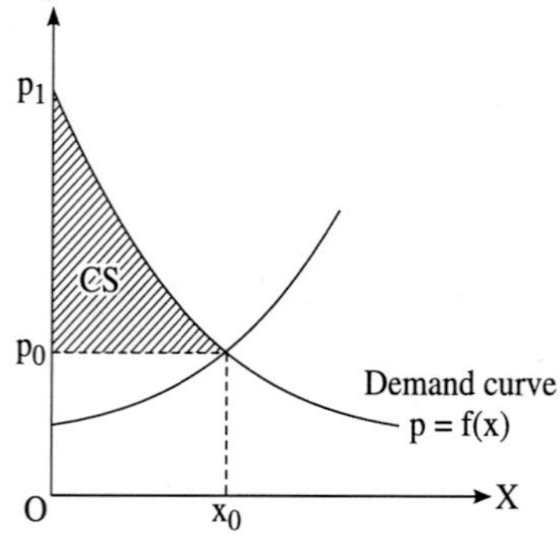

$$\text{Consumer's surplus} = \int_{p_0}^{p_1} \text{(demand function as a function of price)}\, dp,$$

where p_1 is the price at which quantity demanded is zero. It can be obtained by putting $x = 0$ in the demand function.

7.3 PRODUCERS' SURPLUS

Sometimes, it so happens that a producer is willing to sell his product at a certain price but when he actually sells the article he gets more than his expectation. In such cases the *difference between the price* at which the producer would be willing to sell his product and the price he actually receives is called **Producer's Surplus.**

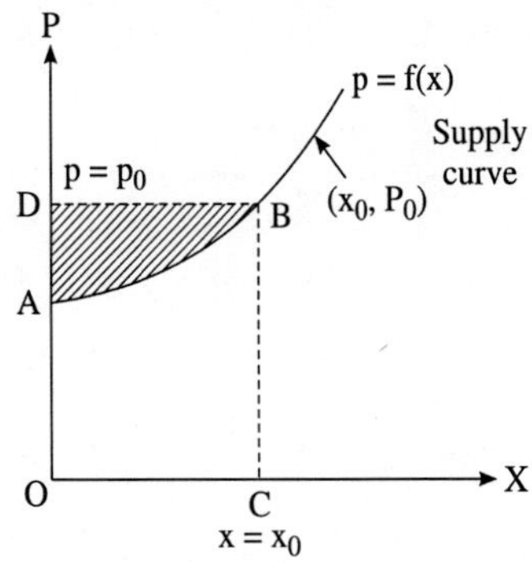

Suppose a producer sells x_0 units at a price p_0, then the amount which he expects is given by the area under the supply curve from $x = 0$ to $x = x_0$ and the amount which he actually receives $= x_0 p_0$.

In the Fig. *OABC* is the area under the supply curve from $x = 0$ to $x = x_0$.

$$\text{Area } OABC = \int_0^{x_0} f(x)\, dx$$

Amount actually received by the producer $= x_0 p_0 =$ Area *ODBC*

$$\textbf{Producers' surplus} = \boldsymbol{x_0 p_0 - \int_0^{x_0} f(x)\, dx}$$

An other way to find producers' surplus by integrating the supply function (as a function of p) with respect to p :

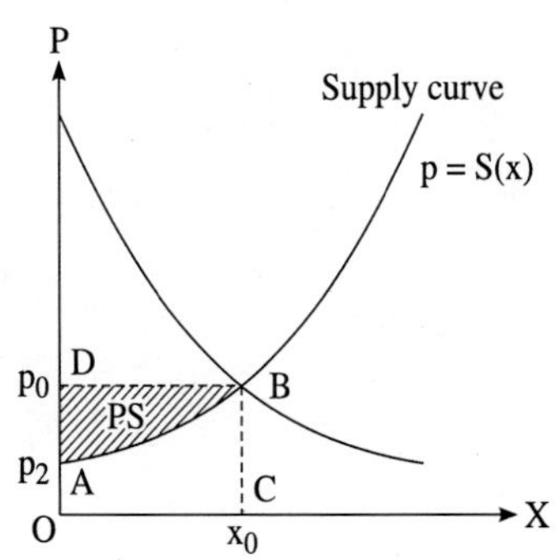

$$PS = \int_{p_2}^{p_0} \text{(supply function as a function of } p) \, dp,$$

where p_2 is the price obtained by setting $x = 0$ in the supply function.

7.4 CONSUMERS' SURPLUS UNDER PURE COMPETITION

Under Pure Competition, the quantity purchased say x_0 is obtained by solving the demand and supply curve equations and by substituting this value of x_0 in the demand equation, the corresponding value of p_0 is obtained. In other words, the point of intersection of demand and supply curves gives the equilibrium price and quantity purchased. Consumer surplus is calculated by the formula given above.

7.5 CONSUMERS' SURPLUS UNDER MONOPOLY

Under monopoly quantity purchased, say x_0, is determined by equating MR and MC functions. Substituting the value of x_0 in demand equation, price is determined and consumers' surplus calculated by usual method.

Example 22. If the demand function is

$$p = 16 - x^2, \text{ find consumer surplus}$$

at $x = 3$. **(Rajasthan Uni. M.A. Eco. 1995)**

Solution : When $x = 3, p = 16 - 9 = 7$

$$\therefore \text{ Consumer surplus} = \int_0^3 (16 - x^2) \, dx - 3 \times 7$$

$$= \left[16x - \frac{x^3}{3}\right]_0^3 - 21 = 48 - 9 - 21 = 18.$$

Example 23. The supply function of a product is $y = 3x^2 + 6$. Find the producer's surplus when 10 units are supplied.

When $x = 10, y = 3 \times 100 + 6 = 306$

Solution : $\therefore$ Producer's surplus $= 306 \times 10 - \int_0^{10} (3x^2 + 6) \, dx$

$$= 3{,}060 - \left[\frac{3x^3}{3} + 6x\right]_0^{10}$$

$$= 3{,}060 - 1{,}000 - 60 = \text{Rs. } 2{,}000$$

Example 24. The supply curve for a commodity is $p = \sqrt{9 + x}$ and the quantity sold is 7 units. Find the producers' surplus. Can you find the consumers' surplus ? If yes find it, if not explain with the help of a diagram why not. **(D.U., B.Com. (Hons.) 1986)**

Solution : The shaded portion represents the Producers' Surplus

Here $$x_0 = 7 = OL$$

$$p_0 = \sqrt{9+7} = \sqrt{16} = 4 = OB$$

Producer's surplus = Area ABC

$$= \text{Area of rectangle } OLAB - \text{Area } OLAC$$

$$= OL \times OB - \int_0^{x_0} \sqrt{9+x}\,dx$$

$$= 28 - \int_0^7 \sqrt{9+x}\,dx$$

$$= 28 - \frac{2}{3}\left|(9+x)^{3/2}\right|_0^7$$

$$= 28 - \frac{2}{3} \times 37 = \frac{10}{3}$$

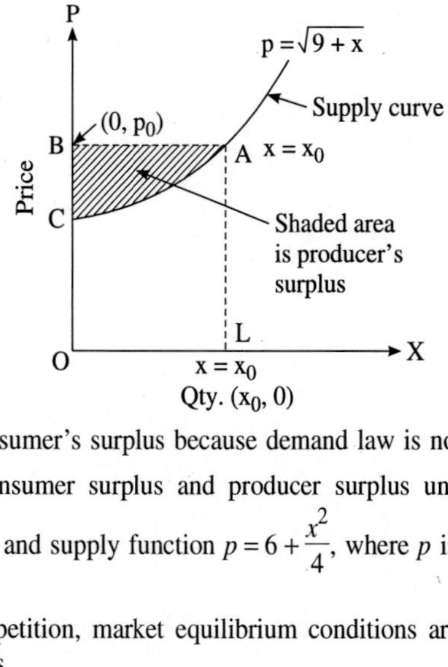

∴ We cannot find Consumer's surplus because demand law is not given.

Example 25. Determine consumer surplus and producer surplus under pure competition for the demand function $p = 36 - x^2$ and supply function $p = 6 + \frac{x^2}{4}$, where p is price and x is quantity.

(D.U., B.Com. (Hons.) 1991)

Solution : Under pure competition, market equilibrium conditions are determined by equating the demand and supply functions

$$\therefore \quad 36 - x^2 = 6 + \frac{x^2}{4} \Rightarrow 144 - 4x^2 = 24 + x^2$$

$$\Rightarrow \quad x^2 = \frac{120}{5} = 24 \Rightarrow x = +2\sqrt{6} = x_0$$

$$\therefore \quad p_0 = 36 - x_0^2 = 36 - (2\sqrt{6})^2 = 12$$

$$\therefore \text{ Consumer's surplus} = \int_0^{2\sqrt{6}} (36 - x^2)\,dx - 12 \times 2\sqrt{6}$$

$$= \left[36x - \frac{x^3}{3}\right]_0^{2\sqrt{6}} - 24\sqrt{6} = (72\sqrt{6} - 16\sqrt{6}) - 24\sqrt{6} = 32\sqrt{6}$$

$$\text{Producer's surplus} = 12 \times 2\sqrt{6} - \int_0^{2\sqrt{6}} \left(6 + \frac{x^2}{4}\right) dx$$

$$= 24\sqrt{6} - \left[6x + \frac{x^3}{12}\right]_0^{2\sqrt{6}}$$

$$= 24\sqrt{6} - (12\sqrt{6} + 4\sqrt{6}) = 8\sqrt{6}$$

Hence, the Consumer's surplus = Rs. $32\sqrt{6}$

and Producer's surplus = Rs. $8\sqrt{6}$.

Example 26. The demand and supply functions under pure competition are $P = 1600 - x^2$ and $P = 2x^2 + 400$ respectively. Find the consumer's and producers' surplus. **(D.U. B.Com. (Hons.) 2000)**

Solution : Under pure competition market equilibrium conditions are determined by Equating supply and demand functions

$$\therefore \quad 1600 - x^2 = 2x^2 + 400$$

[Since production can not be negative, hence negative sign is rejected]

$$\Rightarrow \quad 3x^2 = 1200 \Rightarrow x^2 = 400 \Rightarrow x = 20$$

$$P_0 = 1600 - 400 = 1200$$

$$\text{Consumer's surplus} = \int_0^{20} (1600 - x^2)\, dx - 1200 \times 20$$

$$= \left[1600x - \frac{x^3}{3}\right]_0^{20} - 1200 \times 20$$

$$= 32000 - \frac{8000}{3} - 1200 \times 20 = \frac{8000}{3}$$

$$\text{Producer's Surplus} = 20 \times 1200 - \int_0^{20} (2x^2 + 400)\, dx$$

$$= 2400 - \left[\frac{2x^3}{3} - 400x\right]_0^{20}$$

$$= 24000 - \frac{1600}{3} - 8000 = \frac{32{,}000}{3}.$$

Example 27. Find the Consumer's surplus and Producer's surplus under pure competition for demand function $p = \frac{8}{x+1} - 2$ and supply function $p = \frac{1}{2}(x+3)$, where p is price and x is quantity.

(D.U., B.Com. (Hons.) 1992)

Solution : We have to find C.S. and P.S. under pure competition for the demand function $p_d = \frac{8}{x+1} - 2$ and supply function $p_s = \frac{1}{2}(x+3)$. Under pure competition, market equilibrium conditions are determiend by equating p_d and p_s

$\therefore \qquad \dfrac{8}{x+1} - 2 = \dfrac{1}{2}(x+3) \Rightarrow 8 - 2(x+1) = \dfrac{1}{2}(x+3)(x+1)$

$\Rightarrow \qquad 2(8 - 2x - 2) = (x+3)(x+1)$

$\Rightarrow \qquad 16 - 4x - 4 = x^2 + x + 3x + 3$

$\Rightarrow \qquad x^2 + 8x - 9 = 0$

$\Rightarrow \qquad x^2 + 9x - x - 9 = 0$

$\Rightarrow \qquad x(x+9) - 1(x+9) = 0$

$\Rightarrow \qquad (x-1)(x+9) = 0$

$\Rightarrow \qquad x = 1, -9$

But $x = -9$ is meaningless because quantity cannot be negative

$\therefore \qquad x = 1 = x_0$

Hence, $\qquad p_0 = \dfrac{1}{2}(1+3) = 2$

Now Consumer's surplus $= \displaystyle\int_0^{x_0} p_d\, dx - p_0 x_0 = \int_0^1 \left(\frac{8}{x+1} - 2\right) dx - 2$

$= [8 \log (x+1) - 2x]_0^1 - 2 = 8 \log 2 - 4$

Producer's surplus $= p_0 x_0 - \displaystyle\int_0^{x_0} p_s dx = 2 - \frac{1}{2}\int_0^1 (x+3)\, dx$

$$= 2 - \frac{1}{2}\left|\frac{(x+3)^2}{2}\right|_0^1 = 2 - \frac{7}{4} = \frac{1}{4}$$

Example 28. Find the Producer's surplus and Consumer's surplus under perfect competition, given the demand function

$$x = \frac{25}{4} - \frac{p}{8}$$

and supply function $p = 5 + x$ (where p is price and x is quantity). **(D.U., B.Com. (Hons. 1996)**

Solution : Demand function is

$$x = \frac{25}{4} - \frac{p}{8} \text{ or } p = 50 - 8x$$

and the supply function is

$$p = 5 + x$$

Under perfect competition,

demand = supply

$\therefore \qquad 50 - 8x = 5 + x$

$\Rightarrow \qquad 9x + 45 \text{ or } x = 5$

and hence, $p = 5 + 5 = 10$. Now

$$\text{Consumer's surplus} = \int_0^{x_0} p(x)\,dx - x_0p_0$$

$$= \int_0^{5} (50 - 8x)\,dx - 10 \times 5$$

$$= [50x - 4x^2]_0^5 - 50$$

$$= (250 - 100) - 50 = 100$$

$$\text{and Producer's surplus} = x_0p_0 - \int_0^{x_0} f(x)\,dx$$

$$= 10 \times 5 - \int_0^{5} (5 + x)\,dx$$

$$= 50 - \left[5x + \frac{x^2}{2}\right]_0^5 = 50 - \left(25 + \frac{25}{2}\right) = \frac{25}{2}$$

Example 29. The demand equation of a product is $p - 10e^{-x} = 0$. Find the Consumer's surplus when the market price is $p = 1$. **(D.U., B.Com. (Hons.) 1998)**

Solution : The demand equation is

$$p - 10e^{-x} = 0 \Rightarrow p = 10e^{-x}$$

$$\therefore \text{Consumers' surplus} = \int_0^{x_0} f(x)\,dx - x_0p_0$$

$$= \int_0^{2.3} 10e^{-x}\,dx - 2.3 \times 1$$

$$\left[\because \text{when } p = 1,\ 1 = 10e^{-x} \Rightarrow e^x = 10 \Rightarrow x \log e = \log 10 \Rightarrow x \log_e e = \log_e 10 \Rightarrow x = \log_e 10 \Rightarrow x = \frac{\log_{10} 10}{\log_{10} e} = 2.3\right]$$

$$= [-10e^{-x}]_0^{2.3} - 2.3 = \frac{-10}{e^{2.3}} + 10 - 2.3$$

$$= 9 - 2.3 = 6.7 \qquad [\because e^{2.3} = 10 \text{ Approx.}]$$

Example 30. A monopolist demand function is $x = 210 - 3p$ where x is the quantity demanded when price is Rs. p per unit with the average cost function

$$\text{AC}(x) = x + 6 + \frac{10}{x}$$

Find the Consumer's surplus at the price which monopolist would like to fix.

(D.U. B.Com. (Hons.) 1999)

Solution : Under monopoly

$$MR = MC$$

$$x = 210 - 3p \Rightarrow 3p = 210 - x \Rightarrow p = \frac{210 - x}{3}$$

$$\text{Revenue} = xp = \frac{x(210 - x)}{3}$$

$$MR = \frac{210 - 2x}{3}$$

$$AC = x + 6 + \frac{10}{x}$$

Total cost $= x^2 + 6x + 10$ $\qquad$ [$\because$ Total Cost $= x$ (Average Cost)]

$MC = 2x + 6$

Equating MC and MR, we get

$$\frac{210 - 2x}{3} = 2x + 6 \Rightarrow 210 - 2x = 6x + 18$$

$$\Rightarrow \qquad 8x = 192 \Rightarrow x = 24$$

When $x = 24$, $24 = 210 - 3p \Rightarrow p = 62$.

$\therefore$ Consumer's surplus $= \frac{1}{3}\int_0^{24} (210 - x)\, dx - 24 \times 62$

$$= \frac{1}{3}\left[210x - \frac{x^2}{2}\right]_0^{24} - 24 \times 62 = 1{,}680 - 96 - 1{,}488 = 96.$$

Example 31. Under a monopoly the quantity sold and market price are determined by the demand function. If the demand function for a profit maximizing monopolist is $p = 274 - x^2$ and $MC = 4 + 3x$, find CS.

Solution : Given, demand function, $p = 274 - x^2$ and $MC = 4 + 3x$

$\therefore$ R = Total revenue $= px = (274 - x^2)\, x = 274x - x^3$ and $MR = dR/dx = 274 - 3x^2$

Under monopoly, MR = MC

i.e., $\qquad 274 - 3x^2 = 4 + 3x \Rightarrow 3x^2 + 3x - 270 = 0 \Rightarrow 3(x^2 + x - 90) = 0,$

$$\Rightarrow \qquad x^2 + x - 90 = 0 \qquad [\because 3 \neq 0]$$

$$x^2 + x - 90 = 0 \Rightarrow (x - 9)(x + 10) = 0 \Rightarrow x = 9 \text{ or } x = -10.$$

But $x = -10$ is not possible, therefore $x = 9$

When $x = 9, p = 274 - (9)^2 - 193$

$$\therefore \qquad CS = \int_0^9 (274 - x^2)\, dx - (193)(9) = \left|274x - \frac{x^3}{3}\right|_0^9 - 1{,}737$$

$$= \left[274(9) - \frac{1}{3}(9)^3\right] - 1,737 = 9\left[274 - \frac{81}{3}\right] - 1,737$$

$$= 9[274 - 27] - 1,737 = 2,223 - 1,737 = 486 \text{ units.}$$

Example 32. The demand function of a commodity is $y = 100 - 5x - x^2$ where x is the quantity demanded at price y. Find the Consumers' surplus when the price is Rs. 16.

Solution : When the price is Rs. 16, the number of units bought are given by

$$16 = 100 - 5x - x^2$$

or $$x^2 + 5x - 84 = 0$$

or $$x(x + 12) - 7(x + 12) = 0$$

or $$(x + 12)(x - 7) = 0$$

$\Rightarrow$ $$x = -12 \text{ or } 7$$

Since, the number of units purchased cannot be negative we have $x = 7$ when $y = \text{Rs. } 16$

$$\text{Consumers' surplus} = \int_0^7 (100 - 5x - x^2)\, dx - 16 \times 7$$

$$= \left[100x - \frac{5x^2}{2} - \frac{x^3}{3}\right]_0^7 - 112$$

$$= 700 - \frac{245}{2} - \frac{343}{3} - 112$$

$$= \frac{4,200 - 735 - 686 - 672}{6}$$

$$= \frac{2,107}{6} = \text{Rs. } 351.16 = \text{Rs. } 351 \text{ approx.}$$

Example 33. Given demand function

$$P_d = 25 - Q^2$$

Supply function

$$P_s = 2Q + 1$$

Assuming pure competition find Consumer's surplus and Producer's surplus.

(D.U., B.Com. (Hons.) 1995, Raj. Univ. M.A. Eco. 1998)

Solution : Under pure competition

$$P_d = P_s$$

$\Rightarrow$ $$25 - Q^2 = 2Q + 1 \Rightarrow Q^2 + 2Q - 24 = 0$$

$\Rightarrow$ $$Q^2 + 6Q - 4Q - 24 = 0$$

$\Rightarrow$ $$Q(Q + 6) - 4(Q + 6) = 0$$

$\Rightarrow$ $$(Q - 4)(Q + 6) = 0 \Rightarrow Q = 4 \text{ or } -6$$

Rejecting negative value we get, $Q = 4$ and $P_d = 25 - 16 = 9$

$$\text{Consumers' surplus} = \int_0^4 (25 - Q)^2\, dQ - 4 \times 9$$

$$= \left[25Q - \frac{Q^3}{3}\right]_0^4 - 36$$

$$= 100 - \frac{64}{3} - 36 = \frac{128}{3}$$

$$\text{Producer's surplus} = 36 - \int_0^4 (2Q + 1)\, dQ$$

$$= 36 - \left[\left(\frac{2Q^2}{2} + Q\right)\right]_0^4 = 36 - (16 + 4) = 16$$

Example 34. If the supply function for a particular commodity is $100\,p = (x + 20)^2$ and the market price is Rs. 25, find the producer's surplus by two methods : (a) integrate with respect to x, (b) integrate with respect to p.

Solution : (a) Using integration with respect to x, the producer's surplus is

$$\text{PS} = x_0 \times p_0 - \int_0^{x_0} S(x)\, dx,$$

where (x_0, p_0) is the equilibrium point and $p = S(x)$ is the supply function.

It is given that $p_0 = 25$. Substituting this value in the supply function : $100p = (x + 20)^2$, we obtain $x_0 = 30$. Thus

$$\text{PS} = 25 \times 30 - \int_0^{30} \frac{(x + 20)^2}{100}\, dx$$

$$= 750 - \left[\frac{1}{300}(x + 20)^3\right]_0^{30}$$

$$= 750 - \frac{1}{300}\left[(50)^3 - (20)^3\right] = 360.$$

(b) Using integration w.r.t. P, the producers surplus is

$$\text{PS} = \int_{p_1}^{p_0} [\text{Supply function as a function of } p]\, dp$$

were p_0 is the market price and p_1 is obtained by setting $x = 0$ in the supply function.

We are given $p_0 = 25$. To find, p_1, we set $x = 0$ in the supply function :

$$100p = (x + 20)^2$$

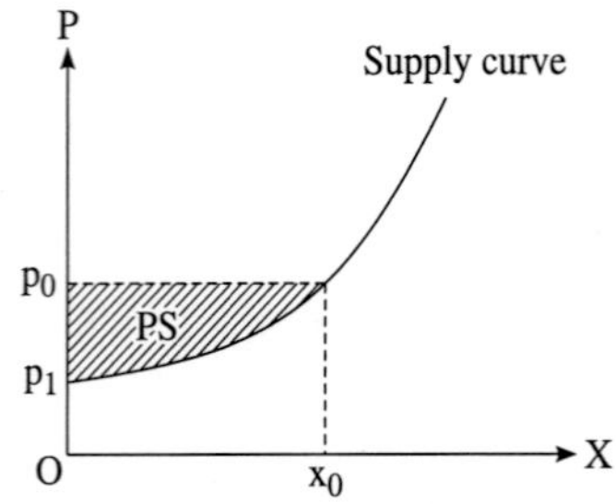

and obtain $p_1 = 4$. Now solving the supply function for x in terms of p, we get

$$x = 10\sqrt{p} - 20$$

Thus,

$$\text{PS} = \int_4^{25} (10\sqrt{p} - 20)\, dp$$

$$= \left(\frac{20p^{3/2}}{3} - 20p\right)_4^{25}$$

$$= \left[\frac{20}{3}(25)^{3/2} - 20(25)\right] - \left[\frac{20}{3}(4)^{3/2} - 20(4)\right]$$

$$= 360 \text{ (on simplification)}$$

which is same as in (Part a).

Example 35. A company suffers a loss of Rs. 121.50 if one of its products does not sell at all. Marginal Revenue and marginal cost function for the product are given by

$$\text{MR} = 30 - 6x,\ \text{MC} = -24 + 3x$$

Determine the total profit function, the break-even points and the total profit between break-even points.

Solution : If P denotes the profit function, then

$$\frac{dp}{dx} = \text{MR} - \text{MC} = (30 - 6x) - (-24 + 3x)$$

$$= 54 - 9x$$

Integrating both sides with respect to P, we have

$$P = \int (54 - 9x)\, dx$$

$$= 54x - \frac{9}{2}x^2 + k,$$

where k is an arbitrary constant. Since, the company suffers a loss of Rs. 121.50 if the product does not sell at all, *i.e.,* $P = -121.50$ when $x = 0$.

Therefore,

$$-121.50 = 54(0) - \frac{9}{2}(0^2) + k \Rightarrow k = -121.50$$

Thus, the total profit function is given by

$$P = 54x - \frac{9}{2}x^2 - 121.50$$

For break-even points, $P = 0$. therefore

$$54x - \frac{9}{2}x^2 - 121.50 = 0$$

$\Rightarrow$ $$x^2 - 12x + 27 = 0 \Rightarrow x^2 - 3x - 9x + 27 = 0$$

$\Rightarrow$ $$x(x-3) - 9(x-3) = 0$$

$\Rightarrow$ $$(x-3)(x-9) = 0$$

$\Rightarrow$ $$x = 3, 9$$

Thus, the break-even points are $x = 3$ and $x = 9$. The total profit between the break-even points is given by

$$\int_3^9 P\,dx = \int_3^8 \left(54x - \frac{9}{2}x^2 - 121.50\right)dx$$

$$= \left| 27x^2 - \frac{3}{2}x^3 - 121.50\,x \right|_3^9$$

$$= \left[27(9)^2 - \frac{3}{2}(9)^3 - 121.50(9)\right] - \left[27(3)^2 - \frac{3}{2}(3)^3 - 121.50(3)\right]$$

$$= \text{Rs. } 162.$$

Example 36. Find the consumption function given that income equals consumption when $Y = 400$, and the marginal propensity to consume (MPC), *i.e.*, $dC/dY = 0.7$, where Y is income and C is consumption.

Solution : Given, $$\text{MPC} = \frac{dC}{dY} = 0.7$$

Integrating both sides, we get

$$C = \int (0.7)\,dY$$

$= 0.7Y + k$, where k is constant of integration.

Since, income equals consumption, when $Y = 400$, *i.e.,* $Y = C$ when $Y = C = 400$, then $400 - 0.7\,(400) + k \Rightarrow k = 120$.

Hence, the consumption function is

$$C = 0.7Y + 120.$$

Example 37. If the 'marginal propensity to save' (MPS) is $S'(Y) = 2 - 0.5Y^{-2/3}$ and aggregate saving = 10 when income is 100 find the saving function.

Solution : Given MPS $= S'(Y) = \dfrac{dS}{dY} = 2 - 0.5\,Y^{-2/3}$, then

$$S = \int (\text{MPS})\, dY = \int [2 - 0.5\, Y^{-2/3}]\, dY$$

$$= 2Y - 1.5Y^{1/3} + k, \text{ where } k \text{ is constant of integration}$$

But $S = 10$, when $Y = 100$.

$$\therefore \quad 10 = 2(100) - (1.5)\,(100)^{\frac{1}{3}} + k$$

$$\Rightarrow \quad k = 10 - 200 + (1.5)\,(4.65) = -183.04.$$

Hence, the saving function is

$$S = 2Y - (1.5)\, Y^{1/3} - 183.04.$$

Example 38. Assume that in 1990 the annual word use of natural gas was 50 trillion cubic feet. The annual consumption of gas is increasing at a rate of 3% compounded continuously. How long will it take to use all available gas, if it is known that in 1990 there were 2,200 trillion cubic feet of the proven reserves? Assume that no new discoveries are made. **(D.U., B.Com. (Hons.) 1991)**

Solution : After t years, 50 trillion cubic feet of gas compounded continuously at the rate of 3% will become

$$\lim_{n \to \infty} 50\left(1 + \frac{3}{100n}\right)^{nt} = 50 \lim_{n \to \infty}\left[\left(1 + \frac{3}{100n}\right)^{\frac{100n}{3}}\right]^{\frac{3}{100}t} = 50\, e^{\frac{3}{100}t}$$

$$= \text{Rate of cosumption of the gas.}$$

Now, according to question

$$\int_0^t 50e^{\frac{3}{100}t}\, dt = 2{,}200$$

$$\Rightarrow \quad 50 \times \frac{100}{3}\left| e^{\frac{3}{100}t} \right|_0^t = 2{,}200$$

$$\Rightarrow \quad e^{\frac{3}{100}t} - 1 = \frac{2{,}200 \times 3}{50 \times 100}$$

$$\Rightarrow \quad e^{\frac{3}{100}t} - 1 = \frac{2{,}200 \times 3}{50 \times 100}$$

$$\Rightarrow \quad e^{\frac{3}{100}t} = 2.32 \Rightarrow (1.03)^t = 2.32 \qquad \left[\because\ e^{\frac{3}{100}} = e^{0.03} = 1.03\right]$$

$$\Rightarrow \quad t = \frac{\log 2.32}{\log 1.03} = 28.6 \text{ years (Approximately).}$$

Example 39. A firm has the current sales of Rs. 50,000 per month. The firm wants to embark on a certain advertising campaign that will increase the sales by 2% per month (compounded continuously) over the period of the campaign which is 12 months. Find the total increase is sales as a result of the campaign. Use calculus. **(D.U., B.Com. (Hons.) 1991)**

Solution : Total increase in sales is given by

$$\int_0^{12} 50{,}000\, e^{0.02t}\, dt - 50{,}000 \times 12 = \frac{50{,}000}{0.02} |e^{0.02t}|_0^{12} - 50{,}000 \times 12$$

$$= 25{,}00{,}000\,(e^{0.24} - 1) - 50{,}000 \times 12$$

$$= 25{,}00{,}000 \times 0.27125 - 6{,}00{,}000$$

$$= 6{,}78{,}122 - 6{,}00{,}000 = \text{Rs. } 78.122.$$

Example 40. A company has a current sales of Rs. 1,000 per month and profit to the company averages 10% of sales. The company's past experience with a certain advertising strategy is that sales increase by 2% per month continuously over the length of the advertising campaign (12 months.) If the advertisement costs Rs. 130 in a year, determine if the company should embark on a similar campaign when the market rate of interest is 12% per annum. **(D.U., B.Com. (Hons.) 1988)**

Solution : Total sales during the advertising campaign

$$= \int_0^{12} (1{,}00)\, e^{0.02t}\, dt = \frac{1000}{0.02} |e^{0.02t}|_0^{12}$$

$$= 50{,}000 - |e^{0.24} - 1| = \text{Rs. } 13{,}565$$

$$\text{Extra Profit due to advertisement} = \frac{10}{100}(14{,}565 - 12{,}000)$$

$$= \text{Rs. } 156.50$$

Cost of advertisement with interest

$$= 130 + \frac{130 + 12 \times 1}{100} = \text{Rs. } 145.60 \text{ which is less than Rs. } 156.50$$

∴ The company should proceed with the advertising campaign.

Example 41. Find by using integral calculus the present value of an annuity of Rs. 100 per annum assumed to be payable continuously for 10 years at the rate of interest 4% per annum compounded continuously. **(D.U. B.Com. (Hos.) 1996)**

Solution : The required present value is given by the formula

$$P = \int_0^N \text{R}e^{-rt}\, dt,$$

where $R = 100$, $r = 0.04$ and $N = 10$. Hence, the required present value $= \int_0^{10} 100e^{-0.04t}\, dt,$

$$= 100 \left| \frac{e^{-0.04t}}{-0.04} \right|_0^{10} = \frac{100}{0.04}(1 - e^{-0.04})$$

$$= 2{,}500\,(1 - 0.67032) = \text{Rs. } 824.20.$$

Example 42. A small business is considering buying an energy-saving device which will reduce its consumption of fuel. The device will cost Rs, 32,000. Engineering estimates suggest that savings from using the device will occur at a rate of $S(t)$ rupees per year, where $S(t) = 20{,}000\, e^{-0.5t}$ and t equals the time measured in years. Determine how long will it take for the firm to recover the cost of device. **(D.U., B.Com. (Hons.) 1994)**

Solution : Let S be the saving after t years

Then $$\frac{dS}{dt} = S(t) = 20{,}000\, e^{-0.5t}$$

On integrating we have

$$S = \int_0^t 20{,}000\, e^{-0.5t}\, dt$$

$$= 20{,}000 \left| \frac{e^{-0.5t}}{-0.5} \right|_0^t$$

$$= -\, 40{,}000\, (e^{-0.5t} - 1)$$

Now device will recover the cost if

$$-40{,}000\, (e^{-0.5t} - 1) = 32{,}000$$

$$\Rightarrow \quad e^{-0.5t} = 0.2 \Rightarrow e^{-0.5t} = 5$$

$$\Rightarrow \quad t = 2\frac{\log 5}{\log_{10} e} = \frac{2 \times 0.6990}{0.4343} \qquad (e = 2.7182)$$

$$= 2 \times 1.609 = 3.22 \text{ years.}$$

Example 43. The purchase price of a car is Rs. 1,75,000. The rate of repairing cost of the car is given by the function

$$C = 8{,}000\, (1 - e^{-0.5t})$$

where t represents the year of use since purchase and C denotes the cost. Determine the cumulative repair cost at the end of 5 years. Also find the equation to find the time in years in which the cumulative repair cost equals the original cost of the car. **(D.U., B.Com. (Hons.) 1995)**

Solution : The rate of repairing cost is given by

$$C = 8{,}000\, (1 - e^{-0.05t}),$$

Cumulative repair cost at the end of 5 years

$$= \int_0^5 8{,}000\, (1 - e^{-0.5t})\, dt = 8{,}000\, |t + 2e^{-0.5t}|_0^5$$

$$= 8{,}000\, [(5 + 2e^{-2.5}) - 2]$$

$$= 8{,}000\, [3 + 2(0.08208)] = \text{Rs. } 25{,}313.28$$

Clearly the cumulative cost at the end of five years is less than the purchase price of the car. Let n be the number of years in which the total repair cost will be equal to the purchase price then n should be more than 5

Now

$$\int_0^n 8{,}000\, (1 - e^{-0.5t})\, dt = 1{,}75{,}000$$

$$\Rightarrow \qquad \left[t + \frac{e^{-0.5t}}{0.5}\right]_0^n = \frac{175}{8} \Rightarrow n + 2e^{-0.5n} - 2 = \frac{175}{8}$$

$$\Rightarrow 8n + 16e^{-0.5n} = 191^* \qquad \ldots(I)$$

Equation (I) gives the required value of $n, x = 24$ years.

Example 44. After an advertising campaign, a product has sales rate $f(t)$ given by $f(t) = 1{,}000\, e^{-0.05t}$ where t is the number of months since the close of the campaign. Find :

(a) The total cumulative sales after 3 months.

(b) The sales during the fourth month.

(c) The total sales as a result of the campaign. **(D.U., B.Com. (Hons.) 1983, 1993)**

(Students should note the solution carefully)

Solution : Total sales over a period of T months after the close of the campaign

$$= \int_0^T f(t)\, dt$$

∴ (a) Total cumulative sales after three months

$$= \int_0^3 f(t)\, dt = \int_0^3 1{,}000\, e^{-0.5t}\, dt = \left[-1{,}000 \frac{e^{-0.5t}}{0.5}\right]_0^3 = -2{,}000\, [e^{-0.5t}]_0^3$$

$$= -2{,}000\, [e^{-1.5} - 1] = 2{,}000\, (1 - e^{-1.5}) = 1{,}5545 \text{ units approx.}$$

(b) Sales during the fourth month

$$= \int_3^4 1{,}000\, e^{-0.5t}\, dt = -1{,}000 \left[\frac{e^{-0.5t}}{0.5}\right]_3^4 \qquad \textbf{(Pl. note)}$$

$$= 2{,}000\, [e^{-1.5} - e^{-2.0}] = 2{,}000\, [0.2231 - 0.1353] = 175.6 \text{ units.}$$

(c) Total sales as result of the campaign

$$= \int_0^\infty 1{,}000\, e^{-0.5t}\, dt = \frac{1{,}000}{0.5} [-e^{-0.5t}]_0^\infty = 200 \text{ units.}$$

Example 45. Calculate the present value of an annuity of Rs. 30,000 per annum, assumed to be payable continuously for 10 years at the rate of interest of 8% p.a. Compounded continuously.

(D.U., B.Com. (Hons.) 2000)

Solution : Present Value $= \int_0^{10} 30{,}000\, e^{-0.08t}\, dt$

$$= 30{,}000 \left[\frac{e^{-0.08t}}{-0.08}\right]_0^{10}$$

---

* The value of n can be found by hit and trial method only.
Here $n = 24$ years approx.

$$= 30{,}000\left[\frac{e^{-0.8t}}{-0.08}+\frac{1}{0.08}\right]=\frac{30{,}000}{0.08}\,[1-e^{-0.8}]$$

$$= \frac{30{,}000}{0.08}\,(1-0.4493)$$

$$= \frac{30{,}000}{0.08}\times 0.55067$$

$$= \text{Rs. } 2{,}06{,}501.6.$$

7.6 INVESTMENT AND CAPITAL FORMATION

If K denotes the Capital, then $K(t)$ denotes the capital stock as a function of time.

$$\frac{d}{dt}[K(t)] = \text{rate of capital formation at time } t$$

$$= \text{rate of net investment flow at time } t = I(t)$$

$$\therefore \qquad \frac{d}{dt}[K(t)] = I(t) \Rightarrow K(t) = \int I(t)\,dt$$

$$= \text{Time path of capital } K$$

Amount of capital formation during the time interval (a, b)

$$[a, b]\int_a^b I(t)\,dt = [K(t)]_a^b = K(b) - K(a).$$

Example 46. Investment function is given by $I(t) = 20e^{0.4t}$, where $I(t) = dK/dt$ and K denotes the capital at time t; and the initial stock is $K(0) = 100$. Find : (i) the time path of capital stock, and (ii) the capital formation during the first five years. **(Delhi Univ., B.Com. (Hons.) 1985 C.C.)**

Solution : (i) The time path of capital stock is

$$K(t) = \int I(t)\,dt = \int 200\,e^{0.4t}\,dt = 500e^{0.4t} + k,$$

where k is the constant of integration. However, since $K(0) = 100$, we have

$$100 = 50e^{0} + k \Rightarrow k = -400$$

Thus, the time path of capital stock is

$$K(t) = 500e^{0.4t} - 400$$

(ii) The capital formation during the first five years

$$= \int_0^5 I(t)\,dt$$

$$= |500e^{0.4t}|_0^5 = 500(e^2 - 1) = 3194.55.$$

Example 47. If the investment flow is given by $I(t) = 3t^{1/2}$ crore of rupess per year, what will be the capital formation in the time period of 5 years, and during the last year of the plan, *i.e.,* 5th year ? Also, find the capital growth equation from the given rate of capital formation if in the initial stock the capital is 20 crores of rupees. **(Delhi Univ., B.Com. (Hons.) 1988 C.C.)**

Solution : Rate of net investment flow = Rate of capital formation

$\therefore$ Capital formation in a period of five years

$$= \int_0^5 3t^{1/2}\, dt$$

$$= \left[\frac{2}{3} \times 3t^{3/2}\right]_0^5 = 2t^{3/2} = 2 \times 5^{2/3} \text{ crores.}$$

Capital formation in the fifth year

$$= \int_4^5 3t^{1/2}\, dt = 3 \times \frac{2}{3}\left[t^{3/2}\right]_4^5 = 2\left[5^{3/2} - 4^{3/2}\right]$$

Capital growth equation is

$$\frac{d}{dt}[K(t)] = I(t)\, dt$$

$$\Rightarrow \qquad K(t) = \int I(t)\, dt + C$$

$$= \int 3t^{1/2}\, dt + C$$

$$= 3\frac{t^{3/2}}{\frac{3}{2}} + C = 2t^{3/2} + C$$

When $t = 0$, $k(t) = 20$

$$\therefore \qquad 20 = C$$

$\therefore$ $K(t) = 2t^{3/2} + 20$ is the required growth equation.

Questions

1. A two wheeler is purchased on instalment basis such that Rs. 1,800 is to be paid each year for 5 years. If interest is charged @ 11% per annum compounded continuously, what would be the cash down price or present value of the two wheeler. **(C.A. (Foundation) May 2000)**

[**Ans.** Rs. 6921.82]

[**Hint :** Present value $= \int_0^5 1800\, e^{-0.11t}\, dt$].

2. Determine the cost of producing 200 cars if the marginal cost (in Rupees per unit) is given by]

$$\text{MC}(x) = \frac{15}{2}x^2 - 4x + 8{,}000$$

What is the cost of production for one car. **(C.A. Foundation May, 2000)**

[**Ans.** Cost of producing 200 cars = Rs. 2,15,20000
Cost of production for one car = Rs. 1,07,600]

[**Hint :** Cost of producing 200 cars $= \int_0^{200} \left(\frac{15}{2}x^2 - 4x + 8{,}000\right) dx$

Cost of producing one car $= \frac{\text{Cost of producing 200 cars}}{200}$].

3. The marginal cost function of a product is given as

$$MC = 6 + 10x - 6x^2$$

The total cost of manufacturing one unit is Rs. 54. Find the total cost function and average cost function.

[**Ans.** Total cost function $= 6x + 5x^2 - 2x^3 + 45$]

4. If the marginal cost function is given by
$MC = 3(3x + 4)^{-1/2}$ and fixed cost is Rs. 2, find the average cost for 7 units of output. [**Ans.** (8/7)]

5. A firm has a marginal revenue function given by $MR = \frac{a}{x+b} - c$, where x is the output and a, b, c are constants. Show that the demand law is given by $p = \frac{a}{x} \log\left(\frac{x+b}{b}\right) - c$.

(Rajasthan Univ. M.A. (Econ.) 1986)

6. For a monopolist, the demand law is $p = 50 - x^2$ and marginal cost $MC = 1 + x^2$. Determine the consumer's surplus at the price which monopolist will like to fix. If there is pure competition in this case, what will be the consumer's surplus at equilibrium prices? **(Delhi Univ. B. Com. (Hons.) 1984 C.C.)**

[**Ans.** $\frac{343}{12}, \frac{343}{3\sqrt{2}}$]

7. When the price of pocket calculators averaged Rs. 400, a compay sold 20 every month. When the price dropped to an average of Rs. 100, 120 were sold every month by the same company. When the price was Rs. 400, 200 calculators were available per month for sale. When the price reached Rs. 100, only 50 remained. Find the demand and supply functions. Also find consumer's and producer's surplus. [**Ans.** C.S. = 12,696; P.S. = 8.464]
[**Hint :** Let the demand function be

$$p = ax + b$$

when

$$p = 400,\ x = 20$$
$$p = 100,\ x = 120$$

find a and b.
Let the supply function be

$$p = mx + c$$

when $p = 400$, $x = 200$, when $p = 100$, $x = 50$.]

8. If the MC of a commodity is equal to half its Average Cost show that the fixed cost is zero. If the cost producing 9 units of a commodity is Rs. 60. Find the cost function. [**Ans.** $C(x) = 20\sqrt{x}$]

9. The rate of change of output Q w.r.t. input x is equal to the product of the input and output. Find the production function if $Q = 4e^2$ when $x = 2$.

[**Ans.** $Q = 4e^{x^2/2}$]

[**Hint :** $\frac{dQ}{dx} = Qx \Rightarrow \frac{dQ}{Q} = x\,dx$
Integrating we get

$$\log_e Q = \frac{x^2}{2} + \log c$$

$$\Rightarrow \qquad \log_e \frac{Q}{c} = \frac{x^2}{2} \Rightarrow Q = ce^{x^2/2}$$

when $x = 2$, $Q = 4e^2$
$\therefore\ 4e^2 = ce^{4/2} = ce^2 \Rightarrow c = 4$
$\therefore$ Production function is $Q = 4e^{x^2/2}$.]

10. The value of machinery depreciates with time according to the relation $\frac{dv}{dt} = -100e^{\left(-\frac{1}{2}t\right)}$ where v is the value of the machinery in thousand, rupees and t the time is years. How much will it depreciate in 3 years.

[**Ans.** Rs. 155.38]

11. (a) The demand function for a commodity is $p = 20 - 3q$ and the supply function on the market is $p = 2q$. Find the Consumer's surplus, under pure competition.

(b) If the supply curve is $p = \sqrt{9 + x}$ and the quantity sold in the market is 7, find Producer's surplus.

[**Ans.** (a) Rs. 234 (b) Rs. $\frac{10}{3}$]

(C.A. Foundation Nov. 1994)

12. The demand and supply function under perfect competition are $y = 16 - x^2$ and $y = 2x^2 + 4$ respectively. Find the market price, Consumers' surplus and Producers' surplus.

[**Ans.** $\frac{16}{3}, \frac{32}{3}$]

13. In $t = 0$, the annual world use of natural gas was 50 trillion cubic feet. The annual consumption of gas is increasing at the rate of 10% continuously. Assuming the rate remains the same in future :

(i) How many cubic feet will be consumed in the next 10 years from $t = 0$?

(ii) How long will it take to use all the available gas if it is known that world reserves are of 2,200 trillions cubic feet in the beginning (assume that no new discoveries are made) ? [**Ans.** 17 years approx.]

(DU, B.Com. (Hons.) 1997)

❑❑❑

8 Basic Mathematics of Finance

8.1 INTRODUCTION

The two terms **moneylender** and **borrower** give rise to the term **interest** in finance. The moneylender can be an individual person or a bank or some other financial institution whereas the borrower can be a single person or a company.

The money lender gives his money to the borrower and at the time of return, charges some extra amount in addition to the sum initially lent. This extra amount charged by the moneylender for the use of his money is called **interest**. The rate of interest is as decided between the lender and the borrower. The interest is charged because of the 'time value of money'.

The amount borrowed is called **principal or capital** and the period for which it is used is called **Time. The rate of interest** is the money payable on the use of unit principal for a unit interval of time. The total sum required to repay the principal as well the interest at the end of a certain period is called **amount**, *i.e.*,

$$\text{Amount} = \text{Sum borrowed [Principal]} + \text{interest.}$$

If the interval of time is one year and the the rate of interest say i is per unit of money per year (or per annum) then for a principal of Rs. P the amount payable at the end of one year = Rs. $P(1 + i)$. If the rate of interest is reckoned on Rs. 100 as principal, then it is called the rate per cent per annum. Thus a rate of interest of 0.06 per unit per annum can also be referred to as a rate of interest of 6 per cent per annum [written as 6% p.a.].

Interest can be reckoned in two ways :

(i) Simple interest.

(ii) Compound interest.

In simple interest the principal remains the same throughout the period the money remains with the borrower. But in compound interest the unpaid interest over any unit of time also earns interest over the subsequent units of time. In compound interest, if the interest is not paid when it is due, the unpaid interest is regarded as an additional loan for the next unit of time (or interval of time) either at the same rate of interest or at some other rate of interest as agreed between the two parties, *i.e.*, lender and the borrower.

In simple interest, interest is calculated at the time of returning the loan by using the formula :

$$\left.\begin{array}{l}\text{Simpler interest}\\ \text{on Rs. } P \text{ for } n \text{ years}\\ \text{at the rate of Re. } i\\ \text{Per rupee per year}\end{array}\right\} = \text{Rs. } Pin$$

If the rate of interest is $r\%$ p.a., then

Simple interest $= \text{Rs.}\ \frac{Prn}{100} = \text{Rs.}\left(\frac{P}{100}\right)rn$ or $P\left(\frac{r}{100}\right)n$

In compound interest the interval of time after which the interest to be calculated is decided before making the loan transaction. The amount due at the end of each interval of time called **conversion period**, becomes the principal for the next interval of time, *e.g.*,

If Rs. 100 are borrowed at 10% p.a. compound interest compounded annually, then

Amount of the end of 1st Year $= 100 + \frac{100 \times 10 \times 1}{100} = \text{Rs. } 110.$

Rs. 110 will become the principal for the second year.

The amount due at the end of 2nd year $= 110 + \frac{110 \times 10}{100} = \text{Rs. } 121.$

Rs. 121 will become the principal for 3rd year and so.

Now $\quad \text{Rs. } 110 = 100\left(1 + \frac{10}{100}\right)$ and

$\text{Rs. } 121 = 110\left(1 + \frac{10}{100}\right) = 100\left(1 + \frac{10}{100}\right)^2$ = amount due at the end of 2nd year.

Likewise we can find the amount due at the end of 3rd year, 4th year and so on.

In general if Rs. P is the principal (borrowed amount or capital), and $r\%$ p.a. the rate of interest compounded annually, then

Amount due at the end of n years $= \text{Rs. } P\left(1 + \frac{r}{100}\right)^n.$

Compound interest for n years $= \text{Rs.}\left[P\left(1 + \frac{r}{100}\right)^n - P\right].$

If $\frac{r}{100} = i$ = rate of interest per rupee per annum, then

Amount due after n years $= \text{Rs. } P(1 + i)^n$*

Compound interest for n years $= \text{Rs. } [P(1 + i)^n - P]$

---

* If P is the principal, $r\%$ p.a. the rate of interest then

Interest for first year $= \frac{P}{100}r = \frac{Pr}{100}$

Amount due after one year $= P + \frac{P}{100}r = P\left(1 + \frac{r}{100}\right)$

In compound interest calculations the principal for 2nd year $= P\left(1 + \frac{r}{100}\right)$

Interest for 2nd year $= \frac{P}{100}\left(1 + \frac{r}{100}\right)r$

Amount due at the end of 2nd year $= P\left(1 + \frac{r}{100}\right) + \frac{P}{100}\left(1 + \frac{r}{100}\right)r = P\left(1 + \frac{r}{100}\right)\left(1 + \frac{r}{100}\right) = P\left(1 + \frac{r}{100}\right)^2$

Similarly amount due at the end of n years $= P\left(1 + \frac{r}{100}\right)^n$

If $\frac{r}{100} = i$, then amount due at the end of n years $= P(1 + i)^n$

If the interest is compounded m times in a year and i is the rate of interest per rupee per annum, then the number of conversion periods in one year = m.

The number of conversion periods in n years = mn

Rate of interest per conversion period $= \frac{i}{m}$

$$= \frac{\text{Nominal rate of interest per year}}{\text{Number of conversion periods in one year}}$$

The amount due at the end of n years = Rs. $P\left[1+\frac{i}{m}\right]^{mn}$

$$\text{Interest for } n \text{ years} = \text{Rs.}\left[P\left(1+\frac{i}{m}\right)^{mn} - P\right]^{*}$$

If compounding is done m times in a year and the rate of interest is $r\%$ p.a. Then

$$\text{Interest for } n \text{ years} = P\left[1+\frac{r}{m}\times\frac{1}{100}\right]^{mn} - P$$

8.2 NOMINAL RATE OF INTEREST AND EFFECTIVE RATE OF INTEREST

The announced or stated rate of interest compounded m times in a year is called the **nominal rate of interest** whereas the actual rate by which money grows annually is called the **effective rate** of interest.

e.g., if a man borrws Rs. 100 at 10% per annum compounded quarterly then the amount due at the end of one year

$$= 10\left[1+\frac{10}{400}\right]^{4} = 10\,(1.1)^4 = 110.38$$

∴ The interest actually paid by the borrower in one year

$$= \text{Rs. } (110.38 - 100) = \text{Rs. } 10.38\% \text{ p.a.}$$

Here 10% p.a. compounded quarterly is the nominal rate of interest and 10.38% p.a. compounded annually is the effective rate of interest. Thus, if a rate of interest $r\%$ p.a. compounded m times a year yields the same amount after a year as the rate of interest $s\%$ p.a. compounded annually, then $r\%$ p.a. is called nominal rate of interest and $s\%$ p.a. is called effective rate of interest.

If $m = 1$, *i.e.*, if the compounding is done annually, then

$$r = s \qquad \text{...(I)}$$

If $m > 1$, *i.e.*, if the compounding is done more than once in a year, then

$$r < s \qquad \text{...(II)}$$

From (I) and (II), we get

$$r \le s$$

i.e., **nominal rate of interest ≤ effective rate of interest.**

* In the compound interest more the number of conversion periods in one year, *i.e.*, lesser the length of the conversion period, more will be the interest for a given time.

For example, if a moneylender lends Rs. 100 to some body at 10% p.a. interest compounded annually the amount due at the end of one year = Rs. 110. If compounding is done half yearly, the amount due at the end of one year $= 100\left(1+\frac{10}{200}\right)^2$ $= 100\left(1+\frac{5}{100}\right)^2$ = Rs. 110.25 and if the compounding is done quarterly the amount due after one year = $100\left(1+\frac{10}{400}\right)^4 = 100\left(1+\frac{5}{200}\right)^4$ = Rs. 110.38.

In brief, the actual interest earned in a year on Rupee one (or Rupees 100) is called the effective rate of interest. The nominal rate of interest merely helps us to decide on proportionate basis the actual rate per conversion period at the end of which interest is convertible.

8.3 CONTINUOUS COMPOUNDING

If the interest is compounded infinite number of times in a year, the interest is said to be **compounded continuously**. If Re. J is the rate of interest per rupee per annum and the compounding is continuous then the amount due at the end of n years for a principal of Rs. P

$$= \lim_{m \to \infty} P\left[1 + \frac{J}{m}\right]^{mn}$$

$$= \lim_{m \to \infty} P\left[\left[\left[1 + \frac{J}{m}\right]^{m/J}\right]^{J}\right]^{n} \qquad [\because [a^m]^n = a^{mn}]$$

$$= \lim_{x \to 0} P\,[[(1+x)^{1/x}]^J]^n \qquad \left[\text{Putting } \frac{J}{m} = x, \text{ when } m \to \infty, \frac{J}{m} \to 0, \text{ i.e., } x \to 0\right]$$

$$= P\,[e^J]^n = Pe^{Jn} \qquad [\because \lim_{x \to 0} (1+x)^{1/x} = e;]$$

Compound interest for ***n* years** $= Pe^{Jn} - P$

$$= P\,[e^{Jn} - 1]$$

If $P = 1, n = 1$ then

Compound interest on Re. 1 for **1 year**

$$= [e^J - 1] = r.$$

Here J is the nominal rate of interest per rupee per annum compounded continuously and r is the corresponding effective rate of interest on Re. 1 per annum.

$\therefore$ Effective rate of interest per rupee per annum corresponding to a given nominal rate of interest per rupee per annum compounded continuously.

$$= [e^{\text{Nominal rate of interest}} - 1]$$

The nominal rate of interest J converted continuously and equivalent to a given effective rate r is called **force of interest**.

8.4 COMPOUND INTEREST CALCULATIONS AT VARYING RATE OF INTEREST

For a principal of Rs. P, if J_1 is the rate of interest per rupee per annum for the first year, J_2 the rate of interest per rupee per annum for 2nd year, ... and J_n the rate of interest per rupee per annum for nth year then,

Amount due after n years $= P\,(1 + J_1)\,(1 + J_2) \ldots (1 + J_n)$

Interest for n years $= P\,(1 + J_1)\,(1 + J_2) \ldots (1 + J_n) - P$

If the rate of interest is r_1 per rupee compound quarterly for n_1 years, r_2 per rupee compounded half yearly for n_2 years and r_3 per rupee compounded annually for n_3 years then

$$\text{Amount due after } (n_1 + n_2 + n_3) \text{ years} = P\left(1 + \frac{r_1}{4}\right)^{4n_1}\left(1 + \frac{r_2}{2}\right)^{2n_2}(1 + r_3)^{n_3}$$

If the rate of interest $r_1\%$ for first n_1 years and $r_2\%$ for next n_2 years and $r_3\%$ for next n_3 years then, amount due

$$= P\left(1+\frac{r_1}{100}\right)^{n_1}\left(1+\frac{r_2}{100}\right)^{n_2}\left(1+\frac{r_3}{100}\right)^{n_3}$$

SOLVED EXAMPLES

Example 1. Ram deposited a sum of Rs. 10,000 in a bank. After 2 years, he withdrew Rs. 4,000 and at the end of 5 yeas he received an amount of Rs. 7,520. Find the rate of simple interest.

(ICWA, Prel. June 1990)

Solution. Total interest received by Ram = 7,520 – 6,000 = Rs. 1,520.

Let $r\%$ p.a. be the rate of interest then simple interest on Rs. 10,000 for 2 years

$$= \frac{10{,}000 \times 2 \times r}{100} = \text{Rs. } 200r$$

In simple interest, interest does not earn interest.

$\therefore$ Principal after 2 years = 10,000 – 4,000 = Rs. 6,000.

Interest on Rs. 6,000 for 3 years $= \frac{6{,}000 \times 3 \times r}{100} = \text{Rs. } 180r.$

Total interest $= 200r + 180r = 380r = 1{,}520 \Rightarrow r = 4\%.$

Example 2. A pressure cooker is available for Rs. 250 each or Rs. 100 cash down payment followed by Rs. 165 after six months. Find the rate of interest charged under the instalment plan.

(ICWA, Prel. June 1993)

Solution. Interest on Rs. (250 – 100) = Rs. 150 for 6 months

$$= 165 - 150 = \text{Rs. } 15$$

Simple interest on Rs. 150 for one year $= 15 \times 2 = \text{Rs. } 30.$

$$\text{Interest on Rs. } 100 = \frac{30}{150} \times 100 = 20\% \text{ p.a.}$$

Example 3. A sum of Rs. 1,000 is invested for 5 years at 12% interest per year. What is the simple interest? If the same amount had been invested for the same period at 10% p.a. compound interest compounded per year, how much more interests would he get? **(ICWA, Prel. June 1987)**

Solution. Simple interest on Rs. 1,000 for 5 years at 12%

$$= \frac{1{,}000 \times 5 \times 12}{100} = \text{Rs. } 600.$$

Compound interest at 10% p.a. on Rs. 1,000 for 5 years

$$= 1{,}000\left(1+\frac{10}{100}\right)^5 - 1{,}000$$

Let $x = 1{,}000\left(1+\frac{10}{100}\right)^5 = 1{,}000 \times \left(\frac{11}{10}\right)^5 = 1{,}000\,(1.1)^5$

$$\log x = \log 1{,}000 + 5 \log 1.1 = 3 + 5 \times 0.0414 = 3.2070$$

$\Rightarrow$ $x = \text{antilog } 3.2070 = 1{,}611$

$\therefore$ Compound interest on Rs. 1,000 at 10% p.a. for 5 years

$$= 1{,}611 - 1{,}000 = 611$$

$$\text{Difference} = 611 - 600 = \text{Rs. } 11$$

Example 4. Find the principal that yields a simple interest of Rs. 40 or a compound interest of Rs. 41 at the same rate of interest per annum in 2 years.

Solution. Let Rs. P be the principal and $r\%$ p.a. the rate of interest, then

$$\frac{P \times 2 \times r}{100} = 40 \Rightarrow Pr = 2{,}000 \qquad \ldots\text{(i)}$$

$$P\left(1+\frac{r}{100}\right)^2 - P = 41 \Rightarrow P\left[1+\frac{r^2}{1{,}0000}+\frac{2r}{100}\right] - P = 41$$

$$\Rightarrow \quad P\left[\left[1+\frac{r^2}{10{,}000}+\frac{2r}{100}\right]-1\right] = 41$$

$$\Rightarrow \quad P\left[\frac{r^2}{10{,}000}+\frac{2r}{100}\right] = 41$$

$$\Rightarrow \quad Pr\left[\frac{r}{10{,}000}+\frac{2}{100}\right] = 41$$

$$\Rightarrow \quad 2{,}000\left[\frac{r}{10{,}000}+\frac{2}{100}\right] = 41 \qquad [\because Pr = 2{,}000 \text{ by (i)}]$$

$$\Rightarrow \quad \frac{r}{10{,}000}+\frac{2}{100} = \frac{41}{2{,}000}$$

$$\Rightarrow \quad \frac{r}{10{,}000} = \frac{41}{2{,}000} - \frac{2}{100} = \frac{1}{2{,}000} \Rightarrow r = \frac{10{,}000}{2{,}000} = 5\% \text{ p.a.}$$

Example 5. Mr. Ram borrowed Rs. 25,000 from a moneylender but he could not repay any amount in a period of 5 years. Accordingly the moneylender demands now Rs. 35,880 from him. At what rate per cent per annum compound interest did the latter lend his money. **(ICWA, June 1987)**

Solution. If $r\%$ p.a. is the rate at which the lender lends his money, then

$$35{,}880 = 25{,}000\left(1+\frac{r}{100}\right)^5.$$

Taking log of both sides, we get

$$\log 35{,}880 = \log 25{,}000 + 5 \log\left(1+\frac{r}{100}\right)$$

$$\Rightarrow \quad 4.55486 - 4.39794 = 5 \log\left(1+\frac{r}{100}\right)$$

i.e.,

$$0.15692 = 5 \log\left(1+\frac{r}{100}\right)$$

$$\Rightarrow \quad \log\left(1+\frac{r}{100}\right) = \frac{0.15692}{5} = 0.31384$$

$$\Rightarrow \quad 1+\frac{r}{100} = 1.0750 \Rightarrow r = 7.5\% \qquad [\text{Anti log } 0.31384 = 1.0750]$$

Example 6. Mr. Brown was given the choice of two payment plans on a piece of property. He may pay Rs. 10,000 at the end of 4 years or Rs. 12,000 at the end of 9 years. Assuming money can be invested annually at 4% per year converted annually, what plan Mr. Brown should choose.

Solution. Time difference between the payment plans = 9 – 4 = 5 years. Rs. 10,000 if invested at 4% p.a. will, at the end of 5 years, become

$$= \text{Rs. } 10{,}000 \times \left(1 + \frac{4}{100}\right)^5$$

$$= \text{Rs. } 10{,}000 \times (1.04)^5 = \text{Rs. } 12{,}166.5. \qquad \ldots\text{(A)}$$

Since Rs. 12,000, *i.e.*, the payment due at the end of 9 years is less than (A), Mr. Brown should choose second plan.

Example 7. A man wishes to borrow Rs. D. He goes to the moneylender and is told that the interest rate is $r\%$ p.a. payable in advance and that Rs. D are to be paid back at the end of one year. Show that the effective rate of interest is $\frac{100r}{100-r}\%$.

Solution. Sum actually received by the man

$$= \left[D - D\left(\frac{r}{100}\right)\right] = \text{Rs. } D\left(1 - \frac{r}{100}\right).$$

If $r'\%$ is the effective rate of interest, then

$$D\left(1 - \frac{r}{100}\right) + D\left(1 - \frac{r}{100}\right)\frac{r'}{100} = D$$

$$\Rightarrow \qquad \left(1 - \frac{r}{100}\right)\frac{r'}{100} = \frac{r}{100} \Rightarrow r' = \frac{100r}{100-r}.$$

Example 8. A lends B Rs. 320. B is to pay interest on whatever amount he has not paid back at the rate of 5% per annum for the first year, 6% of the second year and 7% for the third year. B pays Rs. 100 at the end of first year, Rs. 100 at the end of the second year and enough to pay off completely the debt and the interest at the end of the third year. How much is the last payment.

Solution. Amount due at the end of first year = Rs. $320\left(1 + \frac{r}{100}\right)$ = Rs. 336

Sum paid at the end of first year = Rs. 100

Balance at the end of first year = Rs. (336 – 100) = Rs. 236

Amount due at the end of second year = Rs. $236\left(1 + \frac{6}{100}\right)$ = Rs. 250.16.

Sum paid at the end of second year = Rs. 100

Balance at the end of second year = Rs. (250.16 – 100) = Rs. 150.16

Amount due at the end of third year = Rs. $150.18\left(1 + \frac{7}{100}\right)$

= Rs. 160.67.

Example 9. A man wants to invest Rs. 5,000 for 4 years. He may invest the amount at 10% per annum compound interest, accruing at the end of each quarter of the year or he may invest it at 10.5% per annum compound interest accruing at the end of each year, which investment will give him slightly better return.

Solution. Nominal rate of interest = 10% p.a. accruing at the end of each quarter.

$$\text{Effective rate of interest} = 100\left(1 + \frac{10}{400}\right)^4 - 100 = 1{,}000 \times \left(\frac{41}{40}\right)^4 - 100 = 10.38\% \text{ p.a.}$$

In the second case the effective rate of interest is 10.5% p.a. therefore, the second investment will give better return.

Example 10. An amount of Rs. 820 falls due in 6 months. The creditor agreed to accept immediate payment of half and to defer the payment of the other half for a year. What will be the gain to the creditor by this arrangement if the money is worth 5% p.a.

Solution. Rs. $\frac{820}{2}$ = Rs. 410 paid immediately will after 6 months become

$$= 410 \times \left(1 + \frac{5}{200}\right) = 410 \times 1.025 = \text{Rs. } 420.25.$$

Rs. 410 due after one year will after six months from now become

$$= \frac{410}{\left[1 + \frac{5}{200}\right]} = \frac{410}{1.025} = \text{Rs. } 400.$$

Net gain to the creditor by the new arrangement

$$= \text{Rs. } [(420.25 + 400) - 820 = \text{Re. } 0.25].$$

Example 11. Calculate the nominal rate of interest convertible half-yearly when the effective rate is 6% p.a. **(ICWA, 1985)**

Solution. Let the nominal rate be x% p.a., then

$$100\left(1 + \frac{x}{200}\right)^2 - 100 = 6 \Rightarrow 100\left(1 + \frac{x}{200}\right)^2 = 106$$

$$\Rightarrow \quad \left(1 + \frac{x}{200}\right)^2 = 1.06 \Rightarrow 1 + \frac{x}{200} = \sqrt{1.06} = 1.0296$$

$$\Rightarrow \quad \frac{x}{200} = 0.0296 \Rightarrow x = 5.91\% \text{ p.a.}$$

Example 12. (i) Mr. X took a loan of Rs. 2,000 for 6 months. Lender deducts Rs. 200 as interest while lending. Find the effective rate of interest charged by lender.

(ii) Mr. Y has two investment options—either at 10% per annum compounded semi-annually or 9.5% per annum compounded continuously. Which option is preferable and why?

(D.U., B.Com. (Hons.) 2005C)

Solution. (i) Amount of loan = Rs. 2,000

Amount deducted as interest while lending = Rs. 200

∴ We may treat Rs. 200 as interest on Rs. 1,800 for 6 months

∴ Half yearly rate of interest $= \frac{200}{1800} = \frac{1}{9}$

Let r_e denote the equivalent effective rate of interest

$$\therefore\ r_e = (1 + i)^2 - 1 = \left(\frac{10}{9}\right)^2 - 1 = 0.23456 \text{ approx.}$$

$\therefore\ r_e = 23.46\%$ (approx.)

(ii) r_e corresponding to 10% per annum if the compounding is done semi-annually

$$= \left(1 + \frac{10}{100 \times 2}\right)^2 - 1 = 0.1025 = 10.25\%$$

r_e corresponding to 9.5% per annum if the compounding is done continuously

$$= e^{0.095} - 1 = 0.0996 = 9.96\%$$

Hence the first investment is preferable.

Example 13. What is the effective rate of interest if the nominal rate is 5% p.a. and is convertible quarterly?

Solution. If the nominal rate is 5% convertible quarterly, then the effective rate of interest is

$$100\left(1 + \frac{5}{400}\right)^4 - 100 = 100\,[1.0509] - 100 = 5.09\%.$$

Example 14. Find for each of the following, the amount to which Rs. 100 will accumulate :

(i) at the rate of interest 12% per annum compounded quarterly for 10 years.
(ii) at the force of interest 3% per annum for 3.5 years.
(iii) at the effective rates of interest 3% per annum for 10 years, 4% per annum for 4 years and 5% per annum for 2 years.
(iv) at the rate of interest corresponding to 3% per annum effective rate of discount for 8 years.

(D.U., B.Com. (Hons.) 2004)

Solution. (i) $P =$ Rs. 100, $r = 12\%$. Compounded quarterly for 10 years

$$\therefore \quad A = 100\left(1 + \frac{12}{100 \times 4}\right)^{10 \times 4}$$

Hence amount $= 100\,(1.03)^{40} =$ Rs. 326.20

(ii) $A = 100\, e^{0.03\,(3.5)} =$ Rs. 111 (approx.)

(iii) $A = 100\,(1.03)^{10}\,(1.04)^4\,(1.05)^2 =$ Rs. 173.30 approx.

(iv) Suppose r is the rate of interest corresponding to an effective rate of discount d.

$$\therefore \quad (1 + r)^{-1} = 1 - d \;\Rightarrow\; 1 + r = \frac{1}{1 - d}$$

$$\therefore \quad A = 100\,(1 + r)^9 = 100\left(\frac{1}{1 - 0.l03}\right)^8$$

$$= 100\left(\frac{1}{0.97}\right)^8 = \text{Rs. } 127.60 \text{ approx.}$$

Example 15. Mr. X deposited Rs. 10,000 in a bank for 3 years offering interest at the rate of 6% compounded half-yearly during first year, at the rate of 12% compounded quarterly during second year and at 10% compounded continuously during 3rd year. Find his balance after 3 years.

(D.U., B.Com. (Hons.) 1999)

Solution. Interest per conversion period during first year $= \frac{6}{2} = 3\%$.

Number of conversion periods in the first year = 2.

Interest per conversion period in the 2nd year $= \frac{12}{4} = 3\%$.

Number of conversion periods in the 2nd year.

Amount due at the end of two years

$$= 10{,}000\left(1+\frac{3}{100}\right)^2\left(1+\frac{3}{100}\right)^4$$
$$= 10{,}000\,(1.03)^2\,(1.03)^4 = 10{,}000\,(1.03)^6$$
$$= 10{,}000 \times 1.194052$$
$$= \text{Rs. } 11{,}940.52$$

Calculations for $(1.03)^6$

Let $y = (1.03)^6$

$$\log y = 6 \log 1.03 = 6 \times 0.0128372$$
$$= 0.0770233$$
$$y = \text{antilog } 0.0770233$$
$$= 1.19405$$

In the third year the compounding is done continuously, therefore, the amount due at the end of 3rd year

$= Pe^{Jn}$ where J is the rate of interest per rupee, per annum and n the number of years.

$= 11{,}940 \times e^{0.1}$ [∵ 10% p.a. is equal to 0.1 per rupee p.a.]

Rs. 13,195.7 = Rs. 13,196

Example 16. A person deposited Rs. 4,000 in a bank at 6% compounded continuously. After 3 years, the rate of interest was increased to 7% and after 5 more years, the rate was further increased to 8%. The money was withdrawn at the end of 10 years. Find the amount. **(D.U., B.Com. (Hons.) 2005R)**

Solution. Amount due $= 4000\, e^{0.06 \times 3} \times e^{0.07 \times 5} \times e^{0.08 \times 2}$

$$= 4000 \times e^{0.18} \times e^{0.35} \times e^{0.16}$$
$$= 4000\, e^{0.18 + 0.35 + 0.16}$$
$$= 4000\, e^{0.69} = 4000 \times 1.9937$$
$$= \text{Rs. } 7974.8$$

Example 17. A moneylender charges 'interest' at the rate of 10 paise per rupee per month, payable in advance. What effective rate of interest does he charge per annum? **(D.U., B.Com. (Hons.) 1996)**

Solution. Since the moneylender charges interest at the rate of 10 paise per rupee per month payable in advance, therefore, 10 paise may be treated as interest on 90 paise for one month. Hence actual interest rate per rupee per month is $i = \frac{0.1}{0.9} = \frac{1}{9}$. Thus if r is the effective rate of interest per rupee per annum, then r is given by

$$r = (1+i)^{12} - 1$$
$$= \left(1+\frac{1}{9}\right)^{12} - 1 = \left(\frac{10}{9}\right)^{12} - 1$$

i.e.,
$$1 + r = \left(\frac{10}{9}\right)^{12}.$$

Taking logarithms of both sides, we get

$$\log(1+r) = 12(\log 10 - \log 9) = 12(1 - 0.9542)$$

$$= 12(0.0458) = 0.4596$$

Hence $\qquad 1 + r = \text{Antilog}(0.5496) = 3.545$

$\Rightarrow \qquad r = 2.545$ or 254.5%.

Note : Also refer Ex. 7. Page 8.7.

Example 18. A sum of Rs. 2,000 is invested at a rate of interest of 5% p.a. After 7 years, the rate of interest was changed to 5% p.a. convertible half-yearly. After a further period of 3 years, the rate was again changed to 6% p.a. convertible continuously. What is the accumulated value at the end of 15 years from commencment?

Solution. Number of conversion periods in first 7 years = 7

Rate of interest per conversion period = 5%

Number of conversion periods in next 3 years = 6

Rate of interest per conversion period $= \frac{5}{2} = 2.5\%$.

Accumulated value at the end of 15 years

$$= \text{Rs. } 2{,}000\,(1.05)^7\,(1.025)^6\,e^{0.06 \times 5}$$

$$= \text{Rs. } 2{,}000 \times 1.407 \times 1.16 \times 1.35^*$$

$$= \text{Rs. } 4{,}406.7$$

Example 19. A has taken a loan of Rs. 2,000 at a rate of interest 4% p.a. payable half-yearly. He repaid Rs. 400 after 2 years, Rs. 600 after a further period of 2 years and cleared all outstanding dues at the end of 7 years from the commencement of the transaction. What is the final payment made by him?

Solution. Let the amount paid at the end of 7 years be Rs. x.

Rs. 2,000 at the end of 7 years from now at 4% p.a. payable half-yearly will become Rs. 2,000 $(1.02)^{14}$. Rs. 400 paid at the end of two yearly will, at the end of seven years at 4% p.a. compounded half-yearly will become Rs. 400 $(1.02)^{10}$.

Rs. 600 paid after 4 years will at the end of 7 years at 4% p.a. compounded half-yearly will become 600 $(1.02)^6$, then

$$400(1.02)^{10} + 600(1.02)^6 + x = 2{,}000(1.02)^{14}$$

$$400 \times 1.219 + 600 \times 1.126 + x = 1.319 \times 2{,}000 = 2{,}638$$

$$\Rightarrow \qquad 487.6 + 675.6 + x = 2{,}638$$

$$\Rightarrow \qquad x = \text{Rs. } 1{,}474.8.$$

Example 20. If the annual increase in the population of a state is 25 per thousand. The present number of inhabitants is 2,62,400. What will be the population in 3 years time ? What was it one year ago ?

(D.U., B.Com. (Hons.) 2007E)

* Calculation for $e^{0.06 \times 5} = e^{0.3}$. The value can be seen directly from the tables or calculate as under

let $\qquad y = e^{0.3}$

$$\log y = 0.3 \log e = 0.3 \log 2.71813$$

Solution. P = Present number of inhabitants + 2,62,400

r = rate of increase = 0.25

$\therefore$ Population after 3 years $= P(1+r)^3 = 2{,}62{,}400\,(1.25)^3 = 2{,}82{,}576$

Population one year ago $= P(1+r)^{-1} = 2{,}62{,}400\,(1.25)^{-1}$

$$= \frac{2{,}62{,}400}{1.25} = 2{,}56{,}000$$

Example 21. Three years ago, A, bought industrial share for Rs. 50 and has received on it three yearly dividends of Rs. 5 each, the first one having been paid one year after purchase. He wants to sell the share now and realise a rate of interest of 6% p.a. on the transaction. For how much should he sell the shares ?

Solution. Rs. 5 received at the end of one year will become Rs. $5\,(1.06)^2$ at the end of three years.

Rs. 5 received at the end of two years will become Rs. 5 (1.06) at the end of three years.

Rs. 5 received at the end of three years will remain Rs. 5.

Total dividend at the end of three years

$$= 5\,(1.06)^2 + 5\,(1.06) + 5$$
$$= 5.618 + 5.3 + 5$$
$$= 15.918$$

If interest at 6% p.a. compounded annually is considered, then Rs. 50 at the end of three years will become

$$\text{Rs. } 50\,(1+0.06)^3 = 50\,(1.06)^3 = \text{Rs. } 55.55.$$

He should sell the share for

$$\text{Rs. } (55.55 - 15.918) = \text{Rs. } 46.63.$$

8.5 PRESENT VALUE

The present value (P) of an amount (S) due at a future date is the equivalent amount payable now by considering a specified rate of interest and the time after which the future date falls.*

If Rs. P is the principal [or sum invested now], i the rate of interest per rupee per annum compounded annually then the amount due.

(i) After one year $= P(1+i) = S_1$

(ii) After two years $= P(1+i)^2 = S_2$

(iii) After three years $= P(1+i)^3 = S_3$

....

....

(n) After n years $= P(1+i)^n = S_n$

From the first equation $P = \dfrac{S_1}{1+i} = S_1\,(1+i)^{-1}$

From the second equation $P = \dfrac{S_2}{(1+i)^2} = S_2\,(1+i)^{-2}$ and from the last equation

* The present value of a given sum of money due at the end of a certain period of time is that sum which if invested now at the given rate of interest accumulates to the given sum at the end of the period.

$$P = \frac{S_n}{(1+i)^n} = S_n (1+i)^{-n}.$$

Here Rs. P is the present value of

(i) Rs. S_1 due at the end of 1 year $= \frac{S_1}{1+i} = S_1 (1+i)^{-1}$

(ii) Rs. S_2 due at the end of 2 years $= \frac{S_2}{(1+i)^2} = S_2 (1+i)^{-2}$

....

....

(iii) Rs. S_n due at the end of n years $= \frac{S_n}{(1+i)^n} = S_n (1+i)^{-n}$.

Clearly the present value of a sum due at some future date depends on the rate of interest and the time difference between the present date and the future date under consideration.

If the rate of interest is i per rupee per conversion period then the present value (P) of Rs. S due after m conversion periods

$$= \frac{S}{(1+i)^m} = S (1+i)^{-m}.$$

If V^n denotes the present value of Re. 1 due after n conversion periods then

$$V^n = (1+i)^{-n}.$$

8.6 INTEREST AND DISCOUNT

Interest is calculated on the present value whereas discount is calculated on the future value **(Pl. note)**. When the value of an obligation is known at some future date, the process of finding its value at some earlier date is called **discounting**. The *rate of discount* is the discount on one unit of money payable one period hence.

If an amount of Rs. S due after one period is to be discounted today at the rate of discount Re. d per rupee per period, then the discounted value today

$$= S - Sd = S(1-d).$$

The discounted value two periods before the maturity date $= S(1-d)^2$. Similarly the discounted value of Rs. S due after n periods $= S(1-d)^n$. Discounted value is also called the **present value** under the condition that the interest is paid in advance, *e.g.*, if a man goes to a moneylender and requests for a loan of Rs. 100 and the moneylender says that he will charge an interest at the rate of 10% p.a. payable in advance, then for Rs. 100 the man will actually receive Rs. 90 and will pay Rs.100 at the end of the year. The actual rate of interest paid by the man becomes $\frac{10}{90} \times 100\%$ p.a., *i.e.*, $11\frac{1}{9}\%$ p.a.

Thus corresponding to 10% p.a. rate of discount, the actual rate of interest is $11\frac{1}{9}\%$ p.a. Businessmen get the 'hundi' discounted through the bank before the date of maturity.

The face value of the hundi = [Actual amount to be paid + Interest on the actual amount for the period after which the hundi becomes payable]

Bank will charge interest on the face value while discounting before due date thereby gaining a sum equal to the interest on the [interest on the actual amount] for the period between the date of discounting and the due date. This is **Banker's gain**.

8.7 NOMINAL RATE OF DISCOUNT, EFFECTIVE RATE OF DISCOUNT, FORCE OF DISCOUNT

We have already discussed that if the interest is compounded more man once in a year, then the actual rate of interest per annum earned by the moneylender [or the actual rate of interest per annum paid by the borrower] is the effective rate of interest. Similarly if the discount is converted more than once in a year then the stated rate of discount is called the **nominal rate** of discount and the discount actually received in one year is called the **effective rate** of discount.

Students should note that the effective rate of interest is higher than the nominal rate of interest whereas the effective rate of discount is lower than the nominal rate of discount. **(Important)**

If the nominal rate of discount is 10% p.a. converted quarterly, then the actual rate of discount per year

$$= 100 - 100\left(1 - \frac{10}{400}\right)^4 = 100 - 100\left(\frac{39}{40}\right)^4$$
$$= 100 - [1 - 0.903687] = 9.63\%.$$

If d is the effective rate of discount per rupee per annum and J is the corresponding nominal rate of discount **per rupee** per annum converted m times in a year then

$$1 - d = \left[1 - \frac{J}{m}\right]^m.$$

If the conversion is continuous, *i.e.*, if the nominal rate of discount is converted continuously then

$$1 - d = \lim_{m \to \infty}\left(1 - \frac{J}{m}\right)^m = \lim_{m \to \infty}\left[\left(1 - \frac{J}{m}\right)^{m/J}\right]^J$$
$$= \lim_{x \to 0}\,[(1 - x)^{1/x}]^J \qquad \left[\text{Put } \frac{J}{m} = x \text{ when } m \to \infty, \frac{J}{m} \to 0\right]$$
$$= \lim_{x \to 0}\,[[1 + (-x)^{(-1/x)}]]^{-J} = e^{-J}$$

$$\therefore \qquad d = 1 - e^{-J}$$

Here J is called the **force of discount**.

*The **force of discount** corresponding to a given effective rate of discount d is that nominal rate of discount J which if converted continuously yields d as the effective rate of discount.*

Similarly we can define the force of interest as :

*The **force of interest** corresponding to a given effective rate w is that nominal rate J which if compounded continuously yields w as the effective rate.*

If compounding is continuous then for a principal of Rs. P at **nominal rate of interest** J, the amount due (S) after n years is given by

$$S = Pe^{Jn} \Rightarrow P = Se^{-Jn}$$

i.e., the **present value** of Rs. S due after n years at rate of interest of Re. J per rupee per annum compounded continuously

$$= Se^{-Jn}.$$

8.8 EQUATION OF VALUE

An equation equating the sum of the values on a given date of one set of obligations with the sum of the values on the same date of another set of obligations is called **Equation of Value**.

If a sum C is due at a particular time, then its value at any time before due date is less than C and its value at any time after due date is more than C.

While forming equation of value, any convenient date is chosen and sum of the values of all the payments in one set of obligations is calculated on this date and equated with the sum of the values on this date of the other set of obligations.

[Accumulated value or amount due is calculated as on the chosen date for the payments due before this date and equated with the present value, *i.e.*, the value on the chosen date for the payments due after this date, rate of interest remaining same in both the cases].

By using equation of value it is possible to reduce a series of amounts payable at different times to a single amount payable at one moment of time.

In financial transaction, one set of payments may be made to secure a second set of benefits. The combined value of the payments can be found as at a particular moment and similarly the combined value of the cost of benefits can also be ascertained as at the same moment. Now assuming that there is no profit or loss, the value of payments made must be equal to the value of benefits received. This statement of the equality of payment and benefits is also known as **equation of value.**

Example 22. Calculate the market equilibrium value for an acre of land yielding Rs. 100 per annum after all expenses, indefinitely into the future. The market rate of interest is 10% p.a.

(D.U., B.A. Eco. (Hons.) 1997)

Solution. Present value of an asset = Market equilibrium Price

$$= \frac{100}{0.1} = \text{Rs. } 1{,}000.$$

$$\left[\text{Present value of perpetuity} = \frac{\text{Annual constant amount}}{\text{Rate of interest per rupee per year}}\right]$$

Example 23. How can $\frac{a}{r}$ be taken as the present value of an income stream of Rs. a per annum for ever when interest at $100r$ per cent is compounded yearly ? **(D.U., B.A. Eco. (Hons.) 1996)**

Solution. Sum of the present values

$$= \frac{a}{1+r} + \frac{a}{(1+r)^2} + \ldots$$

$$= \frac{a}{1+r}\left[1 + \frac{1}{1+r} + \frac{1}{(1+r)^2} + \ldots\right]$$

$$= \frac{a}{1+r}\left[\frac{1}{1 - \frac{1}{1+r}}\right] = \frac{a}{r}. \qquad \left[\because \text{Sum of infinite G.P.} = \frac{a}{1-r}\right]$$

Example 24. Suppose that interest rate r is such that the present value of receiving Rs. V_2 in t_2 years from now is the same as receiving Rs. V_1 in t_1 years from now it is given that $t_2 > t_1$.

Assuming interest is compounded annually

(a) Show that $V_2 > V_1$

(b) Show that present value of receiving Rs. V_2 $(t_2 + k)$ years from now is equal to the present value of receiving Rs. V_1 $(t_1 + k)$ years from now for any value of k.

(D.U., B.A. Eco. (Hons.) 2006)

Solution. In this case V_1 and V_2 are the amounts due at the end of t_1 and t_2 years respectively. Then

$$V_2 = \text{Principal}\left(1+\frac{r}{100}\right)^{t_2} \Rightarrow \text{Principal}^* = \frac{V_2}{\left(1+\frac{r}{100}\right)^{t_2}}$$

$$V_1 = \text{Principal}\left(1+\frac{r}{100}\right)^{t_1} \Rightarrow \text{Principal} = \frac{V_1}{\left(1+\frac{r}{100}\right)^{t_1}}$$

As per the question $$\frac{V_2}{\left(1+\frac{r}{100}\right)^{t_2}} = \frac{V_1}{\left(1+\frac{r}{100}\right)^{t_1}} = k \text{ (say)}$$

$$V_2 = \left(1+\frac{r}{100}\right)^{t_2} k,\ V_1 = \left(1+\frac{r}{100}\right)^{t_1} k$$

$$\left(1+\frac{r}{100}\right)^{t_2} > \left(1+\frac{r}{100}\right)^{t_1}$$

$$V_2 > V_1 \qquad \left[\text{Since } t_2 > t_1,\ \left(1+\frac{r}{100}\right)^{t_2} > \left(1+\frac{r}{100}\right)^{t_1}\right]$$

Now $$\frac{V_2}{\left(1+\frac{r}{100}\right)^{t_2}} = \frac{V_1}{\left(1+\frac{r}{100}\right)^{t_1}}$$

$\Rightarrow$ $$\frac{V_2}{\left(1+\frac{r}{100}\right)^{t_2}\left(1+\frac{r}{100}\right)^{k}} = \frac{V_1}{\left(1+\frac{r}{100}\right)^{t_1}\left(1+\frac{r}{100}\right)^{k}}$$

$\Rightarrow$ $$\frac{V_2}{\left(1+\frac{r}{100}\right)^{(t_2+k)}} = \frac{V_1}{\left(1+\frac{r}{100}\right)^{(t_1+k)}}$$

Example 25. In return of a promise to pay Rs. 500 at the end of 10 years, a person agrees to pay Rs. 100 now, Rs. 200 at the end of 6 years and a final payment at the end of 12 years. If the rate of interest is 2% per annum effective, what should the final payment be ?

(D.U., B.Com. (Hons.) 1996, G.G.I.P. Uni. BBA May 2001)

Solution. Let Rs. x be paid at the end of 12 years. We will find the value of each payment at the end of 10 years from now.

Rs. 100 paid now will amount to Rs. 100 $(1+0.02)^{10} = (1.02)^{10}$ at the end of 10 years

Rs. 200 paid at the end of 6 years will become Rs. 200 $(1+0.02)^4$

$= 200\,(1.02)^4$ at the end of 10 years from now.

Since Rs. x are paid at the end of 12 years, its value at the end of 10 years from now.

$$= x\,(1.02)^{-2}$$

$\therefore$ $$(100)\,(1.02)^{10} + 200\,(1.02)^4 + x\,(1.02)^{-2} = 500$$

$\Rightarrow$ $$100\,(1.219) + 200\,(1.08243) + 0.96x = 500$$

* Principal = Present value

$$\Rightarrow \quad 121.9 + 216.48 + 0.96x = 500$$

$$\Rightarrow \quad 0.96x = 500 - 121.9 - 216.48 = 161.62$$

$$\Rightarrow \quad x = \frac{161.62}{0.96} = \text{Rs. } 168.35.$$

Example 26. Under a marriage endowment policy for Rs. 20,000, the sum assured is payable at the end of the fixed term chosen. A claim for a policy of Rs. 20,000 has occurred 12 years before the end of the fixed term and the claimant requests for immediate payment. Presuming that the insurer entertains such requests, what is the sum payable if interest is reckoned at (i) 6% p.a. (ii) 7 p.a. and (iii) 8% p.a. ?

Solution. Here fixed term is not known. We are to find the present value of Rs. 20,000 on the date when exactly 12 years are left for maturity. If the rate of interest is 6%, then immediate payment is done just 12 years before maturity

$$\text{The Payment} = \frac{20{,}000}{\left(1 + \frac{6}{100}\right)^{12}} = \frac{20{,}000}{(1.06)^{12}} \text{ when rate of interest is 6\% p.a.}$$

$$= \frac{20{,}000}{2.012} = 20{,}000 \times 0.4970 = \text{Rs. } 9{,}940.$$

Part II and III pl. try. Ans. Rs. 8,880.20, 7,942.2.

Example 27. In lieu of a single payment of Rs, 1,000 at the present moment a person agrees to receive three equal payments at the end of 3 years, 6 years and 10 years respectively. Assuming a rate of interest of 6% p.a., what should be the value of each of the three payments ?

Solution. Let Rs. x be the value of each payment, then

$$\frac{x}{(1.06)^3} + \frac{x}{(1.06)^6} + \frac{x}{(1.06)^{10}} = 1{,}000$$

$$\frac{x}{(1.06)^3}\left[1 + \frac{1}{(1.06)^3} + \frac{1}{(1.06)^7}\right] = 1{,}000$$

$$\Rightarrow \quad x\,(0.8396)\,[1 + 0.8396 + 0.6650] \Rightarrow x\,(0.8396)\,(2.5046) = 1{,}000$$

$$\Rightarrow \quad 2.1029x = 1{,}000 \Rightarrow x = \text{Rs. } 475.54.$$

Example 28. The amount of Re. 1 in 2 years at a certain nominal rate of interest plus the present value of Re. 1 due in 2 years, at the same nominal rate of discount, both rates convertible half-yearly, is 2.00480032. Find the rate of discount. **(D.U., B.Com. (Hons.) 1996)**

Solution. Let i be the nominal rate of interest as well as discount per rupee per annum, then

$$\left[1 + \frac{i}{2}\right]^4 + \left(1 - \frac{i}{2}\right)^4 = 2.00480032$$

$$^4C_0 + {}^4C_1\left(\frac{i}{2}\right) + {}^4C_2\left(\frac{i}{2}\right)^2 + {}^4C_3\left(\frac{i}{2}\right)^3 + {}^4C_4\left(\frac{i}{2}\right)^4$$

$$+ {}^4C_0 - {}^4C_1\left(\frac{i}{2}\right) + {}^4C_2\left(\frac{i}{2}\right)^2 - {}^4C_3\left(\frac{i}{2}\right)^3 + {}^4C_4\left(\frac{i}{2}\right)^4 \; *$$

$$= 2.00480032$$

* Expanding by Binomial expansion.

$$2 + 12\left(\frac{i^2}{4}\right) + 2\left[\frac{i^4}{16}\right] = 2.00480032$$ [But rate of interest can not be negative]

$$\Rightarrow \quad 2 + 3i^2 + \frac{i^4}{8} = 2.00480032$$ Therefore $i = 4\%$

$$\Rightarrow \quad i^4 + 24i^2 + 16 = 16.0384026$$

$$\Rightarrow \quad i^4 + 24i^2 - 0.0384026 = 0$$

$$\Rightarrow \quad i^2 = \frac{-24 \pm \sqrt{576 + 0.15361024}}{2}$$

$= 0.0016$ [Rejecting –ve value as i^2 cannot be negative]

$$\Rightarrow \quad i = \pm\, 0.04$$

Example 29. The sums of Rs. 5,000. Rs. 12,000 and Rs. 16,000 are due at the ends of 2, 5 and 7 years. It is proposed to replace this series by a single sum of Rs. 32,000 payable at the end of n years. If the rate of interest is 8% p.a. compounded half yearly, determine the value of n.

(D.U., B.Com. (Hons.) 2005C)

Solution. We suppose that the focal date is taken as the current year.

We have $P =$ Rs. 32,000 $\quad r = 8\%$ $\quad$ Time $= n$ years

Since the interest is compound half yearly

$$\therefore \quad 32{,}000\,(1.04)^{-2n} = 5000\,(1.04)^{-4} + 12000\,(1.04)^{-20} + 16000\,(1.04)^{-14}$$

$$\Rightarrow \quad 32\,(1.04)^{-2n} = 5\,(1.04)^{-4} + 12\,(1.04)^{-20} + 16\,(1.04)^{-14}$$

$$= 5\,(0.8548) + 12\,(0.6755) + 16\,(0.5774) = 21.618$$

$$\therefore \quad -2n \log 1.04 = \log 21.618 - \log 32$$

$$\Rightarrow \quad (-2n)\,(0.0170) = 1.3347 - 1.5051 = -0.1704$$

$$\therefore \quad n = \frac{0.1704}{2 \times 0.0170} = 5 \text{ years}$$

Example 30. A person has purchased a bond of the face value of Rs. 100 on which interest is payable yearly at 4% p.a. He received in all three interest payments, the first one falling due one year after purchase. At the end of three yeas the bond has matured for payment at par. If the person has realized an interest yield of 5% p.a. in the transaction, what is the purchase price ?

Solution. The payments received are as under :

(i) Rs. 4 at the end of one year.

(ii) Rs. 4 at the end of two years.

(iii) Rs. (4 + 100) = 104 at the end of three years.

Present value of all the payments at 5% p.a.

$$= \frac{4}{1.05} + \frac{4}{(1.05)^2} + \frac{104}{(1.05)^3}$$

$= 3.81 + 3.628 + 89.84 =$ Rs. $97.277 =$ Purchase price.

Example 31. A has a right to receive an amount of Rs. 1,000 at the end of 12 years from now. This right has been sold to B for a present value calculated at the rate of 8% p.a. The money thus received was invested by A in a deposit account at 9% p.a. payable half-yearly. After 8 years the account had to be closed and A then invested the amount available at 6% p.a. in another bank. How much has A gained or lost in this transaction, as at the end of 12 years ?

Solution. Present value of Rs. 1,000 which is due at the end of 12 years

$$= \frac{1,000}{(1.08)^{12}} = 1,000 \times 0.397 = \text{Rs. } 397$$

With Rs. 397 as principal, the amount due at the end of 8 years at 9% p.a. compounded half-yearly

$$= 397\,(1.045)^{16} = 397 \times 2.0223$$
$$= \text{Rs. } 802.88$$

With Rs. 802.88 as the principal, the amount due at the end of 4 years at 6% p.a.

$$= \text{Rs. } 802.88 \times (1.06)^4 = \text{Rs. } 1,013.62$$
$$\text{Gain} = \text{Rs. } (1,013.62 - 1,000) = \text{Rs. } 13.62.$$

Example 32. A sum of Rs. 1,000 is due at the end of 10 years 6 months. The present interest rates are 7% p.a. but it is expected that there will be a fall in the rates after 6 years bringing down the rate to 6% p.a. Find the present value of the sum of money under these assumptions.

Solution. Present value $= \dfrac{1,000}{(1.03)} \times \dfrac{1}{(1.06)^4} \times \dfrac{1}{(1.07)^6}$

$$= 1,000 \times 0.97087 \times \frac{0.7921}{1.5} = \text{Rs. } 512.684.$$

Example 33. The sums of Rs. 2,000, Rs. 3,000 and Rs. 4,000 are due at the end of 2, 4 and 8 years respectively. It is proposed to replace this series of payments by a single sum of Rs. 9,000 payable at the end of n years. If the rate of interest is 10% per annum effective, find the value of n.

(D.U., B.Com. (Hons.) 1998, G.G.I.P. Uni. B.B.A. May 2001)

Solution. Equating the present values, we get

$$\frac{2,000}{(1.1)^2} + \frac{3,000}{(1.1)^4} + \frac{4,000}{(1.1)^8} = \frac{9,000}{(1.1)^n}$$

$$1,652.89 + 2,049 + 1,866 = \frac{9,000}{(1.1)^n}$$

$$5,567.89 = \frac{9,000}{(1.1)^n}$$

$$\Rightarrow \quad 5,567.89 \times (1.1)^n = 9,000$$

$$\Rightarrow \quad \log 5,567.89 + n \log 1.1 = \log 9,000$$

$$\Rightarrow \quad n \log 1.1 = \log 9,000 - \log 5,567.89$$

$$= 3.9542 - 3.7457 = 0.2085$$

$$\Rightarrow \quad n = \frac{0.2085}{\log 1.1} = \frac{0.2085}{0.0414}$$

$$= 5.036, \text{ years (nearly)}.$$

Example 34. A purchaser of irredeemable debenture gets a dividend of 3 every half year. What should be the purchase price in order yield an effective rate of 8% p.a.

Solution. Nominal rate of interest x% p.a. corresponding to an effective rate of 8% compounded half-yearly is given by

$$\left(1+\frac{x}{200}\right)^2 - 1 = 0.08$$

$$\Rightarrow \qquad \left(1+\frac{x}{200}\right)^2 = 1.08 \Rightarrow 1+\frac{x}{200} = 1.0392$$

$$\Rightarrow \qquad \frac{x}{200} = 0.0392 \Rightarrow x = 7.846.$$

Now we are to find that sum which while earning a nominal rate of interest of 7.846% p.a., yields Rs. 3 every six months. If Rs. P be such sum, then

$$P\left[1+\frac{7.846}{200}\right] - P = 3$$

$$P\left[\frac{7.846}{200}\right] = 3 \Rightarrow P = \frac{3\times 200}{7.846} = \text{Rs. } 76.47$$

$\therefore$ Therefore the purchase price should be Rs. 76.47.

8.9 DEPRECIATION

Depreciation is the reduction in the value of fixed assets such as buildings and equipments through normal use or age. The **depreciated** value of the asset at the end of its useful life is called **scrap value**. The difference between the original cost of the asset and its scrap value is called total **depreciation or wearing value** of the asset. Though there are many methods to calculate depreciation and depreciated value but in this chapter we will confine ourselves to depreciation on reducing balance or diminishing returns.

Let C be the original cost of the machine and $r\%$ p.a. is the rate of depreciation on reducing balance, then

Depreciation for the first year $= \dfrac{Cr}{100}$

Depreciated value after first year $= C - \dfrac{Cr}{100} = C\left(1-\dfrac{r}{100}\right)$

Depreciation for second year $= C\left(1-\dfrac{r}{100}\right)\dfrac{r}{100}$

Depreciated value at the end of second year

$$= C\left(1-\frac{r}{100}\right) - C\left(1-\frac{r}{100}\right)\frac{r}{100}$$

$$= C\left(1-\frac{r}{100}\right)\left(1-\frac{r}{100}\right)$$

$$= C\left(1-\frac{r}{100}\right)^2$$

....

....

Depreciated value at the end of n years $= C\left(1-\dfrac{r}{100}\right)^n$.

Example 35. A machine costing Rs. 80,000 would reduce to Rs. 20,000 in 8 years. Find the rate of yearly depreciation, given that depreciation is calculated using diminishing balances method.

(D.U., B.Com. (Hons.) 2000)

Solution. Let $r\%$ be the rate of yearly depreciation, then

$$80{,}000\left(1-\frac{r}{100}\right)^8 = 20{,}000 \Rightarrow \left(1-\frac{r}{100}\right)^8 = \frac{1}{4}$$

$$\Rightarrow \quad 8\log\left(1-\frac{r}{100}\right) = -\log 4 = -0.6021 \qquad [\because \log 1 = 0]$$

$$\Rightarrow \quad \log\left(1-\frac{r}{100}\right) = -\frac{0.6021}{8} = -0.07526 = \bar{1} + 0.9247$$

$$\Rightarrow \quad \left(1-\frac{r}{100}\right) = \text{antilog of } \bar{1}.9247 = 0.8412$$

$$\Rightarrow \quad \frac{r}{100} = 1 - 0.8412 = 0.1588 \Rightarrow r = 15.88\%$$

Example 36. A machine is purchased for Rs. 10,600. Depreciation is calculated at 8% per annum for the first three years and after that 10% per annum for the next seven years; depreciation being calculated on diminishing value. Find the value of the machine after a period of 10 years.

(G.G.I.P. Univ. BBA May 2001)

Solution. Value of the machine after 10 years.

$$= 10{,}600\left(1-\frac{8}{100}\right)^3\left(1-\frac{10}{100}\right)^7 = 10{,}600\left(\frac{92}{100}\right)^3\left(\frac{90}{100}\right)^7 = \text{Rs. } 3947.9$$

Example 37. A machine depreciates at the rate of 7% of its value at the beginning of a year. If the machine were purchased for Rs. 8,500 what is the minimum number of complete years at the end of which the worth of the machine will be less or equal to half of its original cost price ?

Solution. Let the number of years be n, then as per the conditions given in the question.

$$8{,}500\left(1-\frac{7}{100}\right)^n \le \frac{1}{2}(8{,}500) \Rightarrow \left(1-\frac{7}{100}\right)^n \le \frac{1}{2}$$

$$\Rightarrow \quad (0.93)^n \le \frac{1}{2} = 0.5$$

$$\therefore \quad n\log 0.93 \le \log 0.5 \text{ or } n\times\bar{1}.9685 \le \bar{1}.6990$$

$$\text{or} \quad n\times(-0.0315) \le (-0.3010) \text{ or } n\,(0.0315) \le 0.3010;$$

$$\therefore \quad n \ge \frac{0.3010}{0.0315} * = \frac{3{,}010}{315} = 9.6$$

Hence the minimum number of complete years = 10.

Example 38. The yearly output of a certain mine is found to decrease every year at the rate of 8% of its output in the previous year. The output was worth Rs. 1 crore in the first year. It would be unprofitable to run the mine when the worth of the yearly output falls below Rs. 1 million. For how many years can the mine be run profitably ?

* Change in sign of the terms on both sides of the inequality changes the sign of the inequality.

Solution. Suppose the output falls to Rs. 1 million or below in the nth year then

$$1 \text{ crore}\left(1-\frac{8}{100}\right)^n = 1 \text{ million.}$$

$$\Rightarrow \quad 1,00,00,000\left(1-\frac{8}{100}\right)^n = 10,00,000$$

$$\Rightarrow \quad \left(\frac{92}{100}\right)^n = \frac{1}{10}$$

$$\Rightarrow \quad (0.92)^n = 0.1$$

Taking log of both sides $n \log (0.92) = \log 0.1 \Rightarrow n\,(\bar{1}.9638) = -1$,

i.e., $n\,[-1 + 0.9638] = -1 \Rightarrow -n\,(0.0362) = -1$

$$\Rightarrow \quad n = \frac{1}{0.0362} = 27.6,$$

i.e., the mine will run profitably for 27 years only and the mine will run unprofitably in 28th year.

Note :

I. Note the difference in Ex. 37 and Ex. 38. In Ex. 37 we are to find the minimum number of complete years at the end of which the worth of the machine will be half or less but in Ex. 38 we are to find the year in which the mine will not run profitably.

II. In Ex. 54 we can also, say

$$1 \text{ crore}\left(1-\frac{8}{100}\right)^n \leq 1 \text{ million} \Rightarrow n \geq 27.6$$

i.e., in 28th year the output will be less than 1 million which means that the mine will run profitably for 27 years only.

Example 39. A machine depreciate each year by 10% of its value at the beginning of the year. At the end of the fourth year its value is Rs. 1,31,200. Find its original value.

Given log 1312 = 3.1179, log 90 = 1.9542, antilog 5.3011 = 20,000. **(D.U., B.Com., 1982)**

Solution. If Rs. x is the original value, then the value of the machine at end of 4th year

$$= x\left(1-\frac{10}{100}\right)^4 = x \times \left(\frac{90}{100}\right)^4 = 1,31,200$$

Taking log of both sides, we get

$$\log x + 4\,[\log 90 - \log 100] = \log 1,31,200$$

$$\Rightarrow \quad \log x + 4\,[1.9542 - 2] = 5.1179 \Rightarrow \log x = 5.3011$$

$$\Rightarrow \quad x = \text{antilog } 5.3011 = \text{Rs. } 2,00,000.$$

Example 40. A machine, costing Rs. 20,000, is sold for Rs. 5,000 down and the balance payable in semi annual instalments in the next five years. What is this instalment if interest is :

(i) 4% compounded semi annually :

(ii) 4% compounded annually? **(D.U., B.Com. (Hons.) 2004)**

Solution. (i) Balance = 20,000 – 5,000 = 15,000

Let Rs. P is the equal annual instalment

Then $$15,000 = \frac{P}{1.02} + \frac{P}{(1.02)^2} + \ldots \frac{P}{(1.02)}$$

$$= \frac{P}{1.02}\left(1 + \frac{1}{1.02} + \frac{1}{(1.02)^2} + \dots \frac{1}{(1.02)^9}\right)$$

$$= \frac{P}{1.02}\left[\frac{1 - \left(\frac{1}{1.02}\right)^{10}}{1 - \frac{1}{1.02}}\right] = 8.98258\,P$$

$$P = \frac{15{,}000}{8.98258} = \text{Rs. } 1669.8$$

(ii) In this case 4% is the effective rate of interest

If $r\%$ is the semi annual rate of compounding

Then $$100\left(1 + \frac{r}{100}\right)^2 = 104 \Rightarrow r = 1.98\%$$

$\therefore$ $$15{,}000 = \frac{P}{1.0198}\left[\frac{1 - \left(\frac{1}{1.0198}\right)^{10}}{1 - \frac{1}{1.0198}}\right]$$

$$= \frac{P}{1.0198}(0.1780) = 8.99p$$

$\Rightarrow$ $$P = \frac{15000}{8.99} = 1668.52$$

Interest is compounded annually.

Example 41. A debt of Rs. 3,000 which is due 6 years from now, is instead to be paid off by three payments Rs. 500 now, Rs. 1,500 in 3 years and a final payment of Rs. 475 at the end of n years. If the rate of interest is 6% per annum effective, find the value of n. **(D.U., B.Com. (Hons.) 2005R)**

Solution. Present value of Rs. 3,000 due after 6 years

$$= \frac{3{,}000}{(1.06)^6} = \frac{3{,}000}{1.4185} = 3{,}000 \times 0.7049 = \text{Rs. } 2114.7$$

Let Rs. 475 be paid after n years

Thus $$2114.7 = 500 + \frac{1{,}500}{(1.06)^3} + \frac{475}{(1.06)^n}$$

$$= 500 + 1259.4 + \frac{475}{(1.06)^n}$$

$\Rightarrow$ $$\frac{475}{(1.06)^n} = 2114.7 - 500 - 1259.4 = 355.3$$

$\Rightarrow$ $$355.3\,(1.06)^n = 475 \Rightarrow (1.06)^n = \frac{475}{355.3} = 1.336898$$

$\Rightarrow$ $$n \log 1.06 = \log 1.336898$$

$\Rightarrow$ $$n\,(0.0253) = 0.1261$$

$$\Rightarrow \qquad n = \frac{0.1261}{0.0253} = 4.98 = 5 \text{ approx.}$$

$$= 5 \text{ years approx.}$$

Example 42. The value of a machine depreciates 11% annually. If the present value is Rs. 38,440, find its value three years ago. **(D.U., B.Com. (Hons.) 2007E)**

Solution. Present = Depreciated value = Rs. 38,440

Original cost = Value of the machine three years ago = x

$d = 0.11$ Time = 3 years

$$\therefore \qquad 38{,}440 = x\,(1 - 0.11)^3$$

$$\Rightarrow \qquad x = \frac{38{,}440}{(1-0.11)^3} = \text{Rs. } 54{,}525 \text{ (approx.)}$$

Example 43. A man borrows Rs. 750 from a moneylender and the bill is renewed after every half-year at an increase of 21%. After how much time the amount of bill will become Rs. 7,500 ?

(D.U., B.Com. (Hons.) 2007(R))

Solution. We are given that

P = Rs. 750; R = 21% after every six months; A = Rs. 7,500

Suppose time = t years

We know that : $A = P\,(1 + R)^{2t}$

$$\therefore \qquad 7{,}500 = 750\,(1.21)^{2t}$$

$$\Rightarrow \qquad 10 = (1.21)^{2t} \;\Rightarrow\; \log 10 = 2t \log (1.21)$$

$\therefore\ t = \dfrac{\log 10}{2 \log (1.21)} = 6.04$ approx. (Using $\log 10 = 1$; $\log 1.21 = 0.828$)

$\therefore$ Approx. time = 6.04 years

Example 44. A bond with a face value of Rs. 5,000 matures is 12 years. The nominal rate of interest on bond is 12 p.a. paid annually. What should be the price of the bond so as to yield an effective rate of return equal to 10% per annum ? **(D.U., B.Com. (Hons.) 2007(R))**

Solution. Face value of the bond = Rs. 5,000

Price of bond = P.V. of $\left(\dfrac{12}{100} \times 500\right)$ for 12 years + P.V. of Rs. 5,000

$$= 600\,(a_{12\,!\,0.10}) + 5{,}000\,(1 + 0.10)^{-R}$$

$$= 600 \times 6.814 + 5{,}000 \times 0.319$$

$$= 4088.40 + 1595 = \text{Rs. } 5683.40$$

Example 45. Mr. X wants to make an investment of Rs. 50,000 for six years. He has two alternatives. First alternative fetches him a return of 8% compounded annually and the second alternative fetches him a return of 7.5% compounded semi-annually. Which investment should be go for ?

(D.U., B.Com. (Hons.) 2007(E))

Solution. $P = 50{,}000$, $n = 6$

First Alternative, $r = 0.08$

Amount after six years $= 50{,}000\,(1 + 0.08)^6 = 79320$

Second Alternative, $r = 0.075$ semiannually; $n = 6 \times 2 = 12$

$\therefore$ Amount after six years $= 50{,}000\,(1 + 0.075)^{12} = 77380$

So, he should go for the first alternative.

Example 46. A man requires Rs. 20,00,000 to purchase a house after 5 years. He has an opportunity to invest the funds in an account which can earn 6% p.a. Find how much must be deposited in each quarter so as to have the required amount at the end of 5 years. **(D.U., B.Com. (Hons.) 2006)**

Solution. Suppose the size of each payment deposited at the end of 3 months $= R$

$$\therefore \qquad 20{,}00{,}000 = RS_{n\,!\,i}$$

where $n = 5 \times 4 = 20,\ i = \dfrac{6}{100 \times 4} = 0.015$

We know that $\qquad S_{20\,!\,0.015} = 23.123667$

$$\therefore \qquad R = \frac{20{,}00{,}000}{23.123667} = \text{Rs. } 86491.50 \text{ (approx.)}$$

Example 47. State the relationship between rate of interest and rate of discount both in case of simple and compound interest. Also determine the amount of Rs. 1,000 at the rate of interest corresponding to 5% rate of discounted annually after 5 years. **(D.U., B.Com. (Hons.) 2005C)**

Solution. ***Case of simple interest*** **:** Relationship between rate of interest and rate of discount is given by

$$\frac{A}{1 + rt} = A\,(1 - dt) \qquad \therefore\ d = \frac{r}{1 + rt} \qquad \ldots\text{(i)}$$

Case of compound interest,

$$\frac{A}{(1 + r)^t} = A\,(1 - d)^t \qquad \therefore\ d = \frac{r}{1 + r} \qquad \ldots\text{(ii)}$$

$$\Rightarrow \qquad \frac{5}{100} = \frac{r}{1 + r} \qquad \therefore\ r = \frac{1}{19} \text{ (rate of discount)}$$

$\therefore$ The amount of Rs. 1,000 at $r = \dfrac{1}{19}$ after 5 years

$$1{,}000\,(1 + r)^5 = 1{,}000 \left(\frac{20}{19}\right)^5 = \text{Rs. } 1292.35 \text{ (approx.)}$$

8.10 USE OF CALCULUS IN CALCULATING AMOUNT AND PRESENT VALUES OF ANNUITIES

In the case of continuous compounding, the amount S of an annuity (continuous stream) in which R rupees are paid each year for N years at the rate of interest $r\%$ per annum compounded continuously is given by

$$S = \int_0^N \text{Re}^{(r/100)\,t} = \int_0^N \text{Re}^{it}\,dt$$

Example 48. A bank pays interest @ 6% per annum compounded continuously. If a person places Rs. 1,000 in the bank each year, how much will be in his account after 3 years ?

(D.U., B.Com. (Hons.) 1989)

Solution. The amount S in the account after 3 years is given by

$$S = \int_0^N \text{Re}^{rt}\, dt$$

$$= \int_0^3 1{,}000\, e^{0.06t}\, dt$$

$$= \frac{1{,}000}{0.06} \left| e^{0.06t} \right|_0^3 = \frac{50{,}000}{3} [e^{0.18} - e^0]$$

$$= \frac{50{,}000}{3} (1.1972 - 1) = \text{Rs. } 3{,}286.67$$

Example 49. A bank pays interest at the rate of 6% per annum compounded continuously. Find how much should be deposited in the bank each year in order to accumulate Rs. 6,000 in 3 years ?

(D.U., B.Com. (Hons.) 1992)

Solution. Let Rs. R be deposited in the bank each year for 3 years in order to accumulate Rs. 6,000 at the rate of interest of 6% per annum compounded continuously, then

$$6{,}000 = \int_0^3 \text{Re}^{0.06t}\, dt$$

$$= \frac{R}{0.06} \left| e^{0.06t} \right|_0^3 = \frac{R}{0.06} (e^{0.18} - 1)$$

$$= \frac{R}{0.06} (1.1972 - 1) = \frac{R}{0.06} (0.1972)$$

$$\Rightarrow \qquad R = \frac{36{,}00{,}000}{1{,}972} = \text{Rs. } 1{,}825.56.$$

Example 50. A firm's present annual sales are Rs. 10,000. The sales as continuously increasing at the rate of 20% per annum. Find the cumulative sales of first five years using calculus.

(D.U., B.Com. (Hons.) 1999)

Solution. Cumulative sales $= \int_0^5 10{,}000\, e^{0.2t}\, dt$

$$= 10{,}000 \left[\frac{e^{0.2t}}{0.2} \right]_0^5 = \frac{10{,}000}{0.2} [e - 1]$$

$$= \frac{10{,}000}{0.2} [2.718 - 1]$$

$$= \text{Rs. } 85{,}900.$$

Example 51. Find the present value of an annuity of Rs. 100 per annum, assumed to be payable continuously for 10 years, at the rate of interest of 4% per annum compounded continuously.

(D.U., B.Com. (Hons.) 1996)

Solution. The required present value is given by the formula

$$P = \int_0^N \text{Re}^{-it}\, dt$$

Here $R = 100$, $i = 0.04$ and $N = 10$.

$$\therefore \qquad P = \int_0^{10} 100\, e^{-0.04t}\, dt,$$

$$= 100 \left| \frac{e^{-0.04t}}{-0.04} \right|_0^{10} = \frac{100}{0.04}(1 - e^{-0.04})$$

$$= 2{,}500\,(1 - 0.67032) = \text{Rs. } 824.20.$$

Example 52. Calculate the present value of an annuity of Rs. 30,000 per annum, assumed to be payable continuously for 10 years, at the rate of interest of 8 per cent per annum compounded continuously. **(D.U., B.Com. (Hons.) 2000)**

Solution.

$$\text{Present value} = \int_0^N \text{Re}^{-it}\, dt\ *$$

$$= \int_0^{10} 30{,}000\, e^{-0.08}\, dt$$

$$= 30{,}000 \left[\frac{e^{-0.8t}}{-0.08} \right]_0^{10}$$

$$= 30{,}000 \left[\frac{e^{-0.8t}}{-0.08} + \frac{1}{0.08} \right]$$

$$= \frac{30{,}000}{0.08}[1 - e^{-0.8}]$$

$$= \frac{30{,}000}{0.08}(1 - 0.44933)$$

$$= \frac{30{,}000}{0.08} \times 0.55067$$

$$= \text{Rs. } 2{,}06{,}501.$$

Example 53. A company's initial sales is Rs. 10,000 and is continuously increasing at the rate of 5% per annum. Find :

(i) Cumulative sales of first five years;

(ii) Sale of fifth year. **(D.U., B.Com. (Hons.) 2004)**

Solution. (i) Cumulative sales for the first five years are given by

$$\int_0^5 = 10000\, e^{.05t}\, dt$$

$$= \left| \frac{10000 \times e^{0.05t}}{0.05} \right|_0^5 = 200000\,(e^{0.25} - 1)$$

$$= \text{Rs. } 56805 \text{ (approx.)}$$

$$\text{(ii) Sale of the fifth year} = \int_0^5 10000\, e^{0.05t}\, dt$$

$$= (200000)\, |e^{0.05t}|_4^5 = 200000\,(e^{0.25} - e^{0.20})$$

$$= \text{Rs. } 12524 \text{ (approx.)}$$

* If r% p.a. is the rate of interest then $i = \frac{r}{100}$.

Example 54. What is the present value of a continuous income stream of Rs. 5,000 per year for three years if it is discounted continuously at the rate of 6% per year ? **(D.U., B.Com. (Hons.) 2006C)**

Solution. Let P be the required present value.

Then $$P = \int_0^3 5000\, e^{-0.06t}\, dt$$

$\Rightarrow$ $$P = \frac{5000}{-0.06} |e^{-0.06t}|_0^3 \Rightarrow P = \frac{5000}{-0.06}(e^{-0.18} - 1)$$

$\Rightarrow$ $$P = \frac{5000}{0.06}(1 - 0.83527) = \text{Rs. } 13727.50$$

Example 55. A man wants to deposit a certain amount with a college so that the college can give scholarship of Rs. 500 every year, indefinitely, to the best student of the year. Find the amount to be deposited by the man, if money is worth 6% compounded annually ? **(D.U., B.Com. (Hons.) 2006C)**

Solution. Income per year = Rs. 500 $\quad r = 0.06\%$

Suppose time = n years

$\therefore$ Capital value $= P = \int_0^n 500\, e^{-rx}\, dx = \frac{500}{r}(1 - e^{-rn})$

Since the income is forever $\therefore\ t = \infty$

$\therefore$ $$P = \lim_{t \to \infty} \frac{500}{r}(1 - e^{-rn}) = \frac{500}{0.06} = \text{Rs. } 8{,}333 \text{ (approx.)} \quad (\because \lim_{n \to \infty} e^{-rn} = 0)$$

Example 56. Find out the capital value of a uniform income stream of Rs. R per year for m years, reckoning interest continuously at 100 r% per year. What will be the result if income is forever ? **(D.U., B.Com. (Hons.) 2006(R))**

Solution. Let P be the capital value of a uniform income stream of Rs. R per year for m years.

$\therefore$ $$P = \int_0^m R\, e^{-rx}\, dx = \frac{R}{-r}|e^{-rx}|_0^m$$

$$= \frac{R}{r}(1 - e^{-rm})$$

We now find the value of P when the income is forever.

$$P = \lim_{m \to \infty} \frac{R}{r}(1 - e^{-rm})$$

$$= \frac{R}{r} \lim_{m \to \infty} (1 - e^{-rm})$$

$\therefore$ $$P = \frac{R}{r}(1 - 0) = \frac{R}{r} \qquad \ldots(\because \lim_{m \to \infty} e^{-rm} = 0)$$

The above question can also be stated as :

Deduce that a/r is the present value of an income stream of Rs. a per year for ever, interest being 100 r% yearly compounded continuously.

Questions

1. A bank pays interest @ 6% per annum compounded continuously. If a person places Rs. 1,000 in the bank each year how much will be in his account after 3 years? [**Ans.** Rs. 3,286.67]
(D.U., B.Com. (Hons.) 1989)
2. A bank pays interest at the rate of 6% per annum compounded continuously. Find how much should be deposited in the bank each year in order to accumulate Rs. 6,000 in 3 years? [**Ans.** Rs. 1,825.56]
(D.U., B.Com. (Hons.) 1992)
3. Find out the capital value of a uniform income stream of Rs. R per year for m years recurring interest continuously at $100r\%$ per year. [**Ans.** $\left(\frac{R}{r}\right)$]
(D.U., B.Com. (Hons.) 1984 c.c.)
4. A person invests Rs. 1,000 every year with a company which pays interest at 10% per annum. He allows his deposits to accumulate with the company at compound interest. Find the amount standing to his credit one year after he has made his yearly instalment for the tenth time. [**Ans.** Rs. 17,531]
(ICWA, Dec., 1975)
5. A sum of money was borrowed at 4% p.a. simple interest. At the end of the first year Rs. 3,000 was paid off and the rate of interest on the balance was reduced to 3% p.a. If interest for the second year was $\frac{3}{7}$ of that of the first year, what was the original sum lent out? [**Ans.** Rs. 7,000]
6. What is the amount of a naya paise in 200 years at 6% compound interest given $\log 106 = 2.025059$, $\log 115.127 = 2.0611800$. [**Ans.** Rs. 1,152]
7. Find the effective rate p.a. corresponding to the nominal rate of 8% p.a. covertible quarterly. [**Ans.** 8.243% p.a.]
8. Find the nominal rate p.a. convertible quarterly corresponding to an effective rate of 8% p.a. [**Ans.** 7.7712% p.a.]
9. Find (i) the rate of discount corresponding to a rate of interest 0.06 and (ii) the rate of interest corresponding to a rate of discount of 0.08 per rupee. [**Ans.** (i) 0.566 (ii) 0.08696]
10. What is the present value of Rs. 500 due at the end of 10 years, the rate of interest being 5% for the first 4 years from now and 6% convertible half-yearly for the next 6 years. [**Ans.** Rs. 288.5]
11. Find the accumulated value at the end of 12 years of an amount of Rs. 750 invested at 9% p.a. for the first 6 years and thereafter at 10% convertible half-yearly. [**Ans.** Rs. 2,258.88]
12. Moneylender charges interest at the rate of 5 paise per rupee per quarter, payable in advance. What effective rate of interest does he charge per annum? [**Ans.** 22.8%]
13. A moneylender charges interest at the rate of 10 rupees per 100 rupees per half-year payable in advance. What effective rate of interest does he charge per annum? [**Ans.** 23.5%]
14. To what sum will Rs. 2,000 amount in 8 years if invested at a 6% effective rate for the first four years and 6% compounded semiannually thereafter? [**Ans.** Rs. 3,198.53]
15. A person deposits Rs. 5,000 in a bank which pays an interest of 11% per annum compounded continuously. How much amount will be in his account after 10 years? [**Ans.** Rs. 15,020 (Approx)]
(D.U., B.Com. (Hons.) 1986)
16. A National Savings Certificate costs Rs. 15 and realises Rs. 20 after 10 years. Find the rate of interest involved when it is added (a) yearly (b) continuously. [**Ans.** (a) 2.92% (b) 2.88%]
(D.U., B.A. Eco. (Hons.) 1991)
17. How long will it take for a principal to double if money is worth 6% compounded continuously. [**Ans.** 11.5 years]
18. A person invests money in a bank paying 6% compounded semiannually. If the person expects to receive Rs. 8,000 in 6 years, what is the present value of the investment? [**Ans.** Rs. 5,611.03]

19. A trust fund is being set up by a single payment so that at the end of 20 years there will be Rs. 50,000 in the fund. If interest is compounded continuously at an annual rate of 5% how much money should be paid into the fund initially. [**Ans.** Rs. 18,394]

20. A moneylender charges 'interest' at the rate of 10 paise per rupee per month, payable in advance. What effective rate of interest does he charge per annum?

21. In return of a promise to pay Rs. 500 at the end of 10 years, a person agrees to pay Rs. 100 now, Rs. 200 at the end of 6 years and a final instalment at the end of 12 years. If the rate of interest is 2% per annum effective, what should the final instalment be?

22. How much money is needed to endure a series of lectures costing Rs. 2,500 at the beginning of each year indefinitely, if money is worth

(i) 3% compounded annually; (ii) 12% compounded quarterly.

[**Ans.** (i) 89,833 (ii) 22,420]

(D.U., B.Com. (Hons.) 1999)

23. A machine costing Rs. 80,000 would reduce to Rs. 20,000 in 8 years. Find the rate of yearly depreciation, given that depreciation is calculated using diminishing balances method. [**Ans.** 15.88]

(D.U., B.Com. (Hons.) 1999)

24. Write expression to find the required value in each of the following cases :

(i) To find the amount, A, of the principal, P, invested for n_1 years at a rate of interest of r_1 per annum compounded quarterly and next n_2 years at the rate r_2 per annum compounded annually.

(ii) To obtain the time it will take to triple an amount, P, invested at the rate of r per year compounded quarterly.

(iii) To calculate the effective rate r per annum when the normal rate is i and the compounding is done m times a year.

25. Find, for each of the following, the amount to which Rs. 100 will accumulate :

(i) At the rate of interest 12% per annum compounded quarterly for 10 years.

(ii) At the force of interest 3% per annum for 3.5 years.

(iii) At the effective rates of interest 3% per annum for 10 years, 4% per annum for 4 years and 5% per annum for 2 years.

(iv) At the rate of interest corresponding to 3% per annum effective rate of discount for 8 years.

(v) What constant force of interest would produce the same amount after 16 years at the rates in (iii) above.

[**Ans.** (i) Rs. 326 (ii) Rs. 111.071 (iii) Rs. 173.32 (iv) Rs. 3,09, Rs. 127.6 (v) 0.03437 per rupee]

26. Answer any three of the following :

(a) (i) A certain sum of money is invested at 4% compounded annually. The interest for second year is Rs. 25. Find the interest for the third year.

(ii) A sum of money is put at compound interest for two years at 20% per annum. It would fetch Rs. 482 more, if the interest were payable half yearly than if it were payable yearly. Find the sum.

(b) What should be the monthly sales volume of a company if it desires to earn a 12% annual return convertible monthly on its investment of Rs. 2,00,000? Monthly costs are Rs. 3,000. The investment will have eight years life with no scrap value.

(c) A machine depreciates at the rate of 10% per annum for the first two years and then 7% per annum for the next 3 years, depreciation being calculated on the dimishing value. If the value of the machine be Rs. 10,000 initially, find the average rate of depreciation and the depreciated value of the machine at the end of the fifth year.

(d) Explain the concepts of effective rate of interest, nominal rate of interest and force of interest. Bring out the relationship among them.

(e) Derive, by using calculus, the present value of an annuity of Re. 1 per annum assumed to be payable continuously for n years at the rate of interest 8 per rupee per annum convertible continuously.

[**Ans.** (a) (i) Rs. 26 (ii) Rs. 20,000 (b) Rs. 6,250 (c) 8.214%]

(D.U., B.Com. (Hons.) 2003)

UNIT-II

Business Statistics

1. Introduction to Statistics
2. Preparation of Frequency Distribution
3. Statistical Averages (Measures of Central Tendency)
4. Measures of Variation (Dispersion)
5. Correlation and Regression Analysis
6. Index Numbers
7. Time Series

1 Introduction to Statistics

Coverage Functions, Scope, Distrust and Limitations of Statistics.

1.1 INTRODUCTION AND ORIGIN

The collection of numerical data on various aspects of life is almost as old as civilization itself. Prof. P.C. Mahalanobis said, "*From time immemorial men must have been computing information for peace and war. Statistics in this sense is as old as statecraft.*"

Initially the concept of statistics started with data collection and something like 'Political arithmetic'. The Kings used to collect the information about the people in their State, *i.e.*, they used to collect the data regarding conditions under which people live and work, their health, their incomes and wealth. So this subject developed as 'a King's subject' or 'as a science of Kings'. Kings also used to collect data such as population, man power, wealth in the form of land, buildings and other assets to know the condition of their State. R.A. Fisher regards the science of statistics as "*Mathematics applied to observational data.*" S.S. Wilks in 1951 remarked. "*In a democracy the citizen lives in a world of facts and figures. He makes decisions all of the time on the basis of large or small amounts of information. He carries on in a mass production economy ......* ." Perhaps H.G. Wells was right when he said, "*Statistical thinking will one day be as necessary for efficient citizenship as the ability to read and write.*" Statistics is not a science but a scientific method and scientific methods are essential for better efficiency. Now-a-days practically there is no sphere of life where statistics do not find applications. In words of M. Walker, "*Today, to a very striking degree our culture has become a statistical culture. Even a person who may never have heard of an index number is affected in an intimate fashion by the gyrations of those index numbers which describe the cost of living. Even on the most elementary level it is impossible to understand psychology, sociology, economics, finance or the physical sciences without some general idea of the meaning of an average, of variation, of correlation, of sampling, of how to interpret charts and tables. The state legislatures deal continually with matters in which it is impossible to reach a sound decision without weighing statistical evidence.*"

The word 'statistics' appears to have been derived from the Latin word 'status', Italian word 'statista', German word 'Statistik' or French word 'statistique' all referring to the political State. Achenwall, called the father of statistics, in 1749 defined statistics as "*The political science of the several centuries*". In the early years 'statistics' connoted a collection of facts about the State or the people in the State for administrative or political purposes. Thus, it helped in conducting the census operations and collecting data about the social and economic conditions of different groups of persons living in different parts of the country.

The word 'statistics', until the last quarter of the nineteenth century, was used to signify not only numbers and quantitative information but also facts calculated to illustrate the conditions and prospects of the society. But, by the turn of the twentieth century, the term 'statistics' became identified with quantitative information and today this is almost the exclusive emphasis which is given to the subject.

Lowell J. Reed in 1952 said, "*although statistics is in its infancy in a scientific sense, it is certain to influence profoundly all future scientific thinking.*"

1.2 EMPIRICAL DATA AND QUANTITATIVE* MEASUREMENTS

An empirical investigation consists of collecting information through observations and analysing the information for drawing conclusions. When the information or observations are recorded in quantity, the information is said to have been quantified. For instance, weights, heights, age, number of accidents, production, sales, etc., are phenomena which are measurable in quantity along a scale.

*The systematic and scientific treatment of quantitative measurements** is precisely known as* **statistics**. *The term* **statistics** *covers all fields of study wherein quantitative measurement of observations is possible and need analysis.*

1.3 DEFINITIONS OF STATISTICS

The word statistics is used in two different but inter-related ways :

(i) As a plural noun.

(ii) As a singular noun.

As a plural noun statistics refers to statistical data. As a singular noun it refers to statistical methods. Statistical methods include collection, classification, presentation, analysis and interpretation of data. Therefore, on the whole, statistics refers to statistical data and the method of obtaining and analysing data in order to base decisions on them.

In this sense, the most comprehensive and exhaustive definition has been given by Prof. Horace Secrist. He defines : "*By statistics we mean aggregate of facts affected to a marked extent by multiplicity of causes, numerically expressed, enumerated or estimated according to reasonable standards of accuracy, collected in a systematic manner for a predetermined purpose and placed in relation to each other.*"

According to this definition, statistics in plural sense should have the following characteristics :

(i) They are aggregate of facts.

(ii) They are affected to a marked extent by multiplicity of causes.

(iii) They are numerically expressed.

(iv) They are enumerated or estimated according to a reasonable standard of accuracy.

(v) They are collected in a systematic manner.

(vi) They are collected for a pre-determined purpose.

(vii) They are placed in relation to each other.

---

* Quantitative and qualitative data is further explained at Page 2.13.

** All data are not statistical. Data in quantitative form [numerical form] only can be a statistical data, but all numerical data also are not statistical [Isolated facts expressed numerically do not constitute data. For example, the wage earned by an industrial worker at any one time, taken by itself, is not a statistical data but taken as part of a mass of information, may be a statistical data]. Thus, quantitative data are statistical when they relate to measurement of masses and not statistical when they relate to an individual item or event as a separate entity.

(i) *They are Aggregate of Facts*

Statistics are concerned with "population" of facts rather than with a single fact. Single or unrelated facts do not throw light on any problem. In short "not a datum* but the data are the subject matter of statistics." Suppose a student 'X' is said to be poor in mathematics it is not a case of statistics but if the marks of all the students in a class are recorded, then it becomes statistics. No doubt all statistics are expressed in numbers but all numbers are not statistics.

(ii) *They are Affected to a Marked Extent by Multiplicity of Causes*

Statistics refer to measurement of phenomena in a complex universe. They are related to other measurements. They grow out of a variety of circumstances differing among themselves and are constantly subject to change. They are not traceable to a single cause.

(iii) *They are Numericaly Expressed*

For example, rise in prices may be due to many factors like short supply, untimely rainfall, less production, increase in demand, shortage of power, rise in wages of labourers, etc.

In the words of Prof. H. Secrist, "*The statistical approach to a subject is numerical. Things, attributes and conditions are counted, totalled, divided, subdivided and analysed.*" This means that the facts, to be called statistics, must be numerically expressed. Qualitative expressions like good, poor, young, old, etc., do not form part of statistical studies unless expressed numerically in some form.

For example, saying Aishwarya Rai is Miss World is not a statistics but saying that in 1994 Aishwarya Rai was selected as Miss World is a statistics.

(iv) *They are Enumerated or Estimated according to a Reasonable Standard of Accuracy*

This characteristic of 'statistics' is related with the collection of data. Data can be collected either by enumeration or by estimation. If the sphere of enquiry is limited, then enumeration is resorted to. The data collected by enumeration are more correct. When the sphere or equity is wide, enumeration becomes a difficult job. In such cases estimation is the only way out. According to Dr. A.L. Bowley, "*Great numbers are not counted correctly to a unit, they are estimated.*" It is natural that the estimated figures cannot be absolutely accurate. Prof. Secrist himself has stated that, "*what are reasonable standards depend upon the purpose which the statistics are to serve.*" Prof. Secrist has further said, "*No absolute criterion can be established for all cases, where precision is required, accuracy is necessary, where general impressions are sufficient, appreciable error may be tolerated.*"

(v) *They are Collected in a Systematic Manner*

The collection of facts should be systematic. Haphazardly collected figures are not desirable because they may lead to erroneous conclusions. According to Prof. Secrist, "*Stray and loose bits of quantitative information, hearsay and unrelated material gleaned here and there, from indiscriminate sources having no common basis of selection, even when numerical can be termed statistics only by confusion of terms.*"

(vi) *They are Collected for a Pre-determined Purpose*

Statistics should be collected for a specific, well-defined and pre-determined purpose. This will facilitate the collection of requisite information. For example, if we are to collect statistics of agricultural production, we must decide before hand, the regions, commodities and the period (or periods) for which data are required.

---

* Datum means a piece of information.

(vii) *They are Placed in Relation to Each Other*

Statistics are collected for making comparisons between related phenomena over a specific region or period. For example, the population figures of India, U.K., U.S.A., China, etc., can be compared only if they belong to the same census year. Thus, for a valid comparison, statistics are required to be put in relation to each other. It would be meaningless to compare the weight of horses with those of human beings.

In the absence of the above conditions, numerical expressions of facts cannot be treated as 'statistics'. Therefore, one can say "*all statistics are numerical data but all numerical data are not statistics.*" For example, logarithms are numbers but they cannot be called as statistics because they do not conform to the above stated characteristics.

The other important definitions, defining statistics in the plural sense are :

(i) Webster defines statistics as, "*Statistics are classified facts representing the condition of the people in a state......especially those facts which can be stated in numbers or in tables of numbers or in any tabular or classified arrangement.*" This definition limits the scope of statistics. According to this definition, only those facts are called statistics which are related to the condition of the people in a State. Facts regarding Physics, Chemistry, Sociology and Psychology will not be called statistics. In the present age, statistics are collected in respect of all aspects of human activity. Hence, this definition is inadequate and incomprehensive.

(ii) According to Dr. A.L. Bowley, "*Statistics are numerical statements of facts in any department of enquiry, placed in relation to each other.*" According to this definition statistics have three characteristics, *viz.*, (i) they are numerical statements of facts, (ii) they are concerned with an enquiry and (iii) they are placed in relation to each other for comparison. Other characteristics are not included in this definition. This definition does not speak about different statistical methods also.

(iii) Yule and Kendall state that, "*By Statistics we mean quantitative data affected to a marked extent by a multiplicity of causes.*" This definition is also incomplete as it mentions only two chaacteristics of statistics, *viz.*, (i) they are quantitative and (ii) they are affected by multiplicity of causes.

(iv) According to Connor, "*Statistics are measurements, enumerations or estimates of natural or social phenomena, systematically arranged so as to exhibit their inter-relations*" This definition states that (i) statistics are measurements, estimates or enumerations, (ii) they relate to some natural or social phenomena, (iii) they are systematically arranged to exhibit their inter-relations.

(v) According to Tuttle, "*Statistics are measurements, enumerations or estimates of natural and social phenomena, usually systematically arranged, analysed and presented so as to exhibit important relationships among them.*" This definition also does not include all the characteristics of statistics.

1.4 STATISTICS AS A SINGULAR NOUN, *i.e.*, STATISTICS AS STATISTICAL METHODS

Dr. Bowley has said, "*It will be better to define Statistics as a method.*" Based on this concept, statistics is what statistics does or what statisticians do. In statistical methods, the word 'statistics' has been treated in singular form or a body of knowledge in the area of applied mathematics.

Some of the definitions of statistics under this concept are :

1. *Incomplete and Traditional Definitions.* Dr. Bowley has given a number of incomplete definitions which touch only some of the aspects of statistics.
   (i) "*Statistics may be called the science of counting.*" He himself further observes, "*Great numbers are not counted......they are estimated.*"
   (ii) "*Statistics may rightly be called the science of averages.*"
   (iii) "*Statistics is the device for abbreviating and classifying the statements and making clear the relations.*"
   (iv) "*Statistics is the science of measurement of social organism regarded as a whole in all its manifestations.*" This definition limits the application of the statistical methods in social sciences only. Dr. Bowley himself at another place expresses contrary view that, "*Statistics is not merely a branch of political economy nor it is confined to any one science. Statistics can be applied in varied fields.*" W.I. King while criticising this definition observes that, "*this statement limits the science to only one field—that of man and his activities. Modern Statistics takes into consideration, however, biological, astronomical and physical as well as social phenomena and hence the definitions obviously are too narrow.*"
2. Like Dr. Bowley, Prof. Boddington defines statistics as, "*Statistics is the science of estimates and probabilities.*" This definition is also not complete.
3. "*The subject 'Statistics' is concerned with the collection, presentation, description and analysis of data which are measurable in numerical terms.*" —*P.H. Karmel*
4. In the words of Prof. Lovit, "*Statistics deals with the collection, classification and tabulation of numerical facts as the basis of explanation, description and comparison of phenomena.*"
5. "*Statistics is the science which deals with the methods of collecting, classifying, presenting, comparing and interpreting numerical data collected to throw some light on any sphere of enquiry.*" —*Seligman*
   This definition is short, simple yet comprehensive.
6. "*Statistics or statistical methods may be defined as the collection, presentation, analysis and interpretation of numerical data.*" —*Croxton and Cowden*

The last two definitions are quite exhaustive and appear to be comprehensive. In the light of above definitions statistics as statistical method have the following stages :

(i) Collection of data
(ii) Organisation (classification) of data
(iii) Presentation of data
(iv) Analysis of data
(v) Interpretation of data.

(i) *Collection of data* is the first step in a statistical investigation and forms the backbone of the investigation (or inquiry). Data should be collected for a specific and well-defined purpose very carefully, because faulty data will lead to fallacious conclusions.

(ii) *Organisation of data* is done in three parts :

(1) Editing, (2) Classification, and (3) Tabulation.

To draw meaningful conclusions, the proper organisation of the data is essential. Observed data, without proper organisation will not lead to any conclusion.

(iii) *Presentation of data* : Orderly presentation of data facilitates statistical analysis. Data presented in the form of graph, diagrams, is easily understandable and have greater memorising effect and hence facilitates efficient comparison.

(iv) *Analysis of data* : The next step after collection, organisation and presentation of data is analysis of data. *The main objective of analysis is to prepare data in such a fashion so as to arrive at certain definite conclusions.* The methods most commonly used are measures of **Central Tendency** (called measures of the first order). **Measures of Dispersion** (called measures of the second order). Skewness, Correlation, Regression, Interpolation etc. (Called measures of third order). The analysis of facts based on observation is termed as (a) Scientific analysis, (b) Numerical analysis and (c) Empirical analysis.

(v) *Interpretation of data* : Interpretation or drawing valid conclusions is the last stage in a statistical investigation. The interpretation of analysed data is a difficult task and needs a high degree of skill and experience. If the analysed data are not interpreted carefully, the basic object of enquiry is liable to be defeated.

Therefore, we may give the following complete definition of statistics.

Statistics (as used in the sense of data, i.e., as a plural noun) are numerical statements of facts capable of analysis and interpretation and the science of statistics (i.e., as singular noun) is a study of the principles and methods used in the collection, organisation, presentation, analysis and interpretation of numerical data in any sphere of enquiry.

1.5 EMPIRICAL AND QUANTITATIVE ANALYSIS

Empirical refers to something based on experiment, observation* or experience rather than on theory. **Analysis** refers to detailed examination of statements or observations mainly to draw some conclusions. Therefore **empirical analysis** can be defined *as a method of studying a subject whereby knowledge is acquired as a result of actual experience.*

Quantitative analysis is an attempt to level 'precision' or 'preciseness' to the facts, so that they can be easily compared. In quantitative analysis, numerical data are classified, tabulated and analysed to draw reasonable conclusions. As stated earlier, statistics mainly deals with the analysis of quantitative expressions or data.

In natural sciences like physics, chemistry and biology method of empirical analysis is commonly used and relevant laws and statements can be tested and experimented in laboratories. But economics is a social science wherein human behaviour is the subject of study and it covers the whole world. Empirical analysis is not helpful for the study of economics and other social sciences. The problems of social sciences can be effectively studied through the technique of quantitative analysis. Under quantitative analysis technique the problem can be studied systematically, but the results so obtained cannot be tested in a laboratory and hence they are not so exact and accurate as the results of empirical analysis. On the other hand, the technique of statistical analysis can be applied only when sufficient data concerning the problem under study is available so that proper classification and tabulation for analysis can be done.

Compared to empirical analysis, statistical (quantitative) analysis has much wider applicability, almost in every field of knowledge, may be medical science, physical science, natural science or any other science. As far as the field of economics is concerned, the technique of statistical analysis has become so important that even for the understanding of elementary economics, the knowledge of statistics is essential. Marshall observes, "*statistics are straws out of which I, like every other economist have to make bricks.*"

C.E. Engeberg remarked, "*no economist would attempt to arrive at a conclusion concerning production or distribution of wealth without an exhaustive study of statistics.*"

* Any recording of information whether it be numerical or categorical is observation.

1.6 NEED FOR QUANTITATIVE ANALYSIS (OR IMPORTANCE OF QUANTITATIVE ANALYSIS)

1. Statistical analysis is useful in solving complex problems of modern business and industry.
2. It provides systematic and powerful tools to decision makers to take efficient decisions based on quantitative data.
3. It enables proper allocation of resources.
4. It helps in minimising waiting and servicing cost.
5. It enables the management to decide when to buy and how to buy.
6. It helps in choosing an optimum strategy, *i.e.*, the action which can give maximum benefit at minimum cost.
7. It renders great help in the optimum allocation of resources.
8. Management can know the reactions of the integrated business system through quantitative analysis.

1.7 SCOPE OF STATISTICS

Scope of statistics is studied under the following three heads :

(i) Nature of Statistics
(ii) Subject-matter of Statistics
(iii) Limitations of Statistics.

(i) *Nature of Statistics*

By nature of statistics, we are to determine whether statistics is an art or a science. *Statistics is regarded as an art of applying the science of scientific methods.* In statistics, we not only study different methods of studying a problem but also study how those methods should be applied in different situations.

In the words of Tippett, "*Statistics is both a science and an art. It is a science in that its methods are basically systematic and have general application; and an art in that their successful application depends, to a considerable degree, on the skill and special experience of the statistician and on his knowledge of the field of application.*"

(ii) *Subject-matter of Statistics*

Subject-matter of statistics is divided into two major parts :

(a) Descriptive statistics
(b) Inferential or inductive statistics

(a) Descriptive Statistics : The descriptive statistics consists of those statistical methods which tell us how to describe the characteristics of a body of data. They deal with the collection, tabulation and presentation of data and the calculations of measures which describe the data in various ways. Descriptive statistics can be either univariate or bivariate. Following are included in the univariate descriptive statistics :

(1) Frequency distributions
(2) Measures of central tendency
(3) Measures of variability, *i.e.*, dispersion.

The study of correlation and regression are a part of the bivariate descriptive statistics.

(b) Inferential or Inductive Statistics : Inferential statistics deals with those methods where conclusions about a large group are drawn by studying a part of it. For example, by checking a few grains of rice a housewife makes an estimate whether the whole lot has been cooked.

(iii) *Limitations of Statistics*

Statistics is an extremely useful science, but it has its limitations and shortcomings too. **Newshome** writes, "*It (statistics) must be regarded as an instrument of research of great value, but having severe limitations, which are not possible to overcome and as such, they need our careful attention.*" **Tippet observes**, "*The application of statistical methods to investigation in the technological and indeed in any other field is based on assumptions, is subject to limitations and often leads to uncertain results.*" These limitations should be kept in mind while using or interpreting statistics. These limitations are :

Limitations of Statistics

Deals with quanti-tative charac-teristics only	Deals with averages	Do not study individ-uals	Results are approx-imately correct	Results not always beyond doubt	Only a means and not an end	Misuse possible	Should be used only by experts	Not the only method	Data should be homoge-neous and uniform

(1) Statistics studies only the quantitative aspect of a problem and does not study its qualitative aspects : The science of statistics studies only the quantitative aspect of a problem. The facts which are not capable of being quantitatively measured, like, intelligence, poverty, honesty, etc., cannot be studied unless these attributes are reduced to precise quantitative terms as Prof. **Secrist** observes, "*Some phenomena cannot be quantitatively measured, honesty of resourcefulness, integrity, goodwill, all important in industry as well as in life generally, are not susceptible of direct statistical measurement.*" **Prof. Bowley** also states that, "*They (statisticians) can at best measure only the numerical aspect of a phenomenon while very often they must be contended with measuring not the facts they wish but some allied quantity.*" This limitation of statistics limits the scope of the science.

(2) Statistics deals with averages : According to W.I. King, "*Statistics largely deals with averages and these averages may be made up of individual items radically different from each other.*" Laws of statistics are true on an average. W.I. King states, "*no system which makes a large or complex group intelligible to the mind at a glance can avoid effacing most of the minor irregularities.*"

(3) Statistics does not study individuals : According to Tippett, "*Statistics is essentially totalisation, because it is not concerned with individual values—but only with classes.*" Statistics does not study individual items. It deals with mass phenomena. For example, average wage of factory workers may be quite high but there may be certain workers getting very low wage. This is a serious limitation of statistics. In the words of W.I. King, "*Statistics from their very nature of subject cannot and will never be able to take into account individual cases. When these are important, other means must be used for their study.*" **The study of individual facts lies outside the scope of statistics.**

(4) Statistical results are only approximately correct : Statistics unlike mathematics or accountancy, is a science of estimates. According to Dr. Bowley, "*When observations are extended, many sources of inaccuracy are found to be present and it is very frequently impossible to remove them completely. Statistical results are, therefore, very generally estimates rather than exact statements.*" Statistics fails in the fields where perfect accuracy is desired.

(5) Statistical results are not always beyond doubt : According to **Prof. Connor**, "*Statistics deal only with measurable aspects of things and therefore can seldom give the complete solution to a problem. They provide a basis for judgement but not the whole judgement.*" **Dr. Bowley** has also expressed the same views. According to him, "*Statistical results are essential, when judgement is to be formed on any question that involves numbers, quantities or values, but they must always be brought into relation with personal, political, aesthetic or other non-quantitative considerations that may be of greater importance in deciding on a course of action.*" Too much dependence on statistics may be misleading.

(6) Statistics is only a means and not an end : It analyses the facts and throws light on the real situations. D. Gregory and H. Ward observe, "*statistics cannot run a business or government. Nor can the study of statistics do more than provide a few suggestions or offer a few pointers as to firm's or a government's future behaviour.*" Statistical technique is no doubt essential for processing, analysing and interpreting numerical data. But the methods are by no means perfect and it is very dangerous to jump to conclusions without a clear-cut idea about its limitations.

(7) Misuse of statistics is possible : W.I. King says, "*One of the shortcomings of statistics is that they do not bear on their face the label of their quality. Statistics is like clay and they can be moulded in any manner. Statistics can be misused by deliberately twisting or manipulating data. That is why some people say 'statistics can prove any thing' or 'statistics is an other form of lying.'*"

(8) Statistics should be used only by experts : Yule and Kendall point out, "*statistical methods are most dangerous tools in the hands of an inexpert.*" Only well trained people having good knowledge of statistics should use statistics, otherwise the results are liable to be misleading and dangerous.

(9) Statistical Study is not the only method : According to Croxton and Cowden, "*it must not be assumed that the statistical method is the only method to use in research, neither should this method be considered the best attack for every problem.*" Results obtained by statistical methods should be verified by other methods.

(10) Homogeneity of data : For comparison purpose, data should be homogeneous and uniform. Quantities measured in different units cannot be compared. Since foodgrains and clothes are measured in different units, we can simply compare their value in Rupees but not their quantities.

1.8 FUNCTIONS OF STATISTICS

The functions of the science of Statistics are beautifully summed up by Robert W. Burgess in the following words :

"*The fundamental gospel of statistics is to push back the domain of ignorance, rule of thumb, arbitrary or premature decisions, tradition and dogmatism and to increase the domain in which decisions are made and principles are formulated on the basis of analytical quantitative facts.*"

If we analyse the above statement, we would find that the main functions of the science of statistics are as given below :

(1) To simplify complex facts.
(2) To provide comparative study.
(3) To study relationship between different facts.
(4) To enlarge individual knowledge and experience.
(5) To formulate policies in different fields.
(6) To measure the effects.
(7) To test a hypothesis.

(8) To provide numerical measurements.
(9) To forecast (to indicate trend behaviour).
(10) To classify data.
(11) To measure uncertainty.
(12) To draw valid inferences.

(1) To simplify complex facts : The raw data is often unwieldy and complex. The function of statistical methods is to simplify large mass of numerical data and present them in a more intelligible form.

Dr. Bowley states that, "*the object of a statistical estimate of complex groups is to present an outline to enable the mind to comprehend with a single effort the significance of the whole.*" In the words of W.I. King, "*It is for the purpose of simplifying these unwieldy masses of facts that statistical science is useful. It reduces them to numerical totals or averages which may be abstractly handled like any other mere numbers. It draws pictures and diagrams to illustrate general tendencies and thus in many ways adapt these groups of ideas to the capacity of our intellects.*"

(2) To provide comparative study : The science does not merely mean counting but also comparison. Statistical devices like averages, ratios, percentages, rates, coefficients, standard error, etc., offer the best way of comparison between two phenomena. Boddington states that, "*The object of statistics is to enable comparison to be made between past and present results with a view to ascertaining the reasons for changes which have taken place and the effect of such changes in the future.*" Dr. Bowley observes, "*A chief practical use of statistics is to show relative importance, the very thing which an individual is likely to misjudge. Statistics are almost always comparative.*"

(3) To study relationship between different facts : Another function performed by statistics is to investigate relationship between two or more phenomena. The relationship existing between demand and supply, rainfall and agricultural production can best be measured with the help of statistical methods.

(4) To enlarge individual knowledge and experience : The science of statistics enlarges human knowledge and experience by making it easier for man to understand, describe and measure the effects of his action on the action of others. *Statistical methods have provided such a master key to mankind that we can use them anywhere and can study any problem in its correct perspective and on right lines.*

In the words of Whipple, "*Statistics enables one to enlarge his horizon.*" Dr. Bowley also observes that, "*The proper function of statistics, indeed, is to enlarge individual experience.*"

Bowley further observes, "*a statistical estimate may be good or bad, accurate or reverse, but in almost all cases it is likely to be more accurate than a casual observer's impressions and in the nature of things, can only be disproved by statistical methods.*"

Precise understanding of the concepts like national income, population, agricultural production, industrial production etc. can be possible by using statistical technique.

(5) To formulate policies in different fields : Statistics helps in formulating policies in different fields, specially in social, economic and business fields. Malthus' Theory of Population and Engles' Law of Family Expenditure are due to the statistics collected by these two noted economists. The governmental policies are framed on the basis of statistics. Policies in the fields of taxation, import-export, social welfare, wage, etc. are formulated on the basis of statistical data.

(6) To measure the effects : Not only in formulating a policy but also in measuring the effects of a policy, statistics act as guide. The effects of a change in the Bank Rate or tax-rate, etc., can be thoroughly studied with the help of statistics.

(7) To test a hypothesis : The most important theoretical function of statistics is to test the hypothesis in theory and discover a new theory. According to Prof. J.M. Keynes, "*The function of statistics is first to suggest empirical laws which may or may not be capable of subsequent deductive explanation and secondly to supplement deductive reasoning by checking its results and submitting them to the best of experience.*" The correctness of the laws of the different branches of knowledge can only be tested with the help of statistical data and statistical techniques.

(8) To provide numerical measurements : Facts expressed in numbers are easily understood. It is one of the major functions of statistics to express the facts relating to different phenomena in numbers. For example, if we say that agricultural production of India is increasing it does not carry much meaning but if we give yearwise figures it becomes easier even for a common man to understand index numbers also. In the last few decades, it has been made possible to express even the qualitative phenomena in the numerical forms to give them more concrete shape.

(9) To forecast : Making estimates for future on the basis of past and present behaviour is called **forecasting**. Extrapolation is a statistical method which helps in forecasting. Forecasting can also be done with the help of trend equations. We can forecast food production, growth of population, supply of power etc. on the basis of past records.

(10) To classify data : Statistical methods help us to arrange the data in groups according to their common characteristics. This facilitates analysis and comparisons.

(11) To measure uncertainty : Statistics helps us to take decisions in the face of uncertainty and study the chance of occurrence of certain events. In theory of games we take decision in the face of uncertainty. John I. Griffin observes, "*Statistical knowledge is, however, imperfect and always characterised by uncertainty. Such imperfect and uncertain knowledge is not to be rejected as worthless because statistical methods permit the measurement of the imperfection and uncertainty.*"

(12) To draw valid inferences : In fields of Business and Economics, many times rational and valid conclusions are drawn by collecting and analysing the available facts.

1.9 USES AND IMPORTANCE OF STATISTICS

In words of H.G. Wells, "*Statistical thinking will one day be as necessary for efficient citizenship as the ability to read and write*" are proving true in present-day life. Today there is hardly any sphere of life which is not affected by statistical methods in one way or the other. In fact statistics has become an "arithmetic of human welfare." Commenting on the importance of statistics, Croxton and Cowden write, "*The methods of statistics are useful in ever-widening range of human activities in any field of thought in which numerical data may be had... . Today, there is hardly a phase of endeavour which does not find statistical devices at least occasionally useful.*" Tippet says, "*Statistics affects everybody and touches life at many points.*" Though statistics deals with quantitative data but qualitative data, for example attributes such as defectives, standard, deluxe, new and used can be transformed into number of defectives, proportionals of standard models and so forth. In this manner extensive statistical analysis can be made of qualitative as well as quantitative data.

1.10 IMPORTANCE OF STATISTICS IN ECONOMICS

Statistical data are a powerful aid in economic analysis. The study of every economic problem requires the use of statistical methods just as a doctor needs his stethoscope to examine almost each and every patient.

In the words of Ya-Lun-Chou, "*Economists depend upon statistics to measure economic aggregates such as gross national output, consumption, saving, investment, expenditure and changes in the value of money. They also use statistical method to verify economic theory and to test hypothesis.*" Most of the economic problems can be expressed numerically which help the economist in collecting data and applying statistical methods in formulating policies and arrive at certain decisions. In fact statistical data and statistical methods are the tools and appliances of an economist's laboratory. This led Prof. Marshall, to say, "*Statistics are the straws out of which I like every other economist have to make bricks.*"

Statistics of **consumption** enables us to find out the ways in which people of the different classes of society spend their incomes. Such statistics are very helpful for knowing the standard of living and the taxable capacity of the people.

In **distribution** too, statistics play a vital role. With the help of statistics, the national wealth of the country is estimated and its distribution among the people of the country is found out. Equitable distribution of national income and wealth is another serious problem to be best studied statistically.

In the field of **exchange** we study markets, laws of prices based on supply and demand, cost of production, etc. A systematic study of all these facts cannot be made without statistics. What price should a monopolist charge in order to reap the maximum profit? What shall be the price of a particular commodity if its supply is increased or decreased?—All such questions can best be answered with the help of statistics. Statistics are the foundation-stone of the theory of exchange.

Statistics of **production** helps us to adjust the supply according to demand. Such statistics are the measures of productivity of a country. Thus productivity of different factors of production can be gauged and compared. Statistics of production of an area or a country can be compared with another area or another country. Census of production helps in understanding the economy and taking corrective measures wherever necessary.

Public Finance : Public finance studies the revenue and expenditure activities of the State. The statement which shows the revenue and expenditure of the State is known as '**Budget**'. Budget is a statistical document. Fiscal policy, deficit financing, nature and amount of subsidies, etc., are concepts which are based on statistics.

Monetary Policy : Monetary policy is the policy of central bank of a country. Bank rate, selective credit controls, open market operations, cash reserve ratio, etc., are the chief weapons of monetary policy. All these concepts are statistical in nature.

Macro-economic Concepts : The world-wide recession of 1930's necessitated the study of national economic aggregates. What should be the increase in national income if the investment in the country is increased by Rs. 100 crores? How much consumption will be affected if the national income increases or decreases by some fixed amount?—All these concepts are statistical in nature. The modern economist lays a great stress on all these concepts and theories.

Econometrics : Econometrics is a product of the 20th Century. It is the combination of economics, mathematics and statistics. It is assumed that with the help of econometrics, economics will become an exact science.

Input-output Analysis : Input-output analysis is based on statistical data, which explains the relationship between the input and the output. Modern theory of probability, sampling, correlation and regression, time series, index numbers, etc., are the concepts used in economic analysis.

The Law of Demand and the Elasticity of Demand are based on inductive reasoning. Under inductive reasoning, general laws are propounded by studying data about a particular phenomenon. It is the function of a statistician to collect data about a particular phenomenon. From the data collected by Statisticians, laws are derived by economists.

Modern economists are trying to make Economics as an exact science. In this effort, use of statistics and statistical methods is inevitable. W.A. Spur and C.P. Bonini observe, "*Economists today are no longer content to theorize in abstract term, citing statistics only as needed to buttress their arguments. Instead, they utilise the excellent data now available to build a sound factual foundation for their reasoning.*"

A statistical approach to an economic problem not only leads to its correct description but also indicates lines along which it is to be tackled. Hence, in all types of economic problems statistical approach is essential and statistical analysis is useful. To quote Dr. Bowley, "*No student of Political Economy can pretend to know complete equipment unless he is master of the methods of statistics, knows its difficulties, can see where accurate figures are possible, can criticise the statistical evidence estimates given to him.*" And C.E. Engeberg says, "*No economist would attempt to arrive at a conclusion concerning the production or distribution of wealth without an exhaustive study of statistical data.*" Justice Oliver Vendell Holmes remarked that, "*The black letter man may be the man of the present, but the man of the future is the man of Statistics and the master of Economics.*"

Thus, we find that there is kinship between **Economics and Statistics**.

1.11 STATISTICAL METHODS AND EXPERIMENTAL METHODS

Statistics is not a body of substantive knowledge but a body of methods for obtaining knowledge. Statistics is usually not studied for its own sake, but as a highly valuable tool in the analysis of problems in both the nature and social sciences. That is the reason why A.M. Mood has said, "*Statistics provides tools and techniques for research workers.*"

There are two methods which are employed in conducting research, *viz., experimental method and statistical method.* The *experimental method* is adopted in physical or exact sciences like Physics, Chemistry, etc. Under this method the factors under investigation are isolated and variated according to a pre-determined plan, while other factors having a bearing on the observed phenomena are kept constant. This is what is done in all experiments carried on in the laboratory. However, in the social sciences like Economics, Sociology, Politics, etc., which are less exact sciences, experimentation is seldom possible, because it is not possible to isolate the factors under investigation. H.L. Balsley observes, "*In the social sciences, which deal principally with people, pure experimental method is precluded from use. Whenever experiments may not be made by holding some factors constant, while allowing others to change from case to case and thus come to an exact answer as desired, the answer is sought through a statistical method. For example, if it is sought to study the effect of change of income on the health of the community, it is not possible to keep constant all other factors affecting health and vary income according to pre-determined plan. Thus, in such cases experimental method cannot be applied. The statistical methods and procedures constitute a useful and often indispensable tool for research workers of social sciences, may be like a blind man groping in the dark for a black cat which is not there.*" The statistical methods are useful in an ever-widening range of human activities in any field of thought in which numerical data may be had.

It should not be assumed that statistical method is the only *method of research, neither this method* be considered the best attack for every problem. Just as a carpenter has a number of tools, each appropriate for a particular type of work, similarly researcher can use various techniques which are the tools of his trade and each of which appropriates for a specific type of situation. If a carpenter uses screw driver in lieu of a chisel, the results are not likely to be satisfactory. Similarly, it is important that the investigator should examine the problem under investigation very carefully and *make use of technique appropriate* to it. Just as a carpenter needs to use more than one tool in completing a piece of work, similarly a research worker must make use of not one but several methods. However, it should be kept in mind that the classification of scientific methods into those which are experimental and those which are statistical, like most classifications, is formal and arbitrary and not entirely realistic. Even in exact sciences, where experimentation is possible, statistical methods also help in getting correct results. When a scientist comes to work on a problem in practice, he usually combines elements of both—the *statistical and experimental* approach. Many of the most important of the statistical methods were originated in the fields of physics and astronomy—fields that we usually think of 'exact' sciences. Even in these fields, the scientist has to contend with errors of observation and in addition he usually finds it impossible to record the values of all the variables which are involved. Under such circumstances, the scientist, researching in exact sciences, is forced to combine statistical methods with his experimental procedures. In the words, of J.F. Kennedy and E.S. Keeping, "*Of late years statistics has more and more come to the help of the other sciences, particularly the biological and social sciences, as an aid to the intelligent planning of experiments and as a means of assessing the significance of the results obtained by experiment. In the 'exact' sciences, such as physics and astronomy, with their relatively high precision of measurements, there did not appear to be the same need for statistical methods as in agriculture, medicine, economics and many other fields, where the results of experiments are complicated by a multiplicity of factors beyond the control of the observer. For a long time, in fact, statistics in the physical sciences hardly progressed beyond the calculation of a standard error and the occasional fitting of a curve by least squares. There is now, however, a keen appreciation among physicists of the fundamental role of statistics in the treatment of the complex problems of molecular, atomic and nuclear structure.*" According to George Simpson and Fritz Kafka, "*Statistics is thus a tool of all sciences, indispensable to research and intelligent judgement.*" Thus, experimental method and statistical method both are necessary for research just as the right and left feet are needed for walking. According to B.C. Brookes and W.F.L. Dick, "*for the research worker in any field that is concerned with quantitative results, a knowledge of statistics is useful.*"

1.12 IMPORTANCE OF STATISTICS IN PLANNING

"*Economic planning means making of major economic decisions—what and how much is to be produced and to whom it is to be allocated—by the conscious decisions of a determinate authority on the basis of a comprehensive survey of the economy as a whole.*" It implies a conscious and deliberate action to attain certain pre-determined economic aims. In planning, statistics plays an indispensable role. In order to underline the role of statistics in economic planning we can do no better than to quote Baykon, a Fussian economist, that "*Every act of planning presupposes a preliminary investigation of exiting conditions.*"

Statistics are used in economic planning for the following purposes :

(1) To know the stage of economic development in our country as compared to other countries, statistics are of great help. The problems like over-population, unemployment, lack of industrialisation, low rate of capital formation, etc., which are the characteristics of undeveloped and under-developed countries, can be fully understood only with the help of statistical facts.

(2) Statistical language is essential to understand the effect of determining factors of economic development in the past—what psychological and sociological factors need to be developed for economic progress, how far we have progressed in the technical field, what is the level of productivity and such other questions.

(3) It is necessary to determine priorities in different sectors for development. This can be done by examining the strucure and importance of production, consumption, exchange and distribution sectors of the economy. For determining priorities, statistical methodology is essential.

(4) Economic planning is done to achieve pre-determined objectives and targets. They have to be expressed in quantitative terms. Financial planning is a part of economic planning. Financial resources are mobilised through taxation, public saving, deficit financing, debt-issues, etc. Planing and budgeting are integrated. All these functions can be performed successfully with the help of statistics.

(5) Evaluation of the success of a plan is also done with statistics.

Thus, we see planning without proper and adequate statistics is unconceivable. "Planning without statistics is like a ship without rudder and a compass."

Tippet says, "*Planning is the order of the day and without statistics, planning is inconceivable.*"

1.13 IMPORTANCE OF STATISTICS IN BUSINESS

Business is regarded as a profession of forecasting which requires a high degree of skill. Success in business depends upon precision in forecasting. A businessman must make a proper and scientific analysis of the past record in order to forecast the future business conditions. Business barometers are constructed with the help of existing records to tell us the future course of business events.

Every businessman, irrespective of his nature of business, has to employ statistical techniques in estimating trend of prices, trend of economic activities, etc.

A *producer* estimates probable demand of his goods, analyses the effects of trade cycles and seasonal variations as also of changes in habits and customs of people on the demand of his wares and after taking all these factors into consideration finally takes decision about the quantum of production. Buddington says, "*In order to suceed in any business today, the businessman must study all the factors which enter into production, buying and selling, exporting and importing of goods in which he deals.*"

Buddington further says, "*A successful businessman is one whose estimates most closely approach accuracy.*" Accuracy can be achieved only by proper analysis of past records, with the help of statistical tools. Even the industrialists, farmers and manufacturers take decisions about the volume of future production onthe basis of past records and trends. Again we may quote Ya-Lun-Chou, "*It is not an exaggeration to say that today nearly every decision in business is made with the aid of statistical data and statistical methods.*" We may conclude that industry and business involve risk and when it is risk, it is better to have a calculated risk and the risk can be calculated with the help of statistical tools.

1.14 IMPORTANCE OF STATISTICS IN STATE AND ADMINISTRATION

Importance of statistics in State and administration started its origin as 'science of King's or political arithmetic'. But now, in fact, statistics has become the eyes of the Government administration.

W.M. Harper observes, "*Increasingly, figures have become the basis of rational decisions and events are proving that these decisions based on figures give better results.*" Finance Minister (1) prepares budget by using the data relating to revenue and expenditure, (2) takes decisions for considering new taxes and planning public expenditure on the basis of the data concerning production, import, export, national income, etc.

It is the duty of the Government to promote economic development, reduce inequalities in the distribution of wealth and income and maintain stability and full employment. The Government decides measures to achieve these objectives on the basis of statistical data.

Not only in peace, even during war and periods of crises, natural calamities, etc., statistical records are helpful in framing suitable policies.

Government formulates its policies concerning family planning, nationalisation, establishment of new industries, etc., on the basis of statistical data.

Reports of different commissions and committees appointed by the Government are always substantiated by the statistical facts.

Government makes assessment about the efficiency of different departments like health, education, industry, agriculture, etc., on the basis of statistical data.

1.15 STATISTICS—A SCIENCE OR AN ART

Webster's dictionary defines science as "*accumulated and accepted knowledge that has been systematised and formulated with reference to the discovery of general truths or the operation of general law*". In brief, science is a body of systematised knowledge. It studies cause and effect and tries to find out generalisations which are called *law of that science.* It does not tell whether the results are good or bad. It only describes what are the facts. It is like a lighthouse that gives light to the ships to find out their own ways but does not indicate in which direction they should go. The ships have to find their own direction with the help of the light.

According to Karl Pearson that knowledge which (a) gives mental education to the citizens, (b) throws light on important social problems, (c) gives happiness in practical life, and (d) gives satisfaction to our artistic faculties, may be called science. All these qualities are found in statistics, hence it can be regarded as a science.

Science has two characteristics : one, skeptical attitude of the scientist and second, methodology. A scientist has a skeptical attitude for every phenomenon and therefore is keen to uncover the truth. Care is taken in evaluating the facts scientifically and reasons are examined and accepted on solid grounds. Tolerance limit is decided for probable error, Scientific methodology is kept on a very high standard. The science of Statistics possesses both the characteristics, hence it should be called a **science**.

However, statistics is not a science like physics, chemistry, economics and sociology. It may be called a science of scientific methods. Statistics help other sciences to derive their own laws. Statistical knowledge is not for its own sake but for the sake of other knowledges. This is a difference between science of statistics and other sciences. But it is very useful in the advancement of knowledges. That is why it is said that "*science without statistics bears no fruit, statistics without science has no root.*" This statement indicates the usefulness of statistics in the development of other sciences. According to Tippett, "*As a science, the statistical method is a part of the general scientific method and is based on the same fundamental ideas and processes.*"

Art is practical science. If science is knowledge then art is action.

Art has the following characteristics :

(1) Art is group of actions to solve a problem;
(2) Art does not describe facts, but examines merits and demerits and tells the way to achieve the objectives;
(3) Artist possesses special skill, experience and restraint.

By science we know a thing, through art we create that thing. Statistics is regarded as an art of applying the science of scientific methods. In the science of statistics we do not study different methods of studying a problem only, but we also study how those methods should be applied in different situations. Thus statistics may be regarded as science of scientific methods and art of applying those methods. In the words of Tippett, "*Statistics, is both a science and an art. It is a science in that its methods are basically systematic and have general application; and an art in that their successful application depends, to a considerable degree, on the skill and special experience of the statistician and on his knowledge of the field of application.*"

1.16 TYPES OF STATISTICAL UNITS

Statistical units are of the following kinds :

Statistical Units

Units of Collection and Enumeration	Units of Analysis and Interpretation
(a) Simple Unit	(a) Rate
(b) Composite or Compound Unit	(b) Ratio
(c) Hypothetical Unit	(c) Coefficient

(i) *Units of Collection and Enumeration* : These are the units in terms of which data are collected, such as, for production of cloth, meter; for weight, tonne; for time, an hour, etc. Units of collection and enumeration are of three kinds.

(a) *Simple Unit* : A simple unit is one which prescribes a single determining characteristic, *e.g.*, tonne, a kilometer, etc. Such units are easily definable.

(b) *Composite or Compound Unit* : In certain cases, no single unit will serve. It would be difficult to compare productivities on two bus routes by stating the number of buses operating on each or the miles run or even the number of passengers carried, because the buses might be running half-empty and the passengers might only travel for part of the journey. Compound units overcome this difficulty. They combine two or more simple units. In other words, a composite or compound unit is based upon more than one determining characteristic, such as, a passenger-kilometer, a tonne-kilometer, etc.

Tonne-kilometers are equal to the number of tonnes and number of kilometers carried.

(c) *Hypothetical Unit* : To facilitate comparison, sometimes hypothetical units, such as, horse-power, mule-equivalent-spindles, etc., are used.

(ii) *Units of Analysis and Interpretation* : The units on the basis of which data are compared, are *called the units of analysis* and interpretation. Rate, proportion, coefficient, etc., are such units. With the help of these units the comparison between figures becomes easy.

The units of analysis and interpretation are :

(a) *Rate* : Rate means per thousand. It is used in the analysis of population data.

(b) *Ratio* : The relative values of two homogeneous facts are expressed in terms of ratio. For example, if in a city there are 10,000 men and 8,000 women, then the men-women ratio in that city will be 1,000 : 800.

(c) *Coefficient* : When the rate is expressed per unit instead of per thousand units, it is called a 'coefficient'. This is the unit of interpretation. The formula of computing coefficient is :

$$C = \frac{Q}{N}$$

where C = Coefficient

Q = Quantity

N = Number

If there are 10,000 persons in a city and in a year there are 500 births, then the birth coefficient will be $\frac{500}{10,000} = 0{\cdot}05$.

Coefficient multiplied by the total number of observations, gives the quantity of concerned data, *i.e.*, $Q = C \times N$.

In the above example : $0{\cdot}05 \times 10,000 = 500$.

1.17 DISTRUST OF STATISTICS

Distrust means lack of faith and confidence. Distrust in statistics means lack of faith and confidence in statistical statements and methods. About statistics people often quote derogatory remarks like :

(i) Statistics are lies of the first order.
(ii) An ounce of truth will produce tonnes of statistics.
(iii) Statistics can prove or disprove anything.
(iv) Statistics can prove nothing.
(v) Figures do not lie.
(vi) If figures say so, it cannot be otherwise.
(vii) Statistics is another form of lying.
(viii) There are three kinds of lies—lies, damned lies and statistics.
(ix) Statistics are like bikinis. They reveal what is interesting and conceal what is vital.

It is only the misuse of statistics which results in the its distrust. Prof. Ya-Lun-Chou explains it as, "*Derogatory statements have been inspired by the misuse of statistics on the part of untrained practitioners of statistical methods, ignorant users of statistical results, dishonest people who employ statistics to mislead and confuse rather than to clarify and illuminate.*" King rightly highlights, "*Science of statistics is the most useful servant but only of great value to those who understand its proper use.*" For example, medicines are used for curing diseases, but wrong administration of a medicine may prove poisonous and fatal. For such unpleasant results medicine can not be blamed. Similarly, if the statistical facts are mis-interpreted by the people to serve their ends, we may be dishonest to blame the science of statistics as such. Thus, statistics are like clay of which one can make a **God** or **Devil** as he wishes. The problem of distrust of statistics arises because of its misuse. Jule and Kandall have rightly said, "*Statistical methods are most dangerous tools in the hands of an inexpert.*" Further W.I. King says, "*The science of statistics is a most useful servant but only of great value to those who understand its proper use.*" Statistics should be used for a useful purpose not as an excuse for making wrong interpretations.

According to F.C. Mills, "*Statistics, as a tool, requires proper usage and the results obtained through statistical analysis require intelligent interpretation.*"

From above words of Mills, it is clear that Statistics is useful if used properly. Sir F. Galton has also said, "*Some people hate the very name of Statistics but I find them full of beauty and interest.*

Whenever they are not brutalized but delicately handled by the higher methods and are properly interpreted, their power in dealing with complicated phenomenon is extraordinary." Distrust of statistics can be removed if it is used judiciously and by expert persons having no prejudice and bias. Prof. J.M. Keynes writes, "*If the limitations of statistics are clearly recognised, if they are accurately collected over an adequate range, if they are employed without prejudice and after full enquiry into their true significance and if they are fairly and properly grouped then their value is unique and the statistical method easily makes good its claim to rank as a most effective and reliable instrument of science.*"

Questions

(a) "Statistics affects everybody and touches life at many points. It is both a science and an art". Comment.

Ans. Statistics affects everybody and touches life at many points. Statistics is a science. As science refers to a systematic study of knowledge and studies cause and effect relationship between different variables. In science the degree of precision is very high. Art on the other hand refers to the skill of collecting and handling of data to draw logical inference and arrive at certain conclusions. The degree of precision is not as high as it is in scientific laws.

Statistics deals with such variables whose individual effects cannot be studied in isolation as is possible in natural and physical sciences, which are experimental in nature. So it would be better to call statistics not as a science but as a scientific method. Statistics is an art in as much as it is concerned with the skill of handling facts and figures for the purpose of analysis, interpretation and policy formulation.

According to A.L. Boddington, the ultimate end of statistical research is to enable comparison to be made between past and present results with a view to ascertaining the reasons for changes which have taken place and the effect of such changes in the future. The science of statistics is very closely associated with the progress of human civilization. It helps is assessing the results of the past achievements of human activities and it is also useful for making forecasts about the future course of events.

(b) "Statistics is the Science of Estimates and Probabilities". Explain and illustrate.

Ans. Statistics is the science of estimates and probabilities. Statistics is a very commonly used word but surprisingly it is understood by different people in different sense. Statistics are numerical facts, but statistics is a body of methods for making decisions when there is uncertainty arising from the incomplete of the information available. The science of statistics is the method of judging collective, natural or social phenomenon from the results obtained from the analysis or collection of estimates. Boddington has defined statistics as the science of "estimates and probabilities".

This definition gives expression only to certain methods by which conclusions are derived in this science. No doubt in most of cases statistics are, estimates and probabilities but is should be remembered that the scope of science is not confined merely to these things.

For example : Say there are 100 students in a class and using the formula for arithmetic mean, the average marks obtained are 42. That means all the students of that class obtained round about 42 marks. But it is not true practically since there will be a large variation between the marks of the students. But if the results are derived from advanced techniques of statistics then statistics will not be a science of estimates and probabilities.

(c) "Statistics plays an important role not only in the study of Economics and Commerce, but also in managerial decision-making". Explain briefly.

Ans. Statistics play an important role not in the field of economics and commerce but also in management decision making :

Usefulness in economics : Important phenomena in all branches of economics can be described compared and correlated with the help of statistics. By statistical analysis we can study the manner in which people spend their income over various items of expenditure namely food, clothing, house rent etc. Statistics of production describe the wealth of a nation and compare it year after year showing thereby effect of changing economic policies. Exchange statistics throw light on the commercial development of a nation. Distribution statistics disclose the economic condition of the various class of people.

Usefulness in commerce : Statistics are an aid to business and commerce. When a man enters business and commerce he enters the profession of forecasting. Modern devices have made forecasting more precise and definite. Economic barometers are the gifts of statistical methods and businessmen all over the world make extensive use of them.

A businessman who ignores the effects of booms and depressions can never succeed. Various branches of commerce utilise the services of statistics in different forms. Cost accounting is entirely statistical in outlook and with the help of this technique producers are in a position to decide the price various commodities. So it is extremely important to business and commerce.

The science of statistics is the method of judging collective, natural or social phenomenon from the results obtained from the analysis or enumeration or collection of estimates. In the business enterprise the business manager has the need to make a decision in the face of uncertainty and the essence of modern statistics lies in the development of general principles for dealing wisely with uncertainty.

Modern statistical tools of collection, classification, tabulation, analysis and interpretation of data have been found to be an important aid in making wise decision at various levels of managerial function. The uses to which statistical methods are put in the area are many and varied for example the success of a modern industrial enterprise depends to a significant extent on the accuracy of production programming, sales, quality and inventory control. Statistical tools are relied upon heavily in arriving at correct decisions in all these aspects.

The success of production planning both in short as well as long period depends to a great extent on the quality of sales forecasts and projections. A company's regional offices may make estimates of sales in their respective area. They also take into account seasonal variation. These regional estimates may then be subjected to statistical treatment and final forecasts made. These forecasts may further be recast in the light of statistically estimated variations in aggregate marked variations under the impact of economic programming at the national level.

The variations in sales below or above the forecast may disturb the production schedule. These variations may be seasonal and the policy of establishing production means that the inventory of finished goods would vary rather widely. The upper and lower zones are set, based on statistical study of sales fluctuation and if fluctuations occur outside this zone, changes in production schedule are made.

The determination of aggregate and regional sales potential is based on a statistical study of trends statistical methods help the sales executive in the establishment of the share of responsibility on each regional sales office for the achievement of sales targets.

Statistical methods also come to the aid of quality control. Market Research, Consumer preference study, Trade channel studies and leadership surveys are other methods of sales control which make an extensive use of statistical tools.

Very Short Answer Type Questions

1. Explain statistics as a plural noun.
2. Define empirical analysis.
3. What is inductive statistics?
4. Define descriptive statistics.
5. Explain econometrics.

Short Answer Type Questions

1. Mention four important functions of statistics.
2. Explain statistics in singular and plural sense.
3. What are the two meanings of statistics?
4. Define statistics.

5. Explain the importance of statistics for the Government.
6. What does statistical thinking deal with?
7. Write a note on distrust of statistics.
8. What is the importance of statistics in business?
9. What is the importance of statistics in planning?
10. What is the importance of statistics in economics?
11. "Statistics affects everybody and touches life at many places." Explain.
12. What is statistics of consumption and statistics of production?
13. Explain inferential statistics.
14. Write four limitations of statistics.
15. Write a short note on scope of statistics.

Long Answer Type Questions

1. Explain the limitations and scope of statistics.
2. Explain the functions of statistics.
3. Explain distrust of statistics, how it can be removed?
4. Illustrate the need for empirical analysis in economics with two examples.
5. "Statistical analysis is of vital importance for successful businessmen, economists, administrators and educationists." Discuss with illustrations.
6. "Statistical thinking is as necessary for efficient citizenship as the ability to read and write." Explain this statement in about 200 words.
7. "A well-wrapped statistician is better than Hitler's 'Big-lie' it misleads, yet it cannot be pinned on you." Discuss.
8. The study of an individual fact lies outside the scope of statistics.
9. What is the difference between statistical methods and experimental methods?
10. Statistics is the science of estimates and probabilities. Explain.
11. Comment on the following statements :
    (i) Statistics can prove any thing (ii) Statistics can prove nothing
12. Explain the role of statistics in commerce.
13. Statistics is the science of counting. Explain.
14. The proper function of statistics indeed is to enlarge individual experience.
15. What are the drawbacks of statistics? Give two examples of the misuse of statistics.
16. "All statistics are numerical statements, but all numerical statements are not statistics." Explain.
17. There is hardly any field which does not fall within the scope of statistics. **(Punjab B.Com., Oct. 2000)**
18. Distinguish between units of collection and units of analysis.

State whether the following statements are True or False

1. Statistics is concerned with quantitative data only.
2. Quantitative data lends preciseness to the facts.
3. As a singular noun statistics means statistical data.
4. Any single numerical figure is statistics.
5. All statistics are numerical statements.
6. Statistics can be misused.
7. Statistics do not ignore the quantitative aspects.

8. Statistics deals with collection of data and not its presentation.
9. Statistics helps in forecasting.
10. Statistical analysis is applicable in all investigations wherefrom quantitative data can be obtained.
11. Statistics is the science of analysing quantitative data.
12. Statistics are numerical statements of facts, but all facts numerically stated are not statistics.
13. Statistics is both singular and plural.
14. Statistics affects everybody and touches life at many points.
15. Statistics is always concerned with mass phenomena and never with a singular observation.

ANSWERS

1. T **2.** T **3.** F **4.** F **5.** T **6.** T **7.** F **8.** T **9.** T **10.** T
11. T **12.** T **13.** T **14.** T **15.** T

Fill in the Blanks

1. Statistics as a ...... noun means the same thing as data. (singular, plural)
2. Statistics are ...... statements. (qualitative, quantitative)
3. ...... is the first step in a statistical investigation. (Collection of data, Organisation of data)
4. The nature of statistical inference is .... (deductive, inductive)
5. The statistics as a ...... noun refers to statistical methods. (singular, plural)
6. ...... includes condensation and summarisation of data. (Analysis, Interpretation)
7. ...... facilitates comparison of daa. (Economics, Statistics)

ANSWERS

1. plural **2.** quantitative **3.** Collection of data **4.** inductive
5. singular **6.** Analysis **7.** Statistics

Multiple Choice Questions

Tick the correct answer :

1. Statistics as a singular noun means :
(a) statistical data (b) statistical methods (c) inductive statistics (d) descriptive statistics
2. Statistics as a plural noun indicates :
(a) statistical methods (b) a bundle of lies (c) statistical data (d) Fool's Paradise
3. Distrust of statistics is due to :
(a) misuse of statistics (b) insufficient statistical methods
(c) scope of statistics is limited (d) statistics is considered useless by many people
4. Statistics facilitates :
(a) comparison of data (b) collection of data (c) organisation of data (d) disposal of data
5. Statistics is the science of analysing :
(a) qualitative data (b) quantitative data
(c) any kind of data (d) only very simple facts

ANSWERS

1. (b) **2.** (c) **3.** (a) **4.** (a) **5.** (b)

❑❑❑

2 Preparation of Frequency Distribution

Coverage Classification, Tabulation, Frequency Distributions.

2.1 FREQUENCY DISTRIBUTION

A frequency distribution is any device such as a graph or table that displays the values that a variable can assume along with the frequency of occurrence of these values either individually or as they are grouped into a set of mutually exclusive and exhaustive intervals.

In the words of Croxton and Cowden, frequency distribution is defined as "*a statistical table which shows the set of all distinct values of the variables arranged in order of magnitude either individually or in groups, with their corresponding frequencies side by side.*"

Frequency distribution is a method of organising and condensing data.

2.2 FREQUENCY ARRAY

Frequency array is simply a frequency distribution in which values are arranged in proper order.

Frequency distribution presents data in a relatively compact form and contains information that is adequate for many purposes. It presents raw (unprocessed) data in a more usable and understandable form.

2.3 CLASSIFICATION AND TABULATION

Classification and Tabulation of data forms a part of organisation of data. Classification is preliminary to tabulation and prepares the ground for proper presentation of statistical data. It is only after proper tabulation that statistical analysis and interpretations are possible.

Classification

Classification is the process of arranging the data according to some common characteristic possessed by the facts constituting the data. The facts having a common characteristic are termed as **one class** or **group**. Thus classification is the grouping of related facts into different classes.

According to Conner, "*Classification is the process of arranging things [either actually or notionally] in groups or classes according to their resemblances and affinities and gives expression to the unity of attributes that may subsist amongst a diversity of individuals*".

Secrist has defined classification as : "*classification is the process of arranging data into Sequences and Groups according to their common characteristics or separating them into different related Parts*".

According to Tuttle A.M., "*A classification is a scheme for breaking a category in to a set of parts called classes according to some precisely defined differing characteristics possessed by all the elements of the category*".

Example of Classification :

(i) Letters in a Post Office are sorted and separated according to area. The letters belonging to the same area may be said to be in the same class.

(ii) Some times in hospitals patients are classified according to disease and are admitted to different wards. Patients having the same disease are admitted to the same ward and can be said to form one class.

Thus, by way of Classification a huge heterogeneous data is converted into homogeneous groups or classes.

2.4 PURPOSE OF CLASSIFICATION (OBJECTIVES OF CLASSIFICATION)

The data collected at random is of no use for a statistician unless properly classified and tabulated. *For Proper tabulation, classification is essential.* The main or principal objectives of classification are :

1. to condense the mass of data in such a manner that similarities and dissimilarities are readily apprehended and relationships studied.
2. to facilitate comparison.
3. to have a bird's eye-view of the significant features of the data.
4. to enlight important information while giving less prominence to insignificant items.
5. to utilise the data for tabulation and further statistical analysis.
6. to eliminate unnecessary details contained in the raw data.
7. to present the complex, scattered data in a concise, logical and understandable form.

2.5 CLASS INTERVALS

Class intervals are contiguous non-overlapping intervals selected in such a way that they are mutually exclusive and exhaustive, *i.e.*, each and every value in the set of data can be placed in one and only one of the intervals and the complete data is covered in all the intervals.

2.6 REQUIREMENTS/ESSENTIALS OF A GOOD CLASSIFICATION

(1) Classification should be done in such a way that the entire data is covered and not even a single item is left unclassified, *i.e.*, the Classification should be exhaustive.

(2) Each item in the data should belong only to one class by avoiding overlapping. This can be achieved by defining various classes properly and arranging data in the respective classes. Suppose we want to classify Students on the basis of their family income into various classes, such as poor, middle and rich. These classes must be demarcated in such a way that

there is no overlapping to avoid any confusion in the mind of a research worker as to who is poor and who is rich. The classes should be mutually exclusive.

(3) If the students are classified on the basis of the marks, their intelligence level can be compared. Thus, classification should facilitate comparison.

(4) As far as possible the class intervals should be of equal length, *i.e.*, they should cover equal range of values. If possible these ranges should be multiples of numbers that are easy to work, *e.g.*, 5, 10, 100, etc. This will facilitate constructing, reading and using the distribution.

(5) Classification should conform to the objects of investigation.

(6) It should be flexible but flexibility should not be at the cost of other features.

(7) Items constituting a group (or class) should be homogeneous.

2.7 TYPES [OR KINDS] OF CLASSIFICATION

(i) Quantitative, *i.e.*, in terms of magnitudes.

(ii) Chronological [or Temporal], *i.e.*, on the basis of the time.

(iii) Geographical [or Spatial], *i.e.*, areawise, *e.g.*, cities, districts, etc.

(iv) Qualitative, *i.e.*, according to some attribute—Twofold (Dichotomous) and Manifold.

Methods of Classification

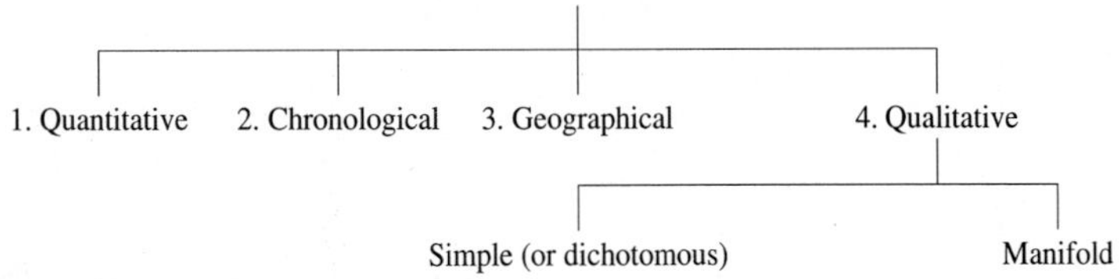

1. Quantitative Classification

If the data are classified on the basis of same characteristics capable of quantitative measurement such as height, weight, age, income, expenditure, marks scored by students in class, etc., the classification is known as quantitative classification. The quantitative characteristics under study is known as variable and hence this classification is sometimes called **'classification by variables'**. For example the students, say 100, in a section can be classified on the basis of percentage marks scored by them.

Percentage Marks	*No. of Students*
0–20	10
20–40	20
40–60	35
60–80	32
80–100	3

In this type of classification thee are two variables. In the above example, the two variables are (i) marks (ii) The frequency, *i.e.*, the number of students in each class or category.

2. Chronological Classification

When the data are classified on the basis of time then it is known as chronological classification. Such series are known as time series because one of the variables is time. If the population of a country, sales of a firm, imports, exports, etc., are classified for a certain number of years it will form a chronological classification. For example,

(i) Population of India

Year	1931	1941	1951	1961	1971	1981	1991	2001
Population (in crores)	27.9	31.8	35.7	43.8	54.6	68.4	84.6	102.7

(ii) Production of TV Sets by a Company *X*

Year	*Quantity*
1992	12800
1993	13500
1994	14200
1995	15800
1996	16300
1997	17000
1998	17500
1999	18300
2000	19000

3. Geographical (Spatial) Classification

In Geographical Classification data are classified on the basis of place, *e.g.*, (i) the production of wheat in different states of the country, (ii) the national income of different countries, etc.

4. Qualitative Classification

If the data are classified on the basis of some attribute or quality [descriptive characteristic] such as sex, literacy, beauty, honesty, intelligence, religion, eductation, colour of hair, etc., the classification is called qualitative classification. In this type of classification, the attribute under study cannot be measured but its presence or absence can be found or felt. This type of classification is called **Simple** or **Dichotomous** or **Two-fold Classification.**

Two-fold and Manifold Classification : When on the basis of presence and absence of an attribute the data are classified into two classes, one possessing that attribute and the other not possessing that attribute, it is called Two-fold Classification, *e.g.*, on the basis of Marital Status, the Whole Population can be divided into two Classes—married and unmarried or on the basis of Sex the Whole Population can be divided into two Classes—males and females.

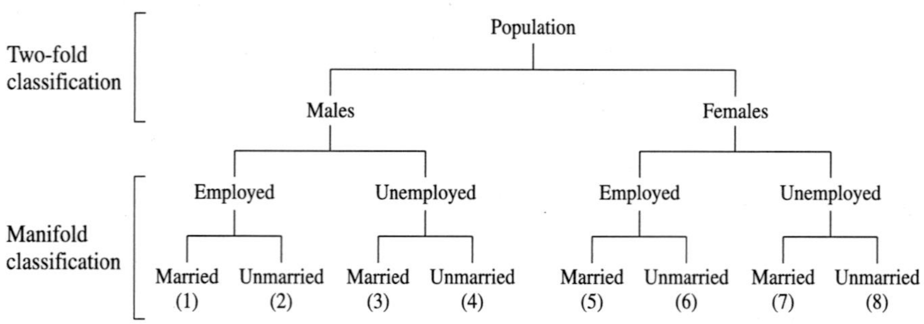

In **manifold classification** each class is divided into two classes on the basis of one attribute. For example, in the above table males and females are first divided on the basis of employment and then on the basis of marital status.

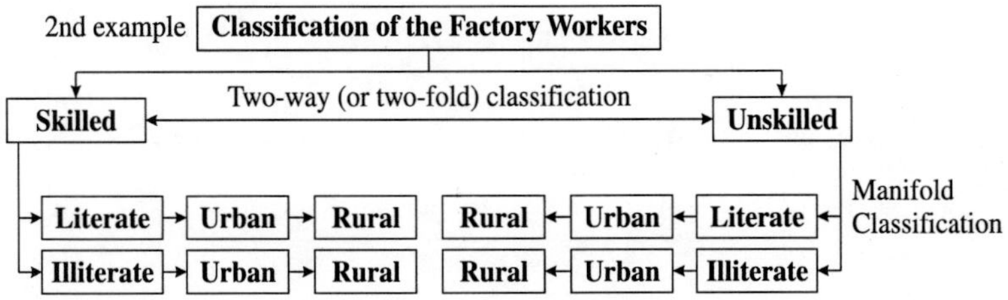

2.8 ENTITY

When we make observations about Persons, Places and Things, we call that which is being observed as entity regardless of type of unit involved.

2.9 VARIABLE

A quantity which can assume a range of numerical values is called a variable and each value within the range is called variable or observation. According to George and Simpson, "*Every frequency distribution involves the Classification of a trait or quality that exhibits differences in magnitude, e.g., prices, wages, number of employees, age, number of units produced or consumed... the trait or quality which varies in amount or magnitude in frequency distribution is called a variable.*"

Variables can be (i) Discrete (ii) Continuous.

A discrete random variable is one which can assume only certain values within an interval. Normally discrete variables remain different from each other by a definite quantity, there are gaps in between the variable values.

A continuous variable is defined as a variable that can assume a large number of values between any two points. A continuous variable (or continuous random variable) has infinitely many values and those values can be associated with points on a continuous scale in such a way that there are no gaps or interruptions. Random variables that represent heights, weights, times and temperatures are usually continuous.

In general, the value of a continuous random variable is found by measuring whereas the value of discrete random variable is determined by counting.

2.10 DIFFERENCE BETWEEN AN ATTRIBUTE AND A VARIABLE

Any characteristic possessed by an object (or individual) may be measurable or non-measurable. Measurable characteristics are quantitative in nature and can be expressed numerically in terms of some units. A measurable characteristic is known a **variable**. *Examples* : height, weight, wage, etc.

* The variable whose value is determined by Chance factors is called a *random variable*. A random variable has a Single numerical value for each outcome of an experiment. The values of the random variable are the numbers we associate with the different outcomes that make up the Sample Space for the experiment, *e.g.*, consider the experiment of randomly selecting three voters. Let the random variable be the number of selected voters who are College graduates. The random variable can then assume the values 0, 1, 2 and 3.

Non-measurable characteristics are those which are qualitative in nature. They cannot be measured but can only be classified under different categories and are termed as **attribute**. *Example* : sex, mother tongue, complexion, honesty, beauty, religion, etc.

2.11 FORMATION OF FREQUENCY DISTRIBUTIONS

The number of times a Value occurs in a Series is called the *frequency of* that value and the arrangement obtained by mentioning the frequency against each value in the series is called *frequency distribution*. For example, in the series 3, 3, 4, 4, 4, 1, 5, 7, 7, 7, 7.

The frequency of 3 is 2, frequency of 4 is 3, etc. and the arrangement

Value	3	4	1	5	7
Frequency	2	3	1	1	4

is called frequency distribution.

There are two types of frequency distributions :

(i) Discrete frequency distribution. (ii) Grouped frequency distribution.

2.12 DISCRETE FREQUENCY DISTRIBUTION

The formation of discrete frequency distribution is quite simple. The number of times a particular value is repeated is noted down and mentioned against that value instead of writing that value repeatedly, *e.g.*, in the distribution 3, 4, 6, 6, 8, 8, 8, 8, 1, 1, 1 we can write

Value	*Frequency*
3	1
4	1
6	2
8	3
1	3

2.13 GROUPED FREQUENCY DISTRIBUTION

In this type of distribution, the various items of a series are classified into groups or classes and the number of items lying in each group or class is put against that group or class. The lowest and the highest values that can be included in the class or group are called **class limits**, the lowest value being the **lower limit** and the highest value being the **upper limit**. The actual lower limit of a class is a value such that no lower value can fall into that class. The actual higher limit of a class is a value such that no higher value can fall into that class.

The width of the class is known as the **class interval**. The number of items falling within the range of the class interval is called the **frequency** of that class.

The systematic presentation of the classes with their respective frequencies is called a *frequency table* or a *frequency distribution*. The number of items falling within a class is known as *class frequency* whereas the total number of items in all the classes is called the *total frequency of the distribution.*

2.14 OPEN-END CLASSES

All open-end class (or unterminated class) is a class in which either the lower limit or the upper limit is missing. In general, the term open is applied to any class of the less than or more than type classification.

If a set of data contain a few values which are much greater than or much smaller than the rest, open-end classes are quite useful in reducing the number of classes required to accommodate the data. However, in practice they are generally avoided because open-end classes make it difficult to calculate certain statistical measures like arithmetic mean.

2.15 FORMATION OF CLASSES

Classes can be formed in two ways :

(i) Exclusive Type, (ii) Inclusive Type

(i) Exclusive Type

When the class intervals are so fixed that the upper limit of one class is the lower limit of the next class, it is known as **exclusive method of classification.**

The following classification illustrates this method.

Marks (Percentage)	*No. of Students*
0–10	15
10–20	17
20–30	22
30–40	30
40–50	39
50–60	45

In this method *highest value of the variable in the class is not included in that class.* To be more elaborate, the classes in the above example can be expressed as

Marks (Percentage)	*No. of Students*
0 and more but less than 10	15
10 and more but less than 20	17
20 and more but less than 30	22
30 and more but less than 40	30
40 and more but less than 50	39
50 and more but less than 60	45

The student getting 10% marks is not included in the class 0–10 but he is included in the class 10–20.

(ii) Inclusive Type

In this method the student getting say 39% marks will be included in the class 30–39 itself.

Marks (Percentage)	*No. of Students*
0–9	5
10–19	8
20–29	7
30–39	13
40–49	25

Note : As far as possible the Inclusive method should not be used. In case the data is given in the Inclusive method it can be converted to the Exclusive method by the following steps :

(1) Find the difference between the upper limit of any class and the lower limit of the next class.

(2) Divide the difference found in (1) by 2.

(3) Subtract the fraction obtained in (2) from the lower limit of all the clases and add the same fraction to the upper limit of all the classes, *e.g.*, in the above example, the difference between the upper limit of first class and lower limit of the second class is $10 - 9 = 1$. Half of this difference $= 1/2 = 0.5$. The above data in the modified way (*i.e.*, Exclusive method) can be written as :

Marks (Percentage)	*No. of Students*
–0.5–9.5	5
9.5–19.5	8
19.5–29.5	7
29.5–39.5	13
39.5–49.5	25

Note : In the grouped frequency distribution it is presumed that frequencies are evenly distributed over class intervals and **mid-value** of each class is a **representative** of that class.

The 'frequencies or observations are evenly distributed' implies that if in the class 10–20 there are 10 values, then first 5 values are included in the class 10–15 and next 5 in the class 15–20.

2.16 CLASS FREQUENCY

The number of items or values or observations belonging to a class is known as the frequency of that class.

2.17 MIDDLE VALUE OF A CLASS OR CLASS MARK

$$\text{Mid-value of a class} = \frac{1}{2}\,[\text{Upper limit of the class} + \text{Lower limit of the class}]$$

$$= \text{Value which is exactly at the middle of the class.}$$

In the class x-y the mid-value $= \dfrac{x+y}{2}$.

$$\frac{\text{Class interval}}{2} \quad \text{or} \quad \left[\frac{1}{2}\text{ width of the class}\right] = \frac{y-x}{2}.$$

$$\text{Upper limit of the class} = \left(\frac{x+y}{2}\right) + \left(\frac{y-x}{2}\right) = y.$$

$$\text{Lower limit of the class} = \left(\frac{x+y}{2}\right) - \left(\frac{y-x}{2}\right) = x.$$

If the classes are of equal width, the difference between any two consecutive mid. points give the class interval (or width of the class). **(Please note)**

$$\text{Upper limit of the class} = \text{Mid-value of the class} + \frac{1}{2}\text{ width of the class.}$$

$$\text{Lower limit of the class} = \text{Mid-value of the class} - \frac{1}{2}\text{ width of the class.}$$

Note : The frequency distribution expressed in different classes by Inclusive or Exclusive method is also called **grouped frequency distribution**.

2.18 CLASS LIMITS AND CLASS BOUNDARIES

Every class in a frequency distribution has 'expressed' class limits and real class limits, also called *class boundaries.* In case of **Exclusive Type** Classification, there is no difference between the class limits and class boundaries. In the case of **Inclusive** Type Classification, class boundaries are obtained by increasing the Upper class limit and decreasing the Lower class limit by one half of the difference between the upper limit of one class and the lower limit of the next class.

Class boundaries are calculated from class limits by the following rule :

$$\text{Lower class boundary} = \text{lower class limit} - \frac{1}{2}d,$$

$$\text{Upper class boundary} = \text{upper class limit} + \frac{1}{2}d,$$

where d = common difference between upper class limit of any class-interval and the lower class limit of the next class-interval. $\frac{1}{2}d$ is called the **correction factor, *e.g.,***

In Exclusive Type of classification d is always zero.

Example : For Class interval 1–10, 11–20, ...

Class boundaries are 0.5, 10.5, 20.5 etc., here $d = 1$
whereas class limits are 1, 10, 20, etc.

2.19 ACTUAL CLASS INTERVAL (OR CLASS WIDTH)

The actual class interval of any class in a frequency distribution is equal to the distance from the lower real limit of that class to the lower real limit of next class.

In Inclusive Type of Classification, two ends of a class-interval are called class-limits of a particular class-interval. The smaller and greater numbers are known as lower and upper class-limits respectively.

In the class interval 1–10, lower class-limit is 1 and upper class-limit is 10.

2.20 INCLUSIVE CLASSIFICATION, EXCLUSIVE CLASSIFICATION AND CONTINUOUS FREQUENCY DISTRIBUTION

Consider the following class intervals

39.5–44.5
44.5–49.5
49.5–54.5
54.5–59.5

Here the class intervals over lap. The value 49.5 belongs to class 49.5–54.5 and not to 44.5–49.5. This brings in continuity of the class intervals and eliminates the defects of overlapping of the classes. As mentioned earlier also this type of classification is called exclusive classification because the item whose value coincides with the upper limit of the class interval is excluded from that interval and is included in the next one. This form of the frequency distribution is known as **continuous frequency distribution**.

If the values are rounded at the time of data collection, the above class intervals can be written as :

40–44
45–49
50–54
55–59 and so on.

Here the value 44 is put in the first class and next value 45 will be put in the next class and no value between 44 and 45 is recorded. This type of distribution has gaps and can not be called **exactly continuous**. However the variable values between 40 and 44, 45 and 49 etc. can vary continuously.

The exclusive type classification is exactly continuously where as inclusive type classification is not exactly continuous. **(Important)**

Illustration : If the class mid-points in a frequency distribution of a group of persons are 125, 132, 139, 146, 153, 160, 167, 174 and 181 lbs, find :

(i) Size of class-interval;
(ii) Class-boundaries;
(iii) Class-limits assuming that the weights are measured to the nearest pound.

(DU B.Com. Hons. 2007)

Solution. If the class intervals are equal the difference between any two consecutive middle values is equal to the class interval.

(i) Class interval = 7.
(ii) Class boundaries are given by

121.5–128.5
128.5–135.5
135.5–142.5
142.5–149.5
149.5–156.5
156.5–163.5
163.5–170.5
170.5–177.5
177.5–184.5

(iii) When the weights are measured to the nearest pound that the class intervals are :

122–128
129–135
136–142
143–149
150–156
157–163
164–170
171–177
178–184

2.21 CUMULATIVE FREQUENCY

Cumulative frequency of any class = Sum of the frequencies of all the preceding classes plus the frequency of that class.

2.22 FREQUENCY DENSITY

Frequency density of a class is defined as the frequency per unit width of the class

$$\text{Frequency density of a class} = \frac{\text{Frequency of the class}}{\text{width of the class}}$$

If the classes in a frequency distribution are not of equal width, we cannot compare the frequencies of the different classes. In this case the frequency densities of the class are used for comparing the concentration of frequencies in the classes.

2.23 RELATIVE FREQUENCY OF A CLASS

Relative frequency of a class is the ratio of the number of observations in that class to the total number of observations in all the classes.

$$\text{Relative frequency of any Class} = \frac{\text{Frequency of that Class}}{\text{Sum of the Frequencies of all the Classes}}$$

2.24 DETERMINATION OF NUMBER OF CLASS FOR A FREQUENCY DISTRIBUTION

An ideal number of classes for any frequency distribution would be that which gives the maximum information in the *clearest fashion*. Ordinarily a frequency distribution should not contain more than 20–25 and not less than 6–8 classes. If the number of classes are very few, there is a danger of obscuring some important characteristics of the data and if the number of classes is too large, it becomes difficult to ascertain them though all the characteristics of the data will be available. Balance is always maintained between the two situations.

Sturges has given a formula to determine the number of classes.

According to this formula, $K = 1 + 3.322 \log N$,

where K is the number of classes and N the total number of observations. If the value of K comes to be a fraction, an approximate round number close to the theoretical value should be taken.

2.25 DETERMINATION OF MAGNITUDE OF CLASS INTERVALS (WIDTH OF CLASSES) IN A GIVEN FREQUENCY DISTRIBUTION

The magnitude of class intervals depends on the range of the data and the number of classes. (*Range is defined as the difference between the highest and the lowest values of the variable*). The magnitude of the class interval should be such that it does not distort or obscure the important characteristics of the data.

Sturges has given a formula for determining width of classes as

$$i = \frac{\text{Range}}{1 + 3.222 \log N} = \frac{\text{Range}}{\text{No. of classes}}$$

SOLVED EXAMPLES

Example 1. The following table shows the time spent by 200 students in watching television programs during in month :

Time (in hours)	*Number of Students*
less than 10	8
10–20	16
20–30	26
30–40	20
40–50	30
50–60	40
60–70	20
70–80	18
80–90	16
more than 90	6

Calculate :

(i) the number of students spending up to 25 hours.

(ii) the number of students spending at least 55 hours.

Solution. The frequency distribution of the given data is shown below :

Times (in hours)	*Mid-Value*	*Frequency*	*Cumulative Frequency*
less than 10	5	8	8
10–20	15	16	24
20–30	25	26	50
30–40	35	20	70
40–50	45	30	100
50–60	55	40	140
60–70	65	20	160
70–80	75	18	178
80–90	85	16	194
more than 90	95	6	200

From the cumulative column it is obvius that

(i) Number of students spending upto 25 hours = 50.

(ii) Number of students spending at least 55 hours, *i.e.*, 55 hours and more = 200 – 100 = 100.

Note : For taking mid-value of the class 'less than 10', the class is assumed to be 0–10 and for the class 'more than 90', the class is assumed to be 90–100.

Example 2. If the class mid-point in a frequency distribution of age of a group of persons are 25, 32, 39, 46, 53 and 60, find :

(i) the size of the class interval.

(ii) the class boundaries.

(iii) the class limits, assuming that the age quoted is the age completed last birthday.

Solution. (i) The size of the class interval

= Difference between the mid-values of any two consecutive classes = 7.

[Since 32 – 25 = 39 – 32 = ... = 60 – 53 = 7]

(ii) Upper limits for different classes

$$25+\frac{7}{2}, 32+\frac{7}{2}, 39+\frac{7}{2}, 46+\frac{7}{2}, 53+\frac{7}{2}, 60+\frac{7}{2}, \textit{i.e.,}$$

28.5, 35.5, 42.5, 49.5, 56.5, 63.5

and corresponding lower limits are

$$25-\frac{7}{2}, 32-\frac{7}{2}, 39-\frac{7}{2}, 46-\frac{7}{2}, 53-\frac{7}{2}, 60-\frac{7}{2},$$

i.e., 21.5, 28.5, 35.5, 42.5, 49.5, 56.5.

∴ The class boundaries (Exclusive type) are :

Class	*Mid-point*
21.5–28.5	25
28.5–35.5	32
35.5–42.5	39
42.5–49.5	46
49.5–56.5	53
56.5–63.5	60

(iii) Assuming the age quoted (X) is the age completed on last birthday then X will be a discrete variable which can take only integral values. Hence, the given distribution can be expressed in an 'inclusive type' of classes with the class interval of magnitude 7, as given in the table below :

Age (on last birthday)	*Mid-point*
22–28	25
29–35	32
36–42	39
43–49	46
50–56	53
57–63	60

2.26 TALLY BARS (OR TALLY MARKS)

The number of times a variable is repeated in a distribution is called the **frequency** of that variable. The number of values contained in a class is called the **frequency** of that class. In order to facilitate counting and determining the frequency, we take the help of *tally bars* (also called *tally marks*). The symbol "|" represents a tallybar (or *tally mark*). We take each item from the data, one at a time and put the tally bar against the class to which it belongs. For the sake of convenience, we record tally bars in the bunches of 5, the fifth one crossing diagonally the other four. For example,

卌 ||| = 8. This method of marking and counting is known as 'Fair and Cross' Method.

Example 3. Using Sturges rules determine the number of classes in which the following data should be divided. Classify the data in equal class intervals.

178	114	110	161	164	155	175	157	108	128
62	30	165	195	87	71	133	151	94	42
141	104	130	167	116	164	156	124	146	149
143	93	103	162	62	79	150	149	113	121
148	204	140	187	40	197	144	184	87	122

Solution. Number of classes or K would be :

$K = 1 + 3.322 \log N$, where N denote the number of items and K the number of classes.

$N = 50$, so $K = 1 + 3.222 \log 50 = 1 + (3.322 \times 1.6990) = 1 + 5.63 = 6.63$ or 7.

The range of the data is $204 - 30 = 174$. If it is divided by the number of classes the magnitude of the class interval would be $\frac{174}{7} = 25$ approximately.

The lowest class would be 30–55 and the next 55–80.

The data would be classified as follows :

Class-Intervals	*Tally bars*	*Frequency*
30–55	|||	3
55–80	||||	4
80–105	~~||||~~ |	6
105–130	~~||||~~ ||||	9
130–155	~~||||~~ ~~||||~~ ||	12
155–180	~~||||~~ ~~||||~~ |	11
180–205	~~||||~~	5

Example 4. Form a frequency distribution from the following data by Inclusive Method taking 4 as the magnitude of class intervals :

10	17	15	22	11	16	19	24	29	18
25	26	32	14	17	20	23	27	30	12
15	18	24	36	18	15	21	28	33	38
34	13	10	16	20	22	29	29	23	31

Solution. Since the minimum value of the variable is 10 which is a very convenient figure for taking the lower limit of the first class and the magnitude of the class intervals is given to be 4, the classes for preparing frequency distribution by the 'Inclusive Method' will be 10–13, 14–17, 18–21, 22–25, ..., 34–37, 38–41, the last class being 38–41, because the maximum value in the distribution is 38. The final frequency distribution along with the tally marks is as under :

Class-Intervals	*Tally bars*	*Frequency*
10–13	~~||||~~	5
14–17	~~||||~~ |||	8
18–21	~~||||~~ |||	8
22–25	~~||||~~ ||	7
26–29	~~||||~~	5
30–33	||||	4
34–37	||	2
38–41	|	1

Example 5. Prepare a statistical table from the following data :

Daily wages of 100 workers (in Rs.) in Factory *A*

88	23	27	28	86	96	94	93	86	99
82	24	24	55	88	99	55	86	82	36
96	39	26	54	87	100	56	84	83	46
102	48	27	26	29	100	59	83	84	48
104	46	30	29	40	101	60	89	46	49
106	33	36	30	40	103	70	90	49	50
104	36	37	40	46	106	72	94	50	60
24	39	49	46	66	107	76	96	46	67
26	78	50	44	43	46	79	99	36	68
29	67	56	99	93	48	80	102	32	51

Solution. The lowest value is 23 and the highest 107. The difference in the highest and the lowest value is 84. If we take a class interval of 10, nine classes would be formed. The first class should be taken as 20–30 instead of 23–33 as per the principles of classification.

Frequency Distribution of the Wages of 100 Workers

Wages (Rs.)	*Tally*	*Frequency*
20–30	~~IIII~~ ~~IIII~~ III	13
30–40	~~IIII~~ ~~IIII~~ I	11
40–50	~~IIII~~ ~~IIII~~ ~~IIII~~ III	18
50–60	~~IIII~~ ~~IIII~~	10
60–70	~~IIII~~ I	6
70–80	~~IIII~~	5
80–90	~~IIII~~ ~~IIII~~ IIII	14
90–100	~~IIII~~ ~~IIII~~ II	12
100–110	~~IIII~~ ~~IIII~~ I	11
Total		100

Example 6. The weights in gm of 54 apples picked out at random from a consignment are as follows :

106	107	76	82	109	107	115	93	187	123	95	123	125	111
92	86	70	126	68	130	129	119	119	115	115	128	100	186
84	99	113	204	111	141	139	123	90	115	115	98	110	78
98	107	81	130	75	85	105	110	80	182	118	82		

Form a grouped frequency table by dividing the variate range into intervals of equal width, each corresponding to 20 gm, in such a way that the mid-value of the first class interval is 70 gm.

Solution. **Frequency Table of the Weight of 54 Apples**

Wages (in gm.)	*Tally*	*Frequency*
60–80	~~IIII~~	5
80–100	~~IIII~~ ~~IIII~~ IIII	14
100–120	~~IIII~~ ~~IIII~~ ~~IIII~~ ~~IIII~~	20
120–140	~~IIII~~ ~~IIII~~	10
140–160	I	1
160–180		–
180–200	III	3
200–200	I	1
		Total 54

Example 7. The following table gives the purchases made by 50 customers at a department store. Form a frequency distribution, taking the class intervals as 0–9.99, 10–19.99, 20–29.99, etc.

Purchases (in Rs.)							
35.00	8.25	19.00	81.40	52.00	41.25	21.25	32.40
60.00	45.50	49.75	43.75	39.10	2.20	25.00	57.50
36.30	65.00	30.00	55.00	70.00	38.25	29.25	52.25
46.50	66.00	36.50	45.25	5.60	56.75	59.00	39.00
33.50	59.00	4.00	17.60	23.50	44.50	39.50	79.00
39.75	20.00	66.50	36.50	47.80	37.25	3.75	50.00
26.00	42.50						

Solution. **Frequency Distribution of the Purchases Made**

Purchases (Rs.)	*Tally*	*Frequency*
0–9.99	~~IIII~~	5
10–19.99	II	2
20–29.99	~~IIII~~ I	6
30–39.99	~~IIII~~ ~~IIII~~ III	13
40–49.99	~~IIII~~ IIII	9
50–59.99	~~IIII~~ III	8
60–69.99	IIII	4
70–70.99	II	2
80–89.99	I	1
		Total 50

Example 8. Given below are the number of tools product by 100 workers in a factory.

43	18	25	18	39	44	19	20	20	26
40	45	38	25	13	14	27	41	42	17
34	31	32	27	33	37	25	26	32	25
33	34	35	46	29	24	31	34	35	24
28	30	41	32	29	28	30	31	30	31
28	31	30	41	32	29	28	30	30	34
31	35	36	29	26	32	36	35	36	37
32	23	22	29	33	37	33	27	24	36
23	42	29	37	29	23	44	41	45	39
21	21	42	22	28	22	15	16	17	28

Form an inclusive frequency distribution showing tally marks in number of class intervals in a descending order. Give also the class boundaries, the class marks and answer the following :

(a) How many workers produced more than 43 tools?

(b) How many workers produced less than 23 tools?

(c) What are the class boundaries of workers in the lower 16% and the upper 20% of the distribution?

Solution. Using the Sturges' formula for determining the number of class intervals, we have

$$\text{Number of class intervals} = 1 + 3.322 \log_{10} N = 1 + 3.322 \log_{10} 100$$

$$= 1 + 3.322 \times 2 = 7.64$$

$$\therefore \qquad \text{Size of classes} = \frac{\text{Highest value} - \text{Lowest value}}{\text{Number of classes}}$$

$$= \frac{46 - 13}{7.64} = \frac{33}{7.64} = 5^* \text{ (approx.)}$$

Hence taking the magnitude of each class interval as 5, we shall get 7 classes. Since the lowest value is 13, which is quite a convenient figure for being taken as the lower limit of the first class, the various classes in descending order by the Inclusive Method would be

43–47, 38–42, ..., 13–17.

Using tally marks, the required frequency distribution is obtained in the following table.

Frequency Distribution of Numbers of Tools Produced by 100 Workers

Class Intervals (inclusive type)	*Tally*	*Number of Tools Produced*
43–47	𝍸 |||	8
38–42	𝍸 𝍸 ||	12
33–37	𝍸 𝍸 𝍸 𝍸	20
28–32	𝍸 𝍸 𝍸 𝍸 𝍸 |||	28
23–27	𝍸 𝍸 𝍸		16
18–22	𝍸 𝍸	10
13–17	𝍸		6
Total		100

The frequency distribution with class boundaries and class marks (the mid-points) is :

Class Boundary (exclusive type)	*Class mark*	*Frequency*
42.5–47.5	45	8
37.5–42.5	40	12
32.5–37.5	35	20
27.5–32.5	30	28
22.5–27.5	25	16
17.5–22.5	20	10
12.5–17.5	15	6

* Here the class interval could have been taken 4 but generally the class interval is taken as 5 or some multiple of 5 to facilitate calculations.

(a) Workers producing more than 43 tools are 8.
(b) Workers producing less than 23 tools are $6 + 10 = 16$.
(c) The class boundaries for lower 16% are 22.5 to 12.5 and for upper 20% they are 47.5 to 37.5.

Example 9. Following the rise in cost of living, the D.A. of the employees of a firm was increased at the following rates :

Rs. 200 for the pay range up to Rs. 1990.
Rs. 250 " " " " from Rs. 2000–3990
Rs. 380 " " " " Rs. 4000–5990
Rs. 450 " " " " Rs. 6000–7990
Rs. 520 " " " " Rs. 8000–9990

No increase in the grade with pay Rs. 10000 or more. What will be the additional amount required to be paid by a department in a year which has 25 employees with the following pay in rupees :

8720, 1640, 6320, 7920, 2850, 1690, 6840, 4850, 5350, 9320, 1230, 7850, 3380, 2190, 5420, 1850, 9920, 5080, 1800, 3800, 4280, 2720, 1360, 1800, 4290.

Solution. As we are given the amount of D.A. according to the pay range of the employees within classes up to 1990, 2000–3990, ..., 8000–9990, we shall convert the given distribution into frequency distribution with these classes.

Frequency Distribution of 25 Employees

Pay Range	*Tally Marks*	*Number of Employees (f)*	*D.A. in Rs. (x)*	*Total Amount (fx)*
up to 1990	~~IIII~~ II	7	200	1400
2000–3990	~~IIII~~	5	250	1250
4000–5990	~~IIII~~ I	6	380	2280
6000–7990	IIII	4	450	1800
8000–9990	III	3	520	1560
Total		$\Sigma f = 25$		$\Sigma fx = 8290$

Additional amount required by the department is $\Sigma fx =$ Rs. 8290.

Example 10. The following are the marks obtained by 30 boys in an examination. Construct a frequency table with class intervals of 10 :

7	18	37	53	24	39
68	40	93	43	11	27
21	19	32	75	52	84
52	29	92	79	45	81
53	8	41	14	26	26

Also give Cumulative Frequency (less than) table.

Solution. **Frequency Table**

Class	*Tally Marks*	*Frequency*	*Less than*	*Cumulative Frequency*
5–15	IIII	4	15	4
15–25	IIII	4	25	8
25–35	~~IIII~~	5	35	13
35–45	~~IIII~~	5	45	18
45–55	~~IIII~~	5	55	23
55–65	–	0	65	23
65–75	I	4	75	24
75–85	IIII	1	85	28
85–95	II	2	95	30
		30		

Example 11. What is frequency distribution?

(i) Prepare a suitable frequency distribution of the weights (in kgs) of 33 students of a class given below :
42, 74, 40, 60, 82, 115, 41, 61, 75, 83, 63, 53, 110, 76, 84, 50, 67, 65, 78, 77, 56, 95, 68, 69, 104, 80, 79, 79, 54, 73, 59, 81, 100.

(ii) Recording the data with the help of tally marks.

(iii) Finding the frequency of each class by counting the frequency tally.

(C.A. Foundation Nov., 1999)

Solution. For the present data, let us take class-intervals of width 10 and jot down tally marks and find frequencies as follows :

Class-Interval	*Tally Marks*	*Frequency*
40–49	III	3
50–29	~~IIII~~	5
60–69	~~IIII~~ II	7
70–79	~~IIII~~ III	8
80–89	~~IIII~~	5
90–99	I	1
100–109	II	2
110–119	II	2
Total		33

Example 12. From the following data relating to wages of 20 workers prepare frequency distribution with a class interval of 5 on exclusive and inclusive basis :

Wages in Rs. : 10, 15, 25, 27, 29, 20, 24, 23, 22, 12, 14, 16, 17, 18, 19, 18, 16, 15, 5, 9.

(M.D.U. B.Com 2004)

Solution. Exclusive class intervals

Class-Interval	*Interval Frequency*
5–10	2
10–15	3
15–20	8
20–25	4
25–30	3

Inclusive class intervals :

Class-Interval	*Interval Frequency*
5–9	2
10–14	3
15–19	8
20–24	4
25–29	3

From the inclusive class intervals we can make exclusive intervals as :

Class-Interval	*Interval Frequency*
4.5–9.5	2
9.5–14.5	3
14.5–19.5	8
19.5–24.5	4
24.5–29.5	3

2.27 TABULATION—MEANING AND DEFINITIONS

Croxton and Cowden say, "Either for one's use or for the use of others, the data must be presented in some suitable form. Usually the figures are arranged in table or represented by graphical devices."

Classification forms a basis for tabulation and is *done prior to tabulation.*

Tabulation may be defined as, "systematic arangement of data in columns and rows". It is designed simply to present data for purposes of analysis and statistical inferences.

Tuttle states about classification as "*The logical listing of related quantitative data in vertical columns and horizontal rows of numbers with sufficient explanatory and qualifying words, phrases and statements in the form of titles, headings and explanatory notes to make clear the full meaning, context and the origin of the data.*"

The above definition brings out the broad structure of a table. It clearly suggests that tabulation gives the data an orderly arrangement and highlights its basic characteristics.

Another definition given below emphasises the use of tabulation as a means of preserving information for future.

Secrist defines, "*Tables are a means of recording in permanent form the analysis that is made through classification and by placing in juxtaposition things that are similar and should be compared.*"

This definition clearly points out that tabulation is a process which gives classification of data a systematic form and is meant for the purpose of making comparative studies.

Bowley refers to tabulation as, "*the intermediate process between the accumulation of data in whatever form they are obtained and the final reasoned out account of the result shown by the statistics.*"

2.28 OBJECTIVES OF TABULATION

As a poet gives beautiful and meaningful presentation to the words by their proper arrangement, the main object of statistical table is to summarise the mass of data in a simple and systematic manner so as to be easily analysed and interpreted for specific requirements. Views of some authorities are as under.

According to D.W. Paden and E.F. Lindquist, "*Tabulation is a medium of communication of great economy and effectiveness for which ordinary prose is inadequate... . In addition to its* function in simple presentation, the statistical table is also a useful tool of analysis.*"

D. Gregory & H. Ward hold the view that, "*Tabulation is the process of condensing classified data in the form of a table, so that it may be more easily understood and so that any comparisons involved may be more readily made.*"

The main objectives of tabulation [Significance of tabulation] are :

(1) To Simplify Complex Data : In tabulation the data are presented systematically in columns and rows in a concise form. Since unnecessary details and repetitions are avoided in tabulation, the data becomes more meaningful and lot of time is saved in its study and understanding.

(2) To Facilitate Comparison : Tabulation facilitates comparison. Comparable figures can be placed is juxtaposed columns for easy comparative study. Since table is divided into various parts and for each part there are totals and subtotals, the relationships between different parts of data can be studied easily.

(3) To Help Reference : When the data are arranged in tables with titles and table numbers, they can be easily identified and made use of as source reference for future studies.

(4) To Economise Space : All unnecessary details and repetitions are avoided in tabulation which results in economy of space without sacrificing quality and usefulness of the data.

(5) To Depict Trend and Pattern of Data : Tabulation of data depicts the trend of the information under study and reveals the patterns within the figures which cannot be understood in a descriptive form of presentation.

(6) To Detect Error and Omissions in the Data : It is only after tabulation that some vital omissions are detected.

(7) To Facilitate Statistical Processing : It is only after classification and tabulation that the data becomes fit for statistical processing.

2.29 ESSENTIAL PARTS OF A TABLE

The following are the essential parts in a good statistical table :

(i) Table number (ii) Title (iii) Head notes
(iv) Captions and stubs (v) Body of the table (vi) Footnote
(vii) Source note

(i) Table number : Numbering of tables in a logical sequence is essential for proper identification and ready reference at the time of need.

* Neiswanger.

(ii) Title : Each table should be given a concise, complete and self explanatory title showing the nature and period of data it covers. Title is given below the table number.

(iii) Head note : It is a supplement to the title and is given below the title if need be felt. It explains the entire table or its major parts.

(iv) Caption and stubs : Captions are the headings for vertical columns and stubs are the headings for horizontal rows.

(v) Body of the table : Body of the table contains figures to be presented to the reader. The data in the body correspond to classification under captions and stubs.

(vi) Footnote : When any item in the table needs some more description, it is given through a footnote. Footnotes are given directly below the body of the table.

(vii) Source note : The source note gives the information about the source from where the data has been taken. Source note is required for reference at the time of need.

2.30 ADVANTAGES (OR FUNCTIONS) OF TABULATION

The tabulated data can perform the following functions :

1. The tabulated data can be more easily understood.
2. The data becomes attractive and leaves a lasting impression.
3. Quick comparison is possible.
4. Tabular arrangement makes summation of items possible.
5. Detection of errors and omissions becomes easier.
6. The repetition of explanatory phrases and headings are avoided.
7. Space is saved because data are placed in rows and columns.
8. Complex data is simplified.
9. Computation of different statistical measures *viz.*, averages, dispersion and correlation is facilitated.
10. Tabulated data are good for references and they make it easy to present the information on graphs and diagrams.
11. The data is presented in an organised and scientific manner.
12. Tabular presentation simplifies the problems.

2.31 ESSENTIALS OF A GOOD TABLE

A good table should possess the following characteristics :

1. The table should be according to the object of statistical investigation.
2. The table should be compact and brief. It should explain only the relevant and desired information needed for further statistical investigation.
3. It should not be overloaded with details.
4. The table should be neat and attractive so that it should appeal to the reader or investigator.
5. The purpose of the table should be clear.
6. Each aspect of the data should be clearly brought out.
7. The table should facilitate comparisons.
8. A table should always make a specific mention of the source of the data, as well as a note about various data gaps.
9. The unit of measurement should be clearly indicated, such as "price in rupees" or "weight in kilograms".

10. There should be no ambiguity in classification.
11. The table should have an appropriate title, captions and stubs, totals, footnotes, etc.
12. The table should show relationship between related items.
13. The table should suit the size of the paper. It should not be a square peg in a round whole.
14. Columns and rows should be numbered to facilitate reference to specific parts of a table.
15. Certain figures which are to be emphasized should be in distinctive type or in a "circle" or a "box" or between thick lines.
16. A table should have miscellaneous column for the data which cannot be grouped in the classification made.

2.32 GENERAL PURPOSE AND SPECIAL PURPOSE TABLES

General purpose tables are also called as reference or repository tables. This type of table provides various types of information on the same subject. These tables are usually appended with a report. They contain massive statistical information which can be used for common purpose without reference to a specific problem. Government publications generally are of this nature which are used as 'Data Bank' by researchers.

Special purpose tables can be called text, summary or analytical tables. Such tables are small in size and designed to highlight a particular set of facts relating to Specific Problem under consideration in a simple and analytical form. They are made to provide the results of analysis in an effective way so that one finds it easy to make comparisons and establish clear relationships.

2.33 ORIGINAL AND DERIVATIVE TABLES

Original tables are called classified tables which contain statistical information in the same form in which they are originally collected.

Derivative tables are derived from original tables. They provide the data in the form of totals, ratios, percentages and other statistical calculations. Such tables can also be derived from general purpose tables.

2.34 SIMPLE TABLE

Simple or one-way table (single tabulation) is the simplest table which shows only one characteristic and takes the form of frequency table, for example :

Marks-wise Distribution of Students of a School

Marks	*No. of Students*
0–10	25
10–20	32
20–30	48
30–40	15
	Total 120

2.35 DOUBLE OR TWO-WAY TABLE (DOUBLE TABULATION)

In such tables two characteristics are shown and a row or column is divided into two parts, as we can divide students into boys and girls as shown below :

Marks and Sex-wise Distribution of Students in a School

Marks	*Students*		*Total*
	Boys	*Girls*	
0–10	8	7	15
10–20	10	2	12
20–30	18	10	28
30–40	4	1	5
Total	40	20	60

2.36 TREBLE OR THREE-WAY TABLE (TREBLE TABULATION)

In such tables three characteristis are shown. Therefore, a row or a column is subdivided into three parts as under :

Distribution of Students in a College
(According to their Sex, Class and Family Status)

Class	*Boys*				*Girls*			
	Rich	*Average*	*Poor*	*Total*	*Rich*	*Average*	*Poor*	*Total*
XI	5	55	10	70	2	8	5	15
B.A. (I)	5	15	15	35	4	4	4	12
B.A. (II)	5	10	15	30	2	3	3	10
B.A. (III)	5	5	20	30	3	2	2	7

The above table can even be called as *manifold table. In higher order table or manifold tabulation* we can increase the number of characteristics, *viz.*, more sections, more subjects, etc.

Such a table is needed when a number of characteristics are to be simultaneously presented. But as more characteristics are studied together, the table becomes more complex and more confusing to the reader.

Example 13. The number of workers in a large factory in 1986 was 540 of which 30% were female and the rest males. In 1988 the strength of the workers increased by 100 females and 200 males. In 1990 the total number of workers increased by 25% over its value in 1988. The female workers were 340. Tabulate the above information.

Solution.

Table

Year	*1986*			*1988*			*1990*		
Category	*M*	*F*	*Total*	*M*	*F*	*Total*	*M*	*F*	*Total*
Workers	378	162	540	578	262	840	710	340	1050

$M \to$ Male, $F \to$ Female

Explanatory notes : In 1986, number of females $= \dfrac{540 \times 30}{100} = 162.$

In 1988 number of females $= 162 + 100 = 262$

In 1990 total number of workers $= 840 \times \frac{125}{100} = 1050.$

Number of female workers in $1990 = 340.$

Number of male workers $= 1050 - 340 = 710.$

Example 14. Present the following information in a suitable tabular form.

In 1990 out of a total of 1,750 workers of a factory, 1,200 were members of a trade union.

The number of women employed was 200, of which 175 did not belong to a trade union. In 1995 the number of union workers increased to 1,580 of which 1,290 were men. On the other hand, the number of non-union workers fell down to 208, of which 180 were men.

In 2000, there were 1,300 employees who belonged to a trade union and 50 who did not belong to a trade union. Of all the employees in 1970, 300 were women of whom only 8 did not belong to a trade union.

Solution. **Table Showing the Sex-wise Distribution of Union and Non-Union Members for 1990, 1995 and 2000.**

Category	*1990*			*1995*			*2000*		
	M	*W*	*Total*	*M*	*W*	*Total*	*M*	*W*	*Total*
Members	1,175	25	1,200	1,290	290	1,580	1,508	292	1,800
Non-members	375	175	550	180	28	208	42	8	50
Total	1,550	200	1,750	1,470	318	1,788	1,550	300	1,850

Notes : $M \rightarrow$ Men, $W \rightarrow$ Women.

Non-members in $1960 = 1750 - 1200 = 550.$

Women members $= 200 - 175 = 25$ and so on.

Example 15. Transforming the ratios into corresponding numbers, prepare a complete table for the following information. Give a suitable title to the table.

In the year 1980, the total strength of students of three colleges, *X*, *Y* and *Z* in a city were in the ratio 4 : 2 : 5. The strength of the college *Y* was 1000. The proportion of girls and boys in all the colleges was in the ratio 2 : 3. The faculty-wise distribution of boys and girls in the faculties of Arts, Science and Commerce was in the ratio 1 : 2 : 2 in all the three colleges.

Solution. **Faculty and Sexwise Distribution of Students in College *X*, *Y* and *Z* During the year 1980**

Colleges	*Sex*	*Faculty*			*Total*
		Arts	*Science*	*Commerce*	
X	Boys	240	480	480	1200
	Girls	160	320	320	800
	Total	400	800	800	2000
Y	Boys	120	240	240	600
	Girls	80	160	160	400
	Total	200	400	400	1000
Z	Boys	300	600	600	1500
	Girls	200	400	400	1000
	Total	500	1000	1000	2500

Note : Strength of college $Y = 1000 \Rightarrow$ strength of college $X = 2000$ and strength of college $Z = 2500$, Girls in college $X = \frac{2}{5} \times 2000 = 800$.

Girls in college X in Arts $= 800 \times \frac{1}{5} = 160$ and so on.

Example 16. In a simple study about coffee habit in two towns, the following information was received :

Town *A* : Females were 40%; Total coffee drinkers were 45% and male non-coffee drinkers were 20%.

Town *B* : Males were 55%, male non-coffee drinkers were 30% and female coffee drinkers were 15%.

Represent the above data in a tabular form.

Solution. **Table Showing the Coffee Drinking Habit of Towns *A* and *B* (In Percentage)**

Attribute	*Town A*			*Town B*		
	Males	*Females*	*Total*	*Males*	*Females*	*Total*
Coffee drinkers	40	5	45	25	15	40
Non-coffee drinkers	20	35	55	30	30	60
Total	60	40	100	55	45	100

Note : Town '*A*' : Females 40%, males = 60%, male non-coffee drinkers = 20%, male coffee drinkers = 60 – 20 = 40%.

Total coffee drinkers = 45%.

∴ Female coffee drinkers = 5%.

Example 17. Present the following information in a suitable tabular form supplying the figures not directly given. In 1975, out of a total of 2,000 workers in a factory, 1,550 were members of a trade union. The number of women employed was 250 out of which 200 did not belong to any union.

In 1974, the number of workers in the union (*i.e.*, union members) was 1725 of which 1,600 were men. The number of non-union workers was 360 of which 155 were women.

Solution. **Table Showing Union and Non-union Workers by Sex**

Union	*1974*			*1975*		
	Males	*Females*	*Total*	*Males*	*Females*	*Total*
Members	1,600	125	1,725	1,500	50	1,550
Non-members	225	155	380	250	200	450
Total	1,825	280	2,105	1,750	250	2,000

Example 18. In a trip organised by a college there were 80 persons each of whom paid Rs. 15.50 on an average. There were 60 students each of whom paid Rs. 16. Members of the teaching staff were charged at a higher rate. The number of servants was 6 (all males) and they were not charged anything. The number of females was 20% of the total of which one was a female staff member.

Tabulate the above information.

Solution. Table Showing the Type of Participants, Sex and Contribution Made

Type of Participants	*Sex*			*Contribution per member Rs.*	*Total Contribution Rs.*
	Males	*Females*	*Total*		
Students	45	15	60	16.00	960
Teaching staff	13	1	14	20.00	280
Servants	6	—	6	—	—
Total	64	16	80	—	1,240

Note : Total contribution = Average contribution × No. of persons who joined the trip.

$$= 15.5 \times 80 = \text{Rs. } 1,240$$

$$\text{Females} = 80 \times \frac{20}{100} = 16.$$

Number of teaching staff members

$$= \text{Total no. of persons} - \text{students} - \text{servants} = 80 - 60 - 6 = 14.$$

Contribution of staff per head

$$= \frac{\text{Total contribution} - \text{Contribution made by the students}}{\text{No. of teaching staff members}} = \frac{1240 - 960}{14} = 20.$$

Example 19. In a sample study about the coffee habits in two towns following data were observed :

Town *A* : 60% people were males.
40% were coffee drinkers, and
26% were male coffee drinkers.

Town *B* : 55% people were males,
35% were coffee drinkers, and
25% were male coffee drinkers.

Tabulate the above observations. [Similar to Example 4].

Example 20. Of the 1,125 students studying in a college during a year, 720 were Hindus, 628 were boys and 440 were science students; the number of Hindu boys was 392, that of boys studying science 205 and that of Hindu students studying science 262; the number of science students among the Hindu boys was 148. Enter these frequencies in a three-way table and complete the table by obtaining the frequencies of the remaining cells.

Solution. Distribution of College Students by Religion and Faculty

Faculty	*Boys*			*Girls*		
	Hindus	*Non-Hindus*	*Total*	*Hindus*	*Non-Hindus*	*Total*
Science	148	57	205	114	121	235
Arts	244	179	423	214	48	262
Total	392	236	628	328	169	497

Example 21. Out of a total number of 1,807 women who were interviewed for employment in a textile factory of Bombay, 512 were from textile areas and the rest from the non-textile areas. Amongst the married women who belonged to textile areas, 247 were experienced and 73 inexperienced, while for non-textile areas the corresponding figures were 49 and 520. The total number of inexperienced women was 1,341 of whom 111 resided in textile areas. Of the total number of women 918 were

unmarried and of these the number of experienced women in the textile and non-textile areas was 154 and 16 respectively. Tabulate.

Solution. **Table Showing the Marital Status of 1,807 Women Residing in Textile and Non-Textile Areas**

	Textile Areas			*Non-Textile Areas*			*Total*		
	M	*U*	*Total*	*M*	*U*	*Total*	*M*	*U*	*Total*
Experienced	247	154	401	49	16	65	296	170	466
Inexperienced	73	38	111	520	710	1,230	593	748	1,341
Total	320	192	512	569	726	1,295	889	918	1,807

$M \rightarrow$ Married; $U \rightarrow$ Unmarried.

Example 22. The following information about whether conditions at different stations were recorded at 8.30 a.m. on Thursday, the 29th August, 1980.

At Ahmednagar station the maximum and minimum temperatures in 24 hrs. were 28°C and 20°C respectively. The rainfall in the past 24 hrs at Ahmednagar was nil. Since June 1 the rainfall was 185 mm which is 105 mm below normal.

Bangalore's minimum and maximum temperatures in 24 hrs for the day were 19°C and 23°C respectively. It had no rainfall in the past 24 hrs and since June 1 the rainfall at Bangalore was 252 mm which is 54 mm below normal.

The minimum temperature at Udaipur ws 21°C and the rainfall in the past 24 hrs at Udaipur was nil. Since June 1 it experienced 434 m of rainfall which is 24 mm below normal.

Panagarh's maximum temperature in 24 hrs was 28°C. It had 4 mm of rain is the past 24 hrs. and since June 1 it had 955 mm of rain.

Calcutta's maximum and minimum day temperature were 30°C and 26°C respectively. It had 3 mm of rainfall in the past 24 hrs. Since June 1 it experienced a rainfall of 1,079 mm which is 154 mm above normal.

Present the above data in a table.

Weather Chart of Different Stations Recorded at 9 A.M. on 9-8-1980

Stations	*In 24 hours*		*Rainfall (m.m.)*			
	Temperature degrees centigrade		*Past 24 hrs.*	*Since June 1*	*Above normal*	*Below normal*
	Max.	*Min.*				
1. Ahmednagar	28	20	0	185		105
2. Bangalore	23	19	0	252		54
3. Udaipur	—	21	0	434		24
4. Panagarh	28	—	4	955		—
5. Calcutta	30	26	3	1079	154	—

The sign (—) indicates that data on the items are not available.

Example 23. Industrial finance in India showed great variation in respect of sources during the First, Second and Third Plans. There were two main sources, *viz.*, internal and external. The former has two source—depreciation and free reserves and surplus. The latter had three sources—capital issue, borrowings and 'other sources'.

During the First Plan internal and external sources accounted for 62% and 38% of the total and in this depreciation, fresh capital and 'other sources' formed 29%, 7% and 10.6% respectively.

During the Second Plan internal sources decreased by 17.3% compared to the First Plan and depreciation was 24.5%. The external finance during the same period consisted of fresh capital 10.9% and borrowings 28.9%.

Compared to the Second Plan, during the Third Plan external finance decreased by 4.4% and borrowings and 'other sources' were 29.4% and 14.9% respectively. During the Third Plan internal finance increased by 4.4% and free reserves and surplus formed 18.6%.

Tabulate the above information with the above details as clearly as possible observing the rules of tabulation.

Table Showing Pattern of Industrial Finance

(in per cent)

Plan	*Sources*						
	Internal			*External*			
	Depre-ciation	*Free Reserves & Surplus*	*Total*	*Capital Issues*	*Borrow-ings*	*Other Sources*	*Total*
First	29	23	62	7	20.4	10.6	38
Second	24.5	20.2	44.7	10.9	28.9	15.5	55.3
Third	30.5	18.6	49.1	6.6	29.4	14.9	50.9

Example 24. In the house of Lok Sabha there were 600 members present. During a discussion on a motion put to vote 400 voted in favour of the resolution. The Govt. members in the house were 380, 65 members belonging to the Opposition voted for the resolution. All the members belong to either of the two groups and there were no absentees. Tabulate the above information.

Solution.

Category	*Voted in favour of Resolution*	*Not voted in favour of Resolution*	*Total*
Govt. members	335	45	380
Opposition members	65	155	220
Total	400	200	600

Notes : Govt. members = 380

Opposition members = 600 – 280 = 220.

No. of opposition members who voted for Resolution = 65.

∴ Govt. members who voted for Resolution = 400 – 65 = 335.

No. of Govt. members not voted for Resolution = 380 – 335 = 45.

Example 25. In a trip organised by a college, there were 100 persons. The average cost works out to Rs. 15.60 per head. There were 80 students each of whom paid Rs. 16. Members of the teaching staff were charged at a higher rate. The number of servants was 6 (all males) and they were not charged. The number of ladies was 20% of the total of which two were lady staff members.

Tabular the above information in proper tabular form.

Solution.

	Students	*Teaching Staff*	*Servants*	*Total*
Males	62	12	6	80
Females	18	2	—	20
			Total	100

Also Refer Ex. 18, page 2.26.

Example 26. (a) Out of the total number of 2,807 women, who were interviewed for employment in a textile factory, 912 were from textile areas and the rest from non-textile areas. Amongst the married women, who belonged to textile areas, 347 were having some work experience and 173 did not have work experience, while for non-textile areas the corresponding figures were 199 and 670 respectively. The total number of women having no experience was 1,841 of whom 311 resided in textile areas. Of the total number of women, 1,418 were unmarried and of these the number of women having experience in the textile and non-textile areas was 254 and 166 respectively.

Tabulate the above information.

Solution. (a) Distribution of number of women interviewed by a textile factory according to their marital status, experience and areas.

	Textile			*Non-Textile*			*Total*		
	Experience	*No Experience*	*Total*	*Experience*	*No Experience*	*Total*	*Experience*	*No Experience*	*Total*
Married	347	173	520	199	670	869	546	843	1389
Unmarried	254	138	392	166	860	1026	420	998	1418
Total	601	311	912	365	1530	1895	966	1841	2807

2.37 DIAGRAMS AND GRAPHS

Diagrams and graphs are nothing but pictorial representation of statistical data. They are geometrical figures like points, bars, squares, rectangles, circles, cubes, etc. pictures, maps or charts.

'Pages of cold figures can be made to speak in clear tones if translated into the diagrammatic language'.

According to M.J. Moroney, "*Cold girues are uninspiring to most people, diagrams help us to see the pattern and shape of any complex situation, just as a map gives us a bird's eye view of the wide stretch of a country, so diagrams help us visualise the whole meaning of a numerical complex data at a single glance.*"

M.J. Moroney further observes, "*Diagrams register a meaningful impession almost before we think*". A diagram is a visual form for presentation of statistical data highlighting their basic facts and relationships. Though diagrams do not add any new meaning to the statistical facts, yet they exhibit the results more clearly.

Bowley says, "*Any list of figures becomes less comprehensible as its length increases. A series of ten numbers can perhaps be easily grasped, of twenty, only with an effort, while a printed list of figures for one hundred successive years leaves hardly any impression on our mind at all; we cannot see the wood for trees.*"

William Playfair writes, "*The advantage proposed by the method is not that of giving a more accurate statement than by figures but it is to give a more simple and permanent idea of the gradual progress and comparative amounts, at different periods, by presenting to the eye a figure, the proportions of which correspond with the amount of the sums intended to be expressed? Linear arithmetic then provides as much information in five minutes as would require whole days to imprint on the memory...by a table of figures.*"

Diagrams are useful because facts can be known at a glance. They are readily intelligible and save a considerable amount of time and energy. Diagrams are more attractive than the figures. Figures when told through diagrams become attractive.

2.38 ADVANTAGES OF DIAGRAMMATIC PRESENTATIONS

1. Diagrams give a **very clear picture** of data. Data with the help of diagrams can be understood and grasped even by a layman in a very short time.
2. **Comparison** between different samples with regard to certain Statistical measures like Mean, Median, Mode etc., can be easily made without using any statistical techniques.
3. This technique can be used **universally** at any place and at any time. This technique is used almost in all the subjects and various fields.
4. Diagrams have **impressive value** also. A common man remains more interested in reading good diagrams than to go through the tabulated data.
5. This technique can be used for **numerical type of statistical analysis.**, *e.g.*, to locate Mean, Median, Mode, etc.
6. It does save not only **time and energy** but is also economical. Preparing even good diagrams is not very costly.
7. These give us **much more information** as compared to tabulation. Technique of tabulation has its own limits.
8. This data is easily remembered. Diagrams have long lasting impression on the human mind.
9. Data can be **condensed** with diagrams. Small pictures some times represent the data which even thousands of words can not express so clearly.
10. Besides being informative, diagrammatic presentation is an entertaining means of data presentation.
11. Diagrammatic presentation helps in *predicting* the behaviour of variables.

2.39 ESSENTIALS OF A GOOD DIAGRAM [OR RULES FOR CONSTITUTING DIAGRAMS]

1. Diagrams should be attractive and should communicate the required information effectively.
2. Diagrams should be accurate. But accuracy should not be bargained for making the diagram attractive and impressive.
3. Diagrams should be proportionate in height and width.
4. Diagrams should be self-explanatory.
5. Diagrams should be simple and easily understandable.
6. Diagrams should suit the size of the paper. It should not look odd with the dimensions of the paper.
7. Vertical diagrams should be preferred over horizontal diagrams.
8. Diagrams must bear proper headings. A heading must be simple, short and informative.
9. Before making a diagram, its scale should be properly determined and indicated on one side of the paper.

10. Diagrams must carry some signs on the nature and classifiation of information. Colours may be used to indicate different classes of the data. Meaning of these signs and colours should be clarified.
11. In diagrammatic presentation of data minimum number of words and calculations should be used.
12. Diagrams must be bordered with bold lines to give them a more attractive look.
13. The construction of diagrams should flow from left to right or from bottom to the top.
14. Data on whose basis the diagrams are drawn should be clearly written.

2.40 LIMITATIONS OF DIAGRAMMATIC PRESENTATION

1. Diagrammatic presentation of data is just an approximation of the actual behaviour of the variables, *i.e.*, diagrams show only an aggregated behaviour of the variables.
2. Only a limited set of data can be presented in the form of a diagram.
3. Diagrammatic presentation of data is a time-consuming process.
4. It is not very easy to arrive at final conclusions after seeing the diagrams. A diagram offers only preliminary conclusions.
5. Diagrams do not present the small differences properly.
6. Only an artist can draw multi-dimensional diagrams.
7. Diagrams can be analysed only mentally and are not amenable to further statistical treatment as the tabular presentation is. Diagrams drawn on false base lines need special care.
8. Diagrams can at best supplement the tabular presentation and are not an alternative to it.
9. Diagrams can be used only for comparative study. They interpret better only when there is another diagram or picture to which it can be compared. Hence diagrams are not of much use when comparison is either not possible or not necessary. Two diagrams drawn on different Scale cannot be compared.
10. Diagrams are capable of being misused easily. A wrong type of diagram, if used, will give fallacious conclusions which may mislead in a very serious way. This happens Specially in Advertisements.

2.41 TYPES OF DIAGRAMS

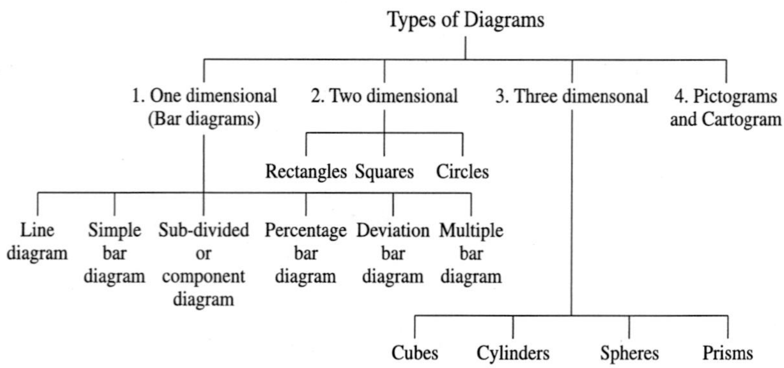

Notes :

1. Bar diagrams are called one dimensional because their height represents the size of the figure. Width is not considered, width is shown merely for atrraction. In any Bar diagram, all bars drawn must be of equal width.
2. In two dimensional diagrams, the area represents the data and the length and breadth both have to be taken into account. Such diagrams are also called *area diagrams or surface diagrams, e.g.*, square, rectangle, circle, pie diagrams etc.
3. Three dimensional diagrams also called *volume diagrams.* These diagrams use length, breadth and height. The given magnitudes are represented by the volumes of the corresponding diagrams. In three dimensional diagrams cubes are the simplest and most commonly used. These diagrams are specially useful if there are wide variations between the smallest and the largest magnitudes to be represented.

Bar Diagrams (or one Dimensional Diagrams)

(i) *Line Diagrams* : Line diagrams consist in drawing verticle lines, length of each line being equal to the frequency. The values of the variate x are presented on a suitable scale on X-axis and the corresponding frequencies are presented on a suitable scale along Y-axis. Line diagrams facilitate comparison though they are not very attractive to see.

(ii) *Simple Bar Diagrams* : Simple bar diagram is a modified form of line diagram. In simple bar diagram equal width is given to all the lines to make them look as rectangles but width carries no meaning and it is only the length of the bar (equal to frequency) that matters. *If the number of items is large, line diagram is preferred to bar diagram to economise space.* Bar diagrams are one dimensional because it is only the length of the bar that matters. A simple bar diagram can be drawn either on horizontal or vertical base, but bars on horizontal base (showing vertical bars), are more common. It is used to represent one variable, *e.g.*, students, prices, etc. Line. diagrams and simple bar diagrams though they are very simple to present but they can present only one variable. For example monthly, quarterly, yearly, production sales, prices, population etc. can be represented by simple bar diagram. To make these diagrams attractive to look at, they are shaded by a single colour.

Limitations of Bar-Diagrams

A bar-diagram has the following limitations :

1. This diagram should be used only when the given figures do not vary much and the proportion between the figures is small.
2. This diagram can be used to represent data on one variable only.

(iii) *Sub-divided or Component Bar Diagrams* : Sub-divided or component bar diagrams and percentage bar diagrams are useful for presenting several items of a variable graphically and enable us to make comparative study of different parts among themselves and study the relationship between each part and the whole. The various components in each bar should be kept in the same order. To distinguish between different components a separate colour or shade is used for each component.

(iv) *Percentage Bar Diagrams* : Sub-divided or component bar diagrams presented graphically on percentage basis give **percentage bar diagrams**. The total for each bar is taken as 100 and the value of each component or part is expressed as percentage of the respective totals. In a percentage bar diagram, all the bars are of equal height and the various segments of the bar representing the different components will vary in height depending on their percentage value to the total. Percentage bar diagrams can be used conveniently to compare two or more sets of data, percentage bars highlight the relative changes in data and thus importance of the various component parts to the whole.

In Sub-divided bar diagrams the component parts are shown in absolute quantities whereas in Percentage bar diagrams the components parts are shown as percentage of the total. In percentage bar diagrams, all bars are of equal height.

(v) *Deviation Bar Diagrams* : **Deviation bar diagrams** *are used to show the net magnitudes of a phenomenon, i.e., net profit, net loss, net exports or imports, etc.* Bars in these diagrams can assume both negative and positive values. Positive valued bars are shown above the base line (*X*-axis) while negative valued bars are drawn below it. Base line is also called line of reference. Deviation bars are some times called as **Bilateral bar diagram**.

(vi) *Multiple Bar Diagrams* : Multiple bar diagrams are used when two or more sets of interrelated variables are to be presented graphically for comparison. A set of simple adjacent bars (one for each variable) is drawn. Different sets of the bars are spaced equally and different bars in the same set are distinguished by different colours/shades.

Comparison can be made either between two or more variables or magnitudes of one variable in two or more aspects.

Note : Different diagrams are explained in the solved examples.

2.42 RULES FOR CONSTRUCTING DIAGRAMS

The following rules be kept in mind while constructing diagrams :

1. *Every diagram should be suitably and briefly titled.* It can be mentioned either at the top of the diagram or below it.
2. *It should suit to the size of the paper.* A diagram should be proportionally matching in terms of its height and width. The scale should be so selected that the diagram suits to the size of the paper.
3. *It should be neat and attractive.*
4. *It should be clearly indexed.* An index be given with the diagram showing representation of different shades and colours so that the items under comparison are easily understood.
5. *Diagrams should contain footnotes and a proper spacing between the figures.*
6. *The details in the diagram should be self-explanatory.* In other words, the nature, time, place and sources of data too should be indicated.

Example 27. Represent the following data on a suitable diagram.

Students	1	2	3	4	5	6
Height in cms.	192	165	171	189	150	168

Solution. Taking students on the *X*-axis at equal distances and their heights on *Y*-axis. The **line diagram** is drawn as under :

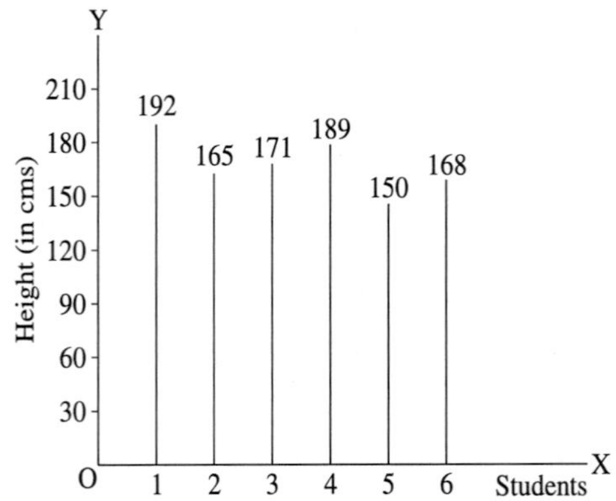

Line Diagram

Example 28. No. of accidents in city in a year is given below :

Month	1	2	3	4	5	6	7	8	9	10	11	12
No. of Accidents	8	12	20	16	10	16	20	14	10	19	16	10

Solution. Represent the data on a suitable diagram :

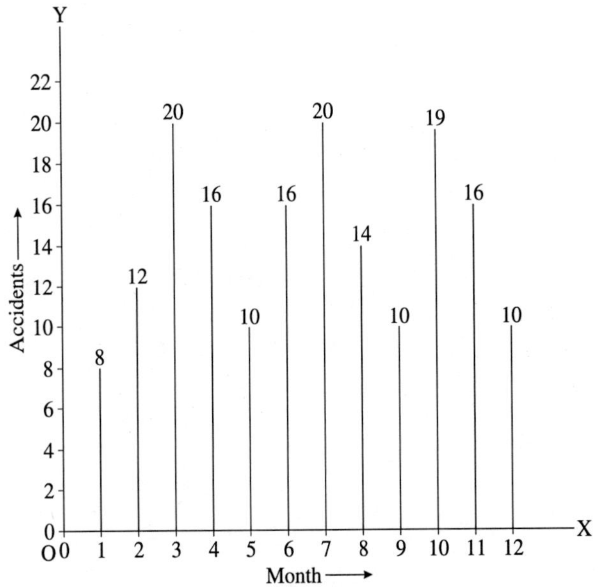

Line Diagram

Example 29. The following table gives the number of Unemployed Engineers in different States of India :

State	*Number of Unemployed Engineers*	*State*	*Number of Unemployed Engineers*
Andhra Pradesh	1559	Maharashtra	240
Delhi	801	Bihar	202
Karnataka	783	Haryana	175
Tamil Nadu	613	West Bengal	137
Kerala	529	Punjab	120
Orissa	433	Gujarat	100
Rajasthan	358		

Represent the above data by a suitable diagram.

Solution. Here, we present the above data by a **simple bar diagram** because the data relate to one variable, *i.e.*, Unemployed Engineers.

Diagram showing the Number of Unemployed Engineers in different States of India.

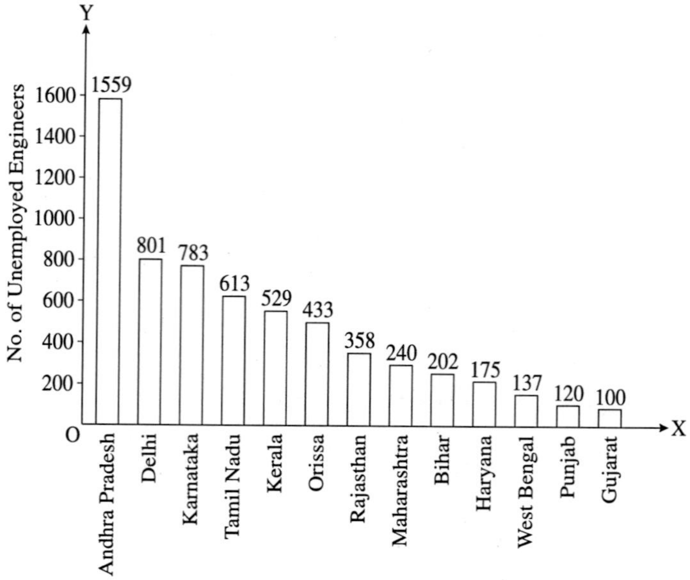

Example 30. Draw a simple bar diagram to represent the following figures relating to manufacturing of machines.

Solution.

Years	1984	1985	1986	1987	1988
No. of Machines	1200	1700	1900	2800	2100

Production of Machines (1984-88)

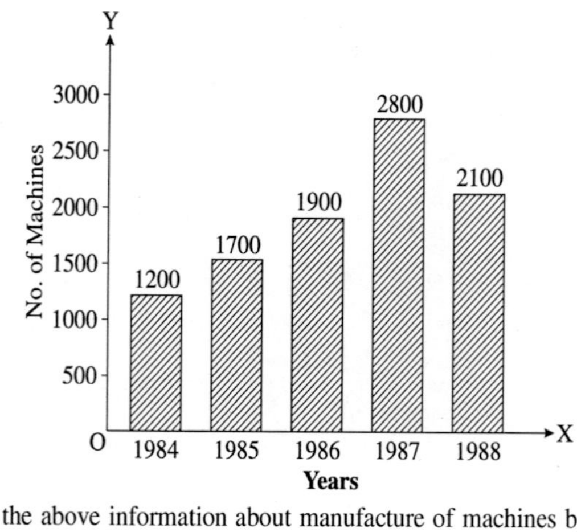

We can present the above information about manufacture of machines by vertical base showing horizontal bars also as under :

Alternative Solution—Vertical base

Production of Machines (1984-88)

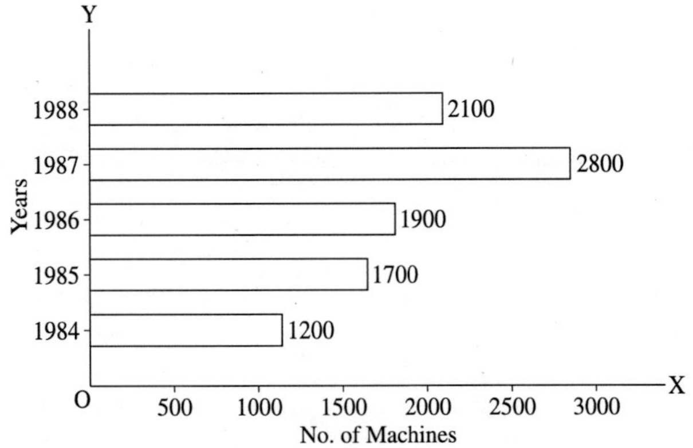

Example 31. The following table gives the result of M.Sc. students of a college for four years :

Year	*Number of Students*		
	Division		
	First	*Second*	*Third*
1983	10	30	50
1984	12	45	70
1985	14	50	60
1986	11	40	75

Represent the above data by a suitable diagram.

Solution. We represent the above data by multiple bar diagram because the values of the variable are inter-related supplied with the inter-related values of the variable.

Diagram Showing the Number of Students Division-wise from 1983 to 1986

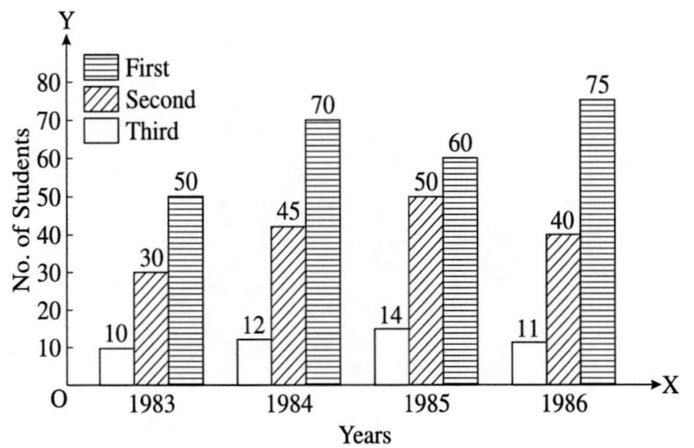

Example 32. Draw a suitable diagram of the following data :

State of Crime in Running Passenger Trains

Year	*Loot*	*Robbery*	*Murder*
1984	321	82	108
1985	386	115	131
1986	352	144	97
1987	285	70	102
1988	245	68	75

Solution. **Multiple Bar Diagram**

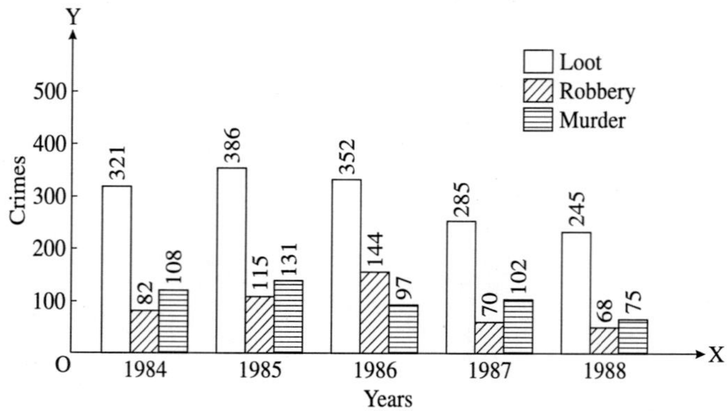

Example 33. The following table shows the monthly expenditure of three families :

Items	*Families*		
	A	*B*	*C*
Food	360	300	200
Clothing	162	140	120
House Rent	126	120	100
Fuel & Light	72	60	40
Misc.	180	130	140

Represent the above data by a suitable diagram.

Solution. **Diagram Showing the Percentage Expenditure on Different Items**

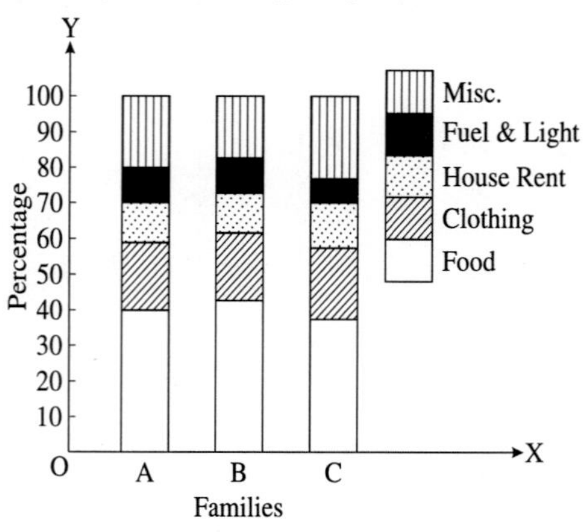

Example 34. The following table gives the number of students sex-wise studying is a degree arts college.

Class	*Number of Students*		
	Boys	*Girls*	*Total*
Pre.	150	75	225
T.D.C. ; Part I	120	65	185
T.D.C. ; Part II	80	50	130
T.D.C. ; Part III	60	40	100

Represent the above data by a suitable diagram.

Solution. Here we employ sub-divided bar diagram to represent the above data. The given data are divided into different components.

Diagram Showing the Number of Students Sex-wise Studying in a Degree Arts College

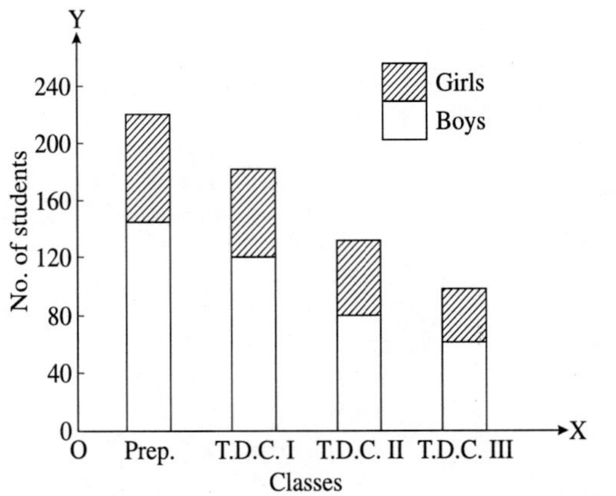

Example 35. The percentage distribution of exports from India of engineering products for two years and 1998-1999 and 1999-2000 are given below :

Region of Destination	*Years*	
	1998-99	*1999-2000*
S.E. Asia	42	23
West Asia	19	22
Africa	16	28
Other regions	23	27
Total	100	100

Draw a suitable diagram to represent the above data.

Solution.

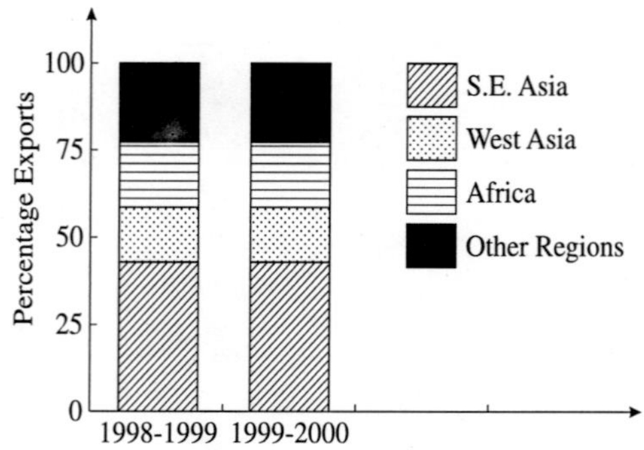

Example 36. Following data relate to the faculty-wise enrolment of students in a college :

Years	:	1993	1994	1995
No. of Arts students	:	95	110	120
No. of Science students	:	160	170	165
No. of Commerce students	:	75	95	110

Represent the data by suitable diagram.

Solution.

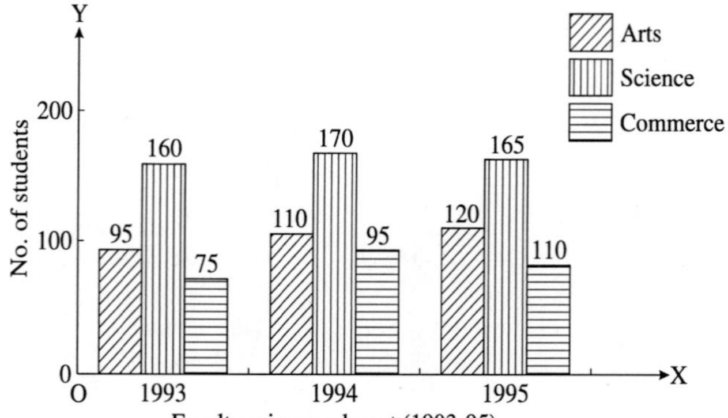

Faculty-wise enrolment (1993-95)

Example 37. Depict the following data by a suitable diagram.

(Balance of Trade = Export – Import)

Year	*Export*	*Import*	*Balance of trade (Millions Rs.)*
1993	95	115	– 20
1994	110	140	– 30
1995	115	96	+ 19
1996	120	100	+ 20

Solution. Deviation bar-diagram representing data on trade balance is shown in the Figure.

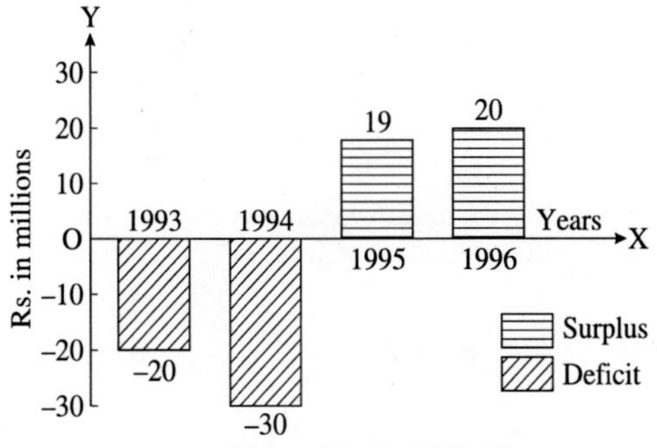

Balance of Trade (1993-96)

Example 38. Following data relate to year-wise enrolment in a college, classified according to faculty.

Year / *Faculty*		*1992*	*1993*	*1994*	*1995*
Arts	:	29.03%	28.79%	29.33%	30.38%
Science	:	48.39%	48.48%	45.34%	41.77%
Commerce	:	22.58%	22.73%	25.33%	27.85%

Represent the data by Percentage bar diagram.

Solution. **Percentage of Students, Faculty-wise and for Each Year**

Year	*Percentage of students*			*Total %*
	Arts	*Science*	*Commerce*	
1992	29.03	48.39	22.58	100.00
1993	28.79	48.48	22.73	100.00
1994	29.33	45.34	25.33	100.00
1995	30.38	41.77	27.85	100.00

Percentage Bar-diagram showing enrolment.

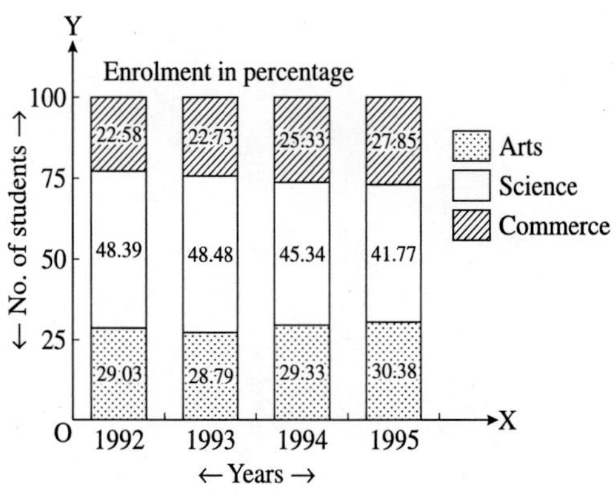

Two-Dimensional Diagrams or Area Diagams

In one-dimensional diagrams we take only the length of the bars into consideration and width is taken simply for the beautification of the diagrams. *However, in* **two-dimensional diagrams** *or* **area diagrams**, *the length as well as the width of the geometric figure is taken into consideration.* Such diagrams are also useful in situations when the proportion between the magnitudes of given values of the variables is quite large. The most popular two-dimensional diagrams are :

1. Rectangle diagrams, 2. Squares and circle diagrams, 3. Pie-diagrams

Rectangle Diagrams

These diagrams are used for two-dimensional comparisons. While drawing the rectangles, we consider the length as well as the width of the bars. *Thus, rectangles used for area diagram vary in heights as well as in the width, so that areas of the rectangles represent the magnitudes of the variable over time or space, or over some other characteristic of variation.** In rectangle diagrams data can be represented in original units or percentages. Rectangles are generally used for representing the total expenditure of two or more families made on various items. These are also used for making comparisons of expenses on various items of production and profit earned per unit in a number of firms. The following example will clarify.

Example 39. The following data represent the expenditure of the two families on various items. Represent the data by a rectangle diagram. (i) In original units, (ii) on percentage basis.

S. No.	*Items*	*Expenditure (Rs.)*	
		Family A	*Family B*
1.	Food	1,200	1,700
2.	Clothing	500	800
3.	House Rent	600	900
4.	Fuel and Electricity	250	300
5.	Miscellaneous	450	800
	Total	3,000	4,500

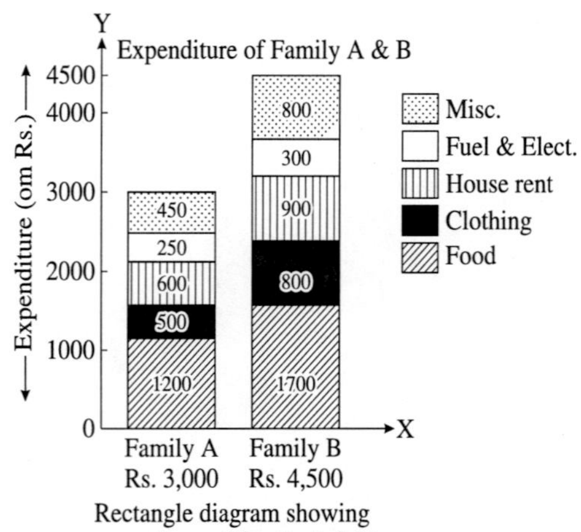

Rectangle diagram showing expenditure of family A & B

* Area of the Rectangle = Length × Width.

Solution. (i) ***In original units*** **:** While drawing the rectangles, the width of the rectangles is kept in proportion to the total expenditures of the families, *i.e.*, 3,000 : 4,500 or 2 : 3 and the height as per the magnitude of the expenditure, *i.e.*, height show the sub-divisions of the expenditures.

(ii) ***On percentage basis*** **:** For representing the data by a rectangle diagram on percentage basis, we express the expenditure on various items as percentage of the total expenditure for both the families as shown in Table below :

Table showing Family Expenditures in percentages

Items	*Expenditure (Rs.)*			
	Family A	*%*	*Family B*	*%*
1. Food	1,200	40.00	1,700	37.78
2. Clothing	500	16.67	800	17.77
3. House Rent	600	20.00	900	20.00
4. Fuel & Electricity	250	8.33	300	6.68
5. Miscellaneous	450	15.00	800	17.77
Total	3,000	100.00	4,500	100.00

Again the widths of the rectangles are kept in proportion to the total expenditure, *i.e.*, 2 : 3 and expenditure is now shown in percentage and not in absolute terms as shown below :

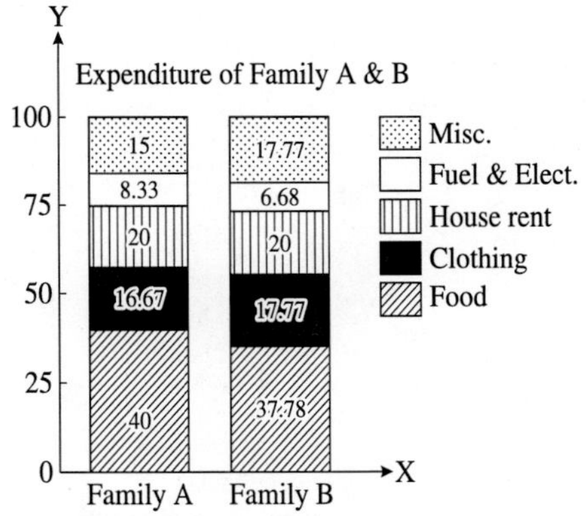

Rectangle diagram showing expenditure of families A & B

Squares and Circle Diagrams

Rectangle diagram is not very useful when the proportion between the magnitudes of the given values is quite large. For example, if the total expenditure of the two families are 45,000 and 3000, then in rectangle diagram the width of the rectangles will be in the ratio 15 : 1 which is an inconvenient ratio to show diagramatically. To overcome this situation, we use **square** or **circle** diagrams.

Steps to Construct Squares

1. Consider the given numerical figures (magnitudes) as different areas.
2. Since the area of square = $(\text{side})^2$, take the square root of each area to find the side of each square.
3. To reduce the size of the sides, we divide each square root by some convenient number.
4. Draw a square on each side calculated in step 3.

Steps to Construct Circles

1. The area of the circle = πr^2.
2. Therefore the area is proportional to the square of the radius.
3. Therefore the radius is proportional to the square root of the area, *i.e.*, the sides of the squares calculated for the construction of squares can represent the radii of different circles. Otherwise radii of circles can be calculated on a different scale also.

The steps will be clear from the solved examples.

Example 40. The following data relate to the plan outlay of a country for three plans.

Five year Plan : I IV VII

Outlay (Rs. '000 crores) : 196 2060 8820

Represent the data by (i) Square diagram (ii) Circle diagram.

Solution. **(i)** ***Square diagram*** **:** As there is a very large difference (or proportion) between the outlays of different plans, square diagram is more suitable. For drawing a square diagram, we first calculate the square roots of the given values as shown in the following table : *

Plan	*Outlay (Rs. '000 crores)*	*Square Root*	*Side of the square (in cms)*
I	196	14.00	0.70
IV	2060	45.39	2.26
VII	8820	93.91	4.70

Here the sides of the squares are in the ratio of the square roots of the outlay. The square diagram is shown in the figure.

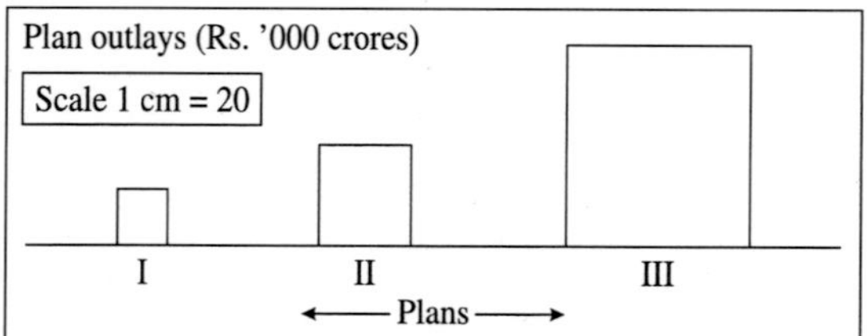

Square diagram showing Plan outlays

(ii) ***Circle diagram*** **:** In this case the radii of the circles will be kept proportional to the magnitude of the different square roots. The circle diagram is shown in the figure.

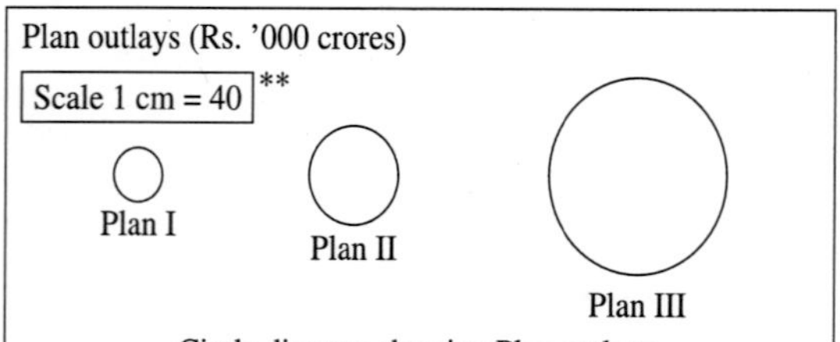

Circle diagram showing Plan outlays.

* Dividing each square root by 20.

** In circles also we can use the scale 1 cm = 20. Changing the scale will not make any difference in comparison.

2.43 PIE DIAGRAM OR ANGULAR DIAGRAM OR SUBDIVIDED CIRCULAR DIAGRAM

A pie diagram is a circle divided into Component Sectors with areas proportional to the size of the corresponding component. Pie diagrams are constructed on percentage basis and not an absolute basis because a series of pie diagrams showing absolute figures would require that larger totals be represented by larger circles which will involve many difficulties. Percentages can be represented by circles equal in size. However, if no comparison is to be made and only one pie diagram is required, we may use any circle.

Pie diagrams are very popular diagrams used for representing breakdown of an aggregate into its components or subdivisions. This diagram is generally used to compare the relationship between various components. For drawing the pie diagram, percentage of each sector is **converted into degrees keeping in view that the whole circle covers 360°**

2.44 STEPS IN THE CONSTRUCTION OF PIE DIAGRAM

1. Convert each component as a percentage of the total.
2. Multiply this percentage by $\frac{360}{100} = 3.6$ to convert into degrees.

   [∵ Complete area of the circle covers 360°, *i.e.*, 100% area covers 360°. Therefore 1% area covers 3.6 degrees.]
3. Starting with the twelve o'clock position on the circle (clockwise) draw the largest component circle.
4. Draw other components in clockwise succession in descending order of magnitude except for catch-all components like 'All others' and miscellaneous which are shown last.
5. Use different colours or shade to distinguish between different components.
6. Explain briefly the different components either within the components in the figure or outside by arrows.

Pie diagrams are often used particularly by newspapers and magazines when national, State or local Government budgets are being determined. The major advantage of the pie diagram is that it is extremely easy to understand.

Example 41. Represent the following data by a pie diagram :

Level of Schools	*Enrolment (in lakhs)*
Elementary	87.02
Junior	36.97
Secondary	11.83
Higher Secondary	4.47
Total	140.29

Solution.

Level of Schools	*Enrolment (in lakhs)*	*Percentage of the Total*	*No. of degrees*
Elementary	87.02	$\frac{87.02}{140.29} \times 100 = 62.03$	$62.03 \times 3.6 = 223°$
Junior	36.97	$\frac{36.97}{140.29} \times 100 = 26.35$	$26.35 \times 3.6 = 95°$
Secondary	11.83	$\frac{11.83}{140.29} \times 100 = 8.43$	$8.43 \times 3.6 = 30°$
Higher Secondary	4.47	$\frac{4.47}{140.29} \times 100 = 3.19$	$3.19 \times 3.6 = 12°$
Total	140.29	100	360°

Pie Diagram Showing Enrolment in Schools

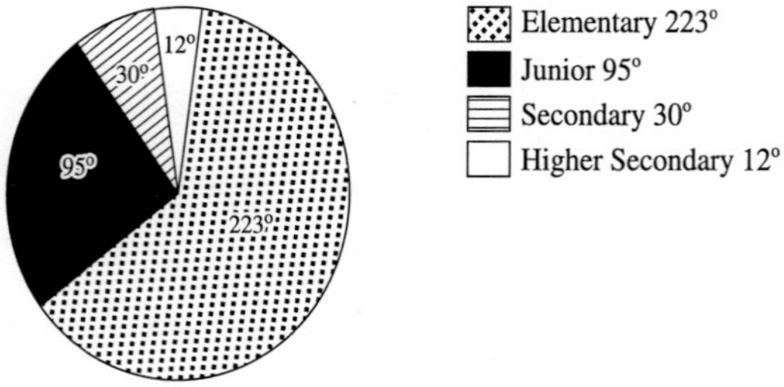

Example 42. Draw a pie diagram to represent the following data :

Average Expenditure Per Month in Rupees

Item	*1978*	*1980*
Food	644	945
Rent	420	525
Clothing	200	322
Medical	80	210
Other items	96	518
Total	1440	2520

Solution. **Computation of Angles**

Item	*1978*		*1980*	
	Expenditure	*Expenditure Angle**	*Expenditure*	*Expenditure Angle*
Food	644	$\frac{644}{1440} \times 360 = 161°$	945	$\frac{945}{2520} \times 360 = 135°$
Rent	420	$\frac{420}{1440} \times 360 = 105°$	525	$\frac{525}{2520} \times 360 = 75°$
Clothing	200	$\frac{200}{1440} \times 360 = 50°$	322	$\frac{322}{2520} \times 360 = 46°$
Medical	80	$\frac{80}{1440} \times 360 = 20°$	210	$\frac{210}{2520} \times 360 = 30°$
Others	96	$\frac{96}{1440} \times 360 = 24°$	518	$\frac{518}{2520} \times 360 = 74°$
Total	1440	360°	2520	360°

* We can calculate the angle either by percentage method or by this method.

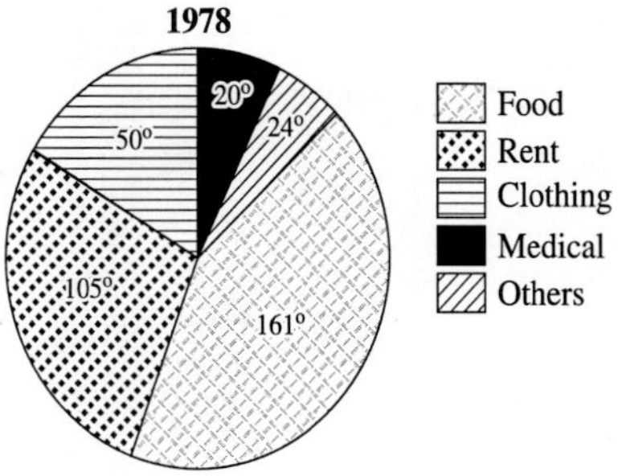

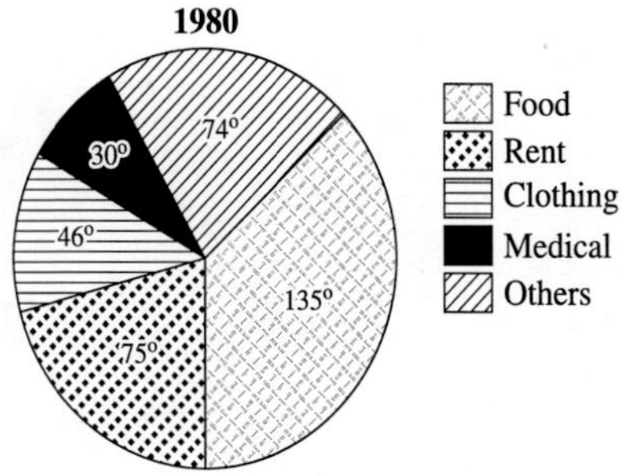

Example 43. The table below gives the ages of drivers of cars involved in total accidents during a certain year. Draw a pie-chart to represent the data :

Age of drivers	:	Under 20	20–40	40–60	Over 60	Total
Per cent of totals	:	15	60	20	5	100

Solution.

Ages	*% of totals*	*Degrees*
Under 20 (A)	15	$15 \times 3.6 = 54$
20–40 (B)	60	$60 \times 3.6 = 216$
40–60 (C)	20	$20 \times 3.6 = 72$
Over 60 (D)	5	$5 \times 3.6 = 18$

1% corresponds to 3.6 degrees.

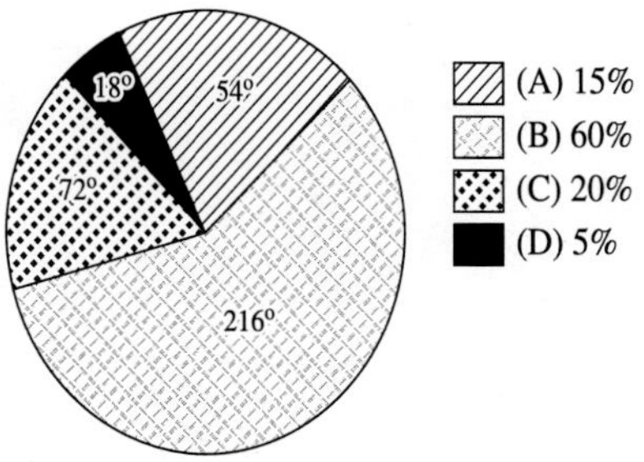

Example 44. Show the following data by means of a suitable diagram :

Production of Tea, Cocoa and Coffee

Tea	3260 Tons
Cocoa	1850 "
Coffee	900 "
Total	6010 "

Solution.

Product	*Production*	*Percentage*	*Angle of degrees**
Tea	3260	$\frac{3260}{6010} \times 100 = 54.24$	$54.26 \times 3.6 = 195°$
Cocoa	1850	$\frac{1850}{6010} \times 100 = 30.78$	$30.78 \times 3.6 = 111°$
Coffee	900	$\frac{900}{6010} \times 100 = 14.975$	$14.975 \times 3.6 = 54°$
	6010		

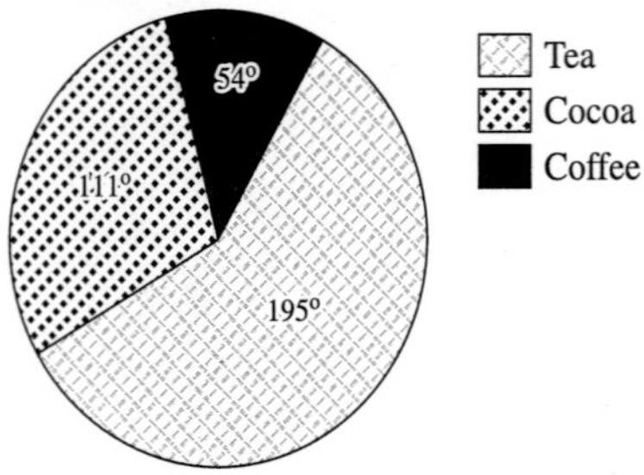

Example 45. Construct a pie diagram with the following data :

Consumers :	*State Electricity Board*	*TISCO*	*CESC*	*Railway*	*Coal*	*Others*
Power Consumed (in KWt. hours) :	7,50,000	3,60,000	6,00,000	1,80,000	4,50,000	6,60,000

Solution.

Consumers	*Power Consumed (in Kwt. hrs.)*	*Expressed in %*	*degrees* $= \% \times 3.6$
SEB	7,50,000	25	90
TISCO	3,60,000	12	43.2
CESC	6,00,000	20	72
RAIL	1,80,000	6	21.6
COAL	4,50,000	15	54
OTHERS	6,60,000	22	79.2
Total	30,00,000	100	360.0

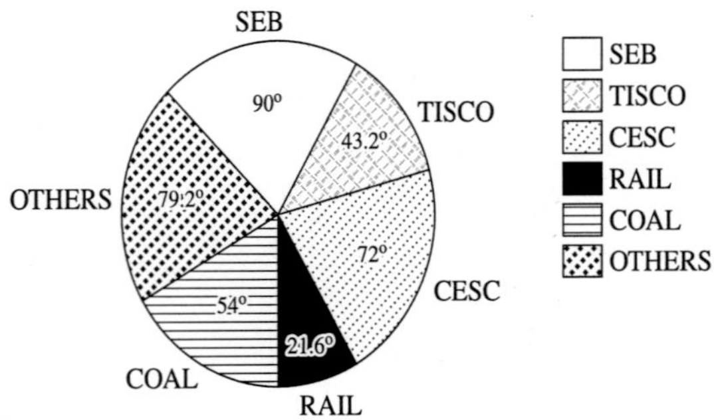

* Multiply each Percentage by 3.6.

Example 46. Show the following data by means of a suitable diagram :

1. Falling objects	8	20%
2. Falls	4	10%
3. Fire	7	$17\frac{1}{2}\%$
4. Traffic	5	$12\frac{1}{2}\%$
5. Machinery	6	15%
6. Cuts and bruises	10	25%
Total	40	100%

Solution.

Accidents due to	*Number*	*Percentage*	*Angle in degrees*
(1) Cuts and bruishes	10	25	$25 \times 3.6 = 90°$
(2) Falling objects	8	20	$20 \times 3.6 = 72°$
(3) Fire	7	$17\frac{1}{2}$	$17\frac{1}{2} \times 3.6 = 63°$
(4) Machinery	6	15	$15 \times 3.6 = 54°$
(5) Traffic	5	$12\frac{1}{2}$	$12\frac{1}{2} \times 3.6 = 45°$
(6) Falls	4	10%	$10 \times 3.6 = 36°$
			Total : 360°

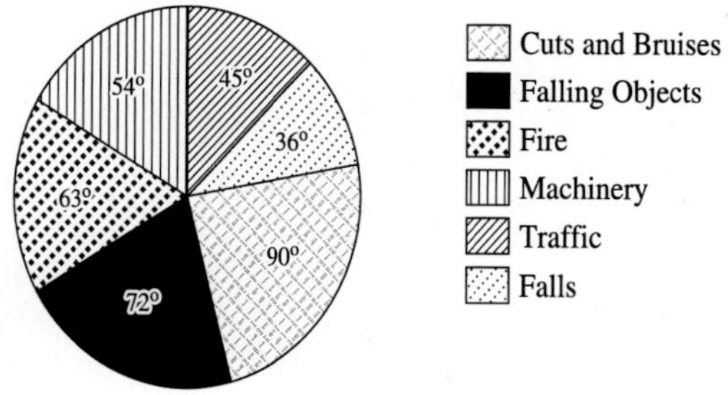

2.45 THREE DIMENSIONAL DIAGRAMS OR VOLUME DIAGRAMS

In case of two dimensional diagrams we used only length and breadth as two dimensions, but in this case we make use of all the three dimensions, *i.e.*, **Length, Breadth and Height**.

Three dimensional diagram may be used where cube roots may be taken easily and there is very large difference between the items. These can be given shape of cube, spheres etc.

Example 47. Present following data by suitable diagram :

University	*I*	*II*	*III*	*IV*
No. of Students (in Thousands)	125	64	27	8

Solution.

Col. I	*Col. II*	*Col. III*	*Col. IV*
University	*Students (000)*	*Cube Root*	*Side [Dividing Col. (III) by 2]*
I	125	5	2.5
II	64	4	2.0
III	27	3	1.5
IV	8	2	1.0

All the values in this question are perfect cubes and hence it is easy to get cube roots. Here also we divided these values by a suitable digit or number to get the suitable size of sides of the different cubes.

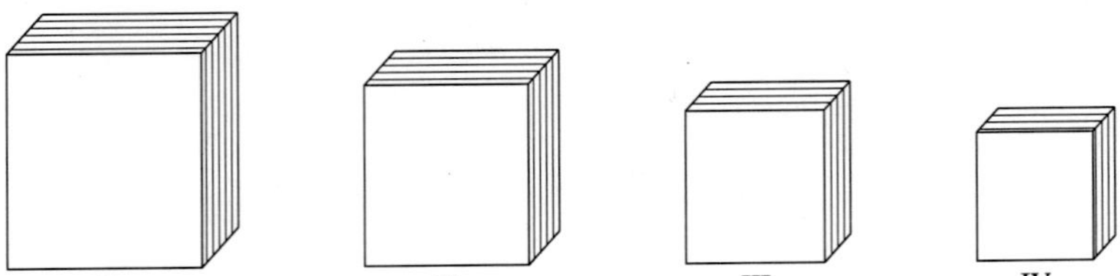

I II III IV

To find the radii of the equivalent spheres we proceed as under. Let r be the radius of the sphere corresponding to a cube whose side is 5 cm, then

$$= \frac{4}{3} \times \pi r^3 = 5 \times 5 \times 5 = 125$$

$$\Rightarrow \qquad r^3 = 125 \times \frac{3}{4} \times \frac{7}{22} = \frac{125}{8} \times \frac{21}{11}.$$

$$\therefore \qquad \text{Radius} = \left[\frac{21}{11}\right]^{1/3} \left(\frac{5}{2}\right).$$

Similarly we find the radii of other corresponding spheres.

2.46 PICTOGRAMS

The technique of presenting data by pictograms was developed by Dr. Otto Neurth who was the resident of Vienna. This method is also known as the Vienna method or Isotype method.

Pictograms are very useful in attracting attention as well as for showing comparisons. They are easily understood and create a lasting impression on the mind. Pictograms are used where the attention of masses is to be drawn, such as exhibitions, fairs etc.

The following table gives the total number of vessels in 1991, 1992, 1993, 1994 and 1995 in a certain factory. Represent the data by means of a pictogram :

Year	*No. of Vessels*
1991	255
1992	359
1993	403
1994	433
1995	488

Solution.

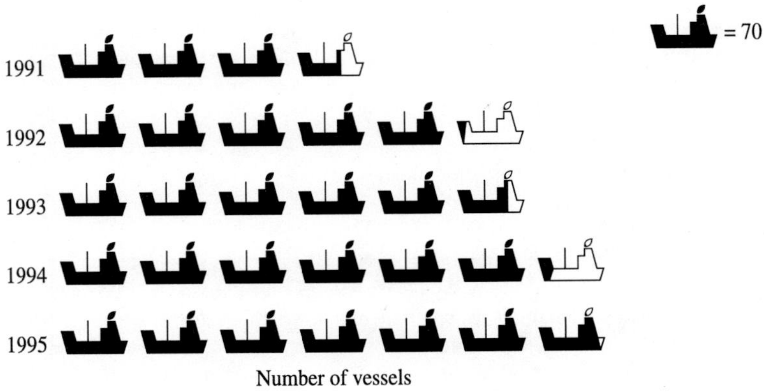

2.47 CARTOGRAMS

The representation of statistical data by maps is called Cartograms. For example, the data showing region-wise measurements of rainfall, region-wise density of population, etc. can be shown on a map of the region. Different type of shades or colours may be used for representing the magnitudes of the variable. A cartogram showing the annual rainfall (in cms.) in India in 1990 is shown in figure.

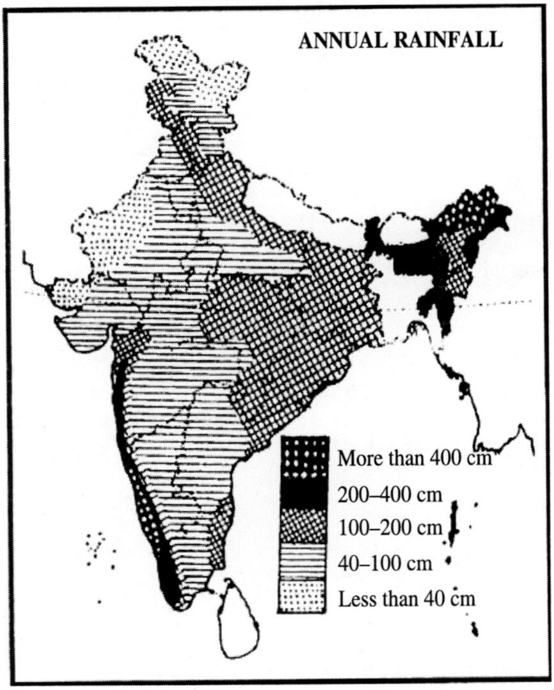

2.48 SELECTION OF A SUITABLE DIAGRAM

There are various methods for depicting statistical data diagrammatically. The selection of a suitable diagram for presenting a particular set of data is a difficult and practical problem. The selection of a suitable diagrammatic technique depends on the following considerations :

(1) nature of data to be represented,
(2) the kind of the persons to whom the characteristics of the data are to be communicated,
(3) the available space for drawing the diagram,
(4) the size of the data, and
(5) the specific purpose of investigation.

The nature of data will help in determining whether the diagram should be one-dimensional and whether simple bar should be made or it should be divided on the basis of percentages or absolute values. In two-dimensional diagrammatic representation whether squares should be made or circles. What is the level of education of the persons to whom the message is to be communicated, is another consideration for selection of method of diagrammatic presentation of statistical facts. To common people, having little knowledge of statistics, cartograms or pictograms may be the most suitable method. To show area, squares or circles are more appropriate.

In practice, statistical data are presented diagrammatically through a number of figures. Bar, rectangle, square, cube, circle, etc. are common media, but bar-diagrams are the most popular method. The kind of bar diagram to be used, depends upon following considerations :

(i) If magnitude of the values is to be specified, simple bar diagram will be most suitable.
(ii) If besides the magnitude of the values, their components are also to be represented, then sub-divided bar diagram can be used.
(iii) If relative changes in the components and their relative importance is to be shown, then bar-diagram based on percentages will be appropriate.

If in a set of data, there are more than three or four components, then bar diagram becomes complex and crowded. In such a case a circular diagram will be suitable to show the relative importance and proportion of the different components. If there is wide gap between the highest and the lowest value in a set of data, three-dimensional diagrams may be used. Pictograms are good for advertising or publicity purposes. Cartograms are employed to show some statistical fact based on geography.

2.49 LIMITATIONS OF DIAGRAMMATIC REPRESENTATION

(1) A diagram presents limited information.
(2) It is subjective in character.
(3) Its interpretation may vary from person to person.
(4) It shows only the approximate values.
(5) The real statistical value is suppressed.
(6) The small differences are generally ignored in the diagrammatic presentation.
(7) Construction of diagrams requires relatively more time and expertise.

2.50 GRAPHS

The comprehension of the different essential features of a distribution become quite easy even for a layman if the data is presented graphically.

According to A.L. Boddington, "*The wandering of line is more powerful in its effect on the mind than a tabulated statement; it shows that is happening and what is likely to take place, just as quickly as the eye is capable of working.*"

According to I.R. Vesselo, "*The simplest and commonest aid to the numerical reading is the graph. This gives a picture of numbers in such a way that the relations between them strike the eye. It is of the greatest value in making numbers clear.*"

But Hubbard Writers holds, "*Graphs are dynamic, dramatic. They may epitomise an epoch, each dot a fact, each slope an event, each curve a history. Wherever there are data to record, inferences to draw or facts to tell, graphs furnish the unrivalled means whose power we are just beginning to realise and apply.*"

2.51 ADVANTAGES OF GRAPHS

A graph is a visual form of presentation of statistical data. Main advantages of graphical presentation are :

(i) Graphs are interesting, attractive and impressive.
(ii) No knowledge of mathematics is required.
(iii) This is the simplest method of presenting the data.
(iv) We can determine the shape of the distribution.
(v) It makes comparison between two or more phenomena easy.
(vi) We can determine the pattern of the distribution.*
(vii) Certain statistical measures like median, quartiles, mode, correlation etc., can be determined graphically.
(viii) It shows any trends that might be present and the way in which the trends change.

2.52 DEFECTS (DISADVANTAGES) OF GRAPHICAL REPRESENTATION

1. Many people are not accustomed to it and they do not attach much importance to it.
2. A curve simply shows tendency and fluctuations, actual values are not known.
3. Graphic presentation may often give misleading impressions. Much depends upon the scale taken. Two different scales may show different fluctuations in the data.
4. Accuracy is rather not possible in a graph.
5. Graphs cannot be quoted in support of some statement.
6. Only one or two characteristics can be depicted in a graph. If more features are shown, the graph becomes difficult to follow.

In spite of these defects, graphic presentation of statistical data is a valuable tool in the hands of a statistician, the utility of which is immense.

2.53 DIAGRAMS AND GRAPHS—SIMILARITIES

1. Both are **simple and attractive**.
2. Both can give **condensed** form to data.
3. Both **help to compare** the variables.

2.54 DIFFERENCE BETWEEN DIAGRAMS AND GRAPHS

1. Graph is drawn on a graph paper whereas a diagram is normally drawn on a plain paper.
2. Graph helps us to study the mathematical relationship whereas diagrams are used for comparison only.

---

* We use Scatter diagram.

3. Data relating to time series and frequency distribution can be depicted by graph only. Diagrams fail to represent this type of data.
4. Graphs are more precise and accurate than the diagrams and are helpful to the Statistician for the study of slopes, rates of change and estimation (Interpolation and extrapolation) wherever possible. Graphs are almost essential for the analysis of economic, business or social data, needed for research work.
5. Construction of graphs is easier as compared to the construction of diagrams.
6. Diagrams are good **even for a Layman** but not the graphs.
7. Value of **Median and Mode** can be determined by graph but not by diagram.

2.55 FALSE BASE LINE

In the function;

$$y = f(x),$$

x is called the independent variable and y the dependent variable. While drawing the graph of $y = f(x)$ the values of x [Independent variable] are shown on X-axis and the values of y [dependent variable] are shown on Y-axis.

One of the fundamental rules for drawing graph is that the vertical axis, *i.e.*, Y-axis must start with zero as origin, but when the gap between zero and the minimum value of the dependent variable is large and the variation in the value of the dependent variable are very small compared with their magnitude, the graph will be mostly on the top of the paper and almost a straightline, hence lot of space will be consumed and minor fluctuations will not be clearly visible on the graph. To overcome this difficulty the concept of **false base line** is used. The portion of the scale between zero and the smallest value of the variable is omitted. False base line is represented by drawing two zig-zag horizontal lines above the base line (x-axis) on the graph. Whenever false base line is used it should be clearly indicated on the graph. The main objectives of using false base line are :

(i) To magnify the minor fluctuations on the graph so that they are clearly visible to the reader.
(ii) To economise in space.

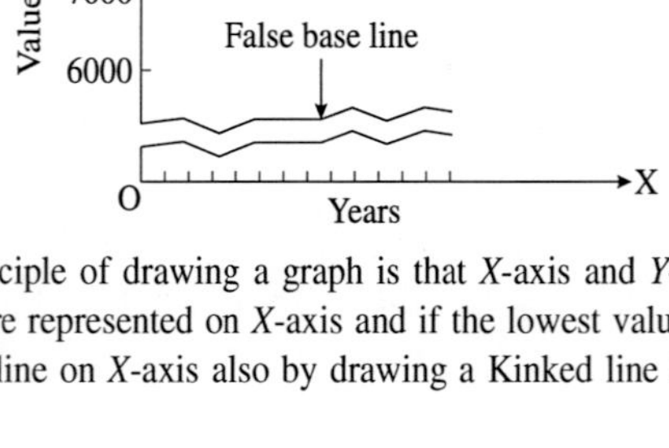

The fundamental principle of drawing a graph is that X-axis and Y-axis both should start from zero. If numerical values are represented on X-axis and if the lowest value is far away from zero, we can also have a false base line on X-axis also by drawing a Kinked line as follows.

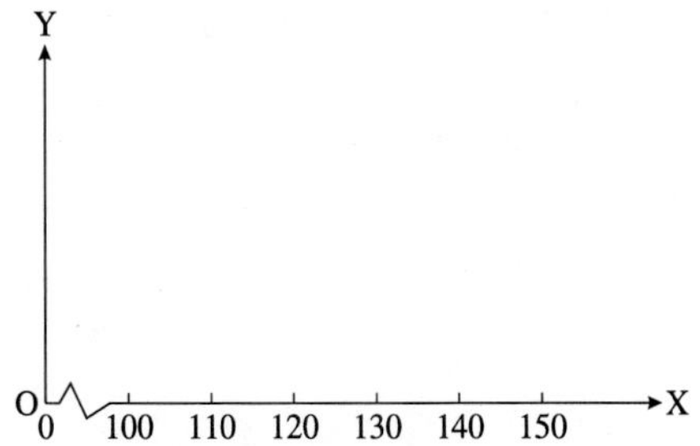

If we do not follow the concept of starting X-axis and Y-axis from zero but instead follow that the X-axis and Y-axis can start from the lowest value or near about from the data given, then there is no usefulness of false base line and it remains only a theoretical concept. (**Please note**)

2.56 TYPES OF GRAPHS

Graphs can be broadly classified in to two main groups :

(i) time series graphs (ii) Frequency graphs.

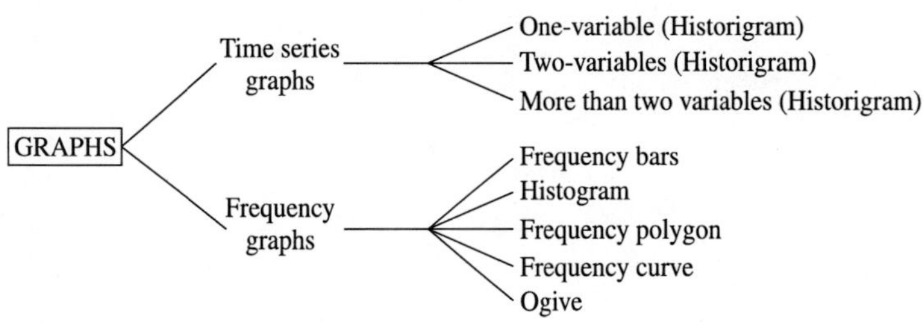

2.57 HISTORIGRAMS

A time series is an arrangement of statistical data in a chronological order. The graph of time series with time on X-axis and dependent variable on Y-axis is called *historigram*. If the actual time series are graphed, the historigram is called *Absolute historigram.* The graph obtained by plotting the index number of the given values is called *Index historigram* and it depicts the percentage change in the values of the phenomenon as compared to some fixed base period.

Notes : In time series, time is always taken as independent variable, variables like the total output of agriculture, the employment, the consumption or the national income may be dependent variables.

2.58 ONE VARIABLE GRAPHS (HISTORIGRAM)

One variable graphs are those graphs in which values of only one variable are shown with respect to different time periods.

Example 48. The following table shows the net revenue earned by Railways during the last nine years. Represent the data diagrammatically.

Year	*Amount*
1996–97	227
1997–98	127
1998–99	403
1999–2000	554
2000–2001	379
2001–2002	270
2002–2003	686
2003–2004	581
2004–2005	721

Solution. Graph of the above data (Historigram) is as under :

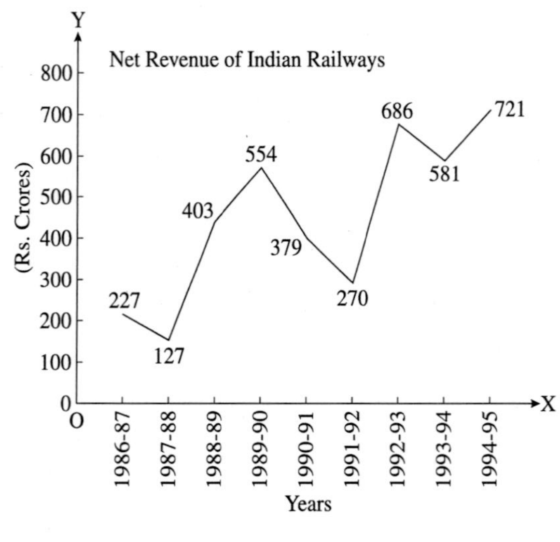

2.59 TWO OR MORE THAN TWO VARIABLES GRAPHS

In these graphs, the values of two or more than two variables are simultaneously shown with respect to same time period. Data on the production and sale of a factory in different months would make a Two Variable Graph or Two Variable Historigram.

Example 49. Following table gives data on the production and sale of a factory (in thousand rupees) between January to June. Make a Two-Variable Historigram.

Month	January	February	March	April	May	June
Production	7.5	10	7.5	12.5	15	17.5
Sale	5	7.5	5	10	12.5	15

Solution. Graph of the above data (Historigram) is as under :

Two-Variable Historigram

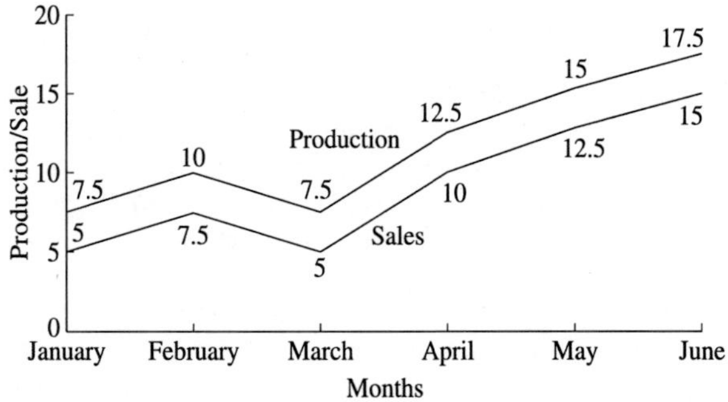

2.60 GRAPHS OF FREQUENCY DISTRIBUTION

Frequency distribution graphs are those graphs which present the frequency distribution of data on graph papers. These graphs are prepared to present those data which are not related to any time period. The data presented in these graphs are in the form of frequency distribution, either grouped or ungrouped. Frequency distribution graphs have better eye appeal than the frequency table. It quickly calls attention to high and low spots in the distribution and offers a vivid picture of characteristics of a given frequency distribution. Frequency distribution can be represented on the graph

(i) in absolute numbers or in percentage form

(ii) in a cumulative form.

In absolute or percentage form we use **histogram** and a **frequency polygon** and for cumulative distribution we use **ogive**.

2.61 LINE FREQUENCY GRAPH

A Line Frequency Graph is that graph which presents a discrete frequency distribution of data on a graph paper. In such a graph, various items or values are shown on the X-axis and their corresponding frequencies on the Y-axis. Frequencies of the values are shown in the form of vertical straight lines.

Graph between the heights of the students and the number of students is shown below. Bold lines show the number of students corresponding to different heights.

Line Frequency Graph

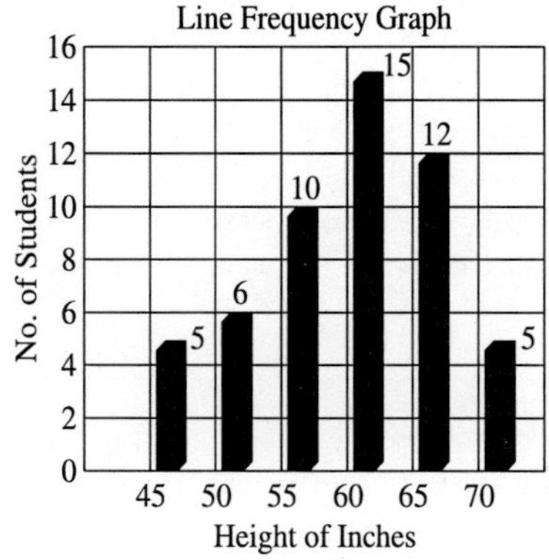

2.62 HISTOGRAM

An Histogram is a graphical representation [graphical picture] of a frequency distribution of a continuous series. It represents the class frequencies in a frequency distribution by verticle rectangles meeting each other from left to right. For each class interval, a rectangle is constructed with the base equal to the width of the class interval and height proportional to the frequency density of that class, so that the area of the rectangle above each class is proportional [or equal] to the frequency of that class. The total area covered under a histogram is proportional to (or equal to) the total frequency*. In histograms the key factor is the area of the rectangles erected on each class interval and not the height of the rectangle.

While constructing the histogram the vertical axis representing the frequency densities begin at zero origin and the horizontal axis representing class intervals can be with any convenient low value. For construction of an histogram, exclusive type of classification is always considered. If the data is given in inclusive type of classification, it is always converted to exclusive type of classification before proceeding to construct an histogram.

Important Note : If we take class intervals on the *X*-axis and corresponding frequency densities on the *Y*-axis then the procedure of drawing the histogram remains the same whether class intervals are equal or unequal. However to give better took to the diagram we may multiply each frequency density **by a constant factor, may be the minimum class interval**.

Following four types are explained in solved examples :

(i) Histogram of Equal Class intervals.

(ii) Histogram when Mid-points are given.

(iii) Histogram of Unequal Class Intervals.

(iv) Histogram when class intervals are given by inclusive method.

2.63 FREQUENCY POLYGON

In case of discrete frequency distribution, frequency polygon is obtained by plotting the frequencies on the vertical axis, *i.e.*, *Y*-axis against the corresponding value of the variable on the horizontal axis, *i.e.*, *X*-axis and joining the points so obtained by straight lines.

In case of grouped frequency distribution, (*i.e.*, continuous frequency distribution), frequency polygon is graphical representation of a frequency distribution in the form of the curve superimposed on a histogram (with equal intervals) by joining the mid-points of the upper horizontal sides of the consecutive rectangles. But we may draw a polygon simply by plotting for each class the value of its mid-point (on *X*-axis) against its frequency (on *Y*-axis) and joining by straight lines the points thus plotted without a histogram. Frequency polygon is drawn under the assumption that the frequencies in a class are evenly distributed throughout the class and the mid-points are the representatives of the class.

Frequency polygon are closed by containing them to the base line. The idea behind this extension is to make the area of the polygon equal to the area under the corresponding histogram. The extension is sometimes shown by dotted lines.

---

* Frequency density of a class $= \dfrac{\text{Frequency of the class}}{\text{Width of the class}}$.

$\therefore$ Frequency density of the class $\times$ Width of the class = Frequency of the class.

2.64 FREQUENCY POLYGON AND HISTOGRAM—COMPARISON

1. Frequency polygon is an improvement over histogram because it provides a continuous curve indicating the gradient of rise and fall in the data. On the other hand, frequency polygon is an approximate curve, but still it is more useful as compared to histogram.
2. In the frequency polygon, it is assumed that the frequency distribution has equal class-width whereas histogram may be used to represent frequency distributions with equal as well as with unequal class width (class interval).
3. In case of frequency polygon, it is assumed that all frequencies in a particular class are concentrated at the mid-point of that class whereas in case of histogram, it is supposed that they are evently spread over the class-interval.
4. Two or more frequency polygons can be drawn on the same graph, *i.e.*, they can be superimposed whereas two histograms cannot be. Hence frequency polygons are preferred for purpose of graphical comparison of frequency distributions. Histogram is a two dimensional figure whereas frequency polygon is a line graph.

2.65 FREQUENCY CURVE

A *frequency curve* for a grouped frequency distribution is a smooth freehand curve drawn by joining the mid-points of the upper horizontal sides of the histograms drawn over that frequency distribution. The area under frequency curve is approximately the same as that under the frequency polygon.

The frequency polygon is not a smooth curve because it is constructed by joining the points by straight lines whereas frequency curve is a smooth freehand curve.

2.66 DIFFERENT TYPES OF CURVES

1. Normal Curve

Normal curve is a bell-shaped symmetrical curve extending to infinity on both sides of X-axis. Normal curves are extremely useful in understanding sampling of data. In normal curves,

Mean = Median = Mode.

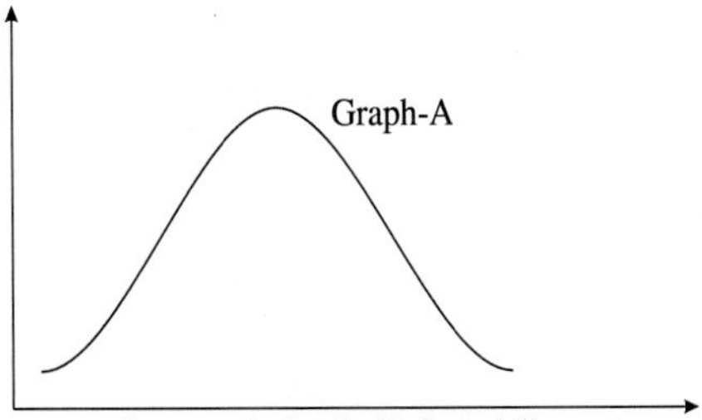

Symmetrical or Bell-shaped Curve
Mean = Median = Mode

2. Skewed Curves

Skewed curves are asymmetrical curves. Skewness can be positive or negative. In positive Skewness the curve leans more towards the right and mean > median > mode and in negative Skewness curve leans more towards the left and mean < median < mode.

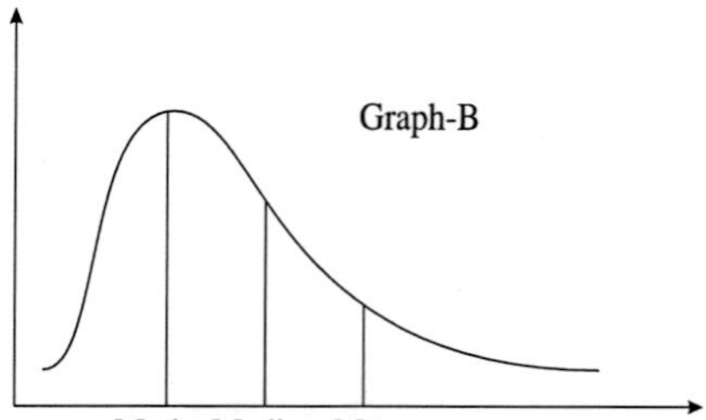

Mode Median Mean

Skewed to the right (positive skewness)

Mean > Median > Mode

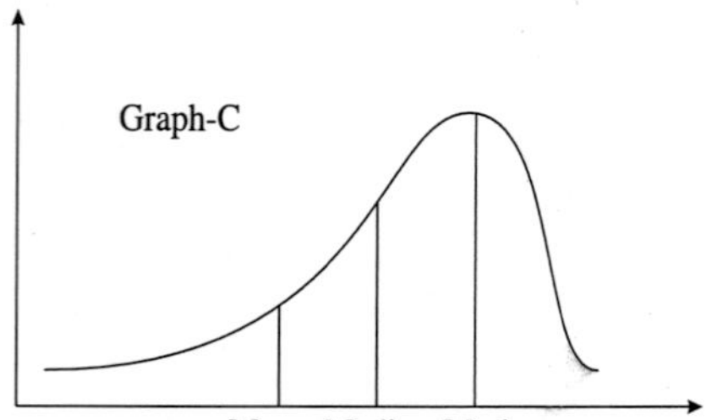

Mean Median Mode

Skewed to the left (negative skewness)

Mean < Median < Mode

3. U-shaped Curve

U-shaped curves are formed if there are two high points in a series both having equals or nearly equal frequency, at the lowest and highest values. U-shaped curve are bi-modal curves having two Peaks one at the lower extreme and one at the upper extreme. For example, unemployment is highest among employable males at earlier working years and at last working years.

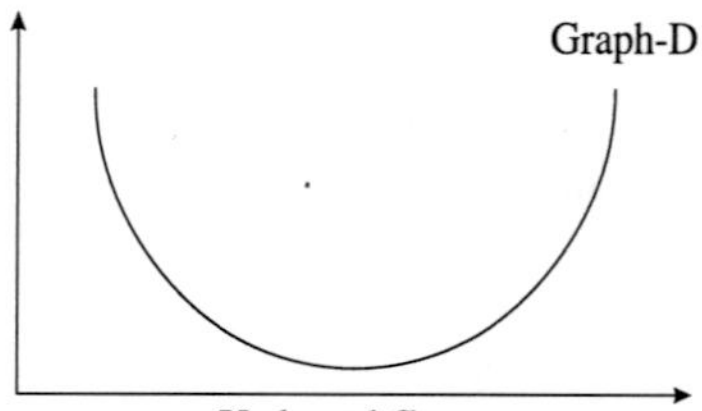

U-shaped Curve

2.67 OTHER BIMODAL CURVES

Graph-E also shows a Bimodal curve. This type of curves are drawn when, in a series, *there are two classes with highest frequencies*. In such series there is no unique modal value.

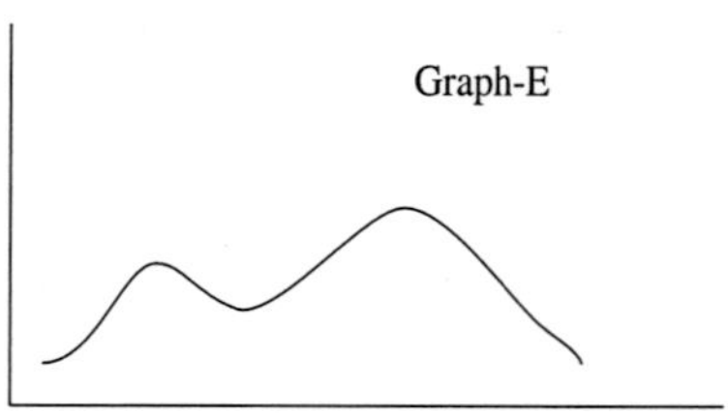

Bimodal Curve

1. J-shaped Curve

A J-shaped curve. This is drawn when, in a distribution *frequencies tend to increase with each class interval. In other words, the curve moves from the low frequencies to the high frequencies*. Thus, highest frequencies are found around the upper end of the curve. Such curves are shaped like the letter 'J' in English language.

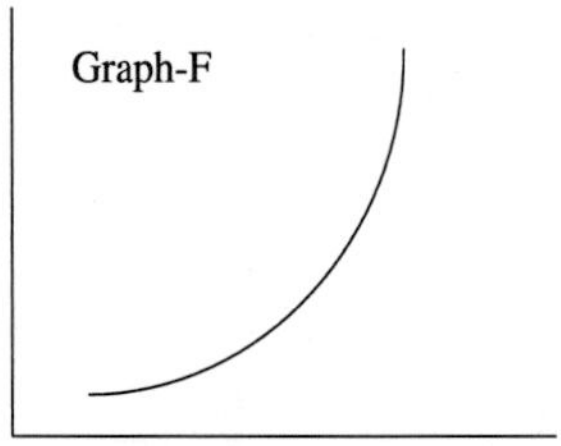

J-shaped Curve

2. Reverse J-shaped Curve

Graph-G shows a reverse J-shaped curve. *This type of curve is drawn when there are highest frequencies corresponding to lowest values in the distribution.* Conversely, there are lowest frequencies corresponding to highest values in the distribution.

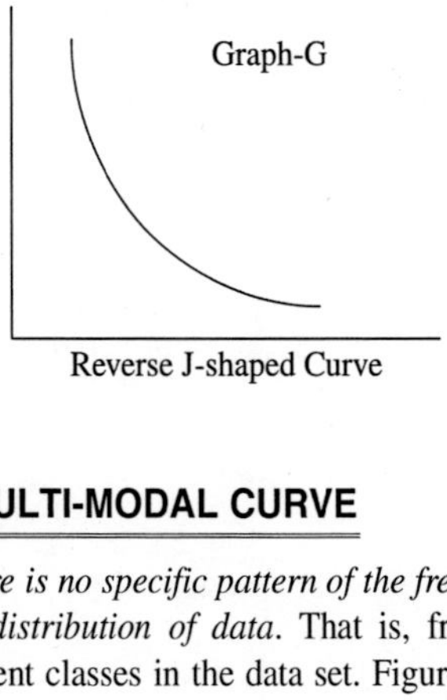

Reverse J-shaped Curve

2.68 MIXED CURVE OR MULTI-MODAL CURVE

These curves are formed *when there is no specific pattern of the frequencies corresponding to different values* (or classes) *in the given distribution of data.* That is, frequencies keep on increasing and decreasing corresponding to different classes in the data set. Figure below shows a mixed curve, also called a multi-modal curve.

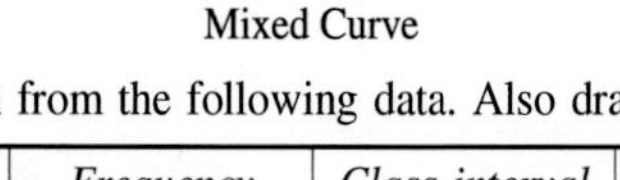

Mixed Curve

Example 50. Draw a histogram from the following data. Also draw the frequency polygon :

Class interval	*Frequency*	*Class interval*	*Frequency*
5–10	3	30–35	13
10–15	5	35–40	11
15–20	9	40–45	7
20–25	15	45–50	4
25–30	18		

Solution. **Construction of Histogram**

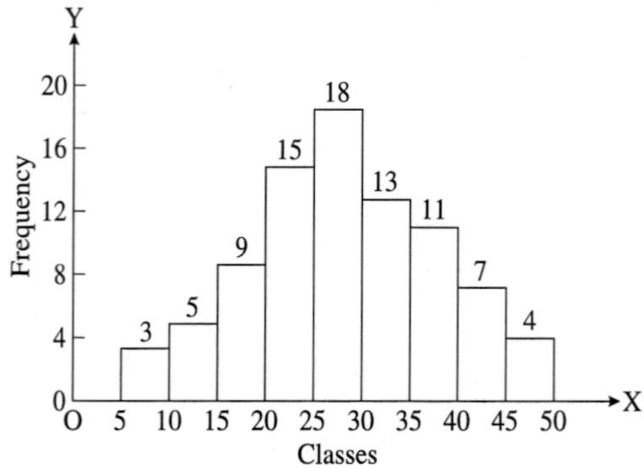

Example 51. Prepare a histogram for the following data. (*Equal Intervals*)

Wages (Rs.)	325–350	350–375	375–400	400–425	425–450
Frequency	30	45	75	60	35

Solution.

Class interval	*Frequency*	*Frequency density*	*Frequency density × Minimum class interval**
325–350	30	$\frac{30}{25} = 12$	$1.2 \times 25 = 30$
350–375	45	$\frac{45}{25} = 1.8$	$1.8 \times 25 = 45$
375–400	75	$\frac{75}{25} = 3$	$3 \times 25 = 75$
400–425	60	$\frac{60}{25} = 2.4$	$2.4 \times 25 = 60$
425–450	35	$\frac{35}{25} = 1.4$	$1.4 \times 25 = 35$

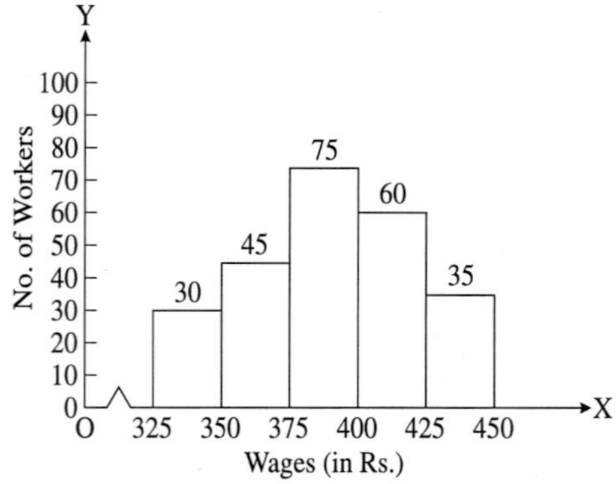

* If the class intervals are equal the minimum class interval is the same as the class interval, so we can multiply the frequency density by class interval. But this multiplication is done just to give large height of the rectangles otherwise it is not essential.

Example 52. Construct a histogram from the following distribution of total marks obtained by 65 students of a school in a Board Examination :

Marks (Mid-points)	*No. of students*
150	8
160	10
170	25
180	12
190	7
200	3

Solution. In this question only mid-points are given :

Respective Class Intervals with Frequency Densities are as under :

Class interal	*Frequency*	*Frequency density*	*Adjusted frequency = Frequency density × Minimum class interval*
145–155	8	$\frac{8}{10}$	$\frac{8}{10} \times 10 = 8$
155–165	10	$\frac{10}{10}$	$\frac{10}{10} \times 10 = 10$
165–175	25	$\frac{25}{10}$	$\frac{25}{10} \times 10 = 25$
175–185	12	$\frac{12}{10}$	$\frac{12}{10} \times 10 = 12$
185–195	7	$\frac{7}{10}$	$\frac{7}{10} \times 10 = 7$
195–205	3	$\frac{3}{10}$	$\frac{3}{10} \times 10 = 3$

Method 1.
[Kinked line method]

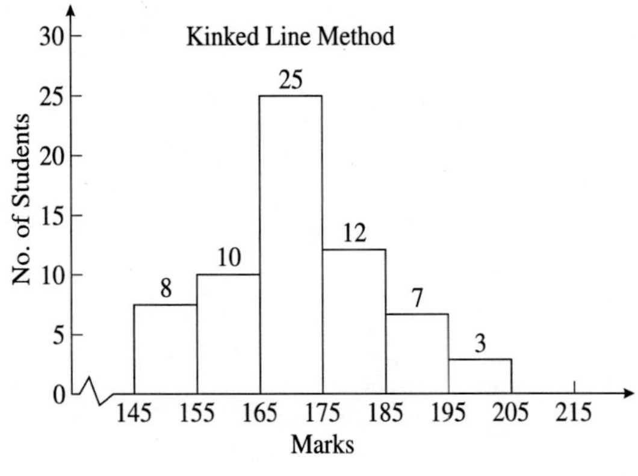

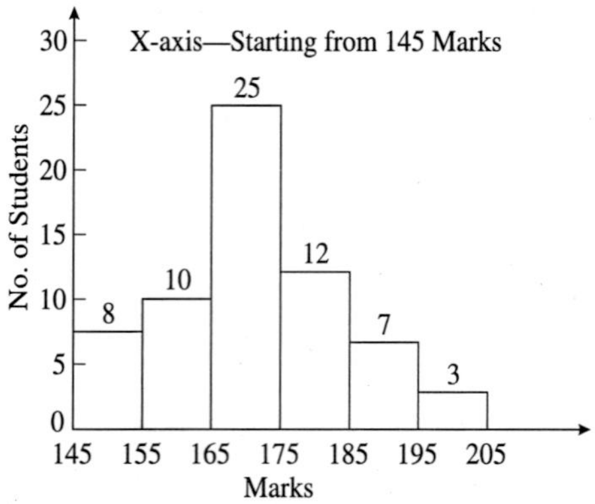

Method 2.
[X-axis starting with the minimum value 145]

Example 53. The following frequency distribution gives the marks of 92 students :

Marks	*No. of students*	*Marks*	*No. of students*
4–8	3	20–28	20
8–12	9	28–40	15
12–16	15	40–56	12
16–20	18		

Draw a histogram with the above data.

Solution. (This is a case of unequal class intervals)

Class interval	*Frequency*	*Frequency density*	*Frequency density × Minimum* class interval*
4–8	3	$\frac{3}{4} = 0.75$	$0.75 \times 4 = 3$
8–12	9	$\frac{9}{4} = 2.25$	$2.25 \times 4 = 9$
12–16	15	$\frac{15}{4} = 3.75$	$3.75 \times 4 = 15$
16–20	18	$\frac{18}{4} = 4.5$	$4.5 \times 4 = 18$
20–28	20	$\frac{20}{8} = 2.5$	$2.5 \times 4 = 10$
28–40	15	$\frac{15}{12} = 1.25$	$1.25 \times 4 = 5$
40–56	12	$\frac{12}{16} = 0.75$	$0.75 \times 4 = 3$

* Frequency Density × Minimum Class Interval = Adjusted Frequency.

Construction of Histogram

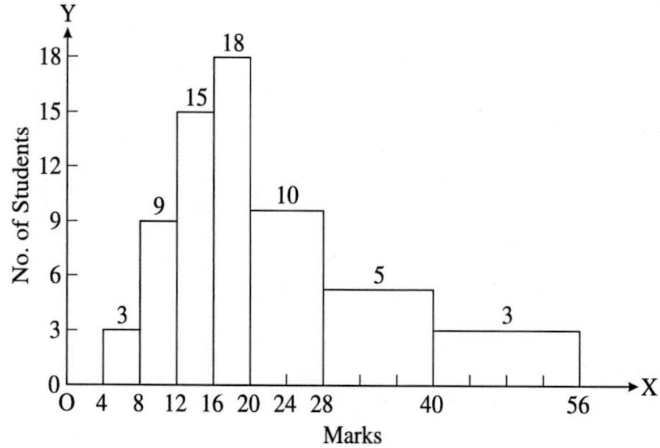

Example 54. Draw a histogram of the following frequency distribution showing the number of boys on the register of primary schools in a certain state :

Age (in years)	2–5	5–11	11–12	12–13	13–14	14–15	15–17
No. of boys	150	3066	497	477	496	143	162

Solution.

Age class	*No. of boys f*	*Frequency density*
2–5	150	150 ÷ 3 = 50
5–11	3066	3066 ÷ 6 = 511
11–12	497	497
12–13	477	477
13–14	496	496
14–15	143	143
15–17	162	162 ÷ 2 = 81

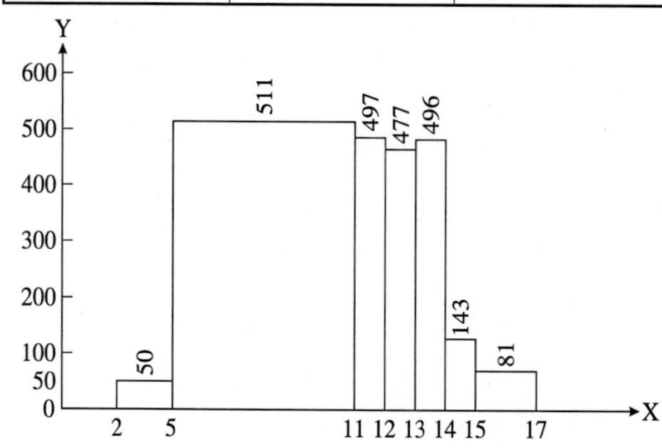

Example 55. Construct a Histogram from the following data :

Marks :	5–9	10–14	15–19	20–24	25–29	30–35
Students :	4	17	25	32	13	6

Solution. In this case when class intervals are given by inclusive method, it is necessary to convert them into exclusive class intervals. The adjusted class intervals will be :

Class interval	*Frequency*	*Frequency density*	*Frequency density* × *Minimum class interval**
4.5–9.5	4	$\frac{4}{5}$	$\frac{4}{5} \times 5 = 4$
9.5–14.5	17	$\frac{17}{5}$	$\frac{17}{5} \times 5 = 17$
14.5–19.5	25	$\frac{25}{5}$	$\frac{25}{5} \times 5 = 25$
19.5–24.5	32	$\frac{32}{5}$	$\frac{32}{5} \times 5 = 32$
24.5–29.5	13	$\frac{13}{5}$	$\frac{13}{5} \times 5 = 13$
29.5–34.5	6	$\frac{6}{5}$	$\frac{6}{5} \times 5 = 6$

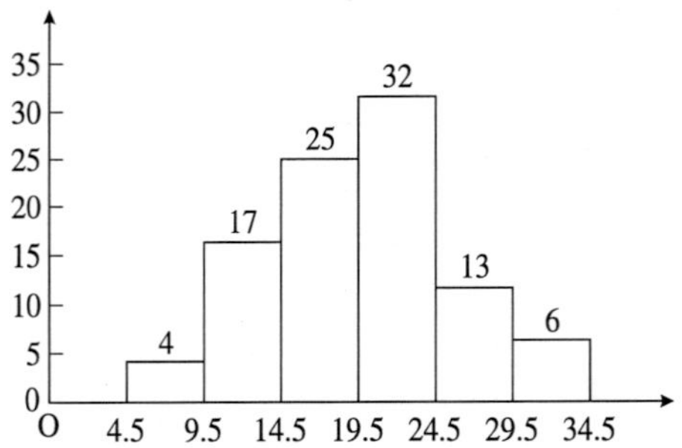

Example 56. Draw a histogram and a frequency polygon from the following frequency distribution table of marks in English :

Class boundaries	:	12.5–17.5	17.5–22.5	22.5–27.5	27.5–32.5	32.5–37.5
Frequency	:	3	4	6	15	19
		37.5–42.5	42.5–47.5	47.5–52.5	52.5–57.5	57.5–62.5
		29	19	15	6	3
		62.5–67.5	67.5–72.5			
		0	1			

* Since the class intervals are equal, we multiply the frequency density by class interval.

Solution.

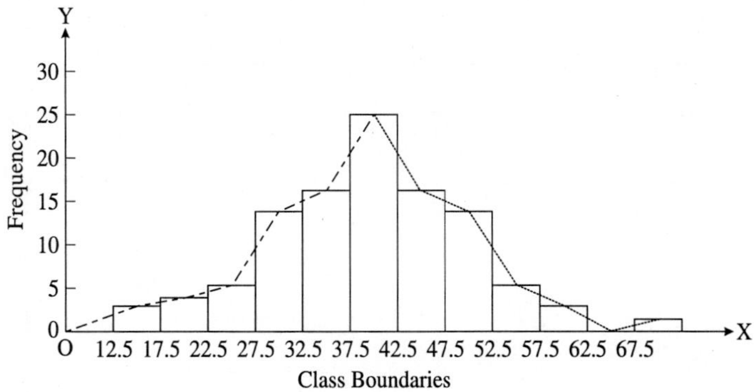

While converting histogram into frequency polygon, there is a loss in the area of rectangles below the mid-point of each rectangle to the left of the mid-point and an addition of equal area to the right of mid-point for rectangles to the left of the maximum rectangle. Similarly, to the right of the maximum rectangle, each rectangle loses area to the right of its mid-point and gains an equal area to the left of its mid-point. In this way, the total area of the frequency polygon remains the same as that of histogram as has been depicted in the figure above.

Example 57. Draw a histogram and a frequency polygon of the distribution of marks :

Weekly wages (in Rs.)	:	1–10	11–20	21–30	31–40	41–50
No. of Workers	:	14	28	36	12	10

Solution.

C.I. Wages (Rs.)	*Class boundaries*	*No. of Workers (frequency)*
1–10	0.5–10.5	14
11–20	10.5–20.5	28
21–30	20.5–30.5	36
31–40	30.5–40.5	12
41–50	40.5–50.5	10

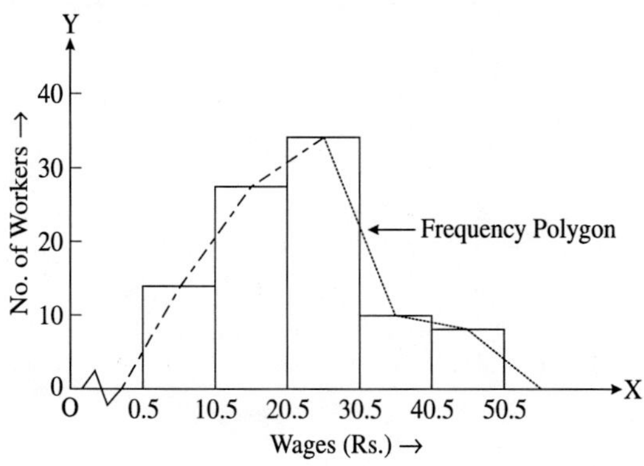

Example 58. Draw a histogram and a frequency polygon for the following data :

Weight (nearest to the pounds) :	102–113	113–124	125–135	135–146	146–157	157–168	168–179
No. of men :	1	1	4	14	12	5	3

Solution.

Wr. (Ib) :	102–113	113–124	124–135	135–146	146–157	157–168	168–179
No. of men (f) :	1	1	4	14	12	5	3

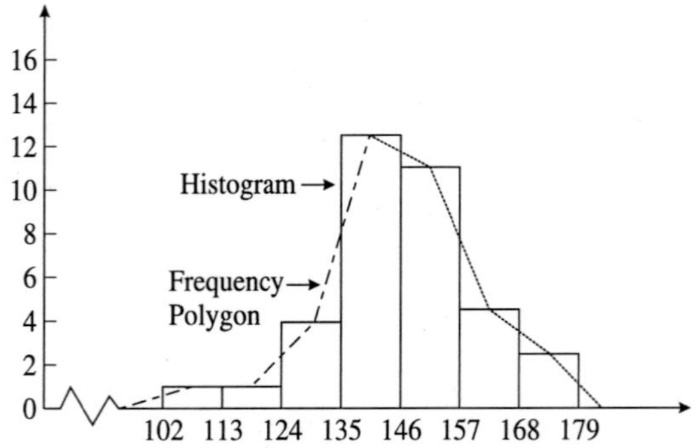

2.69 OGIVES OR CUMULATIVE FREQUENCY CURVE (MORE THAN OGIVE AND LESS THAN OGIVE)

Cumulative frequency of any class is equal to the sum of the frequencies of all the classes preceding that class and its own frequency.

Suppose we have the distribution :

Class interval	*Frequency*
0–10	10
10–20	7
20–30	12
30–40	17
40–50	22

Cumulative frequency of the class

$$0\text{–}10 = 10$$
$$10\text{–}20 = 10 + 7 = 17$$
$$20\text{–}30 = 10 + 7 + 12 = 29$$
$$30\text{–}40 = 10 + 7 + 12 + 17 = 46$$
$$40\text{–}50 = 10 + 7 + 12 + 17 + 22 = 68$$

Total number of values = 68 = Total frequency.

Cumulative frequency of the last class = Total frequency.

All these 68 values are less than 50. If we leave the last class interval, *i.e.*, 44–50, then the remaining (68 – 22) = 46 values are less than 40, 29 values are less than 30 and so on. This data will help us in plotting a '**less than**' ogive.

On the other hand, all these 68 values are more than 0. If we leave the first class interval, then the remaining $(68 - 10) = 58$ values are more than 10, if we leave the first two classes, then $(68 - 10 - 7) = 51$ values are more than 20 and so on. This data will help us in constructing a '**more than**' ogive.

For drawing an Ogive, cumulative frequency (*i.e.*, No. of values) is taken on the *Y*-axis and limits of class intervals on the *X*-axis.

In a "*less than*" type ogive, we plot the upper limit of each class along *X*-axis and in a "*more than*" type ogive, we plot the lower limit of each class along the *X*-axis. Along the *Y*-axis, we plot the cumulative frequencies at the end of each class. Ogive can be drawn even if the class intervals are unequal. If the class intervals are open ended, than

(i) less than ogive can be drawn if lower end is open

(ii) more than ogive can be drawn if upper end is open.

Both the ogives "*less than*" or "*more than*" can be drawn on the same graph. If done so they intersect at a point. The foot of the perpendicular from their point of intersection on the *X*-axis gives the value of the **median**.

Ogives corresponding to different frequency distributions can be drawn on the same graph and are generally less overlapping than the corresponding frequency curves. **So ogives are preferred over frequency curves for comparative study**.

An ogive can be used to locate a number of measures of position, such as median, quartiles, deciles, percentiles and so on. It is also used to find the value of a variable corresponding to a specified cumulative frequency and *vice versa.*

If the class frequencies are large they can be expressed as percentage of the total frequency. The graph of the cumulative *per cent frequency* is called '*percentile curve*'.

Example 59. Draw a less than ogive from the following data :

Wages (in Rs.)	*No. of Workers*
20–40	4
40–60	6
60–80	10
80–100	16
100–120	12
120–140	7
140–160	3

Solution.

Wages (in Rs.) *Less than*	*No. of Workers* *(C.F.)*
40	4
60	10
80	20
100	36
120	48
140	55
160	58

For **less than Ogive curve**, plot the points (40, 4), (60, 10), (80, 20), (100, 36), (120, 48), (140, 55), (160, 58).

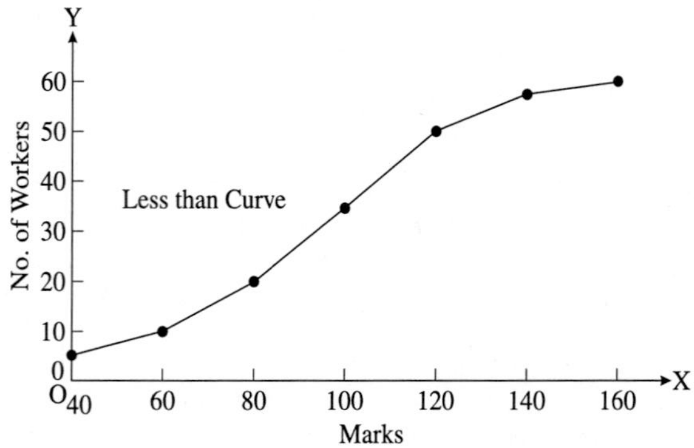

Example 60. On the basis of the following frequency distribution, draw 'less than' and 'more than', cumulative frequency curves :

Marks	*No. of Students (Frequency)*
0–10	2
10–20	3
20–30	5
30–40	8
40–50	4
50–60	3
60–70	5

Solution. (i) 'Less than' Ogive :

Marks (Less than)	*Cumulative Frequency*
10	2
20	5
30	10
40	18
50	22
60	25
70	30

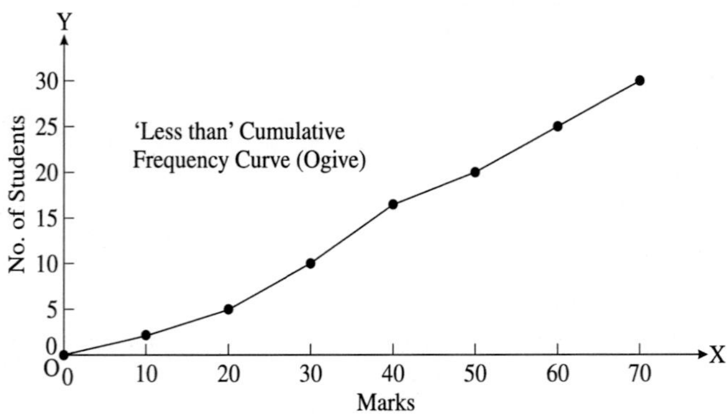

(ii) 'More than' Ogive :

Marks (More than)	*Cumulative Frequency*
0	30
10	28
20	25
30	20
40	12
50	8
60	5

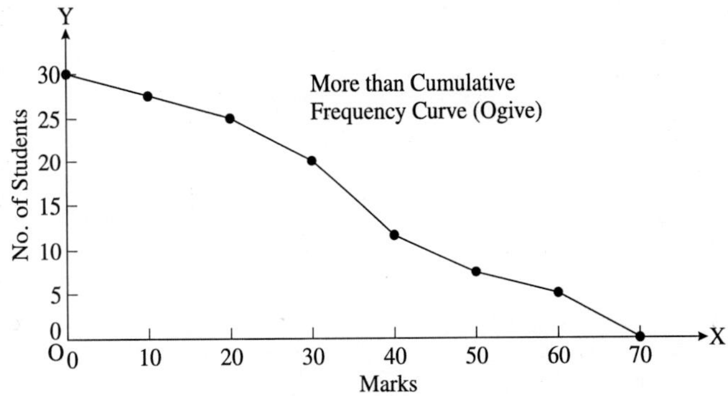

Note : More examples on Ogives have been given in Chapter 4.

Questions

Very Short Answer Type Questions

1. U-shaped curve is symmetrical or asymmetrical. Explain.
2. What is skewness?
3. What is normal curve, write one main characteristic of normal curves?
4. Explain J-shaped curve.
5. What is multimodal curve?
6. What is the difference between uni-modal and bi-modal curves?
7. What is frequency array?

Short Answer Type Questions

1. Dicsuss the utility of diagrammatic presentation.
2. Name the various forms of visual presentation of statistical data.
3. Discuss the rules of constructing diagrams.
4. Name the types of diagrams used for displaying data.
5. What do you mean by one-dimensional diagram?
6. Name different types of one-dimensional diagrams.

7. Mention the situation where simple bar diagram is used.
8. Discuss the situation when multiple bar diagram is suitable.
9. Why do we use a percentage bar diagram?
10. What do you mean by a two-dimensional diagram?
11. 'Two-dimensional diagrams are also called are diagram', why?
12. Name some of the two-dimensional diagrams.
13. Discuss the formation of square and circle diagrams.
14. Define a Pie-diagram. How is it formed? Give example.
15. What do you mean by Pictogram?
16. Define Cartogram and mention its uses.
17. Discuss various limitations of diagrammatical representation.
18. State briefly the purpose served by the graphical representation of a frequency distribution.
19. Define the following terms :
    (i) Histogram (ii) Frequency polygon (iii) Frequency curve (iv) Ogive
20. What do you understand by a cumulative frequency distribution? Point out its special advantages and uses.
21. Name the various ways of presenting a frequency distribution graphically.

Long Answer Type Questions

1. Define classification. Explain inclusive and exclusive type of classification.
2. What are the objectives of tabulation?
3. What are the essential parts of a table?
4. What are the characteristics of a good table?
5. What is the difference between diagrams and graphs?
6. What are the advantages of diagrammatic representation?
7. What are the essentials of a good diagram?
8. What are the limitations of a diagrammatic representation?
9. Explain histogram and historigram.
10. What are the advantages and disadvantages of graphical representation of data?
11. Explain false base line.
12. Explain frequency curve and frequency polygon.

State whether the following statements are True or False

1. Diagrammatic presentation is a tabular form of classified data.
2. Only the length of the bar is considered in a bar diagram.
3. Area diagrams and two-dimensional diagrams are different.
4. Simple bar diagram is used only for one-dimensional comparisons.
5. A multiple bar diagram is used when a comparison is made between values on two or more variables.
6. A sub-divided bar diagram is also known as a component bar diagram.
7. In a percentage bar diagram, the component parts are transformed to percentages of the total.
8. Deviation bar-diagrams are two-dimensional in nature.
9. Pie-diagram is a circle divided into sectors with area equal to the corresponding component.
10. Pictograms and Cartograms are the same.

ANSWERS

1. F **2.** T **3.** F **4.** T **5.** T **6.** T **7.** T **8.** F **9.** T **10.** F

Multiple Choice Questions

Pick up the correct answer :

1. Two dimensional diagrams are :
   (a) linear diagrams (b) volume diagrams (c) area diagrams (d) none of the above
2. Square is a :
   (a) simple bar diagram (b) two dimensional diagram
   (c) volume diagram (d) cumulative frequency graph
3. Ogive is a :
   (a) frequency polygon (b) cumulative frequency curve
   (c) line diagram (d) none of the above
4. Classification is done :
   (a) class limits and class boundaries ae the same (b) class limits are different than class boundaries
   (c) upper class limit of any class is included in that class (d) none of the above
5. In exclusive type classification :
   (a) upper class limit is not included in the class (b) class limits and class boundaries are the same
   (c) class limits and class boundaries are not the same (c) none of the above

ANSWERS

1. (c) **2.** (b) **3.** (b) **4.** (a) **5.** (a) **6.** (c)

Fill in the Blanks

1. Bar diagrams are ...... dimensional diagrams.
2. Pie-diagram is a ...... dimensional diagram.
3. Square diagram is a ...... dimensional diagram.
4. Cartograms are used to give quantitative information on a ....
5. Representation of data by pictures is called ....
6. Multiple bar diagrams is ...... dimensional diagram.
7. Sub-divided bar diagrams are also called ...... bar diagrams.

ANSWERS

1. one **2.** two **3.** two **4.** geographical basis **5.** pictogram
6. one **7.** component

Practice Questions

1. Prepare (a) Histogram (b) Frequency polygon for the following frequency distribution of grades in a final examination :

Class	*Frequency*	*Class*	*Frequency*
10–19	6	50–59	12
20–29	12	60–69	8
30–39	20	70–79	6
40–49	14	80–89	2

2. The following table gives the marks of 100 students in the subject 'English' :

Marks	*Students*	*Marks*	*Students*
0–9	5	40–49	15
10–19	15	50–59	10
20–29	18	60–69	5
30–39	30	70–79	2

Draw 'less than' and 'more than' type ogives. Using these curves find the number of students :

(i) with marks less than 45 (ii) with marks more than 65

(iii) with marks between 45 and 65

[**Hint :** (iii) Marks 45 and 65 on *X*-axis. Let these points be *P* and *Q*. Draw Perpendiculars from *P* and *Q* Cutting the Ogives at points say *M* and *N*. From *M* and *N* Draw Perpendiculars on *Y*-axis cutting the *Y*-axis at points say *A* and *B*. the number of values falling between the points *A* and *B* is the required number of students.]

3. What are different forms of a frequency curve?

4. Prepare histogram for the following cumulative frequency distribution :

Classes	*Frequency*
Less than 10	5
Less than 20	8
Less than 30	15
Less than 40	29
Less than 50	36
Less than 60	45
Less than 70	53
Less than 80	60

5. Represent the following data by a subdivided bar diagram :

Year	*1988*	*1989*	*1990*	*1991*	*1992*
Exports (crores Rs.)	10	110	115	125	122
Import (crores Rs.)	85	89	105	110	115

6. Use pie-diagram to represent the following data :

Items	*Expenditure (Rs.)*
Food	1640
Clothing	600
Rent	660
Medical	250
Others	800
Total	3950

7. Use rectangle diagram to present the following data :

Items	*Expenditure (Rs.)*	
	Family A	*Family B*
Food	1500	1200
Clothing	600	500
Rent	800	700
Education	400	300
Miscellaneous	700	800
Total	4000	3500

8. Draw a suitable diagram to represent the following data :

Table showing crimes

Year	*Murder*	*Robbery*	*Loot*	*Total*
1985	118	85	322	525
1986	140	120	390	650
1987	165	125	420	710
1988	195	153	442	790

[**Hint :** Use either multiple bar diagram or subdivided bar diagram.]

9. Represent the following data by rectangle diagram (i) Using original units, (ii) On percentage basis :

Items of expenditure	*R Family*	*S Family*
Food	400	640
Clothing	250	480
Rent	150	320
Education	340	100
Miscellaneous	160	60
Total	1300	1600

10. Draw (i) Square diagram and (ii) Circle diagram to represent the following data given in the census figures :

Name of the U.T. and States	*Area in Sq. Km.*
Andaman & Nicobar Islands	8293
Delhi	1485
Goa, Daman and Diu	3815
Mizoram	21087

11. Represent the data by a simple bar diagram.

Countries	*Production of Rice (000's tons)*
A	38
B	42
C	29
D	28
E	18
F	11

12. Below are given the wages earned per month in a certain factory. Calculate the number of workers earning more than Rs. 75 per month.

Monthly Income (Rs.)	*No. of Workers*
Upto Rs. 50	50
Upto Rs. 60	150
Upto Rs. 70	300
Upto Rs. 80	500
Upto Rs. 90	700
Upto Rs. 100	800

13. Show the following data of expenditure of an average working class family by a pie diagram.

Item of Expenditure	*Percent of Total Expenditure*
Food	65
Clothing	10
Housing	12
Fuel and lighting	5
Miscellaneous	8

14. Construct a sub-divided bar diagram from the following data, regarding the number of students in 2 colleges.

College	*Arts*	*Science*	*Commerce*	*Agriculture*	*Total*
A	1,200	800	600	600	3,000
B	750	500	300	450	2,000

15. Draw a less than ogive from the following data and estimate the number of employees earning atleast Rs. 63 but less than Rs. 75.

Wages	*No. of Employees*
50–60	8
60–70	10
70–80	16
80–90	14
90–100	10
100–110	5
110–120	2

16. Draw a Pie diagram to represent the following :

Males	*Females*	*Girls*	*Boys*	*Total*
2,000	1,800	4,200	2,000	10,000

17. Draw a blank table to present the following information regarding students of Bangalore University :

1. Faculty : Arts, Science, Commerce
2. Class : Degree, Pre-University
3. Students : Boys, Girls
4. Age : Below 20 years, More than 20 years.

18. Draw up a blank table to show the following divisions :

Year : 1992 1993 1994
Faculty : Arts Science Commerce
Students : Boys, Girls
Medium : English, Hindi

Solution.

Faculty / *Year*	*Arts*				*Science*				*Commerce*			
	Boys		Girls		Boys		Girls		Boys		Girls	
	Hindi	Eng.	Hindi	Eng.	Hindi	Eng.	Hindi	Eng.	Hindi	Eng.	Hindi	Eng.
1992												
1993												
1994												

3

Statistical Averages (Measures of Central Tendency)

Coverage Importance and Requisites of Good Statistical Average : Types of Averages—Arithmetic Mean, Median, Mode, Geometric Mean and Harmonic Mean, Weighted Average, Relationship among Different Averages, Partition Values : Quartiles, Deciles and Percentiles.

3.1 INTRODUCTION

According to Simpson and Kafka, *"A measure of central tendency is a typical value around which other figures congregate or which divides their number in half. Thus an average can be used to describe or represent a whole series of figures involving magnitudes of the same variable. That average is an overall single value which represents the series. Measures of central tendency permits us to compare individual items in the group with it and also permits us to compare different series of figures with regard to their central tendencies."* Averages are derived figures and not the original data.

3.2 DEFINITION OF CENTRAL TENDENCY

(i) According to Clark, *"An average is a figure that represents the whole group."*

(ii) A.E. Wagh defines, *"An average is a single value selected from a group of values to represent them in some way."*

(iii) According to Croxton and Cowden, *"An average is a single value within the range of the data that is used to represent all the values in the series. Since an average is somewhere within the range of the data it is sometimes called a measure of central value."*

According to Lawrence J. Kaplan, *"One of the most widely used sets of summary figures is known as measures of location, which are often referred to as averages, measures of central tendency or central location. The purpose of computing an average value for a set of observations is to obtain a single value which is representative of all the items and which the mind can grasp simply and quickly. The single value is the point of location around which the individual items cluster."*

Crum and Smith say, "An average is sometimes called a 'measure of central tendency' because individual values of the variable usually cluster around it."

Characteristics (Essentials or Requisites or Desiderata) of a Good Average :

An average should be :

(a) rigorously defined,

(b) easy to compute,

(c) capable of simple interpretation,

(d) dependent on all the observed values,

(e) not unduly influenced by one or two extremely large or small values,

(f) should fluctuate relatively little from one random sample to another, and

(g) be capable of mathematical manipulation.

3.3

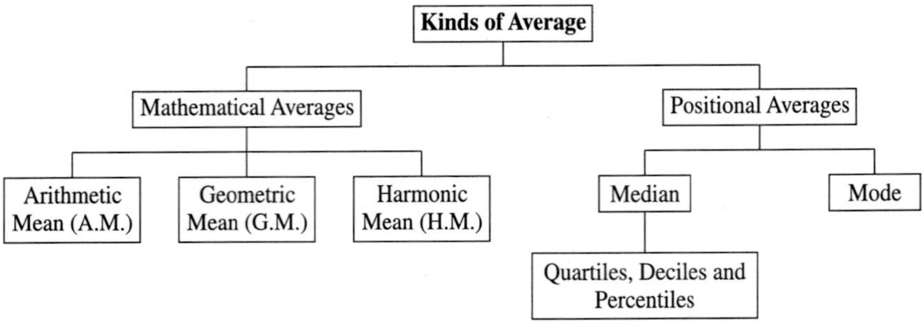

3.4 ARITHMETIC MEAN (A.M.) FOR UNGROUPED DATA

$$\text{A.M.} = \frac{\text{Sum of the Values}}{\text{Number of Values}}$$

$\Rightarrow$ A.M. $\times$ Number of Values = Sum of the Values.

$\Rightarrow$ If each value in any distribution is replaced by the A.M. of that distribution and these A.Ms. are added, the result = sum of the values.

If $x_1, x_2, ..., x_n$ are n values of a variable, then their arithmetic mean (generally denoted by $\bar{x}$) is given by

$$\text{A.M.} = \frac{x_1 + x_2 + x_3 + \ldots + x_n}{n} = \frac{\sum_{i=1}^{n} x_i}{n}.$$

If $f_1, f_2, f_3, \ldots, f_n$ are the respective frequencies of $x_1, x_2, \ldots, x_n$, then

$$\text{A.M.} = \frac{f_1x_1 + f_2x_2 + f_3x_3 + \ldots + f_nx_n}{f_1 + f_2 + f_3 + \ldots + f_n} = \frac{\sum_{i=1}^{n} f_ix_i}{\sum_{i=1}^{n} f_i}.$$

(The number of times a value is repeated in a distribution is called the frequency of that value.)

The A.M. of the values 3, 3, 3, 3, 5, 5, 7, 11, 15, 15

$$= \frac{3+3+3+3+5+5+7+11+15+15}{10} = \frac{70}{10} = 7.$$

Also, the A.M. of these values

$$= \frac{3 \times 4 + 5 \times 2 + 7 \times 1 + 11 \times 1 + 15 \times 2}{4 + 2 + 1 + 1 + 2} = 7$$

(The frequency of 3 is 4, frequency of 5 is 2, etc.)

$$= \frac{\text{Sum of the Products of the values by their respective frequencies}}{\text{Sum of the frequencies}} = \frac{\Sigma fx}{\Sigma f}$$

$\Rightarrow$ $\Sigma fx =$ **Sum of the values** and $\Sigma f =$ **Number of values**.

3.5 COMBINED MEAN* OF TWO GROUPS

If there are two groups G_1, G_2, such that

	Group I (G_1)	**Group II** (G_2)
Arithmetic mean	$\bar{x}_1$	$\bar{x}_2$
Number of values (or observations)	n_1	n_2

Combined mean of group I and group II ($\bar{x}_{12}$)

$$= \frac{\begin{array}{c}\text{Sum of the values}\\ \text{in Group I}\end{array} + \begin{array}{c}\text{Sum of the values}\\ \text{in Group II}\end{array}}{\begin{array}{c}\text{Number of values}\\ \text{in Group I}\end{array} + \begin{array}{c}\text{Number of values}\\ \text{in Group II}\end{array}}$$

$$= \frac{n_1\bar{x}_1 + n_2\bar{x}_2}{n_1 + n_2} = \frac{\text{Sum of the values in both the groups}}{\text{No. of values in both the groups}} \qquad \text{...(I)}$$

If $n_1 = n_2$, *i.e.*, if each group has the same number of values, then

$$\bar{x}_{12} = \frac{\bar{x}_1 + \bar{x}_2}{2} = \text{A.M. of the A.Ms. of the two groups.}$$

Result (I) is applicable for more groups also.

If there are p groups, then their combined arithmetic mean $\bar{x}$ is given by

$$\bar{x} = \frac{n_1\bar{x}_1 + n_2\bar{x}_2 + \ldots\ldots + n_p\bar{x}_p}{n_1 + n_2 + \ldots\ldots n_p}$$

where $\bar{x}_p$ denotes the arithmetic mean of pth group and n_p denotes number of values in the pth group.

If $n_1 = n_2 = n_3 = \ldots\ldots = n_p$ then

$$\bar{x} = \frac{\bar{x}_1 + \bar{x}_2 + \bar{x}_3 + \ldots\ldots + \bar{x}_p}{p}$$

3.6 WEIGHTED A.M.

So far we have given equal importance to each item or value in the series but this concept may be misleading in many situations. Suppose a man purchases mangoes at Rs. 20 per kg., apples at Rs. 30 per kg. and oranges at Rs. 16 per kg. then average rate of purchases will be Rs. $\frac{20 + 30 + 16}{3} = \frac{66}{3}$ = Rs. 22 per kg. provided he purchases only 1 kg. of each fruit but if he purchases different quantities of each fruit, then this result will not hold good.

* Mean ⇒ Arithmetic mean, the word 'arithmetic' is many times omitted.

In such cases, we have to use weighted arithmetic mean $= \frac{\Sigma wx}{\Sigma w}$.

In the above example, if the man purchases 5 kgs. of mangoes, 2 kgs. of apples and 6 kgs. of oranges then the average rate of purchase per kg.

$$= \frac{5 \times 20 + 2 \times 30 + 6 \times 16}{5 + 2 + 6} = \frac{\text{Total amount spent}}{\text{Total quantity purchased}}$$

$$= \frac{100 + 60 + 96}{13} = \frac{256}{13} = \text{Rs. } 19.69 \text{ per kg.*}$$

The weighted mean is particularly useful where the average we are looking for is the *mean of means.* If we have two groups, then to find the combined mean, we weigh the average of each group by the number of items in it, *i.e.,* we find the products of the means by the number of items in the respective groups, add these products and divide by the total number of items in the two groups. Weighted A.M. is also used in the construction of index numbers, comparison of the results of two or more universities where the number of students differ and computation of standardised death and birth rates.

3.7 PROPERTIES OF ARITHMETIC MEAN

(1) *If a constant amount is added or subtracted from each value in the series, mean is also added or subtracted by the same constant amount.*

Let the values be 3, 5, 9, 15, 16.

The A.M. of these values $= 3 + 5 + 9 + 15 + 16 = \frac{48}{5} = 9.6.$

If we add 2 to each value, the values becomes 5, 7, 11, 17, 18.

New mean $= \frac{5 + 7 + 11 + 17 + 18}{5} = \frac{58}{5} = 11.6 = 9.6 + 2,$

i.e., the mean is also added by 2.

Similarly, if the mean age of 5 boys is 10 years, then their mean age after 3 years will become 13 years because in 3 years, the age of each boy will increase by 3 years.

(2) *If each value in a series is multiplied or divided by a constant, the mean is also multiplied or divided by the same constant.*

The mean of the values

3, 5, 9, 15, 16 is 9.6.

If we multiply each value by 3, the values become

9, 15, 27, 45, 48.

The new mean $= \frac{9 + 15 + 27 + 45 + 48}{5} = \frac{144}{5} = 28.8 = 9.6 \times 3,$

i.e., the mean is also multiplied by 3.

(3) *The algebraic sum of the deviations of all the values from their mean* = 0. (**important**)

Let the values be 4, 6, 9, 15, 16.

$$\text{Mean} = \frac{4 + 6 + 9 + 15 + 16}{5} = 10.$$

* **Please note :** The average rate of purchase, in this case, will not be $\frac{20 + 30 + 16}{3} = \text{Rs. } 22$ per kg.

Deviation of 4 from 10 = 4 – 10 = –6.

Deviation of 5 from 10 = 6 – 10 = –4.

Deviation of 9 from 10 = 9 – 10 = –1.

Deviation of 15 from 10 = 15 – 10 = 5.

Deviation of 16 from 10 = 16 – 10 = 6.

Algebraic sum of the deviations = (Sum of the deviations with their signs)

$$= (-6) + (-4) + (-1) + 5 + 6 = 0.$$

If $\bar{x}$ is the mean of the n values, $x_1, x_2, x_3, ..., x_n$, the algebraic sum of the deviations of all the values from $\bar{x}$

$$= (x_1 - \bar{x}) + (x_2 - \bar{x}) + (x_3 - \bar{x}) + ... + (x_n - \bar{x})$$

$$= (x_1 + x_2 + x_3 + ... + x_n) - n\bar{x} = \text{Sum of the values} - n\bar{x}$$

$$= n\bar{x} - n\bar{x} = 0 \qquad [\because \text{Sum of the values} = \text{means} \times \text{number of values.}]$$

In case of frequency distribution, the algebraic sum of the deviations of all the values from their mean

$$= \Sigma f(x - \bar{x}) = \Sigma fx - \bar{x}\Sigma f = \bar{x}\Sigma f - \bar{x}\Sigma f = 0 \qquad \left[\because \bar{x} = \frac{\Sigma fx}{\Sigma f} \Rightarrow \bar{x}\Sigma f = \Sigma fx\right]$$

Notes :

(1) To find the mean

(i) Add the given values, *i.e.,* find the sum of the values.

(ii) Divide the sum by the number of values.

(2) If the mean and the number of values are given but one value is missing in the series, then the missing values = Number of values × mean – sum of known values.

(3) If one or two values have been taken wrongly while calculating the mean, then

Correct mean = $\frac{1}{n}$ [Sum of all the values including wrong values – the value (or values) taken wrongly + correct value (or values)]

(4) The deviation of a value from the mean [or any other arbitrary constant] is repeated as many times as the value is repeated in the distribution. **(imp.)**

For example for the values 3, 3, 3, 4, 7, the mean is 4. The value 3 is repeated three times. Its deviation from 4 [or any other constant] will also be repeated three times.

SOLVED EXAMPLES

Example 1. Find the arithmetic mean of 3, 6, 24, 28. **(ICWA Foundation. Dec., 1999)**

Solution. A.M. $= \frac{\text{Sum of the values}}{\text{Number of values}} = \frac{3 + 6 + 24 + 48}{4} = \frac{81}{4} = 20.25$

Example 2. Calculate the A.M. of the following observations :

32, 35, 36, 37, 39, 41, 43, 47, 48

Solution. A.M. $= \frac{32 + 35 + 36 + 37 + 39 + 41 + 43 + 47 + 48}{9} = \frac{358}{9} = 39.77.$

Example 3. Find the mean from the following data :

x	5	10	15	20	25	30	35	40
f	5	9	13	21	20	15	8	3

Solution. Total Frequency $= \Sigma f = 5 + 9 + 13 + 21 + 20 + 15 + 8 + 3$
$= 94 =$ Number of values

x	f	fx
5	5	25
10	9	90
15	13	195
20	21	420
25	20	500
30	15	450
35	8	280
40	3	120
	$\Sigma f = 94$	$\Sigma fx = 2{,}080$

$\Sigma fx =$ Sum of the products of x values with their respective frequencies.
$=$ Sum of the values $= 2{,}080$.

$$\text{Mean} = \frac{\Sigma fx}{\Sigma f} = \frac{2{,}080}{94} = 22.127.$$

Example 4. The A.M. of the following table is 17 years. Find the value of x.

Age (Years)	8	20	26	29
No. of Persons	3	2	x	1

Solution.
$$\text{A.M.} = \frac{\Sigma fx}{\Sigma f} = \frac{8 \times 3 + 20 \times 2 + 26 \times x + 29 \times 1}{3 + 2 + x + 1} = 17$$

$$\Rightarrow \quad \frac{24 + 40 + 26x + 29}{6 + x} = 17$$

$$\Rightarrow \quad 93 + 26x = 102 + 17x$$

$$\Rightarrow \quad 9x = 9 \Rightarrow x = 1.$$

Example 5. The frequencies of the numbers 3.2, 5.8, 7.9 and 4.5 are respectively x, $x + 2$, $x - 3$, and $x + 6$. If arithmetic mean is 4.876, then determine the value of x.

Solution.
$$\text{A.M.} = \frac{\sum_{i=1}^{n} f_i x_i}{\sum_{i=1}^{n} f_i}$$

$$= \frac{x \times 3.2 + (x + 2) \times 5.8 + (x - 3) \times 7.9 + (x + 6) \times 4.5}{x + x + 2 + x - 3 + x + 6}$$

$$= \frac{3.2x + 5.8x + 7.9x + 4.5x + 2 \times 5.8 - 3 \times 7.9 + 6 \times 4.5}{4x + 5}$$

$$= \frac{21.4x + 14.9}{4x + 5} = 4.876 \text{ (given)}$$

$$\Rightarrow \quad 21.4x + 14.9 = 4 \times 4.876x + 5 \times 4.876$$

$$\Rightarrow \quad 21.4x - 19.504x = 24.38 - 14.9$$

$$\Rightarrow \quad 1.896x = 9.48 \Rightarrow x = \frac{9.48}{1.896} = 5.$$

Example 7. Average marks in Statistics of 10 students of a class was 68. A new student took admission with 72 marks whereas two existing students left the college. If the marks of these students were 40 and 39, find the average marks of the remaining students. **(DU B. Com, 2000)**

Solution. Total marks of 10 students $= 68 \times 10 = 680$

After the entry of one new student and departure of two of existing students, the number of students left $= 9$

Total marks of 9 students $= 680 + 72 - 40 - 39 = 673$

Average marks of the remaining 9 students

$$= \frac{673}{9} = 74.78 \text{ marks (approx.)}$$

Example 8. The average score of boys is 60, that of girls is 70 and that of all the candidates is 64 appearing in Mathematics of annual examination. Find the ratio of number of boys and number of girls there. If the total number of candidates appearing in Mathematics is 150, find the number of boys there. **(ICWA Foundation Dec. 2006)**

Solution. Let n_1 be the number of boys and n_2 the number of girls then

$$\text{Total score of boys} = 60n_1$$

$$\text{Total score of girls} = 70n_2$$

Total score of boys and girls $= (n_1 + n_2)\ 64$

$$\therefore \quad 60n_1 + 70n_2 = 64\ (n_1 + n_2)$$

$$6n_2 = 4n_1 \quad \Rightarrow \quad \frac{n_1}{n_2} = \frac{6}{4} = \frac{3}{2}$$

$$\Rightarrow \quad \frac{n_1}{3} = \frac{n_2}{2} = k \quad \Rightarrow \quad n_1 = 3k, n_2 = 2k$$

$$n_1 + n_2 = 3k + 2k = 150 \quad \Rightarrow \quad k = 30$$

Number of boys $= 3 \times 30 = 90$

Number of girls $= 2 \times 30 = 60$

Example 9. A distribution consists of three components with total frequencies of 200, 250 and 300 having means 25, 10 and 15 respectively. Find the mean of combined distribution.

(DU B.Com, 2006)

Solution.

$$\text{Combined mean} = \frac{n_1\bar{x}_1 + n_2\bar{x}_2 + n_3\bar{x}_3}{n_1 + n_2 + n_3}$$

$$= \frac{200 \times 25 + 250 \times 10 + 300 \times 15}{200 + 250 + 300}$$

$$= \frac{12000}{750} = 16$$

Example 10. (a) Two variables x and y are related by $y = \frac{x-5}{10}$ and each of them has 5 observations. If mean of x is 45, find the mean of y. **(ICWA Foundation Dec., 2006)**

(b) In a series with 100 items, having simply mean of 40, later on it was discovered that two items of 45, 35 were wrongly taken as 35 and 25. Find the correct mean. **(DU B. Com, 2006)**

Solution. (a) $\Sigma y = \Sigma\left(\frac{x-5}{10}\right) = \frac{1}{10}(\Sigma x - 5 \times 5)$

$$= \frac{1}{10}(\Sigma x - 25) = \frac{1}{10}[45 \times 5 - 25] \qquad [\Sigma x = n\bar{x}]$$

$$= \frac{200}{10} = 20$$

$$\frac{\Sigma y}{5} = \text{mean of } y = \frac{20}{5} = 4$$

(b) Sum of 100 items $= 100 \times 40 = 4000$ [Sum of the values $= n\bar{x}$]

Correct sum of 100 items

$$= 4000 - 35 - 25 + 45 + 35 = 4020$$

$$\text{Correct mean} = \frac{4020}{100} = 40.2$$

Example 11. An aeroplane travels distances of d_1, d_2 and d_3 km at speeds v_1, v_2 and v_3 km per hour respectively. Show that the average speed is given by v, where

$$\frac{d_1 + d_2 + d_3}{v} = \frac{d_1}{v_1} + \frac{d_2}{v_2} + \frac{d_3}{v_3}$$

Solution. Time taken to travel distance d_1 km $= \frac{d_1}{v_1}$ hrs.

Time taken to travel distance d_2 km $= \frac{d_2}{v_2}$ hrs.

Time taken to travel distance d_3 km $= \frac{d_3}{v_3}$ hrs.

Total time taken to travel distance $(d_1 + d_2 + d_3)$ kms $= \frac{d_1}{v_1} + \frac{d_2}{v_2} + \frac{d_3}{v_3}$.

$$\text{Average speed } (v) = \frac{\text{Total distance travelled}}{\text{Total time taken}} = \frac{d_1 + d_2 + d_3}{\frac{d_1}{v_1} + \frac{d_2}{v_2} + \frac{d_3}{v_3}}$$

$$\Rightarrow \quad \frac{d_1 + d_2 + d_3}{v} = \frac{d_1}{v_1} + \frac{d_2}{v_2} + \frac{d_3}{v_3}.$$

Example 12. The average marks scored by 10 students are 35. Later the moderator awarded 2 grace marks to 4 students each and 1 grace mark to 2 students each. Find the average marks after moderation.

Solution. Total marks scored by 10 students $= 35 \times 10 = 350$.

Total grace marks $= 4 \times 2 + 2 \times 1 = 10$.

Total marks after moderation $= 350 + 10 = 360$

Average marks after moderation $= \frac{360}{10} = 36$.

Example 13. Show that the weighted arithmetic mean of the squares of n natural numbers whose weights are equal to the corresponding natural numbers is equal to $\frac{n(n+1)}{2}$.

Solution. Weighted A.M. $= \frac{1^2 \times 1 + 2^2 \times 2 + 3^3 \times 3 + \ldots + n^2 \times n}{1 + 2 + 3 + \ldots + n}$

$$= \frac{1^3 + 2^3 + 3^3 + \ldots + n^3}{1 + 2 + 3 + \ldots + n}$$

$$= \left[\frac{n(n+1)}{2}\right]^2 \div \frac{n(n+1)}{2} = \frac{n(n+1)}{2}.$$

[$\because$ Sum of the cubes of first n natural numbers $= \left(\frac{n(n+1)}{2}\right)^2$ and sum of first n natural numbers $= \frac{n(n+1)}{2}$.]

Example 14. An individual purchases three qualities of pencils. The relevant data is given below :

Quality	*Price Per Pencil (Rs.)*	*Money Spent (Rs.)*
A	1	50
B	1.5	30
C	2.00	20

Calculate the average price per pencil.

Solution. Number of pencils purchased of quality $A = \frac{50}{1} = 50.$

Number of pencils purchased of quality $B = \frac{30}{1.5} = 20.$

Number of pencils purchased of quality $C = \frac{20}{2} = 10.$

Average price per pencil $= \frac{\text{Total amount spent}}{\text{Total number of pencils purchased}}$

$$= \frac{50 + 30 + 20}{50 + 20 + 10} = \frac{100}{80} = \text{Rs. } 1.25.$$

Example 15. A man purchased 5 rolls of ribbons each measuring 60 metres at the rate of 4, 6, 10, 12 and 15 metres a rupee respectively. Find the average price of ribbons.

Solution. Cost of the roll purchased at the rate of 4 metres a rupee $= \frac{60}{4} = \text{Rs. } 15.$

Cost of the roll purchased at the rate of 6 metres a rupee $= \frac{60}{6} = \text{Rs. } 10.$

Cost of the roll purchased at the rate of 10 metres a rupee $= \frac{60}{10} = \text{Rs. } 6.$

Cost of the roll purchased at the rate of 12 metres a rupee $= \frac{60}{12} = \text{Rs. } 5.$

Cost of the roll purchased at the rate of 15 metres a rupee $= \frac{60}{15} =$ Rs. 4.

Total length of 5 rolls $= 60 \times 5 = 300$ metres = Total length purchased

Total cost of 5 rolls = Rs. (15 + 10 + 6 + 5 + 4) = Rs. 40.

Average rate $= \frac{300}{40} = 7.5$ metres a rupee.

Example 16. The A.M. of the following distribution is 1.46.

No. of Accidents	0	1	2	3	4	5	Total
No. of Days	46	f_1	f_2	25	10	5	200

Find f_1 and f_2.

Solution.

(x)	(f)	Fx
0	46	0
1	f_1	f_1
2	f_2	$2f_2$
3	25	75
4	10	40
5	5	25
Total	$86 + f_1 + f_2$	$140 + f_1 + 2f_2$

$$\text{Total frequency} = 86 + f_1 + f_2 = 200$$

$$\Rightarrow \quad f_1 + f_2 = 114 \quad \text{...(i)}$$

$$\text{Mean} = \frac{140 + f_1 + 2f_2}{200} = 1.46$$

$$\Rightarrow \quad f_1 + 2f_2 = 152 \quad \text{...(ii)}$$

Subtracting (i) from (ii), we get $f_2 = 38$.

Putting $f_2 = 38$ (i), we get $f_1 = 76$.

Eample 17. Find the mean height (x) (in inches) of 52 students of a school from the following table :*

No. of Students	4	7	10	15	8	5	3
$u = x - 40$ *(in inches)*	–10	–7	–7	0	3	5	8

Solution.

d	f	fd
–10	4	–40
–7	7	–49
–5	10	–50
0	15	0
3	8	24
5	5	25
8	3	24
	$\Sigma f = 52$	$\Sigma fd = -66$

* In this example, assumed mean is 40.

$$\text{Mean} = 40 + \left(-\frac{66}{52}\right) = 40 - \frac{66}{52} = 38.73.$$

Example 18. The arithmetic average of the following distribution is 40. Find the value of k.

x	20	25	30	$k + 10.5$	60
f	2	7	15	10	6

Solution.

x	f	fx
20	2	40
25	7	175
30	15	450
$(k + 10.5)$	10	$10k + 105$
60	6	360
	$\Sigma f = 40$	$\Sigma fx = 1130 + 10k$

$$\text{Mean} = 40 = \frac{\Sigma fx}{\Sigma f}$$

$$\Rightarrow \quad 40 = \frac{1130 + 10k}{40}$$

$$\Rightarrow \quad 40 \times 40 - 1130 = 10k$$

$$\Rightarrow \quad k = \frac{1,600 - 1,130}{10}$$

$$\Rightarrow \quad k = \frac{470}{10} = 47.$$

Example 19. The mean yearly salary of employees of a company was Rs. 24,000. The mean yearly salaries of male and female employees were Rs. 25,000 and Rs. 19,000 respectively. Find out the percentage of males and females employed by the company.

Solution. Let n_1, n_2 denote respectively the number of male and female employees.

Total yearly salary paid to all the employees = Rs. 24,000 $(n_1 + n_2)$

Total yearly salary paid to male employees = Rs. 25,000 (n_1)

Total yearly salary paid to female employees = Rs. 19,000 (n_2)

The total salary paid to all the employees = Total salary paid to male employees + Total salary paid to female employees.

$$\Rightarrow \quad 24,000\,(n_1 + n_2) = 25,000\,(n_1) + 19,000\,(n_2)$$

$$\Rightarrow \quad \frac{n_1}{n_2} = \frac{5}{1} \quad \Rightarrow \quad n_1 = 5n_2.$$

$$\text{Percentage of female employees} = \frac{n_2}{n_1 + n_2} \times 100$$

$$= \frac{n_2}{5n_2 + n_2} \times 100 = 16\frac{2}{3}\%.$$

$$\therefore \text{ Percentage of male employees} = 100 - 16\frac{2}{3} = 83\frac{1}{3}\%.$$

Example 20. The average marks in English obtained by 300 students is 45. If average marks of first and last 100 students are 70 and 20 respectively, then what is the average marks of remaining students.

Solution. Total marks of 300 students $= 300 \times 45 = 13{,}500$.

Total marks of first 100 students $= 100 \times 70 = 7{,}000$.

Total marks of last 100 students $= 100 \times 20 = 2{,}000$.

Average marks of remaining 100 students

$$= \frac{13{,}500 - 7{,}000 - 2{,}000}{100} = 45.$$

Example 21. Fifty students appeared in an examination. The result of the passed students are given below :

Marks	:	40	50	60	70	80	90
No. of Students	:	6	14	7	5	4	4

The marks of all students is 52. Find the average marks of the students who failed in the examination. **(ICWA Foundation Dec., 2006)**

Solution. Total marks of the students who passed

$$= 40 \times 6 + 50 \times 14 + 60 \times 7 + 70 \times 5 + 80 \times 4 + 90 \times 4$$
$$= 240 + 700 + 420 + 350 + 320 + 360 = 2390$$

Total marks of all the 50 students $= 50 \times 52 = 2600$

Total marks of the failed candidates $= 2600 - 2390 = 210$

Number of failed candidates $= 50 - 6 - 14 - 7 - 5 - 4 - 4 = 10$

Average marks of the students who failed $= \frac{210}{10} = 21$

Example 22. 100 students appeared for an examination. The results of those who failed are given below :

Marks	5	10	15	20	25	30
No. of Students	4	6	8	7	3	2

If the average marks of all the 100 students were 68.6, find out the average marks of those who passed.

Solution. Number of students who passed $= 100 - (4 + 6 + 8 + 7 + 3 + 2) = 70$.

Total marks obtained by 100 students $= 100 \times 68.6 = 6{,}860$.

Total marks obtained by the students who failed

$$= 4 \times 5 + 6 \times 10 + 8 \times 15 + 7 \times 20 + 3 \times 25 + 2 \times 30 = 475.$$

Total marks obtained by students who passed $= 6{,}860 - 475 = 6{,}385$.

Average marks of students who passed $= \frac{6{,}385}{70} = 91.2$.

Example 23. The mean salary of 1,000 employees of an establishment was found to be Rs. 541.20. Later on after disbursement of salary it was discovered that the salary of two employees was wrongly entered as Rs. 891 and Rs. 495. The correct salaries were Rs. 591 and 555. Find the correct arithmetic mean.

Solution. Total salary calculated wrongly, of 1,000 employees

$$= 541.20 \times 1{,}000 = \text{Rs. } 5{,}41{,}200.$$

Correct total salary of 1,000 employees

$$= 5{,}41{,}200 - (891 + 495) + (591 + 555)$$

$$= \text{Rs. } 5{,}40{,}960.$$

$$\text{Correct arithmetic mean} = \frac{5{,}40{,}960}{1{,}000} = \text{Rs. } 540.96.$$

Example 24. Two groups of students reported mean weights of 162 kgs. and 148 kgs. respectively. When would be the mean weight of both the groups together be 155 kgs ?

Solution. Let n_1 be the number of students in the first group and n_2 the number of students in the second group, then

	Group I	*Group II*
Numbers of Students	n_1	n_2
Mean	162 kgs.	148 kgs.

Combined mean = 155 kgs.

$$\therefore \quad 155 = \frac{162n_1 + 148n_2}{n_1 + n_2}$$

$$\Rightarrow \quad 155n_1 + 155n_2 = 162n_1 + 148n_2$$

$$\Rightarrow \quad 7n_2 = 7n_1 \Rightarrow n_1 = n_2.$$

$\Rightarrow$ The number of items in the two groups should be the same.

Second Method

$$\text{Combined mean} = \frac{\text{Mean of 1st group} + \text{Mean of 2nd group}}{2}.$$

Since this condition is satisfied in this question, therefore $n_1 = n_2$.

Example 25. The mean weight of 15 wrestlers is 110 kgs. The mean weight of 5 of them is 100 kgs. and another 5 is 125 kgs. What is the mean weight of the remainder ?

Solution. Mean weight of 15 wrestlers = 110 kgs.

Total weight of 15 wrestlers = $110 \times 15 = 1{,}650$ kgs.

Total weight of 5 wrestlers whose mean weight is 100 kgs. = $100 \times 5 = 500$ kgs.

Total weight of 5 wrestlers whose mean weight is 125 kgs. = $125 \times 5 = 625$ kgs.

Total weight of remaining 5 wrestlers = $1{,}650 - 500 - 625 = 525$ kgs.

$$\text{Mean weight of these 5 wrestlers} = \frac{525}{5} = 105 \text{ kgs.}$$

Example 26. Of a batch of 20 students, 4 students failed while 7 students passed with distinction. The remaining students who passed ordinarily scored the following marks :

47, 52, 49, 57, 52, 58, 66, 64, 59

The students who failed scored 23 marks on an average while the students who passed with distinction scored 76 marks on an average. With this information you are required to find the mean marks and the median marks. **(DU B. Com, 2007)**

Solution. Total marks scored by ordinary students

$$= 47 + 52 + 49 + 57 + 52 + 58 + 66 + 64 + 59 = 504$$

Total marks obtained by the students who failed $= 23 \times 4 = 92$

Total marks obtained by the students who passed with distinction $= 76 \times 7 = 532$

$$\text{Mean marks} = \frac{504 + 92 + 532}{20} = \frac{1128}{20} = 56.4$$

For finding the median marks arrange the data as under

23, 23, 23, 23, 47, 49, 52, 52, 57, 58, 59, 64, 66, 76, 76, 76, 76, 76, 76, 76

The average of 58 and 59 $= 58.5$

Hence median marks is 58.5

Eample 27. A group consists of 150 children. The group is divided into three subgroups A, B and C in the ratio of 2 : 5 : 3 respectively. The average age of the children in the subgroup A is 8 years. The average age of the children in the subgroup B is 10 years. The average age of the childen in the subgroup C is 12 years. What is the average of all the 150 children in the group ?

Solution. Number of children in group $A = \frac{2}{2+5+3} \times 150 = 30$

Number of children in group $B = \frac{5}{2+5+3} \times 150 = 75$

Number of children in group $C = \frac{3}{2+5+3} \times 150 = 45$

$$\therefore \text{ Combined mean} = \frac{30 \times 8 + 75 \times 10 + 45 \times 12}{30 + 75 + 45}$$

$$= \frac{240 + 750 + 540}{150} = 10.2 \text{ years.}$$

Example 28. A train runs 25 kms at a speed of 30 kms/hr. Then it travels for 6 minutes at a speed of 100 kms/hr. Lastly it concluded its journey by moving at a speed of 24 kms/hr for a distance of 24 kms. What is the average speed ?

Solution. Time taken to cover 25 kms at a speed of 30 kms per hour $= \frac{25}{30} \times 60 = 50$ minutes.

Time taken to cover 50 kms at a speed of 40 kms/hr $= \frac{50}{40} \times 60 = 75$ minutes.

Distance covered in 6 minutes at a speed of 100 kms/hour $= \frac{100}{60} \times 6 = 10$ kms.

Time taken to cover 24 kms at a speed of 24 kms/hour = 1 hr.

$$\text{Average speed} = \frac{\text{Total distance covered}}{\text{Total time taken}}$$

$$= \frac{25 + 50 + 10 + 24}{\frac{50 + 6 + 60 + 75}{60}} = \frac{109 \times 60}{191} = 34.24.$$

Example 29. The sum of deviations of certain number of items measured from 2.5 is 50 and the sum of deviations of the same series measured from 3.5 is –50. Find the number of observation and their mean.

Solution. Let n be the number of observations and $\bar{x}$ the mean, then

$$\bar{x} = 2.5 + \frac{50}{n} = 3.5 - \frac{50}{n}$$

$$\Rightarrow \quad 2.5 + \frac{50}{n} = 3.5 - \frac{50}{n}$$

$$\Rightarrow \quad \frac{50}{n} + \frac{50}{n} = 3.5 - 2.5 = 1$$

$$\Rightarrow \quad \frac{100}{n} = 1 \Rightarrow n = 100$$

When $n = 100, \bar{x} = 2.5 + \frac{50}{100} = 3.$

Example 30. The mean of 10 numbers is 12.5. The mean of first six is 15 and the last five is 10. Find the sixth number.

Solution. Sum of ten numbers $= 12.5 \times 10 = 125$

Sum of first six numbers $= 6 \times 15 = 90$

Sum of last five numbes $= 10 \times 5 = 50$

Sum of first six numbers + Sum of last five numbers

= Sum of first six numbers + Sixth number + Sum of last four numbers

= Sum of first six numbers + Sum of last four numbers + 6th number

= Sum of ten numbers + 6th number

$\Rightarrow$ 125 + 6th number = 90 + 50

$\Rightarrow$ Sixth number $= 140 - 125 = 15.$

Second Method

Let $\bar{x}$ be the average of last four numbers, then

Sum of ten numbers = Sum of first six numbers + Sum of last four numbers

$$= 6 \times 15 + \bar{x} \times 4 = 125$$

$$\Rightarrow \quad \bar{x} = \frac{125 - 90}{4} = \frac{35}{4}$$

Sum of last four numbers $= \frac{35}{4} \times 4 = 35$

$\therefore$ 90 + Sixth number + 35 = 90 + 50 = 140 [$\because$ Sum of the last five numbers is 50.]

$\Rightarrow$ Sixth number $= 140 - 90 - 35 = 15.$

3.8 ARITHMETIC MEAN FOR CONTINUOUS SERIES OR GROUPED DATA

To find the mean of a large number of values, it may be quite lengthy to add up all the values and divide by the number of values. In such a case, we make groups each containing a range of values and find the number of items falling in each group. For example, if there are 75 students in a class then to find the average marks of students in any subject, we have to add the marks obtained by all the 75 students and divide by 75. In such cases, we divide the marks in small groups. If the total marks are 100, then we can form groups. Let each group contain 10 marks. Then the groups will be 0—10, 10—20, 20—30, 30—40, 40—50, 50—60, 60—70, 70—80, 80—90, 90—100. Then we find the number of students falling in each group which gives the frequency of each group.

In grouped data, **middle value** of each group is the representative of the group, *i.e.,* in grouped data, each item in the group is considered to be equal to the middle value of that group. **(Please note)**

∴ Sum of the value of each group

$$= \text{Middle value of the group} \times \text{Number of values.}$$

Further the procedure is the same as for finding the combined mean of two or more groups. The following examples will make the method clear.

Example 31. Calculate the arithmetic mean of the following frequency distribution :

Class-Interval	*Frequency*
10–20	4
20–40	10
40–70	26
70–120	8
120–200	2
	Total 50

Solution. Mid-value of first class (or group) $= \dfrac{10+20}{2} = 15.$

$$\text{Sum of the values in first class} = 15 \times 4 = 60.$$

$$\text{Mid-value of second class} = \frac{20+40}{2} = 30.$$

$$\text{Sum of the values in the second class} = 30 \times 10 = 300.$$

$$\text{Mid-value of the third class} = \frac{40+70}{2} = 55.$$

$$\text{Sum of the values in third class} = 55 \times 26 = 1{,}430.$$

$$\text{Mid-value of the fourth class} = \frac{70+120}{2} = 95.$$

$$\text{Sum of the values in the fourth class} = 95 \times 8 = 760.$$

$$\text{Mid-value of the fifth class} = \frac{120+200}{2} = 160.$$

Sum of the values in the fifth class = $160 \times 2 = 320$.

$$\text{Mean} = \frac{\text{Sum of the values in all the classes}}{\text{No. of values in all the classes}}$$

$$= \frac{60 + 300 + 1{,}430 + 760 + 320}{4 + 10 + 26 + 8 + 2} = \frac{2{,}870}{50} = 57.4$$

The above steps can be written in the tabular form as :

Class-Interval	*Mid-Values (x)*	*Freq. (f)*	*fx*
10–20	15	4	60
20–40	30	10	300
40–70	55	26	1430
70–120	95	8	760
120–200	160	2	320
70–120	95	$\Sigma f = 50$	$\Sigma fx = 2870$

$$\text{A.M.} = \frac{\Sigma fx}{\Sigma f} = \frac{2{,}870}{50} = 57.4.$$

Example 32. From the data below, find the missing frequency if the Arithmetic Mean is 33.

Loss per Shop	*No. of Shops*
0–10	10
10–20	15
20–30	30
30–40	–
40–50	25
50–60	20

Solution. Let the missing frequency be : f

Loss per Shop	*Mid-Value x*	*No. of Shops*	*fx*
0–10	5	10	50
10–20	15	15	225
20–30	25	30	750
30–40	35	f	$35f$
40–50	45	25	1125
50–60	55	20	1100
		$\Sigma f = 100 + f$	$\Sigma fX = 3250 + 35f$

$$\text{Mean} = \frac{\Sigma fx}{\Sigma f} = \frac{3{,}250 + 35f}{100 + f} = 33.$$

$$\Rightarrow \quad 3{,}250 + 35f = 3{,}300 + 33f$$

$$\Rightarrow \quad 2f = 50 \Rightarrow f = 25.$$

Example 33. From the following data of calculation of arithmetic mean, find the missing item, *i.e.*, find the value of x :

Wages in Rs.	110	112	113	117	(x)	125	128	130
No. of Workers	25	17	13	15	14	8	6	2

Mean wage = 115.86.

Solution.

Wages X	*No. of Workers* f	fX
110	25	2,750
112	17	1,904
113	13	1,469
117	15	1,755
(x)	14	$14x$
125	8	1,000
128	6	768
130	2	260
	$N = 100$	$\Sigma fX = 9{,}906 + 14x$

$$\overline{X} = \frac{\Sigma fX}{\Sigma N} \quad \text{or} \quad 115.86 = \frac{9{,}906 + 14x}{100}$$

$$\Rightarrow \quad 115.86 \times 100 = 9{,}906 + 14x$$

$$\Rightarrow \quad 11{,}586 = 9{,}906 + 14x$$

$$\Rightarrow \quad -14x = 9{,}906 - 11{,}586$$

$$\Rightarrow \quad -14x = -1{,}680$$

$$\Rightarrow \quad x = \frac{1{,}680}{14} = 120.$$

Example 34. In the following frequency distribution the frequency of the class interval (30—40) is not known. Find it out, if the arithmetic mean of the distribution is 28.

Profit per Shop (Rs.)	0–10	10–20	20–30	30–40	40–50	50–60
No. of Shop	12	18	27	(x)	17	6

Solution.

Profit per Shop	*Mid-Points Rs. (m)*	*Frequency* f	fm
0–10	5	12	60
10–20	15	18	270
20–30	25	27	675
30–40	35	x	$35x$
40–50	45	17	765
50–60	55	6	330
		$N = 80 + x$	$\Sigma fm = 2{,}100 + 35x$

$$\overline{X} = \frac{\Sigma fm}{N} \quad \Rightarrow \quad 28 = \frac{2{,}100 + 35x}{80 + x}$$

$$\Rightarrow \quad 28\,(80 + x) = 2{,}100 + 35x$$

$$\Rightarrow \quad 2{,}240 + 28x = 2{,}100 + 35x$$

$$\Rightarrow \quad 28x - 35x = 2{,}100 - 2{,}240$$

$$\Rightarrow \qquad -7x = -140 \Rightarrow 7x = 140 \Rightarrow x = 20.$$

Hence, 20 is the missing frequency.

Short Cut Method : This method is used when we deal with large figures. This method is based on the property of the Mean that the algebraic sum of the deviations of all the values from their mean =0. In this method the following steps are used.

Steps involved :

(1) Take any value (Though not necessary but generally one value among the given values is taken). Call this value as the assumed mean, called as 'A'*.

(2) Calculate the deviations of all the values from this assumed mean 'A' and take mean of these deviation, *i.e.*, calculate $\frac{\Sigma(x-A)}{n}$ or $\frac{\Sigma f(x-A)}{\Sigma f}$ as the case may be.**

(3) Add the mean of the deviation to the assumed mean, *i.e.*,

Required A.M. of the given series $= A + \frac{\Sigma f(x-A)}{\Sigma f} = A + \frac{\Sigma fd}{\Sigma f}$***,

where $d = (x - A)$ are the deviations.

(4) In case there is any common factor in the deviations, then divide each deviation by that common factor (In continuous series, the common, factor in most cases = the class interval), *i.e.*, find $\frac{d}{i} = d'$, where i is the class interval or any other common factor.

(5) Required mean $= A + \frac{\Sigma fd'}{\Sigma f} \times i$. This is called **Step Deviation Method.**

Following examples will make the method clear.

Example 35. If the sum of the deviations of a set of 10 values taken from 35 is 60, find the value of the A.M.

Sol. A.M. $= A + \frac{\Sigma f}{n} = 35 + \frac{60}{10} = 41.$

Note. In this example, the deviations are taken from the value 35, so 35 is assumed mean.

Example 36. Calculate arithmetic mean from the following data :

Class	*Frequency*
20–25	10
25–30	12
30–35	8
35–40	20
40–50	11
45–50	4
50–55	5

* Assumed mean is normally chosen as a figure in the middle of the series (but it is not binding) so that on either side the deviations are not very large.

** If the distribution consists of discrete individual values find $\frac{\Sigma(X-A)}{n}$ and in case of frequency distribution (discrete or continuous) find $\frac{\Sigma f(X-A)}{\Sigma f}$.

*** If the mean of the deviations is positive, the actual mean will be more than the assumed mean and if the mean of the deviations is negative then the actual mean will be less than the assumed mean by an amount equal to the mean of the deviations.

Solution.

Class	*Frequency*	*Mid-Value*	$x - 37.5 = d$	*fd*	$\frac{d}{5} = d'$	*fd′*
20–25	10	22.5	–15	–150	–3	–30
25–30	12	27.5	–10	–120	–2	–24
30–35	8	32.5	–5	–40	–2	–8
35–40	20	37.5	0	0	–1	0
40–45	11	42.5	5	55	0	11
45–50	4	47.5	10	40	2	8
50–55	5	52.5	15	75	3	15
	$N = 70$			$\Sigma fd = -140$		$\Sigma fd' = -28$

$$\text{A.M.} = \overline{X} = A + \frac{\Sigma fd}{N}, \ (A = \text{Assumed mean} = 37.5)$$

$$= 37.5 \times \frac{(-140)}{70} = 37.5 - 2 = 35.5.$$

Notes : (Step deviation method is explained below in Note 2) :

(1) In the above example, Assumed mean = 37.5, mean of the deviations = –2.
∴ The mean of the given series = 37.5 – 2 = 35.5. In this case actual mean is less than the assumed mean.

(2) Here factor 5 is common in the deviations. If we divide each deviation by 5 and define $d' = \frac{d}{5}$ then $\Sigma fd' = -28$. A.M. of the given series $= A + \frac{\Sigma fd'}{\Sigma f} \times 5 = 37.5 - \frac{28}{70} \times 5 = 35.5.$

Example 37. Following table gives the wages paid to 150 workers in a factory. Calculate the arithmetic mean.

Wages (Rs.)	240	250	260	270	280	290	300
No. of Workers	5	15	32	42	15	12	4

Solution. Taking 250 as the assumed mean, the calculations are as under :

Wages X	*No. of Workers* f	$X - A = (X - 250) = d$	*fd*
240	5	–10	–5
250	15	0	0
260	32	+10	+320
270	42	+20	+840
280	15	+30	+450
290	12	+40	+480
300	4	+50	+200
	$N = 125$		$\Sigma fd = 2{,}240$

Mean of the deviations $= \frac{\Sigma fd}{\Sigma f} = \frac{2,240}{125} = 17.92.$

$$= A + \frac{\Sigma fd}{\Sigma f} = \text{Assumed mean} + \text{Arithmetic mean of the deviations}$$

$\therefore$ Mean of the given series $= 250 + 17.92 = 267.92$

Since 10 is common in all the deviations, we can define $d' = \frac{d}{10}$ and follow the method of **Step deviations** as under :

Wages X	*No. of Workers* f	$X - A = (X - 250) = d$	$\frac{X - 250}{10} = (d')$	fd'
240	5	–10	–1	–5
250	15	0	0	0
260	32	+10	+1	+32
270	42	+20	+2	+84
280	15	+30	+3	+45
290	12	+40	+4	+48
300	4	+50	+5	+20
	$N = 125$			$\Sigma fd' = 224$

$$\text{Mean of the given series} = A + \frac{\Sigma fd'}{\Sigma f} \times 10$$

$$= 250 + \frac{224}{125} \times 10 = 267.92.$$

Example 38. From the following frequency table, find the value of x if mean is 23.5 :

Class	50–59	40–49	30–39	20–29	10–19	0–9
Frequency	$x - 4$	$x - 2$	$x + 3$	$x + 5$	$x + 10$	$x - 2$

Solution.

Mid-Value y	$d = \frac{y - 34.5}{10}$	f	fd
54.5	+2	$x - 4$	$2x - 8$
44.5	+1	$x - 2$	$x - 2$
34.5	0	$x + 3$	0
24.5	–1	$x + 5$	$-x - 5$
14.5	–2	$x + 10$	$-2x - 20$
4.5	–3	$x - 2$	$-3x + 6$
Total		$6x + 10$	$-3x - 29$

$$\bar{d} = \frac{3x - 29}{6x + 10}.$$

$$\bar{y} = 34.5 + \frac{\Sigma fd}{\Sigma f} \times 10 = 34.5 - \frac{10\,(3x + 29)}{6x + 10} = 23.5$$

$$\Rightarrow \frac{30x + 290}{6x + 10} = 11 \quad \Rightarrow \quad 30x + 290 = 110 \quad \Rightarrow \quad 36x = 180 \quad \Rightarrow \quad x = 5.$$

Second Method

Mid-Value *Y*	*Frequency* *f*	*fY*
54.5	$x-4$	$54.5x-218$
44.5	$x-2$	$44.5x-89$
34.5	$x+3$	$34.5x+103.5$
24.5	$x+5$	$24.5x+122.5$
14.5	$x+10$	$14.5x+145$
4.5	$x-2$	$4.5x-9$
Total	$6x+10$	$177x+55$

$$\overline{Y}=\frac{177x+55}{6x+10}=23.5$$

$$\Rightarrow \quad 141x+235=177x+55$$

$$\Rightarrow \quad 36x=180$$

$$\Rightarrow \quad x=5.$$

Example 39. Two hundred people were interviewed by a Public Opinion Polling Agency. The frequency distribution gives the Ages of people interviewed.

Age Group	80–89	70–79	60–69	50–59	40–49	30–39	20–29	10–19
Frequency	2	2	6	20	56	40	42	32

Solution.

Age Group	*Frequency*	*Mid-Value* *f*	$x-44.5=d$	$\frac{x-44.5}{10}=d'$	fd'
80–79	2	84.5	40	4	8
70–79	2	84.5	30	3	6
60–69	6	64.5	20	2	12
50–59	20	54.5	10	1	20
40–49	56	44.5	0	0	0
30–39	40	34.5	–10	–1	–40
20–29	42	24.5	–20	–2	–84
10–19	32	14.5	–30	–3	–96
	200				$\Sigma fd'=-174$
Mean $=44.5-\frac{174}{200}\times 10=35.8$					

Notes :

(1) In the above example, even if we change the series in exclusive type, the middle points will remain the same and hence mean will remain unchanged. Therefore, there is no need to change the class intervals to exclusive type.

(2) Writing the class intervals as 10—19, 20—29, ..., etc., also will not affect the mean.

Example 40. Calculate the mean of the following distribution of monthly wages of workers in a factory :

Monthly Wages (Rs.)	*No. of Workers*
100–120	10
120–140	20
140–160	30
160–180	15
180–200	5
	80

Solution.

Calculation of Mean

Monthly Wages (X)	*No. of Workers f*	*Mid-Points x*	fx	$x - 150 = d$	fd	$\frac{d}{20} = d'$	fd'
100–120	10	110	1,100	–40	–400	–2	–20
120–140	20	130	2,600	–20	–400	–1	–20
140–160	30	150	4,500	0	0	0	0
160–180	15	170	2,250	20	300	1	15
180–200	5	190	950	40	200	2	10
	$N = 80$		$\Sigma fx = 11{,}700$		$\Sigma fd = -300$		–15

$$\overline{X} = \frac{\Sigma fx}{\Sigma f} \text{ or } \frac{\Sigma fx}{N} = \frac{11700}{80} = 146.25. \qquad [\because \Sigma f = N]$$

Mean wage of workers is Rs. 146.25.

Short-Cut Method

$$\text{Mean} = A + \frac{\Sigma fd}{\Sigma f} = 150 + \left(\frac{-300}{80}\right)$$
$$= 150 - 3.75 = \text{Rs. } 146.25.$$

Step Deviation Method

$$\text{Mean} = A + \frac{\Sigma fd'}{\Sigma f} \times 20 = 150 - \frac{15}{80} \times 20 = 150 - 3.75 = \text{Rs. } 146.25.$$

3.9 CHARLIER'S ACCURACY CHECK

Prof. Charlier has provided a formula for checking the accuracy of calculations involved in computing mean of a discrete or a continuous frequency distribution. The formula is

$$\Sigma f(d + 1) = \Sigma fd + \Sigma f$$

or

$$\Sigma f(d' + 1) = \Sigma fd' + \Sigma f.$$

In using this check, we need an additional column $\Sigma f(d + 1)$ or $\Sigma f(d' + 1)$ in the table showing calculations involved in the computation of A.M. or Mean.

The point will be clear from the following example :

Example 41. Use Charlier-check to check the accuracy in calculations regarding mean of the following distributions :

Classes	0–10	10–20	20–30	30–40	40–50	50–60
Frequency	12	15	25	36	18	14

Solution.

Calculations in Charlier-Check of Accuracy

Classes	*Mid-Value x*	f	$d' = \frac{x-35}{10}$	fd'	$(d'+1)$	$f(d'+1)$	$d = X-35$	$f(d+1)$	fd
0–10	5	12	–3	–36	–2	–24	–30	–348	–360
10–20	15	15	–2	–30	–1	–15	–20	–285	–300
20–30	25	25	–1	–25	0	0	–10	–225	–250
30–40	35	36	0	0	1	36	0	36	0
40–50	45	18	1	18	2	36	10	198	180
50–60	55	14	2	28	3	42	20	294	280
		$N = \Sigma f =$ 120		$\Sigma fd' =$ –45		$\Sigma f (d'+1) = 75$		$\Sigma f(d+1) = -330$	$\Sigma fd =$ –450

Thus, $\Sigma f(d'+1) = 75$ and $\Sigma fd' + \Sigma f = -45 + 120 = 75.$

$\therefore$ $\Sigma f(d'+1) = \Sigma fd' + \Sigma f$

$$\Sigma f(d+1) = -330,\ \Sigma fd + \Sigma f = -450 + 120 = -330$$

$\Rightarrow \Sigma f(d+1) = \Sigma fd + \Sigma f.$ This also verifies that Charlier's check is satisfied and calculations are correct.]

3.10 GEOMETRIC MEAN (G.M.)

Geometric mean of n positive numbers is the nth root of their product.

If $x_1, x_2, ..., x_n$ are the n values of a variable, then their

$$\text{G.M.} = [x_1 \times x_2 \times \ldots \times x_n]^{\frac{1}{n}}$$

$$\Rightarrow \quad \log(\text{G.M.}) = \frac{1}{n}[\log x_1 + \log x_2 + \ldots + \log x_n]$$

If $f_1, f_2, f_3, ..., f_n$ are the respective frequencies of the values $x_1, x_2, ..., x_n$, then

$$\text{G.M.} = [x_1^{f_1} \times x_2^{f_2} \times ...... \times x_n^{f_n}]^{\frac{1}{f_1 + f_2 + ... + f_n}}$$

$$\Rightarrow \quad \log \text{G.M.} = \frac{1}{f_1 + f_2 + ... + f_n}[f_1 \log x_1 + f_2 \log x_2 + \ldots + f_n \log x_n]$$

i.e., the logarithm of the G.M. of a set of values is equal to the Arithmetic mean of their logarithms.

$$\Rightarrow \quad \text{G.M.} = \text{antilog}\left[\frac{1}{f_1 + f_2 + \ldots + f_n}[f_1 \log x_1 + f_2 \log x_2 + \ldots + f_n \log x_n]\right]$$

In general, Geometric mean of a number of values

$$= \textbf{[Product of the values]}^{\frac{1}{\textbf{Number of Values}}}$$

Note : If one of the values is zero, G.M. cannot be determined.

3.11 COMBINED G.M. OF TWO OR MORE GROUPS

G.M. of n values $x_1, x_2, x_3, \ldots\ldots \times x_n = [x_1 \times x_2 \times x_3 \times \ldots \times x_n]^{\frac{1}{n}}$

$$\Rightarrow \quad [\text{G.M.}]^n = [x_1 \times x_2 \times x_3 \ldots\ldots \times x_n]$$

$$\Rightarrow \quad \text{Product of the values} = [\text{G.M.}]^{\text{Number of values}}$$

If G_1, G_2 be the geometric means of two groups having n_1, n_2 number of items respectively, then the combined Geometric mean G of these two groups is given by

$$G = [[G_1]^{n_1} \times [G_2]^{n_2}]^{\frac{1}{n_1 + n_2}}$$

$$\Rightarrow \quad \log G = \frac{n_1 \log G_1 + n_2 \log G_2}{n_1 + n_2}$$

If $n_1 = n_2$, then $\quad \log G = \dfrac{\log G_1 + \log G_2}{2} = \dfrac{1}{2} \log G_1 G_2$

$$= \log [G_1 G_2]^{\frac{1}{2}} \quad \text{or} \quad G = [G_1 \times G_2]^{1/2},$$

i.e., if G_1, G_2 be the Geometric mean of two series having n items each, then the Geometric mean of the combined series having $2n$ items is $\sqrt{G_1 \times G_2}$.

If there are k groups with respective G.Ms. as $G_1, G_2, G_3, \ldots\ldots, G_k$ and $n_1, n_2, \ldots\ldots\ldots, n_k$ the number of items, then the combined G.M. (G) is given by

$$\log G = \frac{n_1 \log G_1 + n_2 \log G_2 + n_3 \log G_3 + \ldots + n_k \log G_k}{n_1 + n_2 + \ldots + n_k}$$

3.12 THE GEOMETRIC MEAN OF THE RATIO OF TWO VARIABLES IS EQUAL TO THE RATIO OF THEIR GEOMETRIC MEANS

Let $x_1, x_2, \ldots\ldots, x_n$ be the n values of a variable x and $y_1, y_2, \ldots, y_n$ be n values of a variable y. If G_1, G_2 be the respective geometric means of the two series then

$$G_1 = [x_1 \times x_2 \times \ldots \times x_n]^{\frac{1}{n}}, \; G_2 = [y_1 \times y_2 \times \ldots \times y_n]^{\frac{1}{n}}$$

Then the geometric mean of $\dfrac{x_1}{y_1}, \dfrac{x_2}{y_2}, \ldots\ldots, \dfrac{x_n}{y_n}$ is

$$= \left[\frac{x_1}{y_1} \times \frac{x_2}{y_2} \times \frac{x_3}{y_3} \times \ldots \times \frac{x_n}{y_n}\right]^{\frac{1}{n}} = \frac{G_1}{G_2}$$

Notes :

(1) If any one of the values of the variable is zero, the geometric mean becomes zero. But we normally say that if any one of the values is equal to zero. G.M. cannot be determined.

(2) In some cases the G.M. becomes imaginary, *e.g.*, in the series 2, –4, 8, –16, ... The G.M. between 2 and –4 is imaginary. The G.M. between 2 and 8 will normally be taken as 4 which is again misleading. So if the common ratio in a G.P. is negative we cannot find the G.M. accurately.

3.13 GROWTH RATE

If $r\%$ per annum is the rate of interest then the amount P after n years will become

$$= p\left(1+\frac{r}{100}\right)^n$$

If p_0 denotes the present population and $r\%$ per annum the constant rate of growth then the population P_n after n years is given by

$$P_n = P_0\left[1+\frac{r}{100}\right]^n. \qquad \text{...(i)}$$

If the rate of growth varies from year to year say $r_1\%$ in the first years, $r_2\%$ in the 2nd year and ...... $r_n\%$ in the nth year, then the total population after n years

$$= P_0\left[\left(1+\frac{r_1}{100}\right)\left(1+\frac{r_2}{100}\right)......\left(1+\frac{r_n}{100}\right)\right] \qquad \text{...(ii)}$$

Equating (i) and (ii), we get

$$\left[1+\frac{r}{100}\right]^n = \left[\left(1+\frac{r_1}{100}\right)\left(1+\frac{r_2}{100}\right)\cdots\left(1+\frac{r_n}{100}\right)\right]$$

$$\Rightarrow \qquad r = [(100+r_1)\,(100+r_2)\,\ldots\,(100+r_n)]^{\frac{1}{n}} - 100$$

which gives the formula for calculating average percentage growth rate when the varying growth rates for different years are given. In other words the formula states that if the varying rates of growth are given as percentages then the average growth rate can be obtained by subtracting 100 from the G.M. of $(100+r_1)$, $(100+r_2)$, ... $(100+r_n)$.

Similarly in case of *depreciation* on reducing balance, the average rate of depreciation say $r\%$ corresponding to depreciation rate of $r_1\%, r_2\%, ..., r_n\%$ respectively in n year is given by

$$r = 100 - [(100-r_1)\,(100-r_2)\,\ldots\,(100-r_n)^{\frac{1}{n}}$$

$$= 100 - \text{G.M. of } (100-r_1), (100-r_2)\,\ldots\,(100-r_n)$$

Thus, geometric mean is a proper average in finding the average rate of interest on a capital or the average rate of depreciation of a machine in case the compound interest law holds. The geometric mean also comes in when one wants to determine the value of a variable at the middle point of a time interval when the variable changes overtime exponentially. Thus, if the values at two points of time be ar^{t_1} and ar^{t_2}, then its value at the middle of the interval, *i.e.*, at $\frac{t_1+t_2}{2}$ is

$ar^{\frac{t_1+t_2}{2}} = ar^{\frac{t_1}{2}} \times ar^{\frac{t_2}{2}} = \sqrt{(ar^{t_1})\,(ar^{t_2})}$ which is the geometric mean of the values at the terminal points t_1 and t_2 of the interval.

Example 42. Calculate the geometric mean of the following numbers :

70, 10, 500, 75, 8, 250 **(Orissa Board, 1998)**

Solutin. $\text{G.M.} = [70 \times 10 \times 500 \times 75 \times 8 \times 250]^{\frac{1}{6}}$ [∵ The number of terns = 6]

$$\log \text{G.M.} = \frac{1}{6} \log [70 \times 10 \times 500 \times 75 \times 8 \times 250]$$

$$= \frac{1}{6} [\log 70 + \log 10 + \log 500 + \log 75 + \log 8 + \log 250]$$

$$= \frac{1}{6} [1.845 + 1 + 2.69897 + 1.875 + 0.9030 + 2.3979]$$

$$= \frac{1}{6} [9.71991] = 1.619985$$

$$\text{G.M.} = \text{antilog } 1.619985 = 41.6855.$$

Example 43. Find the G.M. of 3, 6, 24, 48. **(ICWA Foundation, June 2001)**

Solution. $\text{G.M.} = [3 \times 6 \times 24 \times 48]^{\frac{1}{4}}$

$$\log \text{G.M.} = \frac{1}{4} \log (3 \times 6 \times 24 \times 48) = \frac{1}{4} \log (20736)$$

$$= \frac{1}{4} (4.316725) = 1.07918$$

$$\text{G.M.} = \text{antilog } (1.07918) = 12^*.$$

Example 44. Calculate the geometric mean of the following data :

3884, 382, 63, 8.0, 4.0, 0.009, 0.003

Solution. Number of observations = 7

$$\text{G.M.} = [3884 \times 382 \times 63 \times 8 \times 4 \times 0.009 \times 0.003]^{\frac{1}{7}}$$

$$= [80760.105]^{\frac{1}{7}}$$

$$\log \text{G.M.} = \frac{1}{7} \log 80760 = \frac{1}{7} \times 4.9072 = 0.701$$

$$\text{G.M.} = \text{antilog } 0.701 = 5.02.$$

Note : In Ex. 44, we can also write

$$\log \text{G.M.} = \frac{1}{7} [\log 3884 + \log 382 + \log 63 + \log 8 + \log 4 + \log 0.009 + \log 0.003]$$

---

* We could have taken $\log \text{G.M.} = \frac{1}{4} [\log 3 + \log 6 + \log 24 + \log 48]$

$$= \frac{1}{4} [0.4771212 + 0.7781512 + 1.38 + 1.6812412]$$

$$= \frac{1}{4} (4.3165) = 1.0791.$$

$$= \frac{1}{7} [3.5892 + 2.5820 + 1.7993 + 0.9030 + 0.6020 - 2.0457 - 2.5228]$$

$$= \frac{1}{7} [4.907] = 0.701.$$

Example 45. Find the geometric mean of the following :

2.156, 1.372, 0.9184, 0.0903, 0.0082, 0.0005, 0.0078, 0.0009.

(Punjab B. Com., April 2000)

Solution. $\text{G.M.} = [2.156 \times 1.372 \times 0.9184 \times 0.0903 \times 0.0082 \times 0.0005 \times 0.0078 \times 0.0009]^{1/8}$

$$\Rightarrow \log \text{G.M.} = \frac{1}{8} \log [(2.156 \times 1.372 \times 0.9184) \times (0.0903 \times 0.0082) \times (0.0005 \times 0.0078 \times 0.0009)]$$

$$= \frac{1}{8} [\log (2.156 \times 1.372 \times 0.9184) + \log (0.0903 \times 0.0082) + \log (0.0005 \times 0.0078 \times 0.0009)]$$

$$= \frac{1}{8} [\log 2.7166566 + \log 0.00074046 + \log 3.51 \times 10^{-9}]$$

$$= \frac{1}{8} [0.4340 - 3.1304948 - 8.4546] = -1.3938989$$

$$\text{G.M.} = \text{antilog } (-1.3938989) = \text{antilog } [\bar{2}.6061] = 0.04037.$$

Example 46. Prove that the product of the ratios of each of the n observations to the G.M. is always unity. **(ICWA Foundation)**

Solution. Let the n observations be

$$x_1, x_2, x_3, \ldots\ldots, x_n$$

Ratios of these n observations to the G.M. are $\frac{x_1}{\text{G.M.}}, \frac{x_2}{\text{G.M.}}, \ldots\ldots, \frac{x_n}{\text{G.M.}}$ respectively.

Product of the ratios

$$= \frac{x_1}{\text{G.M.}} \times \frac{x_2}{\text{G.M.}} \times \frac{x_3}{\text{G.M.}} \times \ldots\ldots \times \frac{x_n}{\text{G.M.}}$$

$$= \frac{x_1 \times x_2 \times x_3 \times \ldots\ldots \times x_n}{(\text{G.M.})^n} = \frac{(\text{G.M.})^n}{(\text{G.M.})^n} = 1.$$

Example 47. If 5, 8, 6 and 1 occur with frequencies 3, 2, 4 and 1 respectively, find the geometric mean. **(ICWA Foundation)**

Solution. $\text{G.M.} = [5^3 \times 8^2 \times 6^4 \times 1]^{\frac{1}{3+2+4+1}} = [125 \times 64 \times 1296 \times 1]^{\frac{1}{10}}$

$$= [10368000]^{\frac{1}{10}}$$

$$\log \text{G.M.} = \frac{1}{10} \log 10368000 = \frac{7.015695}{10} = 0.7015695$$

$$\text{G.M.} = \text{antilog } 0.7015695 = 5.03 = 5 \text{ approx.}$$

Second Method

$$\log \text{G.M.} = \frac{1}{10} [3 \log 5 + 2 \log 8 + 4 \log 6 + \log 1]$$

$$= \frac{1}{10}[3 \times 0.69897 + 2 \times 0.90309 + 4 \times 0.77815 + 0]$$

$$= \frac{1}{10}[2.09691 + 1.806 + 3.1126]$$

$$= \frac{1}{10}[7.0155] = 0.70155$$

$$\text{G.M.} = \text{antilog } 0.70155 = 5.029 = 5 \text{ approx.}$$

Example 48. A machinery depreciates by 40% in the first year, 25% in the second year and 10% p.a. for the next three years. Each percentage being calculated on diminishing value, what is the average percentage of depreciation for the entire period ?

Solution. Average percentage for the entire period

$$= 100 - [(100-40)(100-25)(100-10)(100-10)(100-10)]^{1/5}$$

Let $y = [(100-40)(100-25)(100-10)(100-10)(100-10)]^{1/5}$

$\Rightarrow$ $\log y = \frac{1}{5}[\log 60 + \log 75 + 3 \log 90]$

$$= \frac{1}{5}[1.7782 + 1.8751 + 3 \times 1.9542]$$

$$= \frac{1}{5}[9.5159] = 1.90318$$

$\Rightarrow$ $y = \text{antilog } 1.90318 = 80$

$\therefore$ Average rate of depreciation $= 100 - 80 = 20\%$.

Example 49. The number of divorces per 1000 marriages in a certain state increased from 84 in 1990 to 108 in 2000. Find the annual rate of increase of the divorce rate for the period 1990 to 2000.

Solution. If $r\%$ is the rate of growth, (increase) then

$$108 = 84\left[1 + \frac{r}{100}\right]^{20}$$

$\Rightarrow$ $\log 108 = \log 84 + 20 \log\left[1 + \frac{r}{100}\right]$

or $\log\left[1 + \frac{r}{100}\right] = \frac{\log 108 - \log 84}{20}$

$$= \frac{2.0334 - 1.9243}{20} = \frac{0.1091}{20} = 0.0055$$

$\Rightarrow$ $1 + \frac{r}{100} = \text{antilog } 0.0055 = 1.013 \Rightarrow r = 1.3\%$.

Example 50. The annual rates of growth of output of a factory in 5 years are 5.0, 6.5, 4.5, 8.5 and 7.5 per cent respectively. What is the compound rate of growth of output per annum for the period ?

(D.U., B. Com. 1990)

Solution. If $r\%$ is the compound rate of growth, then

$$r = [(100+5)(100+6.5)(100+4.5)(100+8.5)(10+7.5)]^{1/5} - 100$$

$$= [105 \times 106.5 \times 104.5 \times 108.5 \times 107.5]^{1/5} - 100$$

Let $$y = [105 \times 106.5 \times 104.5 \times 108.5 \times 107.5]^{1/5}$$

$$\Rightarrow \quad \log y = \frac{1}{5}[\log 105 + \log 106.5 + \log 104.5 + \log 108.5 + \log 107.5]$$

$$= \frac{1}{5}[2.0212 + 2.0273 + 2.0191 + 2.0354 + 2.0314]$$

$$= \frac{10.1344}{5} = 2.0269$$

$$\Rightarrow \quad y = \text{antolog } 2.0269 = 106.38$$

Average rate of growth = 106.38 – 100 = 6.38%

Example 51. A sum of money was invested for 5 years. The average rate of interest for the five successive years was 8.50, 10.00, 9.30, 8.76 and 7.88 per cent. Compute the combined average rate of interest for 5 years.

Solution. Let r% be the combined average rate of interest, then

$$\left(1+\frac{r}{100}\right)^5 = \left(1+\frac{8.50}{100}\right)\left(1+\frac{10}{100}\right)\left(1+\frac{9.30}{100}\right)\left(1+\frac{8.76}{100}\right)\left(1+\frac{7.88}{100}\right)$$

$$= (1.085)\,(1.1)\,(1.093)\,(1.0876)\,(1.0788)$$

$$\Rightarrow \quad 5\log\left(1+\frac{r}{100}\right) = \log[1.085 \times 1.1 \times 1.093 \times 1.0876 \times 1.0788]$$

$$= \log(1.5305683)^*$$

$$\Rightarrow \quad \log\left(1+\frac{r}{100}\right) = \frac{1}{5}\log(1.5305683) = \frac{1}{5}(0.1848527) = 0.03697$$

$$\therefore \quad 1+\frac{r}{100} = \text{antilog } 0.03697 = 1.0888$$

$$\Rightarrow \quad \frac{r}{100} = 0.0888 \;\Rightarrow\; r = 8.88\%$$

Example 52. The annual rates of growth achieved by a nation for 5 years are 5%, 7.5%, 2.5%, 5% and 10% respectively. What is the compound rate of growth for the 5-year period ?

(D.U., B.A. Hons. Eco., 1993)

Solution. If r% is the compound rate of growth, then

$$\left(1+\frac{r}{100}\right)^5 = \left(1+\frac{5}{100}\right)\left(1+\frac{7.5}{100}\right)\left(1+\frac{2.5}{100}\right)\left(1+\frac{5}{100}\right)\left(1+\frac{10}{100}\right)$$

$$= \frac{21}{20} \times \frac{43}{40} \times \frac{41}{40} \times \frac{21}{20} \times \frac{11}{10}$$

$$= \left(\frac{21}{20}\right)^2 \times \frac{11}{10} \times \frac{43}{40} \times \frac{41}{40}$$

* We can also write $5\log\left(1+\frac{r}{100}\right) = [\log 1.085 + \log 1.1 + \log 1.083 + \log 1.0876 + \log 1.0788]$

$$5 \log\left(1+\frac{r}{100}\right) = \log 1.336 = 0.1258$$

$$\log\left(1+\frac{r}{100}\right) = \frac{0.1258}{5} = 0.02516$$

$$\Rightarrow \quad 1+\frac{r}{100} = \text{antilog}\, 0.02516$$

$$\Rightarrow \quad 1+\frac{r}{100} = 1.059$$

$$\Rightarrow \quad \frac{r}{100} = 0.059 \Rightarrow r = 5.9\%$$

Example 53. The price of a commodity increased by 10% in first year and by 20% in second year. However, it decreased by 40% in third year. But it again started rising and increased by 5% in the fourth year. Find the average increase or decrease in the price of the commodity over the period of four years. **(D.U. B. Com., 2007)**

Solution. Let Rs. P be the present price of the commodity then the price after 4 years

$$= P\left(1+\frac{10}{100}\right)\left(1+\frac{20}{100}\right)\left(1+\frac{40}{100}\right)\left(1+\frac{5}{100}\right)$$

$$= P \times \frac{11}{10} + \frac{6}{5} \times \frac{6}{10} \times \frac{21}{20}$$

$$= P\left[\frac{8316}{10000}\right]$$

If $r\%$ p.a. be the average rate of interest then

$$P\left(1+\frac{r}{100}\right)^4 = P\left(\frac{8316}{10000}\right)$$

$$\Rightarrow \quad 1+\frac{r}{100} = 0.9549$$

$$\Rightarrow \quad \frac{r}{100} = -0.045 \Rightarrow r = -4.5\%$$

There is an average decrease by 4.5%.

Example 54. If the arithmetic mean of two unequal positive real numbers 'a' and 'b' ($a > b$) be twice as great as their geometric mean, show that $a : b = (2+\sqrt{3}) : (2-\sqrt{3})$. **(ICWA Foundation, June 2001)**

Solution. Arithmetic mean of a and $b = \dfrac{a+b}{2}$

Geometric mean of a and $b = \sqrt{ab}$.

As per the question

$$\frac{a+b}{2} = 2\sqrt{ab} \Rightarrow a+b = 4\sqrt{ab}$$

$$\Rightarrow \quad (a+b)^2 = 16ab$$

$$(a-b)^2 = (a+b)^2 - 4ab = 16ab - 4ab = 12ab$$

$$\Rightarrow \quad (a-b) = \pm\sqrt{12ab} = 2\sqrt{3}\sqrt{ab} \quad [\because a \text{ and } b \text{ are given to be positive numbers}]$$

Now $$a+b=4\sqrt{ab}$$
$$a-b=2\sqrt{3}\sqrt{ab}$$

$\therefore$ $$\frac{a+b}{a-b}=\frac{4\sqrt{ab}}{2\sqrt{3}\sqrt{ab}}=\frac{2}{\sqrt{3}}$$

By componendo and dividendo, we get

$$\frac{2a}{2b}=\frac{2+\sqrt{3}}{2-\sqrt{3}} \Rightarrow a:b=(2+\sqrt{3}):(2-\sqrt{3}).$$

3.14 HARMONIC MEAN (H.M.)

The harmonic mean of a number of values of a variable is a reciprocal of the arithmetic mean of the reciprocals of the given values of the variable. Symbol H.M. is used to denote harmonic mean.

If $x_1, x_2, ..., x_n$ are n values of a variable then

$$\text{H.M.}=\frac{1}{\dfrac{\dfrac{1}{x_1}+\dfrac{1}{x_2}+\ldots+\dfrac{1}{x_n}}{n}}=\frac{n}{\dfrac{1}{x_1}+\dfrac{1}{x_2}+\ldots+\dfrac{1}{x_n}} \qquad ...(i)$$

If $f_1, f_2, ..., f_n$ are the respective frequencies of the values $x_1, x_2, \ldots, x_n$ then

$$\text{H.M.}=\frac{\sum_{i=1}^{n} f_i}{\dfrac{f_1}{x_1}+\dfrac{f_2}{x_2}+\ldots+\dfrac{f_n}{x_n}} \qquad ...(ii)$$

If each observation is divided by a constant. The H.M. is also divided by the same constant. In (i) if each observation is divided by K, then

New $$\text{H.M.}=\frac{n}{\dfrac{1}{x_1/K}+\dfrac{1}{x_2/K}+\ldots+\dfrac{1}{x_n/K}}=\frac{1}{K}\left[\frac{n}{\dfrac{1}{x_1}+\dfrac{1}{x_2}+\ldots+\dfrac{1}{x_n}}\right]$$

$=\dfrac{1}{K}$ times the original H.M.

Similarly, if each observation is multiplied by a constant H.M. is also multiplied by the same constant.

The use of harmonic mean is very limited. It is generally used in averaging speed, price of articles, etc., but here also proper care has to be taken.

Notes :

(1) If time varies *w.r.t.* a fixed distance then harmonic mean determines the average speed but if distance varies, *w.r.t.* a fixed time, then arithmetic mean only determines the average speed, *e.g.,* if a man moves along the sides of a square with speeds v_1, v_2, v_3, v_4 km/hr respectively, then the average speed for the whole journey

$$=\frac{4}{\dfrac{1}{v_1}+\dfrac{1}{v_2}+\dfrac{1}{v_3}+\dfrac{1}{v_4}}.$$

(2) If different types of shares or some other articles are purchased at different rates by making equal investment on each type, then the average price is given by the harmonic mean of different rates. For example, if a family spends Rs. 400 per month for milk and the price of milk per litre during the month of January, February and March is Rs. 5, Rs. 6 and Rs. 7 per litre then in these three months the average price of milk will be

$$= \text{Rs. } \frac{3}{\frac{1}{5}+\frac{1}{6}+\frac{1}{7}} \text{ per litre} = \text{Rs. } 5.89 \text{ per litre.}$$

Otherwise

Quantity of milk purchased in January $= \frac{400}{5}$ litres

Quantity of milk purchased in February $= \frac{400}{6}$ litres

Quantity of milk purchased in March $= \frac{400}{7}$ litres

$$\text{Averge price per litre} = \frac{\text{Total amount spent}}{\text{Total quantity of milk purchased}}$$

$$= \frac{400 \times 3}{\frac{400}{5}+\frac{400}{6}+\frac{400}{7}} = \text{Rs. } 5.89.$$

Example 55. Find the harmonic mean of the numbers $1, \frac{1}{2}, \frac{1}{3}, \frac{1}{4}$, with frequencies 5, 4, 5, 4 respectively.

Solution. Harmonic Mean $= \dfrac{\sum\limits_{i=1}^{n} f_i}{\frac{f_1}{x_1}+\frac{f_2}{x_2}+\ldots+\frac{f_n}{x_n}} = \dfrac{5+4+5+4}{\frac{5}{1}+\frac{4}{1/2}+\frac{5}{1/3}+\frac{4}{1/4}}$

$$= \frac{18}{5+8+15+16} = \frac{18}{44} = 0.4090.$$

Example 56. Find the Harmonic Mean of $1, \frac{1}{2}, \frac{1}{3}, \ldots\ldots, \frac{1}{n}$. **(ICWA Foundation, June 2001)**

Solution. Harmonic Mean $= \dfrac{n}{1+\frac{1}{1/2}+\frac{1}{1/3}+\ldots+\frac{1}{1/n}} = \dfrac{n}{1+2+3+4+\ldots+n}$

$$= \frac{n}{\frac{n(n+1)}{2}} = \frac{2}{n+1} \qquad \left[\because\ 1+2+3+\ldots+n = \frac{n(n+1)}{2}\right]$$

Example 57. The arithemtic mean of two observations is 127.5, and their geometric mean is 60. Find their harmonic mean. Also obtain the observations.

Solution. Let the two observations be a and b,

then $$\text{A.M.} = \frac{a+b}{2},\ \text{G.M.} = \sqrt{ab}$$

$$\text{H.M.} = \frac{2ab}{a+b} = \frac{ab}{\frac{a+b}{2}} = \frac{(\text{GM})^2}{\text{A.M.}}$$

$$= \frac{(60)^2}{127.5} = \frac{60 \times 60}{127.5} = 28.23$$

$$\text{A.M.} = \frac{a+b}{2} = 127.5 \ \Rightarrow\ a+b = 255,$$

$$\text{G.M.} = \sqrt{ab} = 60 \ \Rightarrow\ ab = (60)^2 = 3600$$

$$(a-b)^2 = (a+b)^2 - 4ab = (255)^2 - 4 \times 3600$$

$$= 65025 - 14400 = 50625$$

$\Rightarrow$ $$a - b = \pm\, 225$$

Solving $a + b + 255$ and $a - b = 225$ we get

$$a = 240 \text{ and } b = 15,$$

and solving $a + b = 255$ and $a - b = -225$ we get

$$a = 15, b = 240$$

Hence, the number are 240 and 15.

Example 58. Prove that beween two positive quantities, say a and b.

$$\text{A.M.} \geq \text{G.M.} \geq \text{H.M.}$$

Solution. $\text{A.M.} = \frac{a+b}{2}$, $\text{G.M.} = \sqrt{ab}$, $\text{H.M.} = \frac{2ab}{a+b}$

$$\text{A.M.} - \text{G.M.} = \frac{a+b-2\sqrt{ab}}{2}$$

$$= \frac{(\sqrt{a})^2 + (\sqrt{b})^2 - 2\sqrt{a}\,\sqrt{b}}{2} = (\sqrt{a} - \sqrt{b})^2 \text{ which} > 0$$

[$\because$ The square of a quantity is always positive.]

$\text{A.M.} - \text{G.M.}$ = a positive quantity $\Rightarrow \text{A.M.} > \text{G.M.}$

$\text{A.M.} \times \text{H.M.} = ab = (\text{G.M.})^2$

$\Rightarrow$ $$\frac{\text{A.M.}}{\text{G.M.}} = \frac{\text{G.M.}}{\text{H.M.}} \Rightarrow \text{G.M.} > \text{H.M.}^*$$ [$\because$ A.M. > G.M.]

Hence $\text{A.M.} > \text{G.M.} > \text{H.M.}$

If $a = b$ then $\text{A.M.} = \text{G.M.} = \text{H.M.}$

$\therefore$ $\text{A.M.} \geq \text{G.M.} \geq \text{H.M.}$

Example 59. The geometric mean and harmonic mean of two observations are respectively 18 and 10.8. Find the observations. **(ICWA Foundation, Dec., 2000)**

Solution. Let the two observations be x_1 and x_2,

* Pl. note if $x > y$ then

(i) $x - y > 0$ (ii) $\frac{x}{y} > 1$

then $$\text{G.M.} = (x_1x_2)^{1/2},\ \text{H.M.} = \frac{2x_1x_2}{x_1+x_2}$$

$$(x_1x_2)^{1/2} = 18 \Rightarrow x_1x_2 = 324$$

$$\frac{2x_1x_2}{x_1+x_2} = 10.8 \Rightarrow 2 \times 324 = 10.8 \times (x_1 + x_2)$$

$\Rightarrow$ $$x_1 + x_2 = \frac{2 \times 324}{10.8} = 60$$

$$(x_1 - x_2)^2 = (x_1 + x_2)^2 - 4x_1x_2$$

$$= (60)^2 - 4 \times 324 = 3600 - 1296 = 2304$$

$$x_1 - x_2 = \pm 48.$$

$$x_1 + x_2 = 60 \text{ and } x_1 - x_2 = 48 \text{ gives } x_1 = 54 \text{ and } x_2 = 6$$

$$x_1 + x_2 = 60 \text{ and } x_1 - x_2 = -48 \text{ gives } x_1 = 6 \text{ and } x_2 = 48$$

Hence, the observations are 48 and 6.

Example 60. Verify the relationship A.M. ≥ G.M. ≥ H.M. from the following data :

25, 24, 23, 32, 40, 27, 30, 25, 10, 20.

Solution. $\text{A.M.} = \frac{\text{Sum of the Values}}{\text{No. of Values}}$

$$= \frac{25 + 24 + 23 + 32 + 40 + 27 + 30 + 25 + 10 + 20}{10} = \frac{256}{10} = 25.6$$

$$\text{G.M.} = (25 \times 24 \times 23 \times 32 \times 40 \times 27 \times 30 \times 25 \times 10 \times 20)^{\frac{1}{10}} = 24.29$$

$$\text{H.M.} = \frac{10}{\frac{1}{25} + \frac{1}{24} + \frac{1}{23} + \frac{1}{32} + \frac{1}{40} + \frac{1}{27} + \frac{1}{30} + \frac{1}{25} + \frac{1}{10} + \frac{1}{20}} = 22.63$$

Clearly A.M. > G.M. > H.M.

Example 61. If the A.M. of two observations be 25 and their H.M. is 9, find their G.M.

Solution. Let the two observations be x_1 and x_2. We know

$$\text{A.M.} = \frac{x_1 + x_2}{2},\ \text{G.M.} = \sqrt{x_1x_2},\quad \text{H.M.} = \frac{2x_1x_2}{x_1 + x_2}$$

$$\text{A.M.} \times \text{H.M.} = x_1x_2 = (\text{G.M.})^2$$

$\Rightarrow$ $$(\text{G.M.})^2 = 9 \times 25 = 255 \Rightarrow \text{G.M.} = 15.$$

Example 62. Cities A, B and C are equidistant from each other. A motorist travels from A to B at 30 km/hr, from B to C at 40 km/hr and from C to A at 50 km/hr. Determine his average speed for the entire trip.

Solution. Average speed is the harmonic mean.

$$\text{H.M.} = \frac{3}{\frac{1}{30} + \frac{1}{40} + \frac{1}{50}} = 38.3 \text{ km/hr.*}$$

Example 63. Calculate the arithmetic mean, geometric mean and harmonic mean of the observations 32, 35, 36, 37, 39, 41, 43, 47, and 48 and show that A.M. > G.M. > H.M.

Solution. $\text{A.M.} = \frac{1}{9}(32 + 35 + 36 + 37 + 39 + 41 + 43 + 47 + 48)$

$$= \frac{1}{9} \times 358 = 39.777 = 39.78$$

Calculations for G.M.

X	$log\ x$
32	1.5051
35	1.5441
36	1.5563
37	1.5682
39	1.5911
41	1.6128
43	1.6335
47	1.6721
48	1.6812
Total	14.3644

$$\log \text{G.M.} = \frac{1}{9} \times 14.3644 = 1.5960$$

$\therefore$ $\text{G.M.} = \text{antilog } 1.5960 = 39.45$

$$\text{H.M.} = \frac{9}{\frac{1}{32} + \frac{1}{35} + \frac{1}{36} + \frac{1}{37} + \frac{1}{39} + \frac{1}{41} + \frac{1}{43} + \frac{1}{47} + \frac{1}{48}}$$

$$= \frac{9}{0.03125 + 0.02857 + 0.02778 + 0.02703 + 0.02564 + 0.02439 + 0.02326 + 0.02128 + 0.02083}$$

$$= \frac{9}{0.23004} = 39.13$$

$$39.78 > 39.45 > 39.13, \Rightarrow \text{A.M.} > \text{G.M.} > \text{H.M.}$$

Example 64. (a) The arithmetic mean and the geometric mean of two observations are 110.5 and 70 respectively. Find (i) their harmonic mean and (ii) the two observations.

Solution. $(\text{G.M.})^2 = \text{A.M.} \times \text{H.M.}$ or $70^2 = 110.5 \times \text{H.M}\}$

or $4900 = 110.5 \times \text{H.M.}$

$\Rightarrow$ $\text{H.M.} = \frac{4900}{110.5} = 44.34.$

---

* Average speed $= \frac{\text{Total distance travelled}}{\text{Total time taken}}$. Students may verify by this formula also.

Let a, b be the two observations so that

$$\frac{a+b}{2} = 110.5 \Rightarrow a+b = 221$$

$$\Rightarrow \quad b = 221 - a \quad \text{...(i)}$$

$$\sqrt{ab} = 70 \quad \text{or} \quad ab = 4900 \quad \text{...(ii)}$$

or $$a(221 - a) = 4900 \quad [\because a+b = 221]$$

or $$a^2 - 221a + 4900 = 0$$

or $$a^2 - 196a - 25a + 4900 = 0$$

or $$a(a - 196) - 25(a - 196) = 0$$

or $$(a - 25)(a - 196) = 0 \quad \text{or} \quad a = 25 \quad \text{or} \quad 196$$

From (i), for $$a = 25, b = 221 - 25 = 196$$

$$a = 196, b = 221 - 196 = 25$$

$\therefore$ Hence, the observations are 25, 196.

Example 65. Mr. Ram travels from Delhi to Shimla at a speed of 40 kms per hour. While returning from Shimla to Delhi, he travels at a speed of 50 kms per hour. What is his average speed for the entire journey? **(DU B.Com 2007)**

Solution. If D kms is the distance from Delhi to Shimla the, time taken to travel from Delhi to Shimla

$$= \frac{D}{40} \text{ hrs}$$

Time taken to travel from Shimla to Delhi $= \frac{D}{50}$ hrs

$$\text{Average speed} = \frac{\text{Total distance travelled}}{\text{Total time taken}}$$

$$= \frac{D+D}{\frac{D}{40} + \frac{D}{50}} = \frac{2}{\frac{1}{40} + \frac{1}{50}} = \frac{2 \times 200}{9}$$

$$= 44.44 \text{ km/hr} = \text{H.M. of speeds}$$

Example 66. There are four typists A, B, C, D. Typist A takes 15 minutes, B takes 10 minutes, C takes 6 minutes and D takes 5 minutes respectively to type 1 letter. Then determine the average time required to type a letter by using harmonic mean if

(1) If each typist type 4 letters what is the average time taken to type a letter.

(2) If each typist type for 2 hrs, what is the average time taken to type a letter.

(DU B.Com II 2008)

Solution. (1) Time taken by typist A to type 4 letters $= 15 \times 4 = 60$ minutes

Time taken by typist B to type 4 letters $= 10 \times 4 = 40$ minutes

Time taken by typist C to type 4 letters $= 6 \times 4 = 24$ minutes

Time taken by typist D to type 4 letters $= 5 \times 4 = 20$ minutes

Average time required to type a letter

$$= \frac{\text{Total time taken to type 16 letters}}{\text{Number of letters typed}}$$

$$= \frac{15 \times 4 + 10 \times 4 + 6 \times 4 + 5 \times 4}{16} = \frac{144}{16} = 9^{*} \text{ minutes}$$

(2) Number of letters typed by typist A in 2 hours = 120 minutes $= \frac{1}{15} \times 120$

Number of letters typed by typist B, C and D respectively in 2 hours are

$$\frac{1}{10} \times 120, \ \frac{1}{6} \times 120 \ \text{ and } \ \frac{1}{5} \times 120$$

Average time taken to type one letters

$$= \frac{\text{Total time taken}}{\text{Total number of letters typed}}$$

$$= \frac{4 \times 120}{\frac{120}{15} + \frac{120}{10} + \frac{120}{6} + \frac{120}{5}}$$

$$= \frac{4}{\frac{1}{15} + \frac{1}{10} + \frac{1}{6} + \frac{1}{5}}$$

$$= \frac{4 \times 60}{4 + 6 + 10 + 12} = \frac{240}{32} = 7.5 \text{ HM.}$$

3.15 MEDIAN (GENERAL INTRODUCTION)

According to Simpson and Kafka, *"As distinct from the arithmetic mean, which is calculated from the value of every item in the series, the median is what is called a* ***position average****. The term 'position' refers to the place of a value in a series. The position of the median in a series is such that the number of items (in the series) below it equals the number of items (in the series) above it. Thus, the median is a value in a series which is exceeded by as many values as it exceeds."* *Thus median is the middlemost value.*

According to Connor, *"The median is that value of the variable which divides the group into two equal parts, one part comprising all values greater, and the other, all values less than the median."*

* In the first case Harmonic mean will not be applicable.

According to A.L. Bowley, *"If the numbers of the group are ranked in order according to the measurement under consideration then the measurement of the number most nearly one half is the median."*

To find the median in *ungrouped data* the data is arranged either in **ascending** order of magnitude or in **descending** order of magnitude. After arranging the data is any of two ways the middlemost value gives the median.

If the number of items is **odd**, then $\left(\frac{N+1}{2}\right)$th value gives the median and if the number of items is **even**, then the arithmetic mean of $\frac{N}{2}$th and $\left(\frac{N}{2}+1\right)$th value gives the median value. For example, consider the following series :

2, 10, 5, 7, 15, 17, 21, 13, 4

and 2, 10, 5, 7, 15, 17, 21, 13, 4, 22

After arranging the above series in ascending order, we get

2, 4, 5, 7, 10, 13, 15, 17, 21, ... '*A*'

and 2, 4, 5, 7, 10, 13, 15, 17, 21, 22, ... '*B*'

Series '*A*' contains 9 items, therefore, $\frac{9+1}{2}$ = 5th value, *i.e.,* 15 is the median.

Series '*B*' contains 10 values, therefore, A.M. of $\frac{N}{2}=\frac{10}{2}$ = 5th value and $\left(\frac{N}{2}+1\right)$ = 6th value gives the median. In this case $\frac{10+13}{2}$ = 11.5 is the median, which is a value half way between the two central values.

The above definitions of the median **do not hold** good in situations where a median value is surrounded by neighbouring values which are equal in magnitude to it. For example, in a series of values such as 12, 13, 14, 15, 15, 17 and 18, there is no value which is so located that three values are smaller than it and three are greater than it. However, value 15 is designated as median. Keeping in view such situations, **Croxton and Cowdon** have given a revised definition of median as, *"The median is that value which divides a series so that one half or more of the items are equal to or less than it and one half or more of the items are equal to or greater than it."*

It is clear from the above discussion that the median cannot be located unless the data is arranged **in ascending or descending order** whereas no such arrangement is required to calculate the mean. ***Median*** *is influenced by the* ***position*** *of the items in the array but not by the size and number of the items.*

3.16 MEDIAN FOR GROUPED DATA

In grouped data, the individual items lose their identity, and the middle item cannot be found by counting. Therefore, it is necessary to get inside of a class to find the value that divides the number of all items in half. If we divide the number of frequencies (*N*) in two halves, we find that the middle item falls within a class and to establish this class, **we cumulate the frequencies until we reach the lowest class whose cumulative frequency is greater than** $\frac{N}{2.}$ This class is called the **median class.**

Median class is the lowest class whose cumulative frequencies* contain the value $\frac{N}{2}$ when the class intervals are arranged in ascending order. Clearly the median value is more than the lower limit of the median class and less or equal to the upper limit of the median class. Assuming that all items are evenly distributed over this median class, we proceed towards the upper limit of this class, stopping when we have reached at the frequency equal to $\frac{N}{2}$. This operation brings us to a value within the median class which is presumed to have $N/2$ items on each side of it.

As already stated, median value is, at least, as high as the lower limit of the median class but cannot be higher than the upper limit of the median class.

$$\textbf{Median} = \begin{array}{c}\text{Lower limit of the}\\ \text{median class}\end{array} + \frac{\dfrac{\text{Total frequency}}{2} - \begin{array}{c}\text{Cumulative frequency of the class}\\ \text{preceding the median class}\end{array}}{\text{Frequency of the median class}} \times \begin{array}{c}\text{Class interval of}\\ \text{the median class}\end{array}$$

The steps for finding the median for grouped data are, therefore, as follows :

1. Divide the number of items in the distribution by 2, that is, compute the value $\frac{N}{2}$.
2. Accumulate, *i.e.,* cumulate the frequencies.
3. Find the class whose cumulative frequency is the first to exceed $N/2$. This is the median class.
4. Find the *actual* lower limit of the median class.
5. Then perform the following operations : Subtract from $N/2$ the frequencies we have accumulated before entering the median class. Divide this difference by the frequency of the median class. Multiply the quotient thus obtained by the size of the class interval of the median class.
6. Add the result of the operation in Step 5 to the lower limit of the median class. This sum gives us the median.

The formula for this procedure is

$$\text{median} = l_1 + \left(\frac{\frac{N}{2} - Cf_{-1}}{f_{med}}\right) \times i \qquad ...(1)$$

where l_1 = the lower limit of the median class,**

N = the total frequency,

Cf_{-1} = cumulative frequency of the class preceding the median class,

f_{med} = the frequency of the median class, and

i = the size of the class interval of the median class.

For finding the median for grouped data also, the class intervals should be arranged in ascending order of magnitude.

* Cumulative frequency of any class = Sum of the frequencies of the classes preceding that class + Frequency of that class.

** It may happen that the cumulative frequency of a class equals $\frac{N}{2}$. In such a case, this class is the median class and its upper limit is the median of the distribution.

The median can also be found for grouped data by entering the median class at its upper limit. In this case the formula is

$$\text{median} = l_2 - \left(\frac{\frac{N}{2} - \Sigma f_2}{f_{med}}\right) i \qquad ...(2)$$

where l_2 = the upper limit of the median class,

N = the total frequency,

Σf_2 = the sum of the frequencies accumulated from the highest class to the class immediately above the median class in value, *i.e.,* sum of the frequencies of all the classes succeeding the median class,

f_{med} = the frequency of the median class,

i = the size of the class interval of the median class.

Note : At the initial stage students should avoid using formulae (2) for calculating the median.

3.17 RELATED POSITIONAL MEASURES, *i.e.*, QUARTILES, DECILES AND PERCENTILES

There are other measures which divide a series into equal parts. Of chief importance are the quartiles, the deciles, and the percentiles. Quartiles divide a series into four equal parts, deciles into ten equal parts, percentiles into one hundred equal parts. There are three quartiles 9 deciles and 99 percentiles. The median, it should be clear, is the same value in a series as the second quartile, the fifth deciles, and the fiftieth percentile. For finding quartiles, deciles and percentiles, the data is **arranged in ascending order only.**

In economics and business, the quartiles are more widely applied than deciles or percentiles. The first quartile, Q_1, is a value such that 25% of the items in the series are equal to or smaller (or below) than it, while the third quartile, Q_3, is a value such that 75% of the items in the series are equal to or below it.

First quartile (written as Q_1) for ungrouped data is $\left(\frac{N+1}{4}\right)$th item*. In case $\left(\frac{N+1}{4}\right)$ is a fraction necessary interpolation between the two values between which $\frac{N+1}{4}$ falls should be made, for example, if there are 28, values in a series, then

$$\frac{N+1}{4} = 7.25.$$

Then we have to find the difference between the seventh value and eighth value and find $\frac{1}{4}$ of this difference and add to the 7th value to get the first quartile value. If 7th value is 9 and 8th value is 12, then First Quartile $= 9 + \left(\frac{12-9}{4}\right) = 9.75.$

Third Quartile (written as Q_3) is $3\left(\frac{N+1}{4}\right)$th value. Similar interpolation as explained above may be required in this case also. However, Second Quartile, *i.e.,* Q_2 = Median.

* It should be noted that for finding Quartiles data must be arranged in ascending order. It is only after arranging the data in ascending order, we take $\left(\frac{N+1}{4}\right)$th item.

3.18 QUARTILES FOR GROUPED DATA

In case of grouped frequency distribution, rth Quartile, *i.e.*, Q_r is given by

(i) **Formula for rth Quartile**

$$Q_r = l_1 + \frac{r\left(\frac{N}{4}\right) - c}{f} \times (l_2 - l_1)^* \quad e.g., \frac{3N}{4} \text{ for Third Quartile since } r = 3 \text{ for Third Quartile}$$

Here Q_r = rth quartile. So r can be 1, 2 or 3.

l_1 = the lower limit of the quartile class,

l_2 = the upper limit of the quartile class,

f = the frequency of the quartile class,

c = cumulative frequency of the class preceding the quartile class.

[First Quartile Class is the lowest class whose cumulative frequency contains $\frac{N}{4}$.]

Similarly, rth decile and rth percentile are calculated.

(ii) **Formula for rth Decile**

The formula for the rth decile will be

$$D_r = l_1 + \frac{r\left(\frac{N}{10}\right) - c}{f} \times (l_2 - l_1) \qquad \text{...(iv)}$$

Here, $r = 1, 2, 3, \ldots, 9$ may be replaced in formula (iv) to get the value of any decile value. For example, fourth decile value is obtained by putting $r = 4$ in the above formula, *i.e.*,

$$D_4 = l_1 + \frac{4\left(\frac{N}{10}\right) - c}{f} \times (l_2 - l_1)$$

Other symbols have their usual meaning.

(iii) **Formula for rth Percentile**

$$P_r = l_1 + \frac{r\left(\frac{N}{100}\right) - c}{f} \times (l_2 - l_1) \qquad \text{...(v)}$$

Here $r = 1, 2, 3, \ldots, 99$.

Therefore, for getting 35th percentile, we replace $r = 35$ in the above formula and thus

$$P_{35} = l_1 + \frac{35\left(\frac{N}{100}\right) - c}{f} \times (l_2 - l_1).$$

Example 67. Calculation the median from the following values

18, 16, 14, 11, 13, 10, 9, 20

Solution. Arranging the values in ascending order, we get

9, 10, 11, 13, 14, 16, 18, 20

* $l_2 - l_1$ = Class interval = i.

Number of items = 8.

Since the number of items is even, the A.M. of $\frac{N}{2}$th and $\left(\frac{N}{2}+1\right)$th value is the median, *i.e.*, the A.M. of 4th and 5th value will be median.

4th value = 13, 5th value = 14

A.M. of 13 and 14 $= \frac{13+14}{2} = 13.5.$

$\therefore$ Median = 13.5.*

Example 68. Find the median of the observations

4, 12, 7, 9, 14, 16, 21, 3

Solution. Arranging the data in ascending order, we get

3, 4, 7 9, 12, 14, 16, 17, 21

Since the number of values is odd, *i.e.*, 9, $\frac{N+1}{2}$th value, *i.e.*, $\frac{9+1}{2}$ = 5th *value is the median.*

5th value = 14.

$\therefore$ Median = 14.

Example 69. Find the median of the numbers

25, 24, 23, 32, 40, 27, 30, 25, 20, 10, 15, 45

Solution. Arranging the data in ascending order, we get

10, 15, 20, 23, 24, 25, 25, 27, 30, 32, 40, 45

The number of items in the series is 12 which is an even number.

$\therefore$ The A.M. of 6th and 7th item is the median

6th item = 25.

7th item = 25.

Since 6th and 7th items are equal, therefore, the median is 25.

Example 70. The following table represents the marks obtained by a batch of 20 students in certain class tests in Statistics and Accountancy.

Serial No.	1	2	3	4	5	6	7	8	9	10
Marks in Statistics	53	54	52	32	30	60	47	46	35	28
Marks in Accountancy	58	55	25	32	56	85	44	80	33	72
Serial No.	11	12	13	14	15	16	17	18	19	20
Marks in Statistics	25	42	33	48	72	51	45	33	65	29
Marks in Accountacy	10	42	15	46	50	64	39	38	30	36

Indicate in which subject is the level of knowledge higher ? Give reasons.

Solution. The use of median is recommended when exceptionally large or small values occur in the series. It is also used when one is interested in comparing the average performance of students in different subjects, colleges, etc. In view of this, median is the most suitable measure of central

* If we arrange the data in descending order, the values become 20, 18, 16, 14, 13, 11, 10, 9.
Now 4th value = 14, 5th value = 13, A.M. of 14 and 13 = 13.5 = Median.

tendency in the present example. For determining median marks in the two subjects, the marks are first arranged in an ascending order as follows :

Computation of Median

S. No.	*Marks in Statistics*	*Marks in Accountancy*	*S. No.*	*Marks in Statistics*	*Marks in Accountancy*
1	25	10	11	46	42
2	28	15	12	47	44
3	29	25	13	48	46
4	30	26	14	51	50
5	32	30	15	52	55
6	33	32	16	53	58
7	33	33	17	54	64
8	35	36	18	60	72
9	42	38	19	65	80
10	45	39	20	72	85

Here, we have 20 observations in both the series. So the median will be the A.M. of 10th value and 11th value.

$$\therefore \qquad \text{Median} = \frac{\text{Value of 10th item} + \text{Value of 11th item}}{2}$$

$$\therefore \text{ Median marks (Statistics)} = \frac{45+46}{2} = 45.5.$$

$$\text{Similarly, Median marks (Accountancy)} = \frac{39+42}{2} = 40.5.$$

The comparison of median marks in the two subjects shows that the level of knowledge is higher in Statistics.

Example 71. The marks out of 50 obtained by 100 students in the test are given below. Find the median marks.

Marks	20	35	28	29	33	38	42	45
No. of Students	6	20	24	28	15	4	2	1

Solution. Cumulative frequency table after arranging the marks in ascending order is as under :

Marks Obtained	*Frequency No. of Students*	*Cumulative Frequency*
20	6	6
25	20	26
28	24	50
29	28	78
33	15	93
38	4	97
42	2	99
45	1	100

Total frequency $= \Sigma f = N = 100$ which is even. Hence, median value will be the average of $\frac{N}{2}$th value, *i.e.*, $\frac{100}{2}$th = 50th value and $\left(\frac{N}{2}+1\right)$th = 51st value.

$$50\text{th value} = 28.$$
$$51\text{st value} = 29.$$

$\therefore$ A.M. of 28 and 29 is $\frac{28+29}{2} = 28.5.$

$\Rightarrow$ About 50% of students obtained marks 28.5 or less and the remaining 50% obtained marks 28.5 or more.

Example 72. Find the median of the data :

X	160	150	152	161	156
F	5	8	6	3	7

Solution. Arranging the data in ascending order, we have

x	f	*Cumulative Frequency*
150	8	8
152	6	14
(156)	7	(21)
160	5	26
161	3	29
Total	$N = \Sigma f = 29$	

Total number of observations $N (= 29)$ is odd.

$\therefore$ $\left(\frac{N+1}{2}\right)$th item $= \left(\frac{29+1}{2}\right)$th item = 15th item.

Now, 15th item lies in the row with 21 as cumulative frequency, since $14 < 15 < 21$.

$\therefore$ Median = size of 15th item = 156.

Example 73. Find the median of the following data :

Income (Rs.)	100	150	80	200	250	180
No. of Persons	24	26	16	20	6	30

Solution. Arranging the data in ascending order, we have

Income (in Rs.)	*No. of Persons*	*Cumulative Frequency*
80	16	16
100	24	40
150	26	66
180	30	96
200	20	116
250	6	122
Total	$N = \Sigma f = 122$	

Hence $N\,(=122)$ is even and $\frac{N}{2}=61$.

$\therefore$ Median is the mean of values of $\left(\frac{N}{2}\right)$th and $\left(\frac{N}{2}+1\right)$th items,

i.e., Median = Mean of 61th and 62th value

$$= \text{Rs.}\left(\frac{150+150}{2}\right) = \text{Rs. } 150.$$

Example 74. If the median of the following frequency distribution is 46, find the missing frequencies f_1 and f_2 :

Variable	*Frequency*
10–20	12
20–30	30
30–40	f_1
40–50	65
50–60	f_2
60–70	25
70–80	18
Total	229

Solution. We have the following table :

Variable	*Frequency* (f)	*Cumulative Frequency* (*c.f.*)
10–20	12	12
20–30	30	42
30–40	f_1	$42+f_1$
40–50	65	$107+f_1$
50–60	f_2	$107+f_1+f_2$
60–70	25	$132+f_1+f_2$
70–80	18	$150+f_1+f_2$
Total	$N=\Sigma f=150+f_1+f_2$	

It is given that $N=229 \Rightarrow \frac{N}{2}=114.5.$

$\therefore$ $150+f_1+f_2=229 \Rightarrow f_1+f_2=79$ ...(i)

Now, since the median is given to be 46, hence median class is $40-50$.

$\therefore$ $l_1=40, f=65, Cf_{-1}=42+f_1, i=10.$

$$\therefore \quad \text{Median} = l_1 + \frac{\frac{N}{2}-Cf_{-1}}{f}\times i$$

$$= 40 + \frac{114.5-(42+f_1)}{65}\times 10$$

$$= 40 + \frac{(72.5f_1) \times 2}{13}$$

$$\Rightarrow \quad 46 = 40 + \frac{145 - 2f_1}{13} \qquad [\because \text{ Median} = 46 \text{ (given)}]$$

$$\Rightarrow \quad 46 - 40 = \frac{145 - 2f_1}{13} \quad \Rightarrow \quad 6 \times 13 = 145 - 2f_1$$

$$\Rightarrow \quad 2f_1 = 145 - 78 \quad \Rightarrow \quad 2f_1 = 67$$

$$\Rightarrow \quad f_1 = 33.5 \approx 34.*$$

Hence from (i), $\quad f_2 = 79 - 34 = 45.$

Hence, the missing frequencies are $f_1 = 34$ and $f_2 = 45$.

Example 75. From the following data, calculate the values of the upper and lower quartiles, D_2, P_{30}.

Marks	Below 10	10–20	20–40	40–60	60–80	Above 80
No. of Students	8	10	22	25	10	5

Solution. We make the following cumulative frequency table :

Marks	*No. of Students* *(f)*	*Cumulative Frequency* *(c.f.)*
0–10	8	8
10–20	10	18
20–40	22	40
40–60	25	65
60–80	10	75
80–100	5	80
Total	$N = \Sigma f = 80$	

For Q_1, Lower Quartile :

$$N = 80 \quad \Rightarrow \quad \frac{N}{4} = 20.$$

The cumulative frequency just greater than 20 is 40. So, 20 – 40 is the lower quartile class such that

$$l_1 = 20, f = 22, Cf_{-1} = 18.$$

$$\therefore \quad Q_1 = l_1 + \frac{\frac{N}{4} - Cf_{-1}}{f} \times i$$

$$\Rightarrow \quad Q_1 = 20 + \frac{20 - 18}{22} \times 20, \qquad \text{where } i = 40 - 20 = 20.$$

$$= 20 + \frac{2}{22} \times 20 = 20 + \frac{40}{22} = 20 + 1.82 = 21.82.$$

* Frequency cannot be a fraction, hence we have taken the next whole number.

Hence, the lower quartile is 21.82 marks.

For Q_3, Upper Quartile :

$$N = 80 \Rightarrow \frac{3N}{4} = 60.$$

The cumulative frequency just greater than 60 is 65. So, 40 – 60 is the upper quartile such that

$$l_1 = 40, f = 25, Cf_{-1} = 40.$$

$$\therefore \quad Q_3 = l_1 + \frac{\frac{3N}{4} - Cf_{-1}}{f} \times i$$

$$\Rightarrow \quad Q_3 = 40 + \frac{60 - 40}{25} \times 20$$

$$= 40 + \frac{20}{25} \times 20 = 40 + \frac{400}{25} = 40 + 16 = 56.$$

Hence, the upper quartile is 56 marks.

For D_2, the Second Decile :

$$N = 80 \Rightarrow \frac{2N}{10} = \frac{2 \times 80}{10} = 16.$$

The cumulative frequency just greater than 16 is 18. So 10–20 is the second decile class such that

$$l_1 = 10, f = 10, Cf_{-1} = 8$$

$$\therefore \quad D_2 = l_1 + \frac{\frac{2N}{10} - Cf_{-1}}{f} \times i$$

$$= 10 + \frac{16 - 8}{10} \times 10, \quad \text{where } i = 20 - 10 = 10$$

$$= 10 + \frac{8}{10} \times 10 = 10 + 8 = 18.$$

Hence, the second decile is 18 months.

For P_{30}, 30th Percentile :

$$N = 80 \Rightarrow \frac{30N}{100} = \frac{30 \times 80}{100} = 24.$$

The cumulative frequency just greater than 24 is 40. So 20 – 40 is the 30th percentile class such that

$$l_1 = 20, f = 22, Cf_{-1} = 18$$

$$\therefore \quad P_{30} = l_1 + \frac{\frac{30N}{100} - Cf_{-1}}{f} \times i$$

$$\Rightarrow \quad P_{30} = 20 + \frac{24 - 18}{22} \times 20, \quad \text{where } i = 40 - 20 = 20$$

$$= 20 + \frac{6 \times 20}{22} = 20 + \frac{60}{11} = 20 + 5.46 = 25.46.$$

Hence, the 30th percentile is 25.46 marks.

Example 76. The median of the following observations arranged in ascending order is 42. Find x.

$$22, 24, 33, 37, x+1, x+3, 44, 47, 51, 58$$

Solution. The number of observation is 10 which is even. Hence mean of 5th and 6th value is median.

Mean of 5th and 6th value

$$= \frac{x+1+x+3}{2} = x+2 = 42 \text{ (given)} \Rightarrow x = 40.$$

Example 77. The median of eight observatoins 31, 48, 37, 34, 45, 36, 41 and x is 38 where $37 < x < 41$. Find the value of x. **(ICWA Foundation Dec., 2001)**

Solution. Arranging the data in ascending order, we get

$$31, 34, 36, 37, x, 41, 45, 48$$

Median is the A.M. of 4th and 5th value, *i.e.,*

$$\text{Median} = \frac{37+x}{2} = 38 \Rightarrow x = 39.$$

Example 78. Find the second quartile of the values

69, 75, 72, 71, 73, 74, 76, 75, 70 **(ICWA Foundation, June 2002)**

Solution. Arranging the data in ascending order, we get 69, 70, 71, 72, 73, 74, 75, 75, 76

Second quartile is the same as median.

Hence, second quartile is 73.

Example 79. Find the mising frequencies f_1, f_2 in the following distribution. It is given that the median of the distribution is 41 and the total number of observation is 82.

Class Interval	10–20	20–30	30–40	40–50	50–60	60–70	Total
Frequency	10	f_1	15	20	f_2	11	82

Also, calculate Q_3 and D_7 for the complete frequency distribution.

Solution. We have the following table :

Class Interval	*Frequency* (f)	*Cumulative Frequency* $(c.f.)$	*c.f.*
10–20	10	10	10
20–30	$f_1 = 14$	$10+f_1$	24
30–40	15	$25+f_1$	39
40–50	20	$45+f_1$	59
50–60	$f_2 = 12$	$25+f_1+f_2$	71
60–70	11	$56+f_1+f_2$	82
Total	$N = \Sigma f = 56+f_1+f_2$		

The total frequency is 82.

$\therefore$ $$56+f_1+f_2 = 82 \Rightarrow f_1+f_2 = 26.$$

Median $= 41$. Clearly 41 lies in the class 40—50. So, 40—50 is the median class.

$\therefore$ $$l_1 = 40,\ i = 10, f = 20,\ Cf_{-1} = 25+f_1,\ N = 82.$$

Now, $$\text{Median} = l_1 + \frac{\frac{N}{2} - Cf_{-1}}{f} \times i$$

$$41 = 40 + \frac{41 - (25 + f_1)}{20} \times 10$$

$$\Rightarrow \qquad 41 - 40 = \frac{16 - f_1}{20} \times 10$$

$$\Rightarrow \qquad 1 = \frac{16 - f_1}{2} \Rightarrow 2 = 16 - f_1$$

$$\Rightarrow \qquad f_1 = 16 - 2 = 14.$$

Since, $f_1 + f_2 = 26$, therefore, $f_2 = 26 - 14 = 12$.

Hence, $f_1 = 14$ and $f_2 = 12$.

Third Quartile (Q_3) : $N = 82$

$$\Rightarrow \qquad \frac{3N}{4} = \frac{3 \times 82}{4} = \frac{246}{4} = 61.5.$$

The cumulative frequency just greater than $\frac{3N}{4} = 61.5$ is 71. So 50—60 is the upper quartile class such that

$$l_1 = 50, f = 12, i = 10, Cf_{-1} = 59.$$

$$\therefore \qquad Q_3 = l_1 + \frac{\frac{3N}{4} - Cf_{-1}}{f} \times i = 50 + \frac{61.5 - 59}{12} \times 10$$

$$= 50 + \frac{2.5}{12} \times 10 = 50 + \frac{25}{12} = 50 + 2.1 = 52.1 \text{ approx.}$$

Hence, third quartile is 51.2 approx.

Seventh Decile (D_7) : $N = 82$

$$\Rightarrow \qquad \frac{7N}{10} = \frac{7 \times 82}{10} = \frac{574}{10} = 57.4.$$

The cumulative frequency just greater than $\frac{7N}{10} = 57.4$ is 59. So, 40—50 is the 7th decile class such that

$$l_1 = 40, i = 10, f = 20, Cf_{-1} = 39$$

$$\therefore \qquad D_7 = l_1 + \frac{\frac{7N}{10} - Cf_{-1}}{f} \times Cf_{-1}$$

$$\Rightarrow \qquad D_7 = 40 + \frac{57.4 - 39}{20} \times 10 = 40 + \frac{18.4}{2} = 40 + 9.2 = 49.2.$$

Hence, 7 th decile is 49.2.

Example 80. The mean, median and mode of a group of 25 observations are 143, 144 and 147 respectively. A set of 6 observations is added to this data with values 132, 125, 130, 160, 165 and 157. Find mean, median and mode for the combined group of 31 observations.

Solution. Mean of 25 observations = 143

Sum of 25 observations $= 143 \times 25 = 3575$

Sum of 6 observations which are added $= 132 + 125 + 130 + 160 + 165 + 157$

$= 869$

Sum of 31 observations $= 3575 + 869 = 4444$

Mean of 31 observations $= \frac{4444}{31} = 143.35$

Arranging the 6 observations 132, 125, 130, 160, 165, 157 in ascending order, we get

125, 130, 132, 157, 160, 165

Since three observations are less than 144 and three are more than 144, the median will remain unaffected, *i.e.*, the median will remain 144.

Mode should remain 147.

Example 81. If the quartiles for the following distribution are $Q_1 = 23.125$ and $Q_3 = 43.5$, find missing frequencies and median of the distribution :

Daily Wages	0–10	10–20	20–30	30–40	40–50	50–60
No. of Workers	5	—	20	30	—	10

(ICWA Foundation, June 2001)

Solution. Let f_1 and f_2 be the two missing frequencies, then

Daily Wages	*Number of Workers*	*Cumulative Frequency*
0–10	5	5
10–20	f_1	$5 + f_1$
20–30	20	$25 + f_1$
30–40	30	$55 + f_1$
40–50	f_2	$55 + f_1 + f_2$
50–60	10	$65 + f_1 + f_2$

$$Q_1 = 23.125$$

$$\Rightarrow \quad 23.125 = 20 + \frac{\frac{65 + f_1 + f_2}{4} - (5 + f_1)}{20} \times 10$$

$$\Rightarrow \quad 3.125 = \frac{65 + f_1 + f_2 - 20 - 4f_1}{8} = \frac{45 - 3f_1 + f_2}{8}$$

$$\Rightarrow \quad f_2 - 3f_1 = 3.125 \times 8 - 45$$

$$\Rightarrow \quad 3f_1 - f_2 = 45 - 8 \times 3.125 = 20 \qquad \text{...(i)}$$

$$Q_3 = 43.5$$

$$\Rightarrow \quad 43.5 = 40 + \frac{3\left(\frac{65 + f_1 + f_2}{4}\right) - (55 + f_1)}{f_2} \times 10$$

$$\Rightarrow \quad 0.35f_2 = 195 + 3f_1 + 3f_2 - 220 - 4f_1 = 3f_2 - f_1 - 25$$

$$\Rightarrow \quad 2.65f_2 - f_1 = 25 \quad \Rightarrow \quad 2.65 \times 3f_2 - 3f_1 = 75$$

$$\Rightarrow \quad 7.95f_2 - 3f_1 = 75 \qquad \text{...(ii)}$$

Adding (i) and (ii), we get

$$6.95f_2 = 95 \Rightarrow f_2 = \frac{95}{6.95} = 13.669 = 14 \text{ (Approx.)}$$

$$\therefore \quad 3f_1 - f_2 = 20 \Rightarrow 3f_1 = 34, f_1 = 11 \text{ (Approx.)}$$

3.19 GRAPHICAL METHOD FOR FINDING THE MEDIAN

$$\textbf{Median} = L_1 + \frac{\frac{N}{2} - Cf_{-1}}{f_{med.}} \times i$$

This implies that to find the median we work through the cumulative frequencies. The graph of a cumulative frequency distribution is an Ogive. Ogives are of two types (i) less than type (ii) more than type. Median can be found either by using one Ogive (generally less than type) or by using both the Ogives.

3.20 STEPS TO FIND THE MEDIAN BY USING BOTH THE OGIVES

1. Obtain 'less than' and 'more than' cumulative frequencies for the given grouped data.
2. Draw 'less than' and 'more than' cumulative frequency curves, *i.e.,* 'Ogives'.
3. From the point of intersection of the two Ogives, draw a perpendicular on X-axis.
4. The point where perpendicular touches X-axis is the median value.

3.21 STEPS TO FIND THE MEDIAN BY USING ONE OGIVE

1. Draw less than Ogive [For median, 'more than Ogive' can also be used.]
2. Calculate $\frac{\text{(Total frequency)}}{2}$ and mark on the Y-axis
3. Let the point be P_1. From P_1 draw a line parallel to x-axis and mark the point where it meets the Ogive. Let this point be Q.
4. From Q draw ⊥ on X-axis. Let this ⊥ meet X-axis at M. Then OM is the median value.

3.22 STEPS TO FIND QUARTILES, DECILES AND PERCENTILES GRAPHICALLY

For calculating quartiles, deciles and percentiles graphically we cannot use two Ogives together. Here only 'less than Ogive' should be used. For first and third quartiles we mark the values $\frac{N}{4}$ and $\frac{3N}{4}$ on Y-axis. For rth decile we mark the value $\left(\frac{N}{10}\right)r$ and for rth percentile we mark $\left(\frac{N}{100}\right)r$ on Y-axis. Rest of the steps are the same as explained for drawing median by using 'less than Ogive'.

Example 82. Find graphically the values of median, D_4 and P_{60} from the following data :

Age (in years)	*No. of Persons*
20–25	60
25–30	80
30–35	100
35–40	160

40–45	150
45–50	100
50–55	150
Total	800

Solution. We calculate the cumulative frequencies for **less than Ogive** as follows :

For less than Ogive :

Age (in years)	*No. of Persons*
Less than 25	60
Less than 30	140
Less than 35	240
Less than 40	400
Less than 45	550
Less than 50	650
Less than 55	800

On some chosen scales, we plot the points :

(25, 60), (30, 140), (35, 240), (40, 400), (45, 550), (50, 650), (55, 800) for less than Ogive and join these points by a smooth curve.

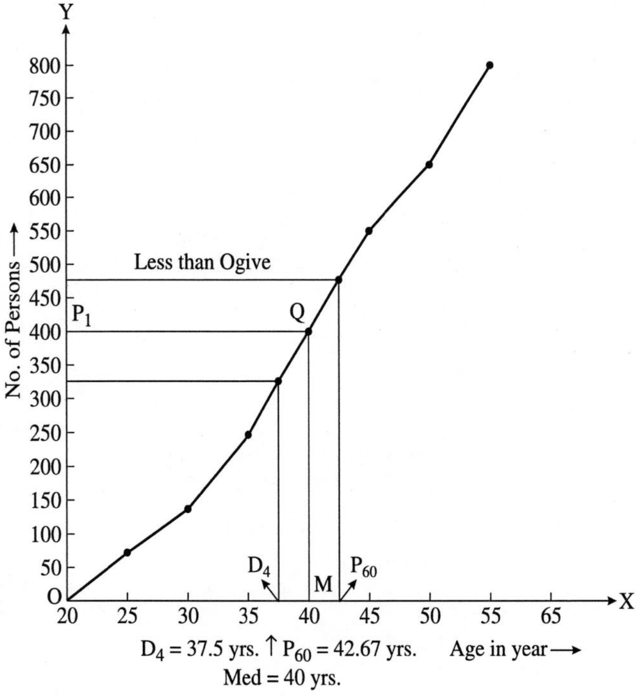

Calculations of Median : $N = 800 \Rightarrow \frac{N}{2} = 400.$

Take 400 on the Y-axis and draw a line parallel to X-axis. Let this line meet the Ogive in M. From M draw perpendicular on the X-axis. The point where this perpendicular meets the X-axis gives the median. It is clear from the above graph that the median is 40 years.

Calculations of Fourth Deciles : $N = 800 \Rightarrow \frac{4N}{10} = \frac{4 \times 800}{10} = 320.$

Take 320 on the Y-axis and repeat the above procedure as for median. It is clear from the above graph that D_4 is 37.5 years.

Calculations of 60th Percentile : $N = 800 \Rightarrow \frac{60N}{100} = \frac{60 \times 800}{100} = 480.$

Take 480 on the Y-axis and repeat the same procedure as for finding the median. It is clear from the above graph that P_{60} is 42.67 years.

Example 83. Draw the 'less than Ogive' and 'more than Ogive' for the following distribution and find the median graphically :

Daily Wages (in Rs.)	*No. of Workers*
50–55	6
55–60	10
60–65	22
65–70	30
70–75	16
75–80	12
80–100	15

Determine the median graphically and verify your result by calculation.

Solution. We calculate the cumulative frequencies for **less than Ogive** and **more than Ogive** as follows :

Table for Less than Ogive

Daily Wages (in Rs.)	*No. of Workers*
Less than 55	6
Less than 60	16
Less than 65	38
Less than 70	68
Less than 75	84
Less than 80	96
Less than 100	111

Table for More than Ogive

Daily Wages (in Rs.)	*No. of Workers*
More than 50	111
More than 55	105
More than 60	95
More than 65	73
More than 70	43
More than 75	27
More than 80	15

On some chosen scales, we plot the points (55, 6), (60, 16), (65, 38), (70, 68), (75, 84), (80, 96), (100, 111) for **less than Ogive** and the points (50, 111), (55, 105), (60, 95), (65, 73), (70, 43), (75, 27), (80, 15) for **more than Ogive** and join these points by smooth curves.

The abscissa of the point of intersection of the two curves is 68. Hence, the median is 68.

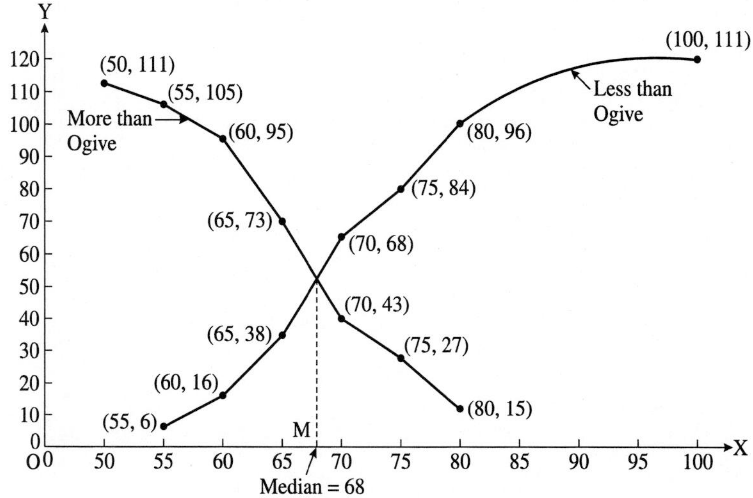

To verify this median by using formula, we first prepare the following cumulative frequency table.

Daily Wages (in Rs.)	*No. of Workers (f)*	*Cumulative Frequency (c.f.)*
50–55	6	6
55–60	10	16
60–65	22	38
65–70	30	68
70–75	16	84
75–80	12	96
80–100	15	111
Total	$N = \Sigma f = 111$	

Here, $$N = 111 \Rightarrow \frac{N}{2} = 55.5.$$

The cumulative frequency just above 55.5 is 68. So, the median class is 65–70 such that

$$l_1 = 65, f_{med} = 30, Cf_{-1} = 38, i = 70 - 65 = 5.$$

$$\text{Median} = l_1 + \frac{\frac{N}{2} - Cf_{-1}}{f} \times i = 65 + \frac{55.5 - 38}{30} \times 5$$

$$= 65 + 2.916 = 67.9 = 68 \text{ (Approx.)}$$

Example 84. Determine the median wage graphically from the following data :

Wages (in Rs.)	*No. of Workers*	*Wages (in Rs.)*	*No. of Workers*
20–40	4	100–120	12
40–60	6	120–140	7
60–80	10	140–160	3
80–100	16		

Solution. Table showing less than and more than cumulative frequencies.

Wages Less than (Rs.)	*No. of Workers*	*Wages More than (Rs.)*	*No. of Workers*
50	4	20	58
60	10	40	54
80	20	60	48
100	36	80	38
120	48	100	22
140	55	120	10
160	58	140	3

Method I. By drawing only one Ogive, say, 'less than type', this is shown by the following graph.

$$\text{Median} = \text{Size of } \frac{58}{2} = 29\text{th item.}$$

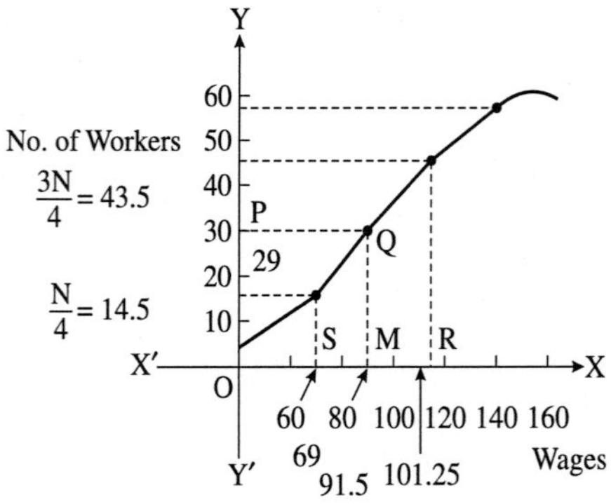

$$\text{Median} = \text{OM} = 91.25$$

$$\text{First Quartile} = \text{OS} = 69$$

$$\text{Third Quartile} = \text{OR} = 101.25$$

Method II. By drawing two cumulative curves – one 'less than' type and another 'more than' type.

From the point where both these curves meet, draw a perpendicular on the *X*-axis and the point where it meets *X*-axis gives the value of the median.

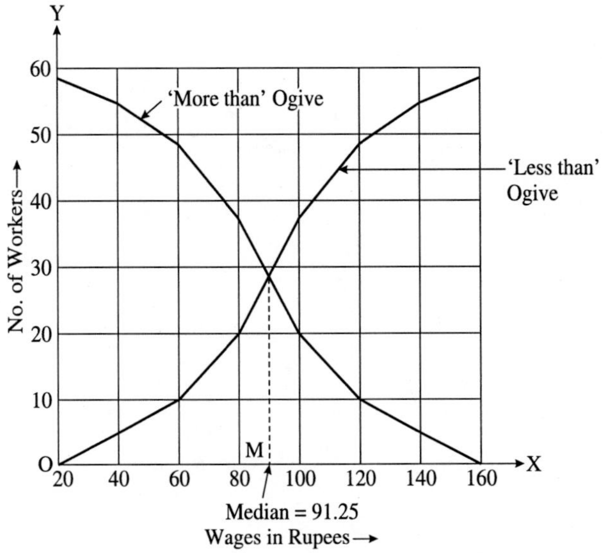

Example 85. Below is given the frequency distribution of marks in Mathematics obtained by 100 students in a class :

Marks	20–29	30–39	40–49	50–59	60–69	70–79	80–89	90–89
No. of Students	8	10	25	31	11	12	2	1

Draw the Ogives (both 'less than' and 'more than' types) from this distribution and use it to determine median (*ie.,* the value of the variable corresponding to cumulative frequency *N*/2).

Solution. Changing the intervals to exclusive type, we get the intervals at 19.5—29.5, 29.5—39.5, etc.

Cumulative Frequency Distribution

Class Boundary	*Cumulative Frequency*	
	'Less than'	*'More than'*
19.5	0	100
29.5	8	92
39.5	18	82
49.5	43	57
59.5	74	26
69.5	85	15
79.5	97	3
89.5	99	1
99.5	100 = *N*	0

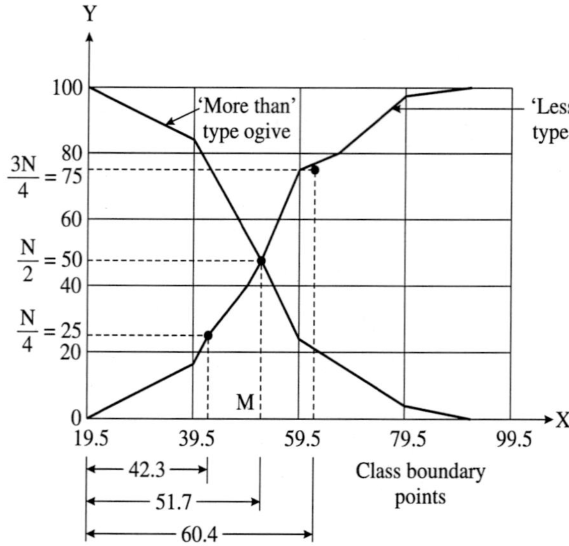

Example 86. The following are the marks obtained by 50 students in statistics :

Marks	*Number of Students*
10 marks or less	4
20 marks or less	10
30 marks or less	30
40 marks or less	40
50 marks or less	47
60 marks or less	50

Draw a curve on the graph paper and show therein :

(i) The range of marks obtained by middle 80% of the students.

(ii) The median.

Also verify your results by direct formula calculations.

Solution. The above data arranged in the form of continuous frequency distribution is as follows :

Marks	*Frequency (f)*	*(Less than) c.f*
0–10	4	4
10–20	6	10
20–30	20	30
30–40	10	40
40–50	7	47
50–60	3	$50 = \Sigma f = N$

Now we plot the less than cumulative frequency against the upper limit of the corresponding class in the above table and join these points by a smooth free hand curve to get Ogive.

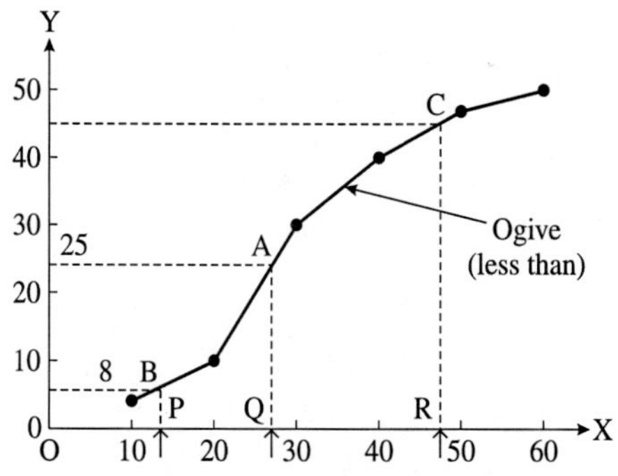

$P_{10} = 11.7$ (Approx.), Median = 27.5, $P_{90} = 47$ (Approx.)

The range of the marks obtained by the middle 80% of the students is given by $P_{90} - P_{10}$. Now P_{90} and P_{10} are obtained graphically as follows :

At the frequencies $\frac{90N}{100} = 45$ and $\frac{10N}{100} = 5$, draw lines parallel to the X-axis meeting the Ogive (less than type) at B and C respectively. Draw BP and CR perpendicular to the X-axis. Then

$$P_{90} = OR = 47 \text{ (Approx.)} \quad \text{and} \quad P_{10} = OP = 11.7 \text{ (Approx.)}$$

Hence required range of marks $= P_{90} - P_{10}$

$$= 47 - 11.7 = 35.3.$$

Further at the frequency $\frac{N}{2} = 25$, draw a line parallel to the X-axis meeting the Ogive at the point A. Draw AQ perpendicular to the X-axis. Then $OQ = 27.5$ is the median marks.

Example 87. Marks in Statistics of 60 students in an examination are given below :

22	47	9	42	31	17	13	15	18	13	2	21
27	38	15	0	33	10	34	29	26	16	25	33
36	10	24	22	26	19	14	16	18	25	21	33
35	25	18	28	25	17	38	10	3	31	24	3
12	16	33	18	26	29	27	28	29	35	26	27

Prepare a frequency table taking class of intervals as 0—7, 8—15, 16—23 and so on. Plot a 'less than Ogive' and from this graph find the median and the quartiles.

Solution.

Marks Intervals	*Frequency*	*Cumulative Frequency* *
0–7	4	4
8–15	10	14
16–23	13	27
24–31	20	57
32–39	11	58
40–47	2	60

* If we do not change to exclusive type of class intervals we can use less than 7, frequency is 4, less than 15, frequency is 12 and so on.

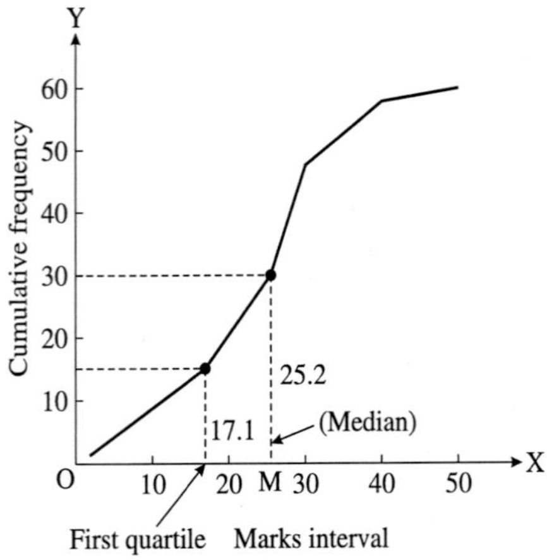

3.23 MODE (INTRODUCTION)

Mode is the most common value, *i.e.,* the value having maximum frequency in the series*. For example,

(i) The retail price paid for a commodity by most of the customers is the modal price.
(ii) In a factory the wage being paid to most of the workers is the modal wage.
(iii) In a grouped frequency distribution the class having the highest frequency is the modal class.

3.24 DEFINITIONS OF MODE

(1) According to Zizek, Mode is *"the value occurring most frequently in a series of items and around which the other items are distributed most densely."*
(2) In words of Croxton and Cowdon, *"The mode of a distribution is the value at the point around which the items tend to be most heavily concentrated. It may be regarded as a most typical of a series of values."*
(3) According to A.M. Tuttle, *"Mode is the value which has the greatest frequency or density in its immediate neighbourhood".*

In case of dispute between the owner of a factory and Trade Union leader, the Trade Union leader will always fight the case on the basis of modal wage whereas the owner will plead his case on the basis of average wage, because due to higher wages of Top Officials, the average wage will always be more than the modal wage.

3.25 MODE FOR UNGROUPED DATA

In case of ungrouped data, the value occurring most frequently is the mode, for example, in the series 2, 4, 5, 5, 5, 8, 9, 10. The mode is 5. Here also it is preferable to arrange the data as discrete frequency distribution. Since, the value 5 is repeated maximum number of times, 5 is the mode.

In the following series, modal value has been circled.

(1) 41, 42, 45, 44, 45, 48, 50, 45, 47, 51, 56

* Mode is also defined as the value in the series most likely to occur.

Arranging the distribution as frequency distribution, we get

Value	*Frequency*
41	1
42	1
(45)	3
44	1
48	1
50	1
47	1
51	1
56	1

(2) Consider the series

27, 28, 26, 35, 40, 37, 38, 28, 27, 25, 30, 28, 40

Arranging as frequency distribution, we get

Value	*Frequency*
25	1
26	1
27	2
(28)	3
30	1
35	1
37	1
38	1
40	2

In the case of a discrete frequency distribution (ungrouped data) mode can be located simply by inspection. Here the value having maximum frequency will represent the mode.

Example 88. Determine mode from the following distribution :

X	10	12	14	16	18	20	22
f	4	6	10	11	21	10	5

Solution. In the above discrete frequency distribution the variate value 18 has the maximum frequency 21. Therefore, 18 is the mode of the given discrete frequency distribution.

3.26 MODE BY GROUPING METHOD

In a uni-modal distribution where the distribution has only one mode and the highest concentration of values is around one value only, we do not face any difficulty in locating the modal value. However, the difficulty is experienced when nearly equal concentration of the frequencies is observed in two or more neighbouring values. In such cases attempt is made to find the value of concentration with the help of **grouping method.** In **grouping method**, the values are first arranged in ascending order and their frequencies are noted. Normally, the grouping table has the following six columns :

Column 1 : Variable values are arranged in ascending order.

Column 2 : Contains the original frequencies marked against each value. The maximum frequency is marked by putting a mark or a circle.

Column 3 : The frequencies are grouped in twos, the totals are indicated against each pair and the highest total is marked.

Column 4 : Leaving the first frequency, the remaining frequencies are grouped in twos and the highest total is marked.

Column 5 : The frequencies are grouped in threes. The highest total is marked.

Column 6 : Leaving the first frequency, the remaining frequencies are grouped in threes and the highest total is marked.

Column 7 : Leaving the first two frequencies, the remaining frequencies are grouped in threes and the highest total is marked.

After completing the grouping table, an **analysis table** is formed for finding the value or the observation which is repeated the highest number of times.

Example 89. The following table gives the measurements of collar sizes of 230 students in a university. Determine the modal size of the collar.

Collar Size (cms)	32	33	34	35	36	37	38	39	40	41
No. of Students	7	14	30	28	35	34	16	14	36	16

Solution. Here, we observe nearly equal concentration of frequencies for the collar sizes 36, 37 and 40. Thus, grouping method is used in locating the modal class. Following the steps listed above in grouping method, we get the following table :

Determining Mode by Grouping Method

Collar Size	*Frequency*					
	Col. 1	*Col.* 2	*Col.* 3	*Col.* 4	*Col.* 5	*Col.* 6
32	7	} 21				
33	14		} 44	} 51		
34	30	} 58			} 72	
35	28		} (63)			} (93)
36	35	} (69)		} (97)		
37	34		} 50		} (85)	
38	16	} 30				} 64
39	14		} 50	} 66		
40	(36)	} 52			} 66	
41	16					

After forming the grouping table, values having highest frequency are counted with the help of an analysis table as shown below :

Analysis Table

Column Number	Collar Size Value contributing to the Highest Frequency									
	32	33	34	35	36	37	38	39	40	41
1										
2					√	√				
3				√	√					
4				√	√	√				
5					√	√	√			
6			√	√	√					
No. of Times	–	–	1	3	5	3	1	–	1	–

From analysis table, we observe that size 36 is repeated the highest number of times. Therefore, modal size is 36 cms.

Example 90. Determine the modal value in the following series.

Value	10	12	14	16	18	20	22	24	26	28	30	32
Frequency	7	15	21	38	34	34	11	19	10	38	5	2

Solution. In this case, equal concentration of values is observed at values 16 and 28. So, we determine modal value by grouping method.

Determining Mode by Grouping Method

Collar Size	*Frequency*					
	Col. 1	*Col.* 2	*Col.* 3	*Col.* 4	*Col.* 5	*Col.* 6
10	7	} 22		} 43		
12	15		} 36		} 74	
14	21	} 59				} **(93)**
16	**(38)**		} **(72)**	} **(106)**		
18	34	} **(68)**			} **(79)**	
20	34		} 45			} 64
22	11	} 30		} 40		
24	19		} 29		} 67	
26	10	} 48				} 53
28	**(38)**		} 43	} 45		
30	5	} 7				
32	2					

Analysis Table

Column Numbers	Values Contributing to the Highest Frequency											
	10	12	14	16	18	20	22	24	26	28	30	32
1				4								
2					4	4						
3				4	4							
4				4	4	4						
5					4	4	4					
6			4	4	4							
Total	–	–	1	4	5	3	1	–	–	1	–	–

Among the highest frequencies in different columns, the value 18 repeats most frequently. Thus, modal value = 18.

3.27 DETERMINATION OF MODE IN A CONTINUOUS SERIES

In a continuous frequency distribution, frequencies are given for various classes and the class having maximum frequency is called the **Modal class**. *In case nearly equal concentration of frequencies is observed in two or more classes,* the **grouping method** *can be used to determine the modal class.* After determining the modal class, the precise value of the mode is obtained by using the following interpolation formula :

$$\textbf{Mode} = l_1 + \frac{f_m - f_{m-1}}{2f_m - f_{m-1} + f_{m+1}} \times i$$

$$= l_1 + \left[\frac{f_m - f_{m1}}{(f_m - f_{m-1}) + (f_m - f_{m+1})}\right] \times i$$

$$= l_1 + \frac{\Delta_1}{\Delta_1 - \Delta_2} \times i,$$

where l_1 = the lower limit of the modal class.

f_m = frequency of the modal class.

f_{m-1} = the frequency of the class preceding the modal class.

f_{m+1} = the frequency of the class succeeding the modal class.

i = size of the modal class, *i.e.,* class interval of the modal class.

Example 91. Find out mode from the following frequency distribution :

*Central Sizes**	1	2	3	4	5	6	7	8	9	10
Frequency	8	6	10	12	20	12	5	3	2	4

Solution. *Please note that in this question central values, i.e., middle values are given. Moreover, the difference between two consecutive central values in the same. We know that if the class intervals are equal then the difference between any two consecutive middle values is the same and is equal to the class interval. Therefore, in this example class interval* = 1.

∴ If l_1 is the lower limit and l_2 the upper limit of the first class, then

$$\frac{l_1 + l_2}{2} = 1 \Rightarrow l_1 + l_2 = 2.$$

Since class interval = 1, we have $l_2 - l_1 = 1$.

Solving $l_1 + l_2 = 2$ and $l_2 - l_1 = 1$, we get

$$l_1 = 0.5 \text{ and } l_2 = 1.5.$$

∴ Class intervals are 0.5 to 1.5, 1.5 to 2.5 and so on.

* Central Size ⇒ Middle Value.

Grouping Table

Class Intervals	*Frequency*					
	(1)	(2)	(3)	(4)	(5)	(6)
0.5–1.5	8	14		24		
1.5–2.5	6		16		28	
2.5–3.5	10	22				42
3.5–4.5	12		32	44		
4.5–5.5*	20	32			37	
5.5–6.5	12		17			20
6.5–7.5	5	8		10		
7.5–8.5	3		5		9	
8.5–9.5	2	6				
9.5–10.5	4					

Since, the class 4.5-5.5 has the highest frequency and the concentration of frequencies in no other lass is nearly equal, the class 4.5-5.5 is the modal class*.

$$\text{Mode} = 4.5 + \frac{20 - 12}{2 \times 20 - 12 - 12} = 5.$$

Example 92. Calculate median and mode of the following distribution :

Wages	10–15	15–20	20–25	25–30
No. of Workers	22	45	67	73
Wages	30–35	35–40	40–45	45–50
No. of Workers	85	190	64	55

Solution.

X	*f*	*c.f.*
10–15	22	22
15–20	45	67
20–25	67	134
25–30	73	207
30–35	85	292
35–40	190	482
40–45	64	546
45–50	55	601

$$\text{Median} = l_1 + \frac{\frac{N}{2} - Cf_{-1}}{f_{med}} \times i = 35 + \frac{305.5 - 292}{190} \times 5 = 35 + 0.35 = 35.35.$$

* Students, if desired, may confirm by analysis table that class 4.5-5.5 is the modal class.

$$\text{Mode} = l_1 + \frac{f_m - f_{m-1}}{2f_m - f_{m-1} - f_{m+1}} \times i = 35 + \frac{190 - 85}{380 - 85 - 64} \times 5$$

$$= 35 + \frac{150}{231} \times 5 = 35 + 2.27 = 37.27.$$

Example 93. compute the mode from the following data :

Class	0–3	3–6	6–10	10–12	12–15	15–18
Frequency	4	8	10	14	16	20
Class	18–20	20–24	24–25	25–28	28–30	30–36
Frequency	24	14	16	11	10	6

Solution. In this question, the class intervals are unequal. For finding the mode, the class intervals have to be equal. So we will first make the class intervals equal by making an interval of 6, *i.e.,* we will make class intervals as

0-6, 6-12, 12-18, — —

[If the class intervals are not equal, they are made equal by combining two or more classes.]

Grouping Table

<table>
<tr><th rowspan="2">Class</th><th colspan="6">Frequency</th></tr>
<tr><th>(1)</th><th>(2)</th><th>(3)</th><th>(4)</th><th>(5)</th><th>(6)</th></tr>
<tr><td>0–6</td><td>$4 + 8 = 12$</td><td rowspan="2">36</td><td></td><td rowspan="3">72</td><td></td><td></td></tr>
<tr><td>6–12</td><td>$10 + 14 = 24$</td><td rowspan="2">60</td><td rowspan="3">98</td><td></td></tr>
<tr><td>12–18</td><td>$16 + 20 = 36$</td><td rowspan="2">74</td><td rowspan="3">111</td></tr>
<tr><td>18–24</td><td>$24 + 14 = 38$</td><td rowspan="2">75</td><td rowspan="3">81</td></tr>
<tr><td>24–30</td><td>$16 + 11 + 10 = 37$</td><td rowspan="2">43</td><td></td></tr>
<tr><td>30–36</td><td>$= 6$</td><td></td><td></td><td></td></tr>
</table>

Analysis Table

Column No.	6–12	12–18	18–24	24–30	30–36
1			4		
2		4	4		
3			4	4	
4			4	4	4
5	4	4	4		
6		4	4	4	
Total	1	3	6	3	1

The mode lies in the class 18-24, whose frequency is 38.

Applying the formula, we get

$$\text{Mode} = l_1 + \frac{f_m - f_{m-1}}{2f_m - f_{m-1} - f_{m+1}} \times i,$$

where $l_1 = 18, f_{m-1} = 36,\ f_m = 38,\ f_{m+1} = 37,\ i = 6.$

$$\therefore\ \text{Mode} = 18 + \frac{38 - 36}{2 \times 38 - 36 - 37} \times 6 = 18 + \frac{2}{3} \times 6 = 22.$$

$\therefore$ Mode $= 22.$

3.28 GRAPHICAL METHOD FOR FINDING THE MODE

Mode is calculated graphically with the help of the curve called **Histogram.**

A histogram consists of a series of continuous vertical rectangles. The class boundaries are plotted along the horizontal axis *OX* and the class frequencies along the vertical axis *OY*. For finding the mode, we consider the distributions with equal **class intervals** throughout. So for plotting a mode, all the rectangles in the histogram will be of equal width and height equal to the frequencies of the respective classes. To locate the mode in the histogram.

(1) Join the top right angular point of the highest rectangle to the top right angular point of the preceding rectangle by a straight line.

(2) Join the top left angular point of the highest rectangle to the top left angular point of the succeeding rectangle by a straight line.

(3) From the point of intersections of these two lines, draw a perpendicular on the horizontal axis *OX*. Let it meet the horizontal axis in *M*.

(4) *OM* is equal to the mode.

Example 94. The monthly profits in ruppes of 100 shops are distributed as follows :

Profit Per Shop	*No. of Shops*	*Profit Per Shop*	*No. of Shops*
0–100	12	300–400	20
100–200	18	400–500	17
200–300	27	500–600	6

Draw a histogram of the data and hence find modal value. Check this value by direct calculation.

Solution. Students may check the value by direct calculation by the formula

$$\text{Mode} = l_1 + \frac{f_m - f_{m-1}}{2f_m - f_{m-1} - f_{m+1}} \times i = 200 + \frac{27-18}{54-18-20} \times 100$$

$$= 200 + \frac{9}{16} \times 100 = 200 + \frac{900}{16} = 200 \times 56.25 = 256.25$$

However graph is as under :

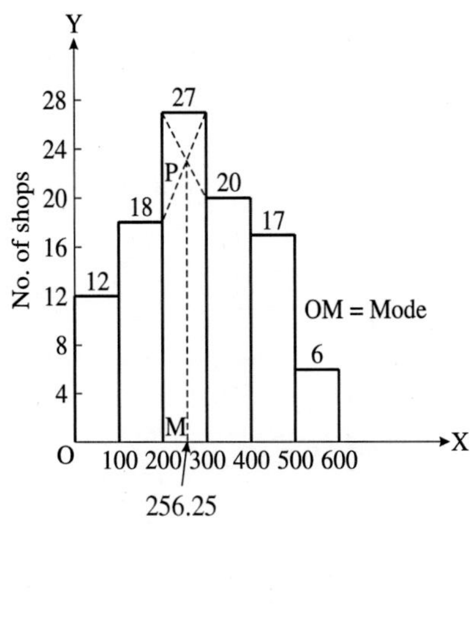

Example 95. Construct a histogram for the following distribution determining the mode graphically :

1–9	9–17	17–25	25–33	33–41	41–49	40–57	57–65
20	30	26	16	10	8	4	4

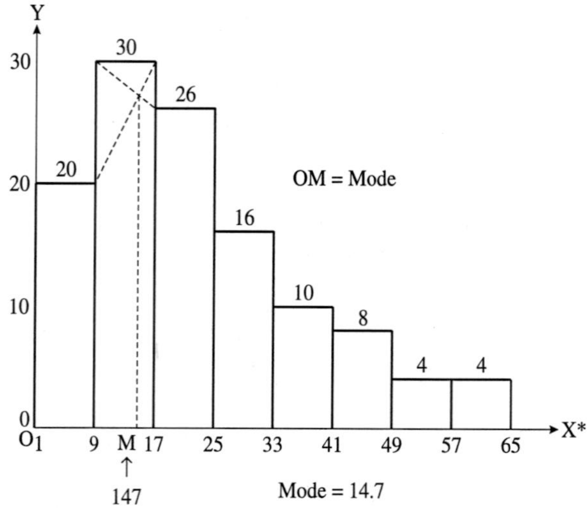

3.29 CALCULATION OF MODE IN SERIES WITH UNEQUAL CLASS INTERVALS

Example 96. Calculate Mode of the following data :

Marks	0–5	5–10	10–20	20–40	40–60	60–80	80–90	90–100
No. of Students	5	7	9	25	30	24	8	6

Solution. This is a series with unequal class intervals. Calculation of Mode requires the class intervals should be equal. There are three classes having class interval as 20. The first three classes can be merged to have a class interval equal to 20. Last two clases can also be merged to have a class interval equal to 20. So we will arrange the whole series with class interval = 20.

Marks	*No. of Students (Frequency)*
0–20	$5+7+9=21$
20–40	25 f_{m-1}
40–60	30 f_m
60–80	24 f_{m+1}
80–100	$8+6=14$

Class 40-60 having maximum frequency is the modal class.

$$\text{Mode} = l_1 + \frac{f_m - f_{m-1}}{2f_m - f_{m-1} - f_{m+1}}$$

$$= 40 + \frac{30-25}{2\times 30 - 25 - 24} \times 20$$

* We can also convert the intervals to exclusive type and draw Histograms.

$$= 40 + \frac{5}{11} \times 20 = 40 + \frac{100}{11}$$

$$= 30 + 9.09 = 49.09.$$

Thus, Mode, $(Z) = 49.09.$

Example 97. Calculate mode of the following series.

Class Interval	10–19	20–29	30–39	40–49	50–59	60–69	70–79
Frequency	10	12	18	30	16	6	8

Solution. For the estimation of mode, inclusive Series are converted into Exclusive Series.

Class Interval	*Exclusive Class Intervals*	*Frequency*
10–19	9.5–19.5	10
20–29	19.5–29.5	12
30–39	29.5–39.5	18 f_{m-1}
40–49	39.5–49.5	30 f_m
50–59	49.5–59.5	16 f_{m+1}
60–69	59.5–69.5	6
70–79	69.5–79.5	8

Class 39.5-49.5 having the maximum frequency is the modal class.

$$\therefore \quad \text{Mode} = Z = l_1 + \frac{f_m - f_{m-1}}{2f_m - f_{m-1} - f_{m+1}} \times i$$

$$= 39.5 + \frac{30 - 18}{2(30) - 18 - 16} \times 10$$

$$= 39.5 + \frac{12}{60 - 34} \times 10 = 39.5 + \frac{12}{26} \times 10$$

$$39.5 + 4.62 = 44.12.$$

Thus, Mode, $(Z) = 44.12.$

Example 98. Calculate mode, given the following data-set.

Mid Value	5	15	25	35	45	55	65	75	85
Frequency	4	5	8	12	16	28	15	3	2

Solution. A series with 'Mid-values' is first to be expanded as a series with class intervals as under :

Mid-Value	*Class Interval**	*Frequency*
5	0–10	4
15	10–20	5
25	20–30	8
35	30–40	12
45	40–50	16 f_{m-1}
55	50–60	28 f_m

* Class width = Difference between any two consecutive mid-values.

65	60–70	15 f_{m+1}
75	70–80	3
85	80–90	2

Class 50-60 is the one with highest frequency. This is the Modal class.

$$\text{Mode} = Z = l_1 + \frac{f_m - f_{m-1}}{2f_m - f_{m-1} - f_{m+1}}$$

$$= 50 + \frac{28 - 16}{2 \times 28 - 16 - 15} \times 10$$

$$= 50 + \frac{12}{56 - 31} \times 10$$

$$= 50 + \frac{12}{25} \times 10$$

$$= 50 + 4.8 = 54.8.$$

Thus, Mode, $(Z) = 54.8.$

Example 99. Find out Mode of the following series :

Wages (Rs.) Less than	200	300	400	500	600	700	800	900
No. of Workers	5	18	38	70	90	95	98	100

Solution. For the calculation of Mode, 'Less than' cumulative frequency distribution is, first to be converted into a normal continuous frequency distribution, as under :

Wages (Rs.)	*No. of Workers (f)*
100–200	5
200–300	$18 - 5 = 13$
300–400	$38 - 18 = 20$ f_{m-1}
$l_1 \rightarrow$ 400–500	$70 - 38 = 32$ f_m
500–600	$90 - 70 = 20$ f_{m+1}
600–700	$95 - 90 = 5$
700–800	$98 - 95 = 3$
800–900	$100 - 98 = 2$

Since the class 400-500 has the highest frequency, this class is the Modal Class. The value of Mode is,

$$\text{Mode} = Z = l_1 + \frac{f_m - f_{m-1}}{2f_m - f_{m-1} - f_{m+1}} \times i$$

$$= 400 + \frac{32 - 20}{2(32) - 20 - 20} \times 100$$

$$= 400 + \frac{12}{64 - 40} \times 100$$

$$= 400 + \frac{12}{24} \times 100 = 450.$$

Example 100. You are given the following incomplete frequency distribution. It is known that the total frequency is 100 and that the median is 44. Estimate by calculation the missing frequencies and find the value of mode.

Value	10–20	20–30	30–40	40–50	50–60	60–70	70–80
Frequency	5	12	–	20	–	10	4

Solution. Let the missing frequencies be f_1 and f_2. The cumulative frequency table is as under :

Cumulative Frequency Table

Value	*Frequency*	*Cumulative Frequency*	Total Frequency
10–20	5	5	$5 + 12 + 20 + 10 + 4 + f_1 + f_2 = 100$
20–30	12	17	$\Rightarrow f_1 + f_2 = 49$
30–40	f_1	$17 + f_1$	
40–50	20	$37 + f_1$	Median Class is 40–50
50–60	f_2	$37 + f_1 + f_2$	
60–70	10	$47 + f_1 + f_2$	
70–80	4	$51 + f_1 + f_2$	
Total Frequency = 100 (given)			

∴ The modal class is 30-40, *i.e.,* the class having the maximum frequency.

$$\text{Mode} = 30 + \frac{25 - 12}{50 - 12 - 20} \times 10$$

$$= 30 + \frac{13}{18} \times 10 = 37.2$$

Example 101. State giving reason, the average to be used in the following situations :

(i) To determine the average size of the shoe sold in a shop.
(ii) To determine the size of agricultural holdings.
(iii) To determine the average wages in an industrial concern.
(iv) To find the per capita income in different cities.
(v) To find the average beauty among a group of girls in a class.
(vi) Average size of readymade garments.
(vii) Average height of a male in a nation.
(viii) Average expenditure of a student in a college.

Solution. (i) Mode (ii) Median (iii) Median (sometimes mode is also used) (iv) Mean (v) Median (vi) Mode (vii) Mode (viii) Mode.

3.30 EMPIRICAL RELATION AMONG MEAN, MEDIAN AND MODE

For a symmetrical distribution

$$\text{Mean} = \text{Median} = \text{Mode}$$

For a moderately asymmetrical (Skewed) distribution.

$$3\ (\text{Mean} - \text{Median}) = \text{Mean} - \text{Mode}$$

$$\Rightarrow \quad \text{Mode} = 3\ \text{Median} - 2\ \text{Mean}$$

Note : If in a distribution one value occurs more frequently than any other value, the distribution is called *unimodal.* If in a distribution two different values have equal and maximum frequencies associated with them, the distribution is known as *bimodal.* When all the values of the observations are non-repetitive or unique, no mode exists.

3.31 SELECTION OF AN AVERAGE

No single average is sufficient to serve each and every purpose. To find a suitable average, the situation has to be analysed thoroughly and carefully. Mean takes into account all the values in the distribution. Hence in cases where we wish to make total value estimates, mean is the most appropriate average, *e.g.,* per capita income of a country.

In case of open ended class intervals mean cannot be found. Hence in such a case, we have to depend on median or mode.

If the data is qualitative, *e.g.,* beauty, honesty, efficiency, intelligence, etc., then only median will be most appropriate average. Regarding agricultural land holdings, psyche and social problems median is used.

In case of business situations, where selection of 'most common', *e.g.,* most common size of shoes for certain age group, most common size of garments, etc., is to be made, then mode is the only average.

Merits, Demerits and Uses of Mean, Median and Mode

3.32 ARITHMETIC MEAN

Merits

1. It is the most commonly and widely used measure of Central tendency.
2. It is rigidly defined. This signifies that whatever method of computation may be adopted, the result will remain the same.
3. It is based on all the observations.
4. It can be subjected to algebraic treatment, *i.e.,* we can find combined mean of two or more groups. This property is extensively used in statistical analysis.
5. It is least affected by fluctuations of sampling.
6. It is very simple to understand and easy to calculate.
7. It can be used for comparison purpose.
8. Algebraic sum of the deviations of all the values from their A.M. is equal to zero. This forms the basis for short-cut method for calculating the mean. **(V. Imp.)**

Demerits

1. It is highly affected by the extreme values. For example in an industry the workers get less pay as compared with top officials. But if we calculate the average wage paid to all employees, it will not represent the workers because the high salary of top officials will increase the average wage much more than what a normal worker gets.
2. The calculated average value may not coincide with any of the given values, for example, the mean of 5, 9, 10 is 8 which is not a value in the series.
3. It cannot be calculated graphically or by inspection.
4. It gives equal weight to high and small observations which distort the average.

5. Arithmetic mean cannot be calculated if even a single observation in the series is missing. Also, in distribution with open-end classes the mean cannot be determined without making certain assumptions regarding the class-end values.
6. For calculating mean in continuous series, two assumptions are made :
   (i) The frequencies are evenly distributed.
   (ii) Middle value of each class is the representative of the class.
   This may not be always true.
7. Sometimes it gives unrealistic conclusions. For example, if there are three classes having 9, 11, 15 students respectively, then the average number of students in each class
$$= \frac{9 + 11 + 15}{3} = \frac{35}{3} = 11.67 \text{ which is absurd.}$$

Uses : Arithmetic mean is very common average and is widely used. It is generally used in all the subjects of studies like Social and Economic studies. It is frequently used in Business and Commerce. for example, 'Average cost of production', 'Average price' and 'Average yield per acre', etc. It is not only used to compare two or more series but also to calculate other statistical measures like mean deviation, standard deviation and coefficient of variation, etc.

Median

Merits

1. Median is the only average which can be used if the data is qualitative. For example, we can find the median of beauty, honesty, intelligence, etc.
2. It is not affected by extreme values. Due to this property it is widely used as a measure of Central tendency in **asymmetrical distribution**.
3. If the class intervals are **open-ended** (*i.e.,* undetermined), median can be calculated unless it falls in the open-ended class interval.
4. Sum of the **absolute deviations** of all the values from the median is **minimum**.
5. It can be located graphically.
6. Even if the data is not complete, *i.e.,* some values are missing, the median can be calculated if the number of items and the middle item is known.
7. In certain cases it can be computed by inspection.
8. It is rigidly defined, *i.e.,* its value remains the same whatever method of computation be applied.
9. If the class intervals are unequal (or changed), the value of the median is not affected. **(Imp.)**

Demerits

1. It cannot be subjected to algebric treatment, *i.e.,* the combined median of two or more groups cannot be calculated.
2. It is affected by Sampling fluctuations more than the mean.
3. If the number of observations are even, it is the A.M. of the two middle values. So in such cases it cannot be precisely expressed.
4. It fails to represent a satisfactory average when there is a great variation among the items of the series.
5. It gives equal weight to all observations but if extreme observations are given more weight, then median as an average become inappropriate.
6. Median $\times$ number of observations $\neq$ sum of the observations.

Uses : It proves very useful when extreme values of the series are either not available or are abnormal.

Mode

Merits

1. Mode is the value which occurs most frequently in the series, *i.e.*, the value having maximum frequency. Hence it is one of the values in the series and not an isolated value.
2. It can be found graphically.
3. It is not affected by extreme value.
4. It is of great use to the manufacturers of garments, shoes, etc., because most common size, *i.e.*, the model size helps them to decide the quantity to be manufactured for each size.
5. It can be found by inspection.
6. It is the most used average in certain situations, *e.g.*, average marks of a class, average number of students in a section and average size of shoes, etc.
7. Trade union leaders use mode to find the average wage.

Demerits

1. It is not rigidly defined. If we calculate mode by different methods, results may be different.
2. Mode can be called the representative average only if the number of values is too large.
3. Series may be bimodal or multimodal, *i.e.*, the mode is not unique.
4. It is most affected by fluctuations of sampling.
5. It is not capable of further algebraic treatment. We cannot find the combined mode.
6. Mode is based only on concentrated values.
7. When multiplied by the number of observations it does not give the sum of the values as mean gives.
8. It is not at all suitable for the cases where weight are given to extreme values.
9. Its computation is not based on all the observations.
10. *Lot of care has to be observed before using mode as a measure of Central tendency because some distributions do not have mode and some other distributions have more than one mode.*

Table Comparing the Various Averages, Arithmetic Mean, Median, Mode, Geometric Mean and Harmonic Mean

	A.M. ($\bar{x}$)	*Median*	*Mode*	*Geometric Mean*	*Harmonic Mean*
Calculation takes into account	All values	Middle value only	Value of highest frequency density or the most frequently occuring value	All values	All values
Exreme values have	Greatest effect	No effect	No effect	Less effect than A.M. but more than harmonic mean	Very small effect
Applied to the same data	Greater than G.M. and H.M.	Between A.M. and mode	May be larger or smaller than A.M. or median	Larger than H.M. but smaller than A.M.	Smaller than G.M. and A.M.
Computation	Computation is possible $\bar{x} = \frac{\Sigma x}{n}$. Also, $\bar{x} = \frac{\Sigma fx}{\Sigma f}$	Not possible in some cases. It is positional average. Interpolated value is taken in many cases.	Not possible in many cases	Possible G.M. $= \sqrt[n]{x_1 . x_2 . \ldots . x_n}$	Possible H.M. $= \frac{n}{\Sigma \frac{1}{x}}$

Open-end class makes :	Indeterminate	Can be calculated	Can be calculated	Indeterminate	Indeterminate
One value is zero	Can be calculated	Can be calculated	We may or may not be able to calculate	Can not be calculated	Can not be calculated
Usage	Most commonly used among all the measures of central tendency. When there is no reason to use any other average it is used.	When representative value is required and distribution is asymmetric	When most frequently occurring value is required (discrete variables)	When rate of growth or decline is required.	Rates of the form distance per unit time and distance is constant.

MISCELLANEOUS EXAMPLES

Example 1. Calculate the median Q_1, Q_3, D_7 and P_{60} from the following data :

Marks	*No. of Students*
Less than 10	4
Less than 20	16
Less than 30	40
Less than 40	76
Less than 50	96
Less than 60	112
Less than 70	120
Less than 80	125

Solution. We calculate the frequency of each class in the following table :

Marks	*Frequency* *(f)*	*Cumulative Frequency* *(c.f.)*
0–10	4	4
10–20	$16-4=12$	16
20–30	$40-16=24$	40
40–50	$76-40=36$	76
50–60	$96-76=20$	96
60–70	$112-96=16$	112
70–80	$120-112=8$	120
70–80	$125-120=5$	125
Total	$N=\Sigma f=125$	

Here, $$\frac{N}{2}=\frac{125}{2}=62.5.$$

The cumulative frequency is just greater than 62.5 is 76. So, 30–40 is the median class such that

$$l_1 = 30, f_{med} = 36, i = 10, C = 40.$$

$$\therefore \quad \text{Median} = l_1 + \frac{\frac{N}{2} - Cf_{-1}}{f_{med}} \times i$$

$$= 30 + \frac{62.5 - 40}{36} \times 10 = 30 + \frac{22.5}{36} \times 10$$

$$= 30 + \frac{225}{36} = 30 + 6.25 = 36.25.$$

Hence, median is 36.25 marks.

First Quartile : Hence, $N = 125 \Rightarrow \frac{N}{4} = \frac{125}{4} = 31.25.$

The cumulative frequency just greater than 31.25 is 40. So 20 – 30 is the Q_1 class such that

$$l_1 = 20, f_{Q_1} = 24, C = 16, i = 10.$$

$$\therefore \quad Q_1 = l_1 + \frac{\frac{N}{4} - Cf_{-1}}{f_{Q_1}} \times i = 20 + \frac{31.25 - 16}{24} \times 10$$

$$= 20 + \frac{15.25}{24} \times 10 = 20 + \frac{152.5}{24}.$$

$$= 20 + 6.35 = 25.35$$

Hence, $Q_1 = 26.35$ marks.

Third Quartile : Here, $N = 125 \Rightarrow \frac{3N}{4} = \frac{3 \times 125}{4} = 93.75.$

The cumulative frequency just greater than 93.75 is 96. So, 40–50 is the Q_3 class such that

$$l_1 = 40, f_{Q_3} = 20, Cf_{-1} = 76, i = 10$$

$$\therefore \quad Q_3 = l_1 + \frac{\frac{3N}{4} - Cf_{-1}}{f_{Q_3}} \times i$$

$$= 40 + \frac{93.75 - 76}{20} \times 10 = 40 + \frac{17.75}{2}$$

$$= 40 + 8.875 = 48.88$$

Hence, $Q_3 = 48.88$ marks.

7th Decile : Here, $N = 125 \Rightarrow \frac{7N}{10} = \frac{7 \times 125}{10} = \frac{175}{2} = 87.5$

The cumulative frequency just greater than 87.5 is 96. So, 40 – 50 is the D_7 class such that

$$l_1 = 40, f_{D_1} = 20, Cf_{-1} = 76, h = 10$$

$$\therefore \quad D_7 = l_1 + \frac{\frac{7N}{10} - Cf_{-1}}{f_{D_1}} \times i$$

$$= 40 + \frac{87.5 - 76}{20} \times 10 = 40 + \frac{11.5}{2} = 45.75$$

Hence, $D_7 = 45.75$ marks.

60th Percentile : Here, $N = 125 \Rightarrow \frac{60N}{100} = \frac{60 \times 125}{100} = 75.$

The cumulative frequency just greater than 75 is 76. So, 30-40 is the P_{60} class such that

$$l_1 = 30, f_{P_{60}} = 36, Cf_{-1} - 40, i = 10$$

$$\therefore \qquad P_{60} = l_1 + \frac{\frac{60N}{100} - Cf_{-1}}{f_{P_{60}}} \times i$$

$$= 30 + \frac{75 - 40}{36} \times 10$$

$$= 30 + \frac{35}{36} \times 10 = 30 + 9.72 = 39.72.$$

Hence, $\qquad P_{60} = 39.72$ marks.

Example 2. A man invested his savings as follows :

Rs. 10,000 in Post Office Savings Bank at 8% p.a.; Rs. 6,000 in National Bank at 7% p.a.; Rs. 3,000 in a Private Firm at 10% p.a.

Find the average rate of interest per cent p.a. **(ICWA Foundation, Dec., 2002)**

Solution. Interest from Post Office S/B $= \frac{10,000 \times 8}{100} =$ Rs. 800

Interest from National Bank $= \frac{6,000 \times 7}{100} =$ Rs. 420

Interest from Private Firm $= \frac{4,000 \times 100}{100} =$ Rs. 400

Total Interest on Rs. $(10,000 + 6,000 + 4,000 = 20,000) =$ Rs. $(800 + 420 = 400) =$ Rs. 1,620

Hence, average percentage rate of interest $= \frac{\text{Total Interest}}{\text{Total amount invested}}$

$$= \frac{1,620}{20,000} \times 100 = 8.10\%$$

Example 3. Find mode and median from the following data :

Class Interval	*Frequency*
0–8	08
8–16	07
16–24	16
24–32	24
32–40	15
40–48	07

Solution. **Assumed Mean (AM) = 20**

Class Interval	*Frequency (f)*	*Mid-Points (m)*	*(d) Deviation (m – AM)*	$d' = \frac{d}{8}$	$f \times d' = fd'$	*Cumulative Frequency c.f.*
0–8	8	4	–16	–2	–16	8
8–16	7	12	–8	–1	–7	15
16–24	16	20	0	0	0	31
24–32	24	28	+8	+1	+24	35
32–40	15	36	+16	+2	+30	70
40–48	7	44	+24	+3	+21	77
	$\Sigma f = 77$				$\Sigma fd' = 52$	

Mode = Since maximum frequency is 24, Mode will lie in the class interval (24–32)

$$\text{Mode} = 24 + \frac{(24 - 16)}{2 \times 24 - 16 - 15} \times 8$$

$$= 24 + \frac{8 \times 8}{48 - 31}$$

$$= 24 + \frac{64}{17} = 24 + 3.76 = 27.76$$

$$\text{Median} = 24 + \frac{38.5 - 31}{4} \times 8 \qquad \left[\because \frac{N}{2} = \frac{77}{2} = 38.5\right]$$

$$= 24 + \frac{7.5}{3} = 24 + 2.5 = 26.5.$$

Hence, Mode = 27.76; Median = 26.5

Example 4. Determine the missing frequency of the class interval 15 – 20 from the following data. The mean of the series is 19.

Class Interval	*Frequency*
5–10	02
10–15	02
15–20	?
20–25	04
25–30	04

Solution.

Suppose the missing frequency is f_1.

Class	*Mid-Value (x)*	*Frequency (f)*	*(x) × (f) = (fx)*
5–10	7.5	2	15
10–15	12.5	2	25
15–20	17.5	f_1	$17.5 f_1$
20–25	22.5	4	90
25–30	27.5	4	110
		Σf or $N = 12 + f_1$	$\Sigma fx = 240 + 17.5 f_1$

$$\overline{X} = \frac{\Sigma fx}{\Sigma f}$$

or $$19 = \frac{240 \times 17.5f_1}{12 + f_1}$$

or $$227 + 19f_1 = 240 + 17.5f_1$$

or $$19f_1 - 17.5f_1 = 240 - 228$$

or $$1.5f_1 = 12$$

or $$f_1 = 8.$$

Thus, the missing frequency of class 15-20 is 8.

Example 5. Find mean, mode and median from the following data :

Wages	*No. of Workers*
Above 0	650
Above 10	500
Above 20	425
Above 30	375
Above 40	300
Above 50	275
Above 60	250
Above 70	100

Solution. **Calculation of Mean and Median**

Wages	*No. of Workers* (f)	*Mid-Value* (m)	$\frac{m-35}{10} = d'$	fd'	*c.f.*
0–10	650 – 500 = 150	5	–3	–450	150
10–20	500 – 425 = 75	15	–2	–150	225
20–30	425 – 375 = 50	25	–1	–50	275
30–40	375 – 300 = 75	35	–0	0	350
40–50	300 – 275 = 25	45	+1	+25	375
50–60	275 – 250 = 25	55	+2	+50	400
60–70	250 – 100 = 150	65	+3	+450	550
70–80	100	75	+4	+400	650
	$\Sigma f = N = 650$			$\Sigma fd' = 275$	

Mean, $$\overline{x} = A + \frac{\Sigma fd'}{\Sigma f} \times I$$

where $A = 35$

$\Sigma fd' = 275$

$\Sigma f = N = 650$

$i = 10.$

$$\overline{x} = 35 + \frac{275 \times 10}{650} = 35 + 4.23 = 39.23.$$

$$\text{Median} = \text{Size of } \frac{N}{2}\text{th item} = \frac{650}{2} = 325\text{th item}$$

Median size lies in class interval 30-40

$$\text{Median} = L_1 + \frac{\frac{N}{2} - Cf_{-1}}{f_{med}} \times i,$$

where $i = 10$, $L_1 = 30$, $\frac{N}{2} = 325$, $Cf_{-1} = 275$, $f = 75$

$$\text{Median} = 30 + \frac{325 - 275}{75} \times 10$$

$$= 30 + 6.67 = 36.67$$

Mode : Since it seems to be a bimodal series, mode can be calculated by using the Empirical formula as under :

$$\text{Mode} = 3\ (\text{median}) - 2\ \text{mean} = 3\ (36.67) - 2(39.23)$$

$$= 110.01 - 78.46 = 31.55.$$

Example 6. Draw a 'less than' and 'more than' Ogive curves from the following data :

Weekly Wages (in Rs.)	*No. of Workers*
100–200	7
200–300	13
300–400	20
400–600	30
600–800	20
800–1100	18

Solution.

'More than' and 'Less than' Cumulative Frequency Table

Weekly Wages (Rs.)	*No. of Workers (f)*	*More than Wages (Rs.)*	*No. of Workers (c.f.)*	*Less than Wages (Rs.)*	*No. of Workers (c.f.)*
100–200	7	100	108	200	7
200–300	13	200	101	300	20
300–400	20	300	88	400	40
400–600	30	400	68	600	70
600–800	20	600	38	800	90
800–1100	18	800	18	1100	108

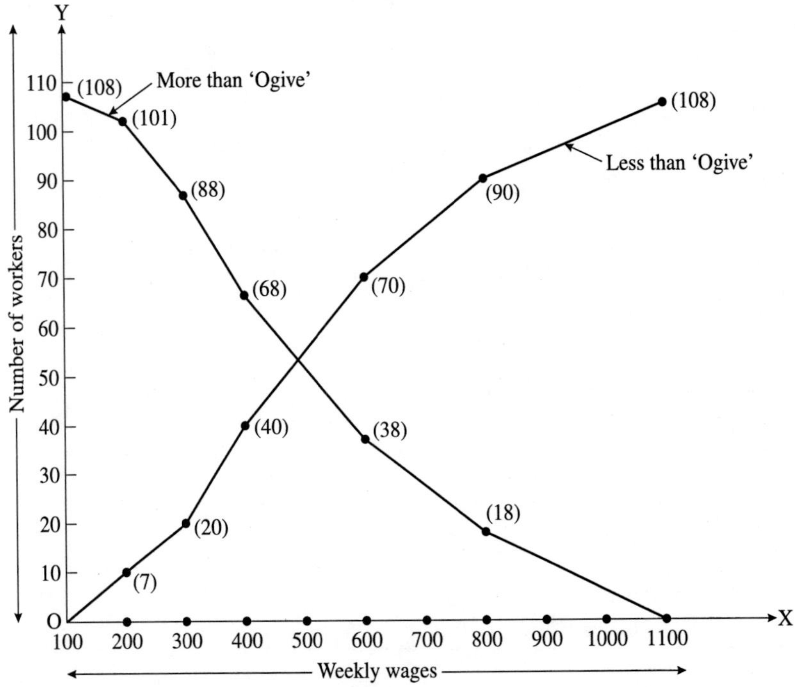

Example 7. The average monthly salary paid to the employees of an establishment is Rs. 9,600. The average salary paid to male and female employees are Rs. 10,000 and Rs. 8,000 respectively.

What is the percentage of male employees in the establishment ?

Solution. Let the total number of employees be 100. Let the male employees be n, then female employees will be $(100 - n)$.

As per the question

$$9{,}600 \times 100 = n \times 10{,}000 + (100 - n)\, 8{,}000$$

$$\Rightarrow \quad 1{,}600 \times 100 = 2{,}000\, n \quad \Rightarrow \quad n = 80$$

$\therefore$ Percentage of male employees = 80

Percentge of female employees = 20

Example 8. Median marks of 50 candidates in mathematics test are 26. Frequencies in the ranges 10-20 and 30-40 are missing in the following tables :

Marks Obtained	0–10	10–20	20–30	30–40	40–50
No. of Candidates	5	–	20	–	7

Determine the missing frequencies.

Solution. Let the missing frequencies be f_1 and f_2.

$$\therefore \quad 5 + f_1 + 20 + f_2 + 7 = 50$$

$$\Rightarrow \quad f_1 + f_2 = 18.$$

Class (Marks)	*No. of Candidates*	*Cumulative Frequencies*
0–10	5	5
10–20	f_1	$5+f_1$
20–30	20	$25+f_1 \rightarrow$ median class
30–40	f_2	$25+f_1+f_2$
40–50	7	$32+f_1+f_2$

$$\text{Median} = 26 = 20 + \frac{25-(5+f_1)}{20} \times 10$$

$$\Rightarrow \quad 6 \times 20 = 200 - 10f_1 \ \Rightarrow \ f_1 = 8$$

$$f_1 + f_2 = 18 \ \Rightarrow \ f_2 = 18 - 8 = 10.$$

∴ The missing frequencies are 8 and 10.

Example 9. An average life of a refrigerator of a certain make is found to be 6.78 years. The following table gives the number of refrigerators in different life groups with frequency missing in (7.0-7.5) years group :

Life in Years	5.5–6.0	6.0–6.5	6.5–7.0	7.0–7.5	7.5–8.0	8.0–8.5
No. of Refrigerators	5	16	13	–	5	4

Find the missing frequency.

Solution. Let f be the missing frequency, then

Life in Years	*Mid-Values (x)*	$x - 7.25 = d$	$d' = \frac{d}{0.5}$	f	fd'
5.5–6.0	5.75	–1.50	–3	5	–15
6.0–6.5	6.25	–1.0	–2	16	–32
6.5–7.0	6.75	–0.5	–1	13	–13
7.0–7.5	7.25	0	0	f	0
7.5–8.0	7.75	0.5	1	5	5
8.0–8.5	8.25	1.0	2	4	8
				$\Sigma f = 43 + f$	$\Sigma fd' = -47$

$$\text{Mean} = 6.78 = 7.25 - \frac{47}{43+f} \times (0.5) \qquad \left[\because \text{Mean} = A + \frac{\Sigma fd'}{\Sigma f} \times i\right]$$

$$\Rightarrow \quad \frac{6.78 - 7.25}{0.5} = \frac{-47}{43+f}$$

$$\Rightarrow \quad \frac{-0.47}{0.5} = \frac{-47}{43+f}$$

$$\Rightarrow \quad 43 + f = 100 \times 0.5$$

$$\Rightarrow \quad f = 50 - 43 = 7.$$

Example 10. Find the median number of workers for industrial establishment in the city of Omega from the following data :

No. of Workers	0–49	50–99	100–149	150–199	200–299	300–399	400–499	500 and above
No. of Establishments	46	59	45	37	31	20	13	9

Solution. Making exclusive type classification, given data can be tabulated as under :

No. of Workers	*No. of Establishments*	*Cumulative Frequency*
–0.5–49.5	46	46
49.5–99.5	59	105
99.5–149.5	45	150 → Median Class
149.5–199.5	37	187
199.5–299.5	31	218
299.5–399.5	20	238
399.5–499.5	13	251
499.5 and above	9	260

$$\Sigma f = 260 = N, \frac{N}{2} = 130.$$

$$\text{Median} = 99.5 + \frac{130 - 105}{45} \times 50$$

$$= 99.5 + \frac{25 \times 50}{45} = 127.28$$

Example 11. Find the median and two quartiles for the following data :

Marks below	10	20	30	40	50	60	70	80
No. of Students	15	35	60	84	96	127	198	250

Solution.

Marks	*No. of Students*	*c.f.*
0–10	15	15
10–20	35	50
20–30	60	110
30–40	84	194
40–50	96	290 → First Quartile Class
50–60	127	417
60–70	198	615 → Median Class
70–80	250	865 → Third Quartile Class

Total frequency $= \Sigma f = N = 865$,

$$\frac{N}{2} = \frac{865}{2} = 432.5, \frac{N}{4} = \frac{865}{4} = 216.25$$

$$\frac{3N}{4} = 648.75.$$

$$\text{Median} = 60 + \frac{432.5 - 417}{198} \times 10 = 70.78$$

$$Q_1 = 40 + \frac{216.25 - 194}{96} \times 10 = 42.3$$

$$Q_3 = 70 + \frac{648.75 - 615}{250} \times 10 = 71.35$$

Example 12. W. Rhodes, the Yorkshire cricketer, scored in 1919, 891 runs or his country with an average of 34.27 runs per innings; in 1920, 949 runs with an average of 28.75; in 1921, 1329 runs with an average of 42.87, and in 1922, 1101 runs with an average of 36.70 runs.

What was the country batting average for the four years ?

Solution. Number of innings in 1919 $= \dfrac{891}{34.27} = 26.$

Number of innings in 1920 $= \dfrac{949}{28.75} = 33.$

Number of innings in 1921 $= \dfrac{1329}{42.87} = 31.$

Number of innings in 1922 $= \dfrac{1101}{36.70} = 30.$

$$\text{Average score for four years} = \frac{\text{Sum of the scores}}{\text{Total number of innings}}$$

$$= \frac{891 + 949 + 1329 + 1101}{26 + 33 + 31 + 30} = \frac{4270}{120} = 35.58 \text{ runs/inning.}$$

Example 13. Mean monthly income of 10 workers in factory *A* is Rs. 4,000 and that of workers in factory *B* is Rs. 3,700. If the mean income of all workers in *A*, and *B* is Rs. 3,800 per month, find the number of workers in *B*.

Solution. Let the number of workers in factory *B* be *x*, then

$$\frac{100 \times 4{,}000 + 3{,}700x}{10 + x} = 3{,}800$$

$$\Rightarrow \quad 40{,}000 + 3{,}700x = 38{,}000 + 3{,}800x$$

$$\Rightarrow \quad 2{,}000 = 100x \quad \Rightarrow \quad x = 20.$$

Example 14. Find the missing frequencies in the following frequency distribution when it is known that the mean = 11.1, total frequency = 60.

Class Limits	*Frequencies*	*Class Limits*	*Frequencies*
9.3–9.7	2	11.3–11.7	14
9.8–10.2	5	11.8–12.2	6
10.3–10.7	–	12.3–12.7	3
10.8–11.2	–	12.8–13.2	1

Solution. Let the missing frequencies be f_1 and f_2.

Class Limit	*Mid-Value* x	$x - 11.5 = d$	*Frequency* f	fd
9.3–9.7	9.5	–2	2	–4
9.8–10.2	10.00	–1.5	5	–7.5
10.3–10.7	10.5	–1	f_1	$-f_1$
10.8–11.2	11.0	–0.5	f_2	$-0.5 f_2$
11.3–11.7	11.5	0	14	0
11.8–12.2	12.00	0.5	6	3
12.3–12.7	12.5	1	3	3
12.8–13.2	13.00	1.5	1	1.5
		$\Sigma d = -2$	$\Sigma f = 31 + f_1 + f_2$	$\Sigma fd = -4 - f_1 - 0.5 f_2$

Since the total frequency = 60, we have

$$31 + f_1 + f_2 = 60 \Rightarrow f_1 + f_2 = 29.$$

$$\text{Mean} = A + \frac{\Sigma fd}{\Sigma f} = 11.5 + \frac{-4 - f_1 - 0.5f_2}{60}$$

$$\Rightarrow \quad 11.1 = 11.5 + \frac{-4 - (f_1 + f_2) + 0.5f_2}{60}$$

$$\Rightarrow \quad 11.1 = 11.5 + \frac{-4 - 29 + 0.5f_2}{60}$$

$$\Rightarrow \quad 11.1 = 11.5 + \frac{-33 + 0.5f_2}{60}$$

$$\Rightarrow \quad -0.4 \times 60 = -33 + 0.5f_2$$

$$\Rightarrow \quad 9 = 0.5f_2 \quad \Rightarrow \quad f_2 = 18$$

$$f_1 + f_2 = 29 \quad \Rightarrow \quad f_1 = 11.$$

Hence, the missing frequencies are 11 and 18.

Example 15. An incomplete frequency distribution is given below :

Class Interval	0–10	10–20	20–30	30–40	40–50	50–60	60–70	Total
Frequency	4	16	–	–	–	6	4	230

Find the three missing frequencies of the table given that median = 33.5 and mode = 34. Also calculate the mean using the empirical relation between mean, median and mode.

Solution. Let the missing frequencies be f_1, f_2, f_3

$$\therefore \quad 4 + 16 + f_1 + f_2 + f_3 + 6 + 4 = 230$$

$$\Rightarrow \quad f_1 + f_2 + f_3 = 200 \qquad ...(i)$$

Cumulative Frequency Table

Class Interval	*Frequency*	*Cumulative Frequencies*
0–10	4	4
10–20	16	20
20–30	f_1	$20 + f_1$
30–40	f_2	$20 + f_1 + f_2$
40–50	f_3	$20 + f_1 + f_2 + f_3$
50–60	6	$26 + f_1 + f_2 + f_3$
60–70	4	$30 + f_1 + f_2 + f_3$

Median = 33.5, therefore median class is 30–40

$$\therefore \quad 33.5 = 30 + \frac{\frac{230}{2} - (20 + f_1)}{f_2} \times 10$$

$$\Rightarrow \quad 0.35f_2 = 115 - 20 - f_1$$

$$\Rightarrow \quad 0.35f_2 = 95 - f_1 \quad \text{...(ii)}$$

$$\text{Mode} = 34$$

$$\therefore \quad 34 = 30 + \frac{f_2 - f_1}{2f_2 - f_1 - f_3} \times 10$$

$$\Rightarrow \quad \frac{f_2 - f_1}{2f_2 - (f_1 + f_3)} = 0.4$$

From (i), $f_1 + f_3 = 200 - f_2$

$$\therefore \quad \frac{f_2 - f_1}{2f_2 - 200 + f_2} = 0.4$$

$$\Rightarrow \quad \frac{f_2 - f_1}{3f_2 - 200} = 0.4$$

$$\Rightarrow \quad f_2 - f_1 = 1.2f_2 - 80 \Rightarrow 0.2f_2 = 80 - f_1 \quad \text{...(iii)}$$

Substracting (iii) from (ii) we get $0.15f_2 = 15 \Rightarrow f_2 = 100$
Substituting $f_2 = 100$ in (ii) we get $f_1 = 60$
Substituting $f_2 = 100, f_1 = 60$ in (i) we get $f_3 = 40$.
Empirical relation between mean, median and mode is

$$\text{mean} - \text{mode} = 3\,(\text{mean} - \text{median})$$

$$\Rightarrow \quad \text{mean} - 34 = 3\,(\text{mean} - 33.5) = 3\,\text{mean} - 100.5$$

$$\Rightarrow \quad 2\,\text{mean} = 100.5 - 34 = 66.5$$

$$\Rightarrow \quad \text{mean} = 33.25.$$

Example 16. Find the A.M. for the following :

Class Interval	–2 – 2	3–7	8–12	13–17	18–22	23–27
Frequency	3277	4096	2048	512	64	3

Solution. Writing the data in exclusive type class intervals, we get

Class Interval	*Mid-Value* x	$d' = \frac{x-5}{5}$	*Frequency* F	fd'
–2.5–2.5	0	–1	3277	–3277
2.5–7.5	5	0	4096	0
7.5–12.5	10	1	2048	2048
12.5–17.5	15	2	512	1024
17.5–22.5	20	3	64	192
22.5–27.5	25	4	3	12
			10,000	$\Sigma fd' = -1$

$$\text{A.M.} = 5 + \frac{\Sigma fd'}{\Sigma f} \times 5 = 5 - \frac{1}{10000} \times 5 = 4.9995.$$

Example 17. In a factory employing 3,000 persons 5 per cent earn less than Rs. 3 per hour, 580 earn from Rs. 3.01 to Rs. 4.5 per hour, 30 per cent earn from Rs. 4.51 to Rs. 6.00 per hour, 500 earn from Rs. 6.01 to Rs. 7.50 per hour, 20 per cent earn from Rs. 7.51 to Rs. 9.00 per hour and the rest earn Rs. 9.01 or more per hour. What is the median wage ?

Solution. Cumulative frequency distribution table by adjusting the class intervals so as to make exclusive type classification.

Class Interval	*Frequency*	*Cumulative Frequency*
Less than 3	$3000 \times \frac{5}{100} = 150$	150
3.005–4.505	580	730
4.505–6.005	$3000 \times \frac{30}{100} = 900$	1630
6.005–7.505	500	2130
7.505–9.005	$3000 \times \frac{20}{100} = 600$	2730
9.005 and above	$3000 - 150 - 580$ $- 900 - 500 - 600 = 270$	3000

Median class $= 4.505 - 6.005$

$$\text{Median} = 4.505 + \frac{1500 - 730}{900} \times 1.5 = 4.505 + \frac{770}{900} \times 1.5 = 5.788.$$

Example 18. For the two frequency distributions given below, the mean calculated from the first was 25.4 and that from the second was 32.5. Find the values of x and y.

Class	*Distribution I Frequency*	*Distribution II Frequency*
10–20	20	4
20–30	15	8
30–40	10	4
40–50	x	$2x$
50–60	y	y

Solution.

Class	*Mid-Value* x	$x - 35 = d$	$\frac{d}{10} = d'$	*Distribution I Frequency* (f_1)	*Distribution II Frequency* (f_2)	f_1d'	f_2d'
10–20	15	–20	–2	20	4	–40	–8
20–30	25	–10	–1	15	8	–15	–8
30–40	35	0	0	10	4	0	0
40–50	45	10	1	x	$2x$	x	$2x$
50–60	55	20	2	y	y	$2y$	$2y$
				$\Sigma f_1 = 45 + x + y$	$\Sigma f_2 = 16 + 2x + y$	$\Sigma f_1d' = x + 2y - 55$	$\Sigma f_2d' = -16 + 2x + 2y$

As per the question

$$25.4 = 35 + \left[\frac{x + 2y - 55}{45 + x + y}\right] \times 10$$

$$\Rightarrow \quad \frac{-9.6}{10} = \frac{x + 2y - 55}{45 + x + y}$$

$$\Rightarrow \quad -432 - 9.6x - 9.6y = 10x + 20y - 550$$
$$\Rightarrow \quad 19.6x + 29.6y - 118 = 0$$
$$\Rightarrow \quad 9.8x + 14.8y - 59 = 0 \qquad ...(i)$$
$$32.5 = 35 + \left[\frac{2x + 2y - 16}{16 + 2x + y}\right] \times 10$$
$$\Rightarrow \quad \frac{-2.5}{10} = \frac{2x + 2y - 16}{16 + 2x + y}$$
$$\Rightarrow \quad -40 - 5x - 2.5y = 20x + 20y - 160$$
$$\Rightarrow \quad 25x + 22.5y = 120$$
$$\Rightarrow \quad 5x + 4.5y = 24 \qquad ...(ii)$$

Equations (i) and (ii) can be written as

$$9.8x + 14.8y - 59 = 0$$
$$5x + 4.5y - 24 = 0$$
$$\frac{x}{-355.2 + 265.5} = \frac{y}{-295 + 235.2} = \frac{1}{44.1 - 74}\ ^{*}$$
$$\Rightarrow \quad \frac{x}{-89.7} = \frac{y}{-59.8} = \frac{1}{29.9}$$
$$\Rightarrow \quad x = \frac{89.7}{29.9} = 3,\ y = \frac{59.8}{29.9} = 2$$
$$\Rightarrow \quad x = 3,\ y = 2.$$

Examploe 19. In the following grouped data, x are the mid-values of the class intervals and c is a constant. If the arithmetic mean of the original distribution is 35.84, find its class intervals.

$x - c$	–21	–14	–7	0	7	14	21	Total
f	2	12	19	29	20	13	5	100

Solution.
$$\text{Mean} = 35.84 = c + \frac{\Sigma f(x - c)}{\Sigma f}$$
$$\Sigma f(x - c) = -21 \times 2 + (-14) \times 12 + (-7) \times 19 + 0 \times 29 + 7 \times 20 + 14 \times 13 + 21 \times 5$$
$$= -42 - 168 - 133 + 140 + 182 + 105 = 84$$
$$\therefore \quad 35.84 = c + \frac{84}{100} = c + 0.84$$
$$\Rightarrow \quad c = 35$$

Let $x_1, x_2, x_3, \ldots$ be the mid-values of first, second, third, etc., classes respectively.

$$x_1 - c = -21 \Rightarrow x_1 - 35 = -21 \Rightarrow x_1 = 14$$
$$x_2 - c = -14 \Rightarrow x_2 - 35 = -14 \Rightarrow x_2 = 21$$
$$x_3 - c = -7 \Rightarrow x_3 - 35 = -7 \Rightarrow x_3 = 28$$

Difference between two consecutive mid-values = 7. Therefore class interval = 7.

* We can solve equations (i) and (ii) by other methods also.

To have the lower limit of the class interval, we substract half the difference between consecutive mid-values from the mid-values and to have the upper limit we add half the difference between the consecutive mid-values to the mid-values,

$\therefore$ The class intervals are :

$$(14 - 3.5) - (14 + 3.5), (21 - 3.5) - (21 + 3.5), (28 - 3.5) - (28 + 2.5), \text{ and so on,}$$

i.e., $$10.5 - 17.5,\ 17.5 - 24.5,\ 24.5 - 31.5, \text{ and so on.}$$

Example 20. Given $\overline{X} = 33$, assumed Mean = 35

Step Deviation	–3	–2	–1	0	+1	+2
Frequency	5	10	25	30	20	10

Find class intervals.

Solution. Let the step deviation be denoted by d'. Then

d'	*Frequency (f)*	fd'
–3	5	–15
–2	10	–20
–1	25	–25
0	30	0
1	20	20
2	10	20
	$\Sigma f = 100$	$\Sigma fd' = -20$

$$\text{Mean} = A + \frac{\Sigma fd'}{\Sigma f} \times i, \text{ where } i \text{ is the class interval,}$$

$$\Rightarrow \quad 33 = 35 - \frac{20}{100} \times i \ \Rightarrow\ -2 = \frac{-20}{100} \times i \ \Rightarrow\ i = 10$$

Let l_1 be the lower limit and l_2 the upper limit of the first interval then

$$l_2 - l_1 = 10 \quad \text{...(i)}$$

and $$\frac{\frac{l_1 + l_2}{2} - 35}{10} = -3 \qquad \left[d' = \frac{x - A}{i}\right]^*$$

$$\Rightarrow \quad l_1 + l_2 = 10 \quad \text{...(ii)}$$

Adding (i) and (ii), we get $2l_2 = 20 \Rightarrow l_2 = 10$

First interval is 0-10, second interval is 10-20, and so on.

Example 21. 20% workers in an industry which employs a total of 2000 men earn less than Rs. 20 per hour, 440 earn from Rs. 20 to 22.4 per hour, 24% earn from 22.5 to 24.9 per hour, 370 earn from Rs. 25 to 27.4 per hour, 12% earn from Rs. 27.5 to 29.9 per hour and the rest earn Rs. 30 or more per hour. Make a frequency table and calculate modal wage.

Solution. **Calculation of Modal Wage**

Earnings Per Hour	*No. of Workers*
Less than Rs. 20	400**
Rs. 20 to Rs. 22.4	440

* x are the mid-values.

** $20\% \text{ of } 2000 = \frac{2000}{100} \times 20 = 400.$

Rs. 22.5 to Rs. 24.9	480
Rs. 25 to Rs. 27.4	370
Rs. 27.5 to Rs. 29.9	240
Rs. 30 and more	70

By inspection mode lies in the class 22.5 to 24.9. Of Course the real limits are 22.45 to 24.95.

$$Z = L_1 + \frac{f_m - f_{m-1}}{2f_m - f_{m-1} - f_{m+1}} \times C$$

$$= 22.45 + \frac{480 - 440}{960 - 440 - 370} \times 2.5$$

$$= 22.45 + 0.67 = \text{Rs. } 23.12.$$

Example 22. For a group of 5000 workers, the weekly wages vary from Rs. 200 to Rs. 800. The wages of 4% of the workers are under Rs. 250 and those of 10% under Rs. 300. 15% of the workers earn Rs. 600 and over and 5% of them got Rs. 700 and over. The Quartile wages are Rs. 400 and Rs. 540 and the sixth Decile is Rs. 500. Put this information in the form of a frequency table. Also evaluate P_2.

Solution. We are given $N = 5000$ and the Range of Wage is Rs. 200 and Rs. 800. Now upto Q_1, D_6 and Q_3, the frequencies will be 1250, 3000 and 3750 workers respectively. The given information in the question is to be presented in the cumulative form first.

Weekly Wages (Rs.)	*Cumulative %*
Less than 250	4
Less than 300	10
Less than 400	25
Less than 500	60
Less than 540	75
More than 600	15
More than 700	5

Now the above cumulative frequency distribution will be converted into a simple frequency distribution.

Weekly Wages (Rs.)	*% of Workers*	*No. of Workers f*	C_f
200–250	4	200	200
250–300	6	300	500
300–400	15	750	1250
400–500	35	1750	3000
500–540	15	750	3750
540–600	10 (Balance)	500	4250
600–700	10	500	4750
700–800	5	250	5000
		$N = 5000$	

$$\frac{2N}{100} = \frac{5000 \times 2}{100} = 100 = N_1$$

$\therefore$ $Cf_{-1} = 0$ **(Note this point)**

$$f = 200;\ L = 200;\ i = 50$$

$$P_2 = L + \frac{N_1 - Cf_{-1}}{f} \times i = 200 + \frac{100 - 0}{200} \times 50$$

$$= 200 + 25 = 225.$$

Example 23. In a factory there are 100 skilled, 250 semi-skilled and 150 unskilled workers. It has been observed that on an average a unit length of a particular fabric is woven by a skilled worker in three hours, by a semi-skilled worker in 4 hours and by an unskilled worker in 5 hours. After a training of two years, the semi-skilled workers are expected to become skilled and the unskilled workers to become semi-skilled. How much less time will be required after two years of training for weaving unit length of fabric by an average worker ?

Solution. Before training

Total number of workers = 100 + 250 + 150 = 500

Total time required to weave (100 + 250 + 150) units length of the fabric

$$= 100 \times 3 + 250 \times 4 + 150 \times 5$$

Average time taken to weave one unit length

$$= \frac{100 \times 3 + 250 \times 4 + 150 \times 5}{500}$$

$$= \frac{300 + 1000 + 750}{500} = \frac{2050}{500} = 4.1 \text{ hr.}$$

After training

Number of skilled workers = 100 + 250 = 350

Number of semi-skilled workers = 150

Time taken to weave (350 + 150) units length of fabric

$$= 350 \times 3 + 150 \times 4 = 1650 \text{ hrs.}$$

Average time taken to weave one unit length of the fabric $= \frac{1650}{500} = 3.3$ hrs.

Time saved due to training to weave one unit length of the fabric = 4.1 – 3.3 = 0.8 hours or 48 minutes.

Example 24. The mean height of 50 students of a college is 5′ – 8″. The height of 30 of these is given in the frequency distribution below. Find the mean height of the remaining 20 students.

Height	5′ 4″	5′ 6″	5′ 8″	5′ 10″	6′ 0″
Frequency	4	12	4	8	2

Solution. Let the variable X_1 denotes the heights of the 30 students in inches.

Let $A = 68''$; i = class interval = 2

Height in inches X_1	*Frequency* f	$d' = \dfrac{X_1 - A}{C}$	fd'
64	4	–2	–8
66	12	–1	–12
68	4	0	0
70	8	+1	+8
72	2	+2	+4
	$\Sigma f = N_1 = 30$		$\Sigma fdx' = -8$

$$\overline{X}_1 = A + \frac{\Sigma fd'}{\Sigma f} \times i = 68 + \left\{\frac{-8}{30}\right\} \times 2$$

$$\overline{X}_1 = 68 - \frac{16}{30} = \frac{2040 - 16}{30} = \left(\frac{2024}{30}\right) = 67.4667 \text{ inches} \quad \text{...(i)}$$

Now $\overline{X}_2$ denotes the mean height of the remaining $N_2 = 50 - 30 = 20$ students. $\overline{X}_{12}$ is the mean height of the 50 students, then we were given that

$$\overline{X}_{12} = 5' - 8'' = 68''$$

$$\overline{X}_{12} = \frac{N_1\overline{X}_1 + N_2\overline{X}_2}{N_1 + N_2}$$

$\Rightarrow$ $$68 = \frac{30\left(\frac{2024}{30}\right) + 20\overline{X}_2}{50}$$

or $$2024 + 20\overline{X}_2 = 68 \times 50$$

$\Rightarrow$ $$20\overline{X}_2 = 3400 - 2024 = 1376$$

$\Rightarrow$ $$\overline{X}_2 = \frac{1376}{20} = 68.8 = 5' - 8.8''.$$

Example 25. From the following data of income consumption calculate the arithmetic mean when it is given that (i) the total income of persons in the highest group is Rs. 435 and (ii) none is earning less than Rs. 20.

Income (Rs.) Below	30	40	50	60	70	80	90
No. of Persons	16	36	61	76	87	95	100

Solution. The data is given in cumulative frequency form. 'Income below Rs. 30' which includes the persons with income less than Rs. 30 is an open end class. Since we are given that none of earning less than 20, this class will be 20-30. By converting from cumulative frequency table to simple frequency table, we get

Income (X)	*No. of Persons* (f)	*Mid-Point* X	fX
20–30	= 16	25	400
30–40	36 – 16 = 20	35	700
40–50	61 – 36 = 25	45	1125
50–60	76 – 61 = 15	55	825

60–70	$87-76=11$	65	715
70–80	$95-87=8$	75	600
80–90	$100-95=5$	85	435 (Given)
	$N=100$		$\Sigma fX=4800$

$$\overline{X}=\frac{\Sigma fX}{N}=\frac{4800}{100}=48.$$

Example. 26. A survey conducted on the effectiveness of a particular weight reduction drug gives the following results. The classification based on the weight of the surveyed people after their using the drug is as follows :

Weight (Kgs)	*Number of People*	
	Before Using Drug	*After Using Drug*
30–40	x	x'
40–50	y	y'
50–60	31	40
60–70	35	20
70–80	11	6
80–90	7	4

The number of people in the first two classes before and after using the drug is not known. You are required to find out the number, given the following additional information.

The medians of the two weights (before and after using drug) are 60.9 Kg and 55.0 Kg.

The people in the first classification after using the drug were 1.5 times as many in number as they were before using it and the people in the second classification after using the drug were twice as many in number as they were before using it.

Solution.

Weight (in Kgs)	*Before Using the Drug (f)*	*Cf*	*After Using the Drug (f')*	*Cf*
30–40	x	x	x'	x'
40–50	y	$x+y$	y'	$x'+y'$
50–60	31	$x+y+31$	40	$x'+y'+40$
60–70	35	$x+y+66$	20	$x'+y'+60$
70–80	11	$x+y+77$	6	$x'+y'+66$
80–90	7	$x+y+84$	4	$x'+y'+70$

Before using the drug

$$60.9=60+\left[\frac{\frac{x+y+84}{2}-(x+y+31)}{35}\right]\times 10$$

$$\Rightarrow \qquad 0.09=\frac{22-x-y}{70} \quad \Rightarrow \quad x+y=15.7$$

$$\Rightarrow \qquad 2x+2y=31.4 \qquad \text{...(1)}$$

After using the drug

$$55 = 50 + \left[\frac{\frac{x' + y' + 70}{2} - (x' + y')}{40} \right] \times 10$$

$$\Rightarrow \qquad \frac{5 \times 80}{10} = 70 - x' - y'$$

$$\Rightarrow \qquad x' + y' = 30 \qquad \text{...(2)}$$

As per question, $x' = 1.5x$, $y' = 2y$

Putting these values in (2), we get

$$1.5x + 2y = 30 \qquad \text{...(3)}$$

Substracting (1) from (3), we get

$$-0.5x = -1.5 \;\Rightarrow\; x = 2.8$$

Putting $x = 2.8$ in (1), we get

$$2 \times 2.8 + 2y = 31.4$$

$$\Rightarrow \qquad 2y = 31.4 - 5.6 = 25.8 \;\Rightarrow\; y = 12.9$$

$$\therefore \qquad x' = 1.5x = 1.5 \times 2.8 = 4.2$$

$$y' = 2y = 2 \times 12.9 = 25.8$$

Example 27. The first and third quartiles of the following data are given to be 25 and 50 marks respectively. From the data given below, find the missing frequencies when total frequency $N = 72$.

Marks	0–10	10–20	20–30	30–40	40–50	50–60	60–70	70–80
Frequency	4	8	?	19	?	10	5	?

Total Frequency = 72 = N **(Punjab University, B.Com. 2006)**

Solution. Let the missing frequencies be f_1, f_2, f_3, then

$$4 + 8 + f_1 + 19 + f_2 + 10 + 5 + f_3 = 72$$

$$\Rightarrow \qquad f_1 + f_2 + f_3 = 72 - 46 = 26$$

Marks	*Frequency*	*Cum frequency*
0–10	4	4
10–20	8	12
20–30	f_1	$12 + f_1$
30–40	19	$31 + f_1$
40–50	f_2	$31 + f_1 + f_2$
50–60	10	$41 + f_1 + f_2$
60–70	5	$46 + f_1 + f_2$
70–80	f_3	$46 + f_1 + f_2 + f_3$

First Quartile is 25, *i.e.,* it lies in the interval 20-30

$$\therefore \qquad 25 = L_1 + \left[\frac{\frac{N}{4} - cf_{-1}}{f} \times i \right]$$

$$= 20 + \frac{18 - 12}{f_1} \times 10$$

$$= 5 = \frac{60}{f_1} \quad \Rightarrow \quad f_1 = 12$$

Third quartile = 50 *i.e.,* it lies in the interval 50-60

$$\therefore \qquad 50 = 50 + \frac{54 - (31 + f_1 + f_2)}{10} \times 10$$

$$\Rightarrow \qquad 23 = f_1 + f_2 = 12 + f_2 \quad \Rightarrow \quad f_2 = 11$$

$$f_1 + f_2 + f_3 = 26 \quad \Rightarrow \quad 12 + 11 + f_3 = 26$$

$$\Rightarrow \qquad f_3 = 3.$$

Example 28. "Every average has its own particular characteristics. It is difficult to say which average is best." Comment briefly. **(D.U. B. Com. (Hons.) Correspondence 2004)**

Solution. The statement is correct. If the data is Qualitative only median can be used. Mode is used to find the most frequently used size. Mode is also used as an average wage by the workers (or trade union leaders). In continuous frequency distribution if the class intervals are unequal, mode cannot be used.

G.M. is used to find the average of ratios or percentages. G.M. is also used to find the average rate of interest against the varying rate of interest in compound interest calculations. If equal distances are travelled at different speed, the average speed is calculated by harmonic mean. (Refer Ex. 29 also)

Example 29. The cost of a machinery has appreciated by 10%, 12% and 8% in three years but in the fourth year it depreciated by 5%. Find out the average increase or decrease in the cost of the machinery during four years. **(D.U. B. Com. (Hons.) Correspondence 2004)**

Solution. If $r\%$ is the average changes then

$$100\left(1 + \frac{r}{100}\right)^4 = 100\left(1 + \frac{10}{100}\right)\left(1 + \frac{12}{100}\right)\left(1 + \frac{8}{100}\right)\left(1 - \frac{5}{100}\right)$$

$$= 100 \times 1.264032$$

$$\Rightarrow \qquad \left(1 + \frac{r}{100}\right)^4 = 1.264032$$

$$\Rightarrow \qquad 1 + \frac{r}{100} = [1.264032]^{\frac{1}{4}} = 1.06032$$

$$\Rightarrow \qquad r = 6.03\% = \text{approx. } 6\% \text{ increase.}$$

Example 30. The mean weight of 50 students in a certain class is 59 kilograms. The mean weight of the boys in the class is 65 kilograms and that of girls is 50 kilograms. Find the number of girls in the class. **(D.U. B.Com. Pass 2008)**

Solution. Let the number of boys be n_1 then the number of girls $= 50 - n_1$. Then

$$50 \times 59 = 65\, n_1 + (50 - n_1)\, 50$$

$$2950 = 65\, n_1 + 2500 - 50\, n_1$$

$$\Rightarrow \qquad 450 = 15\, n_1 \Rightarrow n_1 = 30$$

$\therefore$ The number of girls $= 50 - 30 = 20.$

Example 31. 100 students appeared for an examination. The results of those who failed are given below :

Marks	*No. of Students*
5	4
10	6
15	8
20	7
25	3
30	2
Total	30

If the average marks of all students were 68.6, find out average marks of those who passed.

(D.U. B.Com. (Hons.) 2008)

Solution.

Marks x	*No. of Students f*	*fx*
5	4	20
10	6	60
15	8	120
20	7	140
25	3	75
30	2	60
	Total	475

Total marks obtained by the students who passed

$$= 100 \times 68.6 - 475 = 6880 - 475 = 6385$$

Average marks obtained by the students who passed

$$= \frac{6385}{70} = 91.2.$$

Example 32. Find, the missing frequencies in the following table if the mode is 39 and total students are 70.

Marks	*No. of Students*
0–10	3
10–20	7
20–30	–
30–40	20
40–50	–
50–60	10

(D.U. B.Com. 2008)

Solution. Let the missing frequencies be f_1 and f_2, then

$$f_1 + f_2 = 70 - 40 = 30$$

$$\Rightarrow \qquad f_1 = 30 - f_2$$

Marks	*No. of Students*
0–10	7
10–20	7
20–30	$30 - f_2$
30–40	20
40–50	f_2
50–60	10
Total	70

$$\text{Mode} = 39 = 30 + \frac{20 - (30 - f_2)}{40 - (30 - f_2) - f_2} \times 10$$

$$\Rightarrow \qquad 9 = \frac{(20 - 30 + f_2)\ 10}{10} \Rightarrow 90 = -10 + 10 f_2 = f_2 = 19$$

$$f_1 = 30 - 19 = 11$$

Example 33. Different averages serve different purposes but simple mean is an ideal average which can serve all purposes'. Comment. **(D.U. B. Com. (Hons.) 2003)**

Solution. The statement is not correct. Simple mean, *i.e.,* Arithmetic mean can not serve all purposes. It fails in the following cases :

1. In Compound Interest calculations average rate of interest against varying rate of interest is calculated by Geometric mean.
2. Theoretically G.M. should be used in averaging index numbers.
3. For calculating average speed when equal distances are covered at different speeds, Harmonic mean is used.
4. If different articles are purchased at different rates by spending equal amount on each article, the average rate of purchase is given by Harmonic mean of different rates.
5. Modal wage is considered to be the best average by trade union leaders/workers whereas mean is considered as the best average by factory owners for calculating average wage of workers.
6. In case of open ended (undetermined) class intervals mean cannot be calculated.
7. If the data is qualitative, median is the only average which can be calculated.
8. Manufacturers take the modal size as the average.

There are some other situations also where mean may not be suitable but mean has wide variety of applications. Mean is used in

(i) calculating per capita income of a country.
(ii) mean is used in calculating standard deviations.
(iii) mean is used in correlation as well as regression analysis.
(iv) mean is used in sampling techniques.
(v) mean is used in quality control and many other applications.

Though mean cannot serve all purposes but still mean can be called as a **king of the measures of central tendency.**

Questions

Very Short Answer Type Questions

1. Which is the best average for qualitative data ?
2. Which average cannot be determined in case of open-end class intervals ?
3. For calculating mode, class intervals have to be equal. Comment.
4. If $\frac{\sum_{i=1}^{10} (x_i - 5)}{10} = -1$, find the mean.
5. Which is the best average for the manufacturer of garments ?
6. If there is a dispute between workers and factory owner, which average will be used by workers and which average by factory owner to justify their respective claims ?
7. Why data is arranged only in ascending order for finding quartiles ?

Short Answer Type Questions

1. What purpose is served by an average ?
2. What is similar in mean, median and mode ?
3. Every average has its own peculiar characteristics. Comment.
4. "The figure 2.2 children per adult female was felt to be in some respect absurd and the royal commission suggested that the middle class be paid money to increase the average to a rounder and more convenient number'' (Punch). Commencing on the above statement, discuss the limitations of a arithmetic mean.
5. What are the desiderata (requisites) of a good average ?
6. What is the empirical relationship between mean, median and mode ? When this relation is applied ?
7. Why averages are called as measures of central value ?
8. What are various measures of central tendency ?
9. What is the object of a central value ?
10. Define arithmetic mean. Discuss its merits and demerits.
11. Why is the arithmetic mean most commonly used measure of a central value ?
12. What are the mathematical properties of mean ?
13. Define weighted arithmetic mean. Discuss the importance.
14. Which ones of the measures of central tendency are computational (mathematical) and which ones are positional ?
15. Define median. Discuss its merits and demerits.
16. When would you consider the median to be a better representation of the data ? Discuss.
17. Write a brief note on partititon values.
18. Define mode of a frequency distribution. Discuss its merits and demerits.
19. Define a weighted average. How does it differ from an unweighted average ?
20. 'Median is not affected by extreme values unlike arithmetic mean'. Discuss.

State Whether the Following Statements are True or False

1. Arithmetic mean is always the best measure of central tendency.
2. The median is a computational measure of central tendency.
3. The median of distribution can be determined graphically.
4. The value of mean is least affected by sampling fluctuations.

5. The sum of the absolute deviations from arithmetic mean is least.
6. The median can be obtained for open-ended distributions.
7. Mode is suitable average to know the sale of average size of readymade garments.
8. The arithmetic mean can be computed even if an observation of the series is missing.
9. Average sale of a firm for different years can be better representeed by mode.
10. Arithmetic mean is highly affected by extreme values.
11. Sum of absolute deviations from median is minimum.
12. All positional averages are mathematical averages.
13. Algebraic sum of deviations from arithmetic mean is equal to zero.
14. For a symmetrical distribution, mean = median = mode.
15. For dealing with qualitative data, median is most suitable.
16. The mean of a series remains unaffected if an observation equal to mean is included in the series.
17. Quartiles divide the series into four parts.
18. First quartile divides the series into two equal parts.
19. Percentiles are hundred in number.
20. 50th percentile is median.
21. 5th decile is median.
22. Q_1 equals the 25th percentile.
23. The mean wages of two factories are equal. Does it signify that their workers are equally well off ?
24. 50% observations of a series lie between first and third quartiles.
25. 80% observations of a series lie between first and ninth declies.
26. 75% observations in a distribution are greater than Q_3.
27. 30% observations in a distribution are less than 70th percentile.
28. In calculating the arithmetic mean for grouped data, the calculated mean is the same no matter where the arbitrary mean (Assumed mean) is set.
29. A.M. can be calculated for a distribution with open ends.
30. Median is not affected by extreme observation.
31. The class limits should be exclusive for the determination of median and mode.
32. Mode can be calculated if class intervals are of unequal width.
33. $\Sigma (x_i - \bar{x}) = 50$ for a distribution.
34. G.M. is used to average exponential rates of growth.
35. Mode is the value with maximum frequency in the data.
36. Mean is based on all the observations.
37. If $2u = x$ and the H.M. of x is u, then the H.M. of u is 2. **(CA Found., May, 2002)**

ANSWERS

1. F (In many situations like qualitative data, open-ended class intervals, size of garments, etc., mean cannot be used)
2. F (Median is a positional measure)

3. T **4.** T **5.** F **6.** T **7.** T **8.** F **9.** F (In such case mean should be used)
10. T **11.** T **12.** F **13.** T **14.** T **15.** T **16.** T **17.** T **18.** F
19. F (They are 99 only) **20.** T **21.** T **22.** T **23.** F (Variability may be quite different)
24. T **25.** T **26.** F **27.** F **28.** T (Assumed mean can be any constant)
29. F **30.** T **31.** T **32.** F **33.** F **34.** T **35.** T **36.** T **37.** T

Multiple Choice Questions

1. The algebraic sum of the deviations of 10 observations measured from 15 is 7. Thus the mean is :
(a) 105 (b) 70 (c) 15.7 (d) None of these

2. The algebraic sum of the deviations from mean is :
(a) Maximum (b) Least (c) Zero (d) None of these

3. The sum of squares of deviations from mean is :
(a) Least (b) Zero (c) Maximum (d) None of these

4. An additional observation 15 is included in a series of 11 observations and its mean remains unaffected. The mean of the series was :
(a) 11 (b) 15 (c) 165 (d) 4

5. The median of a series is 10. Two additional observations, 7 and 20, are added to the series. The median of the new series will be :
(a) 9 (b) 20 (c) 7 (d) 10

6. In a batch of 13 students 4 have failed. The marks of the successful candidates were 41, 57, 38, 61, 36, 35, 71, 50 and 40. Find the median marks :
(a) 40 (b) 50 (c) 41 (d) 38

7. From the information given below :

	Factory A	*Factory B*
Mean Daily Wage	Rs. 120	Rs. 80
No. of Workers	80	120

Choose the correct answer :
(a) Factory A pays larger amount as daily wages. (b) Factory B pays larger amount as daily wages.
(c) Both factories pay equal amount as daily wages. (d) It cannot be determined.

8. The arithmetic mean of 9 observations is 100 and that of 6 is 80, the combined mean of all the 15 observations will be :
(a) 100 (b) 80 (c) 90 (d) 92

9. The arithmetic mean of the first N numbers, 1, 2, 3, ..., N is :
(a) $\frac{N}{2}$ (b) $\frac{N+1}{2}$ (c) $\frac{N(N+1)}{2}$ (d) None of these

$\left[\textbf{Hint : } \sum_{i=1}^{n} i = \frac{N(N+1)}{2}\right]$

10. For a symmetrical distribution, Q_1 and Q_3 are 20 and 40 respectively. The value of 50th percentile will be :
(a) 40 (b) 20 (c) 30 (d) None of these

11. For a symmetrical distribution P_{25} and P_{75} are 40 and 60 respectively. The value of median will be :
(a) 50 (b) 40 (c) 60 (d) None of these

12. In a series of 25 observations $\Sigma (x-7) = -8$. Then, mean is equal to :
(a) 7 (b) 8 (c) 6.68 (d) 7.68

13. Total wages paid to the workers in two factories A and B are the same but the modal wage in factory A is larger than the modal wage in factory B. In which factory workers are well off ?
(a) A (b) B c) Equal in both (d) No answer

14. Which of the following would be an appropriate average for determining the average size of readymade garments :
(a) Arithmetic mean (b) Median (c) Mode (d) Geometric mean

15. Most appropriate average to determine the size of oranges on a tree is :
(a) Mode (b) Median (c) A.M. (d) None of these

16. Most appropriate measure for qualitative measurements is :
(a) Mode (b) Median (c) A.M. (d) None of these

17. The most unstable measure of central tendency is :
(a) A.M (b) Median (c) Mode (d) None of these

18. The sum of deviations of observations is zero when measured from :
(a) Median (b) G.M. (c) Mode (d) A.M.

19. The average, most affected by the extreme observations, is :
(a) Mode (b) A.M. (c) G.M. (d) Median

20. The most stable average is :
(a) Mode (b) A.M. (c) Median (d) G.M.

21. For which measure of central tendency the sum of the deviations (taking signs into account) of the items from the measure of central tendency is zero ?
(a) Mode (b) G.M. (c) A.M. (d) Median
(e) None of the above

22. The median of the series 6, 3, 8, 9, 5, 7 is :
(a) 5.5 (b) 6.0 (c) 6.5 (d) 7.0
(e) 7.5

23. Which of the following is false ?
(a) The value of median of a data set gets affected by extreme values that are not representative of the data set.
(b) Median is the point below which 50% of the data items fall.
(c) Median can be calculated from a grouped data with open-ended classes.
(d) Median of a set of ungrouped data can be found out only after arranging them in ascending or descending order.
(e) None of the above

24. Which of the following is not a mathematical average ?
(a) A.M. (b) G.M. (c) H.M. (d) Median
(e) Weighted Arithmetic Mean

25. What is the major assumption that one makes, when computing a mean from a grouped data ?
(a) All values are discrete
(b) No value occurs more than once
(c) Every value in a class is equal to its mid-point
(d) Each class contains exactly the same number of values
(e) None of the above

26. A security analyst studied a hundred companies and obtained the folowing distribution for the dividend declared by the companies during the year 2002 :

Dividend Declared in Percentage	0–8	8–16	16–24	24–32	32–40
No. of Companies	12	20	25	28	15

Which of the following can be best inferred from above mentioned data ?
I. The sum of the deviations of the items from the arithmetic mean is zero.
II. The sum of the squared deviations of the items from the arithmetic mean is less than the sum of the squared deviations of the items from any other value.
III. The given information is not sufficient to compute combined average of any of these two groups.
(a) Only (I) above (b) Only (II) above (c) Only (III) above
(d) Both (I) and (II) above (e) All of the above

27. If the sum of the deviations (taking signs into account) of the items from a measure of central tendency is zero, then that measure is known as :

(a) H.M. (b) G.M. (c) A.M. (d) Median
(e) Mode

28. If the mean, median and mode of a data set are equal then, which of the following is true with regard to the distribution of the data ?

(a) It is symmetrical (b) It is positively skewed (c) It is negatively skewed (d) It is asymmetrical
(e) Insufficient information

29. The 50th percentile is :

I. A.M. II. Median III. Standard Deviations

(a) Only (I) above (b) Only (II) above (c) Only (III) above
(d) Both (I) and (III) above (e) Both (II) and (III) above

30. What is the MODE of the data set {3, 3, 3, 5, 5, 5, 7, 7, 7} ?

(a) 1.3 (b) 2.5 (c) 3.7 (d) 4.3, 5, 7
(e) None of the above

31. For a given set of data set, the values of mode and median were found to 50.25 and 65.50 respectively. The mean of the data set is approximately equal to :

(a) 27.125 (b) 57.875 (c) 59.250 (d) 73.125
(e) 85.325

32. The sum of the squares of the deviations is the least when measured from :

(a) Arithmetic mean (b) Geometric mean (c) Harmonic mean (d) Median
(e) Mode

ANSWERS

1. (c) **2.** (c) **3.** (a) **4.** (b) **5.** (d) **6.** (d) **7.** (c) **8.** (d) **9.** (c) **10.** (c)
11. (a) **12.** (c) **13.** (a) **14.** (c) **15.** (a) **16.** (b) **17.** (c) **18.** (d) **19.** (b) **20.** (b)

21. (c) Arithmetic Mean (Property 3 on Page 5.4)

22. (c) After arranging the data in ascending order we get 3, 5, 6, 7, 8, 9.

Since the number of items is even, A.M. of $\frac{N}{2}$th and $\left(\frac{N}{2}+1\right)$th item is the median, *i.e.,* Median = A.M. of 3rd and 4th items $= \frac{6+7}{2} = 6.5$.

23. (a) Median is only a positional average, it is not infleunced by the extreme values in a set of observations. Hence, when extreme values are present in a set of observations, the median is a more satisfactory measure of the central tendency than the mean.

24. (d) Median is a positional average.

25. (c) In a grouped data middle value of a class is taken as the representative of the class. Therefore, each value is taken as equal to the mid value.

26. (d) I and II, the given information is quite sufficient to find the combined mean of any two groups.

27. (c), **28.** (a), **29.** (b), **30.** (e),
31. (d) Apply mean – Mode = 3 (mean – median) **32.** (a)

Fill in the Blanks

1. Measure of central tendency are called averages of .... (first order, third order)
2. If a constant amount is added to each value in a distribution then the Arithmetic mean .... (remains unchanged, is also added by the same constant amount)
3. If the arithmetic mean of the ages of a group of boys is 20 years, then the arithmetic mean of their ages after 2 years will be .... (22 years, no change)

4. For calculating mode, class intervals have to be .... (equal, no restriction)
5. For finding quartiles, the data should be arranged in .... (ascending order, discending order)
6. Histogram is used to find .... (median graphically, mode graphically)
7. For finding quartiles graphically, we use the diagram called .... (less than Ogive, more than Ogive)

ANSWERS

1. first order, **2.** is also added by the same constant amount, **3.** 22 years, **4.** equal,
5. ascending order, **6.** mode graphically, **7.** less than Ogive.

Practice Questions

A. Questions about Calculating Arithmetic Mean

1. The mean monthly salary paid to all employees in a company is Rs. 1,600. The mean monthly salaries paid to technical and non-technical employees are Rs. 1,800 and Rs. 1,200 respectiely. Determine the percentage of technical and non-technical employees of the company.
[**Ans.** Percentage of technical employees = 66.67 and percentage of non-technical employees = 33.33]

2. The mean marks of 100 students were found to be 40. Later on it was discovered that a score of 53 was misread as 83. Find the corrected mean corresponding to the corrected score. [**Ans.** 39.7 marks]

3. The mean age of a combined group of men and women is 30 years. If the mean age of the group of men is 32 years and that for the group of women is 27 years, find the percentage of men and women in the group.
[**Ans.** 60% men and 40% women]

4. The mean weight of 150 students in a class is 60 kgs. The mean weight of boys in the class is 70 kgs and that of the girls is 55 kgs. Find the number of girls in the class. [**Ans.** 100 girls]

5. In a certain examination, the average grade of all students in Class '*A*' is 68.4 and of students in Class '*B*' is 71.2. If the average of both classes combined is 70, find the ratio of the number of students in Class '*A*' to the number of students in Class '*B*'. [**Ans.** 3/4]

6. The mean weight of 15 students is 110 kgs. The mean weight of 5 of them is 100 kgs and another 5 is 125 kgs. What is the mean weight of the remainder ? [**Ans.** 105 kgs]

7. Fifty students took up a test. The result of those who passed the test is given below :

Marks	4	5	6	7	8	9
No. of Students	10	10	9	6	4	3

If the average of all the 50 students was 5.16 marks, find the average of those who failed. [**Ans.** 1.625]

8. The mean marks in statistics of 100 students of a class was 72. The mean of marks of boys was 75, while their number was 70. Find out the mean marks of girls in the class. [**Ans.** 65 girls]

9. The mean age of a combined group of men and women is 25 years. The mean age of the group of men is 26 years and that of the group of women is 21 years. Find out the percentage of men and women in the group.
[**Ans.** Men 80 and Women 20]

10. The mean age of a combined group of men and women is 30 years. If the mean age of the group of men is 32 and that of group of women is 25, find out the percentage of men and women in the group.
[**Ans.** Men 71.93 and Women 28.57]

11. Average rainfall of a city from Monday to Saturday is 0.3 inch. Due to heavy rainfall on Sunday the average rainfall for the week increased to 0.5 inch. What was the rainfall on Sunday ? [**Ans.** 1.7 inch]

12. Calculate the mean from the following data :

x	10	12	18	16	15	19	18	17
f	8	12	6	3	19	13	17	14

[**Ans.** 15.82]

13. Calculate the mean from the following data :

x	5	10	12	20	25	30	35	40	45	50
f	20	43	75	76	42	45	39	9	8	6

[**Ans.** 21.928]

14. Find the average number of children per family for the two sub-groups separately as well as for the distribution as a whole :

Group 1		*Group 2*	
No. of Children	*No. of Families*	*No. of Children*	*No. of Families*
0	10	4–5	20
1	50	6–7	12
2	60	8–9	4
3	40	10–11	4
	160		40

[**Ans.** 1.8125, 6.1, 2.67]

15. Find the A.M. of first n natural numbers. $\left[\textbf{Ans. } \frac{n+1}{2}\right]$

16. Find the A.M. of first 100 natural numbers. [**Ans.** 50.5]

17. From the following data calculate the G.M. :

15, 250, 15.7, 157, 1.57, 105.7, 10.5, 1.06, 25.7 and 0.257. [**Ans.** 12.75]

18. Calculate the G.M. of the followin data :

0.66, 0.24, 0.60, 1.66, 0.02, 0.26, 1.05. [**Ans.** 0.36]

19. A taxi ride in Delhi costs one rupee for the first km and 60 paise for each additional km. The cost of each km is incurred at the beginning of the km, so that the rider pays for the whole km. What is the average cost for $2\frac{3}{4}$ km ? [**Ans.** Rs. 0.80]

20. Peter travels by car for 4 days. He drove 10 hours each day. He drove first day at the rate of 45 km per hour, second day at the rate of 40 km per hour, third day at the rate of 38 km per hour and fourth day at the rate of 37 km per hour. What is his average speed ? [**Ans.** 40 km]

21. Calculate the G.M. of the following :

375, 0.05, 0.5672, 0.854 [**Ans.** 1.736]

22. The A.M. of two observations is 25 and their G.M. is 15, find their H.M. [**Ans.** 9]

23. If G.M. is 60 and H.M. is 28.24 of the two observations, find their A.M. [**Ans.** 127.478]

24. The geometric mean and harmonic mean of two observations are respectively 18 and 10.8. Find the observations. [**Ans.** 54, 6]

25. In a month of 31 days there are five Mondays, Tuesdays and Wednesdays. Number of accidents on various days is given below. Find average number of accidents in a month.

Day	*Sunday*	*Monday*	*Tuesday*	*Wednesday*	*Thursday*	*Friday*	*Saturday*
Accidents	10	8	6	8	6	8	14

[**Ans.** 8.45]

26. Income of employees in an industrial concern are given below. The total income of 10 employees in the class over Rs. 250 in Rs. 3000. Every employee belonging to the top 25% of the earners is required to pay 1% of his income to workers relief fund. Find the contribution to this fund.

Income (Rs.)	0–50	50–100	100–150	150–200	200–250	250 & over
Frequency	90	150	100	80	70	10

[**Ans.** Rs. 271.20]

27. Twenty boats make 6 transatlantic trips each per year and 10 boats make 4 trips per year. What is the average number of days for a 'turn around', *i.e.,* the time between consecutive departures from the same port ? Take the year as 360 days for convenience. [**Ans.** 67.5]

[**Hint :** Average number of days for a trip $= \frac{360 \times 30}{(20 \times 6 + 10 \times 4)}$]

B. Questions for Calculating Median, Mode, Quartiles, Deciles and Percentiles

1. The marks obtained by 15 students in a class test are given below. Find (i) Mean, (ii) Median, (iii) Lower Quartile and, (iv) Upper Quartile.

6, 9, 10, 12, 18, 19, 23, 23, 24, 28, 37, 48, 49, 53 and 60

[**Ans.** $\overline{X} = 27.93$, Median = 23, $Q_1 = 12$ or $Q_3 = 48$]

2. Calculate mean, median and mode from the following data of the heights in inches of a group of students :

61, 62, 63, 61, 63, 64, 64, 60, 65, 63, 64 65, 66, 64

Now suppose that a group of students whose heights are 60, 66, 59, 68, 67 and 70 inches, is added to the original group. Find mean, median and mode of the combined group. [$\overline{X} = 63.2$, $M = 63.5$; $M_0 = 64$; for the combined group] [**Ans.** $\overline{X} = 63.75$, Median = 64, Mode = 64]

3. The numbers of telephone cells received in 245 successive one minute intervals at an exchange are shown in the following frequency distribution :

Number of Calls	0	1	2	3	4	5	6	7
Frequency	14	21	25	43	51	40	39	12

Calculate the mean, median and mode. [**Ans.** $\overline{X} = 3.763$, Median = 4, Mode = 4]

4. The numbers of fully formed tomatoes on 100 plants were counted with the following results :

2	Plants	Had	0	Tomatoes
5	"	"	1	"
7	"	"	2	"
11	"	"	3	"
18	"	"	4	"
24	"	"	5	"
12	"	"	6	"
8	"	"	7	"
6	"	"	8	"
4	"	"	9	"
3	"	"	10	"

(i) How many tomatoes were there in all ?

(ii) What was the average number of tomatoes per plant ?

(iii) What was the mode or modal number of tomatoes ? [**Ans.** (i) 486, (ii) $\overline{X} = 4.86$, (iii) Mode = 5]

5. Calculate the median, 7th decile and 35th percentile for the following data :

Value	*Frequency*	*Value*	*Frequency*
Less than 10	4	Less than 50	96
Less than 20	16	Less than 60	112
Less than 30	40	Less than 70	120
Less than 40	76	Less than 80	125

[**Ans.** Median = 36.25, 7th Decile = 45.75, 35th Percentile = 31.04]

6. Determine the median wage graphically from the following data :

Wages (in '00 Rs.)	*No. of Workers*
20–40	4
40–60	6
60–80	10
80–100	16
100–120	12
120–140	7
140–160	3

[**Ans.** 9125]

7. Find the mode and median of the following distribution :

Marks	5–10	10–15	15–20	20–25	25–30	30–35	35–40	40–45
No. of Students	5	6	15	10	5	4	2	2

[**Ans.** 18.21, 19.5]

8. A.M. of the following frequency distribution is 67.45. Find f_3.

Height (in inches)	61	64	67	70	73
Frequency	15	54	f_3	81	24

[**Ans.** 113]

9. From the following frequency distribution of 100 items the median was found to be 53.5. Find the missing frequencies :

Variate	20–30	30–40	40–50	50–60	60–70	70–80
Frequency	8	5	2	20	2	5

[**Ans.** 30, 32]

10. Calculate mode and median from the following data :

Age	20–25	25–30	30–35	35–40	40–45	45–50	50–55	55–60
Frequency	50	70	100	180	150	120	70	60

[**Ans.** Mode = 38.82, Median = 40]

11. The following table gives the marks of 100 students in statistics. Draw the Ogive curve and find the median :

Marks	70–80	60–70	50–60	40–50	30–40	20–30	10–20
Frequency	5	6	20	31	22	9	7

[**Ans.** Median = 43.87]

12. Draw the Ogive based on the following data and estimate median and quartiles :

Years Less than	10	20	30	40	50	60
No. of Persons	12	31	44	55	65	70

[**Ans.** Median = 23.1, Q_1= 12.9, Q_3 = 37.7]

13. Compute mean and mode of the following distribution of marks obtained by 125 students in statistics at a certain examination :

Marks	0–10	10–20	20–30	30–40	40–50
No. of Students	3	4	8	10	15
Marks	50–60	60–70	70–80	80–90	90–100
No. of Students	35	20	16	8	6

[**Ans.** Mode = 55.7, $\overline{X}$ = 55.8]

14. Find the median and mode from the following data :

Class	0–3	3–6	6–10	10–12	12–15	15–18
f	4	8	10	14	16	20
Class	18–20	20–24	24–28	28–30	30–36	
f	24	14	11	10	6	

[**Ans.** Median = 17.55, Mode = 18.57]

15. The arithmetic mean of 10 values is 12.5. The A.M. of first 6 values is 15 and that of last five values is 10. Find the 6th value. [**Ans.** 15]

16. The following table gives the frequency distribution of marks obtained by 199 students in an examination :

Marks	0–20	21–30	31–40	41–50	51–60	61–70	71–80
No. of Students	21	19	60	42	24	18	15

Find the median and the percentage of failure if the minimum marks for a pass is 35.

[**Ans.** Median = 40.5, percentage of failure is 35.2]

17. Calculate arithmetic mean and median from the following data :

Marks Less than	80	70	60	50	40	30	20	10
No. of Students	100	90	80	60	32	20	13	5

[**Ans.** $\overline{X}$ = 45, Median = 46.43]

18. Calculate mode from the following data :

Marks above	0	10	20	30	40	50	60	70	80	90	100
No. of Students	80	77	72	65	55	43	28	16	10	8	0

[**Ans.** Mode = 55]

19. From the figures given below, find the mode, median and quartiels. What information could you deduce from them ?

Age	*No. of Persons*	*Age*	*No. of Persons*
20–25	50	40–45	150
25–30	70	45–50	120
30–35	100	50–55	70
35–40	180	55–60	59

[**Ans.** Mode = 38.64, Median = 39.99, Q_1 = 34, Q_3 = 47.08]

20. Calculate the median, mode and seventh percentile of the following distribution :

Age	0–10	10–20	20–30	30–40	40–50	50–60	60–70	Total
Frequency	2	11	20	26	30	10	1	100

[**Ans.** $\overline{X} = 35.5$, Median $= 36.54$, Mode $= 41.67$, $P_7 = 14.57$]

21. Calculate the median of the following data :

Less than	10	20	30	40	50	60	70	80
Frequency	4	16	40	76	96	112	120	125

[**Ans.** Median $= 36.25$]

22. Calculate the median and mode for the distribution of the weights of 150 students from the data given below :

Weight (Kg.)	*Frequency*
30–40	18
40–50	37
50–60	45
60–70	27
70–80	15
80–90	8

[**Ans.** Median $= 54.45$, Mode $= 54.08$]

23. The mean monthly salary paid to all workers in a concern is Rs. 4600. The mean monthly salary paid to skilled and unskilled workers average Rs. 5000 and Rs. 4000 respectively. Find out the percentage of skilled and unskilled workers employed in the concern. [**Ans.** Skilled 60, Unskilled 40]

24. Draw the less than Ogive and estimate the value of the median on the basis of data given below :

Mid-Point	18	25	32	39	46	53	60	Total
Frequency	10	15	32	42	26	12	9	146

[**Ans.** Median $= 38.17$]

25. Draw a cumulative frequency graph of the following distribution showing daily wages of a group of workers and locate first and their quartiles in the graph.

Daily Wages	100	110	120	130	140	150	160
No. of Workers	8	10	12	16	20	25	15

26. (a) Distinguish between simple and weighted averages. State the circumstances under which each should be used.

(b) Calculate the simple and weighted average price per mound of coal purchased by an industry for the half year. Account for the difference between the two.

Month	Jan.	Feb.	March	April	May	June
Price Per Mounds	42.50	51.25	50.00	52.00	44.25	54.00
Mounds Purchased	25	30	40	50	10	45

[**Ans.** (a) $\overline{X} =$ Rs. 49, (b) $\overline{X}_w = 50.36$]

27. Convert the following into an ordinary frequency table and find out mode :

5 students get less than 3 marks
12 " " " 6 "
25 " " " 9 "
30 " " " 12 " [**Ans.** 7.29]

27. The population of five towns A, B, C, D, E was 20,000, 26,000, 23,000, 25,000 and 24,000 respectively. The average income of the residents for the perspective towns was 280, 270, 240, 230 and 300. Find out the average income per head for all the towns combined. [**Ans.** Combined $\overline{X} = 263.47$]

29. A factory has 5 sections employing 105, 184, 130, 93 and 124 workers. The mean earnings in a certain week per workers are Rs. 1384, 1512, 1527, 1819 and 1,422 for the 5 sections. Determine the mean earnings per worker of the whole factory. [**Ans.** Combined $\overline{X} = 1521$]

30. The following frequency distribution shows the marks secured by 885 students in statistics out of 100 at the B.Com. Examination of a university :

Class Interval	*Frequency*
0–10	18
10–20	25
20–30	41
30–40	89
40–50	127
50–60	250
60–70	160
70–80	92
80–90	63
90–100	20

(a) Find the median marks by drawing an 'Ogive'.
(b) Check up the value of the median so found by using the standard formula for finding the median.
[**Ans.** Median = 55.7]

31. In the following data two class frequencies are missing :

Class	100–110	110–120	120–130	130–140	140–150
Frequency	4	7	15	?	40
Class	150–160	160–170	170–180	180–190	190–200
Frequency	?	16	10	6	3

However, it was possible to ascertain that the total number of frequency is 150 and that the median has been correctly found out as 146.25.
You are required to find out with the help of information given :
(a) Two missing frequencies.
(b) Having found missing frequencies calculate arithmetic mean.
(c) Without using the direct formula, find value of the mode.
[**Ans.** (a) Frequency of 130-140 class = 24 and of 150-160 = 25, (b) $\overline{X} = 147.33$, (c) Mode = 144.08]

32. The median marks of 100 students were 45. But, later on it was found that the marks of a student who secured 64 were wrongly taken as 46. Find the correct Median. [**Ans.** 45, *i.e.,* no change in the median value]

33. Find the most suitable average in the following cases :
(i) It is desired to estimate average expenditure of a family.
(ii) It is to be representative expenditure.
(iii) It is desired to study the Pattern of expenditure.
(iv) Average speed of work.
(v) Average income of workers in a firm.
(vi) Average size of oranges on a tree.

(vii) Average marks of a student in a class.
(viii) Average size of Agriculture Holdings.
(ix) Average rate of growth of population in a given period.
(x) Average number of children in a family.
(xi) Average intelligence of students in a class.
(xii) Average size of collars.
(xiii) Average income of a lawyer.
(xiv) Average size of readymade garments.
(xv) Average change in prices.
(xvi) Average level of health.

[**Ans.** (i) A.M. (ii) Mode (iii) A.M. (iv) A.M. (v) Mode (vi) Mode (vii) A.M. (viii) A.M. (ix) G.M. (x) Mode (xi) Median (xii) Mode (xiii) A.M. (xiv) Mode (xv) A.M. (xvi) Mode.]

34. A candidate obtained the following percentages of marks in different subjects in the Half-Yearly Examinations :

English	46%
Statistics	67%
Cost Accountancy	72%
Economics	58%
Income Tax	53%

It is agreed to give double weights to marks in English and Statistics as compared to other subjects. What is the simple and weighted arithmetic mean ? **(D.U., B. Com. 2002)**

[**Hint.** Simple A.M. $= \frac{46+67+72+58+53}{5} = \frac{296}{5} = 58.2\%$

Weighted A.M. $= \frac{46\times 2+67\times 2+72+58+53}{7} = \frac{409}{7} = 58.42\%$]

35. If $x_1, x_2, x_3, \ldots, x_n$ are n values of the variable x and A is any arbitrary value, so that $\Sigma\,(x_i - A)$ is the sum of deviations from A; show that

$\Sigma\,(x_i - A) = n\,(\bar{x} - A)$ where $\bar{x} = \frac{1}{n}\Sigma\, x_i$ is the arithmetic mean of x.

Hence, $\Sigma\,(x_i - A) = 0$ if and only if, $A = \bar{x}$.

[**Hint.** $\Sigma\,(x_i - A) = \Sigma\, x_i - nA$

$$= n\frac{\Sigma\, x_i}{n} - nA = n\bar{x} - nA = n\,(\bar{x} - A)$$

$$\Sigma\,(x_i - A) = 0 \Rightarrow n\,(\bar{x} - A) = 0$$

$$\Rightarrow \bar{x} = A.]$$

❑❑❑

4

Measures of Variation

Coverage Meaning and Significance of Dispersion, Methods of Measuring Dispersion, Range, Quartile Deviation, Mean Deviation, Standard Deviation and Coefficient of Variation, Lorenz Curve.

4.1 INTRODUCTION AND DEFINITION

Measures of central tendency are called **Averages of the first order.** But these average are not *sensitive* to the variability among the data. Two distributors may have the same **mean**, **same median**, **same mode** but the variability among the data in two cases may be quite different. Consider the two groups *A* and *B*. In the table 1.

Table 1

Group A	*Group B*
65	42
66	54
67	58
68	62
71	67
72	77
74	77
77	85
77	93
77	100

Computing the averages, we get

Group A	*Group B*
Mean = 71.5	Mean = 71.5
Median = 72.0	Median = 72.0
Mode = 77	Mode = 77

It is clear that the two groups *A* and *B* have the same values of **mean**, **median** and **mode** but still the careful perusal of data in both the groups show an obvious difference that the values in group *B* are much more widely scattered than the values in group *A*.

Some time the two series may have similar formation as shown in Table 2 but their measures of central tendency may be different, *e.g.,* the series have entirely different average value but have the same formation. Clearly the measures of central tendency do not indicate how the individual values in the distribution differ from each other or from the central value. When the extent of variation of individual values in relation to other values or in relation to a central value is large the measures of central tendency fail to represent the series fully. The measures of dispersion (or variability) coupled with the measures of central tendency give a fairly good idea (but not complete idea) about the nature of the distribution.

Table 2

A	*B*	*C*
10	30	100
11	31	101
12	32	102
13	33	103
14	34	104

[*Groups A, B, C have the same formation, but different Arithmetic Means*]

To have complete idea about the distribution we need the measures of variation, skewness and kurtosis also.

Dispersion is the spread or scatter of values from a measure of central tendency.

A measure of dispersion is designed to state the extent to which individual observations (or items) vary from their average. Here, we shall account only the amount of variation and not the direction.

D.C. Brooks and W.F.L. Dick defines dispersion as "*Dispersion or spread is the degree of the scatter or variation of variable about a central value.*"

A measure of variation or dispersion describes the degree of scatter shown by observations and is usually measured as an average deviation about some central value.

According to George Simpson and Fritz Kafka, "*An average does not tell the full story. It is hardly fully representative of a mass unless we know the manner in which the individual items scatter around it. A further description of the series is necessary if we are to gauge how representative the value is.*"

According to Spiegel, "*The degree to which numerical data tend to spread about an average value is called the variation or dispassion of the data.*"

Measures of dispersion are called Averages of the **Second order** because they are based on the deviations of the different values from the mean or other measures of central tendency which are called average of the **First order.**

The difference between an average and the actual value is called **deviation** from the average value and measures the variability, spread or scatter of the value from that average.

Deviations can be taken from an arbitrary value also.

Concept of dispersion or variation is important and fundamental to statistical theory not as merely supplementary to the average but because the scatter in a distribution may itself be significant. It is useful to determine the nature and cause of the variation in order to control the variation itself. In matters of health, variations in body temperature, pulse beat and blood pressure are basic guides to diagnosis. Prescribed treatment is designed to control their variation. In industrial production efficient operation requires uniformity of quality. In case of heavy variations in quality or even production, causes are sought and remedial measures are applied. In social science the measurement of inequality in the distribution of income and wealth requires the measures of variation.

We can judge the reliability of an average with the help of measures of variation. In this regard William A. Spurr and Charles P. Bonini observe, *"when dispersion is small, the average is a typical value in that it closely represents the individual values*, and it is reliable in that it is a good estimate of the corresponding average in the population. On the other hand when the dispersion is great, the average is not so typical and unless the sample is very large the average may be quite unreliable."* A greater amount of dispersion means lack of uniformity or consistency in the data. In such a case no average will reliably represent the series.

4.2 OBJECTIVES OF DISPERSION

Main objective of measuring dispersion are :

1. To know the average variation of different values of the items from the average of a series.
2. To know about the composition of a series or the dispersion of values on either side of the central tendency.
3. To know the range of values (*i.e.,* difference between the highest and the lowest value).
4. To compare the disparity between two or more series expressed in different units in order to find out the degree of variation.
5. To know whether the central tendency truly represents the series or not. If the dispersion is more, the measures of central tendency do not represent the series.

4.3 IMPORTANCE OF DISPERSION

1. Conclusion drawn from the central tendency alone carries no meaning without knowing the variation of the various items of the series from the Average.
2. Inequalities in the distribution of wealth and income can be measured by dispersion.
3. Dispersion is used to compare and measure concentration of economic power and monopoly in the country.
4. Dispersion is used in output control and price control.

4.4 CHARACTERISTICS OF A GOOD MEASURE OF DISPERSION

1. It should be simple to understand and easy to calculate.
2. It should be rigidly defined.
3. It should be based on all the items of the series.
4. It should not be unduly affected by extreme items.
5. It should be least affected by fluctuations of sampling.
6. It should be amendable to further algebraic treatment.

---

* It closely represents the series.

4.5 MERITS AND DEMERITS OF MEASURES OF DISPERSION

As it has been pointed out earlier, there are different measures of dispersion with their relative merits and demerits. But the merits and demerits common to all types of measures of dispersion are outlined as under :

Merits

1. They indicate the dispersal character of a statistical series.
2. They speak of the reliability of dependability of the average value of a series.
3. They enable the statisticians in making comparison between two or more statistical series with regard to the character of their uniformity or consistency or equitability.
4. They enable one in controlling the variability of a phenomenon under his purview.
5. They facilitate in making further statistical analysis of the series through the devices like co-efficient of skewness, co-efficient of kurtosis, co-efficient of correlation, variance analysis, etc.
6. They supplement the measures of central tendency in finding out more and more information relating to the nature of a series.

Demerits

1. They are liable to misinterpretations and wrong generalisation by a statistician of biased character.
2. They are liable to yield inappropriate results as there are different methods of calculating the dispersion.
3. Excepting one or two, most of the methods of dispersion involve complicated process of computation.
4. They by themselves cannot give any idea about the symmetrical or skewed character of a series.
5. Like measures of central tendency, most of the measures of dispersion do not give a convincing idea about a series to a layman.

4.6 DIFFERENT MEASURES OF DISPERSION

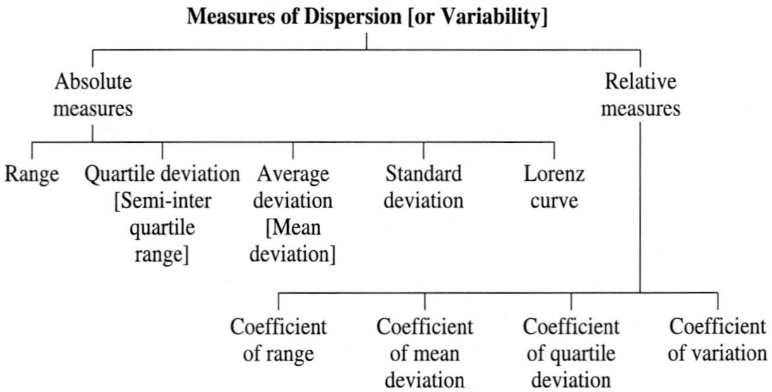

An absolute measure of dispersion is expressed in terms of the units of measurement of the variable. The relative measure of dispersion generally known as coefficient of dispersion is expressed as a pure number independent of the units of measurement of the variable. The main disadvantage of the absolute measures of dispersion is that it can not be used to compare the variability of two distributions expressed in different units. Comparison of distributions with regard to their variability from the central value is done by relative measures of dispersion.

4.7 RANGE

It is defined as difference between the extreme values in the distribution, *i.e.,*

Range = Largest value in the distribution – Smallest value in the distribution.

In case of continuous frequency distributions range is calculated by any one of the following two methods.

(1) By substracting the lower limit of the lowest class from the upper limit of the highest class.

Or

(2) By substracting the mid-value of the lowest class from midvalue of the highest class.

Notes :

(1) In calculations of range only the values of the variable are taken into account and the frequencies are completely ignored.

(2) Open-ended distributions have no range because no highest or lowest values exist in an open-ended class.

(3) Some times, variability of two series is measured by range only though it is a rough measure of variability.

4.8 LIMITATIONS OF RANGE

1. Since, it is based on the smallest and the highest values of the distribution, it is unduly affected by the two extreme values.
2. Not useful for frequency distribution.
3. Since, it is not based on all the observations of the series, it is not affected by the values of the various items comprising the distribution.
4. It is affected by sampling fluctuation. It is extremely sensitive to the size of the sample. As the sample size increases, range also tends to increase though not proportionately. Thus, range of the sample is biased estimate of the variability in the population.
5. It is not amenable to algebraic treatment.
6. It cannot be computed if the distribution is open-ended, because in an open-ended distribution mid-values of the first and the last class cannot be determined.

4.9 MERITS AND USES RANGE

1. It is simplest of the all the measures of dispersion. It is simple to calculate and easy to understand. Wessel, Willet and Simone observe *"Range is still an important concept. Although it does not consider the measure in which values are spread out within the distribution but concerns itself only with top and bottom values, the range is still an extremely useful measure."*
2. It gives a broad picture of the data and includes the limits within which all the items occur.

3. The Weather Bureau of a city always employs range to determine the difference between maximum and minimum temperature and computes the daily mean temperature on the basis of these two readings.
4. Range is also used to study variations in the prices of commodities and movement in the prices of securities.
5. Range has special applications in the quality control measures for industrial products. Due to mass production it may not be possible to check each and every item manufactured, therefore the producers set standards to which their goods must conform. For this purpose range control charts are prepared which helps the production engineer in locating the problem. As long as manufactured products conform to set standards, the production process is assumed to be in control.

4.10 COEFFICIENT OF RANGE

It is a relative measure of dispersion used to compare series in different units.

$$\text{Coefficient of range} = \frac{\text{Max. value} - \text{Min. value}}{\text{Max. value} + \text{Min. value}}$$

$$= \frac{\text{Absolute range}}{\text{Sum of the extreme values}}$$

4.11 INTERQUARTILE RANGE

Interquartile range $(IR) = Q_3 - Q_1 =$ Third quartile (Q_3) – First quartile (Q_1)

In this measure extreme values have no effect. But this measure also cannot be said to be a representative measure because it leaves off 50% of the frequency distribution, *i.e.,* 25% before the first quartile and 25% after the first quartile. However, this measure of dispersion gives more sensible results when data is badly skewed.

4.12 PERCENTILE RANGE (P.R.)

$$P.R. = P_{90} - P_{10}.$$

This measure takes into account two values between which the central 80% of the values occur. This measure also does not make use of all the items in the distribution.

4.13 QUARTILE DEVIATION (Q.D.)

= Half the difference between the upper and the lower quartile.

$= \dfrac{Q_3 - Q_1}{2} =$ **Semi-interquartile range.**

Coefficient of quartile deviation $= \dfrac{Q_3 - Q_1}{Q_3 + Q_1}.$

Q_1, Q_2, Q_3 denote first, second and third quartiles respectively.

For Symmetrical Distributions

$$Q_3 - \text{Median} = \text{Median} - Q_1$$

$$\Rightarrow \quad 2 \text{ median} = Q_3 + Q_1 = Q_3 + Q_1 + Q_1 - Q_1 = (Q_3 - Q_1) + 2Q_1$$

$$\text{Median} = Q_1 + \frac{Q_3 - Q_1}{2} \quad \text{...(i)}$$

$$= Q_1 + \text{quartile deviation.} \quad \left[\because \text{Quartile deviation} = \frac{Q_3 - Q_1}{2}\right]$$

$$2 \text{ Median} = Q_3 + Q_1 + Q_3 - Q_3 = 2Q_3 - (Q_3 - Q_1)$$

$$\Rightarrow \quad \text{Median} = Q_3 - \left(\frac{Q_3 - Q_1}{2}\right) \quad \text{...(ii)}$$

$$= Q_3 - \text{quartile deviation.} \quad \left[\because \text{Quartile deviation} = \frac{Q_3 - Q_1}{2}\right]$$

For Asymmetrical Distributions

$$\text{Median} \neq Q_1 + Q.D.$$

$$\text{Median} \neq Q_3 - Q.D.$$

The greater the difference between these values, the greater would be the dispersion in the distribution.

4.14 MERITS OF QUARTILE DEVIATION

The quartile deviation possesses the following merits :

1. It is simple to understand and easy to compute.
2. It is not affected by extreme values.
3. It can be computed even if the distribution has unequal class intervals.
4. It can be computed in case of open-ended distribution.

4.15 LIMITATIONS OF QUARTILE DEVIATION

The following are the main limitations of quartile deviation :

1. It is not based on all the observations of a series because it does not take frequencies below the lower quartile and above the upper quartile into consideration.
2. It is not amenable to algebraic treatment.
3. It is affected by sampling fluctuations.
4. It is a distance on the scale and is not a measure from average. Therefore, it fails to show variations around an average.

Use

The quartile deviation, as a measure of dispersion, is mainly employed in open-ended distributions. In many situations, we encounter such distribution because of the need to keep certain information confidential.

4.16 MEAN DEVIATION

Mean deviation (also called Average deviation) is defined as the arithmetic mean of the absolute deviations* of all the values from their mean or median or mode

i.e., $$\text{Mean deviation} = \sum_{i=1}^{n} \frac{|x_i - \text{Median}|}{n} \text{ or } \sum_{i=1}^{n} \frac{|x_i - \text{Mean}|}{n}.$$

For a frequency distribution :

$$\text{Mean deviation (M.D.)} = \sum_{i=1}^{n} \frac{f|x_i - \text{Mean}|}{N} \text{ when } N = \Sigma f$$

or $$\text{M.D.} = \sum_{i=1}^{n} \frac{f|x_i - \text{Median}|}{N}.$$

But in practice arithmetic mean is generally used because it is amenable to algebraic treatment. Median at times is not an actual quantity while mode may be ill defined. Although the sum of the absolute deviations is the least from the median (if mean and median are not identical) but on the balance arithmetic mean is generally considered to have greater advantage.

In the words of Clark and Schkade, *"Average deviation is the average amount of scatter of the items in a distribution from either the mean or the median, ignoring the signs of the deviations. The average that is taken of the scatter is an Arithmetic mean which accounts for the fact that this measure is often called the mean deviation."*

Note. Absolute value denotes the value without considering the sign, *i.e.,* taking only positive sign, *e.g.,* the absolute value of $6 - 5 = 1$ and $5 - 6$ is also equal to 1. Algebraic sum implies 'by considering all the signs.'

4.17 STEPS TO CALCULATE MEAN DEVIATION IN INDIVIDUAL VALUES (OR OBSERVATIONS)

In case of individual observations, the following steps are involved in the calculation of mean deviation :

1. Calculate the mean or median of a given series.
2. Write down the deviations (dx_i) of each item (x_i) either from the mean or the median without considering the sign.
3. Sum up the deviations disregarding the signs. This is $= \left(\sum_{i=1}^{n} |dx_i|\right)$.
4. Divide the total of the deviations by the number of observations and the resulting value is the mean deviation.

4.18 STEPS TO CALCULATE MEAN DEVIATION—DISCRETE SERIES

In case of discrete series, the computation of mean deviation involves the following steps :

1. Find the mean or median of the series.
2. Write down the deviations of the various values of the variable from mean or median ignoring signs ($|dx_i|$).

* Deviations when signs of the deviation are ignored.

3. Multiply the deviations by their respective frequencies ($f\,|dx_i|$).
4. Find sum of the products so obtained. This is ($\Sigma f\,|dx|$).
5. Divide the sum of products by the total frequency and the resulting value is the mean deviation. Expressed as a formula form :

$$D. = \frac{(\Sigma f\,|dx|)}{\Sigma f}$$

4.19 CALCULATION OF MEAN DEVIATION—CONTINUOUS SERIES

As regards the calculation of mean deviation in a continuous series, the procedure to be adopted is the same as in the case of discrete series but with a minor difference. Here classes are replaced by mid-values and frequencies are multiplied by deviation of the mid-values from the mean. When the data are in the form of a continuous series, the computation of mean deviation involves the following steps :

1. Write down the mid-values of all classes.
2. Compute the mean or median.
3. Write down the deviations of the mid-values either from mean or from median ($|dx|$) disregarding the signs.
4. Multiply the deviations of each class by their respective class frequencies, *i.e.,* find ($f\,|dx|$).
5. Find the sum of the products obtained in Step 4. *i.e.,* ($\Sigma f\,|dx|$).
6. Divide the total of products by the total frequency and the resulting value is the mean deviation.

4.20 COEFFICIENT OF MEAN DEVIATION

It is a relative measure of dispersion and is computed by the following formulas :

$$\text{Coefficient of Mean deviation} = \frac{\text{Mean deviation}}{\text{Mean}}.$$

(When mean is used as a reference point.)

$$\text{Coefficient of Mean deviation} = \frac{\text{Mean deviation}}{\text{Median}}.$$

(When median is used as a reference point.)

We may also define coefficient of mean deviation as

$$\text{Coefficient of Mean Deviation} = \frac{\text{Mean Deviation}}{\text{Average used}}.$$

4.21 SHORT-CUT METHOD OF FINDING MEAN DEVIATION—INDIVIDUAL SERIES

When mean (or median) is not a whole number but a fraction, the following short-cut formula can be used :

Mean deviation from mean

$$= \frac{[\text{Sum of the values} > \text{Mean}] - [\text{Sum of the values} < \text{Mean}]}{\text{Total Number of Values}}$$

Similarly the mean deviation from the median

$$= \frac{[\text{Sum of the values} > \text{Median}] - [\text{Sum of the values} < \text{Median}]}{\text{Total Number of Values}}.$$

4.22 MEAN DEVIATION IN CASE OF FREQUENCY DISTRIBUTION (SHORT-CUT METHOD)

$$\text{Mean deviation from the mean} = \frac{\Sigma f\,|dx| + (f_a - f_b) \times C}{\Sigma f}$$

$$= \frac{\Sigma f\,|(x - A)| + (f_a - f_b) \times C}{\Sigma f}$$

$\Sigma f\,|dx|$ = Sum of the products of the absolute deviations [considering all deviations positive or negative as positive] and the respective frequencies when the deviations are taken from the assumed mean.

f_a = Sum of the frequencies above the mean = Sum of the frequencies of the values less than the mean.

f_b = Sum of the frequencies below the mean = Sum of the frequencies of the values more than the mean.

C = Difference between the real mean and arbitrary mean.

Note : In the frequency distribution the sum of the frequencies above the mean = Sum of the frequencies of the classes preceding the mean class.

The sum of the frequencies below the mean = Sum of the frequencies of the mean class and the classes succeeding the mean class.

Similarly, we can find the mean deviation from the median.

$$\text{Mean Deviation from the median} = \frac{\Sigma f\,|dx| + (f_a - f_b) \times C}{\Sigma f}.$$

where

$\Sigma f\,|dx|$ = Sum of products of the deviations with their respective frequencies where deviations are taken from the assumed mean.

f_a = Sum of frequencies above the mean.

f_b = Sum of frequencies below the mean.

C = Difference between real mean and arbitrary mean.

Note : The assumed mean or median must be selected from the class in which real mean or median falls.

Mean deviation may also be calculated by the following short-cut method :

$$\text{M.D. from mean} = \frac{\Sigma\, Xf_a - \Sigma\, Xf_a - (\Sigma f_a - \Sigma f_b)\,(\overline{X})}{N}.$$

$\Sigma\, Xf_a$ = Sum of products of midpoint (X) and frequencies above the mean.

$\Sigma\, Xf_b$ = Sum of products of midpoint (X) and frequencies below the mean.

Σf_a = Sum of frequencies of midpoints above the mean.

Σf_b = Sum of frequencies of midpoint below the mean.

N = Total number of observations.

Mean deviation from median :

$$\text{M.D.} = \frac{\Sigma Xf_a - \Sigma Xf_a - (\Sigma f_a - \Sigma f_b) \text{ (medan)}}{N}$$

4.23 STANDARD DEVIATION

The standard deviation is the most important, the most reliable and the most widely used measure of dispersion. The term 'standard' is assigned to this measure of variation probably because of the following reasons :

(i) It is the most commonly used and is the most flexible in terms of variety of applications of all the measures of variation.

(ii) The area under any symmetrical curve rather normal curve remains the same within a fixed number of standard deviations from the mean on either side of it, *e.g.,* in any normal curve area within mean ±1 standard deviation is always 68.27% of the total area and the area is 95.45% of the total area within mean ±2 standard deviations.

(iii) The sum of the squares of the deviations about the mean is the least as compared to the sum of the squares of the deviations about the median or mode, therefore, root mean square deviation about the mean is the least.

It is most important of all the measures of dispersion because it is used in many other statistical operations, *e.g.,* sampling techniques, correlation and regression analysis, finding coefficient of variation, skewness, kurtosis, etc. Standard deviation is also called 'Mean Error' or 'Mean Square Error' or 'Root-Mean Square Deviation'. Unlike the mean deviation, which may be calculated around any average, the standard deviation is always computed around the mean. ***It is the positive square-root of the arithmetic mean of the squared deviations of all the values from the mean.*** It is clear that standard deviation is a measure of the spread in a set of observation. If all the values in a set of observations were identical, the deviation of each value from the mean would be zero. In such a completely uniform distribution, the standard deviation would be **equal to zero**, its minimum value. On the other hand, as items are dispersed more and more widely from the mean, the standard deviation becomes larger and larger.

To quote Yule and Kenndall, *"The standard deviation is the square-root of the arithmetic mean of the squares of all the deviations, deviations being measured from the arithmetic mean of the observation."* The standard deviation is always measured from the mean because of the sum of squared-deviations is minimum from mean compared to the sum of squared deviations from any other value.

Measure of S.D. is very important in many practical circumstances. Manufacturers interested in producing items of consistent quality are very much concerned with S.D. If the mean life of a component is four years and the S.D. is very large, it would correspond to many failures large before four years. Quality control requires consistency and consistency requires a relatively small S.D.

4.24 WHY STANDARD DEVIATION IS PREFERRED TO MEAN DEVIATION

The standard deviation is regarded as the better measure of dispersion compared to mean deviation due to the following reasons :

1. In the calculation of mean deviation, the algebraic signs are ignored which is mathematically illogical, whereas in the calculation of standard deviation, these signs are considered.

2. The standard deviation is amenable to algebraic treatment, *i.e.,* we can find the combined standard deviation of two or more series but it is not possible to compute combined mean deviation.
3. The standard deviation is used in the analysis of statistical series such as in correlation, skewness, sampling or the analysis of variance.

The standard deviation is represented by the Greek letter σ (pronounced as sigma) and is computed with the aid of the following formula :

$$\textbf{Standard Deviation} = \sqrt{\frac{\text{Sum of Squared Deviations of all the values from the Mean}}{\text{Number of Observations (Values)}}}$$

In symbols :

$$\sigma = \sqrt{\frac{\sum_{i=1}^{n} (X_i - \overline{X})^2}{n}}^{*}$$

The above formula may also be written as :

$$\sigma = \sqrt{\frac{\sum_{i=1}^{n} x_i^2}{n}},$$

where $x_i = X_i - \overline{X}$ = deviation of ith value from the mean.

For a frequency distribution :

$$\sigma = \sqrt{\frac{\sum_{i=1}^{N} f_i x_i^2}{\Sigma f}}, \text{ where } x_i = X_i - \overline{X}.$$

The standard deviation is called the **root-mean-square** because it is the square-root of the mean of the squared deviations.

The above formula can be conveniently used when X values are not very large and the mean is a whole number, *i.e.,* when mean is not a fraction.

$$\frac{\Sigma (X - \overline{X})^2}{n} = \frac{\Sigma X^2 + \overline{X}^2 - 2\,\Sigma X\overline{X}}{n}$$

$$= \frac{\Sigma X^2}{n} + \frac{n\overline{X}^2}{n} - 2\overline{X}\frac{\Sigma X}{n}$$

$$= \frac{\Sigma X^2}{n} + \overline{X}^2 - 2\overline{X}^2 = \frac{\Sigma X^2}{n} - \overline{X}^2 = \frac{\Sigma X^2}{n} - \left(\frac{\Sigma X}{n}\right)^2.$$

$\therefore$
$$\sigma = \text{S.D.} = \sqrt{\frac{\Sigma X^2}{n} - \left(\frac{\Sigma X}{n}\right)^2}$$

and
$$\text{variance} = \frac{\Sigma X^2}{n} - \left(\frac{\Sigma X}{n}\right)^2. \qquad [\text{variance} = (\text{S.D.})^2]$$

* If all the values are equal, S.D. = 0 but upper limit cannot be defined. Hence, limits of S.D. are 0 to ∞.

In case frequencies of different values (or observations) are given, then

$$\text{S.D.} = \sqrt{\frac{\Sigma fX^2}{\Sigma f} - \left(\frac{\Sigma fX}{\Sigma f}\right)^2}.$$

4.25 SHORT-CUT METHOD FOR FINDING STANDARD DEVIATION

If the X values are larger or fractional, then the deviation of each value is taken from an arbitrarily assumed value (called assumed mean). We define

$$X - A = d, \text{ where } A \text{ is an Assumed mean.}$$

$$\sigma^2 = \frac{\Sigma (X - \overline{X})^2}{n} = \frac{\Sigma (X - A + A - \overline{X})^2}{n}$$

$$= \frac{\Sigma [(X - A) - (\overline{X} - A)]^2}{n} = \frac{\Sigma (d - \overline{d})^2}{n} = \frac{\Sigma d^2}{n} - \left(\frac{\Sigma d}{n}\right)^2 *$$

$$\Rightarrow \qquad \sigma = \sqrt{\frac{\Sigma d^2}{n} - \left(\frac{\Sigma d}{n}\right)^2}.$$

If frequencies are given, then

$$\sigma = \sqrt{\frac{\Sigma f d^2}{\Sigma f} - \left(\frac{\Sigma f d}{\Sigma f}\right)^2}.$$

4.26 CALCULATIONS OF S.D. FOR CONTINUOUS SERIES

In continuous series X values are the middle points (Middle values) of the class intervals. Other calculations remain the same as described earlier for discrete series.

4.27 STEP DEVIATION METHOD FOR CALCULATING STANDARD DEVIATION

This method can be applied to discrete as well as continuous series but is most common in continuous series. In this method each deviation from the assumed mean is divided by a common factor [in continuous series the common factor is normally taken as class interval because if class intervals are equal then the difference between the consecutive middle values is equal to class interval].

If we define

$$d' = \frac{X - A}{h} = \frac{d}{h},$$

where h is a common factor in d values

[If the class intervals are equal, h is taken as equal to the class interval]

then

$$\text{S.D.} = h\sqrt{\frac{\Sigma d'^2}{n} - \left(\frac{\Sigma d'}{n}\right)^2}$$

and if frequencies are given, then

* When $X - A = d$, $\overline{X} - A = \overline{d}$ mean of deviations.

$$\text{S.D.} = h\sqrt{\frac{\Sigma fd'^2}{\Sigma f} - \left[\frac{\Sigma fd'}{\Sigma f}\right]^2}$$

$$\text{S.D.} = h\sqrt{\frac{\Sigma fd'^2}{\Sigma f} - \left(\frac{\Sigma fd'}{\Sigma f}\right)^2}$$

4.28 COMBINED STANDARD DEVIATIONS OF TWO OR MORE GROUPS

Let A and B be two groups with n_1, n_2 the respective number of values and σ_1, σ_2 the respective standard deviations, then their combined S.D. $\sigma_{1,2}$ is given by

$$\sigma_{1,2} = \sqrt{\frac{n_1(\sigma_1^2 + d_1^2) + n_2(\sigma_2^2 + d_2^2)}{n_1 + n_2}},$$

where $d_1 = \overline{X}_1 - \overline{X}_{1,2}$, $d_2 = \overline{X}_2 - \overline{X}_{1,2}$.

$\overline{X}_1$ = A.M. of group A.

$\overline{X}_2$ = A.M. of group B.

$\overline{X}_{1,2}$ = Combined arithmetic mean of the two groups A and B.

4.29 COMBINED S.D. OF *n* GROUPS

Combined S.D. of n groups

$$= \sqrt{\frac{N_1(\sigma_1^2 + d_1^2) + N_2(\sigma_2^2 + d_2^2) + \ldots + N_n(\sigma_n^2 + d_n^2)}{N_1 + N_2 + \ldots + N_n}}$$

4.30 THE COMBINED S.D. OF TWO GROUPS IS ALSO GIVEN BY THE FOLLOWING FORMULA

$$\sigma = \sqrt{\frac{n_1\sigma_1^2 + n_2\sigma_2^2}{n_1 + n_2} + \frac{n_1 n_2}{(n_1 + n_2)^2}(\overline{X}_1 - \overline{X}_2)^2}.$$

This formula is applicable for finding the combined S.D. of two groups only. **(Please note)**

Notes :

(1) If the two groups have the same number of items, *i.e.*, if $n_1 = n_2$, then their combined S.D. is

$$\sqrt{\frac{\sigma_1^2 + \sigma_2^2}{2} + \frac{1}{4}(\overline{X}_1 - \overline{X}_2)^2}. \qquad \text{[Put } n_1 = n_2 \text{ in 4.30]}$$

(2) If $\overline{X}_1 = \overline{X}_2$, *i.e.*, if both the groups have the same mean, then their combined S.D. is

$$\sqrt{\frac{n_1\sigma_1^2 + n_2\sigma_2^2}{n_1 + n_2}}. \qquad \text{[Put } \overline{X}_1 = \overline{X}_2 \text{ in 4.30]}$$

4.31 THE MAIN CHARACTERISTICS OF STANDARD DEVIATION ARE

1. It is rigidly defined.
2. It is based on all observations.
3. In the case of the values which lie close to the mean, the deviations are small and therefore variance and standard deviations are also small. Variance and standard deviation will be zero when all values are equal.
4. If the same amount is added to or subtracted from all the values, the variance and standard deviations shall remain unchanged, *i.e.,* S.D. is independent of change of origin.
5. If all the values are multiplied by the same quantity, the standard deviations of the changed values get multiplied by the same value, *i.e.,* the S.D. depends on the change of scale.
6. In case a number of samples are drawn from the same population, it may be observed that standard deviation is least affected from sample to sample as compared with other measures of dispersions.

4.32 COEFFICIENT OF VARIATION*

$$\text{Karl Pearson's Coefficient of Variation} = \frac{\text{Standard Deviation}}{\text{Mean}} \times 100\%$$

Coefficient of variation is a relative measure of dispersion and has no units. It is usually expressed as percentage. It is used to compare the variability of two or more distributions. The distribution for which the coefficient of variations is more said to be less stable, less uniform, less consistent, less homogeneous, less equitable but more variable.

The main advantage of the coefficient of variation is that the uniformity, consistency, equitability, stability and homogeneity of entirely different distributions, measured in different units can be compared. The comparison is very helpful for actuaries.

4.33 LORENZ CURVE

Graphically the dispersion is studied by means of the *Lorenz Curve.* Lorenz Curve is a cumulative percentage curve in which the percentage of items (or frequencies) are shown with the corresponding percentage of factors like income, wealth, profits, etc. The curve is also used to study the distribution of land, wages, income, etc., among the population of a country.

To draw the Lorenz Curve, we first find the cumulative totals of both, the values of the items and the frequencies (or number of persons) and then express cumulative values as percentage of the total quantity or total number of persons as the case may be. We now represent the percentage of the cumulative values along the *Y*-axis *OY* and the percentage of the cumulative frequencies along the *X*-axis *OX*. Clearly the range of their percentages of both the axis is from 0 to 100. But on the *X*-axis start the scale from 100 and go up to 0 and on the *Y*-axis start the scale from 0 and go up to 100**. The percentage of cumulative values and the percentage of cumulative frequencies are then plotted on a graph paper to obtain Lorenz Curve for the given data. If the wealth are equally distributed among the people then the Lorenz Curve is a straight line obtained by joining 0 on the *X*-axis with 100 on

* Coefficient of variation is also called coefficient of dispersion or coeficient of variability. Sometimes coefficient of standard deviation is defined as $\frac{\text{S.D.}}{\text{Mean}}$.

** Some authors start the scale on both the axis from 0 to 100.

Y-axis. This line is also known as line of equal distribution. Any point on this line shows the same percentage people Lorenz Curve is not a straight line but a curve like *B* below the line of equal distribution.

The area between the line of equal distribution and the plotted curve gives the extent (shown shaded in the Fig. 4.1) of inequality in the items. The larger the area more is inequality.* If curves of various distributions are shown on the same Lorenz presentation, the curve that is farthest from the

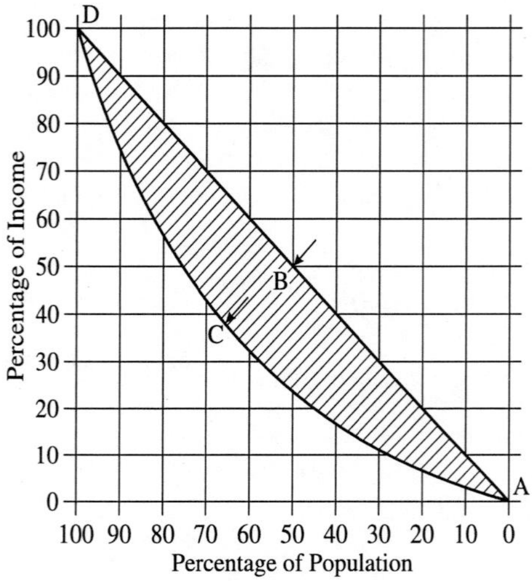

Fig. 4.1. Lorenz Curve.

diagonal line represents greatest inequality.

Mean $\pm$ 1 σ covers 68.27% of the items.
Mean $\pm$ 2 σ covers 95.45% of the items.
Mean $\pm$ 3 σ covers 99.73% of the items.

4.34 RELATION BETWEEN MEASURES OF DISPERSION

For the normal distribution,

(i) Quartile deviation $= \frac{2}{3}$ Standard deviation = 0.67 of S.D.

(ii) Mean deviation $= \frac{4}{5}$ Standard deviation = 0.8 of S.D.

(iii) Arithmetic Mean $\pm$ Q.D. would cover 50% of the items.

(iv) Arithmetic mean $\pm$ Mean deviation would cover 57.51% of the items.

(v) Arithmetic mean $\pm$ S.D. would cover 68.27% of the items.

(vi) For a symmetrical distribution, the following area relationships hold good.

* We can also say that larger the ratio between the area OACDO and the area OADO, the less is dispersion, *i.e.,* inequality and more will be the concentration in the distribution. The ratio is called concentration ratio or Gini's coefficient.

Bienayme-Chebyshev Rule or Theorem

This rule determined by the research of the two mathematicians Bienayme and Chebyshev explains the variability of data about its mean when the distribution of the data set is not known. This rule states

For any set of data (population or sample) and any constant K greater than 1, at least $\left(1-\frac{1}{K^2}\right)\times 100$ per cent of the data must lie within K standard deviations on either side of the mean.

Thus we can be sure that at least $\left(1-\frac{1}{(2)^2}\right)=\frac{3}{4}$ or 75% of the values in any set of data must lie within two standard deviations on either side of the mean; at least $\left(1-\frac{1}{(3)^2}\right)=\frac{8}{9}$ or 88.9% must lie within three standard deviations on either side of the mean.

Chebyshev's theorem applies to **any kind of data regardless of the shape** of the data set and it tells us only 'at least what percentage' of the data must lie between certain limits. Due to this reason the results are usually weak. Further it may be noted that the above formula applies to differences greater than one standard deviation about the mean, *i.e.,* K must be greater than 1. If K = 1, the theorem **fails**.

SOLVED EXAMPLES

Example 1. Given the following numbers : 1, 2, 3, 4, 5.

Calculate (i) Range, (ii) Mean deviation from median, and (iii) Coefficient of variation.

Solution.

(i) Range = Difference between the extreme values = 5 – 1 = 4.

[∵ Highest value = 5, lowest value = 1.]

(ii) Median = Since the number of terms is odd, the median is the value of $\frac{n+1}{2}$th term = value of $\frac{5+1}{2}$ th item = value of the third item = 3 (∵ observations are given in ascending order.)

Obervations (x)	$\lvert x-3\rvert$
1	2
2	1
3	0
4	1
5	2
	$\Sigma\lvert(x-3)\rvert = 6$

M.D. From the medium

$$=\frac{\Sigma\lvert x-3\rvert}{5}=\frac{6}{5}=1.2.$$ [Refer point 4.16]

(iii) A.M. $=\frac{1}{5}(1+2+3+4+5)=\frac{1}{5}\times 15=3.$

∴ A.M. = Median

x	x – mean = x – 3 = d	d^2
1	–2	4
2	–1	1
3	0	0
4	1	1
5	2	4
		$\Sigma d^2 = 10$

$$\sigma = \sqrt{\frac{\Sigma d^2}{n}}* = \sqrt{\frac{10}{5}} = \sqrt{2} = 1.4142.$$

$$\text{C.V.} = \frac{\text{S.D.}}{\text{Mean}} \times 100 = \frac{\sigma}{\text{A.M.}} \times 100$$

[Refer point 4.32]

$$= \frac{1.4142}{3} \times 100 = 47.14.$$

Example 2. If the S.D. of a set of observations is zero then all observations are equal. Comment.

Solution. Yes.

$$(\text{S.D.})^2 = \frac{\Sigma (x - \bar{x})^2}{n}.$$

If S.D. = 0 $\Rightarrow (\text{S.D.})^2 = 0 \Rightarrow \frac{\Sigma (x - \bar{x})^2}{n} = 0 \Rightarrow \Sigma (x - \bar{x})^2 = 0,$

which is possible only if each value is equal to $\bar{x}$.

Example 3. Given x : 9, 7, 5, 11, 1, 5, 7, 3.

Calculate (i) range, (ii) standard deviation. Also calculate these measures with $-x$ and compare them with the corresponding values of x.

Solution.

(i) Arranging the values of x in ascending order

$$x : 1, 3, 5, 5, 7, 7, 9, 11$$

$$\text{Mean} = \frac{1 + 3 + 5 \times 2 + 7 \times 2 + 9 + 11}{8} = \frac{48}{8} = 6.$$

$$\text{Median} = \frac{5 + 7}{2} = 6.$$ [Mean of 4th and 5th value is the median]

$$\text{Range} = 11 - 1 = 10.$$

(ii) $\Sigma x^2 = (1)^2 + (3)^2 + 2 \times (5)^2 + 2 \times (7)^2 + (9)^2 + (11)^2 = 360.$

$$\text{S.D.} = \sqrt{\frac{\Sigma x^2}{n} - \left(\frac{\Sigma x}{n}\right)^2} = \sqrt{\frac{360}{8} - (6)^2} = \sqrt{45 - 36} = 3.$$

---

* When the deviations are taken from the actual mean S.D. = $\sqrt{\frac{\Sigma d^2}{n}}$ where $d = X - \bar{X}$

If x is negative,

$$\text{Range} = -1 - (-11) = 10.*$$
$$\text{S.D.} = 3$$

[$\because$ If x is –ve, Σx^2 and $\left(\frac{\Sigma x}{n}\right)^2$ will remain unchanged.]

$\therefore$ Range and S.D. remains unchanged with negative x.

Example 4. Calculate semi-interquartile range and coefficient of quartile deviation from the following data :

Age (in year)	15–25	25–35	35–45	45–55	55–65	65–75	75–85
No. of persons	3	61	132	153	140	51	3

Solution :

Age	*Freq.*	*cf*
15–25	3	3
25–35	61	64
35–45	132	196
45–55	153	349
55–65	140	489
65–75	51	540
75–85	3	543
	543	

$$\frac{N}{4} = \frac{543}{4} = 135.75, \frac{3N}{4} = 407.25.$$

$$Q_1 = 35 + \frac{135.75 - 64}{132} \times 100 = 40.435.$$

$$Q_3 = 55 + \frac{407.25 - 349}{140} \times 10 = 59.16.$$

$$Q_3 - Q_1 = 18.725, Q_3 + Q_1 = 99.595.$$

$$\text{Semi-interquartile range} = \frac{Q_3 - Q_1}{2} = \frac{18.725}{2} = 9.3625$$

$$\text{Coefficient of quartile deviation} = \frac{Q_3 - Q_1}{Q_3 + Q_1} = \frac{18.725}{99.595} = 0.188.$$

Example 5. Calculate the mean deviation a query from the following table :

Weight (lbs.)	95–105	105–115	115–125	125–135
No. of Students	20	26	38	16

* If x is negative, the given values becomes –1, –3, –5, –5, –7, –7, –9, –11.
Highest value = – 1, Lowest Value = –11, therefore Range = –1 – (–11).

Solution.

Class Interval	*Mid-value*	*Freq.*	$x - 120 = d$	$\frac{d}{10} = d'$	fd'	$x - 115$	$\lvert x - 115\rvert$	$f\lvert x - 115\rvert$
95–105	100	20	–20	–2	–40	–15	15	300
105–115	110	26	–10	–1	–26	–5	5	130
125–125	120	38	0	0	0	5	5	190
125–135	130	16	10	1	16	15	15	240
		100						860

$$\text{Mean} = 120 - \frac{20}{200} \times 10 = 115.$$

$$\text{Mean deviation from the mean} = \frac{\sum_{i=1}^{n} f_i \lvert(x_1 - \bar{x})\rvert}{\Sigma f} = \frac{860}{100} = 8.6.$$

Example 6. Calculate the mean deviation from the mean for the following data :

Marks	0–10	10–20	20–30	30–40	40–50	50–60	60–70
No. of Students	6	5	8	15	7	6	3

Solution. **Computation table**

Class Interval	*Mid-value* (x)	*Frequency* (f)	fx	$\lvert x - \bar{x}\rvert = \lvert x - 34.4\rvert$	$\Sigma f\lvert x - \bar{x}\rvert$
0–10	5	6	30	28.4	170.4
10–20	15	5	75	18.4	92.0
20–30	25	8	200	8.4	67.2
30–40	35	15	525	1.6	24.0
40–50	45	7	315	11.6	81.2
50–60	55	6	330	21.6	129.6
60–70	65	3	195	31.6	94.8
		$\Sigma f = 50 = N$	$\Sigma fx = 1670$		659.2

$$\bar{x} = \frac{\Sigma fx}{N} = \frac{1670}{50} = 33.4.$$

$$\text{Mean deviation} = \frac{\Sigma f\lvert x - \bar{x}\rvert}{\Sigma f} = \frac{659.2}{50} = 13.184.$$

Example 7. The mean of 50 items is 25 and their S.D. is 2. Find the sum and sum of squares of all the items.

Solution. Sum of the item $= 50 \times 25 = 1250$

$$(\text{S.D.})^2 = \frac{\Sigma x^2}{n} - \left(\frac{\Sigma x}{n}\right)^2 = \frac{\Sigma x^2}{n} - (\bar{x})^2$$

$$\Rightarrow 4 = \frac{\Sigma x^2}{n} - 625 \Rightarrow \frac{\Sigma x^2}{n} = 629 \Rightarrow \Sigma x^2 = 629 \times 50 = 31450$$

$\Rightarrow$ Sum of the squares of all the items = 31450.

Example 8. Twenty passengers were found ticketless on a bus. The sum of the squares and the S.D. of the amount found in their pockets were Rs. 2,000, and Rs. 6 respectively. If the total fine imposed on these passengers is equal to the total amount recovered from them and the fine imposed is uniform, what is the amount each one of them has to pay as fine. What difficulties do you visualise if such a system of penalty were imposed. **(Delhi Univ. B.A. Hons., Eco., 1993)**

Soluton. S.D. $= \sqrt{\frac{\Sigma x^2}{n} - \left(\frac{\Sigma x}{n}\right)^2} \Rightarrow 6 = \sqrt{\frac{2000}{20} - \left(\frac{\Sigma x}{20}\right)^2}$

$$\Rightarrow \quad 36 = 100 - \left(\frac{\Sigma x}{20}\right)^2 \Rightarrow \left(\frac{\Sigma x}{20}\right)^2 = 64$$

$$\Rightarrow \quad \frac{\Sigma x}{20} = 8*$$

i.e., each passenger will have to pay Rs. 8 as fine. In such a system, the persons having more money in their pockets may object.

Example 9. Prove that the S.D. is independent of the change of origin and depends on the change of scale.

Solution. S.D. $= \sqrt{\frac{\sum_{i=1}^{n} (x_i - \bar{x})^2}{n}}$

If each value x_1 is increased or decreased by a constant quantity, the mean is also increased or decreased by the same constant quantity and the expression $(x_i - \bar{x})$ remains unaffected. Hence S.D. is independent of the change of origin.

If we multiply (or divide) each value by a constant quantity, the mean is also multiplied or divided by the same constant quantity and hence the expression $\sum_{i=1}^{n} (x_i - \bar{x})$ is also multiplied as divided by the same constant quantity.

Therefore S.D. depends on the change of scale.

Example 10. Find the S.D. of the first n natural numbers.

Solution. The first n natural numbers are 1, 2, 3, 4, ..., n.

$$\text{S.D.} = \sqrt{\frac{\Sigma x^2}{n} - \left(\frac{\Sigma x}{n}\right)^2}$$

$$= \sqrt{\frac{1^2 + 2^2 + 3^3 + \ldots + n^2}{n} - \left[\frac{1 + 2 + \ldots + \ldots + n}{n}\right]^2}$$

$$= \sqrt{\frac{n(n+1)(2n+1)}{6n} - \left(\frac{n(n+1)}{2n}\right)^2} \;**$$

* The fine imposed on the passengers is uniform implies that each of the passengers has to pay the same amount as fine. Since the total fine imposed = Total amount recovered, the fine imposed on each passenger is equal to the A.M. of the amounts available with all the passengers.

** Sum of the squares of first n natural numbers $= 1^2 + 2^2 + 3^3 \ldots + n^2 = \frac{n(n+1)(2n+1)}{6}$

Sum of first n natural numbers $= 1 + 2 + 3 + 4 + \ldots + n = \frac{n(n+1)}{2}$.

$$= \sqrt{\frac{(n+1)(2n+1)}{6} - \left(\frac{n+1}{2}\right)^2}$$

$$= \sqrt{\frac{n+1}{2}} \sqrt{\left[\frac{2n+1}{3} - \frac{n+1}{2}\right]}$$

$$= \sqrt{\frac{n+1}{2}} \sqrt{\frac{4n+2-3n-3}{6}}$$

$$= \sqrt{\frac{n+1}{2}} \sqrt{\frac{n-1}{6}} = \sqrt{\frac{n^2-1}{12}}.$$

Example 11. Answer the following :

(i) The standard deviation of a set of 50 items is 10. Find the standard deviation if every item is increased by 5.

(ii) Find the coefficient of variation if the sum of squared deviations taken from mean 40 of 10 observations is 360.

(iii) If the coefficient of mean deviation is 0.44 and mean deviation from mean is 5.77, find the mean.

Solution.

(i) The standard deviation is independent of the change of origin. Therefore the value of the S.D. will remain unchanged if each value is increased by 5.

(ii) Mean = 40.

Sum of squared deviations taken from mean $\sum_{i=1}^{n} (x_i - \bar{x})^2 = 360.$

$$\therefore \text{ S.D.} = \sqrt{\frac{360}{10}} = 6$$

$$\therefore \text{ C.V.} = \frac{\text{S.D.}}{\text{Mean}} \times 100 = \frac{6}{40} \times 100 = 15.$$

(iii) Coeff. of mean deviation $= \frac{\text{M.D.}}{\text{Mean}} \Rightarrow 0.44 = \frac{5.77}{\text{Mean}} \Rightarrow \text{Mean} = \frac{5.77}{0.44}$

$\Rightarrow$ Mean = 13.11.

Example 12. Find the mean deviation from the mean and the standard deviation from the following sample observations on weight (gms) of a certain product :

19. 22, 20, 21, 18, 23, 21, 22, 20, 21, 21, 22, 21, 18, 21, 26.

Solution. $n = 16, \Sigma x = 336*$ $\therefore \bar{x} = \frac{336}{16} = 21.$

Mean deviation from mean $= \frac{1}{16} \Sigma |x_i - \bar{x}|$

$$= \frac{1}{16} [2 + 1 + 1 + 0 + 3 + 2 + 0 + 1 + 1 + 0 + 0 + 1 + 0 + 3 + 0 + 5]$$

$$= \frac{20}{16} = 1.25.$$

$$\text{S.D.} = \sqrt{\left[\frac{1}{16} \Sigma (x_i - 21)^2\right]}$$

$$= \sqrt{\frac{1}{16} (4 + 1 + 1 + 9 + 4 + 1 + 1 + 1 + 9 + 25)}$$

$$= \sqrt{\left[\frac{1}{16} \times 56\right]} = \sqrt{3.5} = 1.87$$

Example 13. Find the mean deviation about arithmetic mean of the first 10 natural numbers.

Solution. Sum of first 10 natural numbers $= \frac{10\,(10 + 1)**}{2} = \frac{55}{10} = 5.5.$

First 10 natural numbers (x)	$x - \bar{x} = x - 5.5$	$\lvert x - 5.5 \rvert$	
1	–4.5	4.5	Mean deviation from the mean $= \frac{\lvert x_i - \bar{x} \rvert}{n}$ $\frac{25}{10} = 2.5$
2	–3.5	3.5	
3	–2.5	2.5	
4	–1.5	1.5	
5	–0.5	0.5	
6	0.5	0.5	
7	1.5	1.5	
8	2.5	2.5	
9	3.5	3.5	
10	4.5	4.5	
		25	

* (i) $\Sigma x = 19 + 22 + 20 + 21 + 17 + 23 + 21 + 22 + 20 + 21 + 21 + 22 + 21 + 18 + 21 + 26 = 336.$

(ii) Data can also be arranged as a frequency distribution

x	f
19	1
22	3
20	2
21	6
18	2
23	1
26	1

** Sum of first n natural numbers $= \frac{n(n+1)}{2}$.

Example 14. Find the mean deviation about median of the following series :

Marks (x)	5	10	15	20	25
Student (f)	6	7	8	11	8

Solution.

Marks (x)	f	$c.f.$	$\lvert d\rvert$ $\lvert x-15\rvert$	$f\lvert d\rvert$
5	6	6	10	60
10	7	13	5	35
15	8	21	0	0
20	11	32	5	55
25	8	40	10	80
Total	40			230

$$\text{Median} = \text{value of } \frac{N+1}{2}\text{th term} = \text{Value of } \frac{40+1}{2} = 20.5\text{th term} = 15.$$

$$\text{Mean deviation} = \frac{\Sigma f\lvert d\rvert}{\Sigma f} = \frac{230}{40} = 5.75 \text{ marks.}$$

Example 15. Determine the standard deviation for the following data :

Yield (in gm) :	1,216	1,374	1,167
	1,232	1,407	1,453
	1,202	1,372	1,278
	1,141	1,221	1,329

Solution.

(x)	$x - 1200 = d$	d^2
1216	16	256
1232	32	1024
1202	02	4
1141	–59	3481
1374	174	30276
1407	207	42849
1372	172	29584
1221	21	441
1167	–33	1089
1453	253	64009
1278	78	6084
1329	129	16641
	992	195738

$$\text{S.D.} = \sqrt{\frac{\Sigma d^2}{n} - \left(\frac{\Sigma d}{n}\right)^2} = \sqrt{\frac{195738}{12} - \left(\frac{992}{12}\right)^2} \quad \text{[Assumed mean = 1200]}$$

$$= \sqrt{16311.5 - 6833.7} = 97.35.$$

Example 16. Prove that the standard deviation calculated from two values x_1 and x_2 of a variable x is equal to half their difference.

Solution. Sum of the squares of the two values $= \Sigma x^2 = x_1^2 + x_2^2$.

Sum of the two values $= \Sigma x = x_1 + x_2$.

$$\text{S.D.} = \sqrt{\frac{\Sigma x^2}{n} - \left(\frac{\Sigma x}{n}\right)^2}$$

$$= \sqrt{\frac{x_1^2 + x_2^2}{2} - \left(\frac{x_1 + x_2}{2}\right)^2} = \sqrt{\frac{2x_1^2 + 2x_2^2 - x_1^2 - x_2^2 - 2x_1x_2}{4}}$$

$$= \sqrt{\frac{x_1^2 + x_2^2 - 2x_1x_2}{4}} = \sqrt{\left(\frac{x_1 - x_2}{2}\right)^2} = \frac{x_1 - x_2}{2} \text{ or } \frac{x_2 - x_1}{2}.$$

$\therefore$ S.D. $= \sigma = \frac{1}{2}(x_1 - x_2)$ if $x_1 > x_2$ and $\frac{x_2 - x_1}{2}$ if $x_2 > x_1$ since σ is always positive.

Example 17. Calculate the mean deviation from the mean for the following weight distribution :

Weight in gms.	0–5	5–10	10–15	15–20	20–25
No. of items produced	5	8	15	16	6

Solution.

Weight	*Mide-value x*	$\frac{x - 12.5}{5} = d'$	f	fd'	$\lvert x - \bar{x} \rvert = \lvert x - 13.5 \rvert$	$f \lvert x - \bar{x} \rvert$
0–5	2.5	–2	5	–10	11	55
5–10	7.5	–1	8	–8	6	48
10–15	12.5	0	15	0	1	15
15–20	17.5	1	16	16	4	64
20–25	22.5	2	6	12	9	54
			$\Sigma f = 50$	$\Sigma fd' = 10$		$\Sigma f \lvert x - \bar{x} \rvert = 236$

$$\text{Mean} = A + \frac{\Sigma fd'}{\Sigma f} \times i = 12.5 + \frac{10}{50} \times 5 = 13.5.$$

$$\text{Mean deviation about the mean} = \frac{\Sigma f \lvert x - \bar{x} \rvert}{\Sigma f} = \frac{236}{50} = 4.72.$$

Example 18. Average of 2 numbers is 20 and their standard deviation is 5. Find the two numbers.

Solution. Let the two numbers be x_1 and x_2. Then $\frac{x_1 + x_2}{2} = 20 \Rightarrow x_1 + x_2 = 40.$

$$\text{S.D.} = \sqrt{\frac{\Sigma x^2}{n} - \left(\frac{\Sigma x}{n}\right)^2} = 5 \Rightarrow \sqrt{\frac{x_1^2 + x_2^2}{2} - 400} = 5.$$

$$\Rightarrow \quad x_1^2 + x_2^2 - 800 = 50 \Rightarrow x_1^2 + x_2^2 = 850.$$

$$(x_1 + x_2)^2 = x_1^2 + x_2^2 + 2x_1x_2 \Rightarrow 1{,}600 = 850 + 2x_1x_2 \quad [\because (x_1 + x_2)^2 = (40)^2 = 1{,}600]$$

$$\Rightarrow \quad x_1x_2 = \frac{1{,}600 - 850}{2} = 375 \Rightarrow 4x_1x_2 = 1{,}500.$$

$$(x_1 - x_2)^2 = (x_1 + x_2)^2 - 4x_1x_2 = 1600 - 1500 = 100 \Rightarrow (x_1 - x_2) = \pm 10.$$

$$x_1 - x_2 = \pm 10, x_1 + x_2 = 40.$$

$$\Rightarrow \quad x_1 = 25, x_2 = 15 \text{ or } x_1 = 15, x_2 = 25.$$

Hence numbers are 15 and 25.

Example 19. Find the standard deviation by the deviation method for the following data :

Class-interval	0–10	10–20	20–30	30–40	40–50	50–60	60–70
Frequency	6	14	10	8	1	3	8

Solution. Let the Assumed Mean = 45.

Class Interval	*Mid-value* x	*Freq.* f	$X - 45 = d$	$\frac{d}{10} = d'$	d'^2	fd'	fd'^2
0–10	5	6	–40	–4	16	–24	96
10–20	15	14	–30	–3	9	–42	126
20–30	25	10	–20	–2	4	–20	40
30–40	35	8	–10	–1	1	–8	8
40–50	45	1	0	0	0	0	0
50–60	55	3	10	1	1	3	3
60–70	65	8	20	2	4	16	32
			$\Sigma d = -70$			–75	305

$$\text{S.D.} = 10\left[\sqrt{\frac{\Sigma fd'^2}{\Sigma f} - \left(\frac{\Sigma fd'}{\Sigma f}\right)^2}\right] = 10\left[\sqrt{\frac{305}{50} - \left(\frac{-75}{50}\right)^2}\right] = 10\sqrt{6.1 - 2.25} = 19.6.$$

Example 20. The standard deviation of a set of 30 items is 9.5. Find the standard deviation if every item is decreased by 5.

Solution. Since the standard deviation is independent of the change of origin, therefore, the S.D. will remain unchanged.*

Example 21. Find out the standard deviation from the following table giving the wages of 230 persons :

Wages (Rs.)	*No. of Persons*
140–160	12
160–180	18
180–200	35
200–220	42
220–240	50
240–260	45
260–280	20
280–300	8

Solution.

Wage Interval	*Mid-value* x	$x-230$	$\frac{d}{20}=d'$	*Freq.*	fd'	d'^2	fd'^2
140–160	150	–80	–4	12	–48	16	192
160–180	170	–60	–3	18	–54	9	162
180–200	190	–40	–2	35	–70	4	140
200–220	210	–20	–1	42	–42	1	42
220–240	230	0	0	50	0	0	0
240–260	250	20	1	45	45	1	45
260–280	270	40	2	20	40	4	80
280–300	290	60	3	8	24	9	72
				$\Sigma f=230$	$\Sigma fd'=-105$		$\Sigma fd'^2=733$

$$\text{S.D.}=\left[\sqrt{\frac{\Sigma fd'^2}{\Sigma f}-\left(\frac{\Sigma fd'}{\Sigma f}\right)^2}\right]\times i=20\sqrt{\frac{733}{230}-\left(\frac{-105}{230}\right)^2}=34.5.$$

---

* $\text{S.D.}=\sqrt{\frac{\Sigma x^2}{n}-\left(\frac{\Sigma x}{n}\right)^2}$ ...(I), if every item is decreased by 5, then new sum of the squares $=\Sigma(x-5)^2$ and new mean $=\bar{x}-5$.

$$\therefore \text{ New S.D.}=\sqrt{\frac{\Sigma(x-5)^2}{n}-\left[\frac{\Sigma(x-5)}{n}\right]^2}=\sqrt{\frac{\Sigma x^2}{n}+25-10\frac{\Sigma x}{n}-\left[\frac{\Sigma x}{n}-5\right]^2}$$

$$=\sqrt{\frac{\Sigma x^2}{n}-\left(\frac{\Sigma x}{n}\right)^2}$$ which is the same as (I).

Example 22. For two groups of observation the following results were available :

Group I	*Group II*
$\Sigma (x-5) = 8$	$\Sigma (x-8) = -10$
$\Sigma (x-5)^2 = 40$	$\Sigma (x-8)^2 = 70$
$N_1 = 20$	$N_2 = 25$

Find the mean and the S.D. of the 45 observations obtained by combining the two groups.

Solution. For first group $\Sigma (x-5) = 8 \Rightarrow \Sigma x = 20 \times 5 + 8$

$$\Rightarrow \qquad \Sigma x = 108 \Rightarrow \frac{\Sigma x}{20} = \frac{108}{20} \Rightarrow \bar{x} = 5.4*$$

i.e., Mean of the first group, *i.e.,* $\bar{x}_1 = 5.4$.

For the second group

$$\Sigma (x-8) = -10 \Rightarrow \Sigma x = 8 \times 25 - 10 = 190 \Rightarrow \frac{\Sigma x}{25} = \frac{190}{25} = 7.6$$

i.e., Mean of the second group $= \bar{x}_2 = 7.6$

Standard deviation is independent of the change of origin

$\therefore$ Standard deviation of the first group say σ_1 is given by

$$\sigma_1 = \sqrt{\frac{\Sigma (x-5)^2}{20} - \left[\frac{\Sigma (x-5)}{20}\right]^2} = \sqrt{\frac{40}{20} - \left(\frac{8}{20}\right)^2} \; **$$

$$= \sqrt{2 - 0.16} = \sqrt{1.84} = 1.356.$$

Standard deviation of the 2nd group say σ_2 is given by

$$\sigma_2 = \sqrt{\frac{\Sigma (x-8)^2}{25} - \left[\frac{\Sigma (x-8)}{25}\right]^2} = \sqrt{\frac{70}{25} - 0.16} = \sqrt{2.8 - 0.16} = 1.62.$$

Combined S.D. of the two groups

$$\sqrt{\frac{n_1\sigma_2^2 + n_2\sigma_2^2}{n_1 + n_2} + \frac{n_1 n_2}{(n_1+n_2)^2}(\bar{x}_1 - \bar{x}_2)}$$

$$= \sqrt{\frac{20 \times 1.84 + 25 \times 2.64}{45} + \frac{20 \times 25}{(20+25)^2}(5.4 - 7.6)^2}$$

$$= \sqrt{2.284 + 1.195} = 1.865$$

$$\text{Combined mean} = \frac{5.4 \times 20 + 7.6 \times 25}{20 + 25} = \frac{108 + 190}{45} = 6.62$$

Example 23. The first two samples has 50 items with mean 54.4 and standard deviation 8. If the whole group has 150 items with mean 51.7 and S.D. 7.6, find the mean and standard deviation of the second sample.

---

* $\Sigma (x-5) = 8 \Rightarrow$ sum of the deviations of x values from assumed mean $5 = 8$.

$$\therefore \qquad \text{Mean} = 5 + \frac{8}{20} = 5.4.$$

** S.D. of each group can also be found by using the formulae S.D. $= \sqrt{\frac{\Sigma x^2}{n} - \left(\frac{\Sigma x}{n}\right)^2}$

For the first group $\qquad \Sigma (x-5)^2 = 40 \Rightarrow \Sigma x^2 + 25 \times 20 - 5\,\Sigma x = 40.$

Solution.
$$\bar{x}_{12} = \frac{n_1\bar{x}_1 + n_2\bar{x}_2}{n_1 + n_2}$$

$$\Rightarrow \quad 51.7 = \frac{50 \times 54.5 + 100\bar{x}_2}{150}$$

$$\Rightarrow \quad 150 \times 51.7 = 2720 + 100\bar{x}_2$$

$$7{,}755 - 2{,}720 = 100\bar{x}_2$$

$$5{,}035 = 100\bar{x}_2$$

$$\bar{x}_2 = 50.35.$$

$$\sigma_{12}^2 = \frac{n_1(\sigma_1^2 + d_1^2) + n_2(\sigma_2^2 + d_2^2)}{n_1 + n_2}$$

$$(7.6)^2 = \frac{50\,(64 + 7.29) + 100\,(\sigma_2^2 + 1.8225)}{150}$$

$$\left[\begin{array}{l} d_1 = \bar{x}_1 - \bar{x}_{12} = (54.4 - 51.7) = 2.7 \\ \therefore d_1^2 = (2.7)^2, d_2^2 = (50.35 - 51.7)^2 \end{array}\right]$$

$$\Rightarrow \quad 8{,}664 - 3{,}664.5 + 182.2 = 100\sigma_2^2.$$

$$\Rightarrow \quad \sigma_2^2 = 49.7 \Rightarrow \sigma = 7.01.$$

$\therefore$ Second sample has mean 50.35 and S.D. = 7.01.

Example 24. Find the S.D. of the $(2n + 1)$ terms of an A.P.

Solution. Let the A.P. be $a, \quad a + d, \quad a + 2d, \quad \ldots + a + 2nd$

$$\begin{array}{cccc} a & a+d & a+2d & a+2nd \\ \downarrow & \downarrow & \downarrow & \downarrow \\ x_1 & x_2 & x_3 & x_{2n+1} \end{array}$$

Mean
$$(\bar{x}) = \frac{a + (a + d) + (a + 2d) + \ldots + (a + 2nd)}{2n + 1}$$

$$= \frac{2n+1}{2} \frac{[a + a + 2nd]}{2n + 1} = a + nd.$$

[Sum of n terms of an A.P. $= \frac{n}{2}(a + l)$ where l is the last term]

$x_1 - \bar{x} = a - a - nd = -nd;\ x_2 - \bar{x} = (a + d) - (a + nd) = -(n - 1)\,d;$

$x_3 - \bar{x} = (a + 2d) - (a + nd) = -(n - 2)d;$

$x_n - \bar{x} = a + (n - 1)\,d - (a + nd) = -d,\ x_{n+1} - \bar{x} = a + nd - a - nd = 0$

$x_{n+2} - \bar{x} = a + (n + 1)\,d - a - nd = d$

$x_{2n+1} - \bar{x} = a + 2nd - a - nd = nd$

$x_{2n} - \bar{x} = (n - 1)\,d$

$x_{2n-1} - \bar{x} = (n - 2)d$

$$[\text{S.D.}]^2 = \frac{\sum_{i=1}^{n} (x_i - \bar{x})^2}{n} = \frac{2}{(2n + 1)}\,[n^2d^2 + (n - 1)^2d^2 + \ldots + d^2]$$

[$\because$ In this case the number of terms is $2n + 1$]

$$= \frac{2d^2}{2n+1}[1^2 + 2^2 + \ldots + n^2]$$

$$= \frac{2d^2}{2n+1}\left[\frac{n(n+1)(2n+1)}{6}\right] = \frac{n(n+1)d^2}{3}$$

$$\text{S.D.} = \left[\sqrt{\frac{n(n+1)}{3}}\right] d.$$

Example 25. The variable x takes only two values x_1 and x_2 with frequencies f_1 and f_2. If s be the S.D. of x, show that

$$s^2 = f_1 f_2 \left[\frac{x_1 - x_2}{f_1 + f_2}\right]^2$$

Solution. $$\bar{x} = \frac{x_1 f_1 + x_2 f_2}{f_1 + f_2}$$

$$f_1 (x_1 - \bar{x})^2 = f_1 x_1^2 + f_1 \bar{x}^2 - 2 f_1 x_1 \bar{x}\,;\; f_2 (x_2 - \bar{x})^2 = f_2 x_2^2 + f_2 \bar{x}^2 - 2 f_2 x_2 \bar{x}$$

$$s^2 = \frac{\sum_{i=1}^{2} f_i (x_i - \bar{x})^2}{\sum_{i=1}^{n} f_i} = \frac{f_1 (x_1 - \bar{x})^2}{f_1 + f_2} + \frac{f_2 (x_2 - \bar{x})^2}{f_1 + f_2}$$

$$= \frac{f_1 x_1^2 + f_1 \bar{x}^2 - 2 f_1 x \bar{x} + f_2 x_2^2 + f_2 \bar{x}^2 - 2 f_2 x_2 \bar{x}}{f_1 + f_2}$$

$$= \frac{f_1 x_1^2 + f_2 x_2^2 - 2\bar{x}(x_1 f_1 + x_2 f_2) + (f_1 + f_2)\bar{x}^2}{f_1 + f_2}$$

$$= \frac{f_1 x_1^2 + f_2 x_2^2 - 2\bar{x}\left(\frac{x_1 f_1 + x_2 f_2}{f_1 + f_2}\right)(f_1 + f_2) + (f_1 + f_2)\bar{x}^2}{f_1 + f_2}$$

$$= \frac{f_1 x_1^2 + f_2 x_2^2 - \bar{x}^2 (f_1 + f_2)}{f_1 + f_2} = \frac{f_1 x_1^2 + f_2 x_2^2 - \frac{(f_1 x_1 + f_2 x_2)^2}{f_1 + f_2}}{f_1 + f_2} \qquad \left[\because \frac{x_1 f_1 + x_2 f_2}{f_1 + f_2} = \bar{x}\right]$$

$$= \frac{(f_1 x_1^2 + f_2 x_2^2)(f_1 + f_2) - f_1^2 x_1^2 - f_2^2 x_2^2 - 2 f_1 f_2 x_1 x_2}{(f_1 + f_2)^2}$$

$$= \frac{f_1 f_2 (x_1^2 + x_2^2) - 2 f_1 f_2 x_1 x_2}{(f_1 + f_2)^2} = \frac{f_1 f_2 (x_1 - x_2)^2}{(f_1 + f_2)^2} = f_1 f_2 \left[\frac{x_1 - x_2}{f_1 + f_2}\right]^2$$

Second Method

$$\text{SD}^2 = \frac{\Sigma f x^2}{\Sigma f} - \left(\frac{\Sigma f x}{\Sigma f}\right)^2 = \frac{f_1 x_1^2 + f_2 x_2^2}{f_1 + f_2} - \left[\frac{f_1 x_1 + f_2 x_2}{f_1 + f_2}\right]^2$$

$= \frac{1}{(f_1 + f_2)^2}[(f_1 + f_2)(f_1 x_1^2 + f_2 x_2^2) - (f_1 x_1 + f_2 x_2)^2]$ simplify further.

Example 26. In a series of 5 observations, the values of mean and variance are 4.4 and 8.24. If three observations are 1, 2 and 6, find the other two.

Solution. $\bar{x} = \frac{\Sigma x}{n} \Rightarrow \Sigma x = n\bar{x}$

Here $n = 5$; $\bar{x} = 4.4$; $\Sigma x = 5 \times 4.4 = 22.$

Let the two missing items be x_1 and x_2

$\therefore$ $$1 + 2 + 6 + x_1 + x_2 = 22$$

$\Rightarrow$ $$x_1 + x_2 = 22 - 9 \Rightarrow x_1 + x_2 = 13$$

$$\sigma^2 = \frac{\Sigma x^2}{n} - (\bar{x})^2 \Rightarrow 8.24 = \frac{\Sigma x^2}{5} - (4.4)^2$$

$$41.2 = \Sigma x^2 - 19.36 \times 5 \Rightarrow \Sigma x^2 = 138$$

$$\Sigma x^2 = x_1^2 + x_2^2 + 1^2 + 2^2 + 6^2 = x_1^2 + x_2^2 + 41$$

$$x_1^2 + x_2^2 = 138 - 41 = 97$$

$$(x_1 + x_2)^2 = x_1^2 + x_2^2 + 2x_1x_2$$

$$(13)^2 = 97 + 2x_1x_2 \quad \text{or} \quad 169 = 97 + 2x_1x_2$$

$$2x_1x_2 = 169 - 97 \quad \text{or} \quad x_1x_2 = 36$$

$$x_1 + x_2 = 13 \qquad \text{...(i)}$$

$$(x_1 - x_2)^2 = x_1^2 + x_2^2 - 2x_1x_2 = 97 - 2\,(36) = 25$$

$$x_1 - x_2 = 5 \qquad \text{...(ii)}$$

Adding equations (i) and (ii), we get

$$2x_1 = 18 \Rightarrow x_1 = 9$$

Putting the values of x_1 in equation (i)

$$9 + x_2 = 13 \Rightarrow x_2 = 4$$

Thus, the two missing values are $x_1 = 9, x_2 = 4$.

Example 27. The following table gives lengths in cms is 200 pieces :

Length in cms.	8.0–8.5	8.5–9.0	9.0–9.5	9.5–10.0	10.0–10.5	Total
No. of Pieces	15	41	86	43	15	200

Calculate the percentage of items in the interval (Mean – S.D.) and (mean + S.D.).

Solution.

Length in cms.	*Mid-values X*	$x - 9.25 = d$	$d' = \frac{d}{0.5}$	*No. of piece F*	*c.f.*	fd'	d'^2	fd'^2
8.0–8.5	8.25	–1	–2	15	15	–30	4	60
9.5–9.0	8.75	–0.5	–1	41	56	–41	1	41
9.0–9.5	9.25	0	0	86	142	0	0	0
9.5–10.0	9.75	0.5	1	43	185	43	1	43
10.0–10.5	10.25	1	2	15	200	30	4	60
						$\Sigma fd' = 2$		$\Sigma fd'^2 = 204$

$$\text{Mean} = 9.25 + \frac{2}{200} \times 0.5 = 9.255.$$

$$[\text{S.D.}]^2 = (0.5)^2 \left[\frac{204}{200} - \left(\frac{2}{200} \right)^2 \right] \qquad \because \text{S.D.} = \left[\sqrt{\frac{\Sigma fd'^2}{n} - \left(\frac{\Sigma fd'}{n} \right)^2} \times i \right]$$

$$\text{S.D.} = 0.5 \text{ nearly.}$$

$$\text{Mean} - \text{S.D.} = 9.255 - 0.5 = 8.755.$$

$$\text{Mean} + \text{S.D.} = 9.255 + 0.5 = 9.755.$$

8.755 belongs to the interval 8.5 – 9.0 and 9.755 belongs to the interval 9.5-10.0. If the value 8.755 corresponds to n_1th value and 9.755 corresponds to n_2th value, then

$$8.755 = 8.5 + \left(\frac{n_1 - 15}{41} \right) 0.5 \Rightarrow \frac{0.255 \times 41}{0.5} + 15 = n_1 = 35.9$$

and

$$9.755 = 9.5 + \frac{n_2 - 142}{43} \times 0.5 \Rightarrow n_2 = \frac{0.255 \times 43}{0.5} + 142 = 163.93.$$

Example 28. Mr. Basu wants to invest Rs. 10,000 in one of the two companies *A* or *B*. Average return in a year from company *A* is Rs. 4,000 with a standard deviation of Rs. 25, while in company *B* the average return in a year is Rs.5,000 with a standard deviation of Rs. 40.

Which company will you recommend to Mr. Basu for investment ? Justify your answer.

Solution. Coefficient of variation for company $A = \frac{25}{4{,}000} \times 100 = 0.625$

Coefficient of variation for company $B = \frac{40}{5{,}000} \times 100 = 0.8.$

Since the coefficient of variation for company *A* is less, company *A* is more consistent and Mr. Basu should invest in company *A*.

Example 29. State whether the following statements are true or false :

(i) For any data, larger the mean larger is the standard deviation.

(ii) If every item of a distribution of 10 values is 35, then standard deviation is 1.

(iii) Standard deviation can be defined as the mean of the square root of the sum of squared deviations about the man.

(iv) The variance of 10 observations is 4, their S.D. is +2 or –2.

Solution.

(i) Not always. Mean depends on the values whereas S.D. depends on the deviations. If the values are close mean may be more but S.D. will be very small.

(ii) No, S.D. will be zero because if every item is 35, mean will also be 35 and $\sqrt{\frac{\Sigma (x - \bar{x})^2}{n}}$ will be zero.

(iii) No, it is defined as the square root of the mean of the squared deviations about the mean.

(iv) No S.D. is always positive. Therefore the S.D. = +2.

Example 30. Explain the term 'dispersion'. What purpose does a measure of dispersion serve? What are the different measures of dispersion?

Solution. A measure of dispersion is designed to state the extent to which individual observations (or items) vary from their average. Here, we shall account only the amount of variation and not the direction.

Purpose

1. To assess the reliability of an average.
2. To serve as a basis for control of variability.
3. To compare two or more series regarding variability between the items in each series.
4. For facilitating the use of other statistical measures.

Example 31. From the data given below, state which series is more variable.

Values	*Series A*	*Series B*
10–20	10	18
20–30	18	22
30–40	32	40
40–50	40	32
50–60	22	18
60–70	18	10

Solution. Calculation of means and standard deviations :

Mid-value x	$d' = \frac{x-45}{10}$	*Series A*			*Series B*		
		f	fd'	fd'^2	F	fd'	fd'^2
15	–3	10	–30	90	18	–54	+162
25	–2	18	–36	72	22	–44	+88
35	–1	32	–32	32	40	–40	+40
45	0	40	0	0	32	0	0
55	1	22	22	22	18	18	+18
65	2	18	36	72	10	20	+40
Total		140	–40	288	140	–100	348

For Series A

$$\text{Mean } (\bar{x}) = 45 + \frac{\Sigma fd'}{\Sigma f} \times 10 = 45 + \left(\frac{-40}{140}\right) \times 10$$

$$= 45 - \frac{20}{7} = 42\frac{1}{7} = 42.14.$$

$$\text{S.D. } (d') = \sqrt{\frac{1}{N}\Sigma fd' - (d')^2} = \sqrt{\frac{1}{140} \times 288 - \left(-\frac{2}{7}\right)^2}$$

$$\therefore \quad \text{S.D. } (x) = 10 \times \text{S.D. } (d') = 14.06.$$

$$\text{Coefficient of variation} = \frac{14.06}{42.14} \times 100 = 33.36\%$$

For Series B

$$\text{Mean } (\bar{x}) = 45 + \frac{\Sigma fd'}{\Sigma f} \times 10$$

$$= 45 - \frac{100}{140} \times 10 = 45 - 7.14 = 37.86.$$

$$\text{S.D. } (d') = \sqrt{\left[\frac{1}{140} \times 348 - \left(-\frac{100}{140}\right)^2\right]} = \sqrt{[2.4857 - 0.5102]}$$

$$= \sqrt{[+1.9755]} = 1.4055.$$

$$\therefore \quad \text{S.D. } (x) = 10 \times 1.4055 = 14.06$$

$$\text{Coefficient of variation} = \frac{14.06}{37.86} \times 100 = 37.1242 \equiv 37.12\%.$$

Conclusion : Since, the C.V. for series B is larger than that for Series A, the Series B is more variable.

Example 32. The mean and standard deviation of 50 items were found to be 75 and 10 respectively. At the time of checking, it was found that one item 65 was incorrect. Calculate the mean and standard deviation if the wrong item is omitted.

Solution. Sum of items $= 75 \times 50 = 3{,}750$.

If the wrong item is omitted, sum of remaining 49 items $= 3{,}750 - 65 = 3{,}685$.

$$\text{Mean of the 49 items} = \frac{3{,}685}{49} = 75.20.$$

$$(\text{S.D.})^2 = \frac{\Sigma x^2}{x} - \left(\frac{\Sigma x}{x}\right)^2$$

$$\Rightarrow \quad (10)^2 = \frac{\Sigma x^2}{50} - (75)^2 \Rightarrow \frac{\Sigma x^2}{50} = (75)^2 + (10)^2$$

$$= 5{,}625 + 100 = 5{,}725.$$

$$\Rightarrow \quad \Sigma x^2 = 50 \times 5725 = 2{,}86{,}250.$$

Σx^2 for 49 items $= 2{,}86{,}250 - (65)^2 = 2{,}82{,}025$.

$$\text{S.D. for 49 items} = \sqrt{\frac{2{,}82{,}025}{49} - \left(\frac{3{,}685}{49}\right)^2} = \sqrt{5755.6 - 5655.347} = 10.0126.$$

Example 33. The means of the two samples of sizes 500 and 600 were 186 and 175 respectively. The corresponding S.D.s were 9 and 10 respectively. The variable studied was height in centimeter, obtain the mean and variance of combined sample.

Solution. Combined mean $= \dfrac{186 \times 500 + 175 \times 600}{500 + 600}$

$$= \frac{93{,}000 + 1{,}05{,}000}{1{,}100} = \frac{1{,}98{,}000}{1{,}100} = 180.$$

$$\text{Combined S.D. } (\sigma_{12}) = \sqrt{\frac{n_1(\sigma_1^2 + d_1^2) + n_2(\sigma_2^2 + d_2^2)}{n_1 + n_2}}$$

$d_1 = 186 - 180 = 6,\ d_2 = 175 - 180 = -5.$

$$\sigma_{12}^2 = \frac{500\,[(9)^2 + (6)^2] + 600\,[(10)^2 + (-5)^2]}{1,100}$$

$$= \frac{500\,(81 + 36) + 600\,(100 + 25)}{1,100}$$

$$= \frac{5 \times 117 + 6 \times 125}{11} = \frac{585 + 750}{11} = \frac{1,335}{11} = 121.36 = \text{Variance of the combined sample}$$

$$\text{S.D.} = \sigma_{12} = \sqrt{121.36} = 11 \quad \text{(approx.)}$$

Example 34. The coefficient of variation of two series is 60% and 80%. Their standard deviations are 24 and 20. What are their arithmetic means ?

Solution. Let $\bar{x}_1$ and $\bar{x}_2$ be the Arithmetic means, then

$$\frac{\sigma_1}{x_1} \times 100 = 60 \quad \text{...(i)}$$

$$\frac{\sigma_2}{x_2} \times 100 = 80 \quad \text{...(ii)}$$

$$\text{(i)} \Rightarrow \frac{24}{\bar{x}_1} \times 100 = 60 \text{ or } \bar{x}_1 = \frac{24}{0.6} = 40.$$

$$\text{(ii)} \Rightarrow \frac{20}{\bar{x}_2} \times 100 = 80$$

$$\Rightarrow \frac{20}{\bar{x}_2} = 0.8 \Rightarrow \bar{x}_2 = \frac{20}{0.8} = 25.$$

Example 35. The daily earnings of a vendor for a period of 40 days are given below :

Daily Earnings in Rs.	118–126	127–135	136–144	145–153	154–162	163–171	172–180	Total
No. of Days	3	8	9	12	5	4	2	40

Calculate mean, standard deviation and the coefficient of variation.

Solution. Making exclusive type classification, the data can be arranged as under :

Daily Earnings in Rs. (X)	*No. of Days f*	*Mid-Value X*	$A = 149$ $X - 149 = x$	x^2
117.5–126.5	3	122	–27	729
126.5–135.5	5	131	–18	324
135.5–144.5	9	140	–9	81
144.5–153.5	12	149	0	0
153.5–162.5	5	158	9	81
162.5–171.5	4	167	18	324
171.5–180.5	2	176	27	729
			$\Sigma x = 0$	$\Sigma x^2 = 2,268$

$$\text{Mean} = A + \frac{\Sigma x}{n} = 149 + 0 = 149.$$

$$\text{S.D.} = \sqrt{\frac{\Sigma x^2}{n} - \left(\frac{\Sigma x}{n}\right)} = \sqrt{\frac{2,268}{40}} = 7.529.$$

$$\text{Coefficient of variation} = \frac{\text{S.D.}}{\text{Mean}} \times 100 = \frac{7.529}{149} \times 100 = 5.05\%$$

Example 36. The mean and standard deviation of the marks obtained by two groups of students, consisting of 50 each, are given below. Calculate the mean and standard deviation of the marks obtained by all the 100 students :

Group	*Mean*	*Standard Deviation*
1	60	8
2	55	7

Solution. For Group 1 $\bar{x}_1 = 60$ $\sigma_1 = 8$ $n_1 = 50$.

For Group 2 $\bar{x}_2 = 55$ $\sigma_2 = 7$ $n_2 = 50$.

$$\text{Combined Mean : } \bar{x}_{12} = \frac{n_1\bar{x}_1 + n_2\bar{x}_1}{n_1 + n_2}$$

$$= \frac{60 \times 50 + 55 \times 50}{50 + 50} = \frac{5,750}{100} = 57.5.$$

$$\text{Combined Standard Deviation } \sigma_{12} = \sqrt{\frac{n_1\sigma_1^2 + n_2\sigma_2^2 + n_1 d_1^2 + n_2 d_2^2}{n_1 + n_2}}, ^*$$

where $d_1 = \bar{x}_{12} - \bar{x}_1 = 57.5 - 60 = -2.5$ and $d_2 = \bar{x}_{12} - \bar{x}_2 =$ '$57.5 - 55 = 2.5$.

$$\sigma_{12} = \sqrt{\frac{50 \times 8^2 \times 50 \times 7^2 + 50 \times (-2.5)^2 + 50 \times (2.5)^2}{50 + 50}}$$

$$= \sqrt{\frac{3,200 + 2,450 + 312.5 + 312.5}{100}}$$

$$= \sqrt{\frac{3,200 + 2,450 + 312.5 + 312.5}{100}} = \sqrt{62.75} = 7.92.$$

Example 37. Weekly salaries of 100 employees in a firm are given below :

Salaries Rs. per week	400–500	500–600	600–700	700–800
No. of Employees	2	15	21	30
Salaries Rs. per week	800–900	900–1,000	1,000–1,100	
No. of Employees	20	9	3	

Calculate the percentage of employees with salaries, in Rs. per week, in the following range :

(Mean – 2 S.D.), (Mean + 2 S.D.).

* We can apply the other formula also that

$$\sigma_{12}^2 = \frac{n_1\sigma_1^2 + n_2\sigma_2^2}{n_1 + n_2} + (\bar{x}_1 - \bar{x}_2)^2 \frac{n_1 n_2}{(n_1 + n_2)^2}$$

$$= \frac{64 + 49}{2} + (60 - 55)^2 \frac{50 \times 50}{(100)^2} = 62.75$$

Solution.

Salaries in Rs. per week	*Mid-value x*	→ *Frequency*	$x = 750 = d$	$\frac{d}{100} = d'$	fd'	d'^2	fd'^2	*cf*
400–500	450	2	–300	–3	–6	9	18	2
500–600	550	15	–200	–2	–30	4	60	17
600–700	650	21	–100	–1	–21	1	21	38
700–800	750	30	0	0	0	0	0	68
800–900	850	20	100	1	20	1	20	88
900–1,000	950	9	200	2	18	4	36	97
1,000–1,100	1,050	3	300	3	9	9	27	100
		1000			$\Sigma fd' = -10$		$\Sigma fd'^2 = 182$	

$$\text{Mean} = 750 - \frac{10}{100} \times 100 = 740$$

$$\text{S.D.} = 100\sqrt{\frac{\Sigma fd'^2}{n} - \left(\frac{\Sigma fd'}{n}\right)} = 100\sqrt{\frac{182}{100} - \left(\frac{-10}{100}\right)^2}$$

$$= 10\sqrt{182 - 1} = 134.5 \Rightarrow 2 \text{ S.D.} = 134.5 \times 2 = 269.$$

$$\text{Mean} - 2 \text{ S.D.} = 740 - 269 = 471$$

$$\text{Mean} + 2 \text{ S.D.} = 740 + 269 = 1{,}009.$$

If the value 471 corresponds to n_1th value and 1009 corresponds to n_2th value, then

$$471 = 400 + \frac{n_1 - 0}{2} \times 100 \Rightarrow n_1 = \frac{71 \times 2}{100} = 1.4.$$

$$1{,}009 = 1{,}000 + \frac{n_2 - 97}{3} \times 100 \Rightarrow n_2 = \frac{9 \times 3}{100} + 97 = 97.27*.$$

[*c.f.* of the class 1,000 – 1,100 is 100 and 3 is the frequency of this class.]

Required Percentage = 97.27 – 1.4 = 96% approx.

Note : This question can also be solved by drawing a less than ogive.

Example 38. Number of articles produced on 55 old machines and 45 new machines gave the following results :

Machines	*No. of Machines*	*Mean*	*Standard Deviation*
Old	55	61	8
New	45	51	6

Determine whether number of articles produced on new machines are more uniform than those on old machines.

Solution. Coefficient of variation of items produced on old machines

$$= \frac{8}{61} \times 100 = 13.1\%$$

* Median $= L_1 + \frac{\frac{N}{2} - cf_{-1}}{f} \times i$. In median we find $\frac{N}{2}$th value. If we find n_1th value, we will replace $\frac{N}{2}$ by n_1 and change other values accordingly.

Coefficient of variation of items produced on new machines

$$= \frac{6}{51} \times 100 = 11.76.$$

Since the coefficient of variation for the articles produced on new machines is less, the articles produced on new machines are more uniform.

Example 39. If x_1 and x_2 be 2 values of a variable x and each of the remaining $(n-2)$ values be $\frac{1}{2}(x_1 + x_2)$, find the mean and standard deviation of the n values.

Solution.
$$\text{Mean} = \frac{1}{n}\left[x_1 + x_2 + (n-2)\frac{(x_1 + x_2)}{2}\right]$$

$$= \frac{1}{2n}[2(x_1 + x_2) + (n-2)(x_1 + x_2)] = \frac{x_1 + x_2}{2}.$$

$$\text{S.D.} = \sqrt{\frac{\Sigma (x - \bar{x})^2}{n}}$$

$$= \sqrt{\frac{\left(x_1 - \left[\frac{(x_1 + x_2)}{2}\right]^2\right) + \left[x_2 - \left(\frac{x_1 + x_2}{2}\right)^2\right]}{n}}$$

[∵ Deviations of other values are zero.]

$$= \sqrt{\frac{(x_1 - x_2)^2 + (x_2 - x_1)^2}{4n}}$$

$$= \sqrt{\frac{2(x_1 - x_2)^2}{4n}} = \frac{(x_1 - x_2)}{\sqrt{2n}} \text{ or } \frac{x_2 - x_1}{\sqrt{2n}}. \quad [\because (x_1 - x_2)^2 = (x_2 - x_1)^2]$$

Example 40. If the mean of 10 values is 60 and sum of squares of deviations from 50 is 5,000, calculate the standard deviation of these 10 values.

Solution.
$$\text{Mean} = A + \frac{\Sigma d}{n}$$

$$\Rightarrow \quad 60 = 50 + \frac{\Sigma d}{10} \Rightarrow \Sigma d = 100$$

$$\therefore \quad \text{S.D.} = \sqrt{\frac{\Sigma d^2}{n} - \left(\frac{\Sigma d}{n}\right)^2} = \sqrt{\frac{5{,}000}{10} - \left(\frac{100}{10}\right)^2} = \sqrt{500 - 100}$$

Example 41. Find the mean and standard deviation of the 2 values, $(a + b)$ and $(a - b)$.

Solution.
$$\text{Mean} = \frac{a + b + a - b}{2} = a.$$

$$\text{S.D.} = \sqrt{\frac{\Sigma (x - \bar{x})^2}{n}} = \sqrt{\frac{(a + b - a)^2}{2} + \frac{(a - b - a)^2}{2}} = \sqrt{\frac{b^2}{2} + \frac{b^2}{2}} = b.$$

$$\left[\text{Students can also use the formula S.D.} = \sqrt{\frac{\Sigma x^2}{n} - \left(\frac{\Sigma x}{n}\right)^2}.\right]$$

Example 42. A state government decided to give old age pension to people over sixty years of age. The scales of pension were fixed as follows :

Age Group	*Rs. per month*
60–65	2500
65–70	3000
70–75	3500
75–80	4000
80–85	4500

The age of 25 persons who secured the pension rights are given below :

74 62 84 72 83 72 81 64 71 63 61 60 61 67 74 64 79 73 75 76 69 78 66 67 68

Calculate the monthly average pension payable and standard deviation.

Solution.

Classification of Data

Age Group	*Frequency*
60–65	7
65–70	5
70–75	6
75–80	4
80–85	3
	25

Calculation of the Monthly Average Pension Payable and the Standard Deviation

Scale of Pension (Rs.)	$d' = \left(\frac{X - 3500}{500}\right)$	f	fd'	fd'^2
2500	–2	7	–14	28
3000	–1	5	–5	5
3500	0	6	0	0
4000	1	4	4	4
4500	2	3	6	12
		25	–9	49

Here $A = 3500$, $C = 500$, $n = 25$, $\Sigma fd' = -9$ and $\Sigma fd' = 49$

$$\bar{X} = A + \frac{\Sigma fd'}{n} \times C = 3500 - \frac{9}{25} \times 500 = 3320$$

$$\sigma = C \times \sqrt{\frac{\Sigma fd'^2}{n} - \left(\frac{\Sigma fd'}{n}\right)^2} = 500 \times \sqrt{\frac{49}{25} - \left(\frac{-9}{25}\right)^2}$$

$$= 1.353 \times 500 = 676.5$$

Thus, the monthly average pension is Rs. 3320 and standard deviation is Rs. 676.5.

Example 43. The marks obtained by 100 students are distributed as follows :

Marks	0–10	10–20	20–30	30–40	40–50	50–60
No. of Students	10	15	25	18	13	8
Marks	60–70	70–80	80–90	90–100	Total	
No. of Students	5	3	2	1	100	

On scrutiny the following mistakes of tabulation are detected :

Students	*A*	*B*	*C*	*D*	*E*	*F*	*G*	*H*
Correct marks	61	36	54	86	36	21	48	8
Entry marks	16	63	45	68	26	20	18	18

Revise the frequency table and calculate the mean and S.D.

Solution. Revised Frequency table :

Marks	*No. of Students*	*Mid-values*	$x - 45 = d$	$\frac{d}{10} = d'$	fd'	fd'^2
0–10	11	5	–40	–4	–44	176
10–20	12	15	–30	–3	–36	108
20–30	24	25	–20	–2	–48	96
30–40	20	35	–10	–1	–20	20
40–50	13	45	0	0	0	0
50–60	9	55	10	1	9	9
60–70	4	65	20	2	8	16
70–80	3	75	30	3	9	27
80–90	3	85	40	4	12	48
90–100	1	95	50	5	5	25
	$\Sigma f = 100$				$\Sigma fd' = -105$	$\Sigma fd'^2 = 525$

$$\text{Mean} = 45 + \frac{\Sigma fd'}{\Sigma f} \times 10 = 45 \times \frac{105}{100} \times 10 = 34.5$$

$$\text{S.D.} = 10\sqrt{\frac{\Sigma fd'^2}{\Sigma f} - \left(\frac{\Sigma fd'}{\Sigma f}\right)^2} = 10\sqrt{\frac{525}{100} - \left(\frac{105}{100}\right)^2}$$

$$= 10\sqrt{5.25 - 1.1025} = 20.36$$

Example. 44. The mean weight of 150 students is 60 kg. The mean weight of boys is 70 kg with a standard deviation of 10 kg. for the girls the mean weight is 55 kg and the standard deviation is 15 kg. Find the number of boys and the combined standard deviation.

Solution. Let the number of boys $= n_1$, Number of girls $= 150 - n_1$

Total weight of 150 students $= 150 \times 60 = 9{,}000$ kg.

Total weight of boys $= 70n_1$ kg.

$\therefore \quad 9{,}000 = 70n_1 + (150 - n_1)\,55 = 70n_1 + 150 \times 55 - 55n_1$

$\Rightarrow \quad 750 = 15n_1 \Rightarrow n_1 = 50$

$\therefore$ The number of boys = 50, the number of girls = 100

$$\text{Combined S.D.} = \sqrt{\frac{50 \times 100 + 100 \times 225}{150} + \frac{50 \times 100}{(150)^2} \times 225}$$

$$= \sqrt{\frac{5{,}000 + 22{,}500}{150} + 50} = \sqrt{\frac{27{,}500}{150} + 50}$$

$$= \sqrt{\frac{550}{3} + 50} + \sqrt{\frac{700}{2}} = 15.275$$

Example 45. You are incharge of rationing in a state affected by food shortage. The following reports arrive from your local investigators about Daily Caloric Value of food available per adult during current period.

Area	*Mean*	*Standard deviation*
X	2,500	500
Y	2,200	300

The estimated requirement of an adult is taken at 3000 calories daily and the absolute minimum at 1250. Comment on the reported figure and determine which area to your opinion needs more urgent attention. **(DU B. Com. (Hons.), 2002)**

Solution. The limits on the basis of the information given to us would be :

$$\text{Area } X : \overline{X} \pm 3\sigma = 1{,}000 \text{ to } 4{,}000$$

$$\text{Area } Y : \overline{Y} \pm 3\sigma = 1{,}300 \text{ to } 3{,}100$$

some people in area *X* do not get even basic minimum calories, *i.e.*, 1,250 calories so this area needs more attention.

Example 46. An analysis of weekly wages paid to workers in two firms *A* and *B* gives the following results :

	A	*B*
No. of workers	60	80
Average wage	300	350
Standard deviation	40	45

(a) In which firm there is greater variability of wages ?

(b) What are the measures of average wage and standard deviation of the two firms taken together ?

Solution. Coefficient of variation of firm $A = \frac{40}{300} \times 100 = 13.33\%$

Coefficient of variation of firm $B = \frac{45}{350} \times 100 = 12.35\%$

The firm A has greater variability

$$\text{Average wage of two firms taken together} = \frac{300 \times 60 + 350 \times 80}{60 + 80}$$

$$= \frac{18,000 + 28,000}{140} = \text{Rs. } 328.57$$

S.D. of the firms taken together or combined S.D.

$$= \sqrt{\frac{60 \times (40)^2 + 80 \times (45)^2}{60 + 80} + \frac{60 \times 80}{(60 + 80)^2} (300 - 350)^2}$$

$$= \sqrt{\frac{96,000 + 1,62,000}{140} + \frac{4,800}{19,600} \times 2,500}$$

$$= \sqrt{1,842.85 + 612.24} = \sqrt{2,455} = 49.54$$

Example 47. Mean and standard deviation of 100 items is 40 and $\sqrt{11}$ respectively. One item 7 has now been left out as it was incorrect. Find the mean and S.D. of the remaining items. Can we find out each of the remaining items also.

Solution. Sum of the item $= 40 \times 100 = 4,000$

Sum of the items after leaving the item $7 = 4,000 - 7 = 3,993$

$$\text{Mean of the remaining 99 items} = \frac{3,993}{99} = 40.3$$

$$\text{S.D.} = \sqrt{\frac{\Sigma x^2}{n} - \left(\frac{\Sigma x}{n}\right)^2}$$

$$\Rightarrow \quad (\text{S.D.})^2 = \frac{\Sigma x^2}{n} - \left(\frac{\Sigma x}{n}\right)^2$$

$$\Rightarrow \quad 11 = \frac{\Sigma x^2}{n} - 1,600$$

$$\Rightarrow \quad \frac{\Sigma x^2}{n} = 1,600 + 11 = 1611$$

$$\Rightarrow \quad \Sigma x^2 = 100\,(1,611) \qquad [\because n = 100]$$

Sum of the squares of 99 items $= 100\,(1,611) - 49$

$= 1,61,051$

$$(\text{S.D.})^2 \text{ of 99 items} = \frac{1,61,051}{99} - \left(\frac{3993}{99}\right)^2$$

$$= \frac{1,61,051 \times 99\,(3,993)^2}{(99)^2} = 0$$

$$\Rightarrow \quad \sum_{i=1}^{n} (x_i - \bar{x})^2 = 0 \Rightarrow x_i = \bar{x}$$

$\Rightarrow$ Each of the 99 items is equal to the mean, *i.e.*, $= 40.3$

Example 48. The coefficient of variation of wages of male workers and female workers are 55 per cent and 70 per cent respectively while the standard deviations are 22.0 and 15.4 respectively. Calculate the overall wages of all workers given that 80 per cent of the workers are male.

Solution. C.V. of male workers $= \dfrac{\text{S.D.}}{\text{Mean}} \times 100 = 55$

$$\Rightarrow \quad \frac{22}{\text{Mean}} \times 100 = 55 \Rightarrow \text{Mean} = \frac{22}{0.55}$$

$$= 40 = \bar{x}_1 \text{ (say)}$$

C.V. of female workers $= \dfrac{\text{S.D.}}{\text{Mean}} \times 100 = 70$

$$\Rightarrow \quad \frac{15.4}{\text{mean}} = 0.7 \Rightarrow \text{Mean} = \frac{15.4}{0.7} = 22 = \bar{x}_2 \text{ (say)}$$

Mean wage of male workers = 40
Mean wage of female workers = 22
Since 80% are male workers, the combined mean

$$= \frac{80 \times 40 + 20 \times 22}{100} = \frac{3200 + 440}{100} = 36.40$$

$\therefore$ Overall average wages of all the workers = 36.40.

Example 49. Show that the S.D. cannot be less than the mean deviation from the mean.

Solutiom. S.D. $= \sqrt{\dfrac{\Sigma (x_i - \bar{x})^2}{n}}$, mean deviation from the mean $= \dfrac{\Sigma |x_i - \bar{x}|}{n}$

$$(\text{S.D.})^2 = \frac{\Sigma (x_i - \bar{x})^2}{n} = \frac{\Sigma |x_i - \bar{x}|^2}{n} = n\frac{\Sigma |x_i - \bar{x}|^2}{n^2}$$

$$= n\left[\frac{\Sigma |x_i - \bar{x}|}{n}\right]^2 = n\,[\text{Mean deviation}]^2 \qquad [\because \Sigma (x_i - \bar{x})^2 = \Sigma |x_i - \bar{x}|^2]$$

$\Rightarrow (\text{S.D.})^2 \geq [\text{Mean deviation}]^2$

$\Rightarrow$ S.D. $\geq$ Mean deviation from the mean.

Example 50. Comment on the following statement "After Settlement the average weekly wages in a factory had increased from Rs. 800 to Rs. 1,200. S.D. had increased from 200 to 250. After settlement the wage has become higher and more uniform."

Solution. Since the Average Wage has increased from Rs. 800 to Rs. 1,200, the wage has certainly become higher.

Coefficient of variation before settlement $= \dfrac{200}{800} \times 100 = 25\%$

Coefficient of variation after settlement $= \dfrac{250}{1{,}200} \times 100 = 20.8\%$

Since, the coefficient of variation has decreased after settlement, the wages have become more uniform.

Example 51. The mean and S.D. of n observations are $\bar{x}$ and σ_1 respectively. A new observation x_1 is added to the group. Prove that the S.D. of the entire set of $(n+1)$ observation is

$$\sigma_{12} = \sqrt{\frac{n\sigma_1^2}{n+1} + \frac{n}{(n+1)^2}(\bar{x} - x_1)^2}$$

Solution. $$\frac{\Sigma x}{n} = \bar{x}, \quad \text{S.D.} = \sigma_1 = \sqrt{\frac{\sum_{i=1}^{n} (x - \bar{x})^2}{n}}$$

Let m be the mean of the new group, then $m = \dfrac{\Sigma x + x_1}{n+1}$

$$[\text{S.D.}]^2 \text{ of the new group} = \frac{\sum_{i=1}^{n} (x - m)^2}{n+1} + \frac{(x_1 - m)^2}{n+1} \qquad \text{...(i)}$$

Now $$x - m = (x - \bar{x} + \bar{x} - m) = (x - \bar{x}) + (\bar{x} - m)$$

$$\Sigma (x - m)^2 = \Sigma [(x - \bar{x}) + (\bar{x} - m)]^2$$

$$= \Sigma (x - \bar{x})^2 + \Sigma (\bar{x} - m)^2 + 2(\bar{x} - m)\Sigma (x - \bar{x})$$

$$= \Sigma (x - \bar{x})^2 + \Sigma (\bar{x} - m)^2 \qquad [\because \Sigma (x - \bar{x}) = 0]$$

$$\frac{\Sigma (x - m)^2}{n+1} = \frac{n}{n+1}\left[\frac{\Sigma (x - \bar{x})^2}{n}\right] + \frac{\Sigma (\bar{x} - m)^2}{n+1}$$

$$= \frac{n}{n+1}\sigma_1^2 + \frac{1}{n+1}\Sigma\left[\bar{x} - \frac{\Sigma x + x_1}{n+1}\right]^2 *$$

$$= \frac{n}{n+1}\sigma_1^2 + \Sigma\frac{[n\bar{x} + \bar{x} - \Sigma x - x_1]^2}{(n+1)^3}$$

$$= \frac{n}{n+1}\sigma_1^2 + \Sigma\frac{(\bar{x} - x_1)^2}{(n+1)^3} = \frac{n}{n+1}\sigma_1^2 + \Sigma\frac{(\bar{x} - x_1)^2}{(n+1)^3}$$

$$x_1 - m = \left[x_1 - \frac{\Sigma x + x_1}{n+1}\right] = \frac{nx_1 + x_1 - \Sigma x - x_1}{n+1}$$

$$= \frac{n(x_1 - \bar{x})}{(n+1)} \qquad [\because n\bar{x} = \Sigma x]$$

* **Please note :** Mean of the new group $= m = \dfrac{n\bar{x} + x_1}{n+1} = \dfrac{\Sigma x + x_1}{n+1}$

$\therefore$ Standard deviation of the new group

$$= \sqrt{\frac{n}{n+1}\sigma_1^2 + \frac{n(\bar{x}-x_1)^2}{(n+1)^3} + \frac{n^2(x_1-\bar{x})^2}{(n+1)^3}} \qquad \text{[By (i)]}$$

$$= \sqrt{\frac{n}{n+1}\sigma_1^2 + \frac{n(n+1)(x_1-\bar{x})^2}{(n+1)^3}}$$

$$= \sqrt{\frac{n}{n+1}\sigma_1^2 + \frac{n(x_1-\bar{x})^2}{(n+1)^2}}$$

Example 52. The mean of 5 observations is 15 and the variance is 9. If two more observations having value –3 and 10 are combined with these 5 observations, what will be the mean and variance of 7 observations ?

Solution. Total of 5 observations $= 5 \times 15 = 75$

$$\text{Variance} = 9 = \frac{\Sigma x^2}{5} - (15)^2$$

$$\Rightarrow \qquad \Sigma x^2 = (225 + 9) \times 5 = 234 \times 5 = 1{,}170$$

[**Please Note.** variance = $(\text{S.D.})^2$]

$$\sum_{i=1}^{n} x_i^2 = 1170 + 9 + 100 = 1{,}279$$

Sum of the 7 observations $= 75 - 3 + 10 = 82$

Mean of 7 observations $= \frac{82}{7} = 11.7$

$$\text{S.D. of 7 observations} = \sqrt{\frac{\sum_{i=1}^{7} x^2}{7} - \left(\frac{\sum_{i=1}^{7} x_i}{7}\right)}$$

$$= \sqrt{\frac{1279}{7} - (11.7)^2}$$

$$= \sqrt{182.7 - 136.89} = 6.76$$

Example 53. Examine the validity of the following statements :

(1) Mean deviation from the mean is 3 and from the median is 4.

(2) "An average when published should be accompanied by a measure of dispersion for significant interpretation."

(3) If the mean and S.D. of n observations $x_1, x_2, x_3, \ldots, x_n$ be $\bar{x}$ and σ, then the mean and S.D. of $-x_1, -x_2, -x_3, \ldots, -x_n$ will be $-\bar{x}$ and $-\sigma$ respectively.

(4) The S.D. is invariant under change of origin and scale.

(5) The computer showed that the S.D. of 40 observations ranging from 120 to 150 is 35.

(6) After settlement the average monthly wage in a factory had increased from Rs. 3,500 to Rs. 4,000 and the S.D. had increased from 1,000 to 1,250. After settlement the wage has become higher and more uniform.

(7) Ten observations have mean 20 and some standard deviation. If each of these 10 observations is doubled then the S.D of the new observations will be doubled.

(8) The S.D. of 10 items is 4. If each item is increased by 2, the S.D. will become 6.

(9) The S.D. of a distribution is –2.

(10) The S.D. of 5 items is found to be $\sqrt{50}$. What will be the S.D. if the values of all the items are increased by 5 ?

(11) The mean depth of water of a river is 130 cms. Therefore, a man with a height of 165 cms can cross the river safely.

Solution.

(1) No. Because mean deviation from the median is always less than the mean deviation from the mean.

(2) Dispersion means variability. Two distributions may have the same value for a particular average but still there may be significant difference between the distributions because of variability of individual values from the average. To have a fairly good idea about a distribution, average should be accompanied by a measure of dispersion.

(3) No, though mean will be $-\bar{x}$ but S.D. is a always taken as positive and will remain σ.

(4) No. S.D. is not invariant under change of scale.

(5) No. S.D. cannot be more than the range.

(6) $$\text{C.V. before settlement} = \frac{1{,}000}{3{,}500} \times 100 = 28.5\%.$$

$$\text{C.V. after settlement} = \frac{1{,}250}{4{,}000} \times 100 = 31.25\%.$$

Since, the coefficient of variation has increased after settlement the wages have become less uniform.

(7) Yes. Because S.D. depends on the change of scale.

(8) *Not True.* Since the S.D. is independent of the change of origin, it will remain unchanged, *i.e.,* it will remain 4.

(9) No. The S.D. cannot be negative.

(10) Standard deviation is independent of the change of origin so the value will remain the same, *i.e.,* $\sqrt{50}$.

(11) Not necessary at some places the depth may be more than 165 cm.

Example 54. Two workers on the same job show the following results over a long period of time :

	Worker A	*Worker B*
Mean time of completing the job (minutes)	30	25
Standard deviation (minutes)	6	4

(i) Which workers appears to be more consistent in the time he requires to complete the job ?

(ii) Which worker appears to be faster in completing the job ? Explain.

Solution.

$$\text{C.V. for worker } A = \frac{6}{30} \times 100 = 20\%$$

$$\text{C.V. for worker } B = \frac{4}{25} \times 100 = 16\%$$

Since the C.V. is less for worker *B* therefore, worker *B* is more consistent. Since average time for completing the job by *B* is less than the average time for completing the job by *A*, the worker *B* is faster than worker *A*.

Example 55. Following information relating to wages per employee for two factories are given below :

	Factory–*A*	**Factory–*B***
Number of employees	50	100
Average wages per employee per week (Rs.)	120	85
Variance of the wages per employee per week (Rs.)	9	16

(I) Which factory wage distribution is more consistent?

(II) Which factory weekly wage bill is lower? **(D.U. B.Com. Pass 2008)**

Solution. Coefficient of variation of factory '*A*'

$$= \frac{\text{S.D.}}{\text{Mean}} \times 100$$

$$= \frac{3}{120} \times 100 = 2.5\%$$

Coefficient of variation of factory *B*

$$= \frac{\text{S.D.}}{\text{Mean}} \times 100$$

$$= \frac{4}{85} \times 100 = 4.7\%$$

Since C.V. of factory *A* is lower. The wages are more consistent in factory *A*

Weekly wage bill of factory $A = 120 \times 50 = 6000$

Weekly wage bill of factory $B = 100 \times 85 = 8500$

Weekly wage bill of factory *A* is lower.

Example 56. Mean height of male workers is 70 inches and the S.D. is 2.9 inches. The corresponding numbers for female workers are 64 inches and 2.3 inches. If the ratio of male and female workers is 70 : 30, what is the mean and S.D. for male and female workers taken together ?

Solution. Let the number of male workers be $7k$ and the famele workers be $3k$ then combined S.D. is given by

$$\sqrt{\frac{2k\,(2.9)^2 + 3k\,(2.3)^2}{10k} + \frac{21k^2}{(10k)^2}(70-64)^2} \qquad \text{(Refer Points 4.30)}$$

$$= \sqrt{\frac{58.87 + 15.87}{10} + \frac{21 \times 36}{100}} = \sqrt{7.474 + 7.56} = 3.877$$

$$\text{Combined mean} = \frac{7k + 70 \times 3k \times 64}{10k} = \frac{490 + 192}{10} = 68.2.$$

Example 57. You are given the record of daily wages paid to workers in two factories *A* and *B*.

Monthly Wages ('000 Rs.)	*Number of Workers*	
	Factory A	*Factory B*
2–3	15	25
3–4	30	40
4–5	40	60
5–6	60	35
6–7	30	20
7–8	14	15
8–9	7	5

Using the Arithmetic mean and S.D. answer the following points :

(i) Which factory pays higher average wages and by how much ?

(ii) In which factory are wages more uniform ?

Solution.

Monthly Wages	*No. of Workers in Factory A* f	*Mid-value* x	$x - A = d$, *taking* $A = 5.5$	fd	d^2	fd^2	*No. of Workers in Factory B* f	fd	fd^2
2–3	15	2.5	–3	–45	9	135	25	–75	225
3–4	30	3.5	2	–60	4	120	40	–80	160
4–5	44	4.5	–1	–44	1	44	60	–60	60
5–6	60	5.5	0	0	0	0	35	0	0
6–7	30	6.5	1	30	1	30	20	20	20
7–8	14	7.5	2	28	4	56	15	30	60
8–9	7	8.5	3	21	9	63	5	15	45
	Total = 200			$\Sigma fd = -70$		$\Sigma fd^2 = 448$	Total = 200	$\Sigma fd = -150$	$\Sigma fd^2 = 570$

$$\text{Mean monthly wage for Factory } A = 5.5 - \frac{70}{200} = 5.15$$

$$\text{S.D. of Factory } A = \sqrt{\frac{\Sigma fd^2}{\Sigma f} - \left(\frac{\Sigma fd}{\Sigma f}\right)^2}$$

$$= \sqrt{\frac{448}{200} - 0.1225} = 1.455$$

$$\text{C.V. for Factory } A = \frac{1.455}{5.15} \times 100 = 28.25\%$$

Mean monthly wage of Factory $B = 5.5 - \frac{150}{200} = 4.75$

S.D. of Factory $B = \sqrt{\frac{570}{200} + \left(\frac{150}{200}\right)^2} = \sqrt{2.85 - 0.5625} = 1.51$

Coefficient of variation for Factory $B = \frac{1.51}{4.75} \times 100 = 31.78\%$

Since, the C.V. for Factory B is higher, the wages in Factory B are more available.

Factory A pays higher average wage by Re. 0.40 [*i.e.*, $5.15 - 4.75 = 0.40$]

Example 58. For each of the three groups of families we give below the number of families in the group, the mean size and also the S.D. of size :

	Group I	*Group II*	*Group III*
Number	35	50	45
Mean size	6.4	5.9	5.1
S.D. of size	2.1	1.9	1.8

If the three groups are combined, what will be the mean family size and S.D. of family size for the composite group ?

Solution. Combined mean $= \frac{35 \times 6.4 + 50 \times 5.9 + 45 \times 5.1}{35 + 50 + 45}$

$$= \frac{224 + 295 + 229.5}{130} = 5.75 = \bar{x}$$

$\bar{x}_1 - \bar{x} = 64 - 5.75 - 0.65,\ \bar{x}_2 - \bar{x} = 5.9 - 5.75 = 0.15$

$\bar{x}_3 - \bar{x} = 5.1 - 5.75 = -0.65$, where $\bar{x}_1, \bar{x}_2, \bar{x}_3$ are the respective means of Groups I, II and III.

Groups I, II and III

Combined S.D.

$$= \sqrt{\frac{35 \times (2.1)^2 + 50 \times (1.9)^2 + 45 \times (1.8)^2 + 35 \times (0.65)^2 + 50 \times (0.15)^2 + 45 \times (0.65)^2}{35 + 50 + 45}}\ *$$

Combined S.D. $= \sqrt{\frac{154.35 + 180.5 + 145.8 + 14.78 + 1.250 + 19}{130}} = 1.99.$

* Combined S.D. of three groups

$$= \sqrt{n_1 \sigma_1^2 + n_2 \sigma_2^2 + n_3 \sigma_3^2 + n_1 d_1^2 + n_2 d_2^2 + n_3 d_3^2}$$

where $d_1 = \bar{x}_1 - \bar{x}_{1, 2, 3}$, $d_2 = \bar{x}_2 - \bar{x}_{1, 2, 3}$, $d_3 = \bar{x}_3 - \bar{x}_{1, 2, 3}$

Example 59. The following table shows the marks obtained by three students A, B and C in an examination :

Student		*Maximum marks in each subject*				
		800	700	900	600	1000
↓	*Subject* →	S_1	S_2	S_3	S_4	S_5
A		560	553	549	540	500
B		480	420	540	360	600
C		424	427	423	426	420

Determine which student has shown (i) most consistent performance and (ii) most inconsistent performance.

Solution. The given data has been shown below :

Student		*Percentage marks*					*Average % marks*	*Standard Deviation*	*C.V.*
		800	700	900	600	1000			
↓	*Subject* →	S_1	S_2	S_3	S_4	S_5			
A		560	553	549	540	500	70	13.87	19.81
B		480	420	540	360	600	60	0	0
C		424	427	423	426	420	54.8	10.32	18.83

Calculations :

A's average marks $= \dfrac{70+79+61+90+50}{5} = \dfrac{350}{5} = 70^*$

B's average marks $= \dfrac{60+60+60+60+60}{5} = \dfrac{300}{5} = 60$

C's average marks $= \dfrac{53+61+47+71+42}{5} = \dfrac{274}{5} = 54.8$

A's Standard deviation (σ_A)

$$= \sqrt{\frac{0+(9)^2+(-9)^2+(20)^2+(-20)^2}{5}} = \sqrt{\frac{962}{5}} = \sqrt{192.4} = 13.87$$

B's Standard deviation $(\sigma_B) = \sqrt{\dfrac{0}{5}} = 0.^{**}$ $\left[\because \text{ S.D.} = \sqrt{\dfrac{\Sigma(x-\bar{x})^2}{n}}\right]$

C's Standard deviation $= (\sigma_C)$

$$= \sqrt{\frac{(1.8)^2+(6.2)^2+(-7.8)^2+(16.3)^2+(-12.8)^2}{5}}$$

$$= \sqrt{\frac{530.05}{5}} = \sqrt{106.41} = 10.32$$

A's Coefficient of variation (C.V.) $= \dfrac{\sigma}{x} \times 100 = \dfrac{13.87}{70} \times 100 = 19.81\%$

B's Coefficient of variation $= \dfrac{0}{60} \times 100 = 0\%$

* A's percentage marks $= \dfrac{560}{800} \times 100 = 70$

** In case of B average percentage of marks in each subject is 60% then $\Sigma(x-\bar{x}) = 0$.

C's Coefficient of variation $= \dfrac{10.32}{54.8} \times 100 = 18.83\%$

Hence, it is clear from above that

(i) B is most consistent as in his case coefficient of variation is the least.

(ii) A is most inconsistent as in his case the coefficient of variation is the greatest.

Example 60. A collar manufacturer is considering the production of a new style of collar to attract young men. The following statistics of neck circumference are available based upon measurements of a typical group of college students :

Mid-value (in inches)	12.0	12.5	13.0	13.5	14.0	14.5	15.0	15.5	16.0
No. of Students	2	16	36	60	76	37	18	3	2

Compute the standard deviation and use the criterion $\overline{X} + 3\sigma$ where σ is the standard deviation and $\overline{X}$ is the arithmetic mean to determine the largest and smallest size of the collar he should make in order to meet the needs of practically all the customers bearing in mind that collars are worn, on average $\frac{1}{2}$ inches longer than neck size.

Solution.

Calculation of Mean and Standard Deviation

Mid-value in inches (m)	*No. of Students* f	$(m-14)/0.5$ d'	fd'	fd'^2
12.0	2	–4	–8	32
12.5	16	–3	–48	144
13.0	36	–2	–72	144
13.5	60	–1	–60	60
14.0	76	0	0	0
14.5	37	1	37	37
15.0	18	2	36	72
15.5	2	3	9	27
16.0	3	4	8	32
	$N = 250$		$\Sigma fd' = -98$	$\Sigma fd'^2 = 548$

$$\overline{X} = A + \frac{\Sigma fd'}{N} \times C$$

$$A = 14, \Sigma fd' = -98, N = 250, C = 0.5$$

$$\overline{X} = 14 - \frac{98}{250} \times 0.5 = 13.8$$

$$\sigma = \sqrt{\frac{\Sigma fd'^2}{N} - \left(\frac{\Sigma fd'}{N}\right)^2} \times C = \sqrt{\frac{548}{250} - \left(\frac{-98}{250}\right)^2} \times 0.5$$

$$= \sqrt{2.192 - 0.154} \times 0.5$$

$$= \sqrt{2.038} \times 0.5 = 1.43 \times 0.5 = 0.7137$$

Largest and smallest neck size

$$= \overline{X} \pm 3\sigma = 13.8 \pm 3\ (0.7137) = 13.8 \pm 2.1411$$

$$= 11.6589 - 15.9411.$$

Since, collars are worn an average $\frac{1}{2}$ inch longer than the neck size, we should add 0.5 to these limits. Thus, the smallest and largest sizes of collar should be :

$$(11.6589 + 0.5) \text{ and } (15.9411 + 0.5) = 12.1511 \text{ and } 16.4411.$$

Thus, the smallest size of collar should be 12.2 inches long and largest 16.4 inches or 16.5 inches long.

Example 61. Find the missing information from the following :

	Group I	*Group II*	*Group III*	*Combined*
Number	50	?	90	200
Standard deviation	6	7	?	7.746
Mean	113	?	115	116

Solution.

Finding Number of Observations in the Second Group

Let N_1, N_2, N_3 denote number of observations in the 1st, 2nd, 3rd groups respectively.

We are given $N_1 + N_2 + N = 200$

$$N_1 = 50, N_2 = 90 \qquad \therefore N_1 + N_2 = 140$$

$$N_3 = 200 - 140 = 60$$

Finding mean of Second Group

Let $\overline{X}_1, \overline{X}_2, \overline{X}_3$ denote mean of first, second and third groups respectively.

$$\overline{X}_{123} = \frac{N_1\overline{X}_1 + N_2\overline{X}_2 + N_3\overline{X}_3}{N_1 + N_2 + N_3}$$

$$\overline{X}_{123} = 116, N_1 + N_2 + N_3 = 200, \overline{X}_1 = 113, \overline{X}_3 = 115$$

We have to find $\overline{X}_2$

Substituting the given values

$$116 = \frac{50\,(113) + 60\,(\overline{X}_2) + 90\,(115)}{200}$$

$$116 \times 200 = 5650 + 60\,\overline{x}_2 + 10350$$

$$6\overline{X}_2 = 23200 - 16000 = 7200$$

$$\therefore \qquad \overline{X}_2 = 120$$

Finding Standard Deviation of Third Group

$$\sigma_{123} = \sqrt{\frac{N_1\sigma_1^2 + N_2\sigma_2^2 + N_3\sigma_3^2 + N_1d_1^2 + N_2d_2^2 + N_3d_3^2}{N_1 + N_2 + N_3}}$$

$$\sigma_{123} = 7.746, N_1 = 50, \sigma_1 = 6, N_2 = 60, \sigma_2 = 7, N_3 = 90, \sigma_3 = ?$$

$$d_1 = (\overline{X}_1 - \overline{X}_{123}) = (113 - 116) = -3$$

$$d_2 = (\overline{X}_2 - \overline{X}_{123}) = (120 - 116) = 4$$

$$d_3 = (\overline{X}_3 - \overline{X}_{123}) = (115 - 116) = -1$$

Substituting the values,

$$7.746 = \sqrt{\frac{50\,(6)^2 + 60\,(7)^2 + 90\sigma_3^2 + 50\,(-3)^2 + 60\,(4)^2 + 90\,(-1)^2}{200}}$$

$$= \sqrt{\frac{1800 + 2940 + 90\sigma_3^2 + 450 + 960 + 90}{200}}$$

$$\Rightarrow \qquad 200 \times (7.746)^2 = 6240 + 90\sigma_3^2$$

$$\Rightarrow \qquad \sigma_3 = 8.$$

Example 62. The following distribution was obtained by a change of scale and origin :

U	–4	–3	–2	–1	0	1	2	3	4
f	4	8	14	18	20	14	10	6	6

Write down the frequency distribution of X if it is given that $\overline{X} = 59.5$ and the variance of $X = 413$.

Solution. We know that S.D. is independent of the change of origin and depends on the change of scale. Let C be the common factor (or dividing factor) to change the scale.

Computation for Determining Class Groups

U	F	fU	fU^2
–4	4	–16	64
–3	8	–24	72
–2	14	–28	56
–1	18	–18	18
0	20	0	0
1	14	14	14
2	10	20	40
3	6	18	54
4	6	24	96
	$N = 100$	$\Sigma fU = -10$	$\Sigma fU^2 = 414$

We are given variance $X = 413$ and $\overline{X} = 59.5$.

$$\text{Variance} = \left[\frac{\Sigma fU^2}{N} - \left(\frac{\Sigma fU}{N}\right)^2\right] \times C^2$$

$$413 = \left[\frac{414}{100} - \left(\frac{-10}{100}\right)^2\right] \times C^2 \qquad \text{[Pl. note]}$$

$$4.13\, C^2 = 413.$$

$$C^2 = \frac{413}{4.13} = 100 \text{ or } C = 10$$

$$\overline{X} = A + \frac{\Sigma fU}{N} \times C$$

$$59.5 = A - \frac{10}{100} \times 10$$

$$A - 1 = 59.5$$

$$A = 60.5$$

Hence, assumed mean = 60.5 and common factor = 10.

The various classes shall be ascertained as follows :

If m_1 is first mid-point, then $\frac{m_1 - A}{10} = -4$

$\Rightarrow$ $\frac{m_1 - 60.5}{10} = -4$ or $m_1 = 20.5$

If m_2 is 2nd mid-point, then $\frac{m_2 - A}{10} = -3$

$\Rightarrow$ $\frac{m_2 - 60.5}{10} = -3$

$\Rightarrow$ $m_2 = 30.5$, etc.

The mid-points of the various classes shall thus be 20.5, 30.5, 40.5, 50.5, 60.5, 70.5, 80.5, 90.5, 100.5. Since the difference between any two consecutive mid-values is 10, the first class shall be $20.5 - 5$ to $20.5 + 5$, *i.e.*, 15.5 to 25.5, the second class 25.5 to 35.5, etc. The frequency distribution can be written as follows :

Variable	*Frequency*
14.5–25.5	4
25.5–35.5	8
35.5–45.5	14
45.5–55.5	18
55.5–65.5	20
65.5–75.5	14
75.5–85.5	10
85.5–95.5	6
95.5–105.5	6

Example 63. The following table presents the results of students in two colleges *A* and *B* :

Name of Exam.	*College A*		*College B*	
	Appeared	*Passed*	*Appeared*	*Passed*
M.A.	30	25	100	80
M.Com	50	45	120	95
B.A.	200	150	100	70
B.Com	120	75	80	50
Total	400	295	400	295

State which of the two colleges is better, and why ?

Solution. In order to determine the quality of result of the two colleges, weighted mean of the two colleges will be determined.

College *A*

Name of Exam.	*Appeared (X)*	*Passed*	*% Passed (W)*	*XW*
M.A.	30	25	83.33	2,500
M.Com	50	45	90.00	4,500
B.A.	200	150	75.00	15,000
B.Com	120	75	62.50	7,500
Total	400	295		29,500
$\overline{X}_w = \frac{\Sigma XW}{\Sigma W} = \frac{29{,}500}{400} = 73.75$				

College *B*

Name of Exam.	*Appeared X*	*Passed W*	*% Passed*	*XW*
M.A.	100	80	80.0	8,000
M.Com	120	95	79.17	9,500
B.A.	100	70	70.00	7,000
B.Com	80	50	62.50	7,500
Total	400	295		29,500
$\overline{X}_w = \frac{\Sigma XW}{\Sigma W} = \frac{29{,}500}{400} = 73.75$				

As the weighted mean of the two colleges is the same, *i.e.*, 73.65, both the colleges are equal in performance.

Example 64. Three independent distributions each of 100 members and S.D. 4.5 units are located with their A.M. at 12.1, 17.1 and 22.1 units respectively. Find the S.D. of the distribution obtained by combining the three distributions.

Solution. Combined mean $= \frac{100\,(12.1 + 17.1 + 22.1)}{300} = 17.1 = \overline{x}$

If $\overline{x}_1, \overline{x}_2$ and $\overline{x}_3$ be the respective arithmetic means of the three distributions, then

$$\overline{x}_1 - \overline{x} = 12.1 - 17.1 = 5;\ \overline{x}_2 - \overline{x} = 0;\ \overline{x}_3 - \overline{x} = 22.1 - 17.1 = 5.$$

Combined S.D.

$$= \sqrt{\frac{100\,[(4.5)^2 + (-5)^2] + 100\,(4.5)^2 + 100\,[(4.5)^2 + (5)^2]}{300}}$$

$$= \sqrt{\frac{20.25 + 25 + 20.25 + 20.25 + 25}{3}} = \sqrt{\frac{110.75}{3}} = 6.075$$

Example 65. The mean, S.D. and range of a distribution of weights of 12 children are 9 lbs, 2 lbs and 6 lbs respectively. The median of the distribution is the same as the mean. Find the mean and S.D. of group if the lightest and heaviest children are excluded.

Solution. Mean = Median ⇒ The distribution is symmetrical.

In a symmetrical distribution the highest and the lowest values will be at the same distance from the mean but on opposite sides of mean. **(Please note)**

Range = 6 ⇒ Difference between the highest and the lowest value = 6.

∴ The highest value will be (mean + 3) and lowest value = (mean – 3), *i.e.*, the highest and lowest values are 12 and 6 respectively.

$$\text{Sum of the values} = 12 \times 9 = 108$$

Excluding highest and lowest values, the sum of the remaining 10 values = 108 – 12 – 6 = 90.

Mean weight of remaining 10 children $= \dfrac{90}{10} = 9$

$$(\text{S.D.})^2 = \frac{\Sigma x^2}{n} - \left(\frac{\Sigma x}{n}\right)^2$$

Sum of the squares of 12 observations $= n\,[(\text{S.D.})^2 + (\bar{x})^2]$*

$$= 12\,(4 + 81) = 12 \times 85 = 1{,}020.$$

Sum of the squares of value excluding squares of highest and the lowest values

$$= 1020 - (12)^2 - (6)^2 = 840.$$

S.D. of the remaining 10 values $= \sqrt{\dfrac{840}{10} - 81} = \sqrt{3} = 1.732.$

Note : Highest and lowest values ⇒ the weights of heaviest and lightest child respectively.

Example 66. Find Arithmetic mean and mean deviation from the median for the following data :

Class interval	*Frequency*
120–125	10
125–130	25
130–135	30
135–140	50
140–145	40
145–150	30
150–155	15

Solution.

Class interval	*Mid-value X*	*Frequency*	$x - 137.5 = d$	*fd*	*Cumulative frequency*	$\lvert x - \text{median}\rvert = D$	*fD*
120–125	122.5	10	–15	–150	10	16	160
125–150	127.5	25	–10	–250	35	11	275
130–135	132.5	30	–5	–150	65	6	180
135–140	137.5	50	0	0	115	1	50
140–145	142.5	40	5	200	155	4	160
145–150	147.5	30	10	300	185	9	270
150–155	152.5	15	15	225	200	14	210
				$\Sigma fd = 175$			$\Sigma fD = 1305$

---

* $\text{SD}^2 = \dfrac{\Sigma x^2}{n} - \left(\dfrac{\Sigma x}{n}\right)^2 \Rightarrow \dfrac{\Sigma x^2}{n} = (\text{SD})^2 + \left(\dfrac{\Sigma x}{n}\right)^2 \Rightarrow \Sigma x^2 = n\left[(\text{SD})^2 + \left(\dfrac{\Sigma x}{n}\right)^2\right]$

135—140 is the median class.

$$\text{Median} = 135 + \frac{100-65}{50} \times 5 = 135 + \frac{175}{50} = 138.5$$

$$\text{Median deviation from median} = \frac{1305}{200} = 6.525.$$

Example 67. Compute quartile deviation for the following data :

Class Interval of Marks	*Number of Students*
0–10	6
10–20	12
20–30	16
30–40	23
40–50	18
50–60	9
60–70	5

(Punjab, B. Com., 1991)

Solution. Cumulative frequency table is :

Class Interval	*Number of Students*	*Cumulative Frequency*	
0–10	5	6	$N = 89, \frac{N}{2} = 44.5$ $\frac{N}{4} = 22.25, \frac{3N}{4} = 66.75$
10–20	12	18	
20–30	16	34	
30–40	23	57	
40–50	18	75	
50–60	9	84	
60–70	5	89	

First quartile, *i.e.*, Q_1 lies in the class 20-30.

$$\therefore \qquad Q_1 = 20 + \frac{22.25-18}{16} \times 10 = 20 + \frac{4.25}{16} \times 10 = 22.65$$

Third quartile, *i.e.*, Q_3 lies in the class 40-50

$$\therefore \qquad Q_3 = 40 + \frac{60.75-57}{18} \times 10 = 42.08$$

$$\text{Quartile deviation} = \frac{Q_3 - Q_1}{2} = \frac{42.08 - 22.65}{2} = 9.715$$

Example 68. The following is the record of number of bricks laid each day for 10 days by two brick-layers *A* and *B*. Calculate the coefficient of variation in each case and discuss the relative consistency of the two brick-layers :

A	700	675	725	625	650	700	650	700	600	650
B	550	600	575	550	650	600	550	525	625	600

If the figures for A were in every case 10 more and those of B in every case 20 more than the figure given above, how would the answer be affected ?

Calculations for Mean and Standard Deviation

Brick-layer A			*Brick-layer B*		
X	$dx' = \frac{X-700}{25}$	dx'^2	Y	$dy' = \frac{Y-625}{25}$	dy'^2
700	0	0	550	–3	9
675	–1	1	600	–1	1
725	1	1	575	–2	4
625	–3	9	550	–3	9
650	–2	4	650	1	1
700	0	0	600	–1	1
650	–2	4	550	–3	9
700	0	0	525	–4	16
600	–4	16	625	0	0
650	–2	4	600	–1	1
Total	–13	39		–17	51

Common factor $h = 25$

Solution.

Brick-layer A :

$$\overline{X} = A + \frac{\Sigma dx'}{n} \times h = 700 - \frac{13}{10} = 25$$

$$= 700 - 32.5 = 667.5 \text{ bricks per day}$$

$$\sigma_X = 25\sqrt{\frac{\Sigma dx'^2}{n} - \left(\frac{\Sigma dx'}{n}\right)^2}$$

$$= \left[\sqrt{\frac{39}{10} - \left(\frac{-13}{10}\right)^2}\right] \times 25 = [\sqrt{3.9 - 1.69}] \times 25$$

$$= \sqrt{2.21} \times 25 = 1.486 \times 25 = 37.15$$

$$\text{C.V. }(A) = \frac{\sigma_X}{\overline{X}} \times 100 = \frac{37.15}{667.5} \times 100 = 5.56\%$$

Brick-layer B :

$$\overline{Y} = 625 - \frac{17}{10} \times 25 = 582.5 \text{ bricks per day}$$

$$\sigma_Y = \left[\sqrt{\frac{51}{10} - \left(\frac{-17}{10}\right)^2}\right] \times 25 = \sqrt{2.21} \times 25 = 37.15$$

$$C.V.\ (B) = \frac{37.15}{582.5} \times 100 = 6.38\%$$

As the coefficient of variation for brick-layer A is less than that of brick-layer B, brick-layer A is more consistent.

If the figures for A in every case were 10 more and that of B were 20 more, the arithmetic mean of A will increase by 10 and that of B by 20 but the standard deviations of both of them will remain unchanged.

[∵ S.D. is independent of the change of origin]

∴ A.M. of A will be $667.5 + 10 = 677.5$ bricks per day

and A.M. of B will be $582.5 + 20 = 602.5$ bricks per day

$$\therefore \text{ Coefficient of variation of } A \text{ will be } = \frac{37.15}{677.5} \times 100 = 5.48\%$$

$$\text{and coefficient of variation of } B \text{ will be } \frac{37.15}{602.5} \times 100 = 6.16\%$$

After the change also brick-layer A remains more consistent than brick-layer B.

Example 69. Complete the table showing the frequencies with which words of different number of letters occur in the extract reproduced below omitting punctuation marks, treating as the variable the number of letters in each word, and obtain the coefficient of variation of the distribution :

"Her eyes were blue : blue as autumn distance—blue as the blue we see between the retreating mouldings of hills and woody slopes on a sunny September morning : a misty and shady blue, that had no beginning on surface, and was looked into rather than at."

Solution. Here, the variable (X) is the number of letters in each word in the extract given above. We find that in the extract given above there are words with the number of letters arranging from 1 to 10. Hence the variable X takes the values from 1 to 10. The frequency distribution is as follows :

No. of lettes in a word (X)	*Frequency (f)*	$d = X - 6$	*fd*	fd^2
1	2	–5	–10	50
2	8	–4	–32	128
3	9	–3	–27	81
4	10	–2	–20	40
5	5	–1	–5	5
6	4	0	0	0
7	3	1	3	3
8	1	2	2	4
9	3	3	9	27
10	1	4	4	16
Total	$46 = \Sigma f = N$		$\Sigma fd = -76$	$\Sigma fd^2 = 354$

Mean, $\overline{X} = A + \frac{1}{N}\Sigma fd = 6 + \left(\frac{-76}{46}\right) = 4.35$

$$\sigma^2 = \frac{\Sigma fd^2}{N} - \left(\frac{\Sigma fd}{N}\right)^2 = \frac{354}{46} - \left(\frac{-76}{46}\right)^2$$

$$= 7.6956 - 2.7225 = 4.9731$$

∴ S.D., $(\sigma) = \sqrt{4.9731} = 2.23$

$$\text{C.V.} = \frac{\sigma}{\overline{X}} \times 100 = \frac{2.23}{4.35} \times 100 = 51.264\% \quad \text{(approx.)}$$

Example 69(a). The following table shows the results of an analysis relating to the pieces of two shares, X and Y, that were recorded during the year 2000 :

Share	*Arithmetic Mean (Rs.)*	*Standard Deviation (Rs.)*
X	45.00	9.00
Y	62.50	18.75

Which of the two shares, X or Y, have more variability in prices ?

Solution. For determining which share shows greater variability in prices, we have to compare coefficient of variation (C.V.)

$$\text{C.V. (for share } X) = \frac{\text{S.D.}}{\overline{\text{X}}} \times 100 = \frac{9}{45} \times 100 = 20$$

$$\text{C.V. (for share } Y) = \frac{\text{S.D.}}{\overline{\text{Y}}} \times 100 = \frac{18.75}{62.5} \times 100 = 30$$

Since coefficient of variation is higher in shares *Y*, hence share *Y* shows greater variability in prices.

Example 69(b). In a statistical investigation in two villages A and B, the following data was obtained :

Particulars	*Voltage A*	*Voltage B*
No. of Respondents	600	500
Average Income (Rs.)	175	186
Standard Deviation (Rs.)	10	9

(i) What is the average income of the respondents in the village A and B put together ?

(ii) In which village is the variation in income greater ?

Solution. No. of Respondents in village $A\ (n_1) = 600$

No. of Respondents in village $B\ (n_2) = 500$

Average income in village $A\ (\overline{X}_1) = 175$ Rs.

Average income in village $B\ (\overline{X}_2) = 186$ Rs.

Standard deviation in village A $(\sigma_1) = 10$ Rs.

Standard deviation in village B $(\sigma_2) = 9$ Rs.

(i) Combined average income of respondents in village A and B

$$\bar{X}_{12} = \frac{\bar{X}_1 N_1 + \bar{X}_2 N_2}{N_1 + N_2}$$

$$= \frac{175\ (600) + 186\ (500)}{600 + 500}$$

$$= \frac{105000 + 93000}{1100} = \frac{198000}{1100}$$

$$\bar{X}_{12} = 180 \text{ Rs.}$$

(ii) Coefficient of variation in village A

$$\%\ \text{C.V. } X_1 = \frac{\sigma_1}{\bar{X}_2} \times 100$$

$$= \frac{10}{175} + 100 = 5.71\%$$

Coefficient of variation in village B

$$\%\ \text{C.V. } X_2 = \frac{\sigma_2}{\bar{X}_2} \times 100$$

$$= \frac{9}{186} \times 100 = 4.91\%$$

Since the C.V. for village A is more, the variation in income is greater in village A.

Example 70. For a distribution of 100 observations, the sum of the deviations from 4 cm is –11 cm and the sum of the squares of those deviations is 257 sq. cm. Find the coefficient of variation.

Plate note. The solution carefully.

Solution.
$$\text{Mean} = 4 - \frac{11}{100} = \frac{389}{100} = 3.89$$

As the S.D. is independent of the change of origin, we have

$$\text{S.D.} = \sqrt{\frac{\Sigma d^2}{n} - \left(\frac{\Sigma d}{n}\right)^2} = \sqrt{\frac{257}{100} - \left(\frac{-11}{100}\right)^2}$$

$$= \sqrt{2.57 - 0.0121} = \sqrt{2.5579} = 1.599$$

$$\text{Coefficient of variation} = \frac{\text{S.D.}}{\text{Mean}} \times 100 = \frac{1.599}{3.89} \times 100 = 41\%$$

Example 71. The arithmetic mean of runs scored by three batsman Vijay, Subhash and Kumar in the same series of 10 innings are 50, 48 and 12 respectively. The standard deviation of the runs are respectively 15, 12 and 2.

Who is the most consistent of the three ? If one of the three is to be selected, who will be selected ?

Solution. Coefficient of variation of Vijay $= \frac{15}{50} \times 100 = 30\%$

Coefficient of variation of Subhash $= \frac{12}{48} \times 100 = 25\%$

Coefficient of variation of Kumar $= \frac{2}{12} \times 100 = 16.66\%$

Since the coefficient of variation of Kumar is the least, he is more consistent of the three and will be selected.

Example 72. For the following data obtain (i) mean, (ii) median, (iii) mean deviation from median, and (iv) the coefficient of variation :

0, –1, 1, –2, 4, 5,6, 8, 12, 11, 10

Solution. Mean $= \frac{0 - 1 + 1 - 2 + 4 + 5 + 6 + 8 + 12 + 11 + 10}{11}$

$$= \frac{54}{11} = 4.909 = 4.91$$

Writing the given series in ascending order, we get

–2, –1, 0, 1, 4,5, 6, 8, 10, 11, 12

Number of observations = 11

$$\text{Median} = \frac{11+1}{2} = \text{6th observation} = 5$$

Calculations of mean deviation from the median :

$$\text{Mean deviation} = \sum_{i=1}^{n} \frac{|x_1 - \text{median}|}{n}$$

$$= \frac{5 + 6 + 4 + 7 + 1 + 0 + 1 + 3 + 7 + 6 + 5}{11}$$

$$= \frac{45}{11} = 4.1 \text{ (Approx.)}$$

Calculation of Standard Deviation

$$\text{S.D.} = \sqrt{\frac{\Sigma x^2}{n} - \left(\frac{\Sigma x}{n}\right)^2}$$

$$\Sigma x^2 = 0 + 1 + 1 + 4 + 16 + 25 + 36 + 64 + 144 + 121 + 100 = 512$$

$$\therefore \quad \text{S.D.} = \sqrt{\frac{512}{11} - \left(\frac{54}{11}\right)^2} = \sqrt{46.54 - 24.09} = 4.738$$

$$\text{Coefficient of variation} = \frac{\text{S.D.}}{\text{Mean}} \times 100 = \frac{4.738}{4.91} \times 100 = 96.5\%$$

Example 73. The following table gives the frequency distribution of expenditure on education per family per month among middle class family in two towns.

Expenditure (in Rs.)	*No. of Families*	
	Town A	*Town B*
3–6	28	39
6–9	292	284
9–12	389	401
12–15	212	202
15–18	59	48
18–21	18	21
21–24	2	5

(i) Find the arithmetic mean and standard deviations of the expenditure at both towns.
(ii) Find out which of the two towns shows greater variability.

Solution. See next page for table for calculations of mean and standard deviation :

Expenditure (in Rs.)	*Mid-value X*	$x - 13.5 = d$	*Town A*			*Town B*		
			f	fd	fd^2	f	Σfd	Σfd^2
3–6	4.5	–9	28	–252	2,268	39	–351	3,159
6–9	7.5	–6	292	–1.752	10,512	284	–1,704	10,224
9–12	10.5	–3	389	–1,167	3,501	401	–1,203	3,609
12–15	13.5	0	212	0	0	202	0	0
15–18	16.5	3	59	177	531	48	144	432
18–21	19.5	6	18	108	648	21	126	756
21–24	22.5	9	2	18	162	5	45	405
			$\Sigma f = 1,000$	$\Sigma fd = 02,868$	$\Sigma fd^2 = 17,622$	$\Sigma f = 1,000$	$\Sigma fd = -2,943$	$\Sigma fd^2 = 18,585$

$$\text{S.D. (for Town } A) = \sqrt{\frac{\Sigma fd^2}{\Sigma f} - \left(\frac{\Sigma fd}{\Sigma f}\right)^2} = \sqrt{17.622 - (2.868)^2}$$

$$= \sqrt{17.622 - 8.225} = 3.065$$

$$\text{S.D. (for Town } B) = \sqrt{\frac{18585}{1000} - \left(\frac{-2943}{1000}\right)^2} = \sqrt{9.925} = 3.15$$

$$\text{Arithmetic mean of expenditure in Town } A = 13.5 + \left(\frac{-2868}{1000}\right)$$

$$= 13.5 - 2.868 = 10.632$$

Arithmetic mean of expenditure in Town $B = 13.5 - 2.943 = 10.557$

Variability [Coefficient of Variation] for Town $A = \dfrac{3.065}{10.632} \times 100 = 28.82\%$

Coefficient of Variation for town $B = \dfrac{3.15}{10.557} \times 100 = 29.83\%$

$\therefore$ Coefficient of variation for Town B shows greater variability.
Hence, Town B shows greater variability.

Example 74. Examine the validity of the following statements :

(1) The mean and S.D. of a series are 50 and 8 respectively. By adding 2 in each value, New Mean = 52 and New S.D. = 10.

(2) The S.D. of a set of values will be positive or negative according as the given values are positive or negative.

(3) Range is a measure of dispersion which gives us information about scatter of values about the measure of central tendency.

Solution.

(1) No. The new mean will be 52 but new S.D. will remain 8 because S.D. is independent of the change of origin.

(2) No. The S.D. is always positive.

(3) No. Range = Difference between the two extreme values

Example 75. The profit (in Rs. lakhs) earned by 100 companies during 1992-93 are shown below :

Profits	*No. of Companies*	*Profits*	*No. of Companies*
20–30	4	60–70	15
30–40	8	70–80	10
40–50	18	80–90	8
50–60	30	90–100	7

Compute (a) mean (b) Median (c) Standard Deviation. **(AIMA, May, 1993)**

Solution.

Profits	*Mid-value* $= x$	$x = 65 = d$	$\frac{d}{10} = d'$	f	fd'	fd'^2	*Calculative frequency*
20–30	25	–40	–4	4	–16	64	4
30–40	35	–30	–3	8	–24	72	12
40–50	45	–20	–2	18	–36	72	30
50–60	55	–10	–1	30	–30	30	60
60–70	65	0	0	15	0	0	75
70–80	75	10	1	10	10	10	85
80–90	85	20	2	8	16	32	93
90–100	95	30	3	7	21	63	100
				$\Sigma f = 100$	$\Sigma fd' = -59$	$\Sigma fd'^2 = 343$	

$$\text{Mean} = 65 - \frac{59}{100} \times 10 = 59.1$$

$$\text{Median} = 50 + \frac{50 - 30}{30} \times 10 = 50 + \frac{200}{30} = 56.66$$

$$\text{S.D.} = \sqrt{\frac{\Sigma fd'^2}{\Sigma f} - \left(\frac{\Sigma fd'}{\Sigma f}\right)^2} \times 10$$

$$= \sqrt{\frac{343}{100} - \left(\frac{-59}{100}\right)^2} \times 10 = \sqrt{3.43 - 0.3481} \times 10 = 17.5$$

Example 76. Calculate (i) quartile deviation, (ii) standard deviation of wages from the following data :

Daily wages (in Rs.)	35–36	36–37	37–38	38–39	39–40	40–41	41–42
No. of Persons	14	20	42	54	45	18	7

(ICWA Dec., 1993)

Solution.

Daily wages	*Mid-value (x)*	$x - 38.5 = d$	f	*Fd*	fd^2	*Cumulative frequencies*
35–36	35.5	–3	14	–42	126	14
36–37	36.5	–2	20	–40	80	34
37–38	37.5	–1	42	–42	42	76
38–39	38.5	0	54	0	0	130
39–40	39.5	1	45	45	45	175
40–41	40.5	2	18	36	72	193
41–42	41.5	3	7	21	63	200
			$\Sigma f = 200$	$\Sigma fd = -12$	$\Sigma fd^2 = 428$	

$$Q_1 = 37 + \frac{50 - 34}{42} \times 1 = 37 + \frac{16}{42} = 37.38$$

$$Q_3 = 39 + \frac{150 - 130}{45} = 39 + \frac{20}{45} = 39.44$$

$$\text{S.D.} = \sqrt{\frac{\Sigma fd^2}{\Sigma f} - \left(\frac{\Sigma fd}{\Sigma f}\right)^2} = \sqrt{\frac{428}{200} - \frac{144}{40000}} = 1.46$$

$$\text{Quartile deviation} = \frac{Q_3 - Q_1}{2} = \frac{39.44 - 37.38}{2} = 1.03$$

Example 77. The mean, stanard deviation and range of a symmetrical distribution of weights of a group of 20 boys are 40, 5 and 6 Kgs. respectively. Find the mean and standard deviation of the group if the lightest and the heaviest boys are excluded.

Solution. Please note.

(i) Since, the distribution is symmetrical

$$\frac{\text{heaviest value} + \text{lightest value}}{2} = \text{Mean}$$

(ii) Heaviest value – Lightest value = Range.

If L_2 is the heaviest value and L_1 the lightest value then

$$L_2 + L_1 = 2 \times \text{mean} = 40 \times 2 = 80 \qquad \text{...(i)}$$

and

$$L_2 - L_1 = 6 \qquad \text{...(ii)}$$

Adding (i) and (ii), we get $2L_2 = 86$

$\Rightarrow$ $L_2 = 43$

Putting $L_2 = 43$ in (i), we get $L_1 = 80 - 43 = 37$

$\therefore$ Weight of the heaviest boy = 43 kg

Weight of the lightest boy = 37 kg

Sum of the weights of all the boys $= 40 \times 20 = 800$

Sum of the weights of the boys excluding the heaviest and lightest boys

$$= 800 - 80 = 720$$

Mean weight of the remaining 18 boys $= \frac{720}{18} = 40$

$$(\text{S.D.})^2 = \frac{\Sigma x^2}{n} - \left(\frac{\Sigma x}{n}\right)^2$$

$$\Rightarrow \quad 25 = \frac{\Sigma x^2}{20} - (40)^2$$

$$\Rightarrow \quad \Sigma x^2 = (1{,}600 + 25)\ 20 = 1{,}625 \times 20 = 32{,}500$$

Sum of the squares of the weights of the 18 boys excluding heaviest and the lightest boys

$$= 32{,}500 - (43)^2 - (37)^2 = 32{,}500 - 1{,}849 - 1{,}369 = 29{,}282$$

S.D. of the remaining 18 boys

$$= \sqrt{\frac{\Sigma x^2}{18} - \left(\frac{\Sigma x}{18}\right)^2} = \sqrt{\frac{29282}{18} - (40)^2}$$

$$= \sqrt{1626.77 - 1600} = \sqrt{26.77} = 5.174$$

Example. 78. The salaries paid to the managers of a company had a mean of Rs. 20,000 with a standard deviation of Rs. 3,000. What will be the mean and standard deviation if all the salaries are increased by (i) 10%, (ii) 10% of the existing mean ?

Which policy would you recommend if the management does not want to have increased disparities in waves ?

Solution. Increasing all the salaries by 10% $\Rightarrow$ multiplying each salary each by 1.1. Hence the mean is also multiplied by 1.1. Since S.D. depends on change of scale, S.D. is also multiplied by 1.1.

$\therefore$ When all salaries are increased by 10% of the S.D. also increases by 10%.

If each salary is increased by 10% the mean is also increased by 10%.

Increasing all the salaries by 10% of the existing mean $\Rightarrow$ Adding a constant amount to each salary.

Since, S.D. is independent of the change of origin, if will remain unchanged if each salary is increased by 10% of the mean. But mean will increase by 10% of the original mean.

If the management do not want to have increased disparities, it should increase the salary of the workers by 10% of the existing mean.

Example. 79. Initially, there were 9 workers, all being paid a uniform wage. Later 10th worker is added whose wage rate is 20 less than for the other. Compute :

(i) The effect on the mean wage.

(ii) The S.D. of wages for the group of 10 workers.

Solution. Let the mean wage of the 9 workers be Rs. x.

Then the wage of 10th worker = Rs. $(x - 20)$

Average wage of 10 workers $= \frac{9x + x - 20}{10} = x - 2$

$\Rightarrow$ The mean wage is decreased by Rs. 2.

$(\text{S.D.})^2$ for the group of 10 workers

$$= \frac{9(x-x+2)^2}{10} + \left(\frac{x-20-x+2}{10}\right)^{2*} = \frac{36}{10} + \frac{324}{10} = \frac{360}{10} = 36$$

$\Rightarrow$ S.D. for the group of 10 workers $= \sqrt{36} = 6$

Second Method :

$$(\text{S.D.})^2 = \frac{\Sigma fx^2}{n} - \left(\frac{\Sigma fx}{n}\right)^2 = \frac{9x^2 + (x-20)^2}{10} - (x-2)^2$$

$$= 9x^2 + x^2 + 400 - 40x - 10(x^2 - 4x + 4)$$

$$= \frac{10x^2 + 400 - 40x - 10x^2 + 40x - 40}{10} = \frac{360}{10} = 36.$$

$\Rightarrow$ S.D. = 6.

Example 80. In a group of 10 workers, the wage of the lowest paid worker is Rs. 25 less than the average wage of others. Show that the standard deviation of wages is equal to 7.5.

Solution. Let the average wage of the 9 workers = Rs. x then the wage of the lowest paid worker $= x - 25$.

$$\text{Average wage of all the 10 workers} = \frac{9x + x - 25}{10} = x - 2.5$$

$$[\text{S.D.}]^2 = \frac{\Sigma(x-\bar{x})^2}{n} = \frac{9[x-(x-2.5)]^2}{10} + \frac{[x-25-(x-2.5)]^2}{10}$$

$$= \frac{56.25 + 506.25}{10} = \frac{562.5}{10} = 56.25$$

$\therefore$ S.D $= \sqrt{56.25} = 7.5$

Example 81. There are 60 male and 40 female workers in a factory. The standard deviation of their wages (per hour) were computed as Rs. 8 and 11 respectively. The mean wages of the two groups were found to be equal. Compute the combined standard deviation of the wages of all the workers.

Solution. Combined S.D. of two groups

$$= \sqrt{\frac{n_1\sigma_1^2 + n_2\sigma_2^2}{n_1+n_2} + \frac{n_1 n_2}{(n_1+n_2)}(\bar{x}_1 - \bar{x}_2)^2}$$

If $\bar{x}_1 = \bar{x}_2$, then

Combined S.D. of the two groups

$$= \sqrt{\frac{n_1\sigma_1^2 + n_2\sigma_2^2}{n_1+n_2}} = \sqrt{\frac{60 \times 64 + 40 \times 121}{60 + 40}}$$

$$= \sqrt{\frac{3{,}840 + 4{,}840}{100}} = \sqrt{86.8} = 9.31$$

Example 82. Ram appeared in three tests of 20, 50 and 30 marks respectively. He obtained 75% marks in the first and 60% marks in the second test. What should be his percentage of marks in the third test in order that his aggregate is 60%?

Solution. Marks obtained in the first test $= 20 \times 0.75 = 15$

---

* $(\text{S.D.})^2 = \frac{\Sigma f(x-\bar{x})^2}{\Sigma f} = \frac{\Sigma fx^2}{\Sigma f} - \left(\frac{\Sigma fx}{\Sigma f}\right)^2$

Marks obtained in the second test $= 50 \times 0.6 = 30$

Total marks $= 20 + 50 + 30 = 100$

60% of $100 = 60$

$\therefore$ Ram should obtain $60 - 15 - 30 = 15$ marks in the third test

Since 50% of 30 is 15, Ram should obtain 50% marks in the third test.

Example 83. A fund manager is considering investment in the equity shares of one of two companies. The criterion for selecting the company for investment is consistency of return on net worth. The following data have been collected :

Financial Year	*Return on Networth (%)*	
	Modern Industries Ltd. (MIL)	*Pioneer Industries Ltd. (PIL)*
2001–2002	19	20
2000–2001	20	24
1999–2000	16	16
1998–1999	13	15
1997–1998	12	10

You are required to identify the company in which the fund manager should invest.

Solution. Modern Industries Ltd. (MIL) :

$$\text{Mean return on equity } (\bar{x}) = \frac{\Sigma x}{n} = \frac{19 + 20 + 16 + 13 + 12}{5}$$

$$= \frac{80}{5} = 16\%$$

$$\text{Standard deviation (S.D.)} = \left[\frac{\Sigma (x - \bar{x})^2}{n}\right]^{\frac{1}{2}*}$$

$$= \left[\frac{(19-16)^2 + (20-16)^2 + (16-16)^2 + (13-16)^2 + (12-16)^2}{5}\right]^{\frac{1}{2}}$$

$$= 3.16\% \qquad \text{(approx.)}$$

$$\text{Coefficient of variation} = \frac{\text{S.D.}}{\text{Mean}} \times 100 = \frac{3.16}{16} \times 100 = 19.76\% \text{ (approx.)}$$

Pionner Industries Ltd. (PIL) :

$$\text{Mean return on equity } (\bar{x}) = \frac{20 + 24 + 16 + 15 + 10}{5} = 17\%$$

$$\text{Standard deviation (S.D.)} = \left[\frac{\Sigma (x - \bar{x})^2}{n}\right]^{1/2}$$

$$= \left[\frac{(20-17)^2 + (24-17)^2 + (16-17)^2 + (15-17)^2 + (10-17)^2}{5}\right]^{1/2} = 4.73\%$$

* In some higher level books formula for S.D. is used as

$$\text{S.D.} = \left[\frac{\Sigma (x - \bar{x})^2}{n - 1}\right]^{\frac{1}{2}}$$

$$\text{Coefficient of variation} = \frac{\text{S.D}}{\text{mean}} \times 100 = \frac{4.73}{17} \times 100 = 27.83\% \text{ (approx.)}$$

Since, the coefficient of variation for MIL is less than the coefficient of variation for PIL, it can be inferred that profitability of MIL is more consistent than PIL. Hence, investment should be made in MIL.

Example. 84. The mean and standard deviation of the following continuous series are 31 and 15.9 respectively. The distribution after taking step-deviations is as follows :

d (Step-dev.)	–3	–2	–1	0	+1	+2	+3
f (Frequency)	10	15	25	25	10	5	5

Determine actual class intervals. **(GGIPU B. Com., May, 2002)**

Solution.

Step deviations are the deviation obtained by change of origin as well as change of scale.

d	f	d^2	fd^2	fd
–3	10	9	90	–30
–2	15	4	60	–30
–1	25	1	25	–25
0	25	0	0	0
1	10	1	10	10
2	5	4	20	10
3	5	9	45	15
	$\Sigma f = 95$		$\Sigma fd^2 = 250$	$\Sigma fd = -50$

Let A be the assumed mean and i the class interval, then

$$(\text{S.D.})^2 = \left[\frac{\Sigma fd^2}{\Sigma f} - \left(\frac{\Sigma fd}{\Sigma f}\right)^2\right] i^2$$

$$\Rightarrow \quad (15.9)^2 = \left[\frac{250}{95} - \left(-\frac{50}{95}\right)^2\right] i^2 = i^2\,[2.63 - 0.277] = (2.353)\, i^2$$

$$\Rightarrow \quad 252.8 = 2.353 i^2 \Rightarrow i^2 = \frac{252.8}{2.353} = 107.44$$

$$\Rightarrow \quad i = 10.365 = 10 \text{ (Approx.)}$$

$$\text{Mean} = A + \frac{\Sigma fd}{\Sigma f} \times i \Rightarrow 31 = \left[A - \frac{50}{95} \times 10\right]$$

$$\Rightarrow \quad 31 = A - 5.26 \Rightarrow A = 36.26 = 36 \text{ (approx.)}$$

If M_1 is the mid-value of the first group. Then

$$\frac{M_1 - 36}{10} = -3 \Rightarrow M_1 = 6$$

$$\Rightarrow \quad \frac{L_1 + L_2}{2} = 6 \Rightarrow L_1 + L_2 = 12 \qquad ...(1)$$

Since the class interval is 10, we have

$$L_2 - L_1 = 10 \qquad ...(2)$$

Solving (1) and (2), we get

$$L_2 = 11, L_1 = 1$$

$\therefore$ The first class interval is 1 – 11 and the other intervals will be 11-21, 21-31, 31-41, 41-51, 51-61, 61-71.

Example 85. If the IQ of a random sample of 1080 students at large university have a mean score of 120 and a S.D. of 8, use Chebyshev's theorem to determine the interval containing at least 810 of the IQs in the sample. From this interval draw a statistical inference concerning the IQs of all the students at this university. In what range can we be sure that no more than 120 of the IQs lie.

Solution. According to Chebyshev's theorem, $1 - \frac{1}{k^2}$ gives the least fraction of the values lying between k standard deviations of the mean.

$$\therefore \quad 1 - \frac{1}{k^2} = \frac{810}{1080} = \frac{3}{4}$$

$$\Rightarrow \quad 1 - \frac{3}{4} = \frac{1}{k^2}$$

$$\Rightarrow \quad \frac{1}{4} = \frac{1}{j^2} \Rightarrow k^2 = 4$$

$$\Rightarrow \quad k = \pm 2.$$

$$120 \pm 2 \times \text{S.D.} = 120 \pm 2 \times 8 = 104, 136$$

i.e., the interval from 104 and 136 contains at least 3 over 4 (*i.e.*, at least 75%) or at least 810 of the IQs of our sample. From this result we may draw the inference that at least $\frac{3}{4}$ of the IQs for the entire university fall in the interval 104 to 136. In the second part we are to find the range containing no more than 120 of the IQs. By Chebyshev's theorem we can find the range containing $1080 - 120 = 960$ of the IQs.

$$\Rightarrow \quad k = 3^*$$

i.e., at least 960 of the IQs are contained in the range

$$120 \pm 3 \times 8, \textit{ i.e.}, \text{ 96 to 144.}$$

Therefore, no more than 120 of the IQs will fall below 96 and above 144 scores, *i.e.*, maximum 120 of the IQs will fall below 96 and above 144.

Example 86. You are given the position in a factory before and after the settlement of an industrial dispute. Comment on the gains or losses from the point of view of workers and that of management :

	Before	*After*
No. of Workers	6,000	5,800
Mean wages (in Rs.)	440	460
Median wages (in Rs.)	500	480
Standard deviation (in Rs.)	60	52

[D.U.B. Com. (Hons.) Correspondence 2004]

* $1 - k^2 = \frac{960}{1080} = \frac{8}{9} \Rightarrow 1 - \frac{8}{9} = \frac{1}{10^2} \Rightarrow k^2 = 9 \Rightarrow k = \pm 3$

Solution.

Before	*After*
Total amount paid to the workers $= 440 \times 6{,}000$ = Rs. 2,640,000 $\text{C.V.} = \frac{60}{440} = 0.136$	Total amount paid to the workers $= 460 \times 5{,}800$ = Rs. 2,668,000 $\text{C.V.} = \frac{52}{460} = 0.113$

Since, the total amount paid to workers has increased after settlement of dispute. The management has suffered a loss of Rs. 2,668,000 – 2,640,000 = Rs. 28,000.

Since, the C.V. has decreased, the wages have become more uniform which in one respect is gain to the workers.

Since, the median wage has decreased, this is a loss to workers.

Example 87. Measures of dispersion and measures of central tendency and complementary to each other.

Solution. The statement is true. Almost all the measures of dispersion are calculated with the help of measures of central tendency. For example,

$$\text{S.D.} = \sqrt{\frac{\Sigma (\text{X} - \overline{\text{X}})^2}{\text{n}}}, \text{ i.e., S.D. uses A.M.}$$

C.V. also uses A.M.

$$\text{Mean deviation} = \frac{\Sigma |X - \text{Mean}|}{n} \text{ or } \frac{\Sigma |X - \text{Median}|}{n}$$

or
$$\frac{\Sigma |X - \text{Mode}|}{n}$$

Quartile deviation uses Quartiles which are also measures of central tendency.

Any change in the values of the measures of central tendency will change the value of the corresponding measue of dispersion. Therefore, measures of central tendency and measures of dispersion are complementary to each other.

Example 88. The number of marriage licences issued in a certain city during the month of June may be looked upon as a random variable with $\mu = 124$ and S.D. = 7.5. According to Chebyshev's theorem with what probability can we assert that between 64 and 184 marriage licences will be issued during the month of June.

Solution. $124 - 64 = 60$; $184 - 124 = 60$ and $\frac{60}{7.5} = 8 = k$

According to the Chebyshev's theorem we are dealing with the interval within 8 S.D.'s from the mean.

$$\text{Required probability} = 1 - \frac{1}{64} = \frac{63}{64}.$$

Example 89. Suppose that samples of polythene bags from two manufacturers *A* and *B* are tested by a prospective buyer for bursting pressure with the following results :

Bursting pressure (lbs)		5.0–9.9	10.0–14.9	15.0–19.9	20.0–24.9	25.0–29.9	30.0–34.9	Total
Number of bags	*A*	2	9	29	54	11	5	110
	B	9	11	18	32	27	13	110

Which set of bags has the highest average bursting pressure ? Which has more uniform pressure ? If prices are the same, which manufacturer's bags would be preferred by the buyer ? Why ?

[**Hint.** Calculations of Mean and Standard Deviation]

Solution.

Manufacturer A

Bursting pressure (lbs)	*Mid-value X*	$(X-17.45)/5$			fd^2
		f	d	fd	
4.95–9.95	7.45	2	–2	–4	8
9.95–14.95	12.45	9	–1	–9	9
14.95–19.95	17.45	29	0	0	0
19.95–24.95	22.45	54	+1	54	54
24.95–29.95	27.45	11	+2	22	44
29.95–34.95	32.45	5	+3	15	45
		$N = 110$		$\Sigma fd = 78$	$\Sigma fd^2 = 160$

Mean 21; S.D. = 4.875 C.V. = 23.32%

Manufacturer B

Bursting pressure (lbs)	*Mid-value X*	$(X-17.45)/5$			fd^2
		f	d	fd	
4.95–9.95	7.45	9	–2	–18	36
9.95–14.95	12.45	11	–1	–11	11
14.95–19.95	17.45	18	0	0	0
19.95–24.95	22.45	32	1	32	32
24.95–29.95	27.45	27	2	54	108
29.95–34.95	32.45	13	3	39	117
		$N = 110$		$\Sigma fd = 96$	$\Sigma fd^2 = 304$

Mean = 21.81, S.D. = 7.0745, C.V. = 32.44%

If prices are the same, the bags of manufacturer *A* should be preferred by the buyer because they have more uniform pressure.

Example 90. If 10 is subtracted from every item in a data set then the coefficient of variation of the resulting set of data is 20%. If 20 is added to every item of the same data set then the coefficient of variation of the resulting set of data is 10%.

You are required to find out the coefficient of variation of the original set of data.

Solution. Let the average and standard deviation of the original data set be $\bar{x}$ and s

$$\text{Average of all items '}x-10\text{'} = \frac{\Sigma(x-10)}{n} = \frac{\Sigma x}{n} = \frac{n10}{n} = \bar{x} - 10$$

Standard deviation of all items '$x - 10$' = s.

(This is because the value of standard deviation remains the same if each observation in a series is increased or decreased by the same quantity.)

Given : $\frac{s}{\bar{x}-10} \times 100 = 20$

or $100s = 20\bar{x} - 200$

or $20\bar{x} - 100s = 200$ ...(A)

Average of all items '$x + 20$' $= \frac{\Sigma(x+20)}{n} = \frac{\Sigma x}{n} - \frac{n20}{n} = \bar{x} + 20$

Standard deviation of all items '$x + 20$' $= s$

(This is because the value of standard deviation remains the same if each observation in a series is increased or decreased by the same quantity).

Given : $\frac{s}{\bar{x}+20} \times 200 = 10$

or $100s = 10\bar{x} + 200$

or $10\bar{x} - 100s = -200$ ...(B)

Subtracting equation (B) from equation (A), we get :

$$(20\bar{x} - 100s) - (10\bar{x} - 100s) = 200 - (-200)$$

or $10\bar{x} = 400$

or $\bar{x} = \frac{400}{10} = 40$

Putting the value of $\bar{x}$ in equation (B), we get :

$$10(40) - 100s = -200$$

or $400 - 100s = -200$

or $400 + 200 = 100s$

or $100s = 600$

or $s = \frac{600}{100} = 6$

∴ Coefficient of variation of the original data set :

$$= \frac{s}{\bar{x}} \times 100 = \frac{6}{40} \times 100 = 15\%.$$

Example 91. Draw the Lorenz Curve for the data relating to the profit of 50 business firms and show the extent of inequality present in the profits.

Profit ('000 *Rs.*)	10–20	20–30	30–40	40–50	50–60
No. of Firms	5	13	18	10	4

Solution.

Computations for Lorenz Curve

Profit ('000 *Rs.*)	*Mid-Value (m)*	*Frequecy (No. of firms) (f)*	*Cum. Freq.*	*% Cum. Freq. (X)*	*Total value* $(m \times f)$	*Cum. Total Values*	*% Cum. Total Values (Y)*
10–20	15	5	5	10	75	75	4.4
20–30	25	13	18	36	325	400	23.5
30–40	35	18	36	72	630	1,030	60.6
40–50	45	10	46	92	450	1,480	87.1
50–60	55	4	50	100	220	1,700	100.0

Now follow the steps explained earlier one by one and draw the Lorenz Curve taking the percentage cumulative frequency (*X*) on (*X*)-axis and the percentage cumulative total values *Y* on *Y*-axis. Look at figure carefully and study how it is drawn.

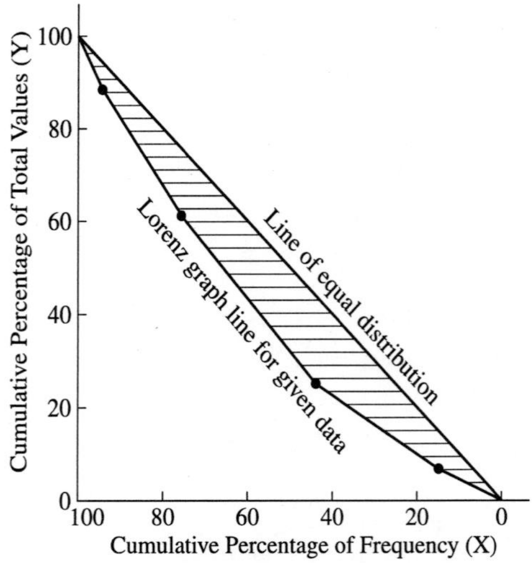

Fig. 4.2 Lorenz Curve showing the inequalities in the profit distribution for 50 business firms.

Example. 92. Draw the Lorenz Graph for the two sets of five workers each whose weekly income figures are given below. Point out which set has greater inequalities in income.

Income of Group A ('00 Rs.)	96	104	103	99	98
Income of Group B ('100 Rs.)	100	270	580	620	430

Solution. In this case frequencies are not given. In fact, each income figure represents the income of one worker. So, the frequencies for various income figures are one each in both Groups *A* and *B*. Arranging the data in ascending order, the computations for two sets are as follows :

Group A			*Group B*			*For Both Sets*		
Income ('00 Rs.)	*Cum. Income*	*% Cum. (Income) (YA)*	*Income ('00 Rs.)*	*Cum. Income*	*% Cum. Income (YB)*	*Freq.*	*Cum. Freq.*	*% Cum. Freq. (X)*
96	96	19.2	100	100	5.0	1	1	20
98	194	38.8	270	370	18.5	1	2	40
99	293	58.6	430	800	40.0	1	3	60
103	396	79.2	580	1,380	69.0	1	4	80
104	500	100.0	620	2,000	100.00	1	5	100

Figure shows that, the line for Group *B* is at a greater distance from line of equal distribution. Compared with the line for workers in Group *A*, it implies that the incomes of the workers in Group *B* have greater inequalities.

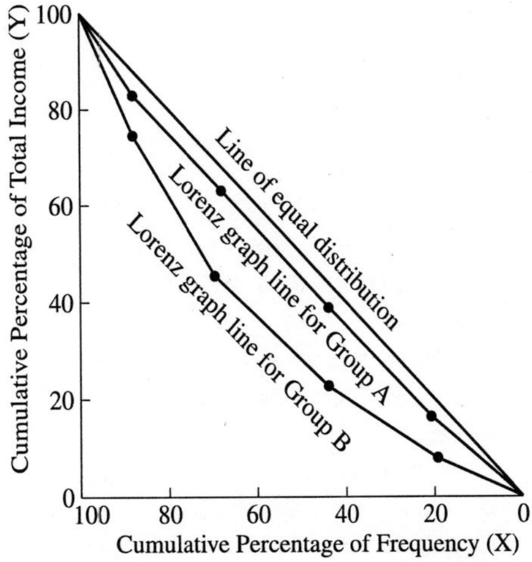

Fig. 4.3. Lorenz Curve showing the income inequalities between two groups of workers.

Questions

Very Short Answer Type Questions

1. What is the relation between Quartile deviation and Standard Deviation ?
2. What percentage of items are covered between Arithmetic mean ± mean deviation ?
3. If the standard deviation of 10 values $x_1, x_2, x_3, \ldots, x_{10}$ is σ, then find the standard deviation of the values $\frac{x_1 - 2}{3}, \frac{x_2 - 2}{3}, \frac{x_3 - 2}{3}, \ldots, \frac{x_{10} - 2}{3}$.

   [**Hint :** S.D. is independent of the change of origin and depends on the change of scale.]
4. Write the formula for finding the combined S.D. of two groups without using the combined mean.
5. Explain whether mean or median should be used for calculating mean deviation.

Short Answer Type Questions

1. What is the importance of dispersion ?
2. Explain Lorenz Curve.
3. Explain the units for measuring standard deviation.
4. Write any two characteristics of standard deviation.
5. Why standard deviation is preferred to other measures of dispersion ?
6. Write two limitations of Quartile deviation.
7. Write two characteristics of a good measure of dispersion.

Long Answer Type Questions

1. Why the word 'standard' is used in standard deviation ?
2. Explain mean deviation and coefficient of mean deviation.
3. What are the merits and demerits of quartile deviation ?

4. Why measures of dispersion are called averages of second order ?
5. Why standard deviation is preferred to mean deviation ?

State whether the following statements are True or False

1. Standard deviation is free from all those defects which the other measures suffer.
2. The variance and the coefficient of variation are the same.
3. Root mean square deviation about arithmetic mean is the least.
4. Reducing each and every item by 5 will reduce standard deviation also by 5.
5. In an ordinary data, standard deviation is less than Quartile deviation.
6. Ranges in the best measure of dispersion.
7. In a symmetrical distribution, semi-interquartile range is one-fourth of the range.
8. Quartile deviation is suitable for open-ended distribution.
9. Coefficient of quartile deviation is an absolute measure.
10. Relative measure of dispersion is a unit free pure number.
11. Standard deviation and mean deviation are relative measures.
12. Absolute measures of variation are used for comparing variability in distributions.
13. Standard deviation can be calculated from any average.
14. Mean deviation is least when deviations are taken from median.
15. Average deviation is the best measure of dispersion.
16. Mean deviation is minimum when deviations are taken from mean.
17. The sum of squared deviation is least when taken from the mean.
18. Measures of dispersion are the averages of second order.
19. The variance is equal to square of S.D.
20. Standard deviation and coefficient of variation are always expressed in the same unit.
21. Range is the difference between two extreme values.
22. M.D. = (4/5) S.D. in a symmetrical distribution.
23. Mean deviation can be negative.
24. Mean and variance of a distribution are 10 and – 15 respectively.

ANSWERS

1. T **2.** F **3.** T **4.** F [S.D. will remain unchanged] **5.** F **6.** F **7.** T
8. T **9.** T **10.** T **11.** F **12.** F **13.** F **14.** T **15.** F **16.** F **17.** T
18. T **19.** T **20.** F **21.** T **22.** T **23.** F **24.** F

2. Fill in the Blanks

1. Mean deviation is .... than standard deviaton.
2. $\overline{X} \pm \sigma$ includes .... per cent of the items.
3. If in a series coefficient of variation is 64 and mean is 10, the standard deviation shall be ....
4. Variance is always .... negative.
5. If each of the 10 values of a set are equal to 5, the standard deviation will be equal to ....

ANSWERS

1. less **2.** 68.26 **3.** 6.4 **4.** None **5.** Zero.

Multiple Choice Questions

1. The measures based on every item of the series :
(a) range (b) standard deviation (c) quartile deviation (d) all of them

2. One of the measures of dispersion which is more useful in case of open-ended distributions :
(a) range (b) mean deviation (c) standard deviation (d) quartile deviation

3. Standard deviation is always computed from :
(a) mean (b) mode (c) median (d) geometric mean

4. Which of the following measures is least affected by extreme items :
(a) quartile deviation (b) range (c) standard deviation (d) mean deviation

5. Mean deviation is :
(a) less than S.D. (b) more than S.D.
(c) not related to S.D. (d) equal to standard deviation

6. Coefficient of variation is given by
(a) $\frac{\sigma}{\overline{X}}$ (b) $\frac{\overline{X}}{\sigma}$ (c) $\frac{\overline{X}}{\sigma} \times 100$ (d) $\frac{\sigma}{\overline{X}} \times 100$

7. Quartile deviation is based on the :
(a) the highest 50% of the items (b) the lowest 25% of the items
(c) the highest 25% of the items (d) middle 50% of the items

8. If 25% of the items are less than 20 and 25% are more than 40, the quartile deviation is :
(a) 20 (b) 30 (c) 40 (d) 10

9. The sum of squares of deviations for 10 observations taken from mean 50 is 25. The coefficient of variation is :
(a) 50% (b) 10% (c) 40% (d) None of these

10. The mean of squared deviation about the mean is called :
(a) S.D. (b) Variance (c) M.D. (d) None of these

11. The mean of absolute deviations from an average is called :
(a) S.D. (b) M.D. (c) Variance (d) Q.D.

12. The range includes :
(a) 100% of the items (b) 50% of the items (c) 25% of the items (d) None of these

13. The interquartile range excludes :
(a) 50% of the values (b) 100% of the values (c) 25% of the values (d) None of these

14. A measure of dispersion is an indicator of the reliability of :
(a) An average (b) Variability (c) Median class (d) None of these

15. Mean $\pm 3\sigma$ covers :
(a) 90% items (b) 99.73% items (c) 95% items (d) None of these

16. For a symmetrical distribution $Q_1 = 25$ and $Q_3 = 45$, the median is :
(a) 20 (b) 25 (c) 35 (d) None of these

17. The percentage of observation lying between D_1 and D_5 is :
(a) 10% (b) 50% (c) 40% (d) 60%

18. The median is :
(a) 4th decile (b) 50th percentile (c) 8th decile (d) None of these

19. The dispersion measure least affected by extreme values is :
(a) S.D. (b) Range (c) M.D. (d) Q.D.

20. Quartile deviation is :
(a) (2/3) σ (b) (4/5) σ (c) (3/2) σ (d) None of these

21. $X = 50, \sigma = 25$; the C.V. is :
(a) 200% (b) 25% (c) 50% (d) None of these

22. If Q.D. is 16, the most likely values of S.D. will be :
(a) 24 (b) 42 (c) 16 (d) None of these

23. If M.D. is 12, the values of S.D. will be :
(a) 15 (b) 12 (c) 24 (d) None of these

24. Which of the following measures is most affected by extreme values :
(a) S.D. (b) Q.D. (c) M.D. (d) Range

25. The percentage of observations lying between P_{10} and P_{90} is :
(a) 100 (b) 80 (c) 10 (d) 90

26. Which of the following is a unit free number :
(a) S.D. (b) Variance (c) M.D. (d) C.V.

27. If every item in a set of data is increased by 10, then which of the following measures will remain unchanged :
(a) Mean (b) Median (c) Mode (d) Variance
(e) None of the above

28. The standard deviation of 30 items in a data set is 5/3. If every item in the data set is multiplied by 3 then, the variance of the group will be :
(a) 5 (b) 8 (c) 15 (d) 25
(e) 41

29. If every observation in a data set is increased by a constant quantity then the coefficient of variation of the resulting set of values will be
(a) Less than the coefficient of variation of the original data set
(b) Greater than the coefficient of variation of the original data set
(c) Equal to the coefficient of variation of the original data set
(d) Equal to the coefficient of variation of the original data set plus the square root of the constant quantity
(e) Equal to the coefficient of variation of the original data set multiplied by the square root of the constant quantity

30. To which of the following distributions is Bienayme and Chebyshev's theorem applicable :
(a) Normal distribution (b) Rectangular distribution
(c) Symmetrical distribution (d) Non-Symmetrical distribution
(e) Any type of distribution

31. Which of the following measures represents the scatter of the values in a data set :
(a) Arithmetic mean (b) Geometric mean (c) Harmonic mean (d) Median
(e) Standard deviation

32. The measure of dispersion which is independent of frequencies of the given distribution is :
(a) Range (b) Standard deviation (c) Mean deviation (d) Quartile deviation
(e) None of these

33. The following data pertain to a sample containing 5 observations :

$$\Sigma x = 15$$
$$\Sigma x^2 = 55$$

The variance of the sample is :
(a) 1.50 (b) 2.00 (c) 2.50 (d) 3.00
(e) 3.50

34. According to Bienayme-Chebyshev theorem, at least what percentage of the observations in a distribution of data wil lie within ± 2 standard deviations of the mean ?
(a) 25% (b) 30% (c) 45% (d) 60%
(e) 75%

35. The mean and standard deviation of a data set are 10 and 3 respectively. If every observation in the data set is increased by 2 then the coefficient of variation of the resulting data set will be :

(a) 20% (b) 25% (c) 30% (d) 35%
(e) 40%

36. Which of the following is a relative measure of dispersion (or variability) :
(a) Variance (b) Standard deviation (c) Range
(d) Coefficient of variation (e) None of these

37. If every item in a data set is decreased by the same quantity then the standard deviation of the resulting data set
(a) Remains the same
(b) Increases by the same quantity by which every data item is increased
(c) Decreases by the same quantity by which every data item is increased
(d) Increases by the square root of the same quantity by which every data item is increased
(e) Decreases by the square root of the same quantity by which every data item is increased

38. The following details are available with regard to a sample of eight observations :

$$\sum_{i=1}^{4} x_i = 64 \quad \sum_{i=2}^{5} x_i = 80 \quad \sum_{i=5}^{8} x_i = 66$$

$$\sum_{i=2}^{4} x_i = 54 \quad \sum_{i=1}^{8} x_i^2 = 2412 \quad x_5 = 26$$

The sample standard deviation is
(a) 6.12 (b) 10.25 (c) 17.36 (d) 42.77
(d) 130

39. The standard deviation of a data set
(a) is expressed in the same unit as the observation in the data set
(b) is expressed in the square of the unit of the observations in the data set
(c) is expressed in the square root of the unit of the observations in the data set
(d) is expressed in a different unit from the unit in which the observations in the data set are expressed
(e) is always expressed as a percentage of the mean of the data set

40. Which of the following measures does not use every observations in the data set :
(a) Variance (b) Coefficient of variation (c) Mode (d) Geometric mean
(e) Arithmetic mean

41. The coefficient of variation cannot be meaningfully used to compare the variability of two or more sets of data, when
(a) the standard deviation is zero for one or more sets of data
(b) the standard deviation is 1 for one or more sets of data
(c) the mean is zero for one or more sets of data
(d) the mean is 1 for one or more sets of data
(e) the mean and standard deviation are equal for one or more sets of data

42. The mean of the squared deviations of all the observations in a data set is equal to the
(a) standard deviation of the data set (b) median of the data set
(c) mode of the data set (d) range of the data set (e) variance of the data set

43. If every item in a data set is multiplied by the same quantity, k, then the standard deviation of the resulting data set
(a) is equal to the variance of the original data set
(b) is equal to the average of the original data set
(c) is equal to the standard deviation of the original data set multiplied by k
(d) is equal to the range of the original data set
(e) is always less than the standard deviation of the original data set

44. If a constant quantity is subtracted from every observations in a data set then the range of the resulting set of values will be equal to

(a) range of the original data set plus the constant quantity
(b) the range of the original data set minus the constant quantity
(c) range of the original data set
(d) range of the original data set multiplied by the constant quantity
(e) range of the original data set divided by the constant quantity

45. If every observations in a data set is increased by a constant quantity then the coefficient of variation of the resulting set of values will be
(a) less than the coefficient of variation of the original data set
(b) greater than the coefficient of variation of the original data set
(c) equal to the coefficient of variation of the original data set
(d) equal to the coefficient of variation of the original data set plus the square root of the constant quantity
(e) equal to the coefficient of variation of the original data set multiplied by the square root of the constant quantity

46. A data set contains the following observations :

10, 12, 8, 14, 15, 7, 17, 20

What is the range of the data set ?
(a) 8 (b) 9 (c) 10 (d) 12
(e) 13

47. The variance of a data set is 9. If every item in the data set is multiplied by 2 then, what is the variance of the resulting data set ?
(a) 9 (b) 12 (c) 36 (d) 72
(e) 324

48. The sum of the observations in a data set containing 10 observations is 120 and the standard deviation of the data set is 3. If 3 is added to every observation in the data set then, what will be the coefficient of variation of the resulting data set ?
(a) 0 (b) 7.5% (c) 20% (d) 25%
(e) 50%

49. The following details are available with regard to a sample of ten observations :
Arithmetic mean of the first four observations = 16
Arithmetic mean of the next six observations = 11

$$\sum_{i=1}^{10} x_i^2 = 1726$$

What is the standard deviation of the sample ?
(a) 1.9 (b) 3 (c) 4 (d) 5
(e) 6

50. Which of the following measures is based only on two observations in a data set :
(a) Arithmetic mean (b) Harmonic mean (c) Range
(d) Mean absolute deviation (e) Standard deviation

51. Which of the following measures is not affected by the presence of extreme values in a data set :
(a) Range (b) Arithmetic mean (c) Standard deviation
(d) Variance (e) Median

52. The coefficient of variation of a data set is 20% and the variance of the data set is 16. What is the mean of the data set ?
(a) 4 (b) 20 (c) 25 (d) 80
(e) 125

53. Which of the following measures will remain unchanged when every observation in the data set is divided by the same quantity :

(a) Range (b) Quartile deviation (c) Standard deviation
(d) Coefficient of variance (e) Mode

54. If every observation in a data set is increased by a constant *k*, then the range of the resulting data set is equal to :
(a) range of the original data set (b) range of the original data set minus *k*
(c) range of the original data set plus *k* (d) *k*
(e) –*k*

55. Which of the following is **true**, if there is no dispersion in a data set :
(a) All the mathematical and positional averages are equal
(b) All the mathematical averages are equal but the positional averages are not equal
(c) All the positional averages are equal but the mathematical averages are not equal
(d) All the mathematical averages are equal to zero
(e) All the positional averages are equal to zero

56. Which of the following measures of dispersion is also called 'root mean square deviation' :
(a) Range (b) Average deviation (c) Mean absolute deviation
(d) Standard deviation (e) None of the above

57. Which of the following measures cannot be combined mathematically :
(a) Standard deviation (b) Arithmetic mean (c) Geometric mean (d) Harmonic mean
(e) Median

58. Which of the following enables us to compare the variability of two or more sets of data :
(a) Variance (b) Standard deviation (c) Coefficient of variation
(d) Mean absolute deviation (e) None of the above

59. If three quantities, *a*, *b* and *c* are in geometric progression then
(a) $a = b + c$ (b) $a = b \cdot c$ (c) $a \cdot b = c$ (d) $b^2 = a \cdot c$
(e) $2b = a + c$

60. The reciprocals of the terms in any harmonic progression are :
(a) odd numbers in multiples of 3 (b) even numbers in multiples of 4
(c) prime numbers (d) in geometric progression
(e) in arithmetic progression

ANSWERS

1. (b),	**2.** (d),	**3.** (a),	**4.** (a),	**5.** (a),	**6.** (d),
7. (d),	**8.** (d),	**9.** (d),	**10.** (b),	**11.** (b),	**12.** (a),
13. (a),	**14.** (a),	**15.** (b),	**16.** (c),	**17.** (c),	**18.** (b),
19. (d),	**20.** (a),	**21.** (c),	**22.** (a),	**23.** (a),	**24.** (d),
25. (b),	**26.** (d),	**27.** (d),	**28.** (d)		

29. (a) $\text{C.V.} = \frac{\text{S.D.}}{\text{Mean}}$. If every observations is increased by a constant, so it will remain unchanged but mean will increase by the same constant. Therefore C.V. will decrease.

30. (e) Bienayme-Chebyshev theorem is applicable to any type of distribution regardless of its shape.

31. (e) The standard deviation represents the scatter of the values in a data set. Arithmetic mean, geometric mean, harmonic mean and median are measures of central tendency.

32. (a) *i.e.*, Range, **33.** (b), **34.** (e), **35.** (b), *i.e.*, 25%,
36. (d), *i.e.*, Coefficient of variation, **37.** (a),

38.
$$\sum_{i=1}^{8} x_i = \sum_{i=1}^{4} x_i + \sum_{i=5}^{8} x_i = 64 + 66 = 130$$
$$\sum_{i=1}^{8} x_i^2 = 2412$$
$$\text{S.D.} = \sqrt{\frac{\Sigma x^2}{n} - \left(\frac{\Sigma x}{n}\right)^2} = \sqrt{\frac{2412}{8} - \left(\frac{130}{8}\right)^2}$$
$$= \sqrt{301.5 - 264} = \sqrt{37.5} = 6.12.$$

Correct answer is (a).

39. (a), **40.** (c), **41.** (c), **42.** (e), **43.** (c), **44.** (c),
45. (a), **46.** (e), **47.** (c),

48. (c) Mean $= \frac{120}{10} = 12$, if 3 is added to each observation, mean = 15.

New coefficient of variation $= \frac{3}{15} \times 100 = 20\%$

49. (a) Sum of the 10 observations $= 16 \times 4 + 11 \times 6 = 130$.

$$\therefore \qquad \text{S.D.} = \sqrt{\frac{1726}{10} - \left(\frac{130}{10}\right)^2}$$
$$= \sqrt{172.6 - 169} = 1.9.$$

50. (c), **51.** (e), **52.** (b),

53. (d) Coefficient of variation $= \frac{\text{S.D.}}{\text{Mean}} \times 100$, if each observation is divided by a constant, S.D. and mean are also divided by the same constant and hence the ratio remains unchanged.

54. (a)

55. (a) If there is no dispersion in the data set, then all the values are equal. If all the values are equal then

Mean = Median = Mode

Also A.M. = G.M. = H.M.

56. (d), **57.** (e), **58.** (e), **59.** (d), **60.** (e).

Practice Questions

1. Find the standard deviation for a set of items 3, 4, 6, 7, 15, 25. [**Ans.** 7.7 Approx.]
2. Find out the standard deviation and variance from the following frequency distribution :

Marks	0–4	4–8	8–12	12–16
No. of Students	4	8	2	1

[**Ans.** S.D. = –3.27, Variance = 10.67]

3. Calculate standard deviation for the following series by direct method and short-cut method using assumed mean as 32.

Serial No.	1	2	3	4	5	6	7	8	9	10
Size	20	22	27	30	31	32	35	40	45	48

[**Ans.** S.D. = 8.73]

4. Calculate the standard deviation and variance from the following frequency distribution by direct and short-cut methods using 14 as assumed mean.

X	10	12	14	16	18	20	22
f	3	5	9	16	8	7	2

[**Ans.** S.D. = 2.97, Variance = 8.8]

5. The profits (in Rs. lakhs) earned by 100 companies during 2000-2001 are shown below. Compute (a) Mean, (b) Variance, and (c) Standard Deviation by using items and their squares.

Profits (Rs. lakhs)	*No. of Companies*
20–30	4
30–40	8
40–50	18
50–60	30
60–70	15
70–80	10
80–90	8
90–100	7

[**Ans.** Mean = 59.10, S.D. = 17.59, Variance = 308.19]

6. Calculate Mean Deviation and Standard Deviation from the following distribution :

Class-Interval	10–20	20–30	30–40	40–50	50–60	60–70	70–80
Frequency	4	8	8	16	12	6	4

[**Ans.** Mean = 45, S.D. = 15.97, Mean Deviation = 12.41]

7. Find the standard deviation of the following distribution :

Class-Interval	0–500	500–1,000	1,000–1,500	1,500–2,000	2,000–3,000
Frequency	90	218	86	41	15

[**Ans.** S.D. = 519.6]

8. The following is the record of goals scored by Team A in a football season :

No. of goals scored in a match	0	1	2	3	4
Number of matches	1	9	7	5	3

For Team B, the average number of goals scored per match was 2.5 with a standard deviation of 1.25 goals. Find which team is more consistent. [**Ans.** Team B is more consistent than Team A]

9. From the data given below, state which series is more variable :

Variable	*Series A*	*Series B*
10–20	10	18
20–30	18	22
30–40	32	40
40–50	40	32
50–60	22	18
60–70	18	10

[**Ans.** S.D. of Series A = 14.06, S.D. of Series B = 14.06, Series B is more Stable.]

Note. In question, the coefficient of variation of Series B is higher than that of Series A. Therefore, Series B is more variable. In this question, you may notice that standard deviation of both the series is same, *i.e.,* 14.06. From this fact we should not conclude that the two series have the same variation. The difference in arithmetic mean has to be taken into account for correct interpretation.

10. The mean and standard deviation of a set of 100 observations were worked out as 40 and 5, respectively. For one of the observations, by mistake, the value is taken as 50 instead of 40. Find the correct mean and variance.

[**Ans.** Correct mean = 39.9, Correct variance 23.99]

11. For a group of 50 male workers, the mean and standard deviation of their daily wages are Rs. 72 and Rs. 9 respectively. For another group of 40 female workers these are Rs. 54 and Rs. 6 respectively. Find the standard deviation for the combined group of 90 workers.

12. Show inequality in wages in two different firms using Lorenz Curve approach, given the following data :

Wages (Rs.)		50–70	70–90	90–110	110–130	130–150
No. of Workers	*A*	20	15	20	25	20
	B	150	100	90	110	50

13. Draw a Lorenz Curve of the data given below :

Income	100	200	400	500	800
No. of Persons	80	70	50	30	20

14. In a small town, a survey was conducted in respect of profits made by retail shops. The following results were obtained :

Profit (+)/Loss (–) (in '000 Rs.)	*No. of Shops*
–4 to –3	4
–3 to –2	10
–2 to –1	22
–1 to 0	28
0 to 1	38
1 to 2	56
2 to 3	40
3 to 4	24
4 to 5	18
5 to 6	10

Calculate (i) the average profit made by a reteail shop, (ii) total profit made by all shops, and (iii) the coefficient of variation of earnings. [**Ans.** (i) 1348, (ii) 3,37,000, (iii) 152.8%]

15. A factory produces two types of electric lamps *A* and *B*. In an experiment relating to their life, the following results were obtained :

Length of Life (in hours)	*No. of Lamps A*	*No. of Lamps B*
500–700	5	4
700–900	11	30
900–1,100	26	12
1,100–1,300	10	9
1,300–1,500	8	6

Compare the variability of the life of the two varieties using coefficient of variation.

[**Ans.** C.V. $(A) = 21.64\%$, C.V. $(B) = 23.41\%$]

16. In two factories *A* and *B*, engaged in the same activity, the average weekly wages and standard deviation are as follows :

Factory	*Average Weekly Wages (Rs.)*	*S.D. of Wages (Rs.)*	*No. of Wage Earners*
A	460	50	100
B	490	40	80

(i) Which factory pays larger amount as weekly wages ?
(ii) Which factory shows greater variability in the distribution of wages ?
(iii) What is the mean and standard deviation of all the workers in these two factories taken together ?

[**Ans.** (i) Factory *A*, (ii) C.V. (*A*) = 10.87%, C.V. (*B*) = 8.16%, (iii) $\overline{X}_{12}$ = Rs. 473.33, $\sigma_{12} = 49.19$]

17. Draw Lorenz Curves from the following data :

Wages (Rs.)	300–400	400–500	500–600	600–700	700–800
No. of Workers					
Factory A	40	30	40	60	40
Factory B	300	200	180	220	100

18. The arithmetic mean and standard deviation of 20 items were found as 20 and 5 respectively. But while calculating an item 13 was misread as 30. Find correct arithmetic mean and standard deviation.

[**Ans.** A.M. = 19.15; $\sigma = 4.66$]

19. The students of the B.Com. Class of a college have obtained the following marks in Statistics out of 100 marks. Calculate the standard deviation of marks obtained.

Student	A	B	C	D	E	F	G	H	I	J
Marks	5	10	20	25	40	42	45	48	70	80

[**Ans.** 23.06]

20. Calculate standard deviation from the following data :

Mid-points	*Frequency*
1	2
2	60
3	101
4	152
5	205
6	155
7	79
8	40
9	1

[**Ans.** 1.57]

21. Compute standard deviation for the following data which relate to the profits of 100 companies :

Profit (Rs. in lakhs)	*No. of Comapnies*
8–10	8
10–12	12
12–14	20
14–16	30
16–18	20
19–20	10

[**Ans.** $\sigma = 2.77$]

22. An analysis of production rejects resulted in the following figures. Calculate mean and standard deviation.

No. of Rejects per Operator	*No. of Operators*
21–25	5
26–30	15
31–35	28
36–40	42
41–45	15
46–50	12
51–55	3

[**Ans.** $\overline{X} = 36.96, \sigma = 6.735$]

23. Two samples of size 40 and 50 have the same mean 53 but different standard deviations 19 and 8 respectively. Find the standard deviation of the combined sample of size 90. [**Ans.** $\sigma_{12} = 14.3$]

24. Find the standard deviation and the coefficient of variation from the following data :

Marks	*No. of Students*
Less than 10	12
Less than 20	30
Less than 30	65
Less than 40	107
Less than 50	202
Less than 60	222
Less than 70	230

[**Ans.** $\sigma = 13.9$, C.V. $= 37.3\%$]

25. You are given the data pertaining to kilowatt hours of electricity consumed by 100 persons in a certain city :

Consumption K. Watt. Hours	*No. of Users*
0 but less than 10	6
10 but less than 20	25
20 but less than 30	36
30 but less than 40	20
40 but less than 50	13

Calculate (i) mean, (ii) standard deviation, and (iii) range within which middle 50% of the consumers fall.

[**Ans.** (i) 25.9, (ii) 10.96, (iii) 34 to 17.6]

26. The number of employees, average wages per employee and variance of the wage per employee for two factories are given below :

	Factory A	*Factory B*
Number of Employees	100	200
Average Wage per Employee (Rs.)	120	200
Variance of the Wages per Employee (Rs.)	16	25

In which factory is there greater variation in the distribution of wages per employee ?

[**Ans.** Factory *A*]

(CA Foundation, May 2000)

27. Calculate the mean deviation from the mean for the following data :

x	5	15	25	35	45	55	65
f	8	12	10	8	3	2	7

[**Ans.** 16]
(CA Foundation, May 2001)

28. Find the S.D. for the following data :

Class-Interval	0–10	10–20	20–30	30–40	40–50	50–60	60–70
Frequency	6	14	10	8	1	3	8

[**Ans.** 19.62]
(C.A. Foundation, Nov., 1998)

29. A variable assumes only two values 0 and 1. Out of 300 observations 1 appears 100 items and the rest are 0. Find the coefficient of variation. **(ICWA Foundation, Dec. 2001)**

[**Hint.** Mean $= \dfrac{100 \times 1 + 0 \times 200}{300} = \dfrac{1}{3} = 0.33$

$$\text{S.D.} = \sqrt{\frac{\Sigma fx^2}{\Sigma f} - \left(\frac{\Sigma fx}{\Sigma f}\right)^2} = \sqrt{\frac{100}{300} - \frac{1}{9}} = \sqrt{\frac{1}{3} - \frac{1}{9}} = \sqrt{\frac{2}{9}} = 0.47$$

$$\text{C.V.} = \frac{\text{S.D.}}{\text{Mean}} \times 100 = \frac{0.47}{0.33} \times 100 = 142.42.]$$

30. The mean and S.D. of 21 observations are 30 and 5 respectively. It was subsequently noted that one of the observations namely 10 was incorrect. Omit it and determine the mean and S.D. of the rest.

[**Ans.** Mean = 3, S.D. = 2.291]
(ICWA Foundation, June 1999)

31. The mean and variance of 10 observations are known to be 17 and 33 respectively. Later it was found that one observation (*i.e.,* 26) is inaccurate and is removed. What is the mean and S.D. of the remaining. [**Ans.** 34.64]
(C.A. Foundation, May 2002)

32. The mean and S.D. of 200 items was found to be 60 and 20 respectively. If at the time of calculations, two items were wrongly recorded as 3 and 67 instead of 13 and 17, find the correct mean, correct S.D. and correct coefficient of variation. [**Ans.** 59.8, 20.09, 33.6%]

33. The following table gives the result of three centres of a Public examination :

Centre	*No. of Examines*	*Mean Score*	*Standard Deviation*
A	200	25	3
B	250	10	4
C	300	15	5

Calculate the mean and S.D. of the combined scores of the three centres. [**Ans.** Mean = 16, S.D. = 7.19]
(ICWA, Dec., 1992)

❑❑❑

5 Correlation and Regression Analysis

Coverage Meaning and Significance of Correlation, Types of Correlation, Methods of Studying Correlation, Scatter Diagram, Karl Pearson's Coefficient of Correlation, Rank Correlation, Coefficient of Determination, Regression Analysis, Meaning and Significance, Difference between Correlation and Regression, Regression Lines, Regression Equations, Coefficients of Regression, Standard Error of the Estimate.

CORRELATION

5.1 INTRODUCTION

So far we have been dealing with the analysis of one variable only. For example, calculations of mean, median, mode, standard deviation, etc., involved only one variable. But the study of *correlation* deals with the degree (strength) of mutual statistical relationship between two or more variables,* *i.e.,* correlation studies, correspondence of movement (going togetherness) between two variables or series of paired items. For example,

(1) If the price increses demand decreases.
(2) If the price increases supply increases.
(3) If income of a family increases the expenditure will increase.
(4) Sale of woollen garments increases as winter approaches.
(5) When the number of students in a college is large, the sale in the canteen is also large, but during vacations when students do not come to college, the sale in the canteen declines.
(6) Cases of lung cancer may increase with increased smoking.

In the above six examples, the two variables move together either in the same direction or in opposite direction. But there are cases where the two variables move independently and there is no tendency of '**going togetherness**' between them. In correlation we do not deal with one series but rather with the *association or relationship between two series,* and we do not measure variation in one series but rather compare *variation in two or more series.* The two series may vary together *in the **same direction**;* or they may vary together in **opposite directions**; or do they not vary together at all.

To measure the association of series though correlation we must have sufficient number of items in the series. If we have only two or three pairs of values, *we cannot generalize* concerning the *way in which they vary together.*** *Moreover, there must not be a blank in one series where there is a value in the other series that is there must be pairing throughout.*

---

* The difference between statistical and functional relationship is that the statistical relation can be felt but can not be measured exactly whereas functional relation is a quantitative relationship and exact measurement can be made. When we deal with one variable the data is said to be univariable. When we deal with two valuables, the data is said to be bivariate.

** If we have only a few pairs of values, then the results of correlation analysis may not be valid.

5.2 DEFINITION OF CORRELATION

Correlation has been defined in different ways. Some of the main definitions are as under :

(1) Correlation measures the closeness of relationship between two variables, more exactly of the closeness of linear relationship.

(2) Acording to Croxton and Cowden, *"When the relationship is of a quantitative nature the appropriate statistical tool for discovering and measuring the relationship and expressing it in a brief formula is known as correlation.'*

(3) In the words of Boddington, *"Whenever some definite connection exists between the two or more groups, classes or series or data there is said to be correlation."*

(4) According to L.R. Connor, *"If two or more quantities vary in sympathy, so that movements in the one tend to be accompanied by corresponding movements in the other(s) then they are said to be correlated."*

5.3 IMPORTANCE (OR UTILITY) OF CORRELATION

(1) The correlation coefficient helps in measuring the extent of relationship between two variables in one figure only.

(2) Existence of relationship between two or more variables enables us to predict what will happen in future, *e.g.,* if the production of wheat has increased, other factors remaining constant, we may expect fall in the price of wheat.

(3) If two variables are closely related, we can estimate the value of one variable given the value of another variable. This is done with the help of Regression Equations.

(4) Correlation facilitates decision-making in business organisations. Expectations about the behaviour of certain variables are also based on correlation analysis.

5.4 KINDS OF CORRELATION

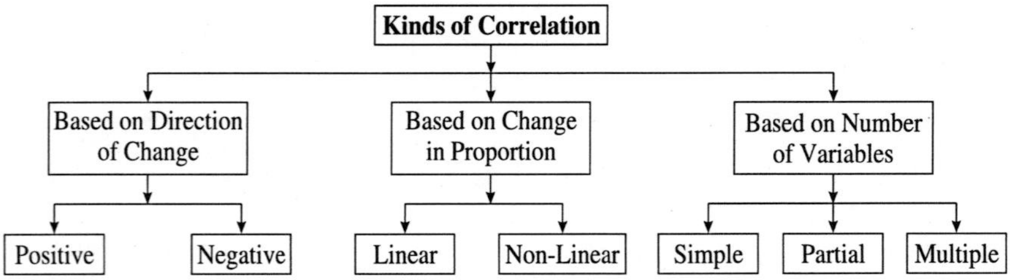

5.5 POSITIVE AND NEGATIVE CORRELATION

If two variables move together in the same direction, the correlation between them is said to be positive. If two variables move together in opposite directions, the correlation between them is said to be negative. If they do not move together at all, then there is no correlation between them.

Example (i) Since price and demand move in the opposite direction, the correlation between them is negative.

(ii) Price and supply move in the same direction, therefore, correlation between them is positive.

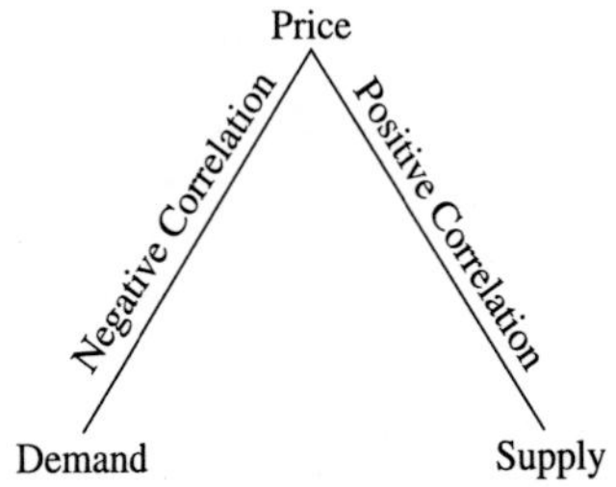

5.6 LINEAR AND NON-LINEAR CORRELATION

To measure the degree of an association between the movements within two or more series, *it is not necessary that there be proportionate change.* For example, the law of demand says 'other factors remaining constant, increase in price of a commodity is followed by a decrease in its demand' but we cannot find any **proportionality** relationship between price and demand.

The correlation is said to be *linear* if there is proportionate change in the value of the two variables, *i.e.,* if x and y are two variables, then in linear correlation $\frac{\Delta y}{\Delta x}$ is always constant, where Δx is a change in the value of x corresponding to a change Δy in the value of y. In linear correlation, the graph drawn with the pair of values of x and y is always a straight line*. If the change in the variables is not proportionate, the correlation is said to be **non-linear** or **curvilinear**. In his book, we will deal only with linear correlation.

Illustrations

(i)

x	1	2	3	4	5
y	2	4	6	8	10

is an example of linear correlation.

(ii)

x	1	2	3	4
y	3	5	8	15

is an example of non-linear correlation.

5.7 CORRELATION BASED ON NUMBER OF VARIABLES

(i) Simple Correlation : Only two variables are involved and the relationship is studied between those two variables.

(ii) Partial Correlation : More than two variables are inolved but the relationship is studied between two variables only, keeping other variables as constant. For example, production of wheat depends on many factors like rainfall, quality of seed, manure etc. and if we study the relation between production of wheat per hectare and quality of seed, keeping rainfall and manure constant, then the correlation is called partial.

(iii) Multiple Correlation : The relationship among two or more variables is studied simultaneously, *e.g.,* in the above example given in partial correlation if we study the relationship between production and all other factors simultaneously, the relationship will be called multiple correlation.

* Correlation between two variables X, Y is said to be linear if corresponding to a change in the value of the independent variable X, the dependent variable Y changes systematically in a straight line manner, *i.e.,* corresponding to a unit change in the value of the independent variable there is a constant change in the value of the dependent variable. This change may be positive (Y increases as X increases) or negative (Y decreases as X increases).

Note. There are so many other cases in which the two variables move together. This correspondence of movement (*i.e.,* going togetherness) of the series of paired items is called *concomitant variation* or *covariation.* The study of covariation leads to correlation analysis.

5.8 SOME IMPORTANT POINTS ABOUT THE STUDY OF CORRELATION ANALYSIS

(1) There should be sufficient number of items in the series. On the basis of two or three pairs of values we cannot generalize their **'going togetherness'.**

(2) In correlation analysis we do not deal with one series but the association or relationship between two or more series.

(3) We do not measure variation in one series only but rather compare variation in two or more series—do the series very together in the same direction, do they vary together in opposite direction or they do not vary together at all.

(4) To measure *Correlation, i.e.,* the degree of association between the movement of two or more series, it is not necessary that there be proportionate change, *i.e.,* it is not necessary that for every unit change in one variable there is a constant change in the other variable.* It is sufficient for the series of paired items to manifest concomitance of movement on the average.

(5) We study only *Linear Correlation.*

(6) Correlation does not necessarily mean cause and effect relationship. If we heat an iron rod, the rod will get more and more hot as we heat it more. Here heating the rod is cause and increase in the hotness of the rod is the effect. But if there is a fear of famine, the demand for essential commodities will increase and the price will also increase. In this case the price and demand move together in the same direction but none is cause for the other. The main cause for this correlation is the fear of famine.

(7) *The sign of r indicates the type of linear relationship, whether positive or negative, and*

(8) *The value of r without regard to sign, indicates the strength of the linear relationship.*

5.9 IS CORRELATION CAUSE AND EFFECT RELATIONSHIP ?

Though the word correlation is used in the sense of mutual dependence of two or more variables yet it is *not at all necessary* that it should always be so. **Even a very high degree of correction between two variables does not necessarily indicate a cause and effect relationship between them.** There can be correlation between two variables due to any one or more of the following reasons :

(1) *There is a cause and effect relationship between two variables.* For example, heat and temperature may be correlated because heat affects temperature. There is a cause and effect relationship between them. However, in social sciences such relationships are rarely found. The reason is that in social sciences there is multiple causation. A variable is affected simultaneously by a large number of factors whose individual effects cannot be isolated. For example, there may be a rise in price due to change in demand, inflation, export policy, and so many other factors. It is almost impossible to find out which factor is the cause (or even the main clause) of rise in prices. Thus in **social sciences correlation rarely indicates cause and effect relationship.**

(2) *Both the correlated variables are being affected by a third variable or by more than one variable.* For example, we may find a high degree of correlation between the prices of rice

* Proportionate change means that for every unit change in the value of one variable there is a constant change in the value of the other variable.

and prices of jute. In reality it may be found that the prices of both these commodities have between affeced by their production which in turn is affected by rainfall. There may be not any real relationship between the prices of rice and prices of jute. The correlation can give misleading conclusions in such cases.

(3) *Related variables might be mutually affecting each other so that neither of them could be designated as a cause or effect.* This situation particularly holds good in the field of economics and business. For example, the demand of a commodity may go down as a result of rise in prices. Here price is the cause an demand is the effect. However, it may be, that the demand of the commodity has gone up due to anticipated shortage in future and has resulted in the price rise. Now demand would be the cause and price would be the effect.

(4) *The correlation may be due to random or chance factors.* Many times correlation is noticed between two variables without any real relationship between them. It may happen due to chance. This generaly happens when a very small sample is chosen from a large universe. For example, a small sample may give us a very high correlation between wages and productivity, leading us to believe that higher wages lead to higher productivity. The real situation in the universe may be just the opposite of it higher wages leading to low productivity or there might be no correlation between them. It is, therefore, **essential that while analysing correlation between two variables, we should not jump to hasty conclusions.**

(5) *There might be a situation of nonsense or spurious correlation between the two variables under study.* One may find a higher degree of correlation between the number of divorces per year and the export of television sets. Obviously there cannot be any relationship between divorces and television exports and that is why it should be understood clearly that **correlation is a relationship between related variables only.**

The above points make it clear that *correlation is only a mathematical relationship and it does not necessarily signify a cause and effect relationship between the variables.*

5.10 MEASURES OF CORRELATION

In case of ungrouped data, the following methods of measuring correlation are normally used :

1. Scatter diagram method.
2. Karl Pearson's coefficient for measuring linear correlation.
3. Method of Rank differences (Spearman's Rank Correlation Coefficient).

(1) Scatter Diagram **(D.U., B.Com. (Hons.), 2006 E)**

Scatter diagram or dot diagram is a graphical representation of pair of numerical values of the two variables. Each pair of values is represented by a dot on the graph. The scatter of points and the direction of the scatter diagram reveals the nature and degree of correlation between two variables. If all the points lie on a straight line having positive slope (*i.e.,* rising line) the correlation is said **to be perfect positive.** In this case, the coefficient of correlation $r = +1$. If all the points lie on a straight line having negative slope (falling line), the correlation is said to be perfect negative. In this case the coefficient of correlation $r = -1$*.

In brief if all the points (*i.e.,* dots) form a line the correlation is said to be **perfect.** In general if low values of one variable go with low values of the other variable, and high values go with the

* The correlation coefficient varies between +1 and –1, *i.e.*, the coefficient of correlation can range from a Perfect Positive Correlation +1 to Perfect Negative Correlation –1.

high values, the path traced by these points runs roughly from the lower left corner to the upper right corner, the relationship is **direct** and is called **positive.** On the other hand, if high values of one variable go with the low values of the other variable, the path runs roughly from the upper left corner to the lower right corner, relationship is **inverse** and is called **negative.**

For example, low sales generally go with small employed labour force and high sales with large employed labour force. Here the correlation is positive. In certain types of jobs high productivity

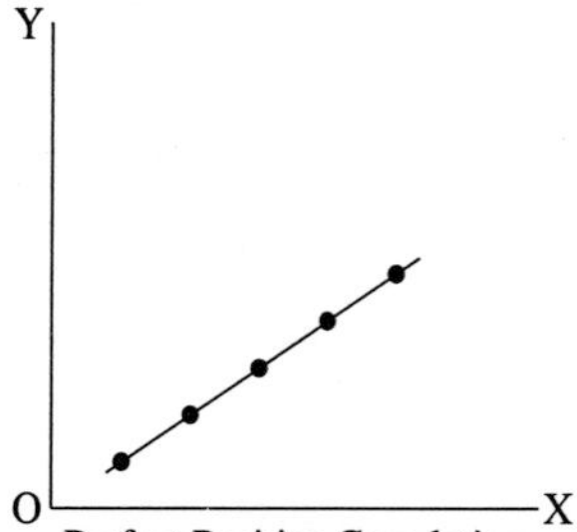

Perfect Positive Correlation
(r = +1)
(a)

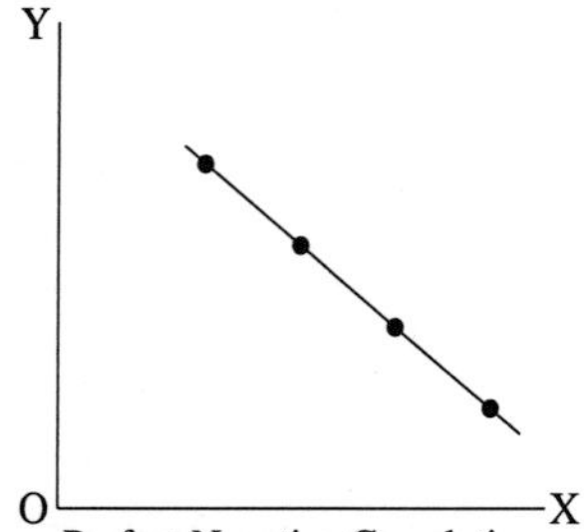

Perfect Negative Correlation
(r = –1)
(b)

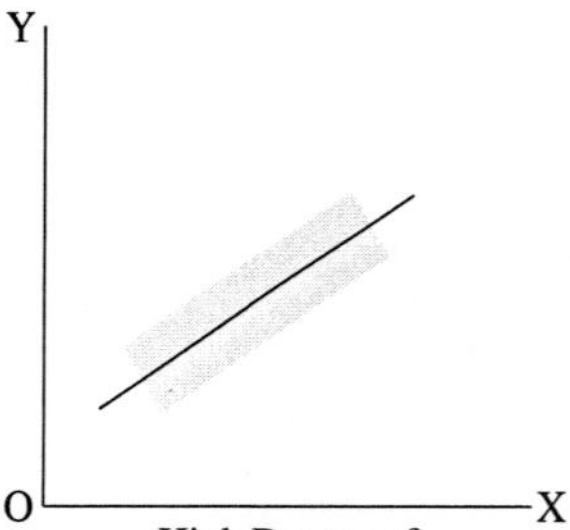

High Degree of
Positive Correlation
(c)

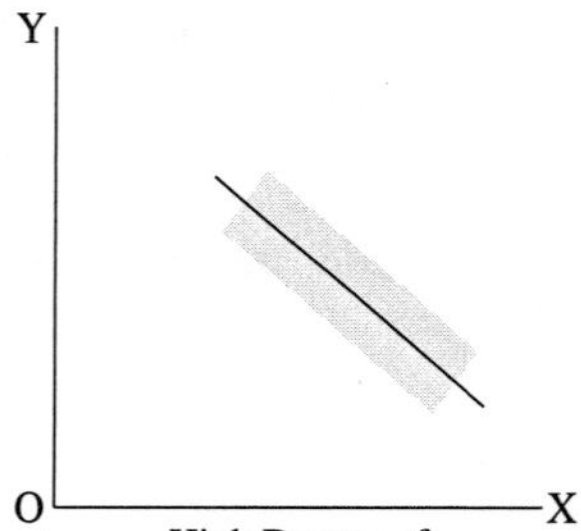

High Degree of
Negative Correlation
(d)

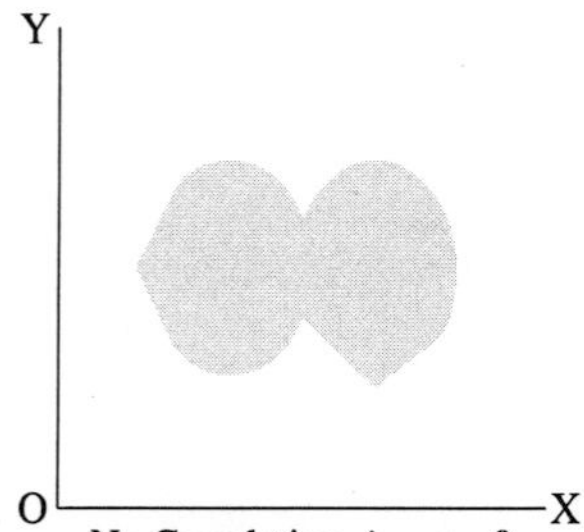

No Correlation, *i.e.*, r = 0
(e)

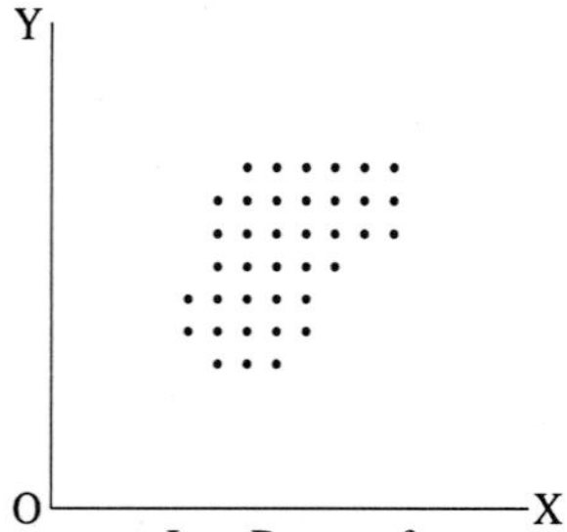

Low Degree of
Positive Correlation
(f)

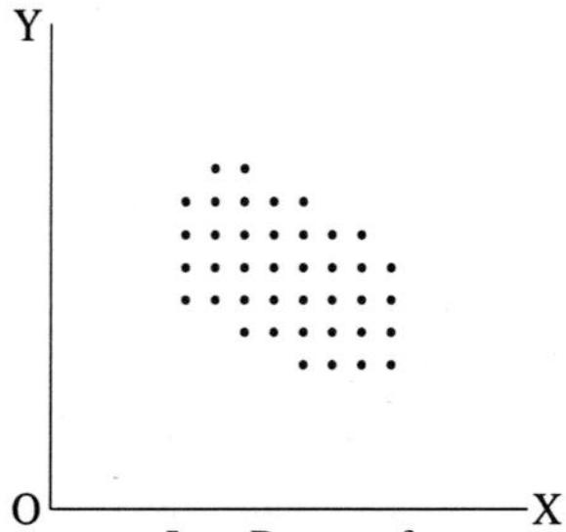

Low Degree of
Negative Correlation
(g)

efficiency is generally achieved with low age and low productivity efficiency with high age. This type of relationship is **inverse** and is termed **negative.** If the independent variable has no influence on the dependent variable, *i.e.,* if one variable does not help to find the value of the other variable (or if the dots do not follow any regular path), the condition is termed as **no correlation** and the correlation coefficient $r = 0$.

Limited Degree of Correlation : In social science, the variables may be correlated, but an increase in one variable need not always be accompanied by a corresponding or equal increase (or decrease) in the other variable. Correlation is said to be *limited positive* when there are unequal changes in the two variables in the *same* direction; and correlation is *limited negative* when there are unequal changes in the *reverse* direction. The limited degree of correlation can be high (between ±0.75 and 1); moderate (between ±0.25 and 0.75), or low (between ±0 and 0.25).

While drawing scatter diagram, there need not be a origin on the *Y*-axis. Both axes can start so as to accommodate the lowest values in the series.

The value of the coefficient of correlation *r* always **lies between −1 and +1**.

Merits and Limitations of the Scatter-Diagram Method

Merits :

1. It is a non-mathematical and easy way of finding the nature of correlation between two variables.
2. By drawing a line of best fit by free hand method through the plotted dots, the method can be used for estimating the missing value of the dependent variable for a given value of independent variable.
3. The shape of scatter-diagram reveals whether the correlation is linear or non-linear which enables us to know the pattern of relationship existing between two variables. Scatter diagram gives an idea whether correlation is positive or
4. The values of the extreme observations do not affect the method.

Demerits : It gives only rough idea of how the two variables are related. The method gives an idea about the direction of correlation and also whether it is high or low. But this method does not give any quantitative measure of the degree or extent of correlation.

Note : If $r = -1$ or $+1$ the correlation is said to be perfectly negative or perfectly positive. An intermediate value of *r* between −1 and +1 indicates the degree of linear relationship between *X* and *Y* whereas its sign tells about the diection of relationship. $r = 0$ means no linear relationship* between the two variables. The correlation coefficient is a pure number independent of the units of measurement. Its value is independent of the change of origin** and state, *i.e.,* if each value of *X* or *Y* (or both) is added to subtracted or (and) multiplied or divided (or multiplied and divided) by a constant either same or different for *X* and *Y*, the coefficient of correlation remains unchanged.

(2) Calculation of Correlation Coefficient (Karl Pearson's Method)***

For ungrouped data, Karl Pearson's coefficient of correlation can be obtained by using any of the following three methods :

(a) Actual mean method (b) Direct method (c) Short-cut method

(a) Actual Mean Method : In this method, we make use of the following formula for computing correlation *i.e.,*

$$r = \frac{\text{cov}(X, Y)}{(\text{S.D. of } X)(\text{S.D. of } Y)} = \frac{\Sigma (X - \bar{X})(Y - \bar{Y})}{n\sigma_X \sigma_Y}$$

* $r = 0$ do not imply no correlation between the two variables but it implies no linear correlation between the two variables.

** Substracting or adding a constant amount from each value is called change of origin. We may change the origin in one series or both the series. Multiplying of dividing each value by a constant is called change of scale.

*** This method measures only linear correlation between the two variables.

$$= \frac{\Sigma (X-\bar{X})(Y-\bar{Y})}{\sqrt{[\Sigma (X-\bar{X})^2 \cdot \Sigma (Y-\bar{Y})^2]}} = \frac{\Sigma xy}{\sqrt{\Sigma x^2}\sqrt{\Sigma y^2}},$$

where $x = X - \bar{X}$, $y = Y - \bar{Y}$,*

where symbols have their usual meanings. *This method is suitable in cases where the mean values $\bar{X}$ and $\bar{Y}$ are whole numbers and not fractional* **(Please note).**

Steps :

1. Calculate arithmetic means of X and Y series.
2. Find out the deviations of X series from mean of X and denote these deviations by x.
3. Square these deviations and obtain the total, *i.e.,* Σx^2.
4. Find out the deviations of Y series from the mean of Y and denote these deviations by y.
5. Square these deviations and obtain the total, *i.e.,* Σy^2.
6. Multiply these calculated deviations of X and Y series and find out the total, *i.e.,* Σxy.

Pl. note Σx and Σy will always be equal to zero.

(b) Direct method : In case mean values of the two series in a bivariate data are fractional values and number of observations and their volume in the two series is not very large, the following simplified form of formula may be used for calculating the value of r

$$r = \frac{\frac{\Sigma XY}{N} - \frac{\Sigma X}{N}\frac{\Sigma Y}{N}}{\sqrt{\frac{\Sigma X^2}{N} - \left(\frac{\Sigma X}{N}\right)^2}\sqrt{\frac{\Sigma Y^2}{N} - \left(\frac{\Sigma Y}{N}\right)^2}} = \frac{N\Sigma XY - (\Sigma X)(\Sigma Y)}{\sqrt{[N\Sigma X^2 - (\Sigma X)^2]}\sqrt{[N\Sigma Y^2 - (\Sigma Y)^2]}}$$

For this formula, we need

N = the number of paired observations.

ΣX = the sum of the observations on X variable.

ΣY = the sum of observations on Y variable.

ΣX^2 = the sum of squares of observations on X variable.

ΣY^2 = the sum of squares of observations on Y variable.

ΣXY = the sum of products of observations on X and Y variables.

Note. *If the correlation between X and Y is r, then the correlation between −X and Y is −r. Here we can also generalise that if one of the variables X or Y is multiplied by a negative constant and the other by a positive constant, the sign of the correlation coefficient changes.*

(c) Short-cut Method : When mean values are fractional and the number of paired observations is large, and the observations have large values, computation of r can be simplified by using deviations of the observations from some suitably chosen constant or constants. The constants for deviations of X and Y can be either same or different. The formula for computing correlation coefficient based on deviations is as under :

$$r = \frac{N\Sigma d_x d_y - (\Sigma d_x)(\Sigma d_y)}{\sqrt{[N\Sigma d_x^2 - (\Sigma d_x)^2]}\sqrt{[N\Sigma d_y^2 - (\Sigma d_y)^2]}}$$

$$= \frac{\frac{\Sigma d_x d_y}{N} - \frac{\Sigma d_x}{N}\frac{\Sigma d_y}{N}}{\sqrt{\frac{\Sigma d_x^2}{N} - \left(\frac{\Sigma d_x}{N}\right)^2}\sqrt{\frac{\Sigma d_y^2}{N} - \left(\frac{\Sigma d_y}{N}\right)^2}}$$

* Instead of using $x = X - \bar{X}$ and $y = Y - \bar{Y}$ sometimes we may use $dx = X - \bar{X}$ and $dy = Y - \bar{Y}$.

$$= \frac{\frac{\Sigma d_x d_y}{N} - \left(\frac{\Sigma (X-A)}{N}\right)\left(\frac{\Sigma (Y-B)}{N}\right)}{\sigma_X \sigma_Y} = \frac{\frac{\Sigma d_x d_y}{N} - \left(\frac{\Sigma X}{N} - \frac{NA}{N}\right)\left(\frac{\Sigma Y}{N} - \frac{NB}{N}\right)}{\sigma_X \sigma_Y}$$

$$= \frac{\frac{\Sigma d_x d_y}{N} - (\overline{X} - A)(\overline{Y} - B)}{\sigma_X \sigma_Y} = \frac{\Sigma d_x d_y - N(\overline{X} - A)(\overline{Y} - B)}{N \sigma_X \sigma_Y}$$

In this formula,

N = the number of paired observations.

$d_x = (X - A)$, the deviations of X taken from assumed mean A.

$d_y = (Y - B)$, the deviations of Y taken from assumed mean B.

Σd_x = the sum of deviations of X values from A.

Σd_y = the sum of deviations of Y values from B.

Σd_x^2 = the sum of squared deviations of X values from A.

Σd_y^2 = the sum of squared deviations of Y values from B.

$\Sigma d_x d_y$ = the sum of the products of deviations d_x and d_y.

$\overline{X}$ = mean of X

$\overline{Y}$ = mean of Y

Steps :

1. Calculate the deviations of X series from an assumed mean and denote them by d_x and find out the total, *i.e.,* Σd_x.
2. Calculate the deviations of Y series from an assumed mean and denote them by d_y and find out the total, *i.e.,* Σd_y.
3. Square the deviations of X series and obtain the total, *i.e.,* Σd_x^2.
4. Square the deviations of Y series and obtain the total, *i.e.,* Σd_y^2.
5. Multiply d_x and d_y and find out the total, *i.e.,* $\Sigma d_x d_y$.

The computation procedure will be more clear from the solved examples.

Notes : (1) Actual Mean Method is used when actual mean is a whole number.

(2) Short-cut Method (method of taking deviations from an arbitrary constant) is used when the actual mean is a fraction and the observations have large values.

Assumptions of the Karl Pearson's Coefficient of Correlation

Karl Pearson's coefficient of correlation is based on the following assumptions :

(i) Linear Relationship : In this method, a linear reelationship between the two variables is assumed. In such a case, the paired observations on the two variables plotted on a scatter-diagram cluster around a straight line.

(ii) Causal Relationship : In studying correlation, we expect a cause and effect relationship between the forces affecting the values in the two series.

Merits and Demerits of Karl Pearson's Coefficient of Correlation

Merits

1. It is important and popular method of measuring the relationship between two variables. It gives a precise and quantitative value indicating the degree of relationship existing between the two variables. The value of r is easily interpretable.
2. It measures the direction as well as the relationship between the two variables.

Demerits :

1. The value of the coefficient is affected by extreme items.
2. It computational procedure is difficult as compared to other methods.
3. It assumes the linear relationship between the two variables.

Properties of Correlation Coefficient

The following are the important properties of the coefficient of correlation :

1. The coefficient of correlation r lies between -1 and $+1$, *i.e.*, $-1 \leq r \leq 1$, *i.e.*, a number between -1 and 1 both inclusive that describes the linear relationship between pairs of quantitative variables.
2. The correlation coefficient is independent of the change of origin and scale.
3. The correlation coefficient is the ratio of two quantities having same units, thus it is a pure number having no units.
4. The value of r does not change if all the values of either variable are converted to a different scale. For example, if the units of X are changed from feet to meters, the value of r does not change.
5. The value of r is not affected by the choice of X or Y. Interchange of all X and Y values will not change the value of r, *i.e.*, the correlation coefficient between X and Y is equal to the correlation coefficient between Y and X.
6. r measures the strength of linear relationship. It is not designated to measure the strength of a relationship that is not linear.
7. If the sign of all the values of one of the variables is changed, the sign of the correlation coefficient changes, *i.e.*, the correlation between X and Y is opposite in sign to the correlation between $-X$ and Y or X and $-Y$. **[Important]***
8. If each value of X and (or) Y a constant amount is added or subtracted, the correlation coefficient remains unchanged, *i.e.*, correlation coefficient is independent of the change of origin.
9. If each value of X and (or) Y is multiplied or divided by a constant, the correlation coefficient remains unchanged, *i.e.*, the correlation coefficient is independent of the change of scale.

Notes :

(1) The change of origin means adding or subtracting a constant amount from given observations of the variables X and Y and the **change of scale** means multiplying or dividing the values of X and Y by some constant amount. The constant amount can be chosen arbitrarily as per convenience. It may be same or different for X and Y.

(2) Property II is very useful for reducing computational work involved in the calculation of coefficient of correlation and forms the basis for short-cut method.

SOLVED EXAMPLES

Exxample 1. Calculate the correlation coefficient between the height of father and height of son from the given data :

Height of Father (in inches)	64	65	66	67	68	69	70
Height of Son (in inches)	66	67	65	68	70	68	72

* If r is the correlation coefficient between X and Y then if each value of X is multiplied by 1 and each value of Y is multiplied by -1 then the correlation coefficient will become $-r$.

Solution. **Computation of *r* (Actual Mean Method)**

Height of Father X	*Height of Son Y*	$(X-\bar{X}) = X-67 = x$	$(Y-\bar{Y}) = Y-68 = y$	$(X-\bar{Y})^2 = x^2$	$(Y-\bar{Y})^2 = y^2$	$(X-\bar{X}) \times (Y-\bar{Y}) = xy$
64	66	–3	–2	9	4	6
65	67	–2	–1	4	1	2
66	65	–1	–3	1	9	3
67	68	0	0	0	0	0
68	70	1	2	1	4	2
69	68	2	0	4	0	0
70	72	3	4	9	16	12
$\Sigma X = 469$	$\Sigma Y = 476$			$\Sigma(X-\bar{X})^2 = 28$	$\Sigma(Y-\bar{Y})^2 = 34$	$\Sigma[(X-\bar{X}) \times (Y-\bar{Y})] = 25$

$$\bar{X} = \frac{\Sigma X}{N} = \frac{469}{7} = 67, \ \bar{Y} = \frac{\Sigma Y}{N} = \frac{476}{7} = 68$$

In this example, $\bar{X}$ and $\bar{Y}$ are not fractional values but whole numbers so actual mean method will be most suitable. for computing r.

$$r = \frac{\Sigma (X-\bar{X})(Y-\bar{Y})}{\sqrt{[\Sigma (X-\bar{X})^2 \cdot \Sigma (Y-\bar{Y})^2]}} = \frac{\Sigma xy}{\Sigma x^2 \Sigma y^2}$$

$$= \frac{25}{\sqrt{[28 \times 34]}} = \frac{25}{\sqrt{952}}$$

$$= \frac{25}{30.85} = 0.81.$$

Example 2. Calculate coefficient of correlation between birth rate and death rate from the following data :

Year	1931	1941	1951	1961	1971	1981	1991
Birth Rate	24	26	32	32	35	30	32
Death Rate	15	20	22	24	27	24	20

Solution. Here mean values $\bar{X}$ and $\bar{Y}$ are fractional values and the volume and number of the paired observations is not large; so either direct or short-cut method can be applied for computing r. Hence, for clear understanding, we make use of both the methods.

Direct Method

Year	*Birth Rate X*	*Death Rate Y*	X^2	Y^2	XY
1931	24	15	576	225	360
1941	26	20	767	400	520
1951	32	22	1024	484	704
1961	33	24	1089	576	792
1971	35	27	1225	729	945
1981	30	24	900	576	720
1991	32	20	1024	400	640
$N = 7$	$\Sigma X = 212$	$\Sigma Y = 152$	$\Sigma X^2 = 6514$	$\Sigma Y^2 = 3390$	$\Sigma XY = 4681$

$$\bar{X} = \frac{\Sigma X}{N} = \frac{212}{7} = 30.285, \; \bar{Y} = \frac{\Sigma Y}{N} = \frac{152}{7} = 21.7$$

$$r = \frac{N \Sigma XY - (\Sigma X)(\Sigma Y)}{\sqrt{[N \Sigma X^2 - (\Sigma X)^2]} \sqrt{[N \Sigma Y^2 - (\Sigma Y)^2]}}$$

Putting the values from the table

$$r = \frac{7 \times 4681 - 212 \times 152}{\sqrt{[7 \times 6514 - (212)^2]} \sqrt{[17 \times 3390 - (152)^2]}}$$

$$= \frac{32767 - 32224}{\sqrt{[654 \times 626]}} = \frac{543.00}{639.85} = 0.85$$

Short-Cut Method

Year	*Birth Rate X*	*Death Reate Y*	$d_x = (X - 30)$	$d_y = (Y - 22)$	d_x^2	d_y^2	$d_x d_y$
1931	24	15	–6	–7	36	49	42
1941	26	20	–4	–3	16	4	8
1951	32	22	2	0	4	0	0
1961	33	24	3	2	9	4	6
1971	35	27	5	5	25	25	25
1981	30	24	0	2	0	4	0
1991	32	30	2	–2	4	4	–4
$N = 7$			$\Sigma d_x = 2$	$\Sigma d_y = -2$	$\Sigma d_x^2 = 94$	$\Sigma d_y^2 = 90$	$\Sigma d_x d_y = 77$

[Formula for Short-cut Method]

$$r = \frac{N \Sigma d_x d_y - (\Sigma d_x)(\Sigma d_y)}{\sqrt{[N \Sigma d_x^2 - (\Sigma d_x)^2]} \sqrt{[N \Sigma d_y^2 - (\Sigma d_y)^2]}}$$

$$= \frac{7 \times 77 - (+2)(-2)}{\sqrt{[7 \times 94 - (2)^2]} \sqrt{[7 \times 90 - (-2)^2]}} = \frac{539 + 4}{\sqrt{[658 - 4]} \sqrt{[630 - 4]}}$$

$$= \frac{543}{\sqrt{[654 \times 626]}} = 0.85$$

The value of r is the same as that obtained by direct formula.

Example 3. The total of the multiplications of deviations of X and $Y = 3044$. No. of pairs of the observations is 10. The total of deviations of $X = -170$. Total of deviations of $Y = -20$. Total of the squares of deviations of $X = 8288$. Total of the squares of the deviations of $Y = 2264$. Find the coefficient of correlation when the arbitrary means of X and Y are 82 and 68 respectively.

(D.U., B. Com. 2001)

Solution. Since the derivations of X and Y are given, the value of the arbitrary means of X and Y are not required in the solution.

$\Sigma d_x = -170, \quad \Sigma d_y = -20, \quad \Sigma d_x d_y = 3044$

$\Sigma d_x^2 = 8288, \quad \Sigma d_y^2 = 2264,$

$$r = \frac{\frac{\Sigma d_x d_y}{n} - \frac{\Sigma d_x}{n}\frac{\Sigma d_y}{n}}{\sqrt{\frac{\Sigma d_x^2}{n} - \left(\frac{\Sigma d_x}{n}\right)}\sqrt{\frac{\Sigma d_y^2}{n} - \left(\frac{\Sigma d_y}{n}\right)^2}}$$

$$= \frac{\frac{3044}{10} - \left(\frac{-170}{10}\right)\left(\frac{-20}{10}\right)}{\sqrt{\frac{8288}{10} - \left(\frac{-170}{10}\right)^2}\sqrt{\frac{2264}{10} - \left(\frac{-20}{10}\right)^2}}$$

$$= \frac{304.4 - 34}{\sqrt{539.8}\sqrt{224.4}} = \frac{270.4}{23.23 \times 14.97} = +0.78.$$

Example 4. Calculate the coefficient of corrlation by Karl Pearson's method from the following data relating to overhead expenses and the cost of production :

Overheads in ('000 *Rs.*)	80	90	100	110	120	130	140	150	160
Cost ('000 *Rs.*)	15	19	16	19	17	18	16	18	15

Solution. **Computation of Correlation Coefficient**

Overheads ('000 *Rs.*) X	*Cost* ('000 *Rs.*) Y	$d_x = (X - \overline{X})$	$d_y = (Y - \overline{Y})$	d_x^2	d_y^2	$d_x d_y$
80	15	–40	–2	1600	4	80
90	19	–30	2	900	4	–60
100	16	–20	–1	400	1	20
110	19	–10	+2	100	4	–20
120	17	0	0	0	0	0
130	18	10	1	100	1	10
140	16	20	–1	400	1	–20
150	18	30	1	900	1	30
160	15	40	–2	1600	4	–80
$\Sigma X = 1080$	$\Sigma Y = 153$	$\Sigma d_x = 0$	$\Sigma d_y = 0$	$\Sigma d_x^2 = 6000$	$\Sigma d_y^2 = 20$	$\Sigma d_x d_y = -40$

$$\overline{X} = \frac{\Sigma X}{N} = \frac{1080}{9} = 120, \; \overline{Y} = \frac{\Sigma Y}{N} = \frac{153}{9} = 17.$$

Since,
$$r = \frac{N\Sigma d_x d_y - (\Sigma d_x)(\Sigma d_y)}{\sqrt{[N\Sigma d_x^2 - (\Sigma d_x)^2]}\sqrt{[N\Sigma d_y^2 - (\Sigma d_y)^2]}}$$

Putting the values from the table, we get

$$r = \frac{9 \times (-40) - 0}{\sqrt{(9 \times 6000 - 0)}\ \sqrt{(9 \times 20 - 0)}} = \frac{-360}{\sqrt{54000 \times 180}} = \frac{-360}{3117.7}*$$

$$= -0.11547$$

Thus, there is a very low degree of negative correlation between overheads and cost.

Example 5. Calculate Karl Pearson's coefficient of correlation from the following data using 20 as the working mean** for price, and 70 as the working mean for demand.

Price	14	16	17	18	19	20	21	22	23
Demand	84	78	70	75	66	67	62	58	60

Solution. In this case, the assumed means for X and Y have been given in the question as 20 and 70 respectively. Therefore we will take the deviations of X from 20 and that of Y from 70.

Computation of Correlation Coefficient

Price X	*Demand Y*	$d_x = (X - 20)$	$d_y = (Y - 70)$	d_x^2	d_y^2	$d_x d_y$
14	84	–6	14	36	196	–84
16	78	–4	8	16	64	–32
17	70	–3	0	9	0	0
18	75	–2	5	4	25	–10
19	66	–1	–4	1	16	+4
20	67	0	–3	0	9	0
21	62	1	–8	1	64	–8
22	58	2	–12	4	144	–24
23	60	3	–10	9	100	–30
$N = 9$		$\Sigma d_x = -10$	$\Sigma d_y = -10$	$\Sigma d_x^2 = 80$	$\Sigma d_y^2 = 618$	$\Sigma d_x d_y = 184$

$$r = \frac{N \Sigma d_x d_y - (\Sigma d_x)(\Sigma d_y)}{\sqrt{[N \Sigma d_x^2 - (\Sigma d_x)^2]}\ \sqrt{[N \Sigma d_y^2 - (\Sigma d_y)^2]}}$$

$$= \frac{9 \times (-184) - (-10)(-10)}{\sqrt{[\{9 \times 80 - (-10)^2\}\ \{9 \times 618 - (-10)^2\}]}}$$

$$= \frac{-1656 - 100}{\sqrt{(720 - 100)\ (5562 - 100)]}} = \frac{-1756}{\sqrt{[620 \times 5462]}} = \frac{-1756}{1840.23} = -0.9542.$$

Thus, there is a high degree of negative correlation between price X and demand Y.

Example 6. Department store gives in service training to its salesman which is followed by a test. It is considering whether it should terminate the services of any salesman who does not do well in the test. The following data gives the test scores and sales made by nine salesman during a certain period.

Test scores	14	19	24	21	26	22	15	20	19
Sales ('00 *Rs.*)	31	36	48	37	50	45	33	41	39

* Since the deviations have been taken from the actual mean of X and actual of Y, we can use the formula $r = \frac{\Sigma d_x d_y}{\sqrt{\Sigma d_x^2}\ \sqrt{\Sigma d_y^2}}$

where $d_x = X - \overline{X}$ and $d_y = Y - \overline{Y}$.

** Working mean means assumed mean.

Calculate the coefficient of correlation between the test scores and the sales. Does it indicate that the termination of services of low test scores is justified ? **(D.U., B. Com. Hons. 1998)**

Solution.

Test X	*Score* $X - 20 = x$	*Sales* Y	$Y - 40 = y$	xy	x^2	y
14	–6	31	–9	54	36	81
19	–1	36	–4	4	1	16
24	4	48	8	32	16	64
21	1	37	–2	–3	1	9
26	6	50	10	60	36	100
22	2	45	5	10	4	25
14	–5	33	–7	35	25	49
20	0	41	1	0	0	1
19	–1	39	–1	1	1	1
				193	$\Sigma x^2 = 120$	$\Sigma y^2 = 346$

$\bar{x} = 20, \qquad \bar{y} = 40$

$$r = \frac{\Sigma xy}{\sqrt{\Sigma x^2}\sqrt{\Sigma y^2}} = \frac{193}{\sqrt{120}\sqrt{346}}$$

$$= \frac{\Sigma xy}{10.95 \times 18.6} = 0.9476.$$

Since there is a high degree of positive correlation between test scores and sales, the persons having low test scores will not be able to make good sales. Hence the termination is justified.

Example 7. Co-efficient of correlation between X and Y for 20 items is 0.3. Mean of X is 15 and that of Y is 20 while standard deviations are 4 and 5 respectively. At the time of calculation one item 27 has wrongly been taken as 17 in case of X series and 35 instead of 30 in case of Y series. Find the correct coefficient of correlation. **(AIMA, 1997)**

Solution. $n = 20$, wrong $r = 0.3$

Wrong sum of X values $= 15 \times 20 = 300$

Correct sum of X value $= 300 - 17 + 27 = 310$

Correct mean of $X = \frac{310}{20} = 15.5$

Wrong sum of Y values $= 20 \times 20 = 400$

Correct sum of Y values $= 400 - 35 + 30 = 395$

Correct mean of $Y = \frac{395}{20} = 19.75$

Wrong S.D. of $X = 4$

$$\therefore \qquad (4)^2 = \frac{\Sigma X^2}{20} - \left(\frac{\Sigma X}{20}\right)^2$$

$$\Rightarrow \qquad 16 = \frac{\Sigma X^2}{20} - 225$$

$\Rightarrow \qquad \Sigma X^2 = (225 + 16)\ 20 = 4820 =$ Wrong sum of the square of X values

Correct sum of the squares of X values

$$= 4820 - (17)^2 + (27)^2$$
$$= 4820 - 289 + 729 = 5260$$

Wrong S.D. of $Y = 5$

$$\therefore \qquad (5)^2 = \frac{\Sigma Y^2}{20} - \left(\frac{\Sigma Y}{20}\right)^2$$
$$= \frac{\Sigma Y^2}{20} - (20)^2$$
$$\Rightarrow \qquad \Sigma Y^2 = 425 \times 20 = 8{,}500$$

$\therefore$ Wrong sum of the squares of Y values $= 8{,}500$

Correct sum of the squares of Y values $= 8{,}500 - (35)^2 + (30)^2$

$$= 8{,}500 - 1225 + 900 = 8{,}175$$

Wrong coefficient of correlation $= 0.3$

$$\therefore \qquad 0.3 = \frac{\frac{\Sigma XY}{n} - \frac{\Sigma X}{n}\frac{\Sigma Y}{n}}{\sigma_X \sigma_Y}$$
$$= \frac{\frac{\Sigma XY}{20} - 15 \times 20}{4 \times 5}$$
$$\Rightarrow \qquad \frac{\Sigma XY}{20} = 0.3 \times 20 + 300 = 306$$
$$\Sigma XY = 306 \times 20 = 6120 = \text{wrong } \Sigma XY$$

Correct $\Sigma XY = 6120 - 17 \times 35 + 27 \times 30 = 6335$

$$\therefore \text{ Correct} \qquad r = \frac{\frac{\Sigma XY}{20} - \frac{\Sigma X}{20}\frac{\Sigma Y}{20}}{\sqrt{\frac{\Sigma X^2}{20} - \left(\frac{\Sigma X}{210}\right)^2}\sqrt{\frac{\Sigma Y^2}{20} - \left(\frac{\Sigma Y}{n}\right)^2}}$$
$$= \frac{\frac{6335}{20} - 15.5 \times 19.75}{\sqrt{\frac{5260}{20} - (15.5)^2} \cdot \sqrt{\frac{8175}{20} - (19.75)^2}}$$
$$= \frac{316.75 - 306.125}{4.7696 \times 4.3229} = \frac{10.625}{20.61} = 0.515.$$

Example 8. Calculate the coefficient of correlation between values of X and Y.

X	21	22	28	32	35	36
Y	18	20	25	30	31	32

If each X value is multiplied by 2 and increased by 3 and each Y value is multiplied by 3 and subtracted by 4, find the new coefficient of correlation. Comment on the result.

Solution. Mean of $X = \frac{174}{6} = 29$, mean of $Y = \frac{156}{6} = 26$.

X	Y	$X - 29 = x$	$Y - 26 = y$*	x^2	y^2	xy
21	18	–8	–8	64	64	64
22	20	–7	–6	49	36	42
28	25	–1	–1	1	1	1
32	30	3	4	9	16	12
35	31	6	5	36	25	30
36	32	7	6	49	36	42
				$\Sigma x^2 = 208$	$\Sigma y^2 = 178$	$\Sigma xy = 191$

$$r_1 = \frac{\Sigma xy}{\sqrt{\Sigma x^2}\sqrt{\Sigma y^2}} = \frac{191}{\sqrt{208}\sqrt{178}} = \frac{191}{\sqrt{37024}} = \frac{191}{192.416} = 0.9926$$

Table for new X and Y values.

X'	Y'	$\bar{X}' - 61$** $= d_x$	$\bar{Y}' - 74$ $= d_y$	d_x^2	d_y^2	$d_x d_y$
45	50	–16	–24	256	576	384
47	56	–14	–18	196	324	252
59	71	–2	–3	4	9	6
67	86	6	12	36	144	72
73	89	12	15	144	225	180
75	92	14	18	196	324	252
				$\Sigma d_x^2 = 832$	$\Sigma d_y^2 = 1602$	$\Sigma d_x d_y = 1146$

$$r_2 = \frac{\Sigma d_x d_y}{\sqrt{\Sigma d_x^2}\sqrt{\Sigma d_y^2}} = \frac{1146}{\sqrt{832}\sqrt{1602}} = \frac{191 \times 6}{\sqrt{4 \times 208}\sqrt{9 \times 178}} = \frac{191}{\sqrt{208}\sqrt{178}}.$$

$r_1 = r_2$

⇒ Coefficient of correlation is independent of the change of origin as well as change of scale.

Example 9. While calculating x and y on computer, the following values figured :

$$N = 30, \quad \Sigma x = 120, \quad \Sigma x^2 = 600$$
$$\Sigma y = 90, \quad \Sigma y^2 = 250, \quad \Sigma xy = 356.$$

On verification it was found that values of two combinations were wrong, while the correct values were :

X	Y		X	Y
6	8		6	10
10	5		8	6

Calculate the correct coefficient of correlation.

* Since the mean of X and mean of Y are whole numbers, the deviations have been taken from the actual mean of X and actual mean of Y.

** $X' = 2X + 3$, $Y' = 3Y - 4$ New $\bar{X} = 61 = \bar{X}'$ and new $\bar{Y} = 74 = \bar{Y}'$.

Solution.

Correct $\Sigma x = 120 - 6 - 10 + 6 + 8 = 118$

Correct $\Sigma y = 90 - 8 - 5 + 10 + 6 = 93$

Correct $\Sigma x^2 = 600 - 36 - 100 + 36 + 64 = 564$

Correct $\Sigma y^2 = 250 - 64 - 25 + 100 + 36 = 297$

Correct $\Sigma xy = 356 - 6 \times 8 - 10 \times 5 + 6 \times 10 + 8 \times 6$

$= 356 - 48 - 50 + 60 + 48 = 366.$

Correct
$$r = \frac{\dfrac{\Sigma xy}{n} - \left(\dfrac{\Sigma x}{n}\right)\left(\dfrac{\Sigma y}{n}\right)}{\sqrt{\dfrac{\Sigma x^2}{n} - \left(\dfrac{\Sigma x}{n}\right)^2}\sqrt{\dfrac{\Sigma y^2}{n} - \left(\Sigma \dfrac{y}{n}\right)^2}}$$

$$= \frac{\dfrac{366}{30} - \dfrac{118}{30} \times \dfrac{93}{30}}{\sqrt{\dfrac{564}{30} - \left(\dfrac{118}{30}\right)^2}\sqrt{\dfrac{297}{30} - \left(\dfrac{93}{30}\right)^2}}$$

$$= \frac{12.2 - 12.193}{\sqrt{18.8 - 15.47}\sqrt{9.9 - 9.61}}$$

$$= \frac{0.007}{1.824 \times 0.538} = 0.00713$$

Example 10. Deviations of two series X and Y from an assumed average are shown below. Calculate coefficient of correlation.

X	+5	–4	–2	+20	–10	0	+3	0	–15	–5
Y	–12	–12	–7	+25	–10	–3	0	+2	–9	–15

Solution.

Deviation of X from an Assumed Mean d_x	*Deviation of X from on Assumed Mean* d_y	d_x^2	d_y^2	d_xd_y
5	5	25	25	25
–4	–12	16	144	48
–2	–7	4	49	14
20	25	400	625	500
–10	–10	100	100	100
0	–3	0	9	0
3	0	9	0	0
0	2	0	4	0
–15	–9	225	81	135
–5	–15	25	225	75
$\Sigma d_x = -8$	$\Sigma d_y = -24$	$\Sigma d_x^2 = 804$	$\Sigma d_y^2 = 1262$	$\Sigma d_xd_y - 897$

$$r = \frac{\frac{\Sigma d_x d_y}{n} - \left(\frac{\Sigma d_x}{n}\right)\left(\frac{\Sigma d_y}{n}\right)}{\sqrt{\frac{\Sigma d_x^2}{n} - \left(\frac{\Sigma d_x}{n}\right)^2}\sqrt{\frac{\Sigma d_y^2}{n} - \left(\frac{\Sigma d_y}{n}\right)^2}}$$

$$= \frac{\frac{897}{10} - \frac{8}{10} \times \frac{24}{10}}{\sqrt{\frac{804}{10} - \left(\frac{-8}{10}\right)^2}\sqrt{\frac{1262}{10} - \left(\frac{-24}{10}\right)^2}}$$

$$= \frac{89.7 - 1.92}{\sqrt{80.4 - 0.64}\sqrt{126.2 - 5.76}} = \frac{87.78}{8.93 \times 10.9745} = \frac{87.78}{98} - 0.8957$$

Example 11. With the help of data relating to density of population and death of six cities, calculate their coefficient of correlation.

Cities	*A*	*B*	*C*	*D*	*E*	*F*
Area (in. sq. miles)	75	90	50	30	60	40
Population	15,000	45,000	20,000	21,000	36,000	12,000
No. of Deaths	150	720	280	420	612	156

Solution.

Density of population per city = population per unit area $= \frac{\text{Population}}{\text{Area}}$.

Death rate per 1,000 $= \frac{\text{No. of deaths}}{\text{Population}} \times 1{,}000$.

In city *A* density of population $= \frac{15{,}000}{75} = 200$/mile.

In city *A* death rate $= \frac{150}{15{,}000} \times 1{,}000 = 10$ per 1,000.

If *X* denotes the density of population and *Y* denotes the death rate, then

City	*Density* *X*	*Death Rate* *Y*	$X - 450 = x$	$Y - 15 = y$	x^2	y^2	xy
A	200	10	–250	–5	62,500	25	1,250
B	500	16	50	1	2,500	1	50
C	400	14	–50	–1	2,500	1	50
D	700	20	250	5	62,500	25	1,250
E	600	17	150	2	22,500	4	300
F	300	13	–150	–2	22,500	4	300
	$\Sigma X = 2{,}700$ $\bar{X} = \frac{2{,}700}{6} = 450$	$\Sigma Y = 90$ $\bar{Y} = \frac{90}{6} = 15$			$\Sigma x^2 = 1{,}75{,}000$	$\Sigma y^2 = 60$	$\Sigma xy = 3{,}200$

$$r = \frac{\Sigma xy}{\sqrt{\Sigma x^2}\sqrt{\Sigma y^2}} = \frac{3200}{\sqrt{1,75,000}\sqrt{60}}$$

$$= \frac{3200}{418.33 \times 7.745} = 0.987$$

Example 12. Data relating to age of the students and their games is given below. Calculate correlation between the age of the students and their playing habits.

Age	15	16	17	18	19	20
No. of Students	250	200	150	120	100	80
Regular Players	200	150	90	48	30	12

Solution.

Age X	*Percentage of Students Having Regular Playing Habits* Y	$X - \bar{X} = x$	$Y - \bar{Y} = y$	x^2	y^2	xy
15	$\frac{200}{250} \times 100 = 80\%$	–2.5	30	6.25	900	–75
16	$\frac{150}{200} \times 100 = 75\%$	–1.5	25	2.25	625	–37.5
17	$\frac{90}{150} \times 100 = 60\%$	–0.5	10	0.25	100	–5
18	$\frac{48}{120} \times 100 = 40\%$	+0.5	–10	0.25	100	–5
19	$\frac{30}{100} \times 100 = 30\%$	1.5	–20	2.25	400	–30
20	$\frac{12}{80} \times 100 = 15\%$	2.5	–35	6.25	1225	–87.5
$\Sigma X = 105$	$\Sigma Y = 300$			$\Sigma x^2 = 17.5$	$\Sigma y^2 = 3350$	$\Sigma xy = -240$

$$\bar{X} = \frac{\Sigma X}{n} = \frac{105}{6} = 17.5,$$

$$\bar{Y} = \frac{\Sigma Y}{n} = \frac{300}{6} = 50$$

$$r = \frac{\Sigma xy}{\sqrt{\Sigma x^2}\sqrt{\Sigma y^2}} = \frac{-240}{\sqrt{17.5}\sqrt{3350}} = \frac{-240}{4.18 \times 57.88} = -0.992.$$

Example 13. The following table gives the distribution of the total population and those who are wholly or partially blind among them. Find out if there is any relation between age and blindness.

Age	*No. of Persons ('000)*	*Blind*
0–10	100	55
10–20	60	40
20–30	40	40
30–40	36	40
40–50	24	36
50–60	11	22
60–70	6	18
70–80	3	15

Solution.

Age Group	Age			Blindness			Product of deviations $d_x d_y$
	Mid Value X	$d_x = \frac{X-45}{10}$	d_x^2	Blind Per Lakh Y	$d_y = Y - 185$	d_y^2	
0–10	5	–4	16	55	–130	16900	520
10–20	15	–3	9	67*	–118	13924	354
20–30	25	–2	4	100	–85	7225	170
30–40	35	–1	1	111	–74	5476	74
40–50	45	0	0	150	–35	1225	0
50–60	55	+1	1	200	15	225	15
60–70	65	+2	4	300	115	13225	230
70–80	75	+3	9	500	315	99225	945
$N = 8$		–4	44		3	157425	2308

It may be noted that frequency of blindness has also been converted into lakhs so as to facilitate comparison with Age.

$$r = \frac{\Sigma d_x d_y - \frac{(\Sigma d_x)(\Sigma d_y)}{n}}{\sqrt{\left[\Sigma d_x^2 - \frac{(\Sigma d_x)^2}{n}\right]\left[\Sigma d_y^2 - \frac{(\Sigma d_y)^2}{n}\right]}}$$

$$= \frac{2308 - \frac{(-4)(+3)}{8}}{\sqrt{\left[44 - \frac{16}{8}\right]\left[157425 - \frac{9}{8}\right]}} = \frac{2309.5}{2571} = +\,0.898$$

Example 14. If two distinct values of the variable are given when is the covariance or coefficient of correlation zero ?

Solution. Let the data be $x : x_1, x_2$

$y : y_1, y_2$

If r, the Coefficient of Correlation of Cov $(x, y) = 0$, we have**

$$\frac{\Sigma xy}{n} - \frac{\Sigma x}{n}.\frac{\Sigma y}{n} = 0$$

---

* $\frac{40}{60{,}000} \times 1{,}00{,}000 = \frac{400}{6} = 66.6 = 67$ approx.

** Covariance $= \frac{\Sigma (x-\bar{x})(y-\bar{y})}{n} = \frac{\Sigma xy}{n} - \left(\frac{\Sigma x}{n}\right)\left(\frac{\Sigma y}{n}\right)$

$r = \frac{\text{cov}(x, y)}{\sigma_x \sigma_y}$

or $$\frac{x_1y_1+x_2y_2}{2}-\left[\frac{x_1+x_2}{2}\right]\cdot\left[\frac{y_1+y_2}{2}\right]=0$$

or $$(2x_1y_1+2x_2y_2)-(x_1y_1+x_2y_1+x_1y_2+x_2y_2)=0$$

or $$(x_1y_1+x_2y_2-x_2y_1-x_1y_2)=0$$

or $$x_1(y_1-y_2)-x_2(y_1-y_2)=0$$

or $$(x_1-x_2)(y_1-y_2)=0$$

i.e., either $x_1=x_2$ or $y_1=y_2$.

Example 15. State with reasons whether the following statements are *true* or *false.*

1. Coefficient of correlation between two variables X and Y is $r=0.6$. Hence the coefficient of correlation between variables X and $\frac{1}{2}Y$ is $r=0.3$.
2. Coefficient of correlation between two variables x and y is $r=\frac{1}{2}$. Hence the coefficient of correlation between variables $x+\frac{1}{2}$ and $y+\frac{1}{2}$ is $r=1$.
3. Coefficient of correlation between two variables X and Y is zero. Hence X and Y are unrelated variables.
4. Coefficient of Correlation between two variables X and Y is 0.8. Hence, the coefficient of correlation between X and $-Y$ is -0.8.
5. Given that the correlation between x and y is 0.5. The correlation between $(2x-4)$ and $(3-2y)$ is -0.5.
6. Karl Pearson's Correlation Coefficient always lies between 0 and 1.

Solution.

1. The statement is false. The coefficient of correlation is independent of the change of scale. Hence the correlation between X and $\frac{Y}{2}$ will remain 0.6 only.
2. Statement is false. The coefficient of correlation between two variables is independent of the change of origin. Hence, the coefficient of correlation between $x+\frac{1}{2}$ and $y+\frac{1}{2}$ will remain same *i.e.,* $\frac{1}{2}$.
3. False $r=0$ indicates Obsence of linear correlation between the two variables.
4. True. When Y change to $-Y$, the expression $(Y-\bar{Y})$ changes to $-(Y-\bar{Y})$.
5. True, because $3-2y$ can be written as $-(2y-3)$.
6. False, $-1\le r\le 1$

Example 16. What is a scatter diagram ? Draw the scatter diagram when (i) $r=+1$, (ii) $r=-1$, and (iii) $r=0$.

Solution. Scatter diagram : The scatter diagram is a diagrammatic representation of ungrouped or simple bivariate data.

If we have n pairs of observations of two variables x and y, we can plot each pair of values (x_i, y_i) as a point on a graph paper. Thus for n pairs, n points will be plotted on the graph paper. This diagrammatic representation of bivariate data is called a **scatter diagram.**

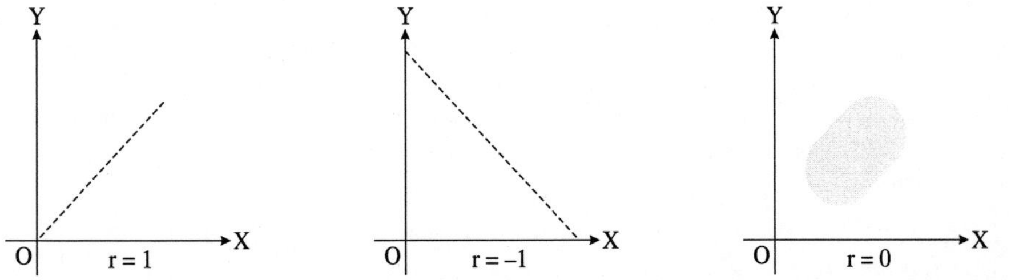

Example 17. In two set of variables x and y with 50 observations each, the following were observed : $\bar{x} = 10$, S.D. of x is 3, $\bar{y} = 6$, S.D. of y is 2, coefficient of correlation between x and y is 0.3. However on subsequent verification it was found that one value of $x = 10$ and one value of $y = 6$ were inaccurate and hence weeded out. With the remaining 49 pairs of values, how is the original value of correlation coefficient affected ?

Solution. Since the values weeded out are $\bar{x}$ and $\bar{y}$; the quantities $\Sigma\,(x - \bar{x})$, $\Sigma\,(y - \bar{y})$, $\Sigma\,(x - \bar{x})^2$, $\Sigma\,(y - \bar{y})^2$ and $\Sigma\,(x - \bar{x})\,(y - \bar{y})$ would remain unaffected. Acordingly $r = \dfrac{\Sigma\,(x - \bar{x})\,(y - \bar{y})}{\sqrt{\Sigma\,(x - \bar{x})^2}\,\sqrt{\Sigma\,(y - \bar{y})^2}}$ will also remain unaffected as 0.3.

Example 18. From the following given sets of observations find the coefficient of correlation :

(a)

X	Y
69	85
70	87

(b)

X	Y
102	50
109	48

Explain why you are not surprised if you find $r = +1$ or -1. If another pair of values be added to (a) or (b), when would you still expect $r = +1$ or -1.

Solution. (a) $\bar{X} = \dfrac{69 + 70}{2} = 69.5$, $\bar{Y} = \dfrac{85 + 87}{2} = 86$

$$r = \frac{\Sigma\,(X - \bar{X})^2\,(Y - \bar{Y})}{\sqrt{\Sigma\,(X - \bar{X})^2\,\Sigma\,(Y - \bar{Y})^2}}$$

$$= \frac{(69 - 69.5)\,(85 - 86) + (70 - 69.5)\,(87 - 86)}{\sqrt{(69 - 69.5)^2 + (70 - (69.5)^2}\,\sqrt{(85 - (86)^2 + (87 - 86)^2}}$$

$$= \frac{(-0.5)\,(-1) + (0.5)\,(1)}{\sqrt{(0.5)^2 + (0.5)^2}\,\sqrt{(1)^2 + (1)^2}} = \frac{+2\,(0.5)}{(0.5)\,\sqrt{2}\,\sqrt{2}} = +1$$

(b) $\bar{X} = \dfrac{102 + 109}{2} = 105.5$, $\bar{Y} = \dfrac{50 + 48}{2} = 49$

$$r = \frac{\Sigma\,(X - \bar{X})\,(Y - \bar{Y})}{\sqrt{\Sigma\,(X - \bar{X})^2\,\Sigma\,(Y - \bar{Y})^2}}$$

$$= \frac{(102 - 105.5)\,(50 - 49) + (109 - 105.5)\,(48 - 49)}{\sqrt{(102 - 105.5)^2 + (109 - 105.5)^2}\,\sqrt{(50 - 49)^2 + (48 - 49)^2}}$$

$$= \frac{-(3.5)\,(1) + (3.5)\,(-1)}{\sqrt{(3.5)^2 + (3.5)^2}\,\sqrt{(1)^2 + (1)^2}} = \frac{-2\,(3.5)}{(3.5)\,\sqrt{2}\,\sqrt{2}} = -1.$$

There is no surprise in the results because whenever only two values are given, the coefficient of correlation would always be either +1 or –1 depending upon whether the straight line perfectly passing through these two points have positive or negative slope.

If another pair of observations be added to data (*a*) or (*b*), the coefficient of correlation would continue to be the same as long as the added pair lies on the straight line passing through the given two points.

Example 19. The following table gives aptitude test scores and productivity indices of 8 randomly selected workers.

Aptitude Score	57	58	59	59	60	61	62	64
Productivity Index	67	68	65	68	72	72	69	71

Calculate the correlation coefficient between aptitude score and productiity index.

(ICWA, Dec., 1992)

Solution. Let X denote the aptitude score and Y the productivity index.

Aptitude Test X	*Productivity Index* Y	$X-\bar{X}=x$	$Y-\bar{Y}=y$	x^2	y^2	xy
57	67	–3	–2	9	4	6
58	68	–2	–1	4	1	2
59	65	–1	–4	1	16	4
59	68	–1	–1	1	1	1
60	72	0	3	0	9	0
61	72	1	3	1	9	3
62	69	2	0	4	0	0
64	71	4	2	16	4	8
$\Sigma X=480$ $\bar{X}=60$	$\Sigma Y=552$ $\bar{Y}=69$	$\Sigma x=0$	$\Sigma y=0$	$\Sigma x^2=36$	$\Sigma y^2=44$	$\Sigma xy=24$

$$\text{Correlation coefficient} = \frac{\frac{\Sigma xy}{n}}{\sqrt{\frac{\Sigma x^2}{n}}\sqrt{\frac{\Sigma y^2}{n}}} = \frac{\frac{24}{8}}{\sqrt{\frac{36}{8}}\sqrt{\frac{44}{8}}}$$

$$= \frac{24}{\sqrt{36}\sqrt{44}} = \frac{2}{\sqrt{11}} = 0.603.$$

Example 20. $r=0.8$, $\Sigma xy=60$, $\sigma_y=2.5$ and $\Sigma x^2=90$.

where x and y are the deviations from their actual means, we have

$$r = \frac{\Sigma xy}{\Sigma x^2 \Sigma y^2} = \frac{60}{90 \Sigma y^2}$$

$$\Rightarrow \qquad 0.8 = \frac{60}{\sqrt{90}\sqrt{\Sigma y^2}} \Rightarrow \sqrt{90} \times 0.8 \sqrt{\Sigma y^2} = 60$$

$$\Rightarrow \qquad 90 \times 0.64 \times \Sigma y^2 = 3600$$

$$\Rightarrow \qquad \Sigma y^2 = \frac{3600}{57.6} = 62.5$$

$$\sigma_y = \sqrt{\frac{\Sigma y^2}{n}} = 2.5 \Rightarrow \frac{\Sigma y^2}{n} = 6.25$$

$$\Rightarrow \qquad \frac{6.5}{n} = 6.35 \Rightarrow n = \frac{62.5}{6.25} = 10$$

Example 21. Calculate the Karl Pearson's coefficient of correlation from the following data :

No.	*Subject*	*Percentage Marks First Term*	*Percentage Marks Second Term*
1.	Hindi	75	62
2.	English	81	68
3.	Economics	70	65
4.	Accounts	76	60
5.	Commerce	77	69
6.	Mathematics	81	72
7.	Statistics	84	76
8.	Costing	75	72

Solution. Computation of Karl Pearson's Coefficient of Correlation

S. No.	*Subject*	*Percentage Marks First Term (X)*	*Percentage Marks Second Term (Y)*	$dx = X - A = X - 77$	$dy = Y - A = Y - 68$	d^2x	d^2y	$dxdy$
1.	Hindi	75	62	–2	–8	4	36	12
2.	English	81	68	+4	0	16	0	0
3.	Economics	70	65	–7	–3	49	9	21
4.	Accounts	76	60	–1	–8	1	64	8
5.	Commerce	77	69	0	+1	0	1	0
6.	Mathematics	81	72	+4	+4	16	16	16
7.	Statistics	84	76	+7	+8	49	64	56
8.	Costing	75	72	–2	+4	4	16	–8
	$N = 8$			$\Sigma dx = +3$	$\Sigma dy = 0$	$\Sigma d^2x = 139$	$\Sigma d^2y = 206$	$\Sigma dxdy = 105$

$$r = \frac{N\Sigma\, dx\, dy - (\Sigma\, dx)\,(\Sigma\, dy)}{\sqrt{N\Sigma\, d^2x - (\Sigma\, dx)^2}\,\sqrt{N\Sigma\, d^2y - (\Sigma\, dy)^2}}$$

$$= \frac{8 \times 105 - (3)\,(0)}{\sqrt{8 \times 139 - (3)^2}\,\sqrt{8 \times 206 - (0)^2}} = \frac{840 - 0}{\sqrt{1112 - 9}\,\sqrt{1648}}$$

$$= \frac{840}{\sqrt{1103}\,\sqrt{1648}} = \frac{840}{\sqrt{1817744}} = 0.623.$$

Examploe 22. Find the Karl Pearson's coefficient of correlation between the age and the playing habits of people from the following information :

Age Group (in years)	*No. of People*	*No. of Players*
15 and less than 20	200	150
20 and less than 25	270	162
25 and less than 30	340	170
30 and less than 35	360	180
35 and less than 40	400	180
40 and less than 45	300	120

Solution. **Calculation of Karl Pearson's Coefficient of Correlation :**

Age Group (in years)	*Mid-point (X)*	*No. of People*	*No. of Players*	*Percent-age of Players*	$dx = \frac{(x-27.5)}{5}$	$dy = \frac{y-50}{5}$	d^2x	d^2y	*dxdy*
15 and less than 20	17.5	200	150	75	–2	+5	4	25	–10
20 and less than 25	22.5	270	162	60	–1	+2	1	4	–2
25 and less than 30	27.5	340	170	50	0	0	0	0	0
30 and less than 35	32.5	360	180	50	+1	0	1	0	0
35 and less than 40	37.5	400	180	45	+2	–1	4	1	–2
40 and less than 45	42.5	300	120	40	+3	–2	9	4	–6
					$\Sigma dx = +3$	$\Sigma dy = +4$	$\Sigma d^2x = 19$	$\Sigma d^2y = 34$	$\Sigma dxdy = -20$

$$r = \frac{N\Sigma\, dx\, dy - (\Sigma\, dx)\,(\Sigma\, dy)}{\sqrt{N\Sigma\, d^2x - (\Sigma\, dx)^2}\,\sqrt{N\Sigma\, d^2y - (\Sigma\, dy)^2}}$$

$$= \frac{6\times(-20) - (+3)\,(+4)}{\sqrt{6\times 19 - (3)^2}\,\sqrt{6\times 34 - (4)^2}}$$

$$= \frac{-120 - 12}{\sqrt{114-9}\,\sqrt{204-16}} = \frac{-132}{\sqrt{105}\,\sqrt{188}}$$

$$= \frac{-132}{\sqrt{19740}} = \frac{-132}{140.499}$$

$$= -\,0.94 \text{ (app.)}$$

Example 23. Calculate coefficient of correlation from the following data and comment on the result :

Experience (x)	**Performance** (y)
16	23
12	22
18	24
4	17
3	19
10	20
5	18
12	21

Solution. **Computation of Karl Pearson's coefficient of Correlation**

Experience (X)	*Performance (Y)*	$dx = X - 10$	$dy = Y - 20$	d^2x	d^2y	$dxdy$
16	23	+6	+3	36	9	+18
12	22	+2	+2	4	4	+4
18	24	+8	+4	64	16	+32
4	17	–7	–3	36	9	+18
3	19	–7	–1	49	1	+7
10	20	0	0	0	0	0
5	18	–5	–2	25	4	+10
12	21	+2	+1	4	1	+2
		$\Sigma dx = 0$	$\Sigma dy = +4$	$\Sigma d^2x = 218$	$\Sigma d^2y = 44$	$\Sigma dxdy = 91$

$$r = \frac{N \Sigma\, dx\, dy - (\Sigma\, dx)\,(\Sigma\, dy)}{\sqrt{N \Sigma\, d^2x - (\Sigma\, dx)^2}\,\sqrt{N \Sigma\, d^2y - (\Sigma\, dy)^2}}$$

$$= \frac{8 \times 91 - (0)\,(4)}{\sqrt{8 \times 218 - (0)^2}\,\sqrt{8 \times 44 - (4)^2}}$$

$$= \frac{728 - 0}{\sqrt{1744 - 0}\,\sqrt{352 - 16}}$$

$$= \frac{728}{\sqrt{1744}\,\sqrt{336}} = \frac{728}{\sqrt{585984}}$$

$$= \frac{728}{765.5} = 0.951$$

Thus, there is high degree of positive correlation between experience and performance.

Example 24. From the following data compute coefficient of correlation between X and Y :

	X Series	*Y Series*
No. of items	15	15
Arithmetic Mean	25	18
Square of deviations from arithmetic Mean	136	138
Summation of products of deviations of X and Y series from their respective arithmetic Mean		122

(DU, B.Com 2005)

Solution. $N = 15, \overline{X} = 25, \overline{Y} = 18, \Sigma x^2 = 136, \Sigma y^2 = 138$

$$\Sigma xy = 122$$

$$r = \frac{\Sigma xy}{\sqrt{\Sigma x^2 \times \Sigma y^2}} = \frac{122}{\sqrt{136 \times 138}} = \frac{122}{\sqrt{18768}} = \frac{122}{136.997} = 0.89$$

Example 25. From the following data, calculate Karl Pearson's coefficient of correlation.

Height of Fathers (inches)	*Height of Sons (inches)*
66	65
68	64
69	67
72	69
65	64
59	60
62	59
67	68
61	60
71	64

(DU B. Com., 2005 (Exert)

Solution.

X *Height of Fathers*	x $(X-\bar{X})$ $(X-66)$	x^2	Y *Height of Sons*	y $Y-64$	y^2	xy
66	0	0	65	1	1	0
68	2	4	64	0	0	0
69	3	9	67	3	9	9
72	6	36	69	5	25	30
65	–1	1	64	0	0	0
59	–7	49	60	–4	16	28
62	–4	16	59	–5	25	20
67	1	1	68	4	16	4
61	–5	25	60	–4	16	20
71	5	25	64	0	0	0
$\Sigma X = 660$		$\Sigma x^2 = 166$	$\Sigma y = 640$		$\Sigma y^2 = 108$	$\Sigma xy = 111$

$$\bar{X} = \frac{\Sigma X}{N} = \frac{660}{10} = 66, \quad \bar{Y} = \frac{\Sigma Y}{N} = \frac{640}{10} = 64$$

$$r = \frac{\Sigma xy}{\sqrt{\Sigma x^2 \times \Sigma y^2}} = \frac{111}{\sqrt{166 \times 108}} = \frac{111}{133.89}$$

Example 26. Calculate Karl Pearson's coefficient of correlation between the variables X and Y using the following data :

X	Y
25	10
40	25
30	40
25	15
10	20
05	40
10	28
15	22
30	15
20	05

(DU B. Com., 2005 Reg)

Solution. **Calculation of Coefficient of Correlation**

X	x $(X-\bar{X})$	x^2	Y	y $(Y-\bar{Y})$	y^2	xy
25	4	16	10	–12	144	–48
40	19	361	25	3	9	57
30	9	81	40	18	324	162
25	4	16	15	–7	49	–28
10	–11	121	20	–2	4	22
05	–16	256	40	18	324	–288
10	–11	121	28	6	36	–66
15	–6	36	22	0	0	0
30	9	81	15	–7	49	–63
20	–1	1	05	–17	289	17
$\Sigma X = 210$	$\Sigma x = 0$	$\Sigma x^2 = 1090$	$\Sigma Y = 220$	$\Sigma y = 0$	$\Sigma y^2 = 1228$	$\Sigma xy = -235$

$$\bar{X} = \frac{\Sigma x}{N} = \frac{210}{10} = 21; \; \bar{Y} = \frac{\Sigma Y}{N} = \frac{220}{10} = 22$$

$$r = \frac{\Sigma xy}{\sqrt{\Sigma x^2 \times \Sigma y^2}} = \frac{-235}{\sqrt{1090 \times 1228}}$$

$$= \frac{-235}{\sqrt{1338520}} = \frac{-235}{1156.94} = -0.203 \text{ (app.)}$$

Example 27. Calculate the coefficient of correlation between x and y.

x	2	4	6	8
y	5	9	13	17

Solution. In the above example $y = 1 + 2x$.* The relationship is linear and the value of r would be +1. The following calculations would illustrate this :

* Pl. note.

Calculations of Coefficient of Correlation

X	*Dev. from assumed av. 5* dx	dx^2	Y	*Dev. from assumed av. 10* dy	dy^2	$dxdy$
2	–3	9	5	–5	25	+15
4	–1	1	9	–1	1	+1
6	+1	1	13	+3	9	+3
9	+3	9	17	–7	49	+21
$n = 4$	$\Sigma dx = 0$	$\Sigma dx^2 = 20$	$n = 4$	$\Sigma dy = +4$	$\Sigma dy^2 = 84$	$\Sigma dxdy = +40$

$$r = \frac{\Sigma\, dx\, dy - \dfrac{(\Sigma\, dx)\,(\Sigma\, dy)}{n}}{\sqrt{\Sigma\, dx^2 - \dfrac{(\Sigma\, dx)^2}{n}}\sqrt{\Sigma\, dy^2 - \dfrac{(\Sigma\, dy)^2}{n}}}$$

$$= \frac{+40\dfrac{(0 \times 4)}{4}}{\sqrt{20 - \dfrac{(0)^2}{4}}\sqrt{84 - \dfrac{(4)^2}{4}}} = \frac{+40}{\sqrt{20 \times 80}} = \frac{+40}{\sqrt{1600}} = \frac{+40}{40} = +1$$

Example 28. Calculate the values of $y = (x - 6)^5$ corresponding to $x = 1, 2, 3, 4$ and 5 and obtain the Correlation. Coefficient between x and y. Explain why the obtained Correlation differs from unity.

Solution. Values of y calculated on the basis of the equation $y = (x - 6)^5$ would be,

X	Y	
1	$(1 - 6)^5 = -3125$	Mean of $X = \frac{15}{5} = 3$
2	$(2 - 6)^5 = -1024$	
3	$(3 - 6)^5 = -243$	Mean of $Y = \frac{-4425}{5} = -885$
4	$(4 - 6)^5 = -32$	
5	$(5 - 6)^5 = -1$	
$\Sigma X = 15$	$\Sigma Y = -4425$	

Calculation of the coefficient of Correlation

X	*Dev. from mean (3)* x	x^2	Y	*Dev. from mean (–885)* y	y^2	xy
1	–2	4	–3125	–2240	5017600	+4480
2	–1	1	–1024	–130	19321	+139
3	0	0	–243	+642	412164	0
4	+1	1	–32	+853	727609	+853
5	+2	4	–1	+884	781556	+1768
$n = 5$	$\Sigma x = 0$	$\Sigma x^2 = 10$	$n = 5$	$\Sigma y = 0$	$\Sigma y^2 = 6958150$	$\Sigma xy = +7240$

Since deviations have been taken from actual arithmetic average, we shall use the Direct formula where,

$$r = \frac{\Sigma xy}{\sqrt{\Sigma x^2}\sqrt{\Sigma y^2}} = \frac{+7240}{\sqrt{10}\sqrt{6958150}} = \frac{+7240}{8341.55} = 0.8679$$

Although y is a function of x as $y = (x-6)^5$ yet it is not the equation of a straight line, *i.e.*, it is not a linear relationship. It is for this season that the **value of *r* is not unity**.

Example 29. The covariance of two perfectly correlated variables X and Y is 96. Determine σ_X and σ_Y if it is known that variance of X and that of Y is in the ratio of 4 : 9. **(D.U. Eco. (Hons.) 2008)**

Solution. Correlation coefficient

$$= 1 = \frac{\text{Cov.}(X, Y)}{\sigma_X \sigma_Y} = \frac{96}{\sigma_X \sigma_Y} \quad \ldots(1)$$

$$\frac{\text{Var. } X}{\text{Var. } Y} = \frac{4}{9} \Rightarrow \frac{\text{SD of } X}{\text{SD of } Y} = \frac{\sigma_X}{\sigma_Y} = \frac{2}{3} \Rightarrow \sigma_X = \frac{2}{9}\sigma_Y \quad \ldots(2)$$

From (1) and (2) we get

$$1 = \frac{96}{\frac{2}{3}\sigma_Y^2} \Rightarrow \frac{2}{3}\sigma_Y^2 = 96$$

$$\Rightarrow \sigma_Y^2 = \frac{93 \times 3}{2} = 144 \Rightarrow \sigma_Y = 12$$

$$\sigma_X = \frac{2}{3}\sigma_Y = \frac{2}{3} \times 12 = 8$$

Example 30. A computer while calculating the coefficient of correlation between two variables X and Y from 25 pairs of observations obtained the following results :

$$N = 25,\ \Sigma X = 125,\ \Sigma X^2 = 650,\ \Sigma Y = 100,\ \Sigma Y^2 = 460,\ \Sigma XY = 508.$$

It was however, discovered at the time of checking that it had copied down two pairs as :

X	Y		X	Y
6	14	While correct value were	8	12
8	6		6	8

Obtain the correct value of correlation coefficient. **(D.U. B.Com (Hons.) 2008)**

Solution. Correct $\Sigma X = 125 - 6 - 8 + 8 + 6 = 125$

Correct $\Sigma Y = 100 - 14 - 6 + 12 + 8 = 100$

Correct $\Sigma X^2 = 650 - 36 - 64 + 64 + 36 = 650$

Correct $\Sigma Y^2 = 460 - 196 - 36 + 144 + 64 = 436$

Correct $\Sigma XY = 508 - 84 - 48 + 96 + 48 = 520$

$$\text{Correct coefficient of correlation} = \frac{\dfrac{\Sigma XY}{n} - \dfrac{\Sigma X}{n}\dfrac{\Sigma Y}{n}}{\sqrt{\dfrac{\Sigma X^2}{n} - \left(\dfrac{\Sigma X}{n}\right)^2}\sqrt{\dfrac{\Sigma Y^2}{n} - \left(\dfrac{\Sigma Y}{n}\right)^2}}$$

$$= \frac{\frac{520}{25} - \frac{125}{25} \times \frac{100}{25}}{\sqrt{\frac{650}{25} - \left(\frac{125}{25}\right)^2} \sqrt{\frac{436}{25} - \left(\frac{100}{25}\right)^2}}$$

$$= \frac{0.8}{1.2} = 0.67$$

5.11 PROBABLE ERROR

Probable error of the coefficient of correlation (P.E.) $= 0.6745\left(\frac{1 - r^2}{\sqrt{n}}\right) = 0.6745$ (S.E.) where r stands for the coefficient of correlation and n for the number of pairs of observation and S.E., *i.e.,* standard error $= \frac{1 - r^2}{\sqrt{n}}$

According to Secrist *"The probable error of the correlation coefficient is an amount which if added to or substracted from the average correlation coefficient produces amounts within which the chances are even that a coefficient of correlation from a series selected at random will fall."*

According to Wheldon, *"probable error defines the limits above and below the size of the coefficient determined within which there is an equal chance that coefficient of correlation similarly calculated from other samples will fall."*

5.12 PROPERTIES OF PROBABLE ERROR

1. In the universe the limits of coefficient of correlation should be $r \pm P.E.$
2. If the value of r is less than 6 times the probable error, there is no evidence of correlation.
3. If the value of r is more than 6 times the probable error it is significant correlation.
4. Probable error as measure for interpreting coefficient of correlation should be used only when the number of pairs of observations is large. If n is small, probable error may give misleading results.

Conditions for the use of probable error :

1. The data must approximate a normal distribution.
2. Sampling has been done in a random manner.
3. The statistical measure for which the probable error is computed must have been calculated from a sample.

Note : Rank correlation and coefficient of concurrent deviations is not mentioned in the syllabus.

REGRESSION

5.13 INTRODUCTION

Regression means stepping back towards the average. In statistics regression analysis is applicable to all those fields where two or more related variables have the tendency to go back to the mean.

According to Blair, *"Regression is the measure of average relationship between two variables and to use it for predicting the most likely value of the dependent variable corresponding to a given known value of the independent variable. However, it may be noted that the regression relation is not*

reversible, *i.e., the regression equation used to predict the value of Y from a given value of X cannot be used to predict the values of X from the given value of Y.* The regression relation is average, irreversible and functional relation.

The two regression equations, *i.e.,* the regression equation of Y on X is used to predict the value of Y from the given value of X and the regression equation of X on Y is used to predict the value of X from the given value of Y. These two equations are derived under different set of conditions. Therefore one regression equation is not sufficient to find the values of both the variables X and Y.

5.14 EXAMPLES OF IRREVERSIBLE RELATION

1. If the income of a family increases the expenditure will increase but if the expenditure increases, there is no guarantee that the income will also increase.
2. If the rainfall is timely and good the crop will be good but if the crop is good there is no guarantee that the rainfall was timely and good.

The existence of relationship between two or more variables constitutes a vital information for decision making in any given situation. For example, it is important for a manufacturer to know how product sales are related to expenses incurred on advertising. Similarly a farmer may like to know the relationship between crop yield and the quantum of fertilizer applied.

5.15 DIFFERENCE BETWEEN REGRESSION AND CORRELATION

The correlation coefficient is merely concerned with determining how strongly (or closely) two variables are linearly related. It is not capable of solving Prediction Problems. Prediction Problems are solved by regression methods. Regression and correlation are basically the twin related aspects of a single problem of examining the relationship between two or more variables. *Correlation precedes regression because if the relationship between the variables is not sufficient strong, there would appear no sound basis for prediction.*

The value of correlation coefficient indicates how well the regression line actually fits the available paired data. If $r = \pm 1$, then the regression line fits the data perfectly. If r is near +1 or –1, the regression line constitutes a very good approximation of the data. But if r is near zero, the regression line fits poorly. When we are to estimate the value of one variable given some value of the other, **we should use the equation of regression line only if r indicates that there is a significant linear correlation. Otherwise we should not use the regression equation for projecting or predicting**.

Correlation coefficient is a symmetrical measure of linear relationship because even if the two variables are interchanged, the value of the correlation coefficient remains unchanged but this symmetry is not present when we deal with the regression equations, *i.e.*, the regression equation used to predict Y by keeping X as independent variable can not be used to predict X by keeping Y as independent variable. The regression represents an irreversible functional relation. Coefficient of correlation is independent of the change of origin and scale whereas coefficient of regression is independent of the change of origin only*.

Coefficient of correlation r satisfies the relation $-1 \le r \le 1$ but the product of two regression coefficients is equal to r^2 which satisfies the relation $0 \le r^2 \le 1$.

* Coefficient of regression becomes independent of the change of scale only if scale of X and scale of Y are changed in the same manner.

Assumptions Underlying Linear Regression

1. For each value of X, there is a group of Y values and these Y values are normally distributed.
2. The means of these normal distributions of Y values will lie on the straight line of regression.
3. The standard deviations of these normal distributions are equal.
4. The Y values are statistically independent. This means that in the selection of a sample, the Y values chosen for a particular X value do not depend on the Y values for any other X value.

5.16 REGRESSION LINES

J.R. Stockton says, "the device used for estimating the value of one variable from the value of the other consists of a line through the points drawn in such a manner as to represent the average relationship between the two variables. Such a line is called the line of **regression**."

As per the method of least squares, the two regression lines are

$$Y - \overline{Y} = b_{YX}(X - \overline{X}) \qquad \ldots(I)$$

and

$$X - \overline{X} = b_{XY}(Y - \overline{Y})$$

where $\overline{X}$ = mean of X values; $\overline{Y}$ = mean of Y values

$$b_{YX} = \text{Regression coefficient of } Y \text{ on } X = r\frac{\sigma_Y}{\sigma_X}.$$

$$b_{XY} = \text{Regression coefficient of } X \text{ on } Y = r\frac{\sigma_X}{\sigma_Y}.$$

$$b_{YX} = r\frac{\sigma_Y}{\sigma_X} = \frac{\Sigma(X-\overline{X})(Y-\overline{Y})}{n\sigma_X\sigma_Y} \times \frac{\sigma_Y}{\sigma_X}$$

$$= \frac{\Sigma(X-\overline{X})(Y-\overline{Y})}{n\sigma_X^2} = \frac{\Sigma xy}{\Sigma x^2}$$

where $x = X - \overline{X}$, $y = Y - \overline{Y}$, *i.e.*, the deviations are taken from the actual mean of x and y.

$$b_{YX} = \frac{\frac{\Sigma XY}{n} - \left(\frac{\Sigma X}{n}\right)\left(\frac{\Sigma Y}{n}\right)}{\frac{\Sigma X^2}{n} - \left(\frac{\Sigma X}{n}\right)^2} \text{ if no deviations are taken}$$

$$= \frac{\frac{\Sigma d_X d_Y}{n} - \frac{\Sigma d_X}{n}\frac{\Sigma d_Y}{n}}{\frac{\Sigma d_X^2}{n} - \left(\frac{\Sigma d_X}{n}\right)^2}$$

where $d_X = X - A$ and $d_Y = Y - B$,* *i.e.*, the deviations are taken from an assumed mean of X and assumed mean of Y.

Likewise $b_{XY} = \frac{\Sigma xy}{\Sigma y^2}$ where $x = X - \overline{X}$, $y = Y - \overline{Y}$

* *A* and *B* may be equal or different.

$$= \frac{\frac{\Sigma XY}{n} - \left(\frac{\Sigma X}{n}\right)\left(\frac{\Sigma Y}{n}\right)}{\frac{\Sigma Y^2}{n} - \left(\frac{\Sigma Y}{n}\right)^2} \text{ if deviations are not taken}$$

$$= \frac{\frac{\Sigma d_X d_Y}{n} - \frac{\Sigma d_X}{n} \frac{\Sigma d_Y}{n}}{\Sigma \frac{d_Y^2}{n} - \left(\frac{\Sigma d_Y}{n}\right)^2} \text{ where } d_X = X - A, d_Y = Y - B.$$

5.17 WHY DO WE NEED TWO REGRESSION LINES TO FIND THE VALUE TWO VARIABLES *X* AND *Y*

Since the regression relation is irreversible one equation is not sufficient to predict the values of two variables X and Y. Moreover the two regression equations are derived under different set of assumptions, therefore, one equation is not sufficient to find X and Y.

5.18 PROPERTIES OF REGRESSION COEFFICIENTS

(1) Both the regression coefficients should be of the same sign :

$$(b_{YX})\,(b_{XY}) = \left[r\frac{\sigma_Y}{\sigma_X}\right]\left[r\frac{\sigma_X}{\sigma_Y}\right] = r^2$$

Since, the square of a quantity is always positive, r^2 is always positive, *i.e.*, the product of two regression coefficients is always positive which is possible only if the two regression coefficients are of the same sign, *i.e.*, either both positive or both negative.

(2) Determination of Sign of Correlation Coefficient from Regression Coefficients : If both the regression coefficients are positive, correlation coefficient is positive and if both the regression coefficients are negative the correlation coefficient is negative.

(3) Change of Origin and Change of Scale : Both the regression coefficients are independent of the change of origin but if the change of scale in X and Y is not identical, they depend on the change of scale. If the change of scale in X and Y is identical, the regression coefficients are independent of the change of scale also.

(4) Both the Regression Coefficients cannot be more than 1 : Since, the product of the two regression coefficients is equal to r^2 which is less than or equal to one, both the regression coefficients cannot be more than 1. Sometimes it may happen that one of the regression coefficients is less than 1 but very close to 1 and the other is more than 1 in such a way that the product of two regression coefficients is more than 1. Such cases are also not possible. The product of two regression coefficients should not be more than 1.

5.19 REGRESSION COEFFICIENTS DENOTE THE RATE OF CHANGE

$$Y - \bar{Y} = b_{YX}(X - \bar{X})$$

$$= (b_{YX})X - (b_{YX})\bar{X}$$

$$\Rightarrow \quad Y = (b_{YX})X + [\bar{Y} - (b_{YX})\bar{X}]$$

Since $\overline{X}$, $\overline{Y}$, b_{YX} are all constants, Equation (I) can be written as

$$Y = [b_{YX}]\, X + C \text{ where } C = [\overline{Y} - b_{YX}\, \overline{X}]$$

$\therefore$ b_{YX} gives the rate of change of Y w.r.t. X.

The equation of the regression line representing regression of X on Y is

$$(X - \overline{X}) = b_{XY}\,(Y - \overline{Y}) \qquad \text{...(II)}$$

Equation (II) can be written in the form

$$X = (b_{XY})\, Y + C$$

Hence, (b_{XY}) gives the rate of change of X w.r.t. Y.

5.20 PROPERTIES OF REGRESSION LINES

(1) The A.M. of X and A.M. of Y lies on the Regression Lines

The equation of two regression lines are

$$Y - \overline{Y} = b_{YX}\,(X - \overline{X}) \quad \text{and} \quad X - \overline{X} = b_{XY}\,(Y - \overline{Y})$$

Clearly the mean of X and mean of Y lies on the regression lines.*

If the regression equations are given, then to find the mean of X and mean of Y, solve the two equations as simultaneous equations in X and Y and the values of X and Y so calculated give the mean of X and mean of Y.

(2) If $r = 0$, the two Regression Lines are perpendicular to each other

The equations of two regression lines are

$$Y - \overline{Y} = b_{YX}\,(X - \overline{X}) = r\frac{\sigma_Y}{\sigma_X}\,(X - \overline{X}) \text{ and } X - \overline{X} = b_{XY}\,(Y - \overline{Y}) = r\frac{\sigma_X}{\sigma_Y}\,(Y - \overline{Y})$$

If $r = 0$, the two equations become

$$Y - \overline{Y} = 0 \text{ or } Y = \overline{Y}$$

and

$$X - \overline{X} = 0 \text{ or } X = \overline{X}$$

From the graph it is clear that $Y = \overline{Y}$ and $X = \overline{X}$ are perpendicular to each other.

(3) If two regression lines are identical, the correlation between the variables is Perfect

The equations of the two regression lines are

$$Y - \overline{Y} = b_{YX}\,(X - \overline{X}) \qquad \text{...(I)}$$

$$X - \overline{X} = b_{XY}\,(Y - \overline{Y}) \qquad \text{...(II)}$$

From (II),

$$Y - \overline{Y} = \frac{X - \overline{X}}{b_{XY}}$$

* In the two regression equation put $X = \overline{X}$, $Y = \overline{Y}$. The two equations are satisfied
$\therefore$ $\overline{X}$, $\overline{Y}$ lies on the regression lines.

If the two regression lines are identical, the value of $Y - \overline{Y}$ calculated from (I) and (II) should be same,

i.e.,
$$b_{YX}(X - \overline{X}) = \frac{(X - \overline{X})}{b_{XY}}$$

$$\Rightarrow \quad [b_{YX}][b_{XY}] = 1 \Rightarrow r^2 = 1 \qquad [\because [b_{YX}][b_{XY}] = r^2]$$

$$\Rightarrow \quad r = \pm 1$$

⇒ The correlation is perfect.

⇒ If the two regression lines are identical the correlation between the two variables is perfect, either perfect positive or perfect negative.

(4) Angle between the two Regression Lines : The two regression equations are

$$Y - \overline{Y} = b_{YX}(X - \overline{X}) = r\frac{\sigma_Y}{\sigma_X}(X - \overline{X}) \text{ and } X - \overline{X} = b_{XY}(Y - \overline{Y})$$

$$\Rightarrow \quad Y - \overline{Y} = \frac{1}{b_{XY}}(X - \overline{X}) = \frac{\sigma_Y}{r\,\sigma_X}(X - \overline{X})$$

If m_1 and m_2 are the slopes of (I) and (II), then

$$m_1 = \frac{r\sigma_Y}{\sigma_X}, \; m_2 = \frac{\sigma_Y}{r\sigma_X}$$

If θ is the angle betweeen the two regression lines, then

$$\tan\theta = \frac{m_1 - m_2}{1 + m_1 m_2} = \frac{\frac{r\sigma_Y}{\sigma_X} - \frac{\sigma_Y}{r\sigma_X}}{1 + \frac{r\sigma_Y}{\sigma_X} \times \frac{\sigma_Y}{r\sigma_X}} = \left[\frac{\sigma_X\sigma_Y}{\sigma_X^2 + \sigma_Y^2}\left(\frac{r^2 - 1}{1}\right)\right]$$

If θ is acute, then $\tan\theta = \left[\frac{\sigma_X\sigma_Y}{\sigma_X^2 + \sigma_Y^2}\left(\frac{1 - r^2}{r}\right)\right]$* $\qquad [\because r^2 \le 1]$

If $r = \pm 1$, $\tan\theta = 0 \Rightarrow \theta = 0 \Rightarrow$ the two lines are identical, coincident or parallel.

If $r = 0$ *i.e.*, if the variables are independent then $\tan\theta = \alpha \Rightarrow \theta = \frac{\pi}{2}$

Two regression lines are perpendicular to each other.

5.21 ERROR OF ESTIMATE : STANDARD ERROR OF ESTIMATE

It may be noted that the estimate of Y, *i.e.*, Y_c derived from a regression equation

$$Y_c - \overline{Y} = b_{YX}(X - \overline{X})$$

will not be exactly the same as the Y-value which are actually observed. The difference between the estimated value of Y, *i.e.*, Y_c and the actual value of Y will depend on the extent of scatter of various points around the line of best fit (Regression Line). The smaller the difference, the greater is the precision of the estimate and vice versa. The square root of the A.M. of the squarred differences measure the error in the estimates called the error of estimates.

* If θ is acute, $\tan\theta$ is positive. Since $0 \le r^2 \le 1$ and S.D. is always positive, $r^2 - 1$ will be either 0 or negative but $1 - r^2$ will be either 0 or positive.

Error of Estimate = $\left|\sqrt{\dfrac{\Sigma (Y - Y_c)^2}{N}}\right|$ = Average absolute amount by which observed Y-value departs from the correspond computed Y_c value.

If Y is the observed value of Y and Y_c, the corresponding value on the least square line (or regression line), the unbiased estimate of the true variance of Y values is

$$\frac{\Sigma (Y - Y_c)^2}{N - 2}$$

Here we lose 2 degrees of freedom because in calculating Y_c from the equation $Y_c = a + bX$, we have to calculate b_{YX}, *i.e.*, the slope of the regression line and

$b_{YX} = \dfrac{\Sigma (X - \overline{X})(Y - \overline{Y})}{\sigma_x^2}$, *i.e.*, to calculate b_{YX} we need the mean of X and mean of Y.

Specifying these two means places two restrictions on the variability of the data and one degree of freedom is lost for each mean.

It is customary to refer $\sqrt{\dfrac{\Sigma (Y - Y_c)^2}{N - 2}}$ as **Standard error of estimate.**

The following easier computational formula is used for standard error of estimate. Standard error of estimate of regression of Y values from Y_c

$$S_{YX} = \sqrt{\frac{\Sigma Y^2 - a\Sigma Y - b\Sigma (XY)}{N - 2}} = \sigma_Y \sqrt{1 - r^2}$$

The standard error of estimate of X on Y

$$S_{XY} = \sqrt{\frac{\Sigma X^2 - a'\Sigma X - b'\Sigma XY}{N - 2}} = \sigma_X \sqrt{1 - r^2}$$

The standard error of estimate measures the variability or scatter of the **observed values** of the dependent variable around the regression line. In the extreme case, when there is no scatter all the Y observations will fall on the regression line and the vertical deviations will all be zero. This will indicate that the true variability of Y is zero and the regression line would provide perfect predictions.

5.22 TOTAL VARIANCE, EXPLAINED VARIANCE, UNEXPLAINED VARIANCE, COEFFICIENT OF DETERMINATION, COEFFICIENT OF ALIENATION

The values of Y given in the original data are the observed values of Y. The values of Y estimated or predicted by the regression equation are called **estimated or predicted values** of Y written as Y_c or Y_e.

$$\text{Total variance} = \frac{\Sigma (Y - \overline{Y})^2}{N} = \frac{\Sigma (Y - \overline{Y} + Y_c - Y_c)^2}{N}$$

$$= \frac{\Sigma [(Y - Y_c) - (\overline{Y} - Y_c)]^2}{N}$$

$$= \frac{\Sigma (Y - Y_c)^2}{N} + \frac{(\overline{Y} - Y_c)^2}{N} - \frac{2\Sigma (Y - Y_c)(\overline{Y} - Y_c)}{N}$$

$$= \frac{\Sigma (Y - Y_c)^2}{N} + \frac{\Sigma (\overline{Y} - Y_c)^2}{N} *$$

* It can be shown that $\Sigma (Y - Y_c)(\overline{Y} - Y_c) = 0$. But the proof is beyond the scope of this work.

= Unexplained Variance + Explained Variance*

$$\Rightarrow \quad \frac{\text{Explained Variance}}{\text{Total Variance}} = \frac{\Sigma(\bar{Y}-Y_c)^2}{\Sigma(Y-\bar{Y})^2} = 1 - \frac{\text{Unexplained Variance}}{\text{Total Variance}} \quad ...(A)$$

$$\frac{\text{Explained Variance}}{\text{Total Variance}} = \frac{\Sigma(\bar{Y}-Y_c)^2}{\Sigma(Y-\bar{Y})^2} = \text{Coefficient of determination, } i.e., \; r^2** \quad ...(B)$$

⇒ Coefficient of determination is proportional to the total variation in Y explained by the regression line.

Since r lies between −1 and 1, r^2 lies between 0 and 1 both inclusive.

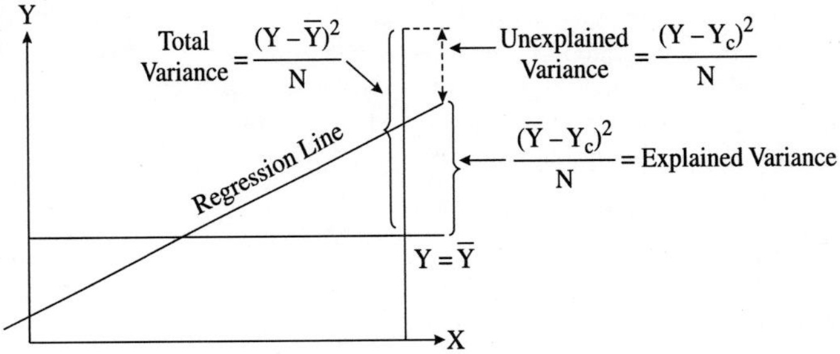

From (A) and (B) we get unexplained variance $= \frac{\Sigma(Y-Y_c)^2}{N} = (1-r^2)$ Total Variance.

Ratio of the unexplained variance to the total variance is called coefficient of **non-determination** and the square root of the coefficient of non-determination is called **coefficient of alineation**,

i.e.,
$$\frac{\text{Unexplained Variance}}{\text{Total Variance}} = \text{Coefficient of non-determination}$$

$$= \frac{\Sigma(Y-Y_c)^2}{\Sigma(Y-\bar{Y})^2}$$

and
$$\text{Coefficient of alination} = \sqrt{\frac{\Sigma(Y-Y_c)^2}{\Sigma(Y-\bar{Y})^2}}$$

To study the true relationship between two variables in regression analysis, it is always recommended to employ the coefficient of determination rather than the coefficient of correlation. Since r is a fraction, *i.e,* $|r| \le 1$, r^2 always satisfies the inequality $0 \le r^2 \le 1$. But the direction of relationship between the two variables is always determined by the coefficient of correlation or slope of the regression lines.

* The amount of vertical scatter of the data points away from the regression line gives an idea about the variance of Y which we are not able to explain with the help of the regression line. This variance is called **unexplained variance.** This unexplained variance (variation) arises when all the data points, *i.e.,* points on the scatter diagram do not lie on the regression line. However, if all the data points lie on the regression line, the unexplained variance is equal to zero and the entire variance in Y is attributed to variaions in X, *i.e.,* when all the data points lie on the regression line, the entire variation (or variance) in Y is *explained variance.* But the regression line does not always explain the total variance.

** Coefficient of determination is also calculated by the formula $r^2 = \frac{a\Sigma Y + b\Sigma XY - n\bar{Y}^2}{\Sigma Y^2 - n\bar{Y}^2}$

Note : IN Predicting the values of Y without any knowledge of X, our best estimate would be $\bar{Y}$, *i.e.*, mean of Y. The difference $(Y - \bar{Y})$ is called an error of estimate. By introducing known values of X into a regression Equation we attempt to reduce this error and the result is $(Y_c - \bar{Y})$ where Y_c is the value of Y calculated from regression Equation. The total improvement based on several estimates is $\Sigma (Y_{ci} - \bar{Y})^2$. The amount of variation explained by the relationship between X and Y is the regression. The coefficient of determination is the ratio of the line of best fit's squared error over that incurred by using $\bar{Y}$ [the total squared error of several predictions using $\bar{Y}$ would be $\Sigma (Y_i - \bar{Y})^2$].

Properties of Coefficient of Determination

(1) As an index of fit it is interpreted as the total proportion of variance in Y explained by X.
(2) As a measure of linear relationship it tells us how well the regression line fits the data.
(3) As an important indicator of the predictive accuracy of the regression Equation, the minimum value of r^2 should be 0.8 otherwise the predictive accuracy is considered low.
(4) The coefficient of determination r^2 is a descriptive measure of the strength of the regression relationship, a measure of how well the regression line fits the data.

Regression and Correlation Correspondence

1. The regression equation of y and x is

$$y = a + bx$$

If $b = 0$, $y = a$ i.e., the regression lines becomes horizontal [Parallel to X-axis], b is the slope of the regression line, therefore if slope is zero the regression line is parallel of X-axis *i.e.*, what ever way the X-may change. There is no change in Y *i.e.*, X and Y are independent and there is no linear correlation between X and Y or $r = 0$ [otherwise also $b = r\dfrac{\sigma_Y}{\sigma_X} = 0 \Rightarrow r = 0$]

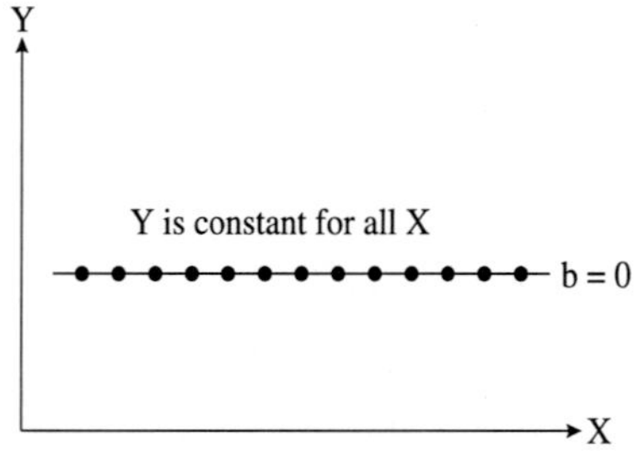

2. When the two variables are *uncorrelated i.e.*, When the correlation between X and Y is zero. Then as X increases Y may increase, or it may decrease, or it may remain constant. There is no *systematic* increase or decrease in the values of Y as X increases. This case is shown in Figure below. As can be seen in the figure, data from this process are not "moving" in any pattern, thus, the line has no direction to follow. With no direction, the slope of the line is, again, zero.

On the other hand, if all observed points are on the regression line, standard error of estimate (computed in regression analysis) is 0 and the correlation is perfect. Therefore, $S_e = 0$ means that $r = 1$ or $r = -1$ *i.e.,* if $S_e = 0$, all the points lie on the regression line and unexplained variance is zero and the coefficient of determination is 1 which implies $r = \pm 1$ i.e., the correlation is perfect.

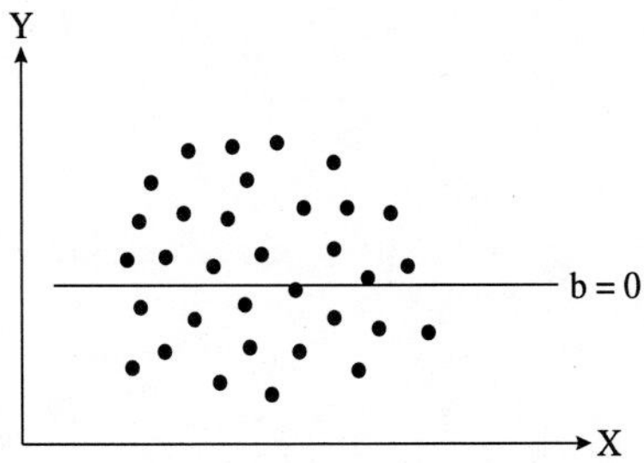

Y is uncorrelated with *X*, *Y* may be either large or small when *X* is large; *Y* may be large or small when *X* is small. There is no systematic trend in *Y* as *X* increases.

Therefore, we can conclude that regression and correlation are the two sides of the same coin.

SOLVED EXAMPLES

Example 1. Calculate the regression equation of *Y* or *X* from the following data :

X	40	38	35	42	30
Y	30	35	40	36	29

(Punjab Univ. B. Com., April 2001)

Solution.

Y	X	$Y-\overline{Y}=y$	$X-\overline{X}=x$	xy	y^2	x^2
30	40	–4	3	–12	16	9
35	38	1	1	1	1	1
40	35	6	–2	–12	36	4
36	42	2	5	10	4	25
29	30	–5	–7	35	25	49
				$\Sigma xy = 22$	$\Sigma y^2 = 82$	$\Sigma x^2 = 88$

Here $n = 5$

$$\Sigma X = 176, \overline{X} = \frac{\Sigma X}{5} = \frac{170}{5} = 34,\ \Sigma Y = 185, \overline{Y} = \frac{185}{5} = 37$$

Since, the mean of *X* and mean of *Y* is a whole number, the deviations have been taken from the actual mean of *X* and actual mean of *Y*.

$$b_{YX} = r\frac{\sigma_Y}{\sigma_X} = \frac{\Sigma xy}{\Sigma x^2} = \frac{22}{88}$$

∴ The regression equation of *Y* on *X* is

$$Y - 37 = \frac{22}{88}(X - 34) \Rightarrow Y = \frac{22}{88}X - \frac{22 \times 34}{88} + 37 = 0.25X - 28.5$$

Example 2. Out of the two regression equations given below which one is the regression of X on Y and which one the regression of Y on X.

$$2X + 4Y = 10 \qquad ...(I)$$

$$4X + 6Y = 8 \qquad ...(II)$$

Solution. To decide which of the two given equations represents the regression of Y on X and which represents the regression of X on Y we follow the following steps :

(1) Assume any one of the equations as regression of Y on X and arrange in the form $Y = a_1 + b_1X$.
Naturally the other equation will represent the regression of X on Y.

(2) Arrange this second equation as $X = a_2 + b_2Y$.

(3) If b_1b_2 is less than or equal to 1 our assumption is correct. If b_1b_2 is greater than 1 then we interchange our assumption $[b_1b_2 = r^2$ and $0 > r^2 \leq 1]$.

Let (I) represent the regression of X on Y and (II) the regression of Y on X then the equations can be written as

$$X = 5 - 2Y,\ b_{XY} = -2 \text{ and } Y = \frac{4}{3} - \frac{2}{3}X,\ b_{YX} = \frac{-2}{3}$$

$$(b_{YX}) \times (b_{XY}) = r^2 = -2 \times \left(-\frac{2}{3}\right) = \frac{4}{3}$$

$\Rightarrow r = 1.15$ which is greater than 1.

Hence, our assumption is wrong.

If (I) represents the regression of Y on X and (II) that of X on Y, then

$$Y = -\frac{1}{2}X + \frac{5}{2} \text{ and } X = \frac{-3}{2}Y + 2$$

$$r^2 = \frac{-1}{2} \times \left(\frac{-3}{2}\right) = \frac{3}{4} \text{ which is possible.}$$

Hence, the equation $2X + 4Y = 10$ represents the regression of Y on X and the other represents the regression of X on Y.

Example 3. For 100 students of a class, the regression equation of marks in Statistics (X) on the marks in Commerce (Y) is $3Y - 5X + 180 = 0$. The mean marks in Commerce is 50 and variance of marks in Statistics is $\frac{4}{9}$th of the variance of marks in Commerce. Find the mean marks in Statistics and the coefficient of correlation between marks in the two subjects. **(CA Foundation, May 2000)**

Solution.

$$3Y - 5X + 180 = 0$$

Since this equation is given to be regression of X or Y, we arrange this equation as under :

$$5X = 3Y + 180 \Rightarrow X = \frac{3}{5}Y + \frac{180}{5} = \frac{3}{5}Y + 36 \qquad ...(I)$$

$$b_{XY} = \frac{3}{5} = r\frac{\sigma_X}{\sigma_Y} \qquad ...(II)$$

Variance of $X = \frac{4}{9}$ variance of Y

$$\Rightarrow \qquad \frac{S.D_X}{S.D_Y} = \frac{\sigma_X}{\sigma_Y} = \frac{2}{3} \qquad \left[\because \frac{\sigma_X^2}{\sigma_Y^2} = \frac{4}{9}\right]$$

$$\therefore \qquad \frac{3}{5} = r \times \frac{2}{3}$$

$$\Rightarrow \qquad r = \frac{3 \times 3}{5 \times 2} = \frac{9}{10} = 0.9$$

Since b_{XY} is positive, r is positive.

Since the mean of X and mean of Y lie on the regression lines, we have

$$\overline{X} = \frac{3}{5}\overline{Y} + 36 \Rightarrow \overline{X} = \frac{3}{5} \times 50 + 36 = 30 + 36 = 66.$$

Example 4. From the following information on values of two variables x and y, find the two regression lines and the correlation coefficient.

$$n = 10, \Sigma x = 20, \Sigma y = 40, \Sigma x^2 = 240, \Sigma y^2 = 410, \Sigma xy = 200 \qquad \textbf{(ICWA, June 2001)}$$

Solution.

$$\text{Regression coefficient of } y \text{ on } x = \frac{\dfrac{\Sigma xy}{n} - \dfrac{\Sigma x}{n}\dfrac{\Sigma y}{n}}{\dfrac{\Sigma x^2}{n} - \left(\dfrac{\Sigma x}{n}\right)^2} = \frac{\dfrac{200}{10} - \left(\dfrac{20}{10}\right)\left(\dfrac{40}{10}\right)}{\dfrac{240}{10} - \left(\dfrac{20}{10}\right)^2}$$

$$= \frac{20 - 2 \times 4}{24 - 4} = \frac{12}{20} = \frac{3}{5} = b_{yx}$$

$$\text{Regression coefficient of } x \text{ on } y = \frac{\dfrac{\Sigma xy}{n} - \dfrac{\Sigma x}{n}\dfrac{\Sigma y}{n}}{\dfrac{\Sigma y^2}{n} - \left(\dfrac{\Sigma y}{n}\right)^2} = \frac{\dfrac{200}{10} - \left(\dfrac{20}{10}\right)\left(\dfrac{40}{10}\right)}{\dfrac{410}{10} - \left(\dfrac{40}{10}\right)^2}$$

$$= \frac{20 - 8}{41 - 16} = \frac{12}{25} = b_{xy}$$

Mean of $x = \overline{x} = \dfrac{20}{10} = 2$, mean of $y = \overline{y} = \dfrac{40}{10} = 4$

The two regression lines are

$$y - \overline{y} = b_{yx}(x - \overline{x}), \text{ i.e., } y - \overline{y} = \frac{3}{5}(x - 2) \Rightarrow y - 4 = 0.6x - 1.2 \Rightarrow y = 0.6x - 2.8$$

$$\text{and } x - \overline{x} = b_{xy}(y - \overline{y}), \text{ i.e., } x - 2 = \frac{12}{25}(y - 4) \Rightarrow x = \frac{12}{25}y + \frac{2}{25}$$

$$b_{yx} \times b_{xy} \Rightarrow \frac{3}{5} \times \frac{12}{25} = r^2 \text{ or } r = \sqrt{\frac{36}{125}} = 0.536$$

Since b_{yx} and b_{xy} are both positive, r will be positive.*

Example 5. The lines of regression of Y on X and X on Y are respectively

$$Y = X + 5 \text{ and } 16X - 9Y = 94.$$

Find the variance of X if the variance of Y is 16. Also find the covariance of X and Y.

---

* $$(b_{yx})(b_{xy}) = r^2 \Rightarrow \sqrt{(b_{yx})(b_{xy})} = \pm r$$

We know that both the regression coefficients must be of the same sign. Therefore r will be +ve if b_{yx} and b_{xy} is +ve and –ve if b_{yx} and b_{xy} are negative.

Solution. Regresson equation of Y on X is : $Y = X + 5 \Rightarrow b_{YX} = 1$

Regression equatin of X on Y is : $16X - 9Y = 94$, *i.e.,* $16X = 9Y + 94$

$$\therefore \qquad X = \left(\frac{9}{16}\right)Y + \frac{94}{16} \Rightarrow b_{XY} = \frac{9}{16}$$

$$\therefore \qquad r^2 = (b_{YX})\,(b_{XY}) = 1 \times \frac{9}{16} \Rightarrow r = \frac{3}{4}$$

$$b_{XY} = r \cdot \frac{\sigma_X}{\sigma_Y} \Rightarrow \sigma_X = \frac{(b_{XY}) \cdot \sigma_Y}{r} = \frac{(9/16) \cdot (4)}{(3/4)} = 3$$

$$r = \frac{\text{cov}\,(X, Y)}{\sigma_X \cdot \sigma_Y} \Rightarrow \text{Cov}\,(X, Y) = (r) \cdot (\sigma_X) \cdot (\sigma_Y) = \frac{3}{4} \times 3 \times \sqrt{16} = 9.$$

Example 6. Find the regression coefficient of y on x from the following regression equations.

$$5x = 22 + y;\; 64x = 24 + 45y$$

It is possible to calculate the standard deviation of y from the given information ? Answer with reason.

Solution. Since we don't know which one of the equations is the regression of x on y let us choose one of the equations arbitrarily.

Let the regression equation of x on y be

$$5x = 22 + y,\; i.e.,\; x = \frac{22}{5} + \left(\frac{1}{5}\right)x; \qquad \left[\therefore\; b_{xy} = \frac{1}{5}\right]$$

Hence the regression equation of y on x is

$$64x = 24 + 45y \text{ or } 45y = 64x - 24$$

or
$$y = \left(\frac{64}{45}\right)x - \frac{24}{45};$$

$$\therefore\; b_{xy} = \frac{64}{45} \;\therefore\; r^2 = (b_{xy})\,(b_{yx}) = \frac{1}{5} \times \frac{64}{45} = \frac{64}{225}$$

Since the condition $r^2 \le 1$ is satisfied our assumption is correct. (If the assumption is not correct, then r^2 will be greater than 1 in which case we have to reverse the assumption and then proceed.)

$$b_{yx} = \frac{64}{45}$$

To find σ_y we require the value of σ_x in which case we could have put the values of r and σ_x in either $b_{xy} = \frac{r\sigma_x}{\sigma_y}$ or $b_{yx} = r\frac{\sigma_y}{\sigma_x}$

Since σ_x is not given, it is not possible to find the value of σ_y.

Example 7. The following sums have been obtained from 100 observation pairs :

$$\Sigma\, x = 12{,}500 \qquad \Sigma\, x^2 = 15{,}85{,}000$$
$$\Sigma\, y = 8{,}000 \qquad \Sigma\, y^2 = 6{,}48{,}100$$
$$\Sigma\, xy = 10{,}07{,}425$$

(i) Find regression of y on x and estimate the value of y, when $x = 130$.

(ii) Compute the correlation coefficient (r) between x and y and state what you learn from the value of r obtained by you.

Solution. Cov. $(x, y) = \frac{\Sigma xy}{n} - \left(\frac{\Sigma x}{n}\right)\left(\frac{\Sigma y}{n}\right) = \frac{107425}{100} - \left(\frac{12500}{100}\right)\left(\frac{8000}{100}\right)$

$= 10074.25 - 10000 = 74.25$

$$\sigma_x = \sqrt{\frac{\Sigma x^2}{n} - \left(\frac{\Sigma x}{n}\right)^2} = \sqrt{\frac{1585000}{100} - \left(\frac{12500}{100}\right)^2} = \sqrt{15850 - (125)^2}$$

$$= \sqrt{15850 - 15625} = \sqrt{225} = 15$$

$$\sigma_y = \sqrt{\frac{648100}{100} - \left(\frac{8000}{100}\right)^2} = \sqrt{6481 - 6400} = \sqrt{81} = 9$$

$$\therefore \quad r = \frac{\text{Cov.}(x, y)}{\sigma_x \sigma_y} = \frac{74.25}{15 \times 9} = 0.55.$$

Regression equation of y on x is $y - \bar{y} = b_{yx}(x - \bar{x})$

$$\Rightarrow \quad y - 80 = 0.55\left(\frac{9}{15}\right)(x - 125)$$

$$\Rightarrow \quad y - 80 = 0.33x - 41.25$$

$$\therefore \quad y = 0.33x + 38.75$$

If $x = 130$, $y = 0.33 \times 130 + 38.75 = 81.65$

Example 8. Given the regression equation of y on x and x on y are respectively $y = 2x$ and $6x - y = 4$ and the second moment of x about the origin is 3. (i) Find the correlation coefficient and (ii) standard deviation of y.

Solution. (i) Regression equation of y on x is $y = 2x \Rightarrow b_{yx} = 2$

Regression equation of x on y is $6x - 4 = 4 \Rightarrow x = \frac{1}{6}y + \frac{4}{6} \Rightarrow b_{xy} = \frac{1}{6}$

$$\therefore \quad r = \sqrt{2 \times \frac{1}{6}} = \sqrt{0.3333} = 0.578$$

(ii) Second raw moment of x [*i.e.,* second moment about origin] = 3 (given)

i.e.,
$$\frac{\Sigma x^2}{n} = 3$$

$$\sigma_x^2 = \frac{\Sigma x^2}{n} - \left(\frac{\Sigma x}{n}\right)^2 = 3 - (\bar{x})^2$$

Solving the two regression equations, we get $\bar{x} = 1, \bar{y} = 2$*

$$\therefore \quad \sigma_x^2 = 3 - 1 = 2$$

$$\therefore \quad \sigma_x = \sqrt{2} = 1.414$$

$$b_{yx} = r\frac{\sigma_y}{\sigma_x}$$

i.e.,
$$2 = 0.578.\frac{\sigma_y}{\sqrt{2}}$$

* By solving the given equation we get the values of x and y. Since $\bar{x}, \bar{y}$ lie on the regression lines, the values of x and y are also the values of $\bar{x}$ and $\bar{y}$.

i.e., $$\sigma_y = \frac{2\sqrt{2}}{0.578} = \frac{2.828}{0.578} = 4.89$$

Example 9. Consider the two regression lines : $3x + 2y = 26$ and $6x + y = 31$.

(i) Find the mean values and correlation coefficient between x and y.* Find the ratio of the variances of the variables.

(ii) Show the regression estimate of y for $x = 0$ is 13 whereas regression estimate of x corresponding to $y = 13$ is 3. Explain the cause of difference.

(AIMA June 1997, ICWA Inter 1998, CA Foundation May 2001)

Solution. (i) Two regression lines intersect at $(\bar{x}, \bar{y})$

Two regression lines are : $$3x + 2y = 26 \quad ...(1)$$
$$6x + y = 31 \quad ...(2)$$

Solving (1) and (2), we get $x = 4, y = 7$.

$\therefore$ $x = 4$ and $y = 7$ (x, y is also the point of intersection of the given lines)

Regression line of y on x is $y = -\frac{3}{2}x + \frac{26}{2}$; $\therefore b_{yx} = -\frac{3}{2}$

Regression line of x on y is $x = -\frac{1}{6}y + \frac{31}{6}$; $\therefore b_{xy} = -\frac{1}{6}$

Correlation coefficient $r_{xy} = -\sqrt{b_{yx} \times b_{xy}}$

$$= -\sqrt{\left(-\frac{3}{2}\right) \times \left(-\frac{1}{6}\right)} = -\frac{1}{2}$$

(r_{xy} is –ve since both the regression coefficients are negative)

$$\frac{b_{yx}}{b_{xy}} = \frac{r\sigma_y}{\sigma_x} \times \frac{\sigma_y}{r\sigma_x} = \frac{\sigma_y^2}{\sigma_x^2} = \frac{3}{2} \times \frac{6}{1} = 9$$

$$= \frac{\text{Variance of } y}{\text{Variance of } x}$$

(ii) Regression estimate of y when $x = 0$ is $y = -\frac{3}{2} \times 0 + 13 = 13$

Regression estimate of x when $y = 13$ is $x = -\frac{1}{6} \times 13 + \frac{31}{6} = 3$.

The difference is due to the fact that the two regression lines are not reversible, *i.e.,* one equation is not sufficient to estimate the values of x and y.

Example 10. Given that $y = mx + 4$ and $x = 4y + 5$ are the two regression lines of y on x and x on y respectively, show that $0 < m \le 0.25$. If actually $m = 0.10$, find the means of x and y and also their correlation coefficient. Can you find their Standard deviations as well?

Solution. Regression equation of y on x is $y = mx + 4 \Rightarrow b_{yx} = m$

Regression equation of x on y is $x = 4y + 5 \Rightarrow b_{xy} = 4$

$\therefore$ $$r^2 = b_{yx} . b_{xy} = 4m.$$

Since $r^2 > 0$ and $r^2 \le 1$ (by definition)

---

* This part of the question can also be stated as 'Find their point of intersection and interpret it'.

$$\therefore \quad r^2 > 0 \Rightarrow 4m > 0 \Rightarrow m > 0$$

$$r^2 \leq 1 \Rightarrow 4m \leq 1 \Rightarrow m \leq 0.25$$

$$\therefore \quad 0 < m \leq 0.25$$

If $$m = 0.10$$

$$y = 0.1x + 4$$

and $x = 4y + 5$ are the regression lines.

Solving the two equations, we get $x = 35, y = 7.5$

Hence $$\bar{x} = 35; \bar{y} = 7.5.$$

Unless any one of the S.D.'s are known (in which case the values can be put in one of the regression coefficients) it is not possible to find the other S.D. Hence the S.D.'s cannot be found.

Example 11. The regression equation of production (x) on capacity utilisation (y) of a certain firm is :

$$5x - 3.83y + 146.80 = 0,$$

the average capacity utilisation of the firm was 70% and the variance of capacity utilisation is (9/16)th of the variance of production. Find the average production and coefficient of correlation between production and capacity utilisation. **(Punjab Univ. B. Com., Sept. 1999)**

Solution. Given : Regression equation of x on y is :

$$5x - 3.83y + 146.80 = 0$$

$$\bar{y} = 70; \ \sigma_y^2 = \frac{9}{16}\sigma_x^2$$

To find : $$\bar{x} = ?; \ r = ?$$

Since $(\bar{x}, \bar{y})$ must satisfy both of the regression equations, we get

$$5\bar{x} - 3.83\bar{y} + 146.80 = 0$$

i.e., $$5\bar{x} = 3.83\,(70) + 146.80 = 0$$

$$\therefore \quad \bar{x} = \frac{-146.80 + 268.10}{5} = 24.26$$

$$5x = 3.83y - 146.80$$

i.e., $$x = \left(\frac{3.83}{5}\right)y - \frac{146.80}{5} \Rightarrow b_{xy} = \frac{3.83}{5} = 0.766$$

i.e., $$r \cdot \frac{\sigma_x}{\sigma_y} = 0.766$$

But since $$\sigma_y^2 = \frac{9}{16}\sigma_x^2 \Rightarrow \frac{\sigma_x^2}{\sigma_y^2} = \frac{16}{9}$$

$$\Rightarrow \quad \frac{\sigma_x}{\sigma_y} = \frac{4}{3}$$

i.e., $$r\left(\frac{4}{3}\right) = 0.766 \Rightarrow r = 0.766 \times \frac{3}{4} = 0.5745.$$

Example 12. The following is the data on the entrance fee and the number of spectators at an entertainment place.

Entrance Fee (in Rs.)	50	55	60	65	70	75	80	85
Number of Spectators (in hundred)	20	17	16	14	13	10	9	5

Obtain the equation of the one litre of regression of the number of spectators on the entrance fee.

Solution. We will have to take y = No. of spectators and x = entrance fee

Hence, $n = 8; \Sigma x = 540; \Sigma y = 104.$

$$\bar{x} = \frac{\Sigma x}{n} = \frac{540}{8} = 67.5 \text{ and } \bar{y} = \frac{\Sigma y}{n} = \frac{104}{8} = 13.$$

$\bar{x}$ is not an integer. Let us take $u = x - 65$, $v = y - 13$.

The calculations are :

x	y	u	v	u^2	uv
50	20	–15	7	225	–105
55	17	–10	4	100	–40
60	16	–5	3	25	–15
65	14	0	1	0	0
70	13	5	0	25	0
75	10	10	–3	100	–30
80	9	15	–4	225	–60
85	5	20	–8	400	–160
540	104	20	0	1100	–405

Here, $\Sigma u = 20; \Sigma v = 0; \Sigma u^2 = 1100; \Sigma uv = -405.$

Coefficient of regression of y on x

$$b_{yx} = \frac{n\Sigma uv - (\Sigma u)(\Sigma v)}{n\,\Sigma u^2 - (\Sigma u)^2}$$

$$= \frac{8 \times (-405) - (20 \times 0)}{8 \times 1100 - (20)^2}$$

$$= \frac{-3240}{8800 - 400}$$

$$= \frac{-3240}{8400} = -0.39.$$

Line of regression of number of spectators on entrance fee, *i.e.,* line of regression of y on x is,

$$y - \bar{y} = b_{yx}(x - \bar{x})$$

i.e., $$y = \bar{y} + b_{yx}(x - \bar{x}) = 13 - 0.39(x - 67.5)$$

$$= 13 - 0.39x + 0.39 \times 67.5$$

$$= 39.325 - 0.39x.$$

Example 13. For a bivariate data, the following calculations were available :

$$\Sigma x = 6, \Sigma y = 9, \Sigma x^2 = 14, \Sigma y^2 = 41, \Sigma xy = 22, n = 3.$$

Obtain the equation of the best fit taking y as a dependent variable.

Solution. If y is a dependent variable, then we are to find the equation of regression line of y on x.

$$\therefore \qquad b_{yx} = \frac{n \Sigma xy - (\Sigma x)(\Sigma y)}{n \Sigma x^2 - (\Sigma x)^2}$$

$$= \frac{(3 \times 22) - (6 \times 9)}{(3 \times 14) - (6)^2} = \frac{66 - 54}{42 - 36} = \frac{12}{6} = 2$$

$$\bar{x} = \frac{\Sigma x}{n} = \frac{6}{3} = 2 \text{ and } \bar{y} = \frac{\Sigma y}{n} = \frac{9}{3} = 3.$$

Let the eqn. of the regression line representing regression of y on x be $y = a + bx$.

Since $\bar{x}, \bar{y}$ lie on it, we have $a = \bar{y} - b\bar{x} = 3 - 2 \times 2 = -1$

Hence, the line of best fit, *i.e.*, the regression of y on x is $y - -1 + 2x$.

Example 14. From the bivariate sample of size 10, following results are obtained for x and y. Find r^2 and check the validity of the assumption : "There is a linear relation between x and y."

$$\Sigma (x - \bar{x})^2 = 120, \Sigma (y - \bar{y})^2 = 180, \Sigma (x - \bar{x})(y - \bar{y}) = -45.$$

Solution. Coefficient of determination $= r^2$

$$= \left[\frac{\Sigma (x - \bar{x})(y - \bar{y})}{n \sigma_x \sigma_y}\right]^2 = \left[\frac{(x - \bar{x})(y - \bar{y})}{\sqrt{\Sigma (x - \bar{x})^2} \sqrt{\Sigma (y - \bar{y})^2}}\right]$$

$$= \left[\frac{-45}{\sqrt{120} \sqrt{180}}\right]^2 = \frac{(-45)^2}{120 \times 180}$$

$$= \frac{2025}{21600} = 0.094.$$

r^2 is near 0, hence the assumption that there is linear relationship between x and y is not valid.

Example 15. A survey was conducted to study the relationship between expenditure on accommodation (x) and expenditure on entertainment (Y) and the following results were obtained :

Expenditure on	Mean	S.D.
Accommodation	173	66
Entertainment	47.8	22
Coefficient of Correlation : 0.57		

Write down the equation of regression of y on x and estimate the expenditure on entertainment if the expenditure on accommodation is 200.

Solution. Here, $\bar{x} = 173; \bar{y} = 47.58; \sigma_x = 66; \sigma_y = 22; r = 0.57$

Coefficient of regression :

$$b_{yx} = r \cdot \frac{\sigma_y}{\sigma_x} = 0.57 \times \frac{22}{66} = 0.57 \times \frac{1}{3} = 0.19.$$

Line of regression of y on x is ,

$$y - \bar{y} = b_{yx} (x - \bar{x})$$

$$\Rightarrow \qquad y = 47.58 = 0.19\,(x - 173)$$

$$\Rightarrow \qquad y = 47.58 + 0.19x - 0.19 \times 173$$

$$\Rightarrow \qquad y = 14.71 + 0.19x$$

When expenditure on accommodation, $x = 200$,

$$y = 14.71 + 0.19 \times 200 = 14.71 + 38 = 52.71.$$

Example 16. From the following results, obtain the two regression equations and estimate the yield, when the rainfall is 29 cms and the rainfall, when the yield is 600 kgs.

	Yield in Kg. (X)	*Rainfall in cms (Y)*
Mean	508.4	26.7
S.D.	36.8	4.6

Coefficient of correlation between yield and rainfall is +0.52. **(CA Foundation, Nov., 2001)**

Solution. Regression equation of X on Y is

$$X - \overline{X} = b_{XY}\,(Y - \overline{Y})$$

$$= \frac{r\,\sigma_X}{\sigma_Y}\,(Y - \overline{Y})$$

$$\Rightarrow \qquad X - 508.4 = 0.52\left[\frac{36.8}{4.6}\right][Y - 26.7]$$

Putting $Y = 29$, we get

$$X = 508.4 + 0.52\left[\frac{36.8}{4.6}\right][29 - 26.7] = 517.968$$

= Yield when rainfall is 29 cms.

Regression equation of Y on X is

$$Y - \overline{Y} = b_{YX}\,(X - \overline{X})$$

$$\Rightarrow \qquad Y - 26.7 = \frac{r\,\sigma_Y}{\sigma_X}\,(X - \overline{X}), \text{ when } X = 600 \text{ kgs.}$$

$$\Rightarrow \qquad Y = 26.7 + \frac{r\sigma_Y}{\sigma_X}\,(600 - 508.4)$$

$$= 26.7 + 0.52\left(\frac{4.6}{36.8}\right)(600 - 508.4) = 32.654.$$

Example 17.

(a) Given

	X Series	*Y Series*
Mean	18	100
Standard Deviation	14	20

Coefficient of correlation between X and Y series = +0.8.

Find the most probable value of Y if X is 70, and most probable value of X if Y is 90.

(b) If two regression coefficients are 0.8 and 0.6 when would be the value of the coefficient of correlation?

Solution. (a) To find out value of X when $Y = 90$, we use the regression equation of X on Y:

$$(X - \overline{X}) = r \frac{\sigma_X}{\sigma_Y} (X - \overline{Y})$$

or $$(X - 18) = 0.8 \frac{14}{20} (90 - 100)$$

or $$(X - 18) = 0.56 (-10)$$

or $$X = 18 - 5.6$$

or $$X = 12.4$$

To find out the value of Y when $X = 70$, we use the regression equation of Y on X :

$$(Y - \overline{Y}) = r \frac{\sigma_Y}{\sigma_X} (X - \overline{X})$$

or $$(Y - 100) = 0.8 \frac{20}{14} (70 - 18)$$

or $$(Y - 100) = 1.144 \times 52$$

or $$Y = 59.48 + 100 = 159.48$$

(b) $r = \sqrt{b_{XY} \times b_{YX}} = \sqrt{0.8 \times 0.6} = \sqrt{0.48} = +0.69.$

Example 18. The following data are given regarding expenditure on advertising and sales of a particular firm :

	Advertisement Expenditure	*Sales*
	(X)	*(Y)*
Mean	10	90
Standard Deviation	3	12

Correlation Coefficient $r_{XY} = 0.8$

(i) Calculate the regression equation of Y on X.

(ii) Estimate the advertisement expenditure required to attain a sales target of Rs. 120 lakhs.

(C.A. Foundation, Nov. 1999)

Sol. (i) Regression equation of Y on X

$$(Y - \overline{Y}) = r \frac{\sigma_Y}{\sigma_X} (X - \overline{X}) \qquad ...(I)$$

Here, $\overline{X} = 10, \overline{Y} = 90, \sigma_X = 3, \sigma_Y = 12, r = 0.8$

Substituting these values in (I), we get

$$(Y - 90) = 0.8 \left(\frac{12}{3}\right) (X - 10)$$

or $$(Y - 90) = 0.8 (4) (X - 10)$$

or $$(Y - 90) = 3.2 (X - 10)$$

or $$(Y - 90) = 3.2X - 32$$

or $$Y = 3.2X + 58$$

(ii) For determining advertisement expenditure for a given sales target, we have to find the regression equation of X on Y.

i.e., $$(X - \overline{X}) = r\frac{\sigma_X}{\sigma_Y}(Y - \overline{Y})$$

$\Rightarrow$ $$X - 10 = 0.8 \times \frac{3}{12}(Y - 90)$$

$\Rightarrow$ $$X - 10 = 0.2Y - 18$$

$\Rightarrow$ $$X = 0.2Y - 8$$

When $Y = 120$, $X = 0.2 \times 120 - 8 = 16$

Hence, the advertisement expenditure required is Rs. 16 lakhs.

Example 19. The heights of a sample of 10 fathers and their eldest sons are given below (to the nearest cm) :

Height of Father (*X*)	170	167	162	193	167	166	169	171	164	165
Height of Son (*Y*)	168	167	166	166	168	165	168	170	165	168

(i) Compute the correlation coefficient *r*.
(ii) Find the regression of *Y* on *X*.
(iii) Compute the coefficient of determination and give your comments.

(CA Foundation, Nov. 2000)

Solution. Since the mean of *X* and mean of *Y* are not whole numbers we take the deviations of *X* and *Y* from arbitrary values.

X	Y	$x = X - 165$	$y = Y - 165$	x^2	y^2	xy
170	168	5	3	25	9	15
167	167	2	2	4	4	4
162	166	–3	1	9	1	–3
163	166	–2	1	4	1	–2
167	168	2	3	4	9	6
166	165	1	0	1	0	0
169	168	4	3	16	9	12
171	170	6	5	36	25	30
164	165	–1	0	1	0	0
165	168	0	3	0	9	0
		14	21	100	67	62

$$\overline{X} = 165 + \frac{14}{10} = 166.4;\ \overline{Y} = 165 + \frac{21}{10} = 167.1$$

$$r = \frac{n\,\Sigma xy - \Sigma y}{\sqrt{\Sigma\,[n\,\Sigma x^2 - (\Sigma x)^2] \times [n\,\Sigma y^2 - (\Sigma y)^2]}}$$

$$= \frac{10 \times 62 - 14 \times 21}{\sqrt{[10 \times 100 - 14^2]\,[10 \times 67 - 21^2]}}$$

$$= \frac{326}{\sqrt{894 \times 229}} = \frac{326}{429.087} = 0.75975 \equiv 0.76$$

The regression equation of *Y* on *X* is given by $Y - \overline{Y} = b_{yx}(X - \overline{X})$

where $b_{yx} = \frac{n\,\Sigma xy - \Sigma x\,\Sigma y}{n\,\Sigma x^2 - (\Sigma x)^2} = \frac{10 \times 62 - 14 \times 21}{10 \times 100 - 14^2} = \frac{326}{804} = 0.405$

The required regression equation is $Y - 167.1 = 0.405\,(X - 166.4)$

$$\Rightarrow \quad Y = 167.1 + 0.405X - 166.4 \times 0.405$$

$$\Rightarrow \quad Y = 0.405X + 99.708$$

Coefficient of determination $= \gamma^2 = (0.76)^2 = 0.58$

Thus, the regression line accounts for 58% of total variation.

Example 20. Point out the inconsistency in the statements :

(i) The regression of y on x is $2y + 3x = 4$ and the correlation coefficient between x and y is 0.8.

(ii) The correlation coefficient of $3x$ and $-2y$ is the same as the correlation coefficient of x and y. **(ICWA, Dec., 1998)**

Solution.

(i) Regression equation y on x is

$$2y + 3x = 4 \Rightarrow y = -\frac{3}{2}x + 2$$

$$\therefore \quad b_{yx} = -\frac{3}{2},$$

Given $\quad r_{xy} = 0.8$ (Given)

The sign of correlation coefficient should be same as the sign of the regression coefficients. So the statement is inconsistent.

(ii) Let $u = 3x$ and $v = -2y$

$$\therefore \quad x = \frac{u}{3} \text{ and } y = \frac{v}{-2}$$

$$r_{uv} = \frac{\text{Cov}\,(u, v)}{\sigma u \sigma_v} = \frac{3 \times (-2)\,\text{Cov}\,(x, y)}{3\sigma_x \times 2\sigma_y}$$

$$= \frac{-\,\text{Cov}\,(x, y)}{\sigma_x \sigma_y} = -r_{xy}$$

So the statement is inconsistent.

Example 21. For a set of 10 pairs of values of x and y, the regression line of x on y is $x - 2y + 12 = 0$, mean and standard deviation of y being 8 and 2 respectively. Later it is known that a pair $(x = 3, y = 8)$ was wrongly recorded and the correct pair detected is $(x = 8, y = 3)$. Find the correct regression line of x on y. **(ICWA, June 1998)**

Solution. Regression line x on y is $x - 2y + 12 = 0 \Rightarrow x = 2y - 12$

Since $\bar{y} = 8$, so $\bar{x} = 2\bar{y} - 12 = 2 \times 8 - 12 = 16 - 12 = 4$

[$\because \bar{x}, \bar{y}$ satisfy the regression equation]

$$b_{xy} = 2 \Rightarrow \frac{\text{Cov}\,(x, y)}{\sigma_y^2} = 2 \Rightarrow \text{Cov}\,(x, y) = 2 \times 4 = 8 \qquad [\because \sigma_y = 2]$$

$$\Rightarrow \quad \frac{1}{10}\Sigma xy - \bar{x}\,\bar{y} = 8 \Rightarrow \Sigma xy = [8 + 8 \times 4]\,10 = 400$$

$$\sigma_y = 2 \Rightarrow \sigma_y^2 = 4 \Rightarrow \frac{1}{n}\Sigma y^2 - (\bar{y})^2 = 4$$

$$\Rightarrow \qquad \frac{1}{10}\Sigma y^2 - 64 = 4 \Rightarrow \Sigma y^2 = 680$$

Corrected values $\bar{x}_c = \frac{4 \times 10 - 3 + 8}{10} = \frac{9}{2},$

$$\bar{y}_c = \frac{8 \times 10 - 8 + 3}{10} = \frac{15}{2}$$

Correct value of $\Sigma xy = 400 - 3 \times 8 + 8 \times 3 = 400$

Correct value of $\Sigma y^2 = 680 - 64 + 9 = +625$

Correct value of $\text{Cov}(x, y) = \frac{\Sigma xy}{10} - \frac{\Sigma x}{10} \times \frac{\Sigma y}{10}$

$$= \frac{400}{10} - \frac{9}{2} \times \frac{15}{2} = \frac{25}{4}$$

Correct value of $\sigma_y^2 = \frac{\Sigma y^2}{10} - \left(\frac{\Sigma y}{10}\right)^2 = \frac{625}{10} - \frac{225}{4} = \frac{25}{4}$

Correct $b_{xy} = \frac{\frac{25}{4}}{\frac{25}{4}} = 1$

Correct regression line x on y is $x - \frac{9}{2} = \left(y - \frac{15}{2}\right)$

$$\Rightarrow \qquad x = y - 3.$$

Example 22. Find the mean of two variables x and y and the correlation coefficient between x and y from two regression lines $x + 6y = 6$ and $2x + 3y = 9$ of a sample. **(ICWA Dec. 1999)**

Solution.

$$4x + 6y = 18 \qquad ...(I)$$

$$x + 6y = 6 \qquad ...(II)$$

Subtracting (II) from (I),

$$3x = 12 \text{ or } x = 4$$

Substituting $x = 4$ in (II),

$$6y = 6 - x = 6 - 4, \text{ i.e., } y = 1/3$$

So, $\bar{x} = 4, \bar{y} = \frac{1}{3}, \bar{x}$ and $\bar{y}$ being means of x and y and lie on the regression lines.

Let $x + 6y = 6$ be the regression line of y on x

i.e., $y = \frac{-1}{6}x + 1, \text{ i.e., } b_{yx} = -1/6$

and $2x + 3y = 9$ be the regression line of x on y

$$\text{i.e., } x = \frac{-3}{2}y + \frac{9}{2} \text{ i.e., } b_{xy} = -\frac{3}{2}$$

where b_{yx} and b_{xy} are the regression coefficients of y on x and that of x on y. Let r_{xy} be the correlation coefficient between x and y.

Now $$b_{yx} \times b_{xy} = \left(-\frac{1}{6}\right)\left(-\frac{3}{2}\right) = \frac{1}{4} < 1, \text{ So } r_{xy}^2 = 1/4$$

i.e., $r_{xy} = -1/2$ since, b_{yx}, b_{xy} and r_{xy} are of same sign.

Example 23. In a partially destroyed laboratory-record of an analysis of correlation data, the following results only are obtained :

Regression equations are $8x - 10 = 64$ and $40x - 18y = 320$.

What are (i) the mean values of x and y, and (ii) the coefficient of correlation between x and y?

(ICWA Dec., 1997)

Solution. The equations are :

$$8x - 10y = 64 \qquad ...(I)$$

$$40x - 18y = 320 \qquad ...(II)$$

(i) Solving the given equations, we get $x = 8, y = 0$, *i.e.,* mean value of $x = 8$, and that of $y = 0$ (as the intersection of regression lines indicates the means of two series).

(ii) For finding correlation coefficient (r) we are to find the values of b_{YX} and b_{XY}. We will find which of the two equations represent the regression of Y on X and X on Y. Let us assume that equation (I) is the regression of Y on X and equation (II) of X on Y.

From :

Equation (I), $y = \frac{8}{10}x - \frac{64}{10}$, *i.e.,* $b_{yx} = \frac{8}{10}$

Equation (II), $x = \frac{18}{40}y + \frac{320}{40}$, *i.e.,* $b_{xy} = \frac{18}{40}$

$$\therefore \qquad r = \sqrt{b_{yx} \times b_{xy}} = \sqrt{\frac{8}{10} \times \frac{18}{40}} = \sqrt{0.8 \times 0.45} = \sqrt{0.36} = 0.6$$

As $|r| < 1$, so our assumption is correct.

$\therefore$ $r = 0.6$. (Since both the regression coefficients as positive, r will be positive.)

Example 24. If $b_{yx} = -2$, $b_{xy} = -\frac{1}{4}$ then r_{xy} is (i) 1, (ii) $\frac{1}{\sqrt{2}}$, (iii) -1, (iv) $-\frac{1}{\sqrt{2}}$

Solution. $$b_{yx} = -2, \; b_{xy} = -\frac{1}{4}$$

$$r_{xy} = -\sqrt{\left(2 \times \frac{1}{4}\right)} = -\sqrt{\frac{1}{2}} \qquad \therefore \text{ Answer : (iv).}$$

Example 25. The following is an excerpt from the correlation-regression study of variables x and y. Obtain the correlation coefficient between x and y and the estimate of y when $x = 8$:

$n = 5$, $\Sigma x = 15$, $\Sigma y = 40$, $\Sigma x^2 = 55$, $\Sigma y^2 = 420$, $\Sigma xy = 150$. **(ICWA, June 1999)**

Solution. $$r_{xy} = \frac{\Sigma xy - \frac{1}{n}\Sigma x \Sigma y}{\sqrt{\left[\left(\Sigma x^2 - \frac{(\Sigma x)^2}{n}\right)\left(\Sigma y^2 - \frac{(\Sigma y)^2}{n}\right)\right]}}$$

$$= \frac{150 - \frac{1}{5} \times 15 \times 40}{\sqrt{\left[\left(55 - \frac{(15)^2}{5}\right)\left(420 - \frac{(40)^2}{5}\right)\right]}}$$

$$= \frac{150 - 120}{\sqrt{[(55-45)(420-320)]}}$$

$$= \frac{30}{\sqrt{10 \times 100}} = \frac{3}{\sqrt{10}} = 0.9487 = 0.95$$

Similarly $b_{xy} = 9.5, a = \bar{y} - b_{yx}\bar{x} = 8 - 9.5 \times 3 = -20.5;$*

$$y_{\text{at } x=8} = -20.5 + 9.5 \times 8 = 55.5.$$

Example 26. For certain data, the following regression equations were obtained :

$$4x - 5y + 33 = 0$$
$$20x - 9y - 107 = 0$$

Estimate y when $x = 20$ and x when $y = 20$.

Solution. Let $$4x - 5y + 33 = 0 \qquad ...(I)$$

be the regression equation of y on x, while

$$20x - 9y - 107 = 0 \qquad ...(II)$$

be the regression equation of x on y.

From (I), $y = \frac{4}{5}x + \frac{33}{5}$, *i.e.*, $b_{yx} = \frac{4}{5}$ and from (II) $x = \frac{9}{10}y + \frac{107}{20}$, *i.e.*, $b_{xy} = \frac{9}{20}$

Now $r = \sqrt{b_{yx} \times b_{xy}} = \sqrt{\frac{4}{5} \times \frac{9}{20}} = \sqrt{\frac{9}{25}} = \frac{3}{5} < 1$, so our assumption is correct.

$\therefore$ when $x = 20, y = \frac{4}{5} \times 20 + \frac{33}{5} = \frac{113}{5} = 22.6$, when $y = 20, x = \frac{9}{20} \times 20 + \frac{107}{20} = \frac{287}{20} = 14.85.$

Example 27. The following are the marks obtained by 8 students in Mathematics and Statistics. Find the regression line of marks in Statistics on marks in Mathematics.

Marks in Mathematics	50	40	60	46	50	48	59	47
Marks in Statistics	30	37	42	32	35	45	40	35

Solution. Mean of $x = 50$, Mean of $y = 37$

Marks in Maths. x	*Marks in Stat.* y	$x - 50 = d_x$	$y - 37 = d_y$	d_x^2	d_y^2	$d_x d_y$
50	30	0	–7	0	49	0
40	37	–10	0	100	0	0
60	42	10	5	100	25	50
46	32	–4	–5	16	25	20
50	35	0	–2	0	4	0
48	45	–2	8	4	64	–16
59	40	9	3	81	9	27
47	35	–3	–2	9	4	6
				310	180	87

* $\bar{y} = \frac{40}{5} = 8, \bar{x} = \frac{15}{5} = 3$

$$\text{S.D. of } x = \sigma_X = \sqrt{\frac{\Sigma d_x^2}{n}} = \sqrt{\frac{310}{8}} = \sqrt{38.75} = 6.225$$

$$\text{S.D. of } y = \sigma_y = \sqrt{\frac{\Sigma d_x^2}{n}} = \sqrt{\frac{180}{8}} = \sqrt{22.5} = 4.74$$

$$b_{yx} = \frac{\Sigma d_x d_y}{n\,\sigma_x^2} = \frac{87}{8 \times 38.15} = 0.2806$$

$\therefore$ Required regression line is $y - \bar{y} = b_{yx}(x - \bar{x})$

$$\Rightarrow \qquad y - 37 = 0.2806\,(x - 50).$$

Example 28. The data in the following table relates the weekly maintenance cost (in Rs.) to the age (in months) of ten machines of similar type in a manufacturing company. Find the least squares regression line of maintenance cost on age and use this to predict the maintenance cost for a machine of this type which is 40 months old.

Machine	1	2	3	4	5	6	7	8	9	10
Age (x)	5	10	15	20	30	30	30	50	50	60
Cost (y)	190	240	250	300	310	335	300	300	300	395

(ICWA, Dec., 1997)

Solution. **Computation of Regression Equation**

	x	y	$x-\bar{x}=d_x$	$y-\bar{y}=d_y$	d_x^2	d_xd_y
	5	190	–25	–107	625	2,675
	10	240	–20	–57	400	1,140
	15	250	–15	–47	225	705
	20	300	–10	3	100	–30
	30	310	0	13	0	0
	30	335	0	38	0	0
	30	300	0	3	0	0
	50	300	20	3	400	60
	50	350	20	53	400	1,060
	60	395	30	98	900	2,940
Total	300	2,970	0	0	3,050	8,550

$$\bar{x} = \frac{300}{10} = 30; \quad \bar{y} = \frac{2970}{10} = 298.$$

Since both means are integers deviations have been taken from actual means.
Regression coefficient of y on x

$$= b_{yx} = \frac{\Sigma d_x d_y}{\Sigma d_x^2} = \frac{8{,}550}{3{,}050} = 2.8033$$

Regression equation of y on x is

$$(Y - \bar{y}) = b_{yx}(x - \bar{x})$$

or $$y - 297 = 2.8033\,(x - 30) = 2.8033x - 84.099$$

or $$y = 2.8033x + 212.901$$

For $x = 40$, $y = 2.8033 \times 40 + 212.901 = 325.033 = \text{Rs. } 325.$

Example 29. Find the regression equation of y on x where y and x are the marks obtained by 10 students are as given below :

x	20	50	55	45	75	35	25	90	10	50
y	20	45	65	40	55	35	15	80	25	50

(C.A. Foundation, May 2002)

Solution. $\Sigma y = 465,\ \dfrac{\Sigma y}{n} = \dfrac{465}{10} = 46.5 = \text{Mean of } y$

$\Sigma x = 430,\ \dfrac{\Sigma y}{n} = \dfrac{430}{10} = 43 = \text{Mean of } x.$

Regression Equation of y on x is $y - \bar{y} = b_{xy}(X - \bar{X})$

y	x	$y - 40 = d_y$	$x - 43 = d_x$	d_xd_y	d_x^2
20	20	–20	–23	460	529
60	45	20	2	40	4
55	65	15	22	330	484
45	40	5	–3	–15	9
75	55	35	12	420	144
35	35	–5	–8	40	64
25	15	–15	–28	420	784
90	80	50	37	1850	1369
10	25	–30	–18	540	324
50	50	10	7	70	49
			$\Sigma dx = 0$	$\Sigma d_xd_y = 4155$	$\Sigma d_x^2 = 3760$

$$(\text{S.D.})^2 = \frac{\Sigma d_x^2}{n} = \frac{3760}{10} = 376$$

$$b_{yx} = \frac{\dfrac{\Sigma d_xd_y}{n} - \dfrac{\Sigma d_x}{n}\dfrac{\Sigma d_y}{n}}{(\text{S.D.})^2} = \frac{\dfrac{4155}{10}}{376} = \frac{415.5}{376}.$$

$\therefore$ The regression equation of y on x is $y - 45.6 = \dfrac{415.5}{376}(x - 43)$.

Example 30. A panel of two judges P and Q graded seven dramatic performances by independently awarding marks as follows :

Performance	1	2	3	4	5	6	7
Marks by P	46	42	44	40	43	41	45
Marks by Q	40	38	36	35	39	37	41

The eighth performance, which Judge Q could not attend, was awarded 37 marks by Judge P. If Judge Q has also been present, how many marks would be expected to have been awarded by him to the eighth performance ?

Solution. **Calculation of Regression Coefficient from the Actual Means**

Marks by P (X)	$X-43=x$	*Marks in Q* (Y)	$Y-38=y$	x^2	y^2	xy
46	+3	40	+2	9	4	+6
42	–1	38	0	1	0	0
44	+1	36	–2	1	4	–2
40	–3	35	–3	9	9	+9
43	0	39	+1	0	1	0
41	–2	37	–1	4	1	+2
45	+2	41	+3	4	9	+6
$\Sigma X=301$	$\Sigma x=0$	$\Sigma Y=266$	$\Sigma y=0$	$\Sigma x^2=28$	$\Sigma y^2=28$	$\Sigma xy=21$

$$\overline{X}=\frac{301}{7}=43;\ \overline{Y}=\frac{266}{7}=38$$

Regression coefficient of X on $Y=b_{xy}$

$$=\frac{\Sigma\, xy}{\Sigma\, x^2} \qquad [\because \text{ The deviations are taken from the actual mean}]$$

$$=\frac{21}{28}=\frac{3}{4}=0.75$$

Regression coefficient of Y on $X=b_{yx}=\dfrac{\Sigma\, xy}{\Sigma\, y^2}=\dfrac{21}{28}=0.75$

$$r^2=b_{xy}\times b_{yx}=0.75\times 0.75=\frac{9}{16}\Rightarrow r=\frac{3}{4}$$

Correlation coefficient $=\dfrac{3}{4}$

Regression equation of Y on X is

$$Y-\overline{Y}=b_{yx}\,(X-\overline{X})\Rightarrow Y-38=0.75\,(X-43)$$

The marks expected to be given by judge Q when judge P gives 37 marks are given by

$$Y=38+0.75\,(37-43)=38-0.75\times 6=38-4.5=33.5.$$

Example 31. (a) Find :

(i) the regression coefficients b_{YX} and b_{XY};

(ii) correlation coefficient r_{XY} and

(iii) the standard error of the estimate of Y on X from the following data :

$$\Sigma XY=350;\ \overline{X}=5;\ \overline{Y}=6;\ \Sigma X=50;\ \Sigma Y=60;$$

$$\text{Var}\,(X)=4;\ \text{Var}\,(Y)=9.$$

(b) Prove that $r_{XY}=r_{UV}$ where $U=(X-A)/h$ and $V=(Y-B)/k$.

Solution. (a)

$$b_{YX} = \frac{\frac{\Sigma XY}{n} - \frac{\Sigma X}{n}\frac{\Sigma Y}{n}}{\sigma_x^2}$$

$$\overline{X} = \frac{\Sigma X}{n} \Rightarrow 5 = \frac{50}{n} \Rightarrow n = 10$$

$\therefore$

$$b_{YX} = \frac{\frac{350}{10} - 5 \times 6}{4} = \frac{35 - 30}{4} = \frac{5}{4}$$

$$b_{XY} = \frac{\frac{350}{10} - 5 \times 6}{9} = \frac{5}{9}$$

$$r = \sqrt{(b_{YX})\,(b_{XY})} = \sqrt{\frac{5}{4} \times \frac{5}{9}} = +\frac{5}{6}$$

$$SE_{YX} = \sigma_y \sqrt{1 - r^2} = 3\sqrt{1 - \frac{25}{36}} = 3\sqrt{\frac{11}{36}} = 1.66$$

(b)

$$\overline{U} = \frac{1}{h}\Sigma\,(X - A) = \frac{1}{h}\,[\Sigma X - nA]$$

$$= \frac{1}{h}\left[\frac{\Sigma X}{h} - \frac{nA}{n}\right] = \frac{1}{h}\,[\overline{X} - A]$$

$$U - \overline{U} = \frac{1}{h}\,[X - \overline{X}]$$

Similarly,

$$V - \overline{V} = \frac{1}{k}\,(Y - \overline{Y})$$

$$\Sigma\,(U - \overline{U})\,(V - \overline{V}) = \frac{1}{hk}\,[\Sigma\,(X - \overline{X})\,(Y - \overline{Y})]$$

$$\sigma_U^2 = \frac{\Sigma\,(X - \overline{X})^2}{h}, \quad \sigma_V^2 = \frac{\Sigma\,(Y - \overline{Y})^2}{k}$$

$$r = \frac{\Sigma\,(U - \overline{U})\,(V - \overline{V})}{n\,\sigma_x\,\sigma_y}$$

$$= \frac{\frac{1}{nhk}\Sigma\,(X - \overline{X})\,(Y - \overline{Y})}{\frac{1}{hk}\sqrt{\frac{\Sigma\,(X - \overline{X})^2}{n}}\sqrt{\frac{\Sigma\,(Y - \overline{Y})^2}{n}}}$$

$$= \frac{\Sigma\,(X - \overline{X})\,(Y - \overline{Y})}{n\,\sigma_x\,\sigma_y} = r_{xy}$$

Example 32. Given $X = 4Y + 5$ and $Y = KX + 4$ are the lines of regression of X on Y and Y on X respectively. If K is positive, prove that it cannot exceed 1/4. If $K = 1/16$, find the means of two variables and the coefficient of correlation between them. **(D.U., B.Com. (Hons.) 2006R)**

Solution. Given : $X = 4Y + 5;\ Y = KX + 4$

So $$b_{XY} = 4 \quad \text{and} \quad b_{YX} = K$$

$$r^2 = (b_{XY})\,(b_{YX}) = 4\,K$$

$r = r^2$, when value of r is equal to 1 *i.e.* maximum value of r^2 is 1

So $$1 = 4\,K \;\Rightarrow\; K \le \frac{1}{4}$$

Given, $$X = 4Y + 5 \qquad \text{...(i)}$$

$$Y = \frac{1}{16}X + 4 \;\Rightarrow\; 16Y = X + 64 \qquad \text{...(ii)}$$

Putting the value of X in (ii),

$$16Y = (4Y + 5) + 64 \;\Rightarrow\; 12Y = 69$$

$\Rightarrow$ $$Y = \frac{69}{12} = 5.75 \;\Rightarrow\; \bar{Y} = 5.75$$

Putting the value of Y in equation (i)

$$X = 4 \times 5.75 + 5 = 23 + 5 = 28 \;\Rightarrow\; \bar{X} = 28$$

Correlation, $$r = \sqrt{(b_{XY})\,(b_{YX})} = \sqrt{4K} = \sqrt{4 \times \frac{1}{16}} = \sqrt{\frac{1}{4}} = \frac{1}{2} = 0.5$$

Example 33. How does 'correlation' differ from 'regression'. State the utility of regression in the field of economic analysis. **(D.U., B.Com. (Hons.) 2006R)**

Solution. Regression is the measure of the average relationship between two or more variables in terms of the original units of the data. Correlation studies covariation of the two variables. It finds out whether the variables under study move in the same direction or in reverse directions. It does not study the nature of relationship. It does not tell about the relative movement in the variables under study and cannot predict the value of one variable by taking into account the value of the other variable.

Utility : Regression helps in estimating or predicting the effect of one variable on the other. Effect of rise in price on the demand of the commodity, effect of increase in publicity expenditure on rise in volume of sales etc., can be studied with the help of regression analysis. Further, with the help of regression studies, we can also calculate the value of coefficient of correlation and coefficient of determination.

Example 34. Given the following data :

X	5	3	7	4	8	2	10	6	8	7	9	15
Y	8	6	8	5	9	6	8	5	11	7	8	15

Obtain the two regression equations and calculate Standard Error of Estimate (SE_{yx}).

(D.U., B.Com. (Hons.) 2007)

Solution.

X	Y	YX	X^2	Y^2
5	8	40	25	64
3	6	18	9	36
7	8	56	49	64
4	5	20	16	25
8	9	72	64	81
2	6	12	4	36
10	8	80	100	64
6	5	30	36	25
8	11	88	64	121
7	7	49	49	49
9	8	72	81	64
15	15	225	225	225
84	96	762	722	854

$$\text{Mean of } X = \frac{\Sigma X}{n} = \frac{84}{12} = 7$$

$$\text{Mean of } Y = \frac{\Sigma Y}{n} = \frac{96}{12} = 8$$

Regression equation of Y on X is $Y - 8 = b_{yx}(X - 7)$

Regression equation of X on Y is $X - 7 = b_{xy}(Y - 8)$

$$b_{yx} = \frac{\frac{\Sigma XY}{n} - \frac{\Sigma X}{n}\frac{\Sigma Y}{n}}{\frac{\Sigma X^2}{n} - \left(\frac{\Sigma X}{n}\right)^2}$$

$$= \frac{\frac{762}{12} - 7 \times 8}{\frac{722}{12} - (7)^2} = \frac{63.5 - 56}{60.16 - 49} = \frac{7.5}{11.16} = 0.672$$

$$b_{xy} = \frac{63.5 - 56}{\frac{\Sigma Y^2}{n} - \left(\frac{\Sigma Y}{n}\right)^2} = \frac{7.5}{\frac{854}{12} - (8)^2} = \frac{7.5}{71.17 - 64} = \frac{7.5}{7.17} = 1.046$$

The equation are :

$$Y - 8 = 0.672\,(X - 7)$$

$\Rightarrow$

$$Y = 0.672X - 4.704 + 8$$
$$= 0.672X + 3.296$$

and

$$X - 7 = 1.046\,(Y - 8)$$

$\Rightarrow$

$$X = 7 + 1.046Y - 1.046 \times 8$$
$$= 1.046Y - 1.368$$

$$SE_{yx} = \sqrt{\frac{\Sigma Y^2 - a\,\Sigma Y - b\,\Sigma XY}{n-2}} = \sqrt{\frac{854 - 3 \times 96 - 0.672 \times 762}{10}} = 2.323 \text{ approx.}$$

Example 35. The equations of two regression lines between two variables are expressed as :

$$2X - 3Y = 0 \quad \text{and} \quad 4Y - 5X - 8 = 0$$

(i) Identify which of the two can be called regression of Y on X and X on Y.
(ii) Find the means of the variables.
(iii) Determine r_{XY}. **(DU Eco. (Hons.) 2008)**

Solution. Equations are :

$$2X - 3Y = 0 \qquad \ldots(1)$$

$$4Y - 5X - 8 = 0 \qquad \ldots(2)$$

Let the first equation be called the regression equation of X on Y, then

$$2X = 3Y \Rightarrow X = \frac{3}{2}Y, \quad i.e.\ b_{XY} = \frac{3}{2}$$

Then the second equation will be regression of Y on X is

$$4Y = 5X + 8$$

$$Y = \frac{5}{4}X + 2 \quad i.e.\ b_{YX} = \frac{5}{4}$$

$$\therefore \qquad r^2 = \frac{3}{2} \times \frac{5}{4} > 1 \qquad \text{which is not possible}$$

Hence our assumption is wrong.

$\therefore$ The regression equation of Y on X is $2X - 3Y = 0$ and that of X on Y is

$$4Y - 5X - 8 = 0$$

Multiplying (1) by (2) and substracting from we get

$$Y = 8, \text{ when } Y = 8, X = 12$$

$$\therefore \qquad \overline{Y} = 8, \ \overline{X} = 12.$$

Example 36. The following table gives the ages (in years) of husbands and wives for 50 newly married couples. Find the two regression equations. Also estimate :

(i) The age of husband when wife is 20.
(ii) The age of wife when husband is 30.
(iii) The coefficient of correlation.
(iv) The standard error of estimates.

Age of Wives	*Age of Husbands*			
	20–25	*25–30*	*30–35*	*Total*
16–20	9	14	–	23
20–24	6	11	3	20
24–28	–	–	7	7
Total	15	25	10	50

(D.U., B.Com. (Hons.) 2007E)

Solution. Let us denote the ages (in years) of husbands by the variable X and that of wives by the variable Y. Let X and Y denote the mid-points of the corresponding classes of X and Y series respectively. Let $A = 27.5$ and $B = 22$.

If we take,

$$u = \frac{X - A}{h} = \frac{X - 27.5}{5}; \qquad v = \frac{Y - B}{k} = \frac{Y - 22}{4}$$

The table for computing the two lines of regression is given :

$$\bar{u} = \frac{\Sigma fu}{N} = \frac{-5}{50} = -0.1 \qquad \bar{X} + A + h\,\bar{u} = 27.5 + 5 \times (-).1) = 27.5 - 0.5 = 27$$

$$v = \frac{\Sigma fv}{N} = \frac{-16}{50} = -0.32 \qquad \bar{Y} = B + k\,\bar{v} = 22 + 4 \times (-0.32) = 22 - 1.28 = 20.72$$

Calculations for Regression Equations

		Age of Husbands →	*20–25*	*25–30*	*30–35*				
		Mid. pt. (*X*)	22.5	27.5	32.5				
Age of Wives ↓	Mid. pt (*Y*)	u → v ↓	–1	0	1	f	fv	fv^2	fuv
16–20	18	–1	9 [9]	14 [0]	[0]	23	–23	23	9
20–24	22	0	6 [0]	11 [0]	3 [0]	20	0	0	0
24–28	26	1	[0]	[0]	7 [7]	7	7	7	7
		f	15	25	10	$N = 50$	$\Sigma fv = -16$	$\Sigma fv^2 = 30$	$\Sigma fuv = 16$
		fu	–15	0	10	$\Sigma fu = -5$			
		fu^2	15	0	10	$\Sigma fu^2 = 25$			
		fuv	9	0	7	$\Sigma fuv = 16$			

$$b_{YX} = \frac{k}{h}\left[\frac{N\Sigma fuv - (\Sigma fu)(\Sigma fv)}{N\Sigma fu^2 - (\Sigma fu)^2}\right]$$

$$= \frac{4}{5}\left[\frac{50 \times 16 - (-5) \times (-16)}{50 \times 25 - (-5)^2}\right]$$

$$= \frac{4}{5}\left[\frac{800 - 80}{1250 - 25}\right]$$

$$= \frac{4}{5} \times \frac{720}{1225} = 0.4702$$

$$b_{XY} = \frac{h}{k}\left[\frac{N\Sigma fuv - (\Sigma fu)(\Sigma fv)}{N\Sigma fv^2 - (\Sigma fv)^2}\right]$$

$$= \frac{5}{4}\left[\frac{800 - 80}{50 \times 30 - (-16)^2}\right]$$

$$= \frac{5}{4} \times \frac{720}{1244} = 0.7235$$

Regression Equation of *Y* on *X*

$$Y - \bar{Y} = b_{YX}(X - \bar{X})$$

$$\Rightarrow \quad Y - 20.72 = 0.4702\,(X - 27)$$

$$= 0.4702X - 12.6954$$

$$Y = 0.4702X + 20.72 - 12.6954$$

$$= 0.4702X + 8.0246$$

Regression Equation of *X* on *Y*

$$X - \bar{X} = b_{XY}(Y - \bar{Y})$$

$$\Rightarrow \quad X - 27 = 0.7235\,(Y - 20.72)$$

$$= 0.7235Y - 14.991$$

$$\Rightarrow \quad X = 0.7235Y + 27.000 - 14.991$$

$$\Rightarrow \quad X = 0.7235Y + 12.009$$

Hence, the most likely age of wife (Y) when the age of husband (X) is 30 years is given by :

$$Y = 0.4702 \times 30 + 8.0246$$
$$= 14.1060 + 8.0246$$
$$= 22.1306$$
$$= 22 \text{ years (approximately)}$$

Hence, the most likely age of husband (X) when the age of wife (Y) is 20 years is given by :

$$X = 0.7235 \times 20 + 12.009$$
$$= 14.470 + 12.009$$
$$= 26.479$$
$$= 26.5 \text{ years (approximately)}$$

Example 37. In fitting of a regression of Y on X to a bivariate distribution consisting of 9 observations, the explained and unexplained variations were computed as 24 and 36 respectively. Find :

(i) The coefficient of determination.

(ii) Standard error of the estimate of Y on X.

(iii) Write properties of regression coefficients.

(iv) Why do we need two regression lines? **(D.U., B.Com. (Hons.) 2007E)**

Solution. (i) Explained variation = 24 Unexplained variation = 36

Total variance = 24 + 36 = 60

$$\text{Coefficient of determination} = \frac{\text{Explained variation}}{\text{Total variation}}$$
$$= \frac{24}{60} = \frac{2}{5} = 0.4$$

(ii) Unexplained variation = 36

$$\text{Standard error of estimate} = \sqrt{\frac{SSE}{n}} = \sqrt{\frac{36}{9}} = 2$$

[Some authors may use the formula $\sqrt{\dfrac{SSE}{n-2}}$ for standard error of estimate]

(iii) **Properties of Regression Coefficients :**

1. Both the regression coefficients must be of the same sign, *i.e.*, either both are positive or both negative.
2. If both the regression coefficients are negative, the correlation coefficient will be negative. If both the regression coefficients are positive, correlation coefficient will be positive.
3. **Change of origin and change of scale :** Both the regression coefficients are independent of the change of origin but if the change of scale in X and Y is not identical, they depend on the change of scale. If the change of scale in X and Y is identical, the regression coefficients are independent of the change of scale also.
4. **Both the regression coefficients can not be more than 1 :** Since the product of the two regression coefficients is equal to r^2 which is less than or equal to one, both the regression coefficients can not be more than 1. Sometimes it may happen that one of the regression coefficients is less than 1 but very close to 1 and the other is more than 1 in such a way that the product of two regression coefficients is more than 1. Such cases are also not possible. The product of two regression coefficients should not be more than 1.

(iv) Since the regression relation is irreversible, one equation is not sufficient to predict the values of two variables X and Y. Moreover the two regression equations are derived under different set of assumptions, therefore, one equation is not sufficient to find X and Y.

Questions

Short Answer Type Questions

1. What is meant by correlation?
2. Discuss the meaning of correlation and distinguish between positive and negative correlation.
3. What do you mean by Scatter diagram? How is Scatter diagram used to determine correlation?
4. Define Karl Pearson's coefficient of correlation. What is intended to measure?
5. Explain the meaning of covariation.
6. Discuss the properties of Karl Pearson's coefficient of correlation.
8. Explain the algebraic properties of coefficient of correlation.
9. What will be your interpretation if,
   (i) $r = 0$ (Ii) $r = +1$ (iii) $r = -1$ (iv) $r = +0.80$
   (v) $r = -0.80$?
10. Distinguish between linear and curvillinear correlation.
12. Write short note on Scatter diagram.
14. Explain the regression coefficients in regression lines.

State whether the Following Statements are True or False :

1. Correlation coefficient establishes cause and effect relationship between the two variables.
2. Coefficient of correlation is a relative measure of association between two variables.
3. Correlation coefficient is an absolute measure of relationship between the two variables.
4. r is exprssed in the units of original data.
5. Correlation coefficient is always positive.
6. $|r|$ is $0 \leq |r| \leq 1$.
7. Scatter diagram helps indetermining the degree of correlation.
8. If $r = +1$ or -1, the correlation will be perfect.
9. Correlation coefficient measures only the degree of linear relationship.
10. Negative correlation means that both the variables have decreasing values.
11. Positive correlation means that both variables have increasing values.
12. If the direction of change in the two variables is the same, the two variables will be positively correlated.
13. If the change in the two variables is in opposite direction, the correlation will be negative.
13. If $X = 2Y$, X and Y will be perfectly correlated.
15. If $X = 20 - Y$, there will be a perfect negative correlation between X and Y.
16. If $X = Y + 2$, $r = +1$.
17. $r = 0$ means absence of correlation.
18. Coefficient of correlation between height and weight is 1.4.
20. Correlation coefficient is independent of change in origin and scale.
21. The correlation coefficient increases with the number of paired observations.
22. Rank correlation can be computed only for quantitative data.
23. The coefficient of correlation has the same unit as the Pearson's r.
24. The coefficient of correlation can be calculated only if both the variables are in the same units.
25. The correlation between the age of applicants for life insurance and premium of insurance is positive.

26. A relation such as (I) $r^2 = 1 - \frac{560}{400}$ (ii) $r^2 = 1 - \frac{650}{500}$ is not possible.

28. If one of the regression coefficients are greater than unity, the other must also be greater than unity.

29. Arithmetic mean of regression coefficients is greater than the correlation coefficient.

30. Correlation coefficient is independent of the change of origin as well as change of scale.

ANSWERS

1. False (Not necessary). Suppose there is a danger of famine, the price will increase and the demand will also increase. But we cannot say that one of the price or demand is the cause and the other is effect because the cause is famine.
2. True, because it gives the relation between two series.
3. False.
4. False. It is free from units.
5. False. It can be positive or negative.
6. True. $-1 \le r \le 1$ but $0 \le |r| \le 1$.
7. False. It gives only rough idea about the correlation between two variables.
8. True.
9. True (Refer Karl Pearson's coefficient of correlation).

10, 11. False, the correlation is positive when both the variables either increase together or decrease together.

12, 13. True.

14, 15. True. In Q. 14, the line has positive slope and in Q. 15 the line has negative slope. In both the cases, the relationship is linear and the Scatter diagram will be a straight line.

16. True. The Scatter diagram will be a straight line with positive slope, therefore, $r = 1$.
17. True.
18. False, there is no fixed correlation between height and weight.
20. True.
21. False.
22. False, because it is used for qualitative data also.
23. Coefficient of correlation has no units.
24. False, correlation between price and demand is negative but they are not measured in the same units.
25. True. As the age increase, the premium of insurance for fresh policies also increases.
26. True because r^2 cannot be negative.
28. False. Both the regression coefficients cannot be greater than unity.
29. AM ≥ GM. Correlation Coefficient = GM of regression coefficients. Otherwise also

$$\text{A.M. of Regression coefficients} = \frac{r\frac{\sigma_y}{\sigma_x} + r\frac{\sigma_x}{\sigma_y}}{2} = \frac{r\,[\sigma_y^2 + \sigma_x^2]}{2\sigma_x\sigma_y}$$

$$= \frac{r\,(\sigma_y - \sigma_x)^2 *}{2\sigma_x\sigma_y} + \frac{r\,2\sigma_x\sigma_y}{2}\sigma_x\sigma_y = r\left[1 + \frac{(\sigma_y - \sigma_x)^2}{2\sigma_x\sigma_y}\right]$$ which is clearly greater than r.

30. True.

Fill in the Blanks

1. Karl Pearson's coefficient of correlation measures .... (linear correlation, non-linear correlation)
2. Maximum positive value of coefficient of correlation is .... (+1, +1.5)
3. Correlation between price and demand is .... (positive, negative)

4. Correlation between price and supply is .... (positive, negative)
5. If the correlation is perfect positive, its value is .... (+1, 0)

ANSWERS

1. linear correlation **2.** +1 **3.** negative **4.** positive **5.** +1.

Multiple Choice Questions

1. The coefficient of correlation between two variables '*X*' and '*Y*' is – 0.70. '*X*' is the independent variable and '*Y*' is the dependent variable in the simple regression relationship between them. What percentage of the variations in the value of '*Y*' is not explained by the regression relationship ?

(a) 30% (b) 49% (c) 51%
(d) 60% (e) 70%

2. The following data pertain to a simple regression relationship :

$$\Sigma (Y - \hat{Y})^2 = 108.16$$

Standard error of estimate = 2.6.
The number of data points is

(a) 12 (b) 13 (c) 15
(d) 16 (e) 18

3. The following data pertain to two variables '*X*' and '*Y*'

Covariance of X and $Y = 160.50$
Variance of $X = 156.25$
Variance of $Y = 275.25$

The coefficient of correlation between X and Y is approximately

(a) 0.57 (b) 0.59 (c) 0.75
(d) 0.78 (e) 0.97

4. Which of the following is true if the coefficient of correlation between two variables '*X*' and '*Y*' is zero ?

(a) Covariance of the variables X and Y is zero (b) Covariance of the variables X and Y is positive
(c) Covariance of the variables X and Y is negative (d) Variance of the variable X is zero
(e) Variance of the variable Y is zero

5. If the dependent variable decreases as the independent variable increases in an estimating equation then, the coefficient of correlation will be in the range of

(a) 0 to –2.00 (b) 0 to –1.00 (c) 0 to 0.50
(d) 0 to 0.25 (e) 0 to 1.00

6. Which of the following represents the proportion of variation in the dependent variable that is explained by the regression line ?

(a) Coefficient of determination (b) Coefficient of correlation
(c) Coefficient of variation (d) Standard error of estimate
(e) None of the above

7. Which of the following is true ?

(a) The coefficient of determination is positive always
(b) The coefficient of determination is negative if the coefficient of correlation is negative
(c) The coefficient of determination is negative if the coefficient of correlation is positive
(d) The coefficient of determination is the square root of the coefficient of correlation
(e) None of the above

8. If the regression equation is a perfect estimator of the dependent variable then which of the following is false ?

(a) The standard error of estimate is zero (b) The coefficient of correlation is zero
(c) The coefficient of determination is 1.00 (d) All the data points fall on the regression line
(e) None of the above

9. The extent to which the observed values of the dependent variable differ from the predicted values according to the regression line is measured by
(a) The slope of the regression line
(b) The intercept of the regression line
(c) The standard error of estimate
(d) The coefficient of correlation
(e) The coefficient of determination

10. If the covaraince between two variables is positive then, the coefficient of correlation between them will be in the range of
(a) 0 to –1.00
(b) 0 to –0.50
(c) 0 to –0.25
(d) 0 to 2.00
(e) 0 to 1.00

11. The variance of variables, X and Y, are 49 and 16 respectively and the covariance between them is 14. What is the coefficient of correlation between X and Y ?
(a) 0.2857
(b) 0.50
(c) 0.875
(d) 0.05
(e) Cannot be determined

12. For a simple linear regression equation $\Sigma (Y - \hat{Y})^2 = 256$ and the standard error of estimate is 4. What is the number of observations?
(a) 4
(b) 12
(c) 16
(d) 18
(e) 24

13. The regression relationship between the variables X and Y is given by $\hat{Y} = 4 + 5X$ and, it explains 64% of the changes in the variable Y. What is the coefficient of correlation between X and Y ?
(s) –0.80
(b) 0.80
(c) 0.08
(d) –0.08
(e) 0.64

14. Which of the following is true with regard to the method of least squares used in simple regression analysis ?
(a) It minimizes the sum of the squared values of the independent variable.
(b) It minimizes the sum of the squared differences between the values of the independent variable and their mean.
(c) It minimizes the sum of the squared differences between the actual values of the dependent variables and the estimated values of the dependent variable as per the regression equation.
(d) It minimizes the sum of the squared differences between the values of the dependent variable and their mean.
(e) It minimizes the sum of the squared values of the dependent variable.

15. If a regression line is represented by $Y = 1.5 - 2.0x$ and the coefficient of determination is 0.49 then, the coefficient of correlation is
(a) –0.70
(b) –0.60
(c) –0.02
(c) 0.50
(e) 0.70

16. If the variance of X and Y are 2.56 and 6.25 respectively and the covariance between them is 3.20 then the coefficient of correlation between X and Y is
(a) 0.10
(b) 0.25
(c) 0.48
(d) 0.68
(e) 0.80

17. The standard error of estimate is developed from the variation of the
(a) Observed values of the dependent variable around a fixed quantity.
(b) Observed values of the dependent variable around the mean of the observed values of the independent variable.
(c) Observed values of the dependent variable around the fitted regression line.
(d) Observed values of the independent variable around their mean.
(e) Observed values of the independent variable around the mean of the observed values of the dependent variable.

18. The Y intercept in the simple linear regression equation (X is the independent variable and Y is the dependent variable) represents the
(a) True value of Y when $X = 0$
(b) Change in average value of Y per unit change in X
(c) Expected value of Y when $X = 0$

(d) Standard deviation of the values of X

(e) Mean of the values of X

19. The slope of the simple linear regression equation (X is the independent variable and Y is the dependent variable) represents the

(a) Mean value of Y when $X = 0$

(b) Change in mean value of Y per unit change in X

(c) True value of Y for a fixed value of X

(d) Variance of the values of X

(e) Variance of the values of Y for a fixed value of X

20. Which of the following is true with regard to simple regression and correlation analysis ?

(a) Correlation analysis provides a measure, which indicates how well an equation for estimating one variable from another, explains the changes in the estimated variable.

(b) If the coefficient of correlation between two variables is close to 1, then there is no correlation between the two variables.

(c) If the slope of a regression equation is positive, then the coefficient of correlation between the variables involved is negative.

(d) If the slope of a regression equation is negative, then the dependent variable decreases as the independent variable decreases.

(e) Correlation analysis necessarily indicates the presence of cause-effect relationship between the variables involved.

21. Which of the following is indicated if the standard error of estimate for a regression equation is zero ?

(a) The standard deviation of the observed values of the dependent variable is zero.

(b) The standard deviation of the independent variable is zero.

(c) The observed value of the dependent variable will always be equal to its estimated value, for a given value of the independent variable.

(d) There is no correlation between the variables involved.

(e) The correlation coefficient between the variables is greater than zero but less than 1.

22. Which of the following is true when the slope of a regression line is negative ?

(a) The correlation coefficient between the dependent and independent variables is 1.

(b) The correlation coefficient between the dependent and independent variables, lies between 0 and 1.

(c) There is a negative correlation between the dependent and independent variables.

(d) The regression line is parallel to the horizontal axis.

(e) The regression line passes through the intersection of the horizontal and vertical axes.

23. which of the following is true with regard to a given coefficient of correlation and its corresponding coefficient of determination ?

(a) The coefficient of determination is always greater than or equal to zero, and less than or equal to 1.

(b) The coefficient of determination is always less than zero.

(c) The coefficient of determination is always equal to 1.

(d) The coefficient of determination always has the same sign as the coefficient of correlation.

(e) The magnitude of coefficient of determination is always higher than the magnitude of coefficient of correlation.

24. Whcih of the following is the graphical plot of the values of the dependent and independent variables, in the context of regression analysis ?

(a) Scatter diagram (b) Frequency polygon (c) Histogram

(d) p chart (e) Ogive

25. r will be positive when :

(a) X increases, Y decreases (b) X decreases, Y increases

(c) Both X and Y are increasing (d) None of these

26. If two variables have the linear relationship $X + Y = 100$, the correlation will be :

(a) -1 (b) $+1$ (c) $+0.80$ (d) 0.20

27. If $X - Y = 50$, the correlation between X and Y will be :

(a) –1 (b) +1 (c) 0.50 (d) None of these

28. The coefficient of correlation :

(a) has no limit (b) is less than 1 (c) is more than –1 (d) None of these

29. If the sum of the product of deviations of X and Y series from their means is zero, the correlation coefficient will be :

(a) 0 (b) 1 (c) –1 (d) None

ANSWERS

1. (c) Proportion of variations in the value of the dependent variable that is explained by the regression relationship $=$ Correlation of determination $= r^2 = (-0.70)^2 = 0.49$

$\therefore$ Proportion of variations in the value of the dependent variable that is not explained by regression relationship $= 1 - r^2 = 1 - 0.49 = 0.51$, *i.e.*, 51%.

2. (e) Standard error of estimate, $S_r = \sqrt{\dfrac{\Sigma (Y - \hat{Y})^2}{n-2}}$ or $S_e^2 = \dfrac{\Sigma (Y - \hat{Y})^2}{n-2}$

$$\therefore (2.6)^2 = \frac{108.16}{n-2} \Rightarrow n - 2 = \frac{108.16}{(2.6)^2} \Rightarrow n = \frac{108.16}{(2.6)^2} + 2 = 18.$$

3. (d) Coefficient of correlation between x and $y = \dfrac{\text{Cov. }(x, y)}{\sigma_x \cdot \sigma_y}$

$$= \frac{160.50}{\sqrt{156.25} \times \sqrt{272.25}} = 0.778 \simeq 0.78.$$

4. (a) Coefficient of correlation between x and $y\ (r) = \dfrac{\text{Cov. }(x, y)}{\sigma_x \cdot \sigma_y}$

Hence, $r = 0 \Rightarrow \text{Cov. }(x, y) = \dfrac{0}{\sigma_x \cdot \sigma_y} = 0.$

5. (b) The dependent variable decreases as the independent variable increases in an estimating equation signifies that there is a negative nature of association (*i.e.*, negative correlation) between the two variables. Hence the correlation coefficient will be negative and will lie in the range of 0 to –1.00 (because correlation coefficient cannot be less than –1).

6. (a) Coefficient of determination represents the proportion of variation in the dependent variable that is explained by the regression line. [Coefficient of correlation signifies the nature of association between two variables. Coefficient of variation states the extent of variability in a set of data points. Standard error of estimate measures the variability of the observed values around the regression line].

7. (a) The coefficient of determination is always positive because it is the square of the coefficient of correlation. [Since the coefficient of correlation is a real number, the coefficient of determination must be positive].

8. (b) If the regression equation is a perfect estimator of the dependent variable, then the following are possible :

(i) The standard error of estimate is zero as $\Sigma (y - \hat{y})^2 = 0$.

(ii) The coefficient of determination is 1.00 as the coefficient of correlation is either +1.00 or –1.00.

(iii) Naturally, all the data points must lie on the regression line.

9. (c) The standard error of estimate for any regression equation is given by

$$S_e = \sqrt{\frac{\Sigma (y - \hat{y})^2}{n-2}}$$

That represents the extent to which the observed values of the dependent variable differ from the predicted values according to the regression line. The factors mentioned in the other alternatives do not represent that fact.

10. (e) The coefficient of correlation is defined as $r = \dfrac{\text{Cov.}(x, y)}{\sigma_x \cdot \sigma_y}$

It is given that, Cov. $(x, y) > 0$ and it is known that the standard deviation is always positive.

$\therefore$ $r > 0$ and hence $0 < r \le 1$ (because r cannot be more than 1).

11. (b) The coefficient of correlation between x and y

$$= \frac{\text{Cov.}(x, y)}{\sqrt{v(x) \cdot v(y)}} = \frac{\text{Cov.}(x, y)}{\sigma_x \cdot \sigma_y} = \frac{14}{\sqrt{49 \times 16}} = 0.5.$$

12. (d) For any simple linear regression equation, the standard error of estimate

$$= \sqrt{\frac{\Sigma (y - \hat{y})^2}{n-2}} = 4.00$$

So, $\dfrac{\Sigma (y - \hat{y})^2}{n-2} = 16$, *i.e.*, $n - 2 = \dfrac{256}{16} = 16 \Rightarrow n = 18.$

13. (b) Coefficient of determination $= r^2 = \dfrac{64}{100} = 0.64$

$\therefore$ Coefficient of correlation $= \sqrt{r^2} = \sqrt{0.64} = 0.8.$

Since the regression equation has a positive slope the coefficient of correlation is positive. (**Please note**)

14. (b) The method of least squares minimizes the sum of the squared differences between the values of the dependent variable and their estimated values as per the regression equation. Hence (a), (b), (d) and (e) are not true with regard to the method of least squares.

15. (a) Here the regression coefficient is negative. Therefore correlation coefficient will also be negative.

In this case correlation coefficient = $\sqrt{\text{Coefficient of Determination}}$.

16. (e) $r = \dfrac{\text{Cov.}(x, y)}{\sigma_x \cdot \sigma_y} = \dfrac{3.20}{\sqrt{2.56} \,.\, \sqrt{6.25}} = 0.8.$

17. (c) The standard error of estimate for any regression equation is given by :

$$S_e = \sqrt{\frac{\Sigma (y - \hat{y})^2}{n-2}} \quad \text{where}$$

y = Observed values of the dependent variable.

$\hat{y}$ = Estimated values from the estimating equation that correspond to each Y values and n = number of data points.

18. (c) The Y intercept of the regression line represents the mean value of Y when $X = 0$.

19. (b) The slope of the simple regression equation represents the change in average value of Y per unit change in X.

20. (a) Correlation analysis provides a measure (coefficient of determination) which indicates, how well an estimating equation explains the changes in the estimated (*i.e.*, dependent) variable.

21. (c) If the standard error of estimate for a regression equation is zero, then it indicate that the observed value of the dependent variable will always be equal to its estimated value, for a given value of the independent variable.

22. (c) When the slope of a regression line is negative, there is a negative correlation between the variables.

23. (a) The coefficient of determination is the square of coefficient of correlation (r). Hence it will always be ≥ 0 and ≤ 1.

24. (a) The graphical plot of the values of the dependent and independent variables in the context of regression analysis, is called scatter diagram.

25. (c), **26.** (a), because slope is negative

27. (b), **28.** (d), **29.** (a).

Practice Questions

1. Calculate coefficient of correlation, using Karl Pearson's formula based on actual mean value of the series given below :

Year	*Index of Industrial Production*	*No. of Unemployed People ('00) Thousand*
1964	100	11.3
1965	102	12.4
1966	104	14.0
1967	107	11.1
1968	105	12.3
1969	112	12.2
1970	103	19.1
1971	94	26.4

[**Ans.** $r = -0.695$]

2. Find out coefficient of correlation between the age of husband and wife, using Karl Pearson's method based on actual mean value of the following series :

Age of Husband	20	23	27	31	35	38	40	42
Age of Wife	18	20	24	30	32	34	36	38

[**Ans.** $r = +9.994$]

3. Information relating to price and quantity of a commodity for five months is given below :

Months	1	2	3	4	5
Price	10	10	11	12	12
Quantity	5	6	4	3	3

Calculate the coefficient of correlation between price and quantity and comment on its sign and magnitude.

[**Ans.** $= -0.96$]

4. Find out coefficient of correlation between X and Y.

X	10	12	8	15	20	25	40
Y	15	10	6	25	16	12	8

[**Ans.** $= -0.18$]

5. Find out coefficient of correlation between values of X and Y given below :

X	70	48	58	55	54	50	60	52
Y	62	47	53	60	55	68	51	48

[**Ans.** 0.25]

6. Find out the coefficient of correlation of the following :

No. of Students	1	2	3	4	5	6	7	8	9	10
Marks in Statistics	20	35	15	40	10	35	30	25	45	30
Marks in Mathematics	25	30	20	35	20	25	25	35	35	30

[**Ans.** 0.76]

7. Find rank correlation coefficient from the following data :

X	8	36	98	25	75	82	92	62	65	35
Y	84	51	91	60	68	62	86	58	35	49

[**Ans.** 0.45]

8. Calculate the rank order correlation between price and supply :

Price	4	6	8	10	12	14	16	18
Supply	10	15	20	25	30	35	40	45

[**Ans.** 1]

9. (i) Compute the coefficient of correlation between the corresponding values of X and Y in the following table :

X	2	4	5	6	8	11
Y	18	12	10	8	7	5

(ii) Multiply each X-value in the table by 2 and add 6. Multiply each value of Y in the table by 3 and subtract 15. Find the coefficient of correlation between the two new sets of values.

Explain why do you or do not obtain the same result as in (i) ? [**Ans.** -0.92]

10. Calculate the correlation coefficient from the following data :

X	12	9	8	10	11	13	7
Y	14	8	6	9	11	12	3

Let each value of X be multiplied by 2 and 6 is added to it. Similarly each value or Y is multiplied by 3 and 2 is substracted from it. What will be the correlation coefficient between the new series of X and Y?

(CA Foundation, May 1997)

[**Ans.** 0.949 in both cases]

11. 10 competitors in a beauty contest are ranked by three judges in the following order. Use Spearman's rank correlation coefficient to determine which pair of judges has the nearest approach to common taste in beauty.

I Judge	1	6	5	10	3	2	4	9	7	8
II Judge	3	5	8	4	7	10	2	1	6	9
III Judge	6	4	9	8	1	2	3	40	5	7

[**Hint.** Find r_{12}, r_{23} and r_{13} by rank correlation method]. [**Ans.** 1st Judge and 3rd Judge]

12. Compute the rank correlation coefficient from the following data :

Section A	115	109	112	87	98	98	120	100	98	118
Section B	75	73	85	70	65	76	82	73	68	80

[**Ans.** 0.73]

13. A work study practitioner made certain observations on the rate of doing work of a worker. His ratings of the worker were :

90 95 115 95 85 110 90 95 95 95

The worker actually performed at the ratings :

95 90 110 100 85 105 95 100 105 95

respectively. How would you place the work study practitioner insofar as his ability in rating is concerned?

[**Ans.** $r = 0.78$. Since there is a high degree of positive correlation between the observation of the practitioner and work actually done by the worker, the ability of the work study practitioner in rating is quite good]

14. The scores of 8 students in an examination in Mathematics and Statistics are given below :

Student No.	1	2	3	4	5	6	7	8
Marks in Mathematics	70	48	58	55	54	50	60	52
Marks in Statistics	62	47	53	60	55	68	51	48

Find (i) Product moment correlation coefficient.
(ii) Rank correlation coefficient and compare the two values. [**Ans.** (i) = 25, (ii) 0.286]

15. Find the Karl Pearson's coefficient of correlation between the age and the playing habits of people from the following information :

Age Group (in Years)	*No. of People*	*No. of Players*
15 and less than 20	200	150
20 and less than 25	270	162
25 and less than 30	340	170
30 and less than 35	360	180
35 and less than 40	400	180
40 and less than 45	300	120

Refer to Ex. 13 page 7.21
(D.U., B.Com. 2002)
[**Ans.** – 0.94]

16. Calculate Spearman's coefficient of correlation between marks assigned to ten students by judges *X* and *Y* in a certain competitive test as shown below :

S. No.	1	2	3	4	5	6	7	8	9	10
Marks by X Judge	52	53	42	60	45	41	37	38	25	27
Marks by Y Judge	65	68	43	38	77	48	35	30	25	50

[**Ans.** 0.46]

17. The coefficient of rank correlation of the marks obtained by 10 students in Mathematis and Statistics was found to be 0.5. It was then detected that the difference in ranks in the two subjects for one particular students was wrongly taken to be 3 in place 7. What should be the correct rank correlation coefficient ? [**Ans.** 0.2576]

18. The regression coefficient of regression of y on x, *i.e.*, $b_{yx} = 1.2$, if $u = \dfrac{x - 100}{20}$ and $v = \dfrac{y - 200}{300}$, find b_{vu}..

[**Ans.** 0.8]

19. In partially destroyed record, for the estimation of the two lines of regression from a bivariate data (*X*, *Y*), the following results were available :

Regression coefficient of regression of Y on $X = -1.6$
Regression coefficient of regression of X on $Y = -0.4$
Standard error of the estimate of Y on $X = 3$
Find
(i) coefficient of correlation between X and Y
(ii) standard deviation σ_X and σ_Y
(iii) standard error of the estimate of X on Y [**Ans.** (i) $r = -0.8$, (iii) $\sigma_X = 2.5$, $\sigma_Y = 5$, (iii) 1.5]

20. A firm knows from its past experience that its monthly average expenses (*X*) on advertisement are Rs. 25,000 with standard deviation of Rs. 25.25. Similarly, its average monthly product sales (*Y*) have been Rs. 45,000 with standard deviation of Rs. 50.50 of sales of its product (Y). Given this information, as also the coefficient of correlation between sales and advertisement expenditure as 0.75.

(i) Estimate the most approximate value of sales against an advertisement expenditure of Rs. 50,000.

(ii) Estimate the most approximate value of advertisement expenditure for achieving a sales target of Rs. 80,000.

[**Ans.** Rs. 82,500, Rs. 48,125]

21. The results of a bivariate analysis are given in the following form :

$$\Sigma x = 30,\ \Sigma y = 60,\ \Sigma x^2 = 190,\ \Sigma y^2 = 760,\ \Sigma xy = 380 \text{ and } n = 10.$$

Determine the regression line of x on y and estimate x for given $y = 8$. **(ICWA, Dec., 2000)**

[**Ans.** $x - 3 = 0.5\ (y - 6), x = 4$ when $y = 1$]

22. Calculate the rank correlation coefficient for the following data :

x	39	65	62	90	82	75	25	98	36	78
y	47	53	58	86	62	68	60	91	51	84

(ICWA, Dec., 2000)

[**Ans.** 0.8182]

23. For a bivariate data, you are given the following information :

$$\Sigma(X - 58) = 46$$
$$\Sigma(Y - 58) = 9$$
$$\Sigma(X - 58)\ (Y - 58) = 1095$$
$$\Sigma(X - 58)^2 = 3086$$
$$\Sigma(Y - 58)^2 = 483$$

Number of pairs of observations = 7. You are required to determine the two regression equations and the coefficient of correlation between X and Y. **(D.U., B.Com. Hons., 2000)**

[**Ans.** $r = 0.9046,\ Y - 59.285 = 0.372\ (X - 64.57),\ X - 64.57 = 2.2\ (Y - 59.285)$]

24. From the following two regression equations calculate the mean values of X and Y and the coefficient of correlation between X and Y.

$$4y - 5x = 0$$
$$5y - x - 63 = 0$$

(D.U., B.Com. Hons., 1999)

[**Ans.** $r = \frac{2}{5}$, Mean of $X = 12$, Mean of $Y = 15$]

25. The percentage marks obtained in graduation and an MBA entrance test of 10 students were as follows :

Graduation	50	52	55	60	62	65	65	66	70	75
Entrance Test	52	50	57	65	62	65	65	65	71	78

From these data, find

(a) the two regression equations, and

(b) the coefficient of correlation between the marks in graduation and those in the entrance test.

[**Ans.** $X_C = 6.182 + 0.886Y,\ Y_C = 1.434 + 0.993X,\ r = 0.938$]

26. The following data relate to marketing expenditure (in Rs. lakh) and the corresponding sales (in Rs. crores) :

Marketing Expenditure	10	12	15	20	23
Sale	10	17	23	21	35

(a) Estimate the marketing expenditure to obtain sales target of Rs. 40 crores.

(b) Also find the coefficient of correlation between marketing expenditure and sales. [**Ans.** Rs. 37, Rs. 0.865]

27. The data for 10 years on sales (Y) and advertisement expenditure (X) of a particular product yielded the following summated values (in Rs. lakh) :

$$\Sigma X = 15,\ \Sigma Y = 110,\ \Sigma XY = 400,\ \Sigma X^2 = 250 \text{ and } \Sigma Y^2 = 3200$$

Find the following :

(a) Regression coefficient b of Y on X and then the Y-intercept a.

(b) X-intercept a', and then the regression coefficient b' of X on Y.

(c) Most approximate value of Y for $X = 5$, and that of X for $Y = 25$. [**Ans.** Rs. 9.451, Rs. 0.118, Rs. 3.151]

28. Find (a) the mean values of variables X and Y, and (b) correlation coefficient between them from the following regression equations :

$$2Y - X - 50 = 0 \quad \ldots(i)$$

$$3Y - 2X - 10 = 0 \quad \ldots(ii)$$

[**Ans.** $\overline{X} = 130$, $\overline{Y} = 90$, $r = +0.866$]

29. If $\overline{X} = 15$, $\overline{Y} = 12$, $\Sigma XY = 1500$, $S_X = 6.4$, $S_Y = 9.0$ and $N = 10$, find the following :

(a) Regression coefficient b of Y on X.

(b) Regression coefficient b' of X on Y.

(c) Correlation coefficient between X and Y. [**Ans.** − 0.73, − 0.37, 0.52]

30. A panel of Judges A and B graded seven debators and independently awarded the following marks :

Debator	1	2	3	4	5	6	7
Marks by A	40	34	28	30	44	38	31
Marks by B	32	39	26	30	38	34	28

An eighth debator was awarded 36 marks by Judge A while Judge B was not present.

If Judge B were also present, how many marks would you expect him to award to the eighth debator assuming that the same degree of relationship exists in their judgement. **(D.U., B.Com. Hons., 1993)**

[**Ans.** 33 marks]

31. A manager of a company is conducting a study examining the relationship bhetween the absentee rate and the number of defectives produced. He sampled production data for 12 weeks and found the following average daily absentee rate (X) and the corresponding number of defectives produced during the week (Y).

X	7.3	6.4	6.2	5.5	6.4	4.7	5.8	7.9	6.7	9.6	10.3	7.2
Y	22	17	9	8	12	5	7	19	13	29	33	18

(a) Develop a linear regression equation that best fits this date.

(b) Calculate the standard error of estimate for this relationship. [**Ans.** (a) $Y = 5.20X - 20.57$, (b) 2.473]

❑❑❑

6 Index Numbers

Coverage Meaning and Significance of Index Numbers, Construction of Index Numbers, Types of Index Numbers, Consumer Price Index Number, Limitations of Index Numbers and Index of Industrial Production.

6.1 INTRODUCTION

Index numbers are one of the most widely used statistical devices. The fact that prices are going up or down, the industrial production is rising or falling, that sales are higher or lower than in a previous period is disclosed by index numbers. They are used to take the pulse of the economy and serve as indicators of inflationary or deflationary tendencies.

Index numbers are basically used in describing relative changes generally expressed in percentage form (without using the word 'percentage') in economic variables like production, wages, price, employment etc. over time. Index numbers can also be used to study the relative changes on the basis of geographic locations or some other characteristics, *e.g.*, incidence of crimes, number of road accidents, etc. Even the number of rooms in the dwelling of a family may be used as an index of the family's economic status. Therefore, index numbers can be called **as devices which measure the relative change in the level of a phenomenon with respect to time, geographical location or some other characteristic.** The phenomenon may be the prices of commodities, the physical quantity of goods produced, marketed or consumed or such concepts as intelligence, beauty or efficiency.

In general, index number are understood to express relative changes in price of a single good but in common practice it is used to represent the variations in a group of related items or series. The series to be grouped may relate to production, prices, unemployment or cost of living index. Such index numbers are described as barometers of economic activity. Thus an **index number** may be defined *as a device which enables us to combine the variations in different series with a view to obtain a figure that faithfully represents the net results of the change in the constituent elements.*

According to Spiegal, *"An index number is a statistical measure designed to show changes in a variable or a group of related variables with respect to time, geographical location or some other characteristic."*

According to Wheldon, *"A index number is a statistical devie for indicating the relative movements of the data where measurement of actual movements is dificult or incapable of being made."*

"Index Numbers are devices for measuring differences in the magnitude of a group of related variables." —*Croxton and Cowdon*

"Index Numbers are used to measure the changes in some quantity which we cannot observe directly." —*Dr. A.L. Broley*

"An index number is a quantity which by reference to a base period shows by its variation, the changes in the magnitudes over a period of time." —*John. I. Griffen*

For calculating index numbers we consider two time periods.

(i) Base Period [Reference Period]

(ii) Current Period.

(i) Base Period is the point of time with which all later changes aree compared, *i.e.,* the period with respect to which changes are measured.

(ii) Current Period is the period of time in which change is measured considering the prices (or other figures) in the base period as standard.

$$\text{Simple Price Index number or price relative} = \frac{P_1}{P_0} = 100 = I_{01}$$

$$= \frac{\text{Current year's price Per unit}}{\text{Base year's price Per unit}} \times 100$$

$$\text{Quantity Index number} = \frac{\text{Current year's quantity produced or consumed}}{\text{Base year's quantity produced or consumed}}$$

Notes.

1. Variables (*i.e.,* price, production, consumption, etc.) are supposed to remain constant over the base period.
2. *Base period and Current period are normally taken as one year, so we call them as Base year and Current year.*
3. When the comparison is made in respect of prices the index number is called *Index number of prices of price index number* or *price relative* or *price index.*
4. If p_1', p_1'', p_1''' ...... are the prices per unit of different commodities in the current year and p_0', p_0'', p_0''' ...... are the prices of the same commodities in the base year and q', q'', q''' ...... are the respective quantities purchased, then

$$\text{Price Index} = \frac{p_1'q' + p_1''q'' + p_1'''q''' + \ldots}{p_0'q' + p_0''q'' + p_0'''q''' + \ldots} \times 100^*$$

i.e., in price index numbers, quantities are used as weights.

If we can consider only the quantities produced or consumed in the Current year and Base year, the index number is called *quantity index number.*** However the sum of the products of quantities purchased with their respective prices per unit in any period is called *True value* (or simply value) for that period. There are two types of weights : quantity weights and price weights. A *quantity weight, symbolized by q,* means *the amount of a commodity produced,* distributed, or consumed *in some time period.* Price is per unit. A value *weight combines price* with quantity *produced,* distributed, or consumed. Value is symbolized by $p \times q$ = Price Per unit × Quantity.

* Quantites for Base year and Current year should not change.

** Price per unit is taken as weight in quantity index numbers.

6.2 SIMPLE AND COMPOSITE INDEX NUMBERS

(Also called single factors index numbers and multiple factor index numbers.)

When the variation in the level of single item is being studied, the index number is called *univariable or simple index number.* For example, the index numbers of sugar production in India during the last 5 years would be an example of univariable index number. If the changes in the average level of various items is being studied together, this index number would be called a *composite index number.* The index number of agricultural production in India would be an example of a composite index number. Sometimes an index number may be taken as a simple index number in a particular situation and the same index number may be used as a composite index number in another situation. For example, an intelligence test score is actually a composite index of intelligence because numerical values for performance on each item of the test are combined to give the test score but in the construcion of an index of general ability we may take an intelligence test score as one of the simple indices to be combined to form a composite index of general ability.

6.3 TYPES OF INDEX NUMBERS

(i) **Price index numbers :** whole sale price index number
(ii) **Quantity index numbers :** index of industrial production
(iii) **Value index numbers :** Index of departmental store sales
(iv) **Special purpose index numbers :** Index of business activity
(v) **Consumer price index number :** Cost of living index number.

(i) Price Index Numbers (Refer Page 6.1)

(ii) Quantity Index Numbers

Quantity index numbers measure, and permit *comparison* of the *phsyical volume* of goods produced, or distributed, or consumed. Though the Price index numbers are more widely used, Production (Quantity) index numbers are highly significant as indicators of the level of output in the economy or in parts of it.

The **direction in which the economy** is moving is shown by how much is being produced and how this level of production compares with previous levels. A quantity index is useful in production decisions because it avoids the effects of inflation and price fluctuations due to competitive market. *Quantity index numbers are frequently used for commodities subject to considerable price variations.* In measuring the standard of living in terms of physical output, quantity index numbers are invaluable.

$$\text{Quantity index number} = \frac{\text{Quantity used in the Current Year}}{\text{Quantity used in the Base Year}} \times 100$$

$$= \frac{q_1}{q_0} \times 100$$

In constructing Quantity index numbers, the problems confronting the statistician are analogous to those involved in price index numbers. We measure changes in quantities, and when we weigh we use prices or values as weights. **Index of industrial production**.

(iii) Value Index Numbers

Value index numbers compare total value (price × quantity) in a given year with total value in the base year.

$$\text{Value index number} = \frac{\text{Sum of the Current Year values}}{\text{Sum of the Base Year values}} \times 100$$

$$= \frac{\Sigma p_1 q_1}{\Sigma p_0 q_0} \times 100$$

These index numbers are not very **popular**.

(iv) Special Purpose Index Numbers

Index of production-workers employment in manufacturing industries, measurement of commercial and industrial efficiency and index number of sales opportunities usually by geographical areas computed by market research organisations are special purpose index numbers.

(v) Consumer Price Index Numbers (Refer page 6.15)

6.4 METHODS OF CONSTRUCTION OF PRICE INDEX NUMBERS

The following are the different methods of construction of index numbers of prices of different commodities :

(i) Simple aggregate of prices
(ii) Weighted aggregate of prices
(iii) Simple arithmetic mean of price relatives
(iv) Simple geometric mean of price relatives
(v) Weighted arithmetic mean of price relatives
(vi) Weighted geometric mean of price relatives
(vii) Family Budget method

(i) Simple Aggregate of Prices

In this method the price for standard unit for each of the items included in the group is aggregated for the base year as well as for the current year separately and then the total price for the current year is expressed as a percentage of total price for the base year.

If $p_1', p_1'', p_1''', \ldots, p_1^{(n)}$ be the price per standard unit of different commodities in the current year and $p'_0, p''_0, p'''_0, \ldots, p_0^{(n)}$ be the corresponding prices in the base year, then price relative is given by

$$I_{01}^{*} = \frac{p_1' + p_1'' + p_1''' + \ldots\ldots + p_1^{(n)}}{p_0' + p_0'' + p_0''' + \ldots\ldots + p_0^{(n)}} \times 100$$

$$= \frac{\Sigma p_1}{\Sigma p_0} \times 100$$

$$= \frac{\text{Sum of the unit prices of all the commodities in the Current Year}}{\text{Sum of the unit prices of all the commodities in the Base Year}} \times 100$$

(ii) Weighted Aggregate of Prices

In this method, the prices in the current year and the base year are both weighted by *same* quantities. The ratio of the aggregate of the weighted prices for the current year to the aggregate of the weighted prices for the base year expressed as a percentage gives the required index number. **Please note that same quantities should be used for base year as well as current year.**

---

* $I_{01} \rightarrow$ Index numbers with 0 as base year are 1 as current year.

$I_{10} \rightarrow$ Index numbers with 1 as base year are 0 as current year.

If $w', w'', ..., w^{(n)}$ are the respective weights, then the *weighted aggregative* price index is given by

$$I_{01} = \frac{p_1'w' + p_1''w'' + ... + p_1^{(n)}w^{(n)}}{p_0'w' + p_0''w'' + ... + p_0^{(n)}w^{(n)}} \times 100$$

$$= \frac{\Sigma p_1 w}{\Sigma p_0 w} \times 100$$

$$= \frac{\text{Sum of the values of all the commodities in the Current year}}{\text{Sum of the values of all the commodities in the Base Year}} \times 100$$

where $p_1', p_1'', p_1''', ...$ and $p_0', p_0'', p_0''', ...$ have the same meaning as in simple aggregate of prices.

Notes. 1. We may use *q* instead of *w*.

2. Same weights should be used in Current Year as well as Base Year.

Index of simple aggregate of prices suffers from the limitations that

(1) It is highly influenced by the high priced items. This limitation severely restricts the usefulness of an index number calculated by the method of simple aggregate of actual prices, *e.g.,* the price of one kg of tea may be many times more than the price of one kg of potatoes.

(2) There is another limitations also that the unit of measurement should be same for all commodities and prices per unit should be for the same unit for all commodities, *e.g.,* if one person uses the unit quintal for one commodity, kg for the other commodity and litre for the third commodity and the other person uses the unit kg for all commodities, then the index number will differ in both the cases.

(3) One limitation of the weighted aggregative index is that it assumes a fixed consumption pattern, *i.e.,* it assumes that the quantities bought in the base year and non-base year (current year) are the same. It may not pose a very serious problem if both the periods are close together but if the time interval is more, this assumption may be unrealistic and the index may lose much of its usefulness.

But still if choice is to be made between weighted and unweighted price index, weighted price index is always preferred because the use of weights assures that relative importance of the items in the index have been taken into consideration.

(iii) Simple Arithmetic Mean of Price Relatives

$$I_{01} = \frac{1}{n}\left[\frac{p_1'}{p_0'} \times 100 + \frac{p_1''}{p_0''} \times 100 + \frac{p_1'''}{p_0'''} \times 100 + ... + \frac{p_1^{(n)}}{p_0^{(n)}} \times 100\right]$$

$$= \frac{100}{n}\left[\frac{p_1'}{p_0'} + \frac{p_0''}{p_0''} + \frac{p_1'''}{p_0'''} ...... + \frac{p_1^{(n)}}{p_0^{(n)}}\right]$$

$$= \frac{100}{n}\Sigma\left(\frac{p_1}{p_0}\right) = \frac{1}{n}\left[\Sigma \frac{p_1}{p_0} \times 100\right]$$

$$= \frac{\text{Sum of the price relatives of all the commodities}}{\text{Total no. of commodities}}$$

where $\frac{p_1'}{p_0'} \times 100, \frac{p_1''}{p_0''} \times 100, \frac{p_1'''}{p_0'''} \times 100, ..., \frac{p_1^{(n)}}{p_0^{(n)}}$

are the price relatives of items 1, 2, 3, ... respectively and n the number of items or commodities included in the group.

In the construction of index numbers by aggregate of prices, it was mentioned that the price of all the quantities must be expressed in the same units but in the simple average of price relatives, this difficulty does not appear because we are comparing price per kg with price per kg and price per tonne with price per tonne but in this method also we find a concealed assumption that in the base year the wholesalers sell an equal number of ruppee worth of each commodity in the basket. This assumption does not correspond to experience.

Suppose in 1995 a wholesaler sells three commodities A, B and C worth Rs. 100 each at the rates of Rs. 2, 4 and 5 per unit of A, B and C respectively, then number of units of A, B and C respectively are 50, 25 and 20. Suppose the price per units of A, B and C in 1998 becomes Rs. 2.5, 5 and 10 respectiely, then the price index for these three items will be $\frac{2.5}{2} \times 100$, $\frac{5}{4} \times 100$ and $\frac{10}{5} \times 100$, *i.e.*, 112.5, 125 and 200. By selling the same number of units in the current year as in the base year, the shopkeeper will get 50×2.5, 25×5 and 20×10, *i.e.*, Rs. 112.5, 125 and Rs. 200 which imply that we have embedded the quantities 50, 25 and 20 in the index. In other words, we have unintentionally weighted the data.

(iv) Simple Geometric Mean of Price Relatives

The index number computed by this method is given by

$$I_{01} = \left[\left\{\frac{p_1'}{p_0'} \times 100\right\}\left\{\frac{p_1''}{p_0''} \times 100\right\}\left\{\frac{p_1'''}{p_0'''} \times 100\right\} \dots \left\{\frac{p_1^{(n)}}{p_0^{(n)}} \times 100\right\}\right]^{\frac{1}{n}}$$

$$= 100 \sqrt[n]{\frac{p_1'}{p_0'} \times \frac{p_1''}{p_0''} \times \frac{p_1'''}{p_0'''} \times \dots \times \frac{p_1^{(n)}}{p_0^{(n)}}}$$

$$\therefore \qquad \log I_{01} = \log 100 + \frac{1}{n} \Sigma \log \frac{p_1}{p_0},$$

where $\frac{p_1'}{p_0'} \times 100$, $\frac{p_1''}{p_0''} \times 100$, $\frac{p_1'''}{p_0'''} \times 1000$, ..., $\frac{p_1^{(n)}}{p_0^{(n)}} \times 100$ are the price relatives of items 1, 2, 3, ..., n respectively.

(v) Weighted Arithmetic Mean of Price Relatives

When the weights $w_1, w_2, \dots\dots, w_n$ are attached to the respective commodities , the index number, calculated as the weighted arithmetic mean of price relatives, is given by

$$I_{01} = \frac{100}{\Sigma w}\left(\frac{p_1'}{p_0'} w_1 + \frac{p_1''}{p_0''} w_2 + \dots + \frac{p_1^{(n)}}{p_0^{(n)}} w_n\right)$$

i.e.,

$$I_{01} = \frac{100}{\Sigma w} \Sigma \frac{p_1}{p_0} w = \frac{1}{\Sigma w} \Sigma \left(\frac{p_1}{p_0} \times 100\right) w,$$

where $\left(\frac{p_1'}{p_0'} \times 100\right) w_1$, $\left(\frac{p_1''}{p_0''} \times 100\right) w_2$, $\left(\frac{p_1'''}{p_0'''} \times 100\right) w_3$, ..., $\left(\frac{p^{(n)}}{p_0^{(n)}} \times 100\right) w_n$ are the weighted price relatives of items 1, 2, 3, ..., n respectively.

(vi) Weighted Geometric Mean of Price Relatives

The index number calculated by this method is not very much in use. It is given by

$$I_{01} = 100\left\{\left(\frac{p_1'}{p_0'}\right)^{w_1} \times \left(\frac{p_1''}{p_0''}\right)^{w_2} \times \ldots\ldots \times \left(\frac{p_1^{(n)}}{p_0^{(n)}}\right)^{w_n}\right\}^{\frac{1}{\Sigma w}},$$

where $w_1, w_2, \ldots\ldots, w_n$ are the weights attached to the respective commodities and $\Sigma w = w_1 + w_2 + w_3 + \ldots\ldots + w_n$.

[If $I_1, I_2, I_3, \ldots\ldots, I_n$ are the price relatives with $w_1, w_2, w_3, \ldots\ldots, w_n$ as respective weights, then

$$I_{01} = \left[(I_1)^{w_1}\,(I_2)^{w_2} \times (I_3)^{w_3} \ldots (I_n)^{w_n}\right]^{\frac{1}{\Sigma w}}$$

$$\log I_{01} = \frac{1}{\Sigma w}[w_1 \log I_1 + w_2 \log I_2 + w_3 \log I_3 + \ldots + w_n \log I_n]$$

$$= \frac{1}{\Sigma w}[\Sigma w \log I],$$

where $$I_1 = \frac{p_1'}{p_0'} \times 100,\ I_2 = \frac{p_1''}{p_0''} \times 100.\ \ldots\ldots].$$

(vii) Family Budget Method of Constructing Consumer Price Index Numbers

Current year's index number or consumer price index number or cost of living index number

$$= \frac{\text{Sum of the products of price relatives and base year values}}{\text{Sum of base year values}}$$

$$= \frac{\Sigma IV}{\Sigma V}$$

where $$I = \frac{\text{Current year price}}{\text{Base year price}} \times 100 = \frac{p_1}{p_0} \times 100 \text{ and } V = p_0 q_0$$

I and V are calculated for different commodities.

6.5 LASPEYRE'S PRICE INDEX NUMBER*

Laspeyre's index number is a weighted aggregate price index number in which base year quantities are used as weights.

---

* Quantity index numbers are obtained by interchanging p and q.

If the quantity index number is prepared by using Laspeyre's method, the base year prices will be used as weights

$$Q_{01} = \frac{\Sigma q_1 p_0}{\Sigma q_0 p_0} \times 100.$$

In Paasche's method current year prices are used as weights

$$Q_{01} = \frac{\Sigma q_1 p_1}{\Sigma q_0 p_1} \times 100$$

Fisher's ideal quantity index number

$$= Q_{01} = \sqrt{\frac{\Sigma q_1 p_0}{\Sigma q_0 p_0} \times \frac{\Sigma q_1 p_1}{\Sigma q_0 p_1}} \times 100$$

If $p_1', p_1'', p_1''', \ldots\ldots, p_1^{(n)}$ denote current year prices per unit for different commodities and $p_0', p_0'', p_0''', \ldots\ldots, p_0^{(n)}$ denote corresponding per unit base year prices and $q_0', q_0'', q_0''', \ldots\ldots, q_0^{(n)}$ be the respective quantities for base year, then Laspeyre's price index number

$$(I_{01}) = \frac{p_1'q_0' + p_1''q_0'' + p_1'''q_0''' + \ldots + p_1^{(n)}q_0^{(n)}}{p_0'q_0' + p_0''q_0'' + p_0'''q_0''' + \ldots + p_0^{(n)}q_0^{(n)}} \times 100$$

$$= \frac{\Sigma p_1q_0}{\Sigma p_0q_0} \times 100$$

$$= \frac{\text{Sum of the amounts paid for base year quantities at current year prices}}{\text{Sum of the amounts paid for base year quantities at base year rices}} \times 100.$$

If the quantity index number is prepared by using Laspeyre's method, the base year prices will be used as weights

$$Q_{01} = \frac{\Sigma q_1p_0}{\Sigma q_0p_0} \times 100.$$

6.6 PAASCHE'S PRICE INDEX NUMBER

Paasche's index number is a weighted aggregate price index number in which current year quantities are used as weights.

If $p_1', p_1'', p_1''', \ldots\ldots, p_1^{(n)}$ denote current year prices per unit for different commodities and $p_0', p_0'', p_0''', \ldots\ldots, p_0^{(n)}$ denote corresponding per unit base year, prices and $q_1', q_1'', q_1''', \ldots\ldots, q_1^{(n)}$ be the respective quantities for current year then Paasche's price index number

$$(I_{01}) = \frac{p_1'q_1' + p_1''q_1'' + p_1'''q_1''' + \ldots\ldots + p_1^{(n)}q_1^{(n)}}{p_0'q_1' + p_0''q_1'' + p_0'''q_1''' + \ldots + p_0^{(n)}q_1^{(n)}} \times 100$$

$$= \frac{\Sigma p_1q_1}{\Sigma p_0q_1} \times 100$$

$$= \frac{\text{Sum of the amounts paid for Current Year quantities at Current year prices}}{\text{Sum of the amounts paid for Current Year quantities at Base Year Prices}} \times 100.$$

In quantity index numbers by Paasche's method current year prices are used as weights.

$$Q_{01} = \frac{\Sigma q_1p_1}{\Sigma q_0p_1} \times 100$$

6.7 DORBISH BOWLEY'S INDEX NUMBER

The Arithmetic Mean of Laspeyre's index number and Paasche's index number is known as *Bowley's index* number.

Thus Bowley's index number (I_{01})

$$= \frac{1}{2} \text{ (Laspeyre's Index number + Paasche's Index number)}$$

$$= \frac{1}{2}\left(\frac{\Sigma p_1q_0}{\Sigma p_0q_0} + \frac{\Sigma p_1q_1}{\Sigma p_0q_1}\right) \times 100$$

6.8 FISHER'S IDEAL INDEX NUMBER

The Geometric mean of Laspeyre's Index number and Paasche's Index number is known as Fisher's Ideal Index number after the name of **Irving Fisher.**

Fisher's Ideal Index Number (I_{01})

$$= \sqrt{\text{Laspeyre's Index number} \times \text{Paasche's Index number}}$$

$$= 100\sqrt{\frac{\Sigma p_1q_0}{\Sigma p_0q_0} \times \frac{\Sigma p_1q_1}{\Sigma p_0q_1}}$$

Fisher's ideal quantity index number

$$= Q_{10} = 100 \times \sqrt{\frac{\Sigma q_1p_0}{\Sigma q_0p_0} \times \frac{\Sigma q_1p_1}{\Sigma q_0p_1}}$$

6.9 COMPARISON BETWEEN LASPEYRE'S INDEX NUMBER AND PAASCHE'S INDEX NUMBER

Laspeyre's index number employs base year quantities as weights which do not change from year to year. But with the passage of time conditions may change and the relative importance of the different items in the index numbers may also change. Keeping in view this aspect it would be advisable to use Paasche's index number where we use current year quantities as weights. But due to changing conditions weights may change in every period and to obtain new quantities (weights) for each period becomes quite expensive and difficult. *Therefore from a practical point of view Laspeyre's index number is often preferred over Paasche's index number.* In general, Laspeyre's price index number is **greater** than Paasche's price index number because Laspeyre's index number uses base year quantities as weights and in times of rising prices, consumption is shifted to cheaper substitutes but in Laspeyre's index numbers weights of expensive as well as cheaper items are not changed so that expensive items retain their higher weights and cheaper items retain their lower weights resulting in an *upward bias* On the other hand, the contrary of this happens in case of Paasche's method and the index has a *downward bias.* This however does not mean that Laspeyre's index number would always be more than the Paasche's index number.

The Laspeyre's price index overestimates the damage done to the consumer by the price changes, *since it neglects the consumer's ability to* react to the changes by altering the *amounts of the products he buys.* By a similar argument, the Paasche's index underestimates the value of price index.

However if the change from the base period to the current period is not substantial both the indices give satisfactory results. Under such circumstances Laspeyre's method becomes more popular as in this index weights are constant and therefore the construction of the index is easier. It should be further noted that if the prices of all the commodities change in the same ratio then these two index numbers will give the same value, as in such a situation weighing system is irrelevant. Similarly if the quantities of all goods change in the same ratio then also the two indices would give the same value of the index as the two weighting systems become identical (Refer Ex. 6 page 6.22).

6.10 FISHER'S INDEX NUMBER IS CALLED IDEAL—WHY

Fisher's index number is called ideal because of the following reasons :

(1) Geometric mean is useful in averaging percentages and ratios. Index numbers indicate percentage changes and Fisher's ideal index number is a Geometric mean between Laspeyre's and Paasche's index numbers.*

(2) It takes into account both Current Year as well as Base Year quantities.

(3) It satisfies certain mathematical tests such as time reversal and factor reversal tests.

(4) It is free from bias upwards as well as downwards (Please note that Laspeyre's index numbers suffer from upward bias whereas Paasche's index suffer from down ward bias.

6.11 MARSHAL-EDGEWORTH'S INDEX NUMBERS (I_{01})

This index number makes use of both the Base Year quantities and the Current Year quantities. (The sum or average of Base Year and Current Year quantities is used as weights.)

Marshall-Edgeworth's Index

$$(I_{01}) = \frac{p_1'(q_0'+q_1') + p_1''(q_0''+q_1'') + p_1'''(q_0'''+q_1''') + \ldots + p_1^{(n)}(q_0^{(n)}+q_1^{(n)})}{p_0'(q_0'+q_1') + p_0''(q_0''+q_1'') + p_0'''(q_0'''+q_1''') + \ldots + p_0^{(n)}(q_0^{(n)}+q_1^{(n)})}$$

$$= \frac{\Sigma p_1(q_0+q_1)}{\Sigma p_0(q_0+q_1)} \times 100 = \frac{\Sigma p_1 \dfrac{(q_0+q_1)}{2}}{\Sigma p_0 \dfrac{(q_0+q_1)}{2}} \times 100 = \frac{\Sigma p_1 q_0 + \Sigma p_1 q_1}{\Sigma p_0 q_0 + \Sigma p_0 q_1} \times 100$$

6.12 KELLY'S METHOD

According to this method ratio of the aggregates with selected weights, not necessarily base year or current year weight, gives the index number with the selected year as base year

$$P_{01} = \frac{\Sigma p_1 q}{\Sigma p_0 q} \times 100,$$

where q refers to quantities of the year selected as base. This formula satisfies time reversal test and circular test.

6.13 SELECTION OF AN AVERAGE FOR THE CONSTRUCTION OF INDEX NUMBERS

Theoretically G.M. is the best average for the construction of index numbers but due to certain difficulties in calculating G.M., practically A.M. is used for constructing index numbers.

* Theoretically, Geometrical mean is considered to be the best average in the construction of index numbers but due to certain problems in calculations, Arithmetic mean is considered for all practical purposes.

6.14 PROBLEMS IN THE CONSTRUCTION OF INDEX NUMBERS

The following problems are mainly faced in the construction of index numbers :

(i) Definition of the Purpose,
(ii) Selection of the Base Period,
(iii) Selection of Items,
(iv) Selection of Sources of Data and Collection of Data,
(v) Selection of Average,
(vi) System of Weighting.

(i) Definition of the Purpose

There are no all purpose index numbers. Therefore, before constructing an index number the specific purpose, *i.e.,* objective for which it is designed must be clearly and rigorously defined. Haberler has rightly said "Different index number are constructed to fulfill different objectives and before setting to construct a particular index number one must clearly define one's objective of study because it is on the objective of the study that the nature and format of the index number depends".

(ii) Selection of the Base Period

Selection of the proper Base Period is an important factor in the construction of index numbers. Base Period is a reference point with which changes in other periods are measured. About the selection of Base Period, following observations may be noted.

Morris Hamburg observers, *"It is desirable that the Base Period be not too far away in times from the present. The further away we move from the Base Period, the dimmer are our recollections of Economic conditions prevailing at that time. Consequently comparisons with these remote periods tend to lose significance and to become rather tenuous in meaning."*

According to George Simpson and Fritz Kafka, *"Since practical decisions are made in terms of index numbers, and economic practices so often are a matter of the short run, we wish to make comparisons between a base which lies in the same general economic framework as the years of immediate interest. Therefore, we choose a base relatively close to the years being studied."* In the fast changing world of today the base year should not be more than a decade old. There is a psychological reason also for taking a recent period as base. T. W. Lewis and R.A. Fox write, *"A fairly recent base is advantageous simply because it keeps the values of the index fairly close to the 100 marks and thereby helps comprehension of changes in the index. If, for example, the base year is too far in the past, current values might be around the 400, 500 or 600 level. If two consecutive values of the index are, say, 573 and 604, further calculation is necessary to appreciate the size of the increase; furthermore there is a psychological effect. If a price index stands at 604 this fact tends to give people the impression that prices are high. A fairly recent base also ensures that comparisons are based on relatively homogeneous quantities."*

(iii) Selection of Items

Selection of items or 'Regimen' or 'Basket'.* In any index number neither it is possible nor necessary to include all the items or commodities. Each index number tries to measure changes pertaining to a particular group.

Selection of items also depends on the purpose of index number. Moreover, items selected should be such as are widely consumed. The items selected for an index number should be relevant, representative, reliable and comparable. In general the larger is the number of items, the lesser will

* The list of items included in the construction of an index number is called 'Regimen' or 'Basket'.

be the chances of error in the average. Since we cannot include a very large number of items. A compromise is always required between the number of items and the reasonable standard of accuracy. We must, however, have manageable number of items and should also aim at reasonable standard of accuracy.

(iv) Selection of Sources of Data and Collection of Data

For sources of data and collection of data we are mainly concerned with the prices. Use of wholesale prices or retail prices depends on the objective of study. Price quotations should be obtained from important markets. In order to ensure better results, it is advisable to take a standard price which implies representative price of a commodity for whole interval under consideration.

(v) Selection of Average

Since index numbers measure the relative changes, Geometric mean* should be the best average but due to certain difficulties in calculations with G.M., for all practical purposes Arithmetic mean is used.

(vi) System of Weighting

According to John I Giffin, *"In simple terms, weighting is designed to give component series an importance in proper relation to their real signifiance."* In order to allow each commodity to have a reasonable influence on the index it is advisable to use a suitable weighting system. In case of an unweighted index number of prices, all commodities are given equal importance. But in actual practice different commodities need a different degree of importance. In an unweighted index number, if price of wheat is doubled and that of tobacco is halved, there may or may not be any change in the index number. But increase in the price of wheat will affect adversely than the benefit which will accrue to the consumer by the reduction in tobacco price. That is the main reason for assigning weights to the various items while constructing index numbers.

The weights may be according to :

(1) The value or quantity produced.
(2) The value or quantity consumed.
(3) The value or quality sold.

For general index number the weights may be according to production and for cost of living index number weights may be according to consumptions. As per the need weights may be selected by Paasche's method, Laspeyre's method or Fisher's method.

6.15 CHARACTERISTICS OF INDEX NUMBERS

The following are the main characteristics of index numbers :

(i) Index Numbers are Specialised Averages.
(ii) Index Numbers as Percentages and Measure of Relative Changee.
(iii) Basis of Comparison.
(iv) Universal Applications.

* (a) Geometric mean is the best measure for measuring the relative changes.
(b) Geometric mean gives more importance to small items and less importance to bigger items. Thus it helps in the evaluation of the real position.
(c) Good index numbers possess the characteristic of reversibility, and that can be had if geometric mean is used.

(i) Index Numbers are Specialised Averages

According to L. R. Connor, *"In its simplest form an index number represents a special case of an average, generally a weighted average compiled from a sample of items judged to be representative of the whole."*

In simple average the data treated have same units of measurement but **index number average variables** having different **units of measurement.** We may find an index number for a group of items including food, clothing, rent, cooking gas etc. which are all measured in different units. **(Imp.)**

(ii) Index Numbers as Percentages and Measures of Relative Change

Index numbers are expressed in percentages but the word or symbol for percentage is never used. They measure relatives changes in price over years with reference to some period called **Base Period** or Base Year. If the index number in 2000 is 150 with reference to 100 in 1995, it indicates that prices have gone up by 50% in 2000 as compared to 1995. Since, the index numbers are quantitative expressions of relative changes, they are expressed in numbers.

(iii) Basis of Comparison

Index numbers are used to make comparisons over different time periods with reference to a base period.

(iv) Universal Applications

They are applied universally to ascertain different types of changes in different sectors of the economy. These changes may be with regard to prices, standard of living of the people, industrial or agricultural production, national or per capita income, exports or imports etc.

6.16 ADVANTAGES OR USES OF INDEX NUMBERS

Some of the main advantages of index numbers are :

(1) Measurement of Change in the Price Level or the Value of Money : A very important use of index numbers is that index numbers measure the value of money during different periods of time. Index numbers can be used to know the impact of the change in the value of money on different sections of the society which helps in finding measures to correct inflationary or deflationary conditions.

(2) Knowledge of the Change in Standard of Living : Index numbers help to ascertain changes in the living standards of people. Money incomes of the people may increase but if the index numbers show a decrease in the value of money, living standards may decline even with increase in money income. Thus index numbers indicate change in real income.

(3) Adjustments in Salaries and Allowances : Cost of living index is a useful guide to the Government and Private Enterprises to make necessary adjustments in salaries and allowances of the workers to enable them to maintain their standard of living. Increase in the cost of living index suggests increase in salaries and allowances of the employees. If the salaries and allowances are increased in equal proportion to the cost of living index, the workers can maintain their original standard of living.

(4) Useful to Business Community : Price index numbers serve as a useful guide to the business community in analysing the business conditions and take useful decisions, with regard to their production keeping in view the trend in prices.

(5) Information regarding Production : Index numbers of production show whether the level of argicultural and industrial production in the economy is increasing or decreasing.

(6) Information regarding Foreign Trade : Index of exports and imports provides useful information regarding foreign trade. Accordingly, export-import policies are formulated.

(7) Useful to Government : It is with the help of index numbers that the Government determines its monetary and fiscal policies and takes concrete steps for the economic development of the country. In other words, with the help of index numbers Government formulates appropriate policies and other programs to increase investment, output, income, employment, trade, price level, consumption, etc.

(8) Useful to Politicians : The analysis of index numbers helps the politicians to know about the actual economic condition of the country. Accordingly, they offer constructive criticism to the Government and make useful suggestions for improvement.

(9) Useful in Contract Escalation : The study of index numbers (*i.e.,* past behaviour or trends) help in adjusting prices in contracts extending over long periods of time.

(10) Index numbers indicate the state of balance between aggregate supply and demand at the primary market level.

(11) Index numbers are used in planning plant construction and production schedules, in purchasing raw materials, in evaluating inventories, in determining production costs and in general investment programmes.

(12) Index numbers are used to forecast future pricing, marketing, and sales policies based on historical price developments.

(13) Index numbers are also used to analyse the changing inter-relationships among individual commodities.

6.17 LIMITATIONS OF INDEX NUMBERS

With regard to the limitations of index numbers, Coulbourn has rightly said, *"In this changing world it is difficult to escape from the theoretical defects and in future, as far as we can see, it will not be possible, from theoretical point of view, to make use of the best method of constructing the index number."*

Main limitations are :

(1) Do not Cover all Commodities : Index numbers measure price changes of certain limited commodities specifically defined for pricing. It is not an overall measure of the 'general price level' or of the 'purchasing power of money'. It does not include all classes of commodities which are factors in the general price level.

(2) Retail Transactions not Covered : The index numbers do not measure the price movements of retail transactions because most of the index numbers are constructed on the basis of wholesale prices. In real life only retail prices are used. Therefore, the index numbers are generally misleading.

(3) International Comparison is not Possible : There is no internationally accepted base to be used by all countries for preparing index numbers. Therefore different countries prepare index numbers on different base periods and the comparison at the international level is not possible.

(4) Effect of Time : With the change in time, general habits of the people, tastes, purchasing capacities, etc. may change. Therefore, the index numbers constructed on old consumption pattern may not give true picture at all times.

(5) Limited Use : Index numbers are always prepared for certain specific objective. The index numbers prepared for one purpose may not be suitable for some other purpose. Index numbers prepared to know about the economic condition of one class of people may not represent another class of people.

(6) Unscientific Weightage : There is no scientific method of according weights to most of the index numbers. Consequently, index numbers do not offer precise conclusions.

6.18 CONSUMER PRICE INDEX (ALSO CALLED COST EXPENDITURE INDEX NUMBER OR COST OF LIVING INDEX NUMBER)

The Consumer Price Index is a statistical measure of changes in retail prices of the goods and services bought by families of city wage earners and clerical workers.* It measures only effect of price changes on the goods. It tells nothing about changes in the kind and amounts of goods and services families buy, or the total amount families spend for living, or the differences in living costs in different places and in different classes of people. John I. Griffin writes, *"The index measures only changes in prices. It tells nothing about changes in the kinds and amounts of goods and services families buy, the total amount families spend for living or the differences in living costs in different places".* The complete name for this index is : *Index of Change in Prices of Goods and Services Purchased by City Wage-Earner and Clerical Worker Families to Maintain their Level of Living.*

The index is concerned with *prices.* These are the retail prices of foods, clothing, house furnishings, fuel, rent and education and other miscellaneous goods and services. The fees paid to doctors and dentists; prices in barber shops and other service establishments; rates charged for transportation, electricity, gas, and other utilities; and so on. They are prices paid by consumers, and they include sales and excise taxes also.

The "market basket" of goods and services priced for the index is representative of the goods and services bought by the above class of families. The main objective of consumer price index number is to know the direction (increase or decrease) in the cost of living of the consumer. Since, the effect of price changes is not uniform on all classes of people, different consumer price index numbers are constructed for different classes of people.

It should be remembered that a cost of living index number tells us about the variations in the cost of living of only one group of persons living in a particular region. By region we mean here an area within which retail prices are almost equal and by groups we mean classes distinguished from each other on the basis of incomes or habits. Thus, there cannot be one cost of living index number for any class of workers of the whole country, because retail prices in different places differ and the pattern of consumption is also not alike in different localities. Similarly, we cannot have a cost of living index number for the whole population of a particular town because groups of persons with different incomes spend on various commodities in different ways and the relative importance of various commodities to all persons is not identical.

Still consumer price-index number in some situations give better information than general price index numbers.

Since general price index numbers fail to measure the exact effect of the change in the price level on the cost of living of people, consumer price index method is employed to overcome this deficiency. This index assists in determining the impact of rise and fall in prices on various classes of consumers.

Consumer price index serves as an effective tool in respect of the following :

(i) Wage negotiations and wage contracts.

(ii) Helps government in the areas like wage policy, price policy, rent control, taxation and other related economic policies.

* Wage earners or salaried clerical workers, include craftsmen, factory workers, labourers, clerks, sales and services workers (except domestic-service workers). Many of the families have two or more wage or salary workers; as a result, average *family incomes* are higher than average *individual earnings.*

(iii) In measuring changes in purchasing power of rupee, real income.*

(iv) In analysing suitability of markets for the goods and services.

6.19 STEPS IN THE CONSTRUCTION OF COST OF LIVING INDEX NUMBERS

(1) Select the class of people for whom the index number is to be constructed.

(2) The goods and services used by the selected class and their importance (weights) are noted to form the index number for the selected class. Weights are considered for goods and services so that their effect on the index number is according to their importance.

6.20 UTILITY OF CONSUMER PRICE INDEX NUMBERS

1. These index numbers are of great use in wage or salary adjustments due to change in price. Automatic adjustment of wages or salaries or dearness allowance are governed in many countries by consumer price index numbers.
2. Such index numbers are used to measure the purchasing power of money and thereby to find real income of a particular group of people.
3. Such index numbers do provide guidelines to the Government for deciding wage policy, price policy, income policy, control of prices and general economic policies.
4. Such index numbers are used in analysing the markets for particular kinds of goods and services.

6.21 METHOD OF CONSTRUCTION OF CONSUMER PRICE INDEX NUMBERS

There are *two* methods of constructing consumer price index numbers :

(i) Consumer Price Index $= \dfrac{\Sigma p_1 q_0}{\Sigma p_0 q_0} \times 100$

This is in fact the Laspeyre's method. This method is the most popular method for constructing consumer price index number.

(ii) *Family Budget Method or Weighted Relatives*

Consumer Price Index $= \dfrac{\Sigma PV}{\Sigma V}$ or $\left[\dfrac{\Sigma IW}{\Sigma W}\right]$

Here, $P = \dfrac{P_1}{P_0} \times 100$ for each item, $I =$ Price relatives,

$V =$ Value weights $(p_0 q_0)$, $W =$ Weights.

This method is the same as the weighted average of price relatives.

---

* Purchasing power of money $= \dfrac{1}{\text{Cost of living index}}$

and Real wage (Real income) $= \dfrac{\text{Money wages}}{\text{Cost of living index}} \times 100$

6.22 TESTS OF ADEQUACY

(i) Unit Test, (ii) Time Reversal Test, (iii) Factor Reversal Test and (iv) Circular Test.

Unit Test

This test requires that the formula should be independent of the unit in which or for which prices and quantities are quoted. All formulae except the simple (unweighted) aggregative index satisfy this test. Because in this test as compared to base year if the unit for current year is changed, the value of the index would change.

Time Reversal Test

It is a test which helps in determining whether a given formula will work in both ways in time forward and backward. This test implies that if the time subscript in any index number formulae be interchanged, then the resulting index number should be the reciprocal of Original index number. Symbolically the product $I_{01} \times I_{10}$ should be equal to 1 (omitting the factor 100 from both the indices) where I_{01} is the price index number with 1 as current year and 0 as base year and I_{10} is the index number with 1 as base year and 0 as current year,

Fisher's Ideal index number formula satisfies this test.

Factor Reversal Test

According to this test the product of a price index the quantity index should be equal to the corresponding value index, *i.e.*, **true value ratio**, omitting the factor 100 from both the indices.

i.e., $$P_{01} \times P_{01} = \frac{\Sigma P_1Q_1}{\Sigma P_0Q_0} = \text{True value ratio}$$

where P_{01} is the price index and Q_{01} is the quantity index with 1 as current year and 0 as base year.

Circular Test

This test is an extension of time reversal test for more than two periods. It is concerned with the measurement of price changes over a period of years when it is desirable to shift the base. According to this test if there are four periods say *a*, *b*, *c* and *d*, then

$I_{ab} \times I_{bc} \times I_{cd} \times I_{da}$ should be equal to 1 where I_{ab} is the price index (without factor 100) for period *b* with period *a* a base and so on

i.e., $$\frac{P_b}{P_a} \times \frac{P_c}{P_b} \times \frac{P_d}{P_c} \times \frac{P_a}{P_d} = 1.$$

If 0 represents the base year and 1, 2, 3, ..., *n* represent *n* different years then the circular test will be satisfied if

$$I_{01} \times I_{12} \times I_{23} \times .... \times I_{(n-1)n} \times I_{n0} = 1.$$

This test is satisfied only by the index number formula based on

(i) Simple geometric mean of price relatives

(ii) Kelly's fixed base method.

6.23 LINK RELATIVE

The figure of a particular year expressed as a percentage of the preceding year's figure is called link relative.

6.24 CHAIN INDEX NUMBERS

In chain index numbers there is no fixed base. A chain base index number is an index number with previous year as base. The figures for each year are first expressed as percentages of the figures for the preceding year. These percentages are then chained together by successive multiplication to form a chain index.

Chain index for current year

$$= \frac{\text{Link relative of current year} \times \text{Chain index of previous year}}{100}$$

Chain index number are useful for long term comparison where as link relatives are used to have a comparison with the immediately preceding year. The fixed base index number compiled from the original data and the chain index numbers compiled from link relatives give the same value of the index provided there is only one commodity whose indices are being constructed. Chain base index for 2000 with 1995 as base is given by

$$P_{1995,\,2000} = P_{1995,\,1996} \times P_{1996,\,1997} \times \ldots\ldots \times P_{1999,\,2000}$$

6.25 CONVERSION OF CHAIN BASE INDEX (CBI) INTO FIXED BASE INDEX (FBI) NUMBER

A chain base index number can be converted into a fixed base index number be the following rule

$$\text{Current year F.B.I.} = \frac{\text{Current Year's C.B.I.} \times \text{Previous Year's F.B.I.}}{100}$$

where F.B.I. stands for Fixed Base Index and C.B.I. stands for Chain Base Index Number.

6.26 CONVERSION OF FIXED BASE INDEX INTO CHAIN BASE INDEX NUMBER

A fixed base index can be converted into a chain base index by the following rule. For the first year the value would be 100.

$$\text{C.B.I.} = \frac{\text{Current year's F.B.I.} \times 100}{\text{Previous year's F.B.I.}}$$

6.27 BASE SHIFTING

Base shifting means changing of the given year of an index number and recasting it into a series based on some recent new base year.

$$\text{New Index of any year} = \frac{\text{Old Index of the year}}{\text{Index Number of the new base}} \times 100$$

$$= \left(\frac{100}{\text{Index of the New Base}}\right) \times \text{(old index of the year)}$$

6.28 SPLICING OF INDEX NUMBERS

Combining two or more series of overlapping index numbers to obtain a single index number on a common base is called *splicing of index numbers.* Splicing of index numbers can be done only if the index numbers are constructed with the same items and have an overlapping year.

Splicing is generally done when an old index number with an old base is being discontinued and a new index with a new base is being started.

6.29 QUANTITY INDEX NUMBERS

Quantity index numbers measure average change in quantity of goods produced or sold. The weights in quantity index numbers are prices.

6.30 VALUE INDEX NUMBERS

$$\text{Value} = \text{Product of Price and Quantity}$$

$$\text{Value index number} = \frac{\text{Sum of the Current Year values}}{\text{Sum of the Base Year values}} \times 100^*$$

These index numbers are not very popular.

SOLVED EXAMPLES

Example 1. A company has three major items that it produces, say items A, B and C. The table below gives prices and units sold in each of three years, *i.e.,* 1991, 1992, 1993.

Item	*1991*		*1992*		*1993*	
	Price	*Qty.*	*Price*	*Qty.*	*Price*	*Qty.*
A	0.50	30	0.75	30	1.00	35
B	2.00	25	4.00	20	3.00	25
C	3.00	20	3.50	25	3.50	20

Use 1991 as the base and answer the following :

(a) The unweighted aggregate price index for 1993.
(b) The unweighted aggregate quantity index for 1993.
(c) The weighted aggregate index for 1993, using Laspeyre's Method to calculate weights.
(d) The weighted aggregate index for 1993, using Paasche's Method to calculate weights.
(e) The weighted average of relatives price index for 1993, using 1991 values as weights.
(f) The unweighted average for relatives quantity index for 1993.

Solution.

Item	*1991*		*1993*		P_0Q_0	P_1Q_0	P_1Q_1	P_0Q_1
	Price	*Qty.*	*Price*	*Qty.*				
	P_0	Q_0	P_1	Q_1				
A	0.50	30	1.00	35	15	30	35	17.50
B	2.00	25	3.00	25	50	75	75.00	50.00
C	3.00	20	3.50	20	60	70	80.00	60.00
	$\Sigma P_0 = 5.50$	$\Sigma Q_0 = 75$	$\Sigma P_1 = 7.50$	$\Sigma Q_1 = 80$	$\Sigma P_0Q_0 = 125$	$\Sigma P_1Q_0 = 175$	ΣP_1Q_1 180	$\Sigma P_0Q_1 = 127.50$

* $$\text{True value ratio} = \frac{\text{Sum of the products of quantities purchased in current year and respective prices per unit in current year}}{\text{Sum of the products of the quantities purchased in base year and respective Prices per unit in base year}}$$

(a) Unweighted aggregate price index for 1993 $= \frac{\Sigma P_1}{\Sigma P_0} \times 100 = \frac{7.50}{5.50} \times 100 = 136$

(b) Unweighted aggregate quantity index for 1993 $= \frac{\Sigma Q_1}{\Sigma Q_0} \times 100 = \frac{80}{75} \times 100 = 107$

(c) Weighted aggregate index for 1993 using Laspeyre's Method

$$= \frac{\Sigma P_1 Q_0}{\Sigma P_0 Q_0} \times 100 = \frac{175}{125} \times 100 = 140$$

(d) Weighted aggregate index for 1993 using Paasche's Method

$$= \frac{\Sigma P_1 Q_1}{\Sigma P_0 Q_1} \times 100 = \frac{180}{127.50} \times 100 = 141.$$

(e) Weighted average of relative price indexes for 1993 using 1991 values as weights

$$= \frac{\left[\Sigma \frac{P_1}{P_0} \times P_0 Q_0\right] \times 100*}{\Sigma P_0 Q_0} = \frac{\left[\frac{1.00}{0.50} \times 15 + \frac{3.00}{2.00} \times 50 + \frac{3.50}{3.00} \times 60\right] \times 100}{(15 + 50 + 60)}$$

$$= \frac{17.500}{125} \times 100 = 140$$

(f) Unweighted average for relative quantity indexes for 1993 $= \frac{\Sigma \frac{Q_1}{Q_0}}{3} \times 100.$

$$= \frac{\left[\frac{35}{30} + \frac{25}{25} + \frac{20}{20}\right]}{3} \times 100 = \frac{317}{3} = 106.$$

Example 2. The cost of living index uses the following weights :

Food : 40, Rent : 15 Clothing : 20, Fuel : 10, Miscellaneous : 15

During the period 1991-2001, the cost of living index raised from 100 to 205.72. Over the same period the percentage rise in prices were : Rent—60, Clothing—180, Fuel—75, Miscellaneous—165. What was the percentage change in the price of food ? **(D.U.B. Com. (Hons) 2005E, 2006E) CA (PE-I) Nove., 2002**

Solution. Let the percentage change in food price $= x$.

Average percentage price change for the whole groups

$$= \frac{\Sigma IW}{\Sigma W}$$

$$= \frac{40x + 60 \times 15 + 180 \times 20 + 75 \times 10 + 165 \times 15}{40 + 15 + 20 + 10 + 15} = 205.72 - 100$$ **(Please note.)**

$$\Rightarrow \frac{40x + 7725}{100} = 105.72 \Rightarrow x = 71.18.$$

* 1991 value mean P_0Q_0. In general value for a commodity for any year = Price per unit × Quantity consumed of that commodity.

Example 3. During a certain period, the cost of living index number goes up from 110 to 200 and the salary of the worker is also raised from Rs. 3250 to Rs. 5000. Does the worker really gain and if so by how much in real terms ?

Solution. The compensate the worker fully, his salary after increase in price index should be

$$= 3250 \times \frac{200}{110} = \text{Rs. } 5909.$$

$\therefore$ If the salary is raised to Rs. 5000, the worker will not gain in real terms but will rather suffer a loss and will not be fully compensated. To compensate the worker fully his pay should be raised to Rs. 5909.

Example 4. The following table gives the index number for different groups together with their respective weights for 2000 (base year = 1995) :

Group	*Group Index No.*	*Group Weight*
Food	130	60
Clothing	280	5
Lighting and fuel	190	7
Rent	300	9
Miscellaneous	200	19

(i) Find out the overall cost of living index number for the year 2000.

(ii) Suppose a person was earning Rs. 1,500 per month in 1995, what should be his salary in 2000, if his standard of living in that year is to be the same as in 1995 ?

Solution. (i) Overall cost of living index in 2000

$$= \frac{130 \times 60 + 280 \times 5 + 190 \times 7 + 300 \times 9 + 200 \times 19}{60 + 5 + 7 + 9 + 19}$$

$$= \frac{7{,}800 + 1{,}400 + 1{,}330 + 2{,}700 + 3{,}800}{100} = \frac{17{,}030}{100} = 170.3$$

(ii) The salary of the man in 2000 should be

$$= \frac{170.3}{100} \times 1500 = \text{Rs. } 2554.5$$

Example 5. The following information relating to works in an industrial town is given :

Item of Consumtion	*Consumer Price Index 2005 (2000 = 100)*	*Proportion of Expenditure on Item*
Food	132	60%
Clothing	154	12%
Fuel and lighting	147	16%
Housing	178	8%
Miscellaneous	158	4%

Average wage per month in 2000 is Rs. 2,000. What should be dearness allowance expressed as % of wages? What should be the average wage per worker per month in 2005 in that town so that the standard of living of worker does not fall below the 2000 level? **(DU B.Com. (Hons.) 2008)**

Solution. Weighted an of price relatives

$$= \frac{132 \times 60 + 154 \times 12 + 147 \times 16 + 178 \times 8 + 158 \times 4}{100}$$

$$= \frac{7920 + 1848 + 2352 + 1424 + 632}{100} = 141.76$$

The average wage in 2005 should be

$$2000 \times \frac{141.76}{100} = 2835.2$$

$$\text{Dearness allowance} = \frac{2835.2 - 2000}{2000} \times 100 = 41.76\%$$

Example 6. An enquiry into the budget of the middle class families of a certain city revealed that on an average the percentage expenses on the different groups were :

Food—45, Rent—15, Clothing—12, Fuel and Light—8, Miscellaneous—20.

The group index numbers for the current year as compared with a fixed base period were respectively 410, 150, 343, 248 and 285. Calculate the cost of living index number for the current year.

Mr. X was getting Rs. 2,400 in the base year and Rs. 4,300 in the current year. State how much he ought to have received as an extra allowance to maintain his former standard of living.

[**Hint :** Cost of living index for the current year $= \dfrac{\Sigma IW}{\Sigma W}$

$$= \frac{410 \times 45 + 150 \times 15 + 343 \times 12 + 248 \times 8 + 285 \times 20}{45 + 15 + 12 + 8 + 20}$$

$$= \frac{18{,}450 + 2{,}250 + 4{,}116 + 1{,}984 + 5{,}700}{100} = 325$$

i.e., if Mr. X is to maintain his base year standard of living his salary in the current year should be

$$= \frac{325}{100} \times 2400 = \text{Rs. } 7{,}800$$

Extra allowance required to be paid = Rs. (7,800 – 4,300) = Rs. 3,500.

Example 7. With regard to Laspeyre's and Paasche's price index number, it is maintained that "if the prices of all the goods change in the same ratio, the two indices will be equal for then the weighting system is irrelevent or if the quantities of the goods change in the same ratio, they will be equal, for then the two weighting systems are the same relatively." —*Karnal.*

Solution. Laspeyre's index number $= \dfrac{\Sigma p_1 q_0}{\Sigma p_0 q_0} \times 100$ ...(I)

Paasche's index number $= \frac{\Sigma p_1 q_1}{\Sigma p_0 q_1} \times 100$ ...(II)

If the prices change in the same ratio, then

$$= \frac{p_1}{p_0} = \text{constant} = k \text{ (say)} \Rightarrow p_1 = kp_0$$

Laspeyre's index number $= \frac{\Sigma p_1 q_0}{\Sigma p_0 q_0} \times 100$

$$= \frac{\Sigma kp_0 q_0}{\Sigma p_0 q_0} \times 100 = k \frac{\Sigma p_0 q_0}{\Sigma p_0 q_0} \times 100 = 100k \quad ...(III)$$

Paasche's index number $= \frac{\Sigma p_1 q_1}{\Sigma p_0 q_1} \times 100 = \frac{\Sigma kp_0 q_1}{\Sigma p_0 q_1} \times 100$

$$= k \frac{\Sigma p_0 q_1}{\Sigma p_0 q_1} \times 100 = 100k \quad ...(IV)$$

From (III) and (IV) we prove that if the prices change in the same ratio, the two index numbers are the same.

If the quantities change in the same ratio, then

$$\frac{q_1}{q_0} = \text{constant} = k' \text{ (say)} \Rightarrow q_1 = k' q_0$$

Paasche's index number

$$= \frac{\Sigma p_1 q_1}{\Sigma p_0 q_1} \times 100 = \frac{\Sigma p_1 k' q_0}{\Sigma p_0 k' q_0} \times 100 = \frac{\Sigma p_1 q_0}{\Sigma p_0 q_0} \times 100 = \text{Laspeyre's index number}$$

The result is proved.

Example 8. Incomplete information obtained from a partly destroyed record on cost of living index is given below :

Group	*Group Index*	*% of Total Expenditure*
Food	134	60
Clothing	140	Not available
Housing	105	20
Fuel & electricity	120	5
Miscellaneous	130	Not available

The cost of living index with % of total expenditure as weight was found to be 127.9. Estimate the weights used for clothing and miscellaneous. **(ICWA June 1999)**

Solution. Let the % total expenditure for clothing $= x$

Let the % total expenditure for Misc. $= y = (100 - 60 - 20 - 5 - x) = 15 - x$

$$\Rightarrow \quad 127.9 = \frac{134 \times 60 + 140 \times x + 105 \times 20 + 120 \times 5 + 130 \times (15 - x)}{100}$$

$$\Rightarrow \qquad 12{,}790 = 8{,}040 + 2{,}100 + 600 + 1{,}950 + 10x = 12{,}690 + 10x$$

$$\Rightarrow \qquad 10x = 100 \Rightarrow x = 10$$

$$\therefore \qquad y = 15 - x = 15 - 10 = 5.$$

Example 9. A textile worker in the city of Mumbai earns Rs. 3,500 per month. The cost of living index for a particular month is given as 136. Using the following information, find out the amount of money he spent on house rent and clothing :

Group	*Expenditure*	*Group Index*
Food	1,400	180
Clothing	?	150
House Rent	?	100
Fuel and lighting	560	110
Miscellaneous	630	80

Solution. Let the expenditure on clothing the house rent = Rs. x and Rs. y respectively.

Then $$1400 + x + y + 560 + 630 = 3500$$

$$\Rightarrow \qquad x + y = 3500 - 1{,}400 - 560 - 630$$

$$\Rightarrow \qquad x + y = 910 \Rightarrow y = 910 - x$$

Then

$$136 = \frac{180 \times 1400 + 150 \times x + 100\,(910 - x) + 110 \times 560 + 80 \times 630}{3{,}500}$$

$$\Rightarrow 3{,}500 \times 136 = 2{,}52{,}000 + 150x + 91{,}000 - 100x + 61{,}600 + 50{,}400$$

$$\Rightarrow \qquad 4{,}76{,}000 - 2{,}52{,}000 - 91{,}000 - 61{,}600 - 50{,}400 = 50x$$

$$\Rightarrow \qquad 50x = 21000 \;\Rightarrow\; x = \text{Rs. } 420$$

$$\therefore \qquad y = 910 - 420 = \text{Rs. } 490$$

$\therefore$ Money spent on Clothing = Rs. 420, and Money spent on House Rent = Rs. 490.

Example 10. The following table gives the cost of living index for 1998 with 1986 as base for different commodity groups :

Food, Clothings, Fuel and Light, Rent and Miscellaneous as 440, 500, 350, 400 and 250 respectively with their weights in order in the ratio 15 : 1 : 2 : 3 : 4.

Obtain, the overall cost of living index number, suppose a person was earning Rs. 4,000 in 1986. What should be his salary in 1998 to maintain the standard of living as in 1986 ? **(ICWA June 2000)**

Solution. Let the weights of the groups, Food, Clothing, Fuel and Light, Rent and Miscellaneous be $15a$, a, $2a$, $3a$ and $4a$ $(a > 0)$ respectively.

Overall index for 1998 with base as 1986

$$\frac{[400 \times 15 + 500 \times 1 + 350 \times 2 + 400 \times 3 + 250 \times 4]}{[15 + 1 + 2 + 3 + 4]}$$

$$= \frac{66{,}000 + 500 + 700 + 1{,}200 + 1{,}000}{25} = 400$$

Salary in 1998 should be $4{,}000 \times \dfrac{400}{100}$ = Rs. 16,000.

Example 11. When the cost of cigarette was increased by 50% a smoker who maintained his former scale of consumption said that the rise of the cigarette price had increased his cost by 5%. What percent of his cost of living was due to buying cigarette before the change of price ?

Solution. Suppose the smoker spends $x\%$ on cigarette before the change of price, then his expenditure on cigarette after change of price $= \frac{3}{2}x\%$

$$\text{Change in expenditure} = \frac{3}{2}x - x = \frac{1}{2}x = 5 \Rightarrow x = 10$$

i.e., 10% of his cost of living was due to buying cigarettes before the change in price.

This question can also be set in the following form :

When the cost of an item was increased by 50%, a man maintaining his former scale of consumption, said that the rise in price of the item has increased his cost of living by 5%. What per cent of his cost of living was due to buying the item before the rise of price ? **(ICWA June 1995)**

Example 12. Net weekly income of an employee was Rs. 800 in 1990. The consumer price index number was 160 in 1990. It rose to 200 in 1998. Calculate the additional D.A. (dearness allowance) to be paid to the employee if he has to be rightly compensated.

Solution. $$\frac{\text{Price in 1998}}{\text{Price in 1990}} = \frac{\text{Price index in 1998}}{\text{Price index in 1990}} = \frac{200}{160} = \frac{5}{4}$$

i.e., $$\text{Price in 1998} = \frac{5}{4} \times \text{Price in 1990}$$

The employee will be rightly compensated if his pay is also increased in the same ratio in which the price has increased.

$\therefore$ The weekly pay of the employee in 1998 should be $800 \times \frac{5}{4} = \text{Rs. } 1000$

Additional dearness allow. per week required to be paid $= 1000 - 800 = \text{Rs. } 200$.

Example 13. The price relatives and weights of a set of commodities are given in the following table

Commodity	A	B	C	D
Price relative	125	120	127	119
Weight	W_1	$2W_1$	W_2	$W_2 + 3$

If the sum of the weights is 40 and the index for the set is 122, find the numerical values of W_1 and W_2. **(D.U.B. Com (Hons) 2007E)**

Solution. $$\text{Index number} = \frac{\Sigma\left(\frac{P_1}{P_0} \times 100\right)W}{\Sigma W}$$

$$122 = \frac{125 \times W_1 + 120 \times 2W_1 + 127 \times W_2 + 119 \times (W_2 + 3)}{W_1 + 2W_1 + W_2 + W_2 + 3} \ * \quad ...(I)$$

$$\Rightarrow \quad 366W_1 + 244W_2 + 366 = 365W_1 + 246W_2 + 357$$

$$\Rightarrow \quad W_1 = 2W_2 - 9 \quad ...(II)$$

$$W_1 + 2W_1 + W_2 + W_2 + 3 = 40 \quad [\because \text{Sum of the weights} = 40]$$

$$\Rightarrow \quad 3W_1 + 2W_2 + 3 = 40$$

$$\Rightarrow \quad 2W_2 = 37 - 3W_1 \quad ...(III)$$

Substituting (II) in (I), we get

$$W_1 = 37 - 3W_1 - 9 \Rightarrow 28 - 4W_1 \Rightarrow W_1 = 7$$

$$\therefore \quad 2W_2 = 37 - 3 \times 7 = 16 \Rightarrow W_2 = 8$$

i.e., $W_1 = 7,\ W_2 = 8.$

Example 14. If the ratio between Laspeyre's index number and Paasche's index number is 28 : 27, find the missing figure in the following table :

Commodity	*Base year*		*Current year*	
	Price	*Quantity*	*Price*	*Quantity*
X	1	10	2	5
Y	1	5	—	2

Solution. Laspeyre's index number $= \dfrac{\Sigma p_1q_0}{\Sigma p_0q_0} \times 100$

Paasche's index number $= \dfrac{\Sigma p_1q_1}{\Sigma p_0q_1} \times 100$

Let the price of commodity *Y* in the current year be Rs. *x*.

Commodity	*Base year*		*Current year*		p_1q_0	p_0q_0	p_1q_1	p_0q_1
	Price	*Qty.*	*Price*	*Qty.*				
X	1	10	2	5	20	10	10	5
Y	1	5	*x*	2	5*x*	5	2*x*	2

$$\Sigma p_1q_0 = 20 + 5x,\ \Sigma p_0q_0 = 15,\ \Sigma p_1q_1 = 10 + 2x,\ \Sigma p_0q_1 = 7$$

As per the question

$$\frac{20 + 5x}{15} \times \frac{7}{10 + 2x} = \frac{28}{27} \Rightarrow \frac{4x}{10 + 2x} = \frac{4}{9}$$

$$\Rightarrow \quad 36 + 9x = 40 + 8x \Rightarrow x = 4.$$

Example 15. Construct the consumer price index number for 2000 on the basis of 1999 from the following data using :

(i) Family Budget method

* (I) can also be written as $122 = \dfrac{125w_1 + 120 \times 2w_1 + 127w_2 + 119(w_2 + 3)}{40}$

(ii) Aggregate expenditure method.

Commodity	*Rice*	*Wheat*	*Pulses*	*Ghee*	*Oil*
Weights	40	20	15	20	5
Price (per unit) 1999 (*Rs.*)	16.00	40.00	0.50	5.12	2.00
Price (per unit) 2000 (*Rs.*)	20.00	60.00	0.50	6.25	1.50

Solution. Constructing Consumer Price Index Number

(i) (Family Budget Method)

Commodity	*Weights (W)*	*Price (Rs.) (per unit) 1999* (p_0)	*Price (Rs.) (per unit) 2000* (p_1)	*Price Relative* $\frac{p_1}{p_0} \times 100$ *(I)*	*Weighted Relatives (IW)*
Rice	40	16.00	20.00	125	5,000
Wheat	20	40.00	60.00	150	3,000
Pulses	15	0.50	0.50	100	1,500
Ghee	20	5.12	6.25	122	2,440
Oil	5	2.00	1.50	75	375
	$\Sigma W = 100$				$\Sigma IW = 12,315$

Cost of Living Index for 2000

$$= \frac{\Sigma IW}{\Sigma W} = \frac{12,315}{100} = 123.15$$

Thus, there is an increase of 23.15% in prices of 2000 with that of 1999.

(ii) (Aggregative Expenditure Method)

Commodity	*Weights* q_0	*Price (Rs.) 1999* p_0	*Price (Rs.) 2000* p_1	p_0q_0	p_1q_0
Rice	40	16.00	20.00	640.00	800.00
Wheat	20	40.00	60.00	800.00	1200.00
Pulses	15	0.50	0.50	7.50	7.50
Ghee	20	5.12	6.25	102.40	125.00
Oil	5	2.00	1.50	10.00	7.50
				$\Sigma p_0q_0 = 1,559.90$	$\Sigma p_1q_0 = 2,140.00$

$$\text{Consumer Price Index for 2000} = \frac{\Sigma p_1q_0}{\Sigma p_0q_0} \times 100$$

$$= \frac{2140}{1559.9} \times 100 = 137.188$$

Thus, there is increase of 37.188 in prices of 2000 with that of 1999.

Example 16. Calculate Fisher's price index from the following data and check whether time Reverse Test is satisfied :

Commodity	*Base year (1990)*		*Current year (1995)*	
	Price (Rs.)	*Quantity (Kg.)*	*Price (Rs.)*	*Quantity (Kg.)*
A	32	50	30	50
B	30	35	25	40
C	16	55	18	50

Solution. Let p_0, q_0 denote the price and quantity for the base year 1990 and p_1, q_1 denote the price and quantity for the current year 1995.

Commodity	*Base year*		*Current year*		p_0q_0	p_1q_0	p_1q_1	p_0q_1
	p_0	q_0	p_1	q_1				
A	32	50	30	50	1,600	1,500	1,500	1,600
B	30	35	25	40	1,050	875	1,000	1,200
C	16	55	18	50	880	990	900	800
					Σp_0q_0 = 3,530	Σp_1q_0 = 3.365	Σp_1q_1 = 3,,400	Σp_0q_1 = 3,600

$$\text{Fisher's ideal index } (I_{0,1}) = \sqrt{\frac{\Sigma p_1q_0}{\Sigma p_0q_0} \times \frac{\Sigma p_1q_1}{\Sigma p_0q_1}} \times 100$$

$$= \sqrt{\frac{3{,}365}{3{,}530} \times \frac{3{,}400}{3{,}600}} \times 100 = 90$$

$$\text{By omitting the factor 100, } (I_{0,1}) = \sqrt{\frac{\Sigma p_1q_0}{\Sigma p_0q_0} \times \frac{\Sigma p_1q_1}{\Sigma p_0q_1}} = 0.9$$

and
$$I_{1,0} = \sqrt{\frac{\Sigma p_0q_1}{\Sigma p_1q_1} \times \frac{\Sigma p_0q_0}{\Sigma p_1q_0}} = \sqrt{\frac{3600 \times 3530}{3400 \times 3365}}$$

$$(I_{0,1}) \times (I_{1,0}) = 1$$

which verifies time reversal Test.

Example 17. Compute the Laspeyre's and Paasche's price index numbers for the year 1997 using the following data concerning four commodities :

		Commodity			
		A	*B*	*C*	*D*
Quantity (kg.)	in 1992	8	10	15	20
	in 1997	6	5	10	15
Price per kg. (Rs.)	in 1992	20	50	40	20
	in 1997	40	60	50	20

Solution.

Commodity	1992 Price p_0	1992 Qty. q_0	1997 Price p_1	1997 Qty. q_1	p_0q_0	p_1q_0	p_1q_1	p_0q_1
A	20	8	40	6	160	320	240	120
B	50	10	60	5	500	600	300	250
C	40	15	50	10	600	750	500	400
D	20	20	20	15	400	400	300	300
					$\Sigma p_0q_0 = 1{,}660$	$\Sigma p_1q_0 = 2{,}070$	$\Sigma p_1q_1 = 1{,}340$	$\Sigma p_0q_1 = 1{,}070$

$$\text{Laspeyre's index number} = \frac{\Sigma p_1q_0}{\Sigma p_0q_0} \times 100 = \frac{2{,}070}{1{,}660} \times 100 = 124.69$$

$$\text{Paasche's index number} = \frac{\Sigma p_1q_1}{\Sigma p_0q_0} \times 100 = \frac{1{,}340}{1{,}070} \times 100 = 125.23$$

Example 18. "For constructing index numbers, the best method on theoretical ground is not best method from practical point of view." Discuss.

Solution. Theoretically, Geometric mean is the best method for the construction of index numbers because

(i) It does not give greater importance to big items.

(ii) Index numbers measure relative changes and for relative change G.M. is preferred.

But due to computational difficulties, Arithmetic mean is preferred in practical calculation.

Example 19. Calculate the index number for 1998 with 1990 as base using the Weighted Average of Price Relatives Method for the following data :

Commodity	Weight	Price 1990	Price 1998
A	2	12	24
B	8	8	12
C	4	15	27
D	5	6	18
E	1	10	12

(D.U.B. Com Pass, 1999)

Solution.

Commodity	Weight w	1990 Price p_0	1998 Price p_1	$p = \frac{p_1}{p_0} \times 100$	p_w
A	2	12	24	$\frac{24}{12} \times 100 = 200$	400
B	8	8	12	$\frac{12}{8} \times 100 = 150$	1200
C	4	15	27	$\frac{27}{15} \times 100 = 180$	720
D	5	6	18	$\frac{18}{6} \times 100 = 300$	1500
E	1	10	12	$\frac{12}{10} \times 100 = 120$	120
	$\Sigma w = 20$				$\Sigma p_w = 3940$

$$P_{01} = \frac{\Sigma Pw}{\Sigma w} = \frac{3940}{20} = 197$$

Example 20. In 1991 for working class people wheat was selling at an average price of Rs. 100 per 10 kg., cloth at Rs. 40 per metre, house rent at Rs. 500 per house and miscellaneous items at Rs. 20 per unit. By 2001 cost of wheat rose by Rs. 50 per 10 kg., house rent by Rs. 300 per house and miscellaneous items doubled the price. The weights of the groups in order were in the ratio 1 : 2 : 4 : 1. The working class cost of living index number for the year 2001 with 1991 as base was 175%. By how much cloth price rose during the period 1991-2001 ?

Solution. In this question we will use weighted Arithmetic mean of price relatives.

$$\text{Per kg Price of Wheat in 1991} = \text{Rs.}\frac{100}{10} = \text{Rs. } 10$$

$$\text{Per kg Price of Wheat in 2001} = \frac{100+50}{10} = \text{Rs. } 15$$

$$\text{Price relative for Wheat in 2001 with 1991 as base} = \frac{15}{10} \times 100 = 150$$

$$\text{Price relative for House} = \frac{500+300}{500} \times 100 = 160$$

Price relative for Miscellaneous items in 2001 with 1991 as base

$$= \frac{2}{1} \times 100 = 200$$

Price relative for Cloth $= \frac{p_1}{40} \times 100$ where Cloth Rs. p_1 per metre is the price of Cloth in 2001.

As per the question

$$175 = \frac{150 \times k + \left[\frac{p_1}{40} \times 100\right] 2k + 160 \times 4k + 200 \times k}{1k + 2k + 4k + k}$$

$$\Rightarrow \quad 175 \times 8 = 150 + 640 + 200 + 5p_1 = 990 + 5p_1$$

$$\Rightarrow \quad 1400 = 990 + 5p_1$$

$$\Rightarrow \quad 5p_1 = 410 \Rightarrow p_1 = \frac{410}{5} = 82$$

$\therefore$ The price of Cloth rose by Rs. $(82 - 40)$ = Rs. 42 per metre.

Example 21. Compute the Weighted Index Numbers for 1997 and 1999 (Based on 1996) by relative method from the following data. Also interpret the computed index numbers :

Years		*Commodities*			
		A	*B*	*C*	*D*
1996	Price	6	8	9	12
	Weight	5	3	1	1
1997	Price	9	10	6	10
1998	Price	12	12	9	15
1999	Price	15	14	12	20

(D.U.B. Com 2001)

Solution.

Comm-odities	1996 p_0	1997 p_1	1998 p_2	Weight w	$P_1 = \frac{p_1}{p_0} \times 100$ (1997)	P_1w	$P_2 = \frac{p_2}{p_0} \times 100$ (1999)	P_2w
A	6	9	15	5	$\frac{9}{6} \times 100$	$\frac{9}{6} \times 100 \times 5 = 750$	$\frac{15}{6} \times 100$	$\frac{15}{6} \times 100 \times 5 = 1250$
B	8	10	14	3	$\frac{10}{8} \times 100$	$\frac{10}{8} \times 100 \times 3 = 375$	$\frac{14}{8} \times 100$	$\frac{14}{8} \times 100 \times 3 = 525$
C	9	6	12	1	$\frac{6}{9} \times 100$	$\frac{6}{9} \times 100 \times 1 = 66.67$	$\frac{12}{9} \times 100$	$\frac{12}{9} \times 100 \times 1 = 133.33$
D	12	10	20	1	$\frac{10}{12} \times 100$	$\frac{10}{12} \times 100 \times 1 = 83.33$	$\frac{20}{12} \times 100$	$\frac{20}{12} \times 100 \times 1 = 166.67$
				$\Sigma w = 10$		$\Sigma P_1w = 1275$		$\Sigma P_2w = 2075$

$$P_{01} = \frac{\Sigma P_1 w}{\Sigma w} = \frac{1275}{10} = 127.50$$

$$P_{02} = \frac{\Sigma P_2 w}{\Sigma w} = \frac{2075}{10} = 207.50$$

Price index in 1997 is 27.5% higher and in 1999 is 107.5% higher as compared to 1996 *i.e.,* in 1999 the over all effect on prices is more than double as compared to 1996.

Example 22. A company spent Rs. 50, Rs. 48, Rs. 18 and Rs. 42 during 1998. The company increased the expenditure to Rs. 100, Rs. 98, Rs. 60 and Rs. 102 during 1999 respectively on four commodities. If the units of four commodities purchased during 1998 and 1999 are identical *i.e.,* 5, 2, 6 and 17, compute the price index for 1999 by the most suitable method. **(D.U.B. Com., 2000)**

Solution. Most suitable method here will be the method of weighted aggregate of prices.

∴ Price index for 1999 with 1998 as base

$$= \frac{\text{Sum of the prices in 1999}}{\text{Weighted sum of the prices in 1998}} \times 100 = \frac{100 + 98 + 60 + 102}{50 + 48 + 18 + 42} \times 100 = \frac{360}{158} \times 100 = 227.84$$

Example 23. The following table gives the annual income of a person and the general price index number for the period 1988 to 1996. Prepare index number to show the changes in the real income of the person.

Year	*Annual Income*	*Price Index Number*
1988	36,000	100
1989	42,000	120
1990	50,000	145
1991	55,000	160
1992	60,000	250
1993	64,000	320
1994	68,000	450
1995	72,000	530
1996	75,000	600

(D.U.B. Com Hon., 2000)

Solution.

Year	*Annual Income in Rs.*	*Price Index No.*	*Real Wage*	*Real Wage Index 1998 = 100*
1988	36,000	100	$\frac{36,000}{100}\times 100 = 36,000$	100
1989	42,000	120	$\frac{42,000}{120}\times 100 = 35,000$	$\frac{35,000}{36,000}\times 100 = 97.22$
1990	50,000	145	$\frac{50,000}{145}\times 100 = 34,482.7$	$\frac{34,482.7}{36,000}\times 100 = 95.78$
1991	55,000	160	$\frac{55,000}{160}\times 100 = 34,375$	$\frac{34,375}{36,000}\times 100 = 95.49$
1992	60,000	250	$\frac{60,000}{250}\times 100 = 24,000$	$\frac{24,000}{36,000}\times 100 = 66.67$
1993	64,000	320	$\frac{64,000}{320}\times 100 = 20,000$	$\frac{20,000}{36,000}\times 100 = 55.56$
1994	68,000	450	$\frac{68,000}{450}\times 100 = 15,111.1$	$\frac{15,111.1}{36,000}\times 100 = 41.9$
1995	72,000	530	$\frac{72,000}{530}\times 100 = 13,589.9$	$\frac{13,589.9}{36,000}\times 100 = 37.74$
1996	75,000	600	$\frac{75,000}{600}\times 100 = 12,500$	$\frac{12,500}{36,000}\times 100 = 34.72$

Example 24. In calculating a certain cost of living index number the following weights were used : Food—15, Clothing—3, Rent—4, Fuel and Light—2, Miscellaneous—1. Calculate the index for a date when the average percentage increases in prices of items in the various groups over the base period were 32, 54, 48, 78 and 58 respectively.

Suppose a business executive was earning Rs. 20,500 in the base period, what should be his salary in the curent period if his standard of living is to remain the same ?

Solution. **Calculation of Cost of Living Index**

Group	*Average % Increase in Price*	*Group Index I**	*Weight W*	*IW*
Food	32	132	15	1980
Clothing	54	154	3	462
Rent	47	147	4	588
Fuel and Light	78	178	2	356
Miscellaneous	58	158	1	158
			$\Sigma W = 25$	$\Sigma IW = 3544$

$$\text{Cost of living index} = \frac{\Sigma\, IW}{\Sigma\, W} = \frac{3544}{25} = 141.76$$

For maintaining the same standard, the business executive should get,

$$\frac{20,500 \times 141.76}{100} = \text{Rs. } 29060.8$$

* Group index has been obtained by adding 100 to the average increase in price, *i.e.*, for Food, it is $100 + 32 = 132$

Example 25. Prove that whereas Fisher's Ideal index number satisfy time interval test and factor reversal test Laspeyre's index number and Paasche's index number do not satisfy these tests.

Solution. Laspeyre's Price index number

$$P_{0,1} = \frac{\Sigma p_1 q_0}{\Sigma p_0 q_0}* \quad \text{...(I)}$$

By omitting the factor 100

Interchanging 0 and 1, we get

$$P_{1,0} = \frac{\Sigma p_0 q_1}{\Sigma p_1 q_1}$$

$$P_{0,1} \times P_{1,0} = \frac{\Sigma p_1 q_0}{\Sigma p_0 q_0} \times \frac{\Sigma p_0 q_1}{\Sigma p_1 q_1} \neq 1 \quad \text{...(II)}$$

Interchanging p and q in (I), we get

$$Q_{0,1} = \frac{\Sigma p_0 q_1}{\Sigma p_0 q_0}$$

$$P_{0,1} \times Q_{0,1} = \frac{\Sigma p_1 q_0}{\Sigma p_0 q_0} \times \frac{\Sigma p_0 q_1}{\Sigma p_0 q_0} \neq \frac{\Sigma p_1 q_1}{\Sigma p_0 q_0} \quad \text{...(III)}$$

(II) and (III) prove that Laspeyre's index numbers do not satisfy time reversal and factor reversal test.

$$\text{Paasche's index number} = P_{0,1} = \frac{\Sigma p_1 q_1}{\Sigma p_0 q_1} \quad \text{...(IV)}$$

Interchanging 0 and 1, we get $P_{1,0} = \dfrac{\Sigma p_1 q_1}{\Sigma q_0 p_1} = \dfrac{\Sigma p_1 q_1}{\Sigma p_1 q_0}$

$$P_{0,1} \times P_{1,0} = \frac{\Sigma p_1 q_1}{\Sigma p_0 q_1} \times \frac{\Sigma p_0 q_0}{\Sigma p_1 q_0} \neq 1 \quad \text{...(V)}$$

Interchanging p and q in (IV), we get

$$Q_{0,1} = \frac{\Sigma p_1 q_1}{\Sigma p_1 q_0}, \; Q_{1,0} = \frac{\Sigma p_1 q_1}{\Sigma p_0 q_1}$$

$$P_{0,1} \times Q_{0,1} = \frac{\Sigma p_1 q_1}{\Sigma p_0 q_1} \times \frac{\Sigma p_1 q_1}{\Sigma p_1 q_0} \neq \frac{\Sigma p_1 q_1}{\Sigma p_0 q_0} \quad \text{...(VI)}$$

Note for provide the tests of adequacy factor 100 has to be ignored.

(V) and (VI) prove that Paasche's index number do not satisfy time reversal test and factor reversal test

Fisher's ideal Price Index $P_{0,1}$ is given by

$$P_{0,1} = \sqrt{\frac{\Sigma p_1 q_0}{\Sigma p_0 q_0} \times \frac{\Sigma p_1 q_1}{\Sigma p_0 q_1}} \text{ (Omitting the factor 100)} \quad \text{...(VII)}$$

* For proving time reversal test and factor reversal test factor 100 is omitted.

Interchanging 0 and 1 in (VII), wet

$$P_{1,0} = \sqrt{\frac{\Sigma p_0q_1}{\Sigma p_1q_1} \times \frac{\Sigma p_0q_0}{\Sigma p_1q_0}} \qquad ...(VIII)$$

$$p_{0,1} \times p_{1,0} = 1$$

⇒ Fisher's index number satisfies time reversal test.

Interchaning p and q, in (VII) we get Fisher's Quantity Index

$$Q_{0,1} = \sqrt{\frac{\Sigma q_1p_0}{\Sigma q_0p_0} \times \frac{\Sigma q_1p_1}{\Sigma q_0p_1}} = \sqrt{\frac{\Sigma p_0q_1}{\Sigma p_0q_0} \times \frac{\Sigma p_1q_1}{\Sigma p_1q_0}} \qquad ...(IX)$$

Multiplying (VII) and (IX), we get

$$P_{0,1} \times Q_{0,1} = \sqrt{\frac{\Sigma p_1q_0}{\Sigma p_0q_0} \times \frac{\Sigma p_1q_1}{\Sigma p_0q_1} \times \frac{\Sigma p_0q_1}{\Sigma p_0q_0} \times \frac{\Sigma p_1q_1}{\Sigma p_1q_0}}$$

$$= \sqrt{\frac{(\Sigma p_1q_1)^2}{(\Sigma p_0q_0)^2}} = \frac{\Sigma p_1q_1}{\Sigma p_0q_0} = \text{the true value ratio.}$$

This shows that Fisher's ideal index Number satisfies Factor Reversal Test.

Example 26. Calculate price index numbers from the following data by :

(i) Laspeyre's Method (ii) Paasche's Method (iii) Fisher's Method

Commodity	*Base Year*		*Current Year*	
	Price	*Quantity*	*Expenditure*	*Quantity*
A	6	50	560	56
B	2	100	240	120
C	4	60	360	60
D	10	30	288	24
E	8	40	432	36

(DU Eco. (Hons.) 2008)

Solution.

Commodity	*Base Year*		*Current Year*		P_1Q_0	P_0Q_0	P_1Q_1	P_0Q_1
	Price P_0	*Quality* Q_0	*Price* P_1	*Quality* Q_1				
A	6	50	$\frac{560}{56} = 10$	56	500	300	560	336
B	2	100	$\frac{240}{120} = 2$	120	200	200	240	240
C	4	60	$\frac{360}{60} = 6$	60	360	240	360	240
D	10	30	$\frac{288}{24} = 12$	24	360	300	288	240
E	8	40	$\frac{432}{36} = 12$	36	1440	320	432	288
					2860	1360	1880	1344

$$\text{Laspeyre's index number} = \frac{\Sigma P_1 Q_0}{\Sigma P_0 Q_0} \times 100$$

$$= \frac{2860}{1360} \times 100 = 210.3$$

$$\text{Paasche's index number} = \frac{\Sigma P_1 Q_1}{\Sigma P_0 Q_1} = \frac{1880}{1344} \times 100 = 139.9$$

Fisher's index number.

Example 27. Using Fisher's form find price index number of the following data :

Commodity	*Unit*	*Base Year*		*Current Year*	
		Price (Rs.)	*Total Value (Rs.)*	*Quantity*	*Total Value (Rs.)*
A	Kg.	12.5	125	12	156
B	Kg.	14	112	9	135
C	Metre	11	88	9	108
D	Kg.	13	78	6	90

Solution. $p_0, p_1 \Rightarrow$ prices of base and current year;

$q_0, q_1 \Rightarrow$ quantity consumed of base and current year.

Commodity	*Base Year*		*Current Year*		p_0q_0	p_0q_1	p_1q_0	p_1q_1
	p_0	q_0	p_1	q_1				
A	12.5	10	13	12	125	150	130	156
B	14	8	15	9	112	126	120	135
C	11	8	12	9	88	99	96	108
D	13	6	15	6	78	78	90	90
					403 $= \Sigma p_0q_0$	453 $= \Sigma p_0q_1$	436 $= \Sigma p_1q_0$	489 $= \Sigma p_1q_1$

$$\text{Base year quantity} = \frac{\text{Total value for Base Year}}{\text{Price per unit in Base Year}}$$

$$\text{Current year price/unit} = \frac{\text{Total value for Current Year}}{\text{Quantity for Current Year}}$$

Hence, Price Index by Fisher's method $= \sqrt{\frac{\Sigma p_1 q_0}{\Sigma p_0 q_0} \times \frac{\Sigma p_1 q_1}{\Sigma p_0 q_1}} \times 100$

$$= \sqrt{\frac{436}{403} \times \frac{489}{453}} \times 100 = \sqrt{1.0819 \times 1.0795} \times 100$$

$$= \sqrt{1.1679} \times 100 = 108.07\%$$

Example 28. Calculating price index number of the year 1996 with 1986 as base year from the following data using :

(i) Laspeyre's
(ii) Paasche's and
(iii) Fisher's formulae :

Commodity	*Unit*	*1986*		*1996*	
		Price (Rs.)	*Value (Rs.)*	*Quantity Consumed*	*Value (Rs.)*
A	Kg.	10	1,500	160	1,760
B	Kg.	12	1,080	100	1,300
C	Metre	15	900	60	960
D	Packets	9	450	40	480

Solution. Table for calculation of Index number :

Commodity	p_0	q_0	p_0q_0	p_1	q_1	p_1q_1	p_0q_1	p_1q_0*
A	10	150	1,500	11	160	1,760	1,600	1,650
B	12	90	1,080	13	10	1,300	1,200	1,170
C	15	60	900	16	60	960	900	960
D	9	50	450	12	40	480	360	600
Total			3,930			4,500	4,060	4,380

(i) Laspeyre's Index Number

$$= \frac{\Sigma p_1 q_0}{\Sigma p_0 q_0} \times 100 = \frac{4,380}{3,930} \times 100 = 111.4$$

(ii) Paasche's Index Number

$$= \frac{\Sigma p_1 q_1}{\Sigma p_0 q_1} \times 100 = \frac{4,500}{4,060} \times 100 = 110.84$$

(iii) Fischer's Index Number

$$= \sqrt{\text{Laspeyre's} \times \text{Paasche's Index Number}}$$
$$= \sqrt{111.45 \times 110.84} = \sqrt{12,353.188} = 111.14$$

* $p_0, q_0 \rightarrow$ Price and quantity in 1986
$p_1, q_1 \rightarrow$ Price and Quantity in 1996

Example 29. Calculate index number of prices for 1995 on the basis of 1990 from the data given below :

Commodity	*Weight*	*Price per unit 1990 (Rs.)*	*Price per unit 1995 (Rs.)*
A	40	16	20
B	25	40	50
C	20	12	15
D	15	2	3

If the weights of commodities *A*, *B*, *C*, *D* are increased in the ratio 1 : 2 : 3 : 4 what will be increase in index number ? **(ICWA June, 1998)**

Solution.

Commodity	*Weight* (w)	*Price of 1990* (p_0)	*Price of 1995* (p_1)	$\frac{p_1}{p_0} \times 100 = I_{01}$	*Increased wt* (w')
A	40	16	20	$\frac{20}{16} \times 100 = 125$	44
B	25	40	50	$\frac{50}{40} \times 100 = 125$	30
C	20	12	15	$\frac{15}{12} \times 100 = 125$	26
D	15	2	3	$\frac{3}{2} \times 100 = 150$	21

Index number of price for 1995 on the basis of 1990 :

$$= \frac{\Sigma \frac{p_1}{p_0} w}{\Sigma w} \times 100 = \frac{40 \times 125 + 25 \times 125 + 20 \times 125 + 15 \times 150}{40 + 25 + 20 + 15} = 128.75$$

The weights of commodities *A*, *B*, *C* and *D* are increased in the ratio 1 : 2 : 3 : 4, *i.e.*, new weights are :

$$40\left(1 + \frac{1}{10}\right); 25\left(1 + \frac{2}{10}\right) = 30; 20\left(1 + \frac{3}{10}\right) = 26$$

and

$$15\left(1 + \frac{4}{10}\right) = 21$$

So the new index number is

$$= \frac{44 \times 125 + 30 \times 125 + 26 \times 125 + 21 \times 150}{44 + 30 + 26 + 21}$$

$$= \frac{5{,}500 + 3{,}750 + 3{,}250 + 3{,}10}{121} = \frac{15{,}650}{121} = 129.33$$

So increase in index number is $129.33 - 128.75 = 0.58\%$

Example 30. The table below gives detail of price and consumption of 5 commodities for 1995 and 1997. Using an appropriate formula arrive at an index number for 1997 prices with 1995 as base.

Commodity	Price per unit 1995 (Rs.)	Price per unit 1997 (Rs.)	Consumption value 1995 (Rs.)
Rice	40	48	800
Wheat	25	27	400
Oil	95	105	760
Fresh	110	120	1,100
Milk	80	100	480

(ICWA Dec. 1998)

Solution. Since the base year consumption value (p_0q_0) is given the appropriate formulae is Laspeyre's Index Number.

Commodity	p_0	q_0	p_1	p_1q_0	p_0q_0
Rirce	40	$\frac{800}{40}=20$	48	960	800
Wheat	25	$\frac{400}{25}=16$	27	432	400
Oil	95	$\frac{760}{95}=8$	105	840	760
Fish	110	$\frac{1,100}{110}=10$	120	1,200	1,100
Milk	80	$\frac{480}{80}=6$	100	600	480
Total				4,032	$\Sigma p_0q_0 = 3540$

$$\left[q_0 = \frac{p_0q_0}{p_0} = \frac{\text{Consumption value 1995 (Rs.)}}{\text{Price per unit 1995 (Rs.)}}\right]$$

$$\Sigma\, p_0q_0 = 3,540$$

$$I_{01} = \frac{4032}{3540} \times 100 = 113.9$$

Example 31. Find Laspeyre's and Fisher's price index numbers for the following data for the year 1990 with base year 1985 :

Commodity	1985		1996	
	Price Rs./Kg.	Quantity Kg.	Price Rs./Kg.	Quantity Kg.
Wheat	4.0	20	5.5	20
Sugar	5.0	10	7.0	9
Edible oil	10.5	5	14.5	4
Vegetables	4.5	4	8.0	4

(ICWA Dec., 1994)

Solution. Laspeyre's index number $= \dfrac{\Sigma\, p_1q_0}{\Sigma\, p_0q_0} \times 100$

$$= \frac{5.5 \times 20 + 7 \times 10 = 14.5 \times 5 + 8.0 \times 4}{20 \times 4 + 10 \times 5 + 5 \times 10.5 + 4 \times 4.5} \times 100$$

$$= \frac{110 + 70 + 72.5 + 32}{80 + 50 + 52.5 + 18} \times 100$$

$$= \frac{284.5}{200.5} \times 100 = 141.89 = 141.9$$

$$\text{Paasche's index number} = \frac{\Sigma p_1 q_1}{\Sigma p_0 q_1} \times 100$$

$$= \frac{20 \times 5.5 + 9 \times 7 + 4 \times 14.5 + 4 \times 8.0}{4 \times 20 + 5 \times 9 + 10.5 \times 4 + 4.5 \times 4} \times 100$$

$$= \frac{110 + 63 + 58 + 32}{80 + 45 + 42 + 18} \times 100 = \frac{263}{185} \times 100 = 142.16$$

$$\text{Fisher's Index Number} = \sqrt{\frac{284.5}{200.5} \times \frac{263}{185}} \times 100 = 142.029$$

Example 32. Find Fisher's and Marshall-Edgeworth index number for the following data :

Commodity	*1989*		*1990*	
	Price (Rs.)	*Qty. (Kg.)*	*Price (Rs.)*	*Qty. (Kg.)*
A	20	8	40	9
B	50	10	60	10
C	40	15	60	12
D	10	15	151	15

(ICWA June, 1996)

Solution.

Commodity	*1989*		*1990*		p_0q_0	p_0q_1	p_1q_0	p_1q_1
	p_0	q_0	p_1	q_1				
A	20	8	40	9	160	180	320	360
B	50	10	60	10	500	500	600	600
C	40	15	60	12	600	480	900	720
D	10	15	15	15	150	150	225	225
Total	—	—	—	—	1410	1310	2045	1905

$$\text{Fisher's Index Number} = \sqrt{\frac{\Sigma p_1 q_0}{\Sigma p_0 q_0} \times \frac{\Sigma p_1 q_1}{\Sigma p_0 q_1}} \times 100$$

$$= \sqrt{\frac{2045}{1410} \times \frac{1905}{1310}} \times 100$$

$$= 1.4523 \times 100 = 145.23$$

Marshall-Edgeworth Index number

$$= \frac{\Sigma p_1 q_0 + \Sigma p_1 q_1}{\Sigma p_0 q_0 + \Sigma p_0 q_1} \times 100 = \frac{2045 + 1905}{1410 + 1310} \times 100$$

$$= \frac{3950}{2720} \times 100 = 1.4522 \times 100 = 145.22$$

Example 33. From the information given below find Laspeyres' and Paasche's price index number for year II with year I as base :

Commodity	*Price/Unit (in Rs.)*		*Commodity (in kg.)*	*Total Value (in Rs.)*
	Year I	*Year II*	*Year I*	*Year II*
A	24.09	12.30	5	184.50
B	20.66	10.17	8	122.04
C	18.60	16.42	16	295.56
D	9.45	10.21	25	189.00

Verify whether the above index numbers satisfy the time reversal Test. **(ICWA June 1994)**

Solution. Quantity in kg of commodity A in year II $= \frac{184.50}{12.30} = 15$

Quantity in kg of commodity B in year II $= \frac{122.04}{10.17} = 12$

Quantity in kg of commodity C in year II $= \frac{295.56}{16.42} = 18$

Quantity in kg of commodity D in year II $= \frac{189.00}{10.21} = 18.5$

Laspeyre's index number $= \frac{\Sigma p_1 q_0}{\Sigma p_0 q_0} \times 100$

$$= \frac{12.30 \times 5 + 10.17 \times 8 + 16.42 \times 16 + 10.21 \times 25}{24.09 \times 5 + 20.66 \times 8 + 18.60 \times 16 + 9.45 \times 25} \times 100$$

$$= \frac{61.5 + 81.36 + 262.72 + 255.5}{120.45 + 165.28 + 297.6 + 236.25} \times 100$$

$$= \frac{661.08}{819.58} \times 100 = 80.66$$

Paasche's index numbers $= \frac{\Sigma p_1 q_1}{\Sigma p_0 q_1} \times 100$

$$= \frac{184.50 + 122.04 + 295.56 + 189.00}{24.09 \times 15 + 20.66 \times 12 + 18.60 \times 18 + 9.45 \times 18.51} \times 100$$

$$= \frac{791}{361.35 + 247.92 + 334.8 + 174.92}$$

$$= \frac{791}{1119} \times 100 = 70.68$$

Further students may check that time reversal test is not verified.

Example 34. From the following data, find the index numbers for the current year and the base year based on each other and show that the G.M. makes it reversible but the Arithmetic mean does not.

Commodity	*Base year*	*Current year*
A	25	55
B	30	45

Solution. Price Index of commodity *A* for current year w.r.t. Base year

$$= \frac{55}{25} \times 100 = 220$$

Price index of commodity *B* for current year w.r.t. Base year

$$= \frac{45}{30} \times 100 = 150$$

Price index of commodity *A* for Base year w.r.t. Current year

$$= \frac{25}{55} \times 100 = 45.45$$

Price index of commodity *B* for Base year w.r.t. Current year

$$= \frac{30}{45} \times 100 = 66.67$$

AM of P_{01} for commodities *A* and $B = \frac{220 + 150}{2} = 185$

AM of P_{10} for commodities *A* and $B = \frac{45.45 + 66.67}{2} = \frac{112.12}{2} = 56.06$

$P_{01} \times P_{10}$ (for AM) $= \frac{185 \times 56.06}{100 \times 100} = 1.0371$ ...(I)

GM of P_{01} for commodities *A* and $B = \sqrt{220 \times 150} = 181.66$

GM of $P_{1,0}$ for commodities *A* and $B = \sqrt{45.45 \times 66.67} = 55.05$

$$P_{01} \times P_{10} \text{ for GM} = \frac{181.0 \times 55.05}{100 \times 100} = 1.000 \qquad \text{...(II)}$$

(I) and (II) prove the required.

Example 35. Prepare price index numbers for 2000 with 1991 as base year for the following data by using (i) Laspeyre's, (ii) Paasche's and (iii) Fisher's method. (Correct up to 4 places of decimal)

Year	*Article*							
	I		*II*		*III*		*IV*	
	P	*Q*	*P*	*Q*	*P*	*Q*	*P*	*Q*
1991	12.50	9	9.63	4	7.75	6	5.00	5
2000	12.75	9	7.75	6	8.80	10	6.50	7

(*P* Price; *Q* : Quantity)

With the help of above data prove that the Time Reversal Test is satisfied by Fisher's formula, but not necessarily by the Laspeyre's and Paasche's index numbers.

Solution. Let P_0, Q_0 denote respectively the price and quantity in the base year (1991) and P_1, Q_1 denote respectively in price and quantity in the current year (2000).

Article	*Year 1991*		*Year 2000*					
	Price (P_0)	*Qty.* (Q_0)	*Price* (P_1)	*Qty.* (Q_1)	P_0Q_0	P_0Q_1	P_1Q_0	P_1Q_1
I	12.50	9	12.75	9	112.50	112.50	114.75	114.75
II	9.63	4	7.75	6	38.52	57.78	31	46.5
III	7.75	6	8.80	10	46.5	77.5	52.8	88
IV	5.00	5	6.50	7	25	35	32.5	45.5
					$\Sigma P_0Q_0 = 222.52$	$\Sigma P_0Q_1 = 282.78$	$\Sigma P_1Q_0 = 231.05$	$\Sigma P_1Q_1 = 294.75$

$$\text{Laspeyre's index number } (I'_{01}) = \frac{\Sigma P_1Q_0}{\Sigma P_0Q_0} \times 100$$

$$= \frac{231.05}{222.52} \times 100 = 103.83$$

$$\text{Paasche's index number } (I''_{01}) = \frac{\Sigma P_1Q_1}{\Sigma P_0Q_0} \times 10$$

$$= \frac{294.75}{282.78} \times 100 = 104.23$$

$$\text{Fisher's ideal index number } (I'''_{01}) = \sqrt{\frac{\Sigma P_1Q_0}{\Sigma P_0Q_0} \times \frac{\Sigma P_1Q_1}{\Sigma P_0Q_1}} \times 100$$

$$= \sqrt{\frac{231.05}{222.52} \times \frac{294.75}{282.78}} \times 100 = 104.03$$

Time Reversal test will be satisfied by Laspeyre's index number if

$$\frac{\Sigma P_1Q_0}{\Sigma P_0Q_0} \times \frac{\Sigma P_0Q_1}{\Sigma P_1Q_1} = 1$$

As per the given data

$$\frac{\Sigma P_1Q_0}{\Sigma P_0Q_0} \times \frac{\Sigma P_0Q_1}{\Sigma P_1Q_1} = \frac{231.05}{222.52} \times \frac{282.78}{294.75} = 0.996 \neq 1$$

Time reversal test will be satisfied by Paasche's index number if

$$\frac{\Sigma P_1Q_1}{\Sigma P_0Q_1} \times \frac{\Sigma P_0Q_0}{\Sigma P_1Q_0} = 1$$

But as per the given data

$$\frac{\Sigma P_1Q_1}{\Sigma P_0Q_1} \times \frac{\Sigma P_0Q_0}{\Sigma P_1Q_0} = \frac{294.75}{282.78} \times \frac{222.52}{231.05} = 1.0038486 \neq 1$$

Time reveresal test will be satisfied by Fisher's ideal index number if

$$\sqrt{\frac{\Sigma p_1q_0}{\Sigma p_0q_0} \times \frac{\Sigma p_1q_1}{\Sigma p_0q_1}} \times \sqrt{\frac{\Sigma p_0q_1}{\Sigma p_1q_1} \times \frac{\Sigma p_0q_0}{\Sigma p_1q_0}} = 1$$

$$\sqrt{\frac{\Sigma p_1q_0}{\Sigma p_0q_0} \times \frac{\Sigma p_1q_1}{\Sigma p_0q_1}} = \sqrt{\frac{231.05}{222.52} \times \frac{294.75}{282.78}} \quad ...(I)$$

$$\sqrt{\frac{\Sigma p_0q_1}{\Sigma p_1q_1} \times \frac{\Sigma p_0q_0}{\Sigma p_1q_0}} = \sqrt{\frac{282.78}{294.75} \times \frac{222.52}{231.05}} \quad ...(II)$$

I × II = 1 which implies that Fisher's ideal index number satisfies time reversal test.

Example 36. An index is at 100 in 1991. It rises 4% in 1992, falls 6% in 1993, falls 4% be 1994 and rises 3% in 1995. Calculate the index number for 5 years with 1993 as the base.

Solution. First we will have to calculate the index at the old base of 1991 and then construct the new series with 1993 as the base. This is done in the following table

Index Number with the new base of 1993

Year	*Old Index (1991 = 100)*	*New Index* (1993 = 100)*
1991	100	$\frac{100}{97.76} \times 100 = 102.32$
1992	$100 + 4 = 104$	$\frac{100}{97.76} \times 104 = 106.40$
1993	$104 \times \frac{94}{100} = 97.76$	$= 100.00$
1994	$97.76 \times \frac{96}{100} = 93.85$	$\frac{100}{97.76} \times 93.85 = 93.00$
1995	$93.85 \times \frac{103}{100} = 96.66$	$\frac{100}{97.76} \times 96.66 = 98.88$

Example 37. The price index of cosmetics was 110 in 1996 with base as 1990 and 120 in 1997 with 1996 as base. It further increased by 30% in 1998 in relation to the price index of 1997 and decreased by 10% in 1999 as compared to its level in 1998. Find the index for 1999 with 1990 as base.

Solution. Price index for 1996 with 1990 as base

$$= I_{90,\,96} = \frac{P_{96}}{P_{90}} \times 100 = 110 \Rightarrow \frac{P_{96}}{P_{90}} = 1.1$$

Similarly $$\frac{P_{97}}{P_{96}} = 1.2, \quad \frac{P_{98}}{P_{97}} = 1.3, \quad \frac{P_{99}}{P_{98}} = 0.9$$

∴ Price index for 1999 with 1990 as base

$$= \frac{P_{1999}}{P_{1990}} \times 100$$

$$= \frac{P_{96}}{P_{90}} \times \frac{P_{97}}{P_{96}} \times \frac{P_{98}}{P_{97}} \times \frac{P_{99}}{P_{98}} \times 100$$

$$= 1.1 \times 1.2 \times 1.3 \times 0.9 \times 100 = 154.5$$

Example 38. Between 1980 and 1990, the price of a commodity increased by 60% while the production decreased by 30%. By what percentage did the value index of production of commodity change in 1990 with respect to its value in 1980.

Solution. Value index for one commodity $= \dfrac{P_1Q_1}{P_0Q_0} \times 100$

$$= \left(\frac{P_1}{P_0}\right)\left(\frac{Q_1}{Q_0}\right) \times 100 = 1.6 \times 0.7 \times 100* = 112$$

$\Rightarrow$ Value index increased by 12%.

Example 39. Given that $\Sigma p_1 q_1 = 250, \Sigma p_0 q_0 = 150$, Paasche's index number $= 150$ and Dorbish-Bowley's index number = 145. Find Fischer's index number. **(D.U., B.Com. (Hons) 2005)**

Solution. Paasche's index number $= \dfrac{\Sigma p_1 q_1}{\Sigma p_0 q_1} \times 100 = 150$

$$\Rightarrow \quad \frac{250}{\Sigma p_0 q_1} = \frac{150}{100}$$

$$\Rightarrow \quad \Sigma p_0 q_1 = \frac{250 \times 100}{150} = \frac{500}{3}$$

Dorbish-Bowley's index number $= \dfrac{L+P}{2}$

where L = Laspeyre's index number
P = Laasche's index number

$$\Rightarrow \quad 145 = \frac{L+150}{2} \quad \Rightarrow \quad L = 140$$

Laspeyre's index number $= \dfrac{\Sigma p_1 q_0}{\Sigma p_0 q_0} \times 100 = 140$

$$\Rightarrow \quad \frac{\Sigma p_1 q_0}{150} = \frac{140}{100}$$

$$\Rightarrow \quad \Sigma p_1 q_0 = \frac{140 \times 150}{100} = 210$$

Fisher's index number $= \sqrt{\dfrac{\Sigma p_1 q_0}{\Sigma p_0 q_0} \times \dfrac{\Sigma p_1 q_1}{\Sigma p_0 q_1}} \times 100 = \sqrt{\dfrac{210}{150} \times \dfrac{250}{\frac{500}{3}}} \times 100 = 144.91$

Example 40. The data below show the percentage increase in prices of a few selected food items and weights attached to each of them, calculate index number for the food group :

Food Item	*Weightage*	*Percentage Increase in Price*
Rice	33	180
Wheat	11	202
Dal	8	115
Ghee	5	212
Oil	5	175
Spices	3	517
Milk	7	260
Fish	9	426
Vegetables	9	332
Refreshmentns	10	279

* Price increase by 60% $\Rightarrow \dfrac{P_1}{P_0} = \dfrac{160}{100} = 1.6$

Using the above food index and the information given below, calculate the consumer price index number :

Group	*Index*	*Weight*
Food	–	60
Clothing	310	5
Fuel and Light	220	8
Rent and Rates	150	9
Miscellaneous	300	18

(D.U., B.Com. (Hons.) 2006R)

Solution. **Calculation of Index Number for Good Group**

Food Item	*W*	*I*	*IW*
Rice	33	280	9240
Wheat	11	302	3322
Dal	8	215	1720
Ghee	5	312	1560
Oil	5	275	1375
Spices	3	617	1851
Milk	7	360	2520
Fish	9	526	4734
Vegetables	9	432	3888
Refreshments	10	379	3790
	100		34000

$$\therefore \quad P_{01} = \frac{\Sigma IW}{\Sigma W} = \frac{34000}{100} = 340$$

Calculation of Consumer Price Index

Group	*Index* *I*	*Weight* *W*	*IW*
Food	340	60	20400
Clothing	320	5	1550
Fuel and Light	220	8	1760
Rent and Rates	150	9	1350
Miscellaneous	300	18	5400
		$\Sigma W = 100$	$\Sigma IW = 30460$

$$\therefore \quad CPI = \frac{\Sigma IW}{\Sigma W} = \frac{30460}{100} = 304.6$$

Example 41. An enquiry into the budgets of middle class families in a city gave the following information :

Item	% Expenditure	Prices (2004)	Prices (2005)
Food	35	140	170
Rent	15	80	90
Clothing	20	110	130
Fuel	10	60	80
Misc.	20	90	110

What change do you notice in the cost of living in 2005 as compared to 2004?

(D.U., B.Com. Pass 2006R)

Solution. **Computations of Cost of Living Index**

Items of Expenditure	*2004* P_0	*2005* P_1	W	$\frac{P_1}{P_0} \times 100$ P	PW
Food	140	170	35	$\frac{170}{140} \times 100 = 121.43$	4250.05
Rent	80	90	15	$\frac{90}{80} \times 100 = 112.50$	1687.5
Clothing	110	130	20	$\frac{130}{110} \times 100 = 118.18$	2363.6
Fuel	60	80	10	$\frac{80}{60} \times 100 = 133.33$	1333.3
Misc.	90	110	20	$\frac{110}{90} \times 100 = 122.22$	2444.4
			$\Sigma W = 100$		$\Sigma PW = 12078.85$

$$\text{Cost of living index} = \frac{\Sigma PW}{\Sigma W}$$

$$= \frac{12078.85}{100} \times 120.79$$

Example 42. Compute : (i) Laspeyre's; (ii) Paasche's; (iii) Fisher's; (iv) Bowley's, index numbers from the following table :

Items	p_0	q_0	p_1	q_1
A	10	4	12	6
B	15	6	20	4
C	2	5	5	3
D	4	4	4	4

(D.U., B.Com. 2006E)

Solution. **Calculation of Various Indices**

Items	p_0	q_0	p_1	q_1	p_0q_0	p_1q_1	p_1q_0	p_0q_1
A	10	4	12	6	40	72	48	60
B	15	6	20	4	90	80	120	60
C	2	5	5	3	10	15	25	6
D	4	4	4	4	16	16	16	16
					$\Sigma p_0q_0 = 156$	$\Sigma p_1q_1 = 183$	$\Sigma p_1q_0 = 209$	$\Sigma p_0q_1 = 142$

(i) Laspeyre's index number

$$P_{01} = \frac{\Sigma p_1q_0}{\Sigma p_0q_0} \times 100 = \frac{209}{156} \times 100 = 133.97$$

(ii) Pasche's index number

$$P_{01} = \frac{\Sigma p_1q_1}{\Sigma p_0q_1} \times 100 = \frac{183}{142} \times 100 = 128.87$$

(iii) Fisher's index number

$$P_{01} = \sqrt{\frac{\Sigma p_1q_0}{\Sigma p_0q_0} \times \frac{\Sigma p_1q_1}{\Sigma p_0q_1}} \times 100 = \sqrt{\frac{209}{156} \times \frac{183}{142}} \times 100$$

$$= \sqrt{\frac{38247}{22152}} \times 100 = \sqrt{1.726} \times 100 = 1.31 \times 100 = 131$$

(iv) Bowley's index number

$$P_{01} = \frac{\dfrac{\Sigma p_1q_0}{\Sigma p_0q_0} + \dfrac{\Sigma p_1q_1}{\Sigma p_0q_1}}{2} \times 100 = \frac{\dfrac{209}{156} + \dfrac{183}{142}}{2} \times 100$$

$$= \frac{1.34 + 1.29}{2} \times 100 = \frac{2.63}{2} \times 100 = 131.5$$

(v) Marshal-Edgeworth index number

$$P_{01} = \frac{\Sigma (q_0 + q_1)\, p_1}{\Sigma (q_0 + q_1)\, p_0} \times 100 = \frac{\Sigma q_0p_1 + \Sigma q_1p_1}{\Sigma q_0p_0 + \Sigma q_1p_0} \times 100$$

$$= \frac{209 + 183}{156 + 142} \times 100 = \frac{392}{298} \times 100 = 131.54$$

Example 43. An enquiry into the budgets of the middle class families in a city provide the following information :

Commodity	*Quantity*	*Price*	
		2004	*2005*
A	20	8	12
B	40	6	7.50
C	100	5	5.50
D	25	12	15
E	10	10	18

What change in the cost of living has taken place in 2005 as compared to 2000?

(D.U., B.Com. 2007E)

Solution.

Commodity	*Quantity* (q)	*Price*			
		2000 (P_0)	*2005* (P_1)	P_1q	P_0q
A	20	8	12	240	160
B	40	6	7.50	300	240
C	100	–	5.50	550	500
D	25	12	15	375	300
E	10	10	18	180	100
				1645	1300

Cost of living index in 2005 $= \frac{1645}{1300} \times 100 = 126.5$

Change (increase) in index $= 26.5\%$.

Example 44. State the 'factor reversal test'. Show that Fisher's method for constructing index numbers satisfies it.

The price index of cosmetics was 110 in 2001 with base as 1995 and 120 in 2002 with 2001 as base. It further increased by 30% in 2003 in relation to the price index of 2002 and decreased by 10% in 2004 as compared to its level in 2003. Find the index for 2004 with 1995 as base.

(D.U., B.Com. (Hons.) 2005E)

Solution. Factor reversal test is said to be satisfied if the product of price index number and quantity index number is equal to true value ratio.

Fisher price index number

$$P_{01} = \sqrt{\frac{\Sigma p_1q_1}{\Sigma p_0q_1} \times \frac{\Sigma p_1q_0}{\Sigma p_0q_0}} \quad \text{[without taking the factor 100]} \quad \ldots\text{(i)}$$

Fisher quantity index number

$$Q_{01} = \sqrt{\frac{\Sigma p_1q_1}{\Sigma q_0p_1} = \frac{\Sigma p_0q_1}{\Sigma p_0q_0}} \quad \ldots\text{(ii)}$$

Multiplying (i) and (ii), we get

$$P_{01} \times Q_{01} = \frac{\Sigma p_1q_1}{\Sigma p_0q_0}$$

$$\frac{P_{2001}}{P_{1995}} \times 100 = 110 \Rightarrow \frac{P_{2001}}{P_{1995}} = 1.1$$

Similarly, $\frac{P_{2002}}{P_{2001}} = 1.2, \quad \frac{P_{2003}}{P_{2002}} = 1.3, \quad \frac{P_{2004}}{P_{2003}} = 0.9$

Price index for 2004 with 1995 as base $= \frac{P_{2004}}{P_{1995}} \times 100$

$$= \frac{P_{2001}}{P_{1995}} \times \frac{P_{2002}}{P_{2001}} \times \frac{P_{2003}}{P_{2002}} \times \frac{P_{2004}}{P_{2003}} \times 100$$

$$= 1.1 \times 1.2 \times 1.3 \times 0.9 \times 100 = 154.4$$

Example 45. The price index and quantity index of a commodity were 120 and 110 respectively in 2005 with base 2004. Find its value index number in 2005 with base 2004. **(D.U., B.Com. 2007R)**

Solution.
$$\frac{p_1}{p_0} \times 100 = 120 \Rightarrow \frac{p_1}{p_0} = 1.2 \Rightarrow p_1 = 1.2\, p_0$$

$$\frac{q_1}{q_0} \times 100 = 110 \Rightarrow \frac{q_1}{q_0} = 1.1 \Rightarrow q_1 = 1.1\, q_0$$

$$\therefore \text{ Value index } = \frac{p_1 q_1}{p_0 q_0} \times 100 = 1.2 \times 1.1 \frac{p_0 q_0}{p_0 q_0} \times 100 = 132.$$

Example 46. Given that $\Sigma p_1 q_1 = 250$, $\Sigma p_0 q_0 = 150$, Paasche's index number = 150 and Dorbish and Bowley's index number = 145. Find out :

(i) Fisher's ideal index number; and

(ii) Marshal-Edgeworth's index number. **(D.U., B.Com. (Hons.) 2007R)**

Solution. Paasche's index number $= \frac{\Sigma p_1 q_1}{\Sigma p_0 q_1} \times 100 = 150$

$$\Rightarrow \quad \Sigma p_0 q_1 = \frac{\Sigma p_1 q_1}{1.5} = \frac{250}{1.5} = 166.67$$

Dorbish and Bowley's index number

$$= \frac{1}{2}\left[\frac{\Sigma p_1 q_1}{\Sigma p_0 q_1} + \frac{\Sigma p_1 q_0}{\Sigma p_0 q_0}\right] 100 = 145$$

$$\Rightarrow \quad \left[\frac{250}{1} \times \frac{1.5}{250} + \frac{\Sigma p_1 q_0}{150}\right] = \frac{290}{100}$$

$$\Rightarrow \quad 1.5 + \frac{\Sigma p_1 q_0}{150} = 2.9$$

$$\Rightarrow \quad \frac{\Sigma p_1 q_0}{150} = 1.4$$

$$\Rightarrow \quad \Sigma p_1 q_0 = 210$$

(i) Fisher's index number $= 100 \sqrt{\frac{\Sigma p_1 q_1}{\Sigma p_0 q_1} \times \frac{\Sigma p_1 q_0}{\Sigma p_0 q_0}}$

$$= 100 \sqrt{\left[1.5 \times \frac{210}{150}\right]} = 100\sqrt{[1.5 \times 1.40]} = 145$$

(ii) Marshal-Edgeworth's index number

$$= \frac{\Sigma p_1 q_0 + \Sigma p_1 q_1}{\Sigma p_0 q_0 + \Sigma p_0 q_1} \times 100$$

$$= \left[\frac{210 + 250}{150 + 166.67}\right] 100 = \frac{460}{316.67} \times 100 = 145.26$$

Example 47. Incomplete information obtained from a partially destroyed records on cost of living analysis is given below :

Group	*Group Index*	*Percent (%) of Total Expenditure*
Food	268	60
Clothing	280	Not available
Housing	210	20
Fuel and Electricity	240	5
Miscellaneous	260	Not available

The cost of living index with percent of total expenditure as weight was found to be 255.8. Estimate the missing weights. **(D.U., B.Com. (Hons.) 2005R)**

Solution. Let the weight for clothing be x_1 and for miscellaneous be x_2

Then $$60 + x_1 + 20 + 5 + x_2 = 100$$

$$x_1 + x_2 = 15$$

$$255.8 = \frac{268 \times 60 + 280x_1 + 210 \times 20 + 240 \times 5 + 260x_2}{100}$$

$$\Rightarrow \quad 25580 = 16080 + 280x_1 + 4200 + 1200 + 260x_2$$

$$= 21480 + 280x_1 + 260x_2$$

$$\Rightarrow \quad 4100 = 260\,(x_1 + x_2) + 20x_1$$

$$= 260 \times 15 + 20x_1 \qquad [\because x_1 + x_2 = 15]$$

$$= 3900 + 20x_1$$

$$\Rightarrow \quad 200 = 20x_1 \Rightarrow x_1 = 10$$

$$\therefore \quad x_2 = 15 - 10 = 5$$

$$\therefore \quad x_1 = 10, x_2 = 5$$

Example 48. The price of a commodity increased by 10% in the first year and by 20% in the second year. However, it decreased by 40% in the third year. But it again started rising and increased by 5% in the fourth year. Find the average increase or decrease in the price of the commodity over the period of four years. **(D.U., B.Com. 2007R)**

Solution. Let r% be the average increase or decrease in the price

then $$\left(1 + \frac{r}{100}\right)^4 = \left(1 + \frac{10}{100}\right)\left(1 + \frac{20}{100}\right)\left(1 - \frac{40}{100}\right)\left(1 + \frac{5}{100}\right)$$

$$= 1.1 \times 1.2 \times 0.6 \times 1.05 = 0.8316$$

$$\Rightarrow \quad \left(1 + \frac{r}{100}\right) = (0.8316)^{1/4} = 0.955$$

$$\Rightarrow \quad \frac{r}{100} = 0.955 - 1 = -0.045$$

$$\Rightarrow \quad r = -4.5\%$$

i.e., On an average there is a decrease of 4.5% in the price of the commodity.

Example 49. "In the construction of index numbers the advantages of geometric mean are greater than those of arithmetic mean". Discuss.

Solution.

(i) In the index numbers we deal with ratios and relative changes and G.M. gives equal weights to equal ratios of change because G.M. of ratios is equal to ratio of GMs, *e.g.*, if the price of one commodity is doubled and that of the other is halved, then GM is not affected but AM will change.

(ii) GM gives more importance to small numbers and less importance to large numbers and is therefore not unduly affected by extreme values.

(iii) Good index numbers possess the characteristic of reversibility which can be achieved by using G.M.

Example 50. The table below gives you the money wages of cost of living index based on 1998 :

Year	1998	1999	2000	2001	2002	2003	2004
Wages	65	70	75	80	90	100	120
Cost of Living Index	100	110	120	130	150	200	250

Calculate (i) Real wages (ii) Index number of real wages. **(D.U., B.Com. (Hons.) 2006E)**

Solution.

Year	*Wages (Rs.)*	*Cost of Living Index*	*Real Wages (Rs.) = (Wages/CLI) × 100*	*Real Wage Index (1998 = 100)*
1998	65	100	65.00	100.00
1999	70	110	63.64	$\frac{100}{65} \times 63.64 = 97.91$
2000	75	120	62.50	96.15
2001	80	130	61.54	94.68
2002	90	150	60.00	92.31
2003	100	200	50.00	76.92
2004	120	250	48.00	73.85

Example 51. Construct the cost of living index number by

(i) Aggregative expenditure method.

(ii) Family budget method for the year 2004 taking 2000 as base.

Commodity	*Quantity*	*Unit*	*Price (Rs.)*	
			2000	*2004*
Wheat	2 Qnts.	Qnt	56	75
Rice	25 kg.	Qnt	100	120
Sugar	10 kg.	Qnt	80	120
Pure Ghee	5 kg.	kg	10	10
Veg. Ghee	5 kg.	kg	3	5
Oil	25 kg.	Qnt.	200	200
Clothing	25 met.	Meter	4	5
Fuel	4 Qnts.	Qntl	8	10
Rent	One	House	20	25

(Kurukshetra University, B.Com. 2005)

Solution. Current year index number by aggregative expenditure method

$$=\frac{\Sigma p_1q_0}{\Sigma p_0q_0}\times 100$$

Current year index number by family budget method

$$=\frac{\Sigma IV}{\Sigma V}$$

where $I=\frac{p_1}{p_0}\times 100,\quad V=q_0p_0$

$$\frac{\Sigma p_1q_0}{\Sigma p_0q_0}\times 100=\frac{75\times 2+120\times\frac{1}{4}+120\times\frac{1}{10}+10\times 5+5\times 5+200\times\frac{1}{4}+5\times 25+10\times 4+25}{50\times 2+10\times\frac{1}{4}+80\times\frac{1}{10}+10\times 5+3\times 5+200\times\frac{1}{4}+4\times 25+8\times 4+20}\times 100$$

$$=\frac{150+30+12+50+25+50+125+40+25}{100+25+8+50+15+25+100+32+20}\times 100$$

$$=\frac{507}{375}\times 100=135.2$$

$\Sigma p_0q_0=375=V$

Commodity	$\frac{p_1}{p_0}\times 100=I$	$p_0q_0=V$	IV
Wheat	(75/50) 100 = 150	100	15000
Rice	(120/100) 100 = 100	25	2500
Sugar	(120 = 80) 100 = 150	8	1200
Pure Ghee	(10/10) 100 = 100	50	5000
Veg. Ghee	(5/3) 100 = 166.66	15	2500
Oil	(200/200) 100 = 100	25	2500
Clothing	(5/4) 100 = 125	100	12500
Fuel	(10/8) 100 = 125	32	5000
Rent	(25/20) 100 = 125	20	2500
		375	47700

$$\text{Index number}=\frac{47700}{375}=127.2$$

Example 52. The price paid and quantities purchased by a household in base and current years are given below. Calculate the additional dearness allowance to be given to the household so as to fully compensate it for the price rise, using both the Laspeyre's and Paasche's index numbers.

Commodity	*Base Period*		*Current Period*	
	Price	*Quantity*	*Price*	*Quantity*
A	30	10	40	8
B	12	20	15	18

Solution. Laspeyre's Index Number

$$= \frac{\Sigma p_1 q_0}{\Sigma p_0 q_0} \times 100$$

$$= \frac{40 \times 10 + 20 \times 15}{30 \times 10 + 12 \times 20} \times 100 = \frac{400 + 300}{300 + 240} \times 100 = \frac{700}{540} \times 100 = 129.63$$

Additional dearness allowance to be paid = 29.63%

$$\text{Paasche's index number} = \frac{\Sigma p_1 q_1}{\Sigma p_0 q_1} \times 100$$

$$= \frac{40 \times 8 + 14 \times 18}{30 \times 8 + 12 \times 18} \times 100 = \frac{320 + 270}{240 + 216} \times 100$$

$$= \frac{590}{456} \times 100 = 129.386$$

Additional dearness allowance to be paid = 29.386%.

Example 53. Owing to change in prices the consumer price index of the working class in a certain area rose in a month by one quarter of what it was before to 225. The index of food became 252 from 198, that of clothing from 185 to 205, that of fuel and lighting from 175 to 195 and that of miscellaneous from 138 to 212. The index rent, however, remained uncharged at 150. It was known that the weight of clothing, rent, fuel and lighting were the same. Find out the exact weight of all the groups. **(Burdwan M. Com. 2000), (D.U., B.Com. (Hons.) 2007E)**

Solution. Suppose the weights of the different groups are as under :

Food X Rent Z

Fuel and Lighting Z Clothing Z

Miscellaneous Y

Therefore, the weighted index in the beginning of the month was :

	Index	*Weights*	
	I	*W*	*IW*
Foot	198	X	$198X$
Clothing	185	Z	$185Z$
Fuel and Lighting	175	Z	$175Z$
Rent	150	Z	$150Z$
Miscellaneous	138	Y	$138Y$
		$\Sigma W = X + Y + 3Z$	$\Sigma IW = 198X + 138Y + 510Z$

$\therefore$ $$\text{Index number} = \frac{198X + 138Y + 510Z}{X + Y + 3Z} \quad \text{...(I)}$$

Similarly the weighted index at the end of the month was :

	Index	*Weights*	
	I	*W*	*IW*
Foot	252	X	$252X$
Clothing	205	Z	$205Z$
Fuel and Lighting	195	Z	$195Z$
Rent	150	Z	$150Z$
Miscellaneous	212	Y	$212Y$
		$\Sigma W = X + Y + 3Z$	$\Sigma IW = 252X + 212Y + 550Z$

$\therefore$ $$\text{Index number} = \frac{252X + 212Y + 550Z}{X + Y + 3Z} \quad \text{...(II)}$$

The weighed index at the end of the month was 225 (given). This index is a rise from the previous index by one quarter. Therefore, the index at the beginning of the month was 4/5 of 225 = 180.

$\therefore$ $$180 = \frac{198X + 138Y + 510Z}{X + Y + 3Z} \quad \text{...(By I)}$$

$\Rightarrow$ $$180X + 180Y + 540Z = 198X + 138Y + 510Z$$

$\Rightarrow$ $$18X - 42Y - 30Z = 0 \quad \text{...(III)}$$

Similarly at the end of the month, we have

$$225 = \frac{252X + 212Y + 550Z}{X + Y + 3Z} \quad \text{...(By II)}$$

$\Rightarrow$ $$225X + 225Y + 675Z = 252X + 212Y + 550Z$$

$\Rightarrow$ $$27X + 13Y + 125Z = 0 \quad \text{...(IV)}$$

Let the total weight be equal to 100 then

$$X + Y + 3Z = 100 \quad \text{...(V)}$$

Multiplying (V) by 18 and subtracting from (III), we get

$$-60Y - 84Z = -1800$$

$\Rightarrow$ $$60Y + 84Z = 1800 \quad \text{...(VI)}$$

Multiplying (V) by 27 and subtracting from (IV) we get

$$-40Y - 206Z = -2700$$

$\Rightarrow$ $$40Y + 206Z = 2700 \quad \text{...(VII)}$$

Multiplying (VI) by 2 and (VII) by 3 and subtracting, we get

$$450Z = 4500 \Rightarrow Z = 10$$

when $Z = 10$, $Y = \frac{1800 - 840}{60} = \frac{960}{60} = 16$ (From VI)

Putting the values of Y, Z in (IV), we get

$$27X - 13 \times 16 - 125 \times 10 = 0$$

$\Rightarrow$ $$27X = 208 + 1250 = 1458$$

$\Rightarrow$ $$X = 54.$$

Example 54. In the working class consumer price index number in a particular town, the weights corresponding to different groups of items were as follows :

Food 55, Fuel 15, Clothing 10, Rent 8, Misc. 12. In October, 1979, the dearness allowance was fixed by a mill of that town at 182 per cent of the worker's wages which fully compensated for the rise in the prices of food and rent but did not compensate for anything else. Another mill of the same town paid dearness allowance of 46.5 per cent which compensated for the rise in fuel and miscellaneous groups. It is known that the rise in food is double that of fuel and the rise in the miscellaneous group is doubled that of rent. Find the rise in food, fuel, rent & miscellaneous groups.

(D.U.B. Com. Hons., 2002)

Solution.

(i) Since nothing is mentioned about the rise in clothing group it is presumed that it remained unaffected so that the current index for the clothing group is the same as in base year, *i.e.*, 100.

(ii) Let us suppose that the percentage *rise* in fuel is X. Then the percentage *rise* in food would be $2X$. Let us further suppose that the percentage rise in rent is Y so that the percentage rise in miscellaneous group is $2Y$.

(iii) Since the first mill gave a D.A. of 182%, the Current Index (I) would be 100 + 182 or 282. Since the mill fully compensated the workers for rise in price of food and rent only, the group index for food would be $(100 + 2X)$ and for rent $(100 + Y)$. The other group indices will be taken as 100 each.

(iv) **Calculation of Consumer Price Index (First Mill)**

Group	Index	Weights	
	I	W	IW
Food	$100 + 2X$	55	$5500 + 110x$
Fuel	100	15	1500
Clothing	100	10	1000
Rent	$100 + Y$	8	$800 + 8Y$
Miscellaneous	100	12	1200
		$\Sigma W = 100$	$\Sigma IW = 10{,}000 + 110X + 8Y$

$$\text{Index Number} = \frac{\Sigma IW}{\Sigma W} = \frac{10{,}000 + 110X + 8Y}{100} = 282$$

or $$10{,}000 + 110X + 8Y = 28200$$

or $$110X + 8Y = 18200 \qquad ...(i)$$

Consumer Price Index (Second Mill)

Group	Index	Weights	
	I	W	IW
Food	100	55	5500
Fuel	$100 + X$	15	$1500 + 15X$
Clothing	100	10	1000
Rent	100	8	800
Miscellaneous	$100 + 2Y$	12	$1200 + 24Y$
		$\Sigma W = 100$	$\Sigma IW = 10{,}000 + 15X + 24Y$

$$\text{Index Number} = \frac{\Sigma IW}{\Sigma W} = \frac{10{,}000 + 15X + 24Y}{100} = 146.5$$

or $$10{,}000 + 15X + 24Y = 146{,}50$$

or $$15X + 24Y = 4650 \quad \text{...(ii)}$$

Multiplying equation (i) by 3, we have

$$330X + 24Y = 54{,}600 \quad \text{...(iii)}$$

Subtracting (ii) from (iii), we get

$$315X = 49{,}950$$

$$X = 158.57$$

Subtracting the value of X in (ii), we get

$$24Y = 4650 - (15 \times 158.57)$$

$$24Y = 4650 - 2378.55 = 2271.45$$

$$Y = \frac{2271.45}{24} = 94.64$$

Hence the percentage increase in different groups is as follows :

Group	*% Increase*
Food	$2X = 2 \times 158.57 = 317.14$
Fuel	$X = 158.57$
Rent	$Y = 94.64$
Miscellaneous	$2Y = 2 \times 94.64 \times 189.28$

Example 55. Expenditure of a family on 3 times are in the ratio of 2 : 5 : 3. The prices of these commodities rise by 30 per cent, 20 per cent and 40 per cent respectively. By what per cent has total expenditure increased ?

Solution. The price of a commodity rise by 30 per cent means that with suitable time as base, index number of a commodity in current year is 100 + 30, *i.e.,* 130,

Item	*Index No. (I)*	*Weight (W)*	*IW*
A	130	2	260
B	120	5	600
C	140	3	420
		10	1280

$$\text{Price Index No.} = \frac{\Sigma IW}{\Sigma W} = \frac{1280}{10} = 128$$

The total expenditure has increased by (128 – 100) = 28%.

LIST OF FORMULAE

1. Price Relative $= P_{01} = \frac{p_1}{p_0} \times 100.$

   Quantity Relative $= Q_{01} = \frac{q_1}{q_0} \times 100$

2. **Unweighted Index Numbers :**

(a) Simple Aggregate Price Index $= P_{01} = \dfrac{\Sigma p_1}{\Sigma p_0} \times 100$

(b) Simple Average of Price Relatives $= P_{01} = \dfrac{\Sigma \dfrac{p_1}{p_0} \times 100}{N}$

3. **Weighted Index Numbers :**

(a) *Weighted Aggregate Method :*

Simple Aggregative with fixed weight $= P_{01} = \dfrac{\Sigma p_1 w}{\Sigma p_0 w} \times 100$

Laspeyre's Price Index $= P_{01} = \dfrac{\Sigma p_1 q_0}{\Sigma p_0 q_0} \times 100$

Paasche's Index number $= \dfrac{\Sigma p_1 q_1}{\Sigma p_0 q_1} \times 100$

Marshall-Edgeworth's Index $= P_{01} = \dfrac{\Sigma p_1 (q_0 + q_1)}{\Sigma p_0 (q_0 + q_1)} \times 100$

Fisher's Ideal Index $= P_{01} = \sqrt{\left[\dfrac{\Sigma p_1 q_0}{\Sigma p_0 q_0} \times \dfrac{\Sigma p_1 q_1}{\Sigma p_0 q_1}\right]} \times 100$

(b) *Weighted Average of Price Relatives :*

Weighted Average of Price Relatives $= P_{01} = \dfrac{\Sigma \left(\dfrac{p_1}{p_0} \times 100\right) W}{\Sigma W}$

Weighted Average of Price Relative (base year values as weights)

$$= P_{01} = \frac{\Sigma \left(\dfrac{p_1}{p_0} \times 100\right) p_0 q_0}{\Sigma p_0 q_0} = \frac{\Sigma p_1 q_0}{\Sigma p_0 q_0} \times 100.$$

Weighted Average of Price Relatives (current year values as weights)

$$= P_{01} = \frac{\Sigma \left(\dfrac{p_1}{p_0} \times 100\right) p_1 q_1}{\Sigma p_1 q_1}$$

Questions

Short Answer Type Questions

1. Define an index number stating its utility.
2. Index numbers are economic barometers.
3. What are the problems faced in the construction of index numbers ?
4. Write critical note on uses and limitations of index numbers.
5. Distinguish between aggregate type and average type index formula.

6. How will you choose a base year for constructing an index number ?
7. Discuss various used in the construction of an index number.
8. Explain Laspeyre's and Paasche's price index numbers.
9. What is Fisher's ideal index number ? Why is it so called ?

Long Answer Type Questions

1. Explain consumer price index number.
2. Explain index of industrial production.
3. How do you compare Laspeyre's index number and Paasche's index number ?
4. Why Fisher's ideal index number is called 'Ideal' ?
5. What are the problems in the construction of index numbers ?
6. Explain different types of index numbers.

State whether the following statements are True or False :

1. Index numbers are specialised averages.
2. Index number is a statistical measure representing changes in a variable or group of related variables.
3. Index numbers are not expressed in percentages.
4. Index numbers measure changes which are not directly measurable.
5. Index numbers are economic barometers.
6. Index numbers cannot be used in forecasting.
7. Index numbers are used in computing the real income or wages.
8. The ideal average used in the construction of index numbers is the arithmetic mean.
9. Base year used in index numbers should be a normal year.
10. Weighted index numbers are better and more representative than simple index numbers.
11. Geometric mean is the most suitable average in the construction of index numbers.
12. Price relatives are formed to study price changes over time.
13. Quantity relatives are used to measure changes in quantum or volume of production or consumption of a commodity.
14. Laspeyre's formula is a weighted aggregate index with base year quantity weights.
15. Paasche's formula is a weighted aggregate index with base year quantity weights.
16. Fisher's index formula is the arithmetic mean of Laspeyre's and Paasche's formula.

ANSWERS

1. T **2.** T **3.** F **4.** T **5.** T **6.** F **7.** T **8.** T **9.** T **10.**. T
11. T **12.** T **13.** T **14.** T **15.** F **16.** F

Fill in the Blanks

1. Consumer price index number ....
(measures changes in the retail prices, measures changes in wholesale prices)
2. If the price index increases by 20% the cost of Campa Cola, which at present is Rs. 10, will ....
(increase by Rs. 2, increase by Rs. 12)
3. Index numbers are .... (simple averages, specialised averages)
4. Index of industrial production is a .... (price index, quantity index)
5. Index numbers indicate .... (relative changes, absolute changes)

ANSWERS

1. measures changes in retail price **2.** Rs. 2 **3.** specialised averages,
4. quantity index **5.** relative changes

Multiple Choice Questions

Choose the correct answer :

1. For comparing yearly changes in price level, the suitable index to be used is :
(a) F.B.I. with average price as base (b) F.B.I.
(c) C.B.I. (d) None of these

2. The weights used in Laspeyre's price index are denoted as :
(a) q_0 (b) q_1 (c) p_0 (d) p_1

3. The weights used in Laspeyre's quantity index are denoted as :
(a) q_0 (b) q_1 (c) p_0 (d) p_1

4. The weights used in Paasche's price index are denoted as :
(a) q_0 (b) q_1 (c) p_0 (d) p_1

5. The weights used in Paasche's quantity index are denoted as :
(a) q_0 (b) q_1 (c) p_0 (d) p_1

6. Weighted aggregate index formula using base year quantities as weight is called :
(a) Laspeyre's price index (b) Paasche's price index (c) Bowley's price index (d) Fisher's price index

7. Weighted aggregate index formula using the average of base year and current year's quantities as weights is called :
(a) Laspeyre's price index (b) Fisher's price index
(c) Marshall-Edgeworth's index (d) Bowley's index

8. The geometric mean of Laspeyre's and Paasche's indices is :
(a) Fisher's ideal index (b) Bowley's index
(c) Marshall and Edgeworth's index (d) None of these

9. Weighted average of relatives if base year value is taken as weights gives :
(a) Fisher's index (b) Laspeyre's index (c) Paasche's index (d) Bowley's index

10. The formula for simple average of price relative is :
(a) $\frac{1}{n}\Sigma\frac{p_1}{p_0}\times 100$ (b) $\frac{\Sigma p_1}{\Sigma p_0}\times 100$ (c) $\frac{1}{n}\Sigma\frac{q_1}{q_0}\times 100$ (d) None

ANSWERS

1. (c) **2.** (a) **3.** (c) **4.** (b) **5.** (d) **6.** (a) **7.** (c) **8.** (a) **9.** (b) **10.** (a)

Practice Questions

1. A worker earned Rs. 9000 per month in 1995. The cost of living index increased by 60% between 1995 and 1999. How much extra income should the worker have earned in 1999 so that he could buy same quantities as in 1995 ? [**Ans.** Rs. 5400]

2. A worker earned Rs. 8000 per month in 1992. The cost of living index increased by 79% between 1992 and 1995. How much extra income should the worker have earned in 1995 so that he could buy the same quantities as in 1992 ? [**Ans.** 5,600]

3. From the following data calculate price index numbers for 2000 with 1990 as base by (i) Laspeyre's method, (ii) Paasche's method, (iii) Marshal-Edgeworth's method, and (iv) Fisher's ideal method.

Commodities	1990		2000	
	Price	*Quantity*	*Price*	*Quantity*
A	20	8	40	6
B	50	10	60	5
C	40	15	50	15
D	20	20	20	25

[**Ans.** (i) 124.67, (ii) 121.76, (iii) 122, (iv) 123.2]

4. The following table gives the cost of living index numbers for different commodity groups together with respective weights for 2000 (Base = 1980) :

Group	*Food*	*Clothing*	*Fuel and lighting*	*Rent*	*Misc.*
Group index	425	475	300	400	250
Group weight	62	4	6	12	16

Obtain the overall cost of living index number. Suppose a person was earning Rs. 6000 in 1980. What should be his salary in 2000 if his standard of living in that year is to be the same as in 1980 ?

[**Ans.** 388.50, Rs. 23,310]

5. From the following data, calculate Fisher's ideal index number. Check whether this index number satisfies both the tests :

Item	*Base Year*		*Current Year*	
	Price (Rs.)	*Quantity (Kg.)*	*Price (Rs.)*	*Quantity (Kg.)*
A	8	50	10	56
B	2	100	2	120
C	4	60	6	60
D	10	30	12	24
E	8	40	12	36

6. The price relative and weight of a set of commodities are given in the following table :

Commodity	C_1	C_2	C_3	C_4
Price Relative	120	127	125	119
Weight	$2w_1$	w_2	w_1	$w_2 + 3$

If the index for the set is 122 and the sum of the weights is 40, find w_1 and w_2. **(ICWA June 1997)**

[**Ans.** $w_1 = 7, w_2 = 8$]

7. The following table gives the cost of living index numbers for different groups with the respective weights for the year 2000 (base year 1985) :

Group	*Cost of living index*	*Weight*
Food	430	60
Clothing	480	5
Lighting and Fuel	320	7
Rent	390	10
Miscellaneous	300	18

(i) Calculate the overall cost of living index number.

(ii) Suppose, a person was earning Rs. 7000 p.m. in 1985. What should be his salary in 2000, if his standard of living in 2000 is to be the same as in 1985 ? [**Ans.** 397.4, Rs. 27,818]

(IWA Dec., 1995)

8. A textile worker in the city of Ahmedabad earns Rs. 750 per month. The cost of Living Index for January, 1986 is given as 160. Using the following data, find out the amount spent on (i) Food (ii) Rent.

Group	*Expenditure (Rs.)*	*Group index*
Food	?	190
Clothing	125	181
Rent	?	140
Fuel and lighting	100	118
Miscellaneous	75	101

(DU B. Com. Hons., 1993)

9. A textile worker in the city of Bombay earns Rs. 350 per month. The cost of living index for that particular month is given as 136. Using the following data find out the amounts he spent on house rent and clothing :

Group	*Expenditure*	*Group index*
Food	140	180
Clothing	?	150
House rent	?	100
Fuel and lighting	56	110
Miscellaneous	63	80

10. Calculate Fisher's price index from the following data and check whether it satisfies the Time Reversal test :

Commodity	*Base year (1990)*		*Current year (2000)*	
	Price (Rs.)	*Quantity (Kg.)*	*Price (Rs.)*	*Quantity (Kg.)*
A	32	50	30	50
B	30	35	25	40
C	16	55	18	50

[**Ans.** 90]

11. An index is at 100 in 1991. It rises 5% in 1992, falls 6% in 1993, falls 5% in 1994, rises 4% in 1995 and 7% in 1996. Calculate the index number for all these years with 1991 as base. **(CA Foundation Nov. 1999)**

(Refer Ex. 35 Page 8.40)

12. The following table gives the prices of some food items in the base year and current year and the quantities sold in the base year. Calculate the weighted index number by using the weighted average of price relatives :

Items	*Base Year Qnaitities (Units)*	*Base Year Price (in Rs.)*	*Current Year Price (in Rs.)*
A	7	18.00	21.60
B	6	3.00	4.65
C	16	7.50	9.00
D	21	2.50	2.25

[**Ans.** 11.6]

13. Compute Fisher's ideal index number from the data given below and check whether the time reversal test is satisfied :

Commodity	*Base Year*		*Current Year*	
	Price	*Quantity*	*Price*	*Quantity*
A	2	7	6	6
B	3	6	2	3
C	4	5	8	5
D	5	4	2	4

(CA Foundation, May 2001)
[**Ans.** 145]

14. Calculate the cost of living index number from the following data :

Group/Commodities	*Weights W*	*Group/Commodity Index Number I*
Food	71	370
Clothing	3	423
Fuel etc.	9	469
House rent	7	110
Miscellaneous	10	279

Sol. Required index number

$$= \frac{71 \times 370 + 3 \times 423 + 9 \times 469 + 7 \times 10 + 10 \times 279}{71 + 3 + 9 + 7 + 10}$$

$$= \frac{26270 + 1269 + 4221 + 770 + 2790}{100} = \frac{35320}{100} = 353.2$$

15. Prices paid and quantities consumed of three commodities during two time periods are :

Commodity	*Time Period I*		*Time Period II*	
	p_1	q_1	p_2	q_2
A	10	2	15	1
B	15	2	10	3
C	25	4	15	4

(i) Keeping the quantity mix of period I as weights, what percentage change in price has occurred between the two periods ?

(ii) What is the perecentage change in prices if quantity mix of periods II is used as weights ?

(iii) What is the percentage change in quantities between two periods when prices of period I are taken as weights ?

(iv) What percentage change in the value of consumption has occurred ?

[**Hint.**(i) Calculate $\frac{\Sigma p_2 q_1}{\Sigma p_1 q_1} \times 100$ (ii) Calculate $\frac{\Sigma p_2 q_2}{\Sigma p_1 q_2} \times 100$

(iii) Calculate $\frac{\Sigma q_2 p_1}{\Sigma q_1 p_1} \times 100$ (iv) Calculate $\frac{\Sigma p_2 q_2}{\Sigma p_1 q_1} \times 100$]

16. Calculate Paasche's Quantity Index and Laspeyre's Price Index for the following data :

Commodities	*Quantity (Units)*		*Value (in Rs.)*	
	q_0 *(1995)*	q_1 *(1999)*	*1995*	*1999*
A	100	150	500	900
B	80	100	320	500
C	60	72	120	360
D	30	33	360	297

(D.U., B. Com., 2002)

[**Hint :** p_0 for commodity $A = \frac{500}{100} = \text{Rs. } 5$

p_0 for commodity $B = \frac{320}{80} = \text{Rs. } 4$

p_0 for commodity $C = \frac{120}{60} = \text{Rs. } 2$

p_0 for commodity $D = \frac{360}{30} = \text{Rs. } 12$

Per unit price in 1995

Similarly

p_1 for commodity $A = \frac{900}{150} = \text{Rs. } 6$

p_1 for commodity $B = \frac{500}{100} = \text{Rs. } 5$

p_1 for commodity $C = \frac{360}{72} = \text{Rs. } 5$

p_1 for commodity $D = \frac{297}{33} = \text{Rs. } 9$

Per unit price in 1999

Paasche's Quantity Index Number

$$= \frac{\Sigma q_1p_1}{\Sigma q_0p_1} \times 100 = \frac{900 + 500 + 360 + 297}{600 + 400 + 300 + 270} \times 100$$

Laspeyre's Price Index Number

$$= \frac{\Sigma p_1q_0}{\Sigma p_0q_0} \times 100 = \frac{600 + 400 + 300 + 270}{500 + 320 + 120 + 360} \times 100]$$

17. The price relative (in %) and weight of three commodities are given below :

Commodities	*A*	*B*	*C*
Price relative	115	125	120
Weight	W_1	W_2	$W_1 + W_2$

If the sum of weights is 20 and price index number for the commodities is 119%, find the numerical values of the weights. **(CA Foundation, May, 2002)**

[**Hint :** $119 = \frac{115W_1 + 125W_2 + 120\,(W_1 + W_2)}{W_1 + W_2 + W_1 + W_2}$

and $W_1 + W_2 + W_1 + W_2 = 20$

$\Rightarrow$ $W_1 + W_2 = 10$] [**Ans.** $W_1 = 7, W_2 = 3$]

❑❑❑

7

Time Series

Coverage Meaning, Significance and Components of Time Series, Measurement of Trend—Graphic, Semi-Average, Moving Average and Least Square Method, Measurement of Seasonal Variations and Forecasting.

7.1 DEFINITION OF TIME SERIES

A time series is a statistical data that are collected, observed or recorded at regular intervals of time.* The term time series applies, for example, to the data recorded periodically showing the total annual sales of retail stores, the total quarterly value of construction contracts awarded, the total amount of unfilled orders in durable goods, industries at the end of each month, weekly earnings of workers in an industrial town, hourly temperature in a particular city.

Some of the important definitions of time series, given by different experts are as under :

"*A time series may be defined as a sequence of repeated measurement of a variable made periodically through time.*" —Cecil H. Mayers

"*A time series may be defined as a collection of readings belonging to different time period of some economic variable or composite of variables. Such as production of steel, per capita income, gross national product, price of tobacco or index of industrial production.*" —Ya-lun Chou

"*A time series consist of statistical data which are collected, recorded or observed over successive increments.*" —Patterson

"*A time series is a set of observations taken at specified times, usually at equal intervals. Mathematically a time series is defined by the value $y_1, y_2, \ldots$ of a variable y (temperature, closing price of share etc.) at time $t_1, t_2, \ldots$. Thus, y is a function of t symbolised by $y = F(t)$.*" —Spiegel

Thus it is clear from above definition that time series consist of data which are arranged chronologically. It establishes a relationship between two variables in which one of the variable is independent variable, *i.e.*, the time and other variable y is the dependent variable whose value changes with regard to time variable, *e.g.*, total agricultural production in different years.

The analysis of time series is useful not only for individual business concerns but it is equally important for the central and state governments for devising suitable future growth strategies. A close

* The intervals of time may not be always equal.

understanding of time series data on major national aggregates such as population, national income, capital formation, etc., provides crucial information about the strength or weakness of strategies adopted in the past. This analysis also helps in setting future growth targets.

Mathematically a time series is defined by the values $X_1, X_2, X_3, \ldots, X_n$ of the variable X at times $t_1, t_2, \ldots, t_n$. Symbolically

$$X = f(t), \text{ i.e, } X \text{ is a function of time.}$$

Objective of Time Series Analysis

By the analysis of time series, we mean the decomposition of y_1 in to various components that affect this value in a given period. There are two main objectives of analysis of any series :

(a) to study of past behaviour of data
(b) to forecast for future.

Significance of Time Series Analysis

Time series is an important statistical technique which is widely used by business executives, financial analysists, economist, sociologists, research workers, policy administrators in the government etc. because of the following reasons :

(1) It helps to understand the past behaviour of data and enables the users to forecast the future behaviour.
(2) It enables the users to compare the actual performance with the expected performance and analyse the different causes of variation.
(3) It is helpful in future operations.
(4) It enables to evaluate the current accomplishments or performances and ascertain the causes in case of poor performance.
(5) It is useful in planning, state administration business, social sciences and other areas of human knowledge.

7.2 COMPONENTS OF TIME SERIES

A time series consists of the following four components :

1. Secular trend (long-term variation or secular variations)
2. Seasonal variations
3. Cylical variations or cyclic variations
4. Irregular variations.

1. Secular trend (Long-term variations)

Secular trend refers to long-term variations where the variable tends to increase or decrease over a long period of time. The steady increase in the cost of living index recorded by the consumer price index is an example of secular trend. Rising trend refers to gradual increase where as declining trend refers to gradual decreases over a period of time. Trend does not include short-term variations.

Secular trend is the smooth, regular and long-term movement of a series showing continuous growth, stagnation or decline over a long period of time. Therefore, the secular trend or long-term variations is the result of long-term forces that gradually operate on the time series variable.

According to A.E. Waugh, secular trend, is "*that irreversible movement which continues, in general, in the same direction for a considerable period of time*". This definition consists of two parts : (i) movement in same direction, which implies that if the values are increasing (or decreasing) in successive periods, the tendency continues; and (ii) a considerable period of time. There is no specific period which can be called as a long period. Long periods are different for different situations.

The total period required for emergence of a trend depends on :

(i) The type of event forming the series.

(ii) The length of the time unit (year, month, hour, etc.) against which the data are recorded.

The total period or time period to know the trend can be as short as a few days or as long as fifty years depending on the nature of the event under consideration. In case of population or output trends long period could be even 10 years but for daily demand trend of vegetables it could be a month.

Examples of secular trend :

(1) Long-term changes in productivity

(2) Increase in the rate of capital formation

(3) Growth of population

(4) Technological innovations

(5) Transformation in socio-economic set-up

(6) 5-year weather pattern described by the Ministry of Agriculture.

Objectives of Measuring Trend : There are four main objectives of measuring trend in a time series data :

(i) To study past growth or decline of the series. On ignoring the short term fluctuations, trend describes the basic growth or decline tendency of the data.

(ii) Assuming that the same behaviour would continue in future also, the trend curve can be extended for forecasting future.

(iii) In order to analyse the influence of other factors, the trend may first be measured and then eliminated from the observed values.

(iv) Trend values of two or more time series can be used for their comparison.

2. Seasonal Variations

Seasonal variations involve patterns of change within a year that tend to be repeated from year to year. Seasonal variations are **short-term** periodic movements.

The time interval of occurrence of seasonal variations may vary from a few hours to a few weeks or a few months. To note the seasonal variations, the data must be recorded at least quarterly, monthly, weekly or daily depending on the nature of the variable under consideration.

The forces responsible for seasonal variations are :

(i) varying seasons

(ii) weather conditions

(iii) changing social customs and ceremonies.

Being regular and periodic, seasonal variations are useful in forecasting future.

Examples :

(i) Going down of the prices of grains during harvest season.

(ii) Effect of a festival on the sale of a departmental store.

(iii) Agriculture production.

(iv) Production of construction material like bricks, sometimes stops during rainy season.

(v) Sale of construction material going down during rainy season.

3. Cyclical Variations

Cyclic variations are periodic movements in the time series around the trend line. These are the movements in time series that like seasonal variations are recurrent but that unlike seasonal variation occure in cycles of longer than one year. These variations do not follow any regular pattern and move in somewhat unpredictable manner. These are upswings and downswings in the time series that are observable over extended periods of time. Long cycles extending upto 50 years have also been observed. These cycles get mixed up with the trend line and are difficult to distinguish. Cycles of 8 or 9 years duration called ***trade cycles*** are more common. Cycles occurring after a time interval of 3 to 4 years are known as ***minor cycles***.

The most common example is cyclical variations is business cycle. The four phases of a business cycle are :

(i) Boom or Prosperity, (ii) Recession,
(iii) Depression and (iv) Recovery.

The period of business cycle is not uniform and further cyclical variations are mixed up with erratic and irregular variations which are difficult to isolate. Due to this reason, **it is difficult to identify the exclusive effect of cyclical variations.**

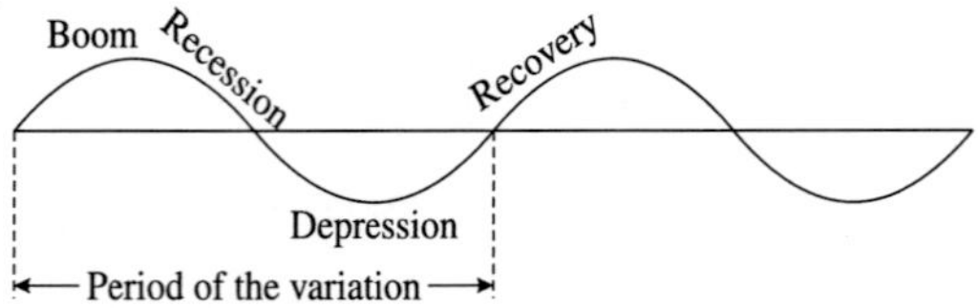

The cyclical variations will take positive or negative signs depending on whether they are above or below the trend line. The positive values of cyclical variations during upswing cancel the negative values during down swing so that the net effect over the period of the cycle will be zero.

4. Irregular or Random Variations

The *irregular component* of the time series is the residual or "catchall" factor that accounts for the deviations of the actual time series values from what we would expect from the trend, cyclical and seasonal components. It accounts for the random variability in the time series. The irregular component is caused by the short term, unanticipated and nonrecurring factors that affect the time series. Since this component accounts for the random variability in the time series, it is unpredictable. We cannot attempt to predict its impact on the time series in advance.

Random variations are unexpected or episodic variations both in terms of amplitude and the length of time interval of occurrence. These are caused by random forces and sporadic causes.

Examples : Variations such as strikes, lockouts, floods, wars, famines, storms, earthquakes, etc.

7.3 ANALYSIS OF TIME SERIES (MODELS OF DECOMPOSITION)

Analysis of time series essentially involves decomposition of the time series into its four components. The main purpose of the analysis of time series is to estimate and separate the four types of variations and to bring out the relative impact of each on the overall behaviour of the time series.

There are two **models** of decomposition of time series :

(1) The additive model

(2) The multiplicative model

The choice between the two models depends on the assumption that we make about the nature of relationship among the four components.

Additive Model

This model is used when it is assumed that the four components are independent of one another, *i.e.*, when the pattern of occurrence and the magnitude of movement in any particular component are not affected by other components, under this assumption the magnitude of time series (Y_t) at any time t is the sum of the separate influence of its four components, *i.e.*,

$$Y_t = T_t + C_t + S_t + R_t$$

where T_t = Trend variations

C_t = Cyclical variations

S_t = Seasonal variations

R_t = Random variations

The units of measurements are the same for all the four components.

When the time series data are **recorded against years**, the seasonal component vanish and in that case

$$Y_t = T_t + C_t + R_t$$

Multiplicative Model

This model is used where it is assumed that the forces giving rise to the four types of variations are independent. According to this model

$$Y_t = T_t\, C_t\, S_t\, R_t$$

or $$\log Y_t = \log T_t + \log C_t + \log S_t + \log R_t$$

Points of difference between the two models : In the additive model there is no interaction of forces that give rise to the four components because it is assumed that the values of an individual component neither affect, nor are affected by the values of the other components whereas in the multiplicative model the forces that generate the four components are interdependent and the time series is the commutative result of the interaction of all the concerned forces. Generally multiplicative model is used because most of the business and economic time series data are the result of interaction of a large number of forces. Since the forces responsible for one type of variation are also responsible for the others, the multiplicative model is often more suited for the purpose of decomposition.

7.4 MEASUREMENT OF TREND

Methods of measuring trend :

(1) Free hand method or graphical method

(2) Semi-average method

(3) Method of least squares

(4) Moving average method

(1) Free Hand Method or Graphical Method

Original data are plotted on a graph paper with time as X-axis and the quantity of measurement on Y-axis. Trend line is drawn free hand in such way that the line must pass roughly mid-way through each peak and closely represent the time series.

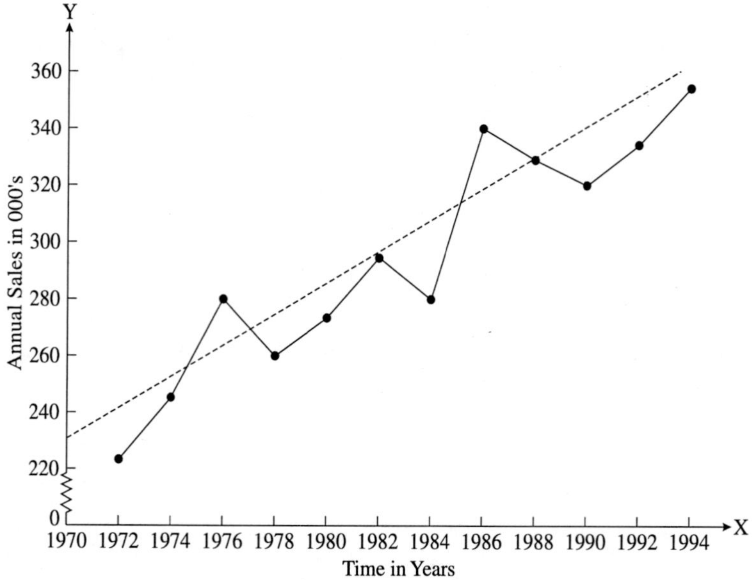

Illustration. Determine the trend of the following time series data by graphical method.

Years	1989–90	1990–91	1991–92	1992–93	1993–94	1994–95
Fert. Imports	2005	2759	2041	1132	1355	3624
Years	1995–96	1996–97	1997–98	1998–99	1999–2000	2000–01
Fert. Imports	3399	2310	984	1608	3114	2758

Solution.

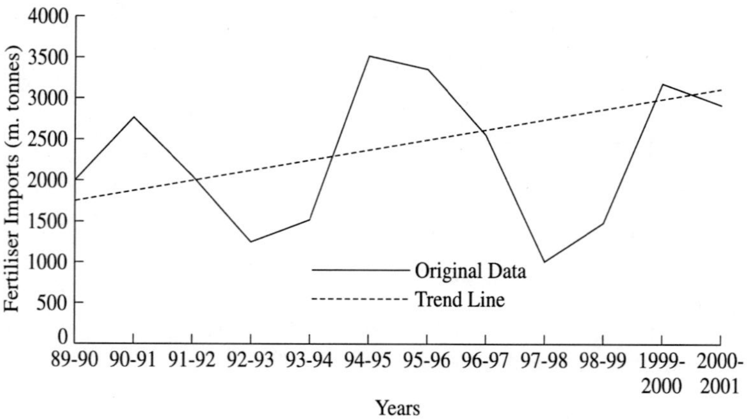

The dotted line, in the figure, shows the required trend line. This line can be extended to get the predicted values for future.

Merits :

(i) It is a simple method of estimating trend which requires no mathematical calculations.

(ii) It is a flexible method as compared to rigid mathematical trends and therefore, a better representative of the trend of the data.

(iii) If the observations are relatively stable, the trend can easily be approximated.

Demerits :

(i) It is a subjective method. The values of trend, obtained by different statisticians would be different and hence, not reliable.

(ii) Prediction made on the basis of this method is not of much value.

(2) Semi-Average Method

Steps :

(i) Divide the series into two equal parts.

(ii) Take average of each part separately.

(iii) Plot the average of each part against the middle of the time period covered by the respective parts.

(iv) Join the plotted points.

Note : If the number of time periods is even, we can have two parts easily but if the number of time periods is odd, the normal practice is to ignore the middle period and divide the resulting series in to two halves. If after dividing the series into two equal parts, each part contains odd number of time periods, middle point in each part will correspond to one time period [different for each part] but if each part contains even number of time periods we have to take the centre of two middle values for each parts. **Time period is taken on *X*-axis.**

Illustration 1. Compute trend by the semi-average method of the following data :

Year	*Sales (Lakhs of Rs.)*	*Semi-total*	*Semi-average*
1991	38		
1992	40		
1993	46	224	44.8
1994	49		
1995	51		
1996	55		
1997	61		
1998	63		
1999	69	345	69.0
2000	72		
2001	80		

These two semi-averages **are plotted in the middle of the respective time spans**. Thus 44.8 is plotted against 1993; and 69.0 against 1999. These two points are then connected by a straight line as shown the following fig.

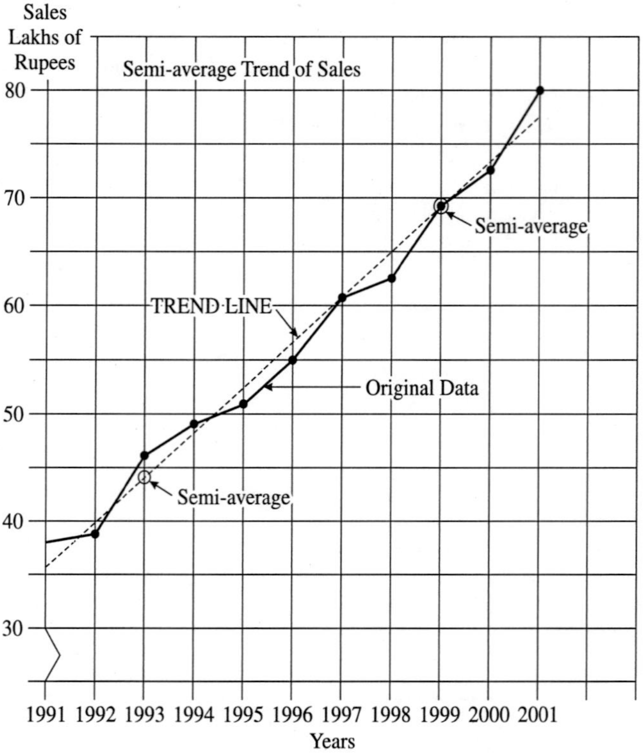

Illustration 2. The sale of commodity in tonnes varied from January 2000 to December 2000 in the following manner :

280	300	280	280	270	240
230	230	220	200	210	200

Fit a trend by the method of semi-average.

Solution.

	Sales	*Semi-average*		*Sales*	*Semi-average*
January	280		July	230	
February	300		August	230	
March	280		September	220	
		275			215
April	280		October	200	
May	270		November	210	
June	240		December	200	

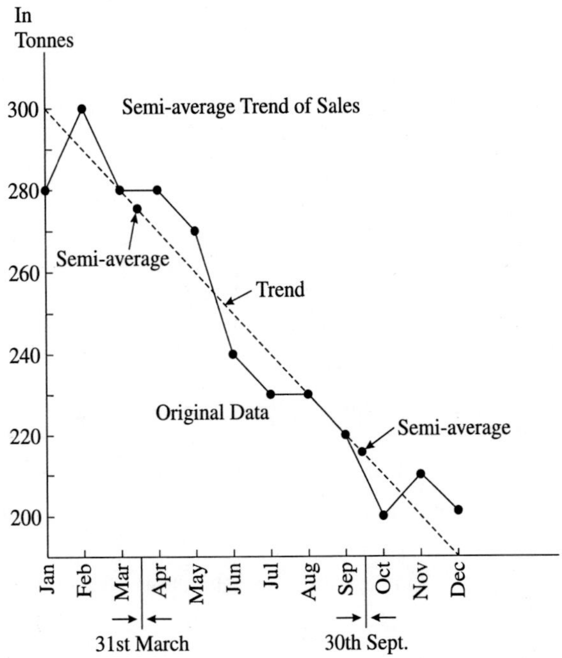

Merits :

(i) It is simple method of measuring trend.

(ii) It is an objective method because anyone applying this method to given data would get identical trend values.

Demerits :

(i) This method can give only a linear trend of the data irrespective to whether it exists or not.

(ii) This is only a crude method of measuring trend, since we do not know whether the effects of other components is completely eliminated or not.

(3) Method of Least Squares

1. Fitting a Straight Line (Line of Best Fit) : The equation of a straight line is of the form $Y = a + bX$ where a and b are constants. By the method of **least squares**, the normal equations to find the values of a and b are

$$\Sigma Y = na^* + b\,\Sigma X$$

$$\Sigma XY = a\,\Sigma X + b\,\Sigma X^2$$

X is taken as time period and is treated as an independent variable.

If the values of X are large, we redefine X by taking deviations from some constant value, *i.e.*, we take $x = X - A$

Calculations are further simplied if we take $A = \overline{X}$ in that case $\Sigma x\ \Sigma(X - \overline{X}) = 0$ [$\because$ The algebraic sum of the deviations of all the values from their mean $= 0$].

* In the euation $Y = a + bX$, a is a constant and will be added as many times as X and Y occurs. If n values of Y, *i.e.*, $Y_1, Y_2, Y_3, \ldots, Y_n$ are taken then a will be added n times *i.e.* na.

(i) If the number of time periods are odd, the deviations are taken from the middle value, *i.e.*, middle year which is $\left(\frac{n+1}{2}\right)$th value and the deviations are taken from $\left(\frac{n+1}{2}\right)$th value so that Σx becomes equal to zero and the value of a is determined from the equation :

$$\Sigma Y = na + b\,\Sigma x$$

by putting $\Sigma x = 0$, *i.e.*, $\Sigma y = na$ gives the value of a and of b is determined from the equation

$$\Sigma xY = a\,\Sigma x + b\,\Sigma x^2$$

by putting $\Sigma x = 0$, *i.e.*, from the equation $\Sigma xY = b\,\Sigma x^2$.

(ii) If the number of time periods are even then, we define

$$x = \frac{X - (\text{Middle Year})*}{\frac{1}{2}\,(\text{Interval between two middle years})} = \frac{(X - \text{Middle Year})}{0.5\,(\text{Interval between two middle years})}$$

e.g., If we have the years 1990, 1991, 1992, 1993, 1994, the middle year is 1992 and if the years are 1989, 1990, 1991, 1992, 1993, 1994, then the two middle years are 1991 and 1992.

How ever if the values are 1960, 1970, 1980, 1990, the two middle values are 1970 and 1980 and the middle value, *i.e.*, mean of two middle values is 1975 and $\frac{1}{2}$ [Interval between middle values] $= 5$.

When the deviations are taken from an arbitrary value the equation of fitted straight line or trend equation will be

$$Y = a + b\,(X - A)$$

where A is the value from which the deviations are taken. If A is the actual mean of X then $Y = a + b\,(X - \overline{X})$.**

2. To Find the Parabolic Trend of the Type :

$$Y = a + bX + CX^2$$

In this case the normal equations by the method of least squares are

$$\Sigma Y = na + b\,\Sigma X = c\,\Sigma X^2$$

$$\Sigma XY = a\,\Sigma X + b\,\Sigma X^2 + c\,\Sigma X^3$$

$$\Sigma X^2 Y = a\,\Sigma X^2 + b\,\Sigma X^3 + c\,\Sigma X^4$$

3. Exponential Curves for Trend Values : The equation of the exponential curve is of the

* Middle year is the A.M. of $\frac{N}{2}$th year and $\left(\frac{N}{2}+1\right)$th year.

** The following points may be notes for straight line trend

(i) In the equation $y = a + bX$, b indicates the yearly increase and decrease.

(ii) Monthly trend equation is $y = \frac{a}{12} + \frac{b}{144}$

(iii) Quarterly trend equation is $y = \frac{a}{4} + \frac{b}{16}$

(iv) For the year of origin the middle of the year is taken as origin, *i.e.*, 1st July of the year will be taken as origin, for month 15th of month will be taken as origin for quarter the middle of quarter will be taken as origin.

following form :

$$Y = AB^X$$

Taking the logarithm of both sides, we get

$$\log Y = \log A + X \log B$$

To obtain the values of the constants A and B the two normal equations by the method of least squares are

$$\Sigma \log Y = N \log A + \log B \, \Sigma X$$

$$\Sigma (X \log Y) = \log A \, \Sigma X + \log B \, \Sigma X^2$$

When deviations are taken from the middle year so that $\Sigma(X - \bar{X}) = \Sigma x = 0$ the above equations are simplified and assume the following form :

$$\Sigma \log Y = N \log A \quad \Rightarrow \quad \log A = \frac{\Sigma \log Y}{N}$$

$$\Sigma (x \cdot \log Y) = \log B \, \Sigma x^2 \quad \Rightarrow \quad \log B = \frac{\Sigma (x \log Y)}{\Sigma x^2}$$

To fit an equation of the above type we need the value of x, Y, $\Sigma x \log Y$, Σx^2 and values of A and B along with their logs.

To get these values the following steps are necessary.

(i) x = Deviation of the X-series from its mid-value. Find these time deviations and square these values and total the squares to get the value of Σx^2.

(ii) Obtain the logarithms of the value of Y.

(iii) Multiply $\log Y$ with corresponding values of x, total these products to get $\Sigma (x \log Y)$.

(iv) Obtain the value of $\log A$ by dividing $\Sigma \log Y$ by N.

(v) Obtain the value of $\log B$ by dividing $\Sigma (x \log X)$ by Σx^2. This would be the **slope of the line** or the **value of growth**.

The values of A and B can be obtained by taking antilogarithms.

The required equation would be $Y = AB^x = AB^{(X - \bar{X})}$.

The following examples would illustrate the above rules.

SOLVED EXAMPLES

Example 1. Calculate the trend values by the method of least squares from the data given below and estimate the sales for the year 2001.

Year	1992	1993	1994	1995	1996
Sales of T.V. (in '000)	12	18	20	23	27

Solution. **Calculation of Trend Values by Least Squares Method**

Years X	*Sales* Y	$X - 1994$ x	xY	x^2	*Trend Values* Y_c
1992	12	–2	–24	4	13.0
1993	18	–1	–18	1	16.5
1994 (Mid-value)	20	0	0	0	20.0
1995	23	+1	+23	1	23.5
1996	27	+2	+54	4	27.0
$N = 5$	$\Sigma Y = 100$	$\Sigma x = 0$	$\Sigma xY = 35$	$\Sigma x^2 = 10$	

Equation of Straight line = $Y_c = a + b\,(X - 1994)$*

Since $\Sigma x = 0$, Value of $a = \dfrac{\Sigma Y}{N} = \dfrac{100}{5} = 20$

$$\text{Value of } b = \frac{\Sigma xY}{\Sigma x^2} = \frac{25}{10} = 3.5$$

Equation of Straight line = $Y_c = a + b\,(X - 1994)$ or $Y_c = 20 + 3.5x$

When $X = 1992$, $Y_c = 20 + (3.5 \times -2) = 13$

When $X = 1993$, $Y_c = 20 + (3.5 \times -1) = 16.5$

When $X = 1994$, $Y_c = 20$

When $X = 1995$, $Y_c = 20 + (3.5 \times 1) = 23.5$

When $X = 1996$, $Y_c = 20 + (3.5 \times 2) = 27.0$

Now for 2001 the value of x would be 7 and $Y_c = 20 + (3.5 \times 7) = 44.5$

Thus the likely sale of T.Vs. in 2001 is 44.5 thousands.

Example 2. Given below are the figures of production (in thousand maunds) of a Sugar factory :

Year	1991	1992	1993	1994	1995	1996	1997
Production ('000) *maunds*	40	45	46	42	47	50	46

Fit Straight line trend by the least squares method and tabulate the trend.

Solution. **Calculation of Trend Values by Least Square Method**

Year t	*Production* (*'000 Maunds*) Y	$t - 1994$ X	X^2	XY	*Trend Values* Y_c
1991	40	–3	9	–120	42.035
1992	45	–2	4	–90	43.071
1993	46	–1	1	–46	44.107
1994	42	0	0	0	45.143
1995	47	+1	1	+47	46.179
1996	50	+2	4	+100	47.215
1997	46	+3	9	+138	48.250
$N = 7$	$\Sigma Y = 316$	$\Sigma X = 0$	$\Sigma X^2 = 28$	$\Sigma XY = +29$	$\Sigma Y_c = 316$

$$Y_c = a + bX = a + b\,(t - 1994)$$

$$a = \frac{\Sigma Y}{N} = \frac{316}{7} = 45.143$$

$$b = \frac{\Sigma XY}{\Sigma X^2} = \frac{29}{28} = 1.036$$

$$Y_c = 45.143 + 1.036X$$

When $X = -3$, $Y_c = 45.143 + [1.036 \times (-3)] = 42.035$, similarly we can calculate trend values for other years.

* Pl. note that the values of Y_c are not the same as the values of Y.

Example 3. The following are the annual profits in thousands in a certain business :

Year	1991	1992	1993	1994	1995	1996	1997
Profit (thousands)	60	72	75	65	80	85	95

By the method of least squares fit a straight line. Using that estimate find the profit for the year 2001.

Solution. **Fitting a Straight Line Trend by Least Squares**

Year t	*Profits* (000) Y	$t - 1994$ X	XY	X^2	Y_c
1991	60	–3	–180	9	
1992	72	–2	–144	4	
1993	75	–1	–75	1	
1994	65	0	0	0	$Y_c = 76 + 4.857X$
1995	80	+1	+80	1	
1996	85	+2	+170	4	
1997	95	+3	+285	9	
$N = 7$	$\Sigma Y = 532$	$\Sigma X = 0$	$\Sigma XY = 136$	$\Sigma X^2 = 28$	

The equation of Straight line trend is $Y_c = a + bX$.

By the method of least squares the two equations are :

$$\Sigma y = na + b\,\Sigma X$$

$$\Sigma XY = a\,\Sigma X + b\,\Sigma X^2$$

Since $\Sigma X = 0$, $a = \dfrac{\Sigma Y}{N}$ and $b = \dfrac{\Sigma XY}{\Sigma X^2}$. Therefore

$$a = \frac{532}{7} = 76 \quad \text{and} \quad b = \frac{136}{28} = 4.857$$

The trend equation would be $Y_c = 76 + 4.857X$

For $t = 2001$, the value of Y would be

$$Y_c = 76 + 4.857\,(2001 - 1994) = 109.99$$

Example 4. The sales of a commodity (in '000 of Rs.) are given below :

Year	1999	2000	2001	2002	2003	2004	2005
Sales	82	86	81	86	92	90	99

(i) Using the method of least squares, fit a straight line trend equation to the data.

(ii) What is the average annual change in the sales?

(iii) Obtain the trend values for the year 1999-2005 and show that the sum of difference between the actual and the trend values is equal to zero.

(iv) What are the expected sales for the year 2010?

Solution. (i) **Calculation of Trend Value**

Year	*Sales (in '000 of Rs.) (Y)*	*X (X – 2002)*	X^2	*XY*	*Trend Values* $Y_c = 88 + 2.5X$
1999	82	–3	9	–246	88 + 2.5 (–3) = 80.5
2000	86	–3	4	–172	88 + 2.5 (–2) = 83.0
2001	81	–1	1	–81	88 + 2.5 (–1) = 85.5
2002	86	0	0	0	88 + 2.5 (0) = 88.0
2003	92	1	1	+92	88 + 2.5 (1) = 90.5
2004	90	2	4	+180	88 + 2.5 (2) = 93.0
2005	99	3	9	+297	88 + 2.5 (3) = 95.5
$N = 7$	$\Sigma Y = 616$	$\Sigma X = 0$	$\Sigma X^2 = 28$	$\Sigma XY = 70$	616

The equation of the straight line trend is :

$$Y = a + bX$$

$$\Sigma Y = Na + b\,\Sigma X \quad \ldots(i)$$

$$\Sigma XY = a\,\Sigma X + b\,\Sigma X^2 \quad \ldots(ii)$$

As the origin is middle year, *i.e.*, 2002 hence $\Sigma X = 0$

∴ Our equations will be :

$$a = \frac{\Sigma Y}{N} = \frac{616}{7} = 88$$

$$b = \frac{\Sigma XY}{\Sigma X^2} = \frac{70}{28} = 2.5*$$

(ii) From (*) it is obvious that the trend values increase by a constant amount '*b*' units every year. Thus the average annual change in sale is '*b*' units, *i.e.*, $2.5 \times 1{,}000$ = Rs. 2,500.

(iii) Sum of difference between the actual (*Y*) and trend value (*Y*) is equal to zero

$$\Sigma(Y - Y_c) = 0 \quad \text{or} \quad 616 - 616 = 0$$

(iv) Expected sales for the year 2010

X will be +8 (*i.e.*, 2010 – 2002)

$$Y_{2010} = 88 + 2.5\,(8) = 88 + 20 = 108$$

Example 5. Fit a straight line trend by the method of least squares to the following data. What would be the predicted earnings for the year 2008?

Year	*Earnings (Rs. in crores)*
1998	60
1999	70
2000	72
2001	63
2002	74
2003	79
2004	72
2005	80

(D.U., B.Com. Pass, 2008)

Solution.

Year x	*Earnings* y	$x = \dfrac{X - 2001.5}{0.5}$	xy	x^2
1998	60	–7	–420	49
1999	70	–5	–350	25
2000	72	–3	–210	9
2001	63	–1	–63	1
2002	74	+1	74	1
2003	79	3	237	9
2004	72	5	360	25
2005	80	7	560	49
	$\Sigma y = 570$	$\Sigma x = 0$	181	168

$$\Sigma y = na + b\Sigma x \Rightarrow 570 = 8a \Rightarrow a = 71.25$$

$$\Sigma xy = a\,\Sigma x = b\,\Sigma x^2 \Rightarrow 181 = 168b \Rightarrow b = \frac{181}{168}$$

The regression equation is

$$y = a + b\,\frac{x - 2001}{0 - 5}$$

$$= 71.25 + \frac{181}{168}(x - 2001)^2$$

$$y_{2008} = 71.25 + \frac{181}{168}(2008 - 2001{\cdot}5)^2$$

$$= 71.25 + 9.8 = 99.25$$

Example 6. Calculate trend values by the method of least square from the data below and estimate the sales for 2003.

Year	1996	1997	1998	1999	2000
Sales (*Rs. lakhs*)	70	74	80	86	90

Solution. **Calculation of Trend Values by Least Squares Method**

Year t	*Sales* Y	$t - 1998$ X	XY	X^2	*Trend Values* Y_c
1996	70	–2	–140	4	69.6
1997	74	–1	–74	1	74.8
1998	80	0	0	0	80.0
1999	86	+1	+86	1	85.2
2000	90	+2	+180	4	90.4
$N = 5$	$\Sigma Y = 400$	$\Sigma X = 0$	$\Sigma XY = 52$	$\Sigma X^2 = 10$	$\Sigma Y_c = 400$

Equation of trend line is $Y_c = a + bX \rightarrow Y_c = a + (t - 1998)$
As per the method of least squares the two Equations

$$\Sigma Y = na + b\,\Sigma X$$

$$\Sigma XY = a\,\Sigma X + b\,\Sigma X^2$$

Since $\Sigma X = 0$,
$$a = \frac{\Sigma Y}{N} = \frac{400}{5} = 80$$

$$b = \frac{\Sigma XY}{\Sigma X^2} = \frac{52}{10} = 5.2$$

The equation of Straight line, *i.e.*, trend line would be $Y_c = 80 + 5.2X$. The value of Y when $t = 2003$ or in terms of deviation $X = +5$

$$Y_{2003} = 80 + (5.2 \times 5) = 80 + 26 = 106$$

$$\text{Trend value for } 1996 = 80 + (1996 - 1998) \times 5.2 = 69.6$$

Similarly trend values for 1997, 1998, etc., have been calculated.

Example 7. Use the method of least square to fit a trend line for the income from the following data. Estimate the income for the year 1993 :

Year	1985	1986	1987	1988	1989	1990	1991	1992
Income ('000 *Rs.*)	85	90	110	100	95	120	125	115

Solution.

Year X	*Income* Y	$x = \frac{X - 1988.5}{1/2}$	xY	Σx^2
1985	85	–7	–595	49
1986	90	–5	–450	25
1987	110	–3	–330	9
1988	100	–1	–100	1
1989	95	+1	95	1
1990	120	+3	360	9
1991	125	+5	625	25
1992	115	+7	805	49
$n = 8$	$\Sigma Y = 840$	$\Sigma x = 0$	$\Sigma xY = 410$	$\Sigma x^2 = 168$

Let the trend equation be

$$Y = a + bx \text{ where } x = 2\,(X - 1988.5)$$

and a, b are determined from the normal equations

$$\Sigma Y = na + b\,\Sigma x \qquad \text{...(I)}$$

$$\Sigma xY = a\,\Sigma x + b\,\Sigma x^2 \qquad \text{...(II)}$$

From (I), $a = \frac{840}{8} = 105$

From (II), $b = \frac{410}{168} = 2.44$ $\quad [\because \Sigma x = 0]$

Trend equation is $Y = 105 + 4.88\,(X - 1988.5)$.

Example 8. Fit a straight line trend to the following data by Least Squares Method :

Year	*Production*
1991	18
1992	21
1993	23
1994	27
1995	16

Specify the year of origin. Estimate production for the years 1998 and 2000. **(D.U. B.Com 2002)**

Solution.

Year X	*Production* Y	$X - 1995 = x$	xY	x^2
1991	18	–4	–72	16
1992	21	–2	–42	4
1993	23	0	0	0
1994	27	2	54	4
1995	16	4	64	16
	$\Sigma y = 105$	$\Sigma x = 0$	$\Sigma xY = 4$	$\Sigma x^2 = 40$

Let the straight line be

$$Y = a + bx \quad \text{where } x = X - 1995$$

The equations as per method of least squares are :

$$\Sigma Y = na + b\,\Sigma x \quad \ldots(I)$$

$$\Sigma xY = a\,\Sigma x + b\,\Sigma x^2 \quad \ldots(II)$$

From (I) $\quad 105 = 5a \Rightarrow a = 21$

From (II) $\quad 4 = 40b \Rightarrow b = \frac{4}{40} = 0.1$

$\therefore$ The equation of the required line is

$$Y = 21 + 0.1x, \text{ i.e., } Y = 21 + 0.1\,(X - 1995)$$

To find the year of origin put $x = 0$

i.e., $\quad X - 1995 = 0 \Rightarrow X = 1995$

i.e., July 1995 is the year of origin.

Production Y_{1998} for 1998 is given by

$$Y_{1998} = 21 + 0.1\,(1998 - 1995) = 21.3$$

and $\quad Y_{2000}$ = Production for 2000

$$= 21 + 0.1\,(2000 - 1995) = 21 + 0.5 = 21.5$$

Example 9. The trend of annual sales of a company is described by the following equation

$$Y_c = 15 + 0.5X$$

[Origin 1995; X–unit = 1 year; Y–unit = annual production]

Convert the equation to (i) a monthly trend equation, (ii) quarterly trend.

Solution. (i) Monthly trend equation would be

$$Y_c = \frac{15}{12} + \frac{0.5}{144}X \quad \text{or} \quad Y_c = 1.25 + .00347X$$

(ii) The quarterly trend equation will be

$$Y = \frac{a}{4} + \frac{b}{4 \times 4} X$$

Example 10. The equation for yearly sales (in '000 Rs.) for a commodity with year 2000 as origin is $Y = 91.6 + 28.8X$. Determine the trend equation to given monthly trend value with January, 2001 as origin and calculate the trend for March, 2001.

Solution. (i) The annual trend equation $Y = 91.6 + 28.8X$ when reduced to monthly trend equation would be

$$Y_c = \frac{91.6}{12} + \frac{28.8}{144} X = 7.63 + 0.2X$$

Now the origin is 1st July, 2000; X–unit = 1 month and Y–unit = average monthly sales.

(ii) We have now to shift the origin from 1st July, 2000 to 15 January, 2001, *i.e.*, $6\frac{1}{2}$ months ahead or forward.

The new trend equation would be

$$Y_c = 7.63 + 0.2\,(X + 6.5) = 7.63 + 0.2X + 1.3 = 8.93 + 0.2X$$

Now the origin is January, 2001, *i.e.*, (15th January); X–unit = one month, Y–unit = monthly sales.

(iii) The trend value for March 2001, would be taking $X = 2$.

Substituting $X = 2$ in the monthly equation we, get

$$Y\,(\text{March 2001}) = 8.93 + (0.2 \times 2) = 8.93 + 0.40 = 9.33\ (000\ \text{rupees})\ \text{or Rs. } 9{,}330.$$

Example 11. The production of a commodity during 1993/98 is given below. Fit the second degree parabola to these data and estimate the production for the year 2000 :

Year	1993	1994	1995	1996	1997	1998
Production (000 *tonnes*)	14	12	13	15	18	20

(D.U., B.Com. Hons., 2001)

Solution. Second degree parabola curve can be drawn with the following equation :

$$Y_c = a + bX + cX^2$$

To find the values of constants a, b and c, we solve the following equations :

$$\Sigma Y = na + b\,\Sigma X + c\,\Sigma X^2 \quad \text{...(i)}$$

$$\Sigma XY = a\,\Sigma X + b\,\Sigma X^2 + c\,\Sigma X^3 \quad \text{...(ii)}$$

$$\Sigma X^2 Y = a\,\Sigma X^2 + b\,\Sigma X^3 + c\,\Sigma X^4 \quad \text{...(iii)}$$

Table for Computing *a*, *b* and *c*

Year t	*Y (Period, 000)*	$t - 1995 = X$	X^2	X^3	X^4	XY	X^2Y
1993	10	–2	4	–8	16	–20	40
1994	12	–1	1	–1	1	–12	12
1995	13	0	0	0	0	0	0
1996	15	1	1	1	1	15	15
1997	18	2	4	8	16	36	72
1998	20	3	9	27	81	60	180
	88	3	19	27	115	79	319

Take base year = 1995

Substituting the table values in three equations :

$$88 = 6a + 3b + 19c \quad \text{...(iv)}$$

$$79 = 3a + 19b + 27c \quad \text{...(v)}$$

$$319 = 19a + 27b + 115c \quad \text{...(vi)}$$

Multiplying equation (iv) by 9 and subtracting from equaion (vi)

$$19a + 27a + 115c = 319$$

$$54a + 27b + 171c = 792$$

$$-35 \qquad -56c = -473$$

or $$35a + 56c = 473 \quad \text{...(vii)}$$

Multiplying equation (vi) by 19 and equation (v) by 27 and substracting (v) from(vi)

$$361a + 513b + 218c = 6061$$

$$81a + 513b + 729c = 2133 \quad \text{...(viii)}$$

$$280a \qquad +1456c = 3928$$

Multiplying equation (vii) by 26 and substracting it from equation (viii)

$$280a + 1456c = 3928$$

$$910a + 1456c = 12298$$

$$-630a \qquad = -8370$$

or $$a = \mathbf{13.285}$$

Substituting the value of a in equation (vii)

$$35a + 56c = 473$$

or $$35 \times 13.285 + 56c = 473$$

or $$56c = 473 - 464.975$$

or $$56c = 8.025$$

$$c = \mathbf{0.143}$$

Substituting the values of a and c in equation (iv)

$$6a + 3b + 18c = 88$$

$$6 \times 13.285 + 3b + 19 \times 19 \times 0.143 = 88$$

$$3b = 88 - 79.71 - 2.717$$

$\Rightarrow$ $$3b = 5.573b$$

or $$= \mathbf{1.857}$$

$\therefore\ Y_c = 13.285 + 1.857X + 0.143X^2$

To estimate the value for the year 2000, substitute $X = 5$* in the regression equation :

$$Y_c = 13.285 + 1.857 \times 5 + 0.143 \times 5^2$$

$$= 13.285 + 9.285 + 3.575$$

$$= 26.145 \text{ or } \mathbf{26145\ tonnes.}$$

Note : We can also solve this example by defining $X = \dfrac{t - 1995.5}{0.5} = 2\,(t - 1995.5)$, by this substitution, the solution will be simplified.

* For the year 2000, 2000 – 1995 = 5

Example 12. Fit a parabolic trend $y = a + bX + cX^2$ to the following data where y denotes the output in thousand unit.

Year	1991	1992	1993	1994	1995	1996	1997	1998	1999
Y	2	6	7	8	10	11	12	20	9

Also compute the trend values. Estimate the value for 2000. **(D.U., B.Com. Hons., 2000)**

Solution. $X - 1995 = x$.

Year X	y	x	xy	x^2	x^3	x^4	x^2y	$y_c = a + bx = cx^2$
1991	2	–4	–8	16	–64	256	32	$10.02 + 0.85(-4) + [-0.27(-4)^2] = 2.3$
1992	6	–3	–18	9	–27	81	54	$10.02 + 0.85(-3) + [-0.27(-3)^2] = 5.04$
1993	7	–2	–14	4	–8	16	28	$10.02 + 0.85(-2) + [-0.27(-2)^2] = 7.24$
1994	8	–1	–8	1	–1	1	8	$10.02 + 0.85(-1) + [-0.27(-1)^2] = 8.9$
1995	10	0	0	0	0	0	0	$10.02 + 0.85(0) + [-0.27(0)^2] = 10.02$
1996	11	1	11	1	1	1	11	$10.02 + 0.85(1) + [-0.27(1)^2] = 10.60$
1997	11	2	22	4	8	16	44	$10.02 + 0.85(2) + [-0.27(2)^2] = 10.64$
1998	10	3	30	9	27	81	90	$10.02 + 0.85(3) + [-0.27(3)^2] = 10.14$
1999	9	4	36	16	64	256	144	$10.02 + 0.85(4) + [-0.27(4)^2] = 9.1$
	74		51	60	0	708	411	

$$\Sigma(Y) = na + b\,\Sigma(x) + c\,\Sigma(x^2)$$

$$\Sigma xY = a\,\Sigma(x) + b\,\Sigma(x^2) + c\,\Sigma(x^3)$$

$$\Sigma x^2Y = a\,\Sigma(x^2) + b\,\Sigma(x^3) + c\,\Sigma(x^4)$$

$$74 = 9a + 60c \quad \ldots(I)$$

$$51 = 60b \quad \ldots(II)$$

$$411 = 60a + 708c \quad \ldots(III)$$

From (II), $b = \frac{60}{51} = 0.85$

From (I) and (III), $9a + 60c = 74 \quad \ldots(IV)$

$$60a + 708c = 411 \quad \ldots(V)$$

Multiplying (IV) by 20 and (V) by 3, we get

$$180a + 1200c = 1480$$

$$180a + 2124c = 1233$$

$$-942c = 247 \quad \ldots\text{By substracting}$$

$$c = -0.27$$

Putting the value of c in equation (IV)

$$9a + 60c = 74$$

$\Rightarrow$ $9a + 60\,(-0.27) = 74$

$\Rightarrow$ $9a - 16.2 = 74$

$\Rightarrow$ $9a = 90.2$

$\Rightarrow$ $a = 10.02$

Hence the required trend equation is

$$Y_c = 10.02 + 0.85x - 0.27x^2$$

$$x = X - 1995$$
$$= 2000 - 1995 \text{ when } X = 2000 = 5$$

and value for 2000 $\quad Y_c = a + bx + cx^2$

$$= 10.02 + 0.85(5) + [-0.27(5)^2]$$
$$= 10.02 + 0.85(5) + [-0.27(25)]$$
$$= 10.02 + 4.25 + (-6.75) = 7.52.$$

Example 13. Compute the quadratic trend of the form $Y = a + bX + cX^2$ for the data showing the production of an item, over a 7 years period :

Year	1995	1996	1997	1998	1999	2000	2001
Production (in '000 Nos.)	10	11	12	9	10	13	11

(CA Foundation, Modified, Nov., 1996)

Solution.

Year x	X $X = (x - 1998)$	Y	XY	X^2	X^2Y	X^3	X^4
1995	–3	10	–30	9	90	–27	81
1996	–2	11	–22	4	44	–8	16
1997	–1	12	–12	1	12	–1	1
1998	0	9	0	0	0	0	0
1999	1	10	10	1	10	1	1
2000	2	13	26	4	52	8	16
2001	3	11	33	9	99	27	81
$N = 7$	ΣX	ΣY	ΣXY	ΣX^2	ΣX^2Y	ΣX^3	ΣX^4

$$\Sigma Y = na + b\,\Sigma X + c\,\Sigma X^2$$
$$\Sigma XY = a\,\Sigma X + b\,\Sigma X^2 + c\,\Sigma X^3$$
$$\Sigma X^2Y = a\,\Sigma X^2 + b\,\Sigma X^3 + c\,\Sigma x^4$$

Hence $76 = 7a + 28c$, $5 = 28b$ and $307 = 28a + 196c$

Solving these 3 equations, we get

$$a = \frac{75}{7} = 10.71, \quad b = \frac{5}{28} = 0.178 \approx 0.18 \text{ and } c = \frac{3}{84} = 0.0357 \approx 0.036$$

Hence the quadratic trend is

$$Y = a + bX + cX^2$$
$$= 10.71 + 0.18X + 0.036X^2$$

where $X = 0$ for 1998.

Example 14. What is multiplicative model and additive model in time series. Given below is the quarterly trend equation for sales (in Rs. thousand) of a commodity :

$$Y_c = 130 + 1.8X$$

Origin, first quarter of 2002
1 unit of $X = 1$ quarter
Y unit = average quarterly sales in Rs. thousand
Convert the above equation to annual trend equation and estiamte the sales for the year 2006.

(D.U., B.Com. (Hons.) 2005E)

Solution. Additive model $\qquad Y = T + S + C + I$

Multiplicative model $\qquad Y = TSCI$

where T = Trend, S = Seasonal variation, C = Cyclic variation, I = Irregular variation.

Trend equation is $\qquad Y_c = 130 + 1.8X$

First quarter of 2002 is origin ⇒ Middle of first quarter of 2002 is origin.

To convert this equation to annual trend, we should take the mid of 2002, *i.e.*, 30-06-2002 or 01-07-2002 as origin, *i.e.*, we have to shift the origin by 1.5 quarter.

∴ The trend equation with 30-06-2002 as origin is

$$Y_c = 130 + 1.8(X + 1.5) = 132.7 + 1.8X$$

Annual trend equation is $\qquad Y_c = 132.7 \times 4 + 1.8 \times 16X$

$$= 530.8 + 28.8X$$

Expected sales for 2006 $\qquad Y_{2006} = 530.8 + 28.8 \times 4$

$$= 530.8 + 115.2 = 646$$

Example 15. Given below is a trend equation :

$$Y_C = 372 + 288X$$

(Origin 2006, X unit = 1 year, Y unit = Annual sales)
Convert the above equation :

(i) To monthly trend equation with January, 2007 as origin and estimate sales for March, 2007.
(ii) To quarterly trend equation with first quarter, 2007 as origin and estimate sales for third quarter of 2007.

(D.U., B.Com. (Hons.) 2008)

Solution. Monthly trend equation :

$$Y_C = \frac{372}{12} + \frac{288}{144}X = 31 + 2X$$

Monthly trend equation with January, 2007 as origin is

$$Y_C = 31 + 2\,(X + 6.5)$$

Sales for March, 2007 is

$$Y_C = 31 + 2\,(2 + 6.5) = 31 + 17 = 48$$

Quarterly trend equation is

$$Y_C = \frac{372}{4} + \frac{288}{16} X = 93 + 18X$$

Quarterly trend equation with first quarter of 2007 as origin is

$$Y_C = 93 + 18\,(X + 2.5)$$

Sales for 3rd quarter of 2007 is

$$Y_C = 93 + 18X\,(2 + 2.5) = 174$$

Example 16. Fit a trend function $Y = A \,.\, B^X$ to the following data :

X	Y
1	2.08
2	6.74
3	23.1
4	45.27
5	138

(D.U., B.Com. (Hons.) 2005R)

Solution. $Y = AB^X \;\Rightarrow\; \log Y = \log A + X \log B$

Y	$\log Y$	X	$X - 3 = x$	x^2	$x \log Y$
2.08	0.3180	1	–2	4	– 0.636
6.74	0.8286	2	–1	1	– 0.8286
23.1	1.3636	3	0	0	0
45.27	1.6558	4	1	1	1.6558
138	2.1399	5	2	4	4.2798
	6.3059	15		10	4.471

As per the method of least squares, the two equations are :

$$\Sigma \log Y = n \log A + \log B\, \Sigma x$$

and

$$\Sigma x \log Y = \log A\, \Sigma x + \log B\, \Sigma x^2$$

i.e.,

$$6.3059 = 5 \log A \Rightarrow \log A = \frac{6.3059}{5} = 1.26118$$

$$4.471 = 10 \log B \Rightarrow \log B = 0.4471$$

$$\log A = 1.26118 \Rightarrow A = 18.2465$$

$$\log B = 0.4471 \Rightarrow B = 2.8$$

∴ The required equation is $Y = (18.2465)\,(2.8)^{(X-3)}$

Example 17. Fit a non-linear trend of the form $Y = a + bX + CX^2$ in the following data :

X	0	1	2	3	4
Y	1.0	1.5	1.5	2.5	3.5

(CA Foundation June, 1993)

Solution. A table is constructed to find ΣX, ΣX^2, ΣX^3, ΣX^4, ΣXY and ΣX^2Y.

X	Y	X^2	X^3	X^4	XY	X^2Y
0	1	0	0	0	0	0
1	1.5	1	1	1	1.5	1.5
2	1.5	4	8	16	3	6.0
3	2.5	9	27	81	7.5	22.5
4	3.5	16	64	256	14	56.0
$\Sigma X = 10$	$\Sigma Y = 10$	$\Sigma X^2 = 30$	$\Sigma X^3 = 300$	$\Sigma X^4 = 354$	$\Sigma XY = 26$	$\Sigma X^2Y = 86$

The normal equations are :

$$\Sigma Y = Na + b\,\Sigma X + c\,\Sigma X^2$$

$$\Sigma XY = a\,\Sigma X + b\,\Sigma X^2 + c\,\Sigma X^3$$

$$\Sigma X^2Y = a\,\Sigma X^2 + b\,\Sigma X^3 + c\,\Sigma X^4$$

$$\therefore \quad 10 = 5a + 10b + 30c \qquad \ldots(I)$$

$$26 = 10a + 30b + 100c \qquad \ldots(II)$$

$$86 = 30a + 100b + 354c \qquad \ldots(III)$$

Solving these equations, we get

$$(II) - (I) \times 2 \quad : 6 = 10b + 40c \qquad \ldots(IV)$$

$$(III) - (II) \times 3 \quad : 8 = 10b + 54c \qquad \ldots(V)$$

$$(V) - (IV) \quad : 2 = 14c \Rightarrow c = \frac{1}{7} = 0.143$$

From (IV),
$$b = \frac{6 - \frac{40}{7}}{10} = \frac{2}{70} = \frac{1}{35} = 0.028$$

From (I),
$$a = \frac{10 - \frac{2}{7} - \frac{30}{7}}{5} = \frac{38}{35} = 1.086$$

Thus, $a = 1.086$; $b = 0.0286$; $c = 0.143$

Hence the non-linear trend is given by,

$$Y = 1.086 + 0.0286X + 0.143X^2$$

Example 18. The following table gives the total expenditure of the Govt. of India during 1994–95 to 2000–01 :

Year	1994–95	1995–96	1996–97	1997–98	1998–99	1999–200	2000–01
Expenditure (in crores)	177.2	185.0	224.9	254.0	304.9	359.9	438.8

Fit an exponential trend curve $y = ab^X$ to the data. **(CA Foundation, Nov., 2000)**

Solution. We have to fit curve $Y = ab^X$

Taking logarithm of both the sides, we get,

$$\log Y = \log a + X \log b$$

Year	*Mid-Value* X	*Expenditure* Y	$X - 1997.5 = x$	x^2	*Log Y*	$x \log Y$
1994–95	1994.5	177.2	–3	9	2.2485	–6.7455
1995–96	1995.5	185.0	–2	4	2.2672	–4.5344
1996–97	1996.5	224.9	–1	1	2.352	–2.352
1997–98	1997.5	254.0	0	0	2.4048	0
1998–99	1998.5	304.9	1	1	2.4842	2.4842
1999–2000	1999.5	359.9	2	4	2.5562	5.1124
2000–01	200.5	438.8	3	9	2.6423	7.9269
		1944.7	0	28	16.9552	1.8916

Normal equations by the method of least squares are

$$\Sigma \log Y = n \log a + \log b\, [\Sigma\, x]$$

$$\Sigma x \log Y = \log a\, [\Sigma x] + \log b\, [\Sigma x^2]$$

$$\log a = \frac{\Sigma \log Y}{n} = \frac{16.9552}{7} = 2.4222 \Rightarrow a = \text{Antilog } 2.4222 = 264.36$$

$$\log b = \frac{1.8916}{28} = 0.06755 \qquad \Rightarrow b = \text{Antilog } 0.06755 = 1.168$$

Thus exponential trend curve is given by $Y_c = 264.36 \times 1.168^x$ when $x = X - 1997.5$.

Example 19. The following table gives the profits of a concern for 5-years ending 2001.

Year	1997	1998	1999	2000	2001
Profit ('000 *Rs.*)	1.6	4.5	13.8	40.2	125.0

Find out the trend values by using the equation $Y_c = AB^X$.

Solution. **Trend Values by the Equation $Y_c = AB^X$**

Year t	*Profits* (*'000 Rs.*) Y	$t - 1999 = X$	$\log Y$	X^2	$X \log Y$	*Trend Values* Y_c
1997	1.6	–2	0.2041	4	–0.4082	1.56
1998	4.5	–1	0.6532	1	–0.6532	4.63
1999	13.8	0	1.1399	0	0.0000	13.79
2000	40.2	+1	1.6042	1	1.6042	41.05
2001	125.0	+2	2.0969	4	4.1938	122.21
$N = 5$		$\Sigma X = 0$	$\Sigma \log Y = 5.6983$	$\Sigma X^2 = 10$	$\Sigma (X \log Y) = 4.7366$	

The two equation by the method of least squares are

$$\Sigma \log Y = N \log A + \log B\, \Sigma x$$

$$\Sigma x \log Y = \log A\, \Sigma x + \log B\, (\Sigma x^2)$$

Since $\Sigma X = 0$; the value of $\log A = \dfrac{\Sigma \log Y}{N} = \dfrac{5.6983}{5} = 1.1397$

The value of $\log B = \dfrac{\Sigma (X \log Y)}{\Sigma X^2} = \dfrac{4.7366}{10} = 0.47366$

Therefore A = Anti log 1.1397 = 13.79, B = Anti log 0.47366 = 2.977

The equation $Y_c = A \,.\, B^X$, now would be

$$Y_c = 13.79 \times (2.977)^X$$

Now when $t = 1979$, $X = -2$, the value of $Y_c = 13.79 \times (2.977)^{(-2)} = \dfrac{13.79 \times 1}{(2.977)^2} = \dfrac{13.79}{8.8625} = 1.56$

Similarly when $X = -1$ the value of $Y_c = 13.79 \times (2.977)^{-1} = \dfrac{13.79}{2.977} = 4.63$

When $X = 1$ the value of $Y_c = 13.79 \times (2.97)^1 = 41.05$

When $X = 2$ the value of $Y_c = 13.79 \times (2.977)^2 = 13.79 \times 8.8625 = 122.21$

The above values have been shown in the last column of the table. The value of Y_c when $X = 0$ is 13.79 which is the value of A.

(4) Method of Moving Averages

The method is based on the principle that the total effect of periodic variations at different points of time in its cycle gets completely neutralised in the period of cyclic variations. In this method we take the average of a definite number of terms (time period) in a time series in succession, each time dropping one value at the beginning and taking up the new value (in sequence) from the remaining entries in the series. This tends to smooth out short term irregularities in the data series. The period of moving average say 3 years or 4 years or 5 years must coincide with the actual period or the multiples of the actual period of the cyclical variations so that the **cyclic variations are eliminated.**

This method do not always yield a straight line trend. However, if the pattern of variation in the time series is consistent and uniform both in terms of amplitude and the interval of occurrence, the moving average method does yield nearly a straight line trend. This method helps in eliminating cyclic, seasonal and random fluctuation in the time series data. The period of moving average can be (i) old, (ii) even.

(i) Odd period of moving averages : When odd period of moving averages is use for computing trend values moving totals and moving averages are placed against the center of the period covered.

(ii) Even period of moving averages : In this case the moving totals and moving averages do not correspond to any year but fall between two time periods, therefore the process of *centering* the values is adopted. In this process two periods moving average of the previously computed moving averages is taken.

Disadvantages of Moving Average Method

(i) An apparent disadvantage of the moving-average technique is that one must recompute the average of the last N observations each time a new demand (observation) become available.

(ii) The use of simple moving averages is not an appropriate forecasting method when there is a trend in the series because moving average lags behind the trend.

Consider a demand process in which there is a definite trend. For example, suppose that the observed demand is 2, 4, 6, 8, 10, 12, 14, 16, 18, 20, 24. Consider the one step-ahead MA(3) and MA(6) forecasts for this series.

Period	*Demand*	*MA(3)*	*MA(6)*
1	2		
2	4	4	
3	6	6	
			7
4	8	8	
			9
5	10	10	
			11
6	12	12	
			13
7	14	14	
			15
8	16	16	
			17
9	18	18	
			19
10	20	20	
11	22	22	
12	24		

* In fact we take the average of the most recent N observations. The moment a new observation becomes available, the oldest observation is ignored and new observation is taken into account.

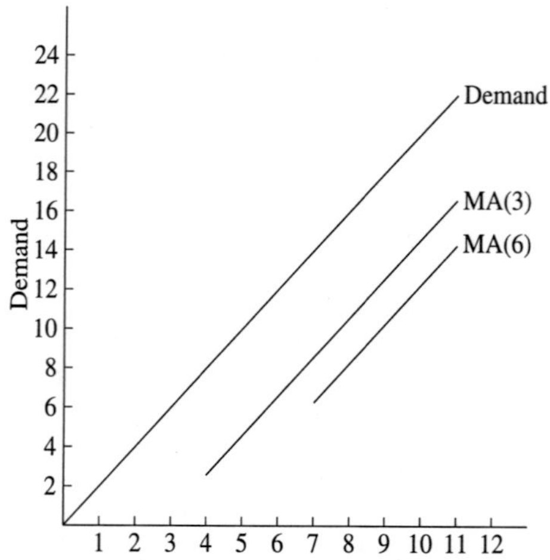

The demand and the forecasts for the respective periods are shown in the figure above.

Notice that both the MA(3) and the MA(6) forecasts lag behind the trend. Furthermore, MA(6) has a greater lag.

(iii) Moving averages are useful if we can assume that the market demands will stay fairly steady over time.

SOLVED EXAMPLES

Example 1. What is a trend in a time series? The following table gives the annual sales (in Rs. '000) of a commodity :

Year	*Sales*	*Year*	*Sales*
1990	710	1996	644
1991	705	1997	783
1992	680	1998	781
1993	687	1999	805
1994	757	2000	872
1995	629		

Determine the trend by calculating the 5-yearly moving average. **(ICWA QM June, 1995)**

Solution. **Calculation of Trend by 5-Yearly Moving Average**

Year	*Sales*	*5-Yearly Moving Total*	*5-Yearly Moving Average*
1990	710		
1991	705		
1992	680	3539	$\frac{3539}{5} = 707.8$
1993	687	3458	$\frac{3458}{5} = 691.6$
1994	757	3397	$\frac{3397}{5} = 679.4$
1995	629	3500	$\frac{3500}{5} = 700$
1996	644	3594	$\frac{3594}{5} = 718.8$
1997	783	3642	$\frac{3642}{5} = 728.4$
1998	781	3885	$\frac{3885}{5} = 777$
1999	805		
2000	872		

Example 2. Calculate 5-yearly moving averages for the following data :

Year	1991	1992	1993	1994	1995	1996	1997	1998	1999	2000
Value ('000 *Rs.*)	123	140	110	98	104	133	95	105	150	135

(CA Foundation Dec., Modified, 1993)

Solution. **Computation of 5-Yearly Moving Averages**

Year	*Value ('000 Rs.)*	*5-Yearly Moving Total ('000 Rs.)*	*5-Yearly Moving Average ('000 Rs.)*
1991	123	—	—
1992	140	—	—
1993	110	575	115
1994	98	585	117
1995	104	540	108
1996	133	535	107
1997	95	587	117.4
1998	105	618	123.6
1999	150	—	—
2000	135	—	—

Example 3. From the following data, calculate the trend values using 4-yearly moving average :

Year	1989	1990	1991	1992	1993	1994	1995	1996	1997
Value	506	620	1036	673	588	696	1116	738	663

(CA Foundation Nov., 2001)

Solution.

Year 1	*Value* 2	*4-Yearly Moving Total* 3	*2-Yearly Moving Total of Column (3)*	*4-Yearly Centred Moving Total*
1990	506			
1990	620			
		2835		
1991	1036		5752	$\frac{5752}{8}=719$
		2917		
1992	673		5910	$\frac{5910}{8}=738.75$
		2993		
1993	588		6066	$\frac{6066}{8}=758.25$
		3073		
1994	696		6211	$\frac{6211}{8}=776.375$
		3178		
1995	1116		6351	$\frac{6351}{8}=793.8$
		3213		
1996	738			
1997	663			

Example 4. Calculate 5-yearly weighted moving averages for the following data, using the weights 1. 1, 3, 2, 1 respectively :

Year	1	2	3	4	5	6	7	8	9	10
Coded Sales	40	33	72	81	76	68	91	97	98	97

(CA Foundation May, 1998)

Solution. *

Year (I)	*Sales (II)*	*5-Yearly Weighted Moving Total (III)*	*5-Yearly Weighted Moving Average*
1	40		
2	33		
3	72	$(40\times1+33\times1+72\times3+81\times2+76\times1)=527$	65.88
4	81	$(33\times1+72\times1+81\times3+76\times2+68\times1)=568$	71.00
5	76	$(72\times1+81\times1+76\times3+68\times2+91\times1)=608$	76.00
6	68	$(81\times1+76\times1+68\times3+91\times2+87\times1)=630$	78.750
7	91	$(76\times1+68\times1+91\times3+87\times2+90\times1)=681$	85.125
8	87	$(68\times1+91\times1+87\times3+98\times2+97\times1)=713$	89.125
9	98		
10	97		

* Weighted Average $=\frac{\Sigma W_x}{\Sigma W}$, *e.g.*, $\frac{527}{8}=65.88$, etc.

Example 5. For the following data, verify that the 5-yearly weighted moving average with weights 1, 2, 2, 2, 1 respectively is equivalent to 4-yearly centred moving average :

Year	1989	1990	1991	1992	1993	1994	1995	1996	1997	1998	1999
Sales (Rs. in Lakhs)	5	3	7	6	4	8	9	10	8	9	9

(CA Foundation May, Modified, 1996)

Solution. 5-yearly weighted moving average would be as under :

Year	*Sales*	*5-Yearly Moving Total with Weights* 1, 2, 2, 2, 1 *Respectively*	*5-Yearly Weighted Moving Averages*
1989	5	—	
1990	3	—	
1991	7	41	$\frac{41}{8} = 5.125$
1992	6	45	$\frac{45}{8} = 5.625$
1993	4	52	$\frac{52}{8} = 6.500$
1994	8	58	$\frac{58}{8} = 7.250$
1995	9	66	$\frac{66}{8} = 8.250$
1996	10	71	$\frac{71}{8} = 8.875$
1997	8	72	$\frac{72}{8} = 9.000$
1998	9	—	
1999	9	—	

4-yearly centred moving average would be as under :

Year 1	*Sales* 2	*4-Yearly Moving Total* (3)	*2-Yearly Moving Total of Column* (3)	*4-Yearly Centred Moving Average*
1989	5	—		—
1990	3			—
		21		
1991	7		41	$\frac{41}{8} = 5.125$
		20		
1992	6		45	$\frac{45}{8} = 5.625$
		25		
1993	4		52	$\frac{52}{8} = 6.500$
		27		
1994	8		58	$\frac{58}{8} = 7.250$
		31		
1995	9		66	$\frac{66}{8} = 8.250$
		35		
1996	10		71	$\frac{71}{8} = 8.875$
		36		
1997	8		72	$\frac{72}{8} = 9.000$
		36		
1998	9			—
1999	9			—

Hence both the series are same.

Example 6. For the following series of observations, verify that the 6-yearly centred mvoing average is equivalent to a 7-yearly weighted moving average with weights 1, 2, 2, 2, 2, 2, 1 respectively.

Year	1990	1991	1992	1993	1994	1995	1996	1997	1998	1999	2000
Sales (in Lakhs)	2	4	3	6	7	9	4	6	7	8	10

Solution. **Calculations of 6-Yearly centred Moving Average**

Year	*Annual Sale (in Lakhs)*	*6-Yearly Moving Total*	*2-Yearly Moving Total of Col. 3*	*6-Yearly Centred Moving Average Col. 4 + 12*
1990	2			
1991	4			
1992	3			
		31		
1993	6		64	$\frac{64}{12} = 5.33$
		33		
1994	7		68	$\frac{68}{12} = 5.66$
		35		
1995	9		74	$\frac{74}{12} = 6.16$
		39		
1996	4		80	$\frac{80}{12} = 6.66$
		41		
1997	6		85	$\frac{85}{12} = 7.083$
		44		
1998	7			
1999	8			
2000	10			

Table for 7-Yearly Weighted Moving Average

Year	*Annual Sale (in Lakhs)*	*7-Yearly Weighted Moving Total*	*7-Yearly Weighted Moving Average Col. 3 ÷ Sum of Weights*
1990	2		
1991	4		
1992	3		
1993	6	$2 \times 1 + (4 + 3 + 6 + 7 + 9) \times 2 + 4 \times 1$	$\frac{64}{12} = 5.33$
1994	7	$4 \times 1 + (3 + 6 + 7 + 9 + 4) \times 2 + 6 \times 1 = 68$	$\frac{68}{12} = 5.66$
1995	9	$3 \times 1 + (6 + 7 + 9 + 4 + 6) \times 2 + 7 \times 1 = 74$	$\frac{74}{12} = 6.16$
1996	4	$6 \times 1 + (7 + 9 + 4 + 6 + 7) \times 2 + 8 \times 1 = 80$	$\frac{80}{12} = 6.66$
1997	6	$7 \times 1 + (9 + 4 + 6 + 7 + 8) \times 2 + 10 \times 1 = 85$	$\frac{85}{12} = 7.083$
1998	7		
1999	8		
2000	10		

From the last columns of the above tables the result is proved.

Questions

Short Answer Type Questions

1. What characteristic component of time series should each of the following be associated

(i) A fire in factory delaying production for three weeks

(ii) An era of prosperity

(iii) Sales of a textile firm during depawali

(iv) A need for increased wheat production due to constant increase in population

(v) The increase in day temperature from winter to summer **(D.U., B.Com. 2001)**

[**Ans.** (i) irregular variations (ii) cyclic variations (iii) seasonal variations (iv) secular trend (v) seasonal variations]

Unsolved Questions

1. Fit a straight line trend to the following data by least square method and estimate the sale for the year 2006.

Year	1999	2000	2001	2002	2003	2004
Sales (in 000's)	70	80	90	100	95	114

(D.U., B.Com. 2007E)

[**Ans.** (i) $\hat{y} = 92.5 + \frac{269}{70}\left[\frac{x - 2001.5}{0.5}\right]$, (ii) 127]

2. Below are given the figures of sales in thousand quintals of a firm operating in a sugar industry.

Year	*Sales*
1997	70
1999	90
2001	100
2003	130
2005	170

(i) Fit a straight line trend to these figures using least square method.

(ii) Estimate the sales of the firm for the year 2008.

(iii) What is the annual increase in the expected sales of the firm. **(D.U., B.Com. 2007R)**

[**Ans.** (i) $\hat{y} = 112 + 12\,(X - 2001)$ (ii) 196 (iii) annual increase 12]

3. Determine the trend values of the following by semi-average method

Year	1980	1981	1982	1983	1984	1985	1986	1987
Sales (lakh Rs.)	39.2	39.1	38.94	38.6	38.1	38.1	38.1	37.7

(D.U., B.Com. 2004)

[**Ans.** 38.96, 38]

❑❑❑

QUESTION PAPERS

B.Com. Pass II Year 2008 (Regular)

Part–A

(Business Mathematics)

Q.1. A salesman has the following record of sales during three months for three items A, B, C which have different rates of commission :

Months	*Sales of Units*			*Total Commission Drawn (in Rs.)*
	A	*B*	*C*	
January	90	100	20	800
February	130	50	40	900
March	60	100	30	850

Find out the rates of commission on items A, B and C.

Solution. Let the commission per unit on items A, B and C is Rs. x, y and respectively. Then the equation in the matrix form is

$$\begin{bmatrix} 90 & 100 & 20 \\ 130 & 50 & 40 \\ 60 & 100 & 30 \end{bmatrix}\begin{bmatrix} x \\ y \\ z \end{bmatrix} = \begin{bmatrix} 800 \\ 900 \\ 850 \end{bmatrix}$$

Let $$P = \begin{bmatrix} 90 & 100 & 20 \\ 130 & 50 & 40 \\ 60 & 100 & 30 \end{bmatrix} = \begin{bmatrix} a_{11} & a_{12} & a_{13} \\ a_{21} & a_{22} & a_{23} \\ a_{31} & a_{32} & a_{33} \end{bmatrix}$$

$$\text{Cofactor of } a_{11} = A_{11} = 1500 - 4000 = -2500$$

$$A_{12} = -(3900 - 2400) = -1500$$

$$A_{13} = 13000 - 3000 = 10000$$

$$A_{21} = -(3000 - 2000) = -1000$$

$$A_{22} = 2700 - 1200 = 1500$$

$$A_{23} = -(9000 - 6000) = -3000$$

$$A_{31} = 4000 - 1000 = 3000$$

$$A_{32} = -(3600 - 2600) = -1000$$

$$A_{33} = 4500 - 13000 = -8500$$

Determinant $$A = 90 \times (-2500) + 100(-1500) + 20(10000)$$

$$= -225000 - 150000 + 200000$$

$$= -175000$$

$$A^{-1} = -\frac{1}{175000}\begin{bmatrix} -2500 & -1000 & 3000 \\ -1500 & 1500 & -1000 \\ 10000 & -3000 & -8500 \end{bmatrix}$$

$$\therefore \quad \begin{bmatrix} x \\ y \\ z \end{bmatrix} = -\frac{1}{175000}\begin{bmatrix} -2500 & -1000 & 3000 \\ -1500 & 1500 & -1000 \\ 10000 & -3000 & -8500 \end{bmatrix}\begin{bmatrix} 800 \\ 900 \\ 850 \end{bmatrix}$$

$$= -\frac{1}{175000}\begin{bmatrix} -2500 \times 800 - 1000 \times 900 + 3000 \times 850 \\ -1500 \times 800 + 1500 \times 900 - 1000 \times 850 \\ 10000 \times 800 + 3000 \times 900 - 8500 \times 850 \end{bmatrix}$$

$$= -\frac{1}{175000}\begin{bmatrix} -350000 \\ -700000 \\ -1925000 \end{bmatrix} = \begin{bmatrix} 2 \\ 4 \\ 11 \end{bmatrix}$$

$\Rightarrow \quad x = 2, y = 4, z = 1$

Rate of commission per unit on item

$$A = \text{Rs. } 2$$
$$B = \text{Rs. } 4$$
$$C = \text{Rs. } 11$$

Q.2. A firm produce chairs, tables and cupboards, each requiring three types of raw material—timber, nails and varnish. You are given below, the units of different raw materials required for producing one unit of each product :

Product	*Timber (c.ft.)*	*Nails (dozen)*	*Varnish (litres)*
Chair	0.7	2	1
Table	1	4	1.5
Cupborad	3.2	6	2

If the firm produces 300 units of each product, find the quantity of each raw material using matrix algebra.

Solution. The given matrix can be written as :

	Chair	*Table*	*Cupboard*
Timber (c.ft.)	0.7	1	3.2
Nails (dozen)	2	4	6
Varnish (litres)	1	1.5	2

$$\begin{bmatrix} 300 \\ 300 \\ 300 \end{bmatrix} \begin{matrix} \text{Chair} \\ \text{Table} \\ \text{Cupboard} \end{matrix}$$

$$= \begin{bmatrix} 0.7 \times 300 + 1 \times 300 + 3.2 \times 300 \\ 2 \times 300 + 4 \times 300 + 6 \times 300 \\ 1 \times 300 + 1.5 \times 300 + 2 \times 300 \end{bmatrix}$$

$$= \begin{bmatrix} 21 + 300 + 960 \\ 600 + 1200 + 1800 \\ 300 + 450 + 600 \end{bmatrix} = \begin{bmatrix} 1281 \\ 3600 \\ 1350 \end{bmatrix} \begin{matrix} \text{Timber} \\ \text{Nails} \\ \text{Varnish} \end{matrix}$$

$\therefore$ We need 1281 cu. ft. of wood, 3600 dozens of nails and 1350 litres of varnish.

Q.3. The total cost $C(x)$ of a firm is :

$$C(x) = 1500 + 30x + x^2$$

where x is the output. Determine :

(i) The average cost;

(ii) The marginal cost;
(iii) The marginal cost when 20 units are produced;
(iv) The actual cost of producing twenty first unit 4.

Solution. (i) Average cost $= \frac{C}{x} = \frac{1500}{x} + 30 + x$

(ii) Marginal cost $= \frac{dC}{dx} = 30 + 2x$

(iii) Marginal cost when $x = 20$ is

$$30 + 2 \times 20 = 70$$

(iv) $C(21) - C(20) = 1500 + 30 \times 21 + (21)^2 - [1500 + 30 \times 20 + (20)^2]$

$$= 30 + 441 - 400 = 71.$$

Q.4. If the demand function for a certain commodity is :

$$q = 500 - 2p \ (0 \le p \le 250)$$

then determine where the demand is elastic, inelastic and of unit elasticity with respect to price.

Solution. Elasticity of demand

$$e_d = -\frac{dq/q}{dp/p} = -\frac{p}{q}\frac{dq}{dp}$$

$$= -\frac{p}{q}(-2) = \frac{2p}{500 - 2p}$$

For the demand to be elastic $e_d > 1$

$\Rightarrow$ $$\frac{2p}{500 - 2p} > 1 \Rightarrow 2p > 500 - 2p \Rightarrow p > 125$$

For the demand to be inelastic $e_d < 1$

$\Rightarrow$ $$\frac{2p}{500 - 2p} < 1 \Rightarrow 2p < 500 - 2p \Rightarrow 4p < 500 \Rightarrow p < 125.$$

For the demand to have unit elasticity $e_d = 1$

$\Rightarrow$ $$\frac{2p}{500 - 2p} = 1 \Rightarrow 2p = 500 - 2p \Rightarrow 4p = 500 \Rightarrow p = 125$$

Q.5. The marginal cost of production is found to be :

$$C'(q) = 1000 - 20q + q^2$$

where q is no. of units produced.

The fixed cost of production is Rs. 2,000. Find the cost function if the manufacturer fixes the price per unit at Rs. 3,400. Find the profit function and the sales volume that yields the maximum profits. Also obtain profit at this sales volume.

Solution. $$\frac{dc}{dq} = 1000 - 20q + q^2$$

$$C = \int (1000 - 20q + q^2)\, dq$$

$$= 1000q - \frac{20q^2}{2} + \frac{q^3}{3} + k$$

$$= 1000q - 10q^2 + \frac{q^3}{3} + k$$

when $q = 0$, $C = 2000$, therefore $k = 2000$.

Cost function is

$$C = 1000q - 10q^2 + \frac{q^3}{3} + 2000$$

Total revenue = $34000q$

Profit function $(P) = 34000q - 1000\,q + 10q^2 - \frac{q^3}{3} - 2000$

$$= 2400q + 10q^2 - \frac{q^3}{3} - 2000$$

$$\frac{dP}{dq} = 2400 + 20q - \frac{3q^2}{3}$$

$$\frac{dP^2}{dq^2} = + 20 - \frac{6q}{3}$$

$$\frac{dP}{dq} = 0 \Rightarrow 2400 + 20q - \frac{3q^2}{3} = 0$$

$$q^2 - 20q - 2400 = 0$$

$\Rightarrow$ $$q = \frac{20 \pm \sqrt{400 + 9600}}{2} = \frac{20 \pm 100}{2} = 60 \text{ or } -40$$

Since q cannot be negative

$\therefore$ $$q = 60$$

when $q = 60$

$$\text{Profit} = 2400 \times 60 + 10 \times 60 \times 60 - \frac{60 \times 60 \times 60}{3} - 2000$$

$$= 144000 + 36000 - 72000 - 2000$$

$$= \text{Rs. } 1{,}06{,}000$$

Q.6. If the marginal revenue function for output X is given by :

$$MR = \frac{6}{(x+2)^2} + 5$$

Find the total revenue function and the demand equation.

Solution. $$MR = \frac{6}{(x+2)^2} + 5$$

$$R = \int \left[\frac{6}{(x+2)^2} + 5 \right] dx$$

$$= 6 \frac{(x+2)^{-2+1}}{-2+1} + 5x + c$$

$$= \frac{-6}{x+2} + 5x + c$$

when $x = 0, R = 0$

$$\therefore \qquad \frac{-6}{2} + c = 0 \Rightarrow c = 3$$

$$\therefore \qquad R = 5x - \frac{6}{x+2} + 3$$

$$\frac{R}{x} = \text{price per unit}$$

$$\therefore \qquad \text{Price per unit} = 5 - \frac{6}{x(x+2)} + \frac{3}{x}.$$

This is the demand equation here x = units demanded.

Q.7. Attempt any two parts :

(a) Find for each of the following the amount to which Rs. 100 will accumulate :

(i) at the rate of interest 12% compounded quarterly for 10 years.

(ii) at the effective rate of interest 3% per annum for 10 years, 4% per annum for 4 years and 5% per annum for 2 years.

(b) An asset costing Rs. 2,000 will depreciate to a scrap value of Rs. 160 in 10 years. Find the rate of depreciation.

(c) Mr. X deposited Rs. 10,000 in a bank for 3 years offering interest at the rate of 6% compounded half yearly during first year, at the rate of 12% compounded quarterly during second year and at 10% compounded continuously during third year. Find his balance after 3 years.

(d) A machine costing Rs. 50,000 depreciates at a constant rate of 8%. What is the depreciation charge for 8th year?

Solution. (a) (i) Amount $= 100\left(1 + \frac{3}{100}\right)^{40}$

$$= 100\,(1.03)^{40} = 400 \times 3.261933$$

$$= 326.2$$

(ii)

$$100\,(1.03)^{10} = 134.4$$

$$100\,(1.04)^{4} = 117$$

$$100\,(1.05)^{2} = 110.25$$

(b)
$$2000\left(1-\frac{r}{100}\right)^{10}=160$$

$$\left(1-\frac{r}{100}\right)^{10}=\frac{160}{2000}=\frac{16}{200}=\frac{2}{25}=0.08$$

$$\left(1-\frac{r}{100}\right)=0.7768 \Rightarrow 1-0.7768=\frac{r}{100} \Rightarrow r=22.32\%$$

(b) Balance after three years $= 10000\,(1.03)^2\,(1.03)^4\,e^{0.1}$

$$= 10000\,(1.03)^2\,(1.03)^4 \times 1.105$$

$$= 10000 \times 1.0609 \times 1.1255 \times 1.105 = \text{Rs. } 3196$$

(d) Depreciation charge for 8th year

= Depreciated value at the end of 7 years – depreciated value at the end of 8 years

$$= 50000\left(1-\frac{8}{100}\right)^7 - 50000\left(1-\frac{8}{100}\right)^8$$

$$= 50000\left(1-\frac{8}{100}\right)^7\left[1-\left(1-\frac{8}{100}\right)\right]$$

$$= 50000 \times \frac{8}{100}\left(1-\frac{8}{100}\right)^7 = 50000 \times 0.08 \times (0.92)^7$$

= Rs. 2231.4

❑

Part–B

(Business Statistics)

Q.1. (i) The mean of a certain number of items is 20. If an observation 25 is added to the data, the mean becomes 21. Find the number of items in the original data.

(ii) Are the seasonal variations present in a time series of year-wise sales of a firm? Give reason for your answer.

(iii) The 25th percentile and 75th percentile of a distribution of marks of the students of a class are 30 and 70 respectively. On the basis of these information, find the coefficient of 'Quartile Deviation' of the distribution of marks.

(iv) Why the mean deviation from median is less than the mean deviation from arithmetic mean?

Solution.

(i) Let the number of items in the original series be n.
Sum of the values in the original data = $20n$.
Sum of the values when are more item is added = $20n + 25$

$$\text{Mean of } (n+1) \text{ items} = \frac{20n+25}{n+1} = 21$$

$$\Rightarrow \quad 20n + 25 = 21n + 21$$

$$\Rightarrow \quad n = 4$$

(ii) Yes, seasonal variations are present in the year-wise sales of a firm because of following reasons.
Almost each type of business activity is affected by seasonal variations to a greater or lesser degree. Therefore seasonal variations are regarded as normal phenomenon recurring evey year. Year wise sales of a firm are also affected by seasonal variations because the intensity of the whether also determines the volume of sales.

(iv) 25th percentile = 1st quartile = Q_1
75th percentile = 3rd quartile = Q_3

$$\text{Coefficient of quartile deviation} = \frac{Q_3 - Q_1}{Q_3 + Q_1} = \frac{70-30}{70+30} = 0.4$$

(iv) While calculating mean deviation, the signs of the deviations are ignored and the deviations of all the values are taken from the median. Since the sum of the deviations of all the items from the median in minimum when signs are ignored. The mean deviation from median is less than the mean deviation from arithmetic mean or any other average.

Q.2. (a) Four typists take 15, 10, 6 and 5 minutes respectively to type a letter. Determine the average time required to type a letter by using 'harmonic mean' if :

(i) four letters are to be typed by each typists.

(ii) each typist works for two hours.

(b) Fifty students took a test. The result of those who passed the test is given below :

Marks	*No. of Students*
4	8
5	11
6	8
7	6
8	4
9	3

If the average marks for all the fifty students were 5.16, find the average marks of those who failed.

Solution. (a) (i) Average time taken $= \dfrac{15 \times 4 + 10 \times 4 + 6 \times 4 + 5 \times 4}{4 + 4 + 4 + 4}$

$$= \frac{36}{4} = 9 \text{ mintues}$$

(ii) Number of letters types in 2 hrs, *i.e.*, 120 minutes by typist A $= \dfrac{1}{15} \times 120$

Number of letters types in 2 hrs by typist B $= \dfrac{1}{10} \times 120$

Number of letters types in 2 hrs by typist C $= \dfrac{1}{6} \times 120$

Number of letters types in 2 hrs by typist D $= \dfrac{1}{5} \times 120$

Average time taken to type a letter

$$= \frac{\text{Total time}}{\text{Total number of letters}} = \frac{120 \times 4}{120\left[\dfrac{1}{15} + \dfrac{1}{10} + \dfrac{1}{6} + \dfrac{1}{5}\right]}$$

$$= \frac{4 \times 60}{4 + 6 + 10 + 12} = \frac{240}{32} = 7.5 \text{ minutes.}$$

Q.3. (a) Represent the following data by means of a histogram :

Monthly Income (Rs.)	*No. of Workers*
3,000–3,500	7
3,500–4,000	19
4,000–4,500	27
4,500–5,000	15
5,000–6,000	12
6,000–8,000	12
8,000–10,000	8

(b) The value of a machine is depreciated by 30% in first year, 13% in second year and by 5% in each of the following three years. Determine the average rate of depreciation for the entire period by using an appropriate measure of central value.

Solution. (a)

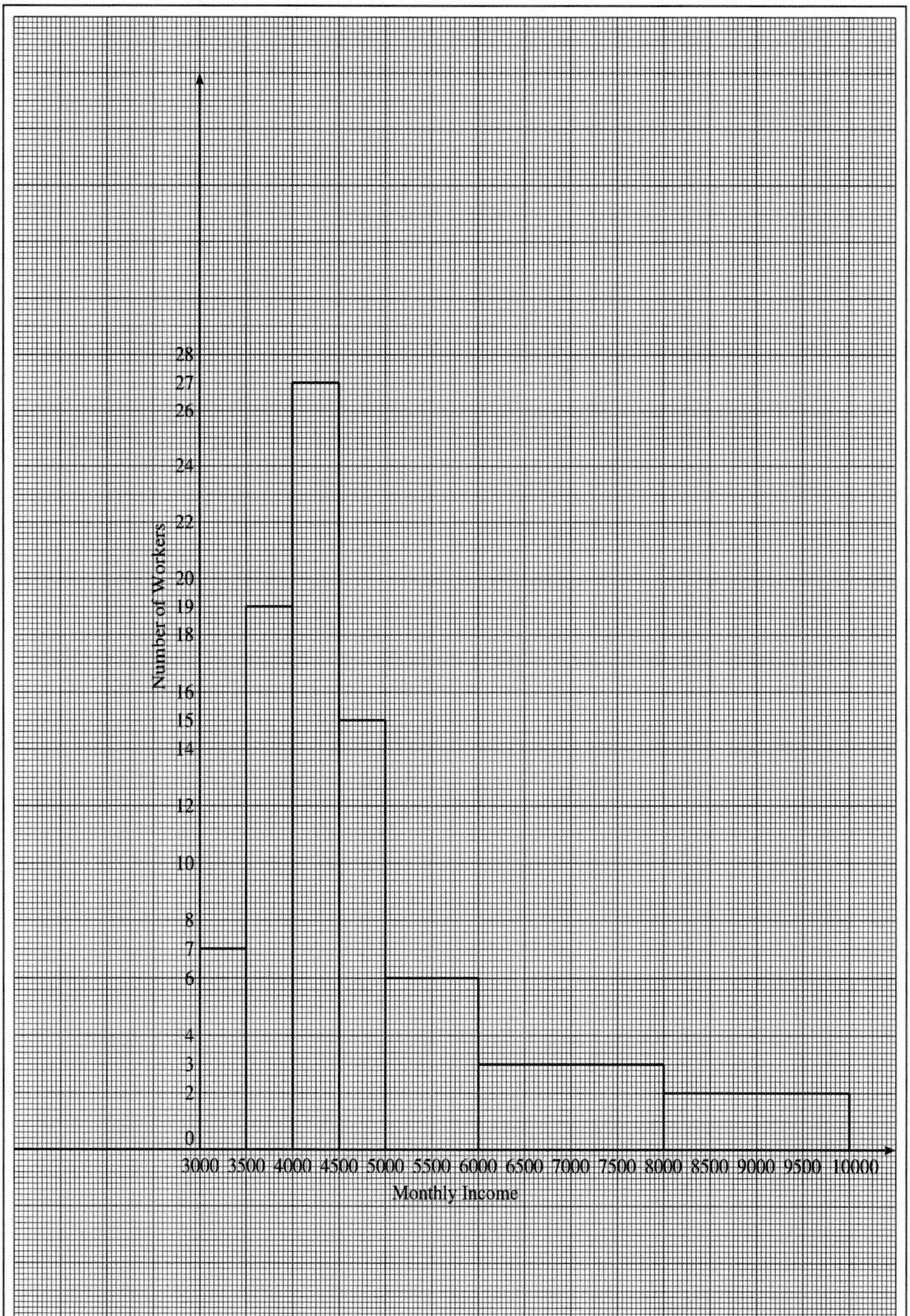
Number of Workers
28
27
26
24
22
20
19
18
16
15
14
12
10
8
7
6
4
3
2
0
3000 3500 4000 4500 5000 5500 6000 6500 7000 7500 8000 8500 9000 9500 10000
Monthly Income

(b) Let $r\%$ be the average rate of depreciation, then

$$\left(1-\frac{r}{100}\right)^5=\left(1-\frac{30}{100}\right)\left(1-\frac{13}{100}\right)\left(1-\frac{5}{100}\right)^3$$

$$=0.7\times 0.87\times (0.95)^3=0.522$$

$$\left(1-\frac{r}{100}\right)=(0.522)^{1/5}=0.8780$$

$$r=1-0.8780=0.122 \Rightarrow r=12.2\%.$$

Q.4. (a) Two workers on the same job, show the following results over a long period of time :

	Workers A	*Worker B*
Mean time of completing the job (in minutes)	30	25
Median time of completing the job (in minutes)	25	30
Mean deviation from median (in minutes)	10	9

(i) Which worker appears to be more consistent in the time required to complete the job?
(ii) Which worker appears to be slower in completing the job?

(b) Given :
Covariance between X and $Y=16$
Variance of $X=25$
Variance of $Y=16$
(i) Calculate coefficient of correlation between X and Y.
(ii) If arithmetic means of X and Y are 20 and 30 respectively, find regression equation Y on X.
(iii) Estimate Y when $X=30$.

Solution. (a) Mean deviation $=0.7979\,\sigma$

$$\therefore \qquad \sigma=\frac{\text{Mean deviation}}{0.7979}$$

$$\sigma_A=\text{S.D. for worker } A=\frac{10}{0.7979}=12.53$$

$$\sigma_B=\text{S.D. for worker } B=\frac{9}{0.7979}=11.27$$

$$\text{C.V. for worker } A=\frac{12.53}{30}\times 100=41.76$$

$$\text{C.V. for worker } B=\frac{11.27}{25}\times 100=45.08\%$$

Since the C.V. for the worker A is less, worker A is more consistent.
Since the average time for worker A is more worker A appears to be slower.

(b) Cov. $(X, Y)=16$

$$\sigma_X^2=25,\ \sigma_X=5$$

$$\sigma_Y^2=16,\ \sigma_Y=4$$

$$r_{(XY)} = \frac{\text{Cov. }(X, Y)}{\sigma_X \sigma_Y} = \frac{16}{5 \times 4} = 0.8$$

Regression equation of Y on X is

$$Y - \overline{Y} = b_{YX}(X - \overline{X})$$

$$\Rightarrow \quad Y - 30 = \frac{\text{Cov. }(X, Y)}{\sigma_X^2}(X - 20)$$

$$= \frac{16}{25}(X - 20)$$

$$\Rightarrow \quad 25Y - 750 = 16X - 300$$

$$\Rightarrow \quad 16X - 25Y + 400 = 0$$

when $X = 30$

$$Y - 30 = \frac{16}{25}(30 - 20) = \frac{160}{25} = 6.4$$

$$\Rightarrow \quad Y = 36.4.$$

Q.5. (a) (i) How is 'Standard Deviation' as a measure of dispersion superior to 'Mean Deviation'?

(ii) Distinguish between 'absolute' and 'relative' measure of dispersion.

(b) The data about the sales and advertisement expenditure of a firm are given below :

	Sales (in crores of Rs.)	*Advertisement Expenditure (in crores of Rs.)*
Mean	40	6
Standard deviation	10	1.5

Coefficient of correlation = + 0.9.

(i) Estimate the likely sales for a proposed advertisement expenditure of Rs. 10 crores.

(ii) What should be the advertisement expenditure if the firm proposes a sales target of 60 crores?

Solution. (a)

(1) The following points makes the S.D. a superior measure of dispersion as computed to mean deviation :

(i) In mean deviation signs of the deviations are ignored which is illogical where as in S.D. algebraic signs are not ignored.

(ii) S.D. is amenable to algebraic treatment, *i.e.*, we can find the combine S.D. of two or more series but we cannot find the combined mean deviation.

(iii) The S.D. is used in the analysis of statistical series such as in correlation, skewness, sampling or the analysis of variance.

(2) (i) Absolute measures of dispersion are expressed in terms of the units of measurement of the variable whereas relative measures are expressed as pure numbers independent of the units of measurement of the variable.

(ii) Absolute measures of dispersion cannot be used to compare the variability of two distribution expressed in different units whereas relative measures can be used to compare the variability of two distribution even if they are expressed in different units.

(b) Let the sales be represented by Y and advertisement expenditure by X, then

$$\bar{Y} = 40, \quad \bar{X} = 6$$

$$\sigma_Y = 10, \quad \sigma_X = 1.5, \quad r = 0.9$$

Regression equation of Y on X is

$$Y - \bar{Y} = b_{YX}(X - \bar{X})$$

$$Y - 40 = r\frac{\sigma_Y}{\sigma_X}(X - 6) = 0.9\left(\frac{10}{1.5}\right)(10 - 6)$$

$$= \frac{9}{1.5} \times 4 = 24$$

$$Y = 64 \text{ crores}$$

Regression equation of X on Y is

$$X - \bar{X} = b_{XY}(Y - \bar{Y})$$

$$X - 6 = r\frac{\sigma_X}{\sigma_Y}(Y - 40)$$

$$\Rightarrow \quad X - 6 = 0.9\left(\frac{1.5}{10}\right)(60 - 40) = 1.35 \times 2 - 2.7$$

$$X = 8.7 \text{ crore.}$$

Q.6. (a) What are the components of time series?

(b) An enquity into the budgets of the middle class families in a city in India gave the following information :

	Food	*Rent*	*Clothing*	*Fuel*	*Others*
Expenses	35%	15%	20%	10%	20%
Price in 2006	150	50	100	20	60
Price in 2007	174	60	125	25	90

What change in the cost of living of 2007 has taken place as compared to 2006?

Solution. (a) **Components of time series :** "A time series consists of statistical data which are collected, recorded or observed over successive increments" **(Patterson)**.

(i) Trends or secular movements : These are changes that have occurred as a result of general tendency of data to increase or decrease. The concept of trend does not include short range oscillations but rather steady movements over a long time. Secular trend movements are attributable to factors such as change in population technological progress and large scale shifts in consumer tastes.

(ii) Seasonal variations : The changes that take place during a period of 12 months as a result of change in climate, weather conditions, festivals etc., such changes are marked as seasonal variations. When data are expressed annually there is no seasonal variation. However, monthly or quarterly data, frequently exhibit strong seasonal movements and considerable interest is attached to devising a pattern of average seasonal variations. For example, if we observe the sales of a bookseller, we find that for the quarter July-September, sales are maximum. If we know by how much the sales of this quarter are usually above or below the previous quarter for seasonal reasons, we shall be able to answer, whether this was due to upward or downward tendency or simply because this quarter has generally seasonal higher sales.

(iii) Cyclical variations : The changes that have taken place as a result of booms or depressions, such changes are classified as cyclical variations. Business cycles are the most important type of fluctuation in economic data. But, because successive cycles vary widely in timing, amplitude and pattern, they are most difficult to measure. Residual method, Reference cycle analysis method, Direct method and Harmonic analysis method are few methods to measure cyclical variations.

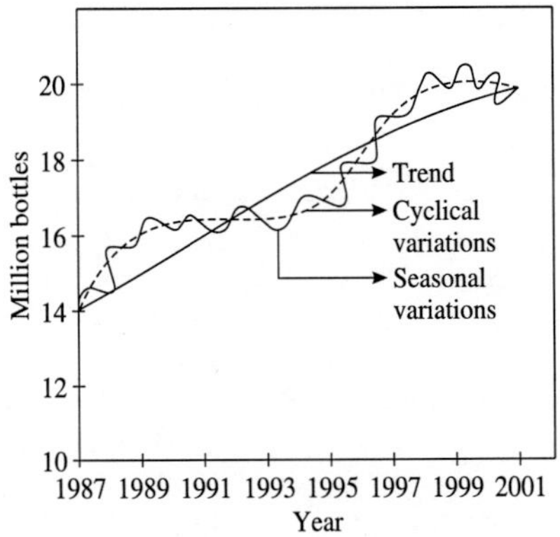

(iv) Irregular or erratic variations : The changes that have taken place as a result of such forces that could not be predicted like floods, earthquakes, famines etc. Such changes are classified under irregular or erratic variations. These variations are also called as episodic variations.

If may be noted that any or all of these components may be present in any particular series.

The graph gives the sales of Pepsi for the years 1987–2001.

(b)

Commodity	*Base Year 2006*	*Current Year 2007*	*Price Relative*	*Weight*
Food	150	174	$\frac{174}{150} \times 100 = 116$	35
Rent	50	60	$\frac{60}{50} \times 100 = 120$	15
Clothing	100	125	$\frac{125}{100} \times 100 = 125$	20
Fuel	20	25	$\frac{25}{20} \times 100 = 125$	10
Others	60	90	$\frac{90}{60} \times 100 = 150$	20

$$\text{Price relatives} = \frac{116 \times 35 + 120 \times 15 + 125 \times 20 + 125 \times 10 + 150 \times 20}{35 + 15 + 20 + 10 + 25}$$

$$= \frac{4060 + 1800 + 2500 + 1250 + 3000}{100}$$

$$= \frac{12610}{100} = 126\%$$

Change is 26.1%.

Q.7. (a) Explain the 'time reversal' test and show that Fisher's Ideal Index Number statisfies it.

(b) Fit a straight line trend to the following data by least squares method :

Year	Sales (Rs. in lakhs)
1999	18
2001	21
2003	23
2005	27
2007	16

(i) Estimate sales for the year 2010.

(ii) What is the annual increase/decrease in the trend values of sales?

Solution. (a) **Time Reversal Test :** It is a test to determine whether a given method will work both ways in time, forward and backward. The test is that the formula for calculating the index number should be such that it will give the same ratio between one point of comparison and the other, no matter which of two is taken as base. In other words when data for any two years are treated by same method, but with base reversed, the two index numbers secured should be reciprocals of each other so that their product is unity,

$$P_{01} \times P_{10} = 1$$

where P_{01} is index for time "1" on time "0" as base and P_{10} is index for time "0" on time "1" as base. Fisher's ideal index satisfy the test :

(i) The Fisher's ideal formula :

$$P_{01} = \sqrt{\frac{\Sigma p_1 q_0}{\Sigma p_0 q_0} \times \frac{\Sigma p_1 q_1}{\Sigma p_0 q_1}} \qquad P_{10} = \sqrt{\frac{\Sigma p_0 q_1}{\Sigma p_1 q_1} \times \frac{\Sigma p_0 q_0}{\Sigma p_1 q_0}}$$

$$P_{01} \times P_{10} = \sqrt{\frac{\Sigma p_1 q_0}{\Sigma p_0 q_0} \times \frac{\Sigma p_1 q_1}{\Sigma p_0 q_1}} \times \sqrt{\frac{\Sigma p_0 q_1}{\Sigma p_1 q_1} \times \frac{\Sigma p_0 q_0}{\Sigma p_1 q_0}} = \sqrt{1} = 1$$

$\Rightarrow$ Fisher's ideal index number satisfies time reversal test.

(b)

Years (X)	Sales (Rs. in lakhs) (Y)	$[X - 2003 = x]$	x^2	xy
1999	18	–4	16	–72
2001	21	–2	4	–42
2003	23	0	0	0
2005	27	2	4	54
2007	16	4	16	64
$\Sigma Y =$	105	0	40	4

The two equations are

$$\Sigma y = na + b\,\Sigma x$$

$$\Sigma xY = a\,\Sigma x + b\,\Sigma x^2$$

Putting the values we get

$$105 = 5a \quad \Rightarrow a = 21$$

$$4 = 40b \quad \Rightarrow \quad b = \frac{1}{10}$$

The trend equation

$$Y = 21 + \frac{1}{10}(X - 2003)$$

Sales for 2010 is

$$Y = 21 + \frac{1}{10}(2010 - 2003)$$

$$= 21 + \frac{7}{10} = 21.7 \text{ lakh}$$

Annual increase $= \frac{1}{10}$ lakh.

❑❑❑